T0140507

Lecture Notes on Data Engineering and Communications Technologies **169**

Series Editor

Fatos Xhafa, *Technical University of Catalonia, Barcelona, Spain*

The aim of the book series is to present cutting edge engineering approaches to data technologies and communications. It will publish latest advances on the engineering task of building and deploying distributed, scalable and reliable data infrastructures and communication systems.

The series will have a prominent applied focus on data technologies and communications with aim to promote the bridging from fundamental research on data science and networking to data engineering and communications that lead to industry products, business knowledge and standardisation.

Indexed by SCOPUS, INSPEC, EI Compendex.

All books published in the series are submitted for consideration in Web of Science.

Jemal H. Abawajy · Zheng Xu ·
Mohammed Atiquzzaman · Xiaolu Zhang
Editors

Tenth International Conference on Applications and Techniques in Cyber Intelligence (ICATCI 2022)

Volume 2

Editors
Jemal H. Abawajy
Faculty of Science, Engineering and Built Environment
Deakin University
Geelong, VIC, Australia

Mohammed Atiquzzaman
School of Computer Science
University of Oklahoma
Norman, OK, USA

Zheng Xu
School of Computer Engineering and Sciences
Shanghai Polytechnic University
Shanghai, China

Xiaolu Zhang
Department of Information Systems and Cyber Security
The University of Texas at San Antonio
San Antonio, TX, USA

ISSN 2367-4512 ISSN 2367-4520 (electronic)
Lecture Notes on Data Engineering and Communications Technologies
ISBN 978-3-031-28892-0 ISBN 978-3-031-28893-7 (eBook)
https://doi.org/10.1007/978-3-031-28893-7

This Springer imprint is published by the registered company Springer Nature Switzerland AG
The registered company address is: Gewerbestrasse 11, 6330 Cham, Switzerland

Foreword

The 10th International Conference on Applications and Techniques in Cyber Intelligence (ATCI 2022), building on the previous successes online conference (2021 and 2020 due to COVID-19), Huainan, China (2019), Shanghai, China (2018), Ningbo, China (2017), Guangzhou, China (2016), Dallas, USA (2015), Beijing, China (2014), and Sydney, Australia (2013), is proud to be in the tenth consecutive conference year. ATCI 2021 has moved online due to COVID-19.

The purpose of ATCI 2022 is to provide a forum for presentation and discussion of innovative theory, methodology and applied ideas, cutting-edge research results, and novel techniques, methods, and applications on all aspects of cyber and electronics security and intelligence. The conference establishes an international forum and aims to bring recent advances in the ever-expanding cybersecurity area including its fundamentals, algorithmic developments, and applications.

Each paper was reviewed by at least two independent experts. The conference would not have been a reality without the contributions of the authors. We sincerely thank all the authors for their valuable contributions. We would like to express our appreciation to all members of the Program Committee for their valuable efforts in the review process that helped us to guarantee the highest quality of the selected papers for the conference.

We would like to express our thanks to the strong support of the general chairs, publication chairs, organizing chairs, program committee members, and all volunteers.

Our special thanks are due also to the editors of Springer Thomas Ditzinger and Suresh Dharmalingam for their assistance throughout the publication process.

Jemal Abawajy
Zheng Xu
Mohammed Atiquzzaman
Xiaolu Zhang

Organization

General Chairs

Hui Zhang	Tsinghua University, China
Liang Wang	Chinese Academy of Sciences, China

Online Conference Organizing Chairs

Xianchao Wang	Fuyang Normal University, China
Shibing Wang	Fuyang Normal University, China

Program Chairs

Jemal Abawajy	Deakin University, Australia
Zheng Xu	Shanghai Polytechnic University, China
Mohammed Atiquzzaman	University of Oklahoma, USA
Xiaolu Zhang	The University of Texas at San Antonio, USA

Publication Chairs

Mazin Yousif	T-Systems International, USA
Vijayan Sugumaran	Oakland University, USA

Publicity Chairs

Kewei Sha	University of Houston, USA
Neil. Y. Yen	University of Aizu, Japan
Shunxiang Zhang	Anhui University of Science and Technology, China

Program Committee Members

William Bradley Glisson	Sam Houston State University, USA
George Grispos	University of Nebraska at Omaha, USA
V. Vijayakumar	VIT Chennai, India
Aniello Castiglione	Universit di Salerno, Italy
Florin Pop	University Politehnica of Bucharest, Romania
Neil Yen	University of Aizu, Japan
Xianchao Wang	Fuyang Normal University & Tech., China
Feng Wang	Fuyang Normal University & Tech., China
Jia Zhao	Fuyang Normal University & Tech., China
Xiuyou Wang	Fuyang Normal University & Tech., China
Gang Sun	Fuyang Normal University & Tech., China
Ya Wang	Fuyang Normal University & Tech., China
Bo Han	Fuyang Normal University & Tech., China
Xiuming Chen	Fuyang Normal University& Tech., China
Xiangfeng Luo	Shanghai Univ., China
Xiao Wei	Shainghai Univ., China
Huan Du	Shanghai Univ., China
Zhiguo Yan	Fudan University, China
Abdulbasit Darem	Northern Boarder University, Saudi Arabia
Hairulnizam Mahdin	Universiti Tun Hussein Onn, Malaysia
Anil Kumar K. M.	JSS Science & Technology University, Mysore, Karnataka, India
Haruna Chiroma	Abubakar Tafawa Balewa University, Bauchi, Nigeria
Yong Ge	University of North Carolina at Charlotte, USA
Yi Liu	Tsinghua University, China
Foluso Ladeinde	SUNU, Korea
Kuien Liu	Pivotal Inc., USA
Feng Lu	Institute of Geographic Science and Natural Resources Research, Chinese Academy of Sciences, China
Ricardo J. Soares Magalhaes	University of Queensland, Australia
Alan Murray	Drexel University, USA
Yasuhide Okuyama	University of Kitakyushu, Japan
Wei Xu	Renmin University of China, China
Chaowei Phil Yang	George Mason University, USA
Hengshu Zhu	Baidu Inc., China
Morshed Chowdhury	Deakin University, Australia
Elfizar	University of Riau, Indonesia
Rohaya Latip	Universiti Putra, Malaysia

The 10th International Conference on Applications and Techniques in Cyber Intelligence (ATCI2022)

18 June 2022
ATCI 2022 has moved online due to COVID-19.
http://atci.com.cn/

Program Book

Conference Program at a Glance

Saturday, June 18, 2022, Tencent Meeting Online Link		
10:00–10:10	Opening ceremony by conference PC Chair	Tencent Meeting
10:10–10:50	Keynote 1: Kim-Kwang Raymond Choo	Tencent Meeting
10:50–11:30	Keynote 2: Jemal Abawajy	Tencent Meeting
Saturday, June 18, 2022, Tencent Meeting Online Link		
13:00–18:00	Session 1	Tencent Meeting
	Session 2	Tencent Meeting
	Session 3	Tencent Meeting
	Session 4	Tencent Meeting
	Session 5	Tencent Meeting
	Session 6	Tencent Meeting
	Session 7	Tencent Meeting

Please download Tencent Meeting.
We will send an online conference link to your email.
https://meeting.tencent.com/sg/en/index.html

ATCI 2022 Keynotes

Kim-Kwang Raymond Choo holds a Ph.D. in information technology from Queensland University of Technology, Australia. Prior to starting his Cloud Technology Endowed Professorship at UTSA, Dr. Choo spent five years working for the University of South Australia and five years working for the Australian Government Australian Institute of Criminology. He was also a visiting scholar at INTERPOL Global Complex for Innovation between October 2015 and February 2016 and a visiting fulbright scholar at Rutgers University School of Criminal Justice and Palo Alto Research Center (formerly Xerox PARC) in 2009.

In April 2017, he was appointed an honorary commander, 502nd Air Base Wing, Joint Base San Antonio-Fort Sam Houston, USA. He is also a fellow of the Australian Computer Society, a senior member of IEEE, and a co-chair of IEEE Multimedia Communications Technical Committee's Digital Rights Management for Multimedia Interest Group.

Jemal Abawajy is a faculty member at Deakin University and has published more than 100 articles in refereed journals and conferences as well as a number of technical reports. He is on the editorial board of several international journals and edited several international journals and conference proceedings. He has also been a member of the organizing committee for over 60 international conferences and workshops serving in various capacities including the best paper award chair, general co-chair, publication chair, vice-chair, and program committee. He is actively involved in funded research in building secure, efficient, and reliable infrastructures for large-scale distributed systems. Toward this vision, he is working in several areas including pervasive and networked systems (mobile, wireless network, sensor networks, grid, cluster, and P2P), e-science and e-business technologies and applications, and performance analysis and evaluation.

Contents

Cyber Intelligence for Industrial, IoT, and Smart City

Cyber Intelligence for CV Process and Data Mining

Cyber Intelligence for Health and Education Informatics

Cyber Intelligence for Industrial, IoT, and Smart City

D_ Modeling and Simulation of Intelligent Building Robot Based on Star Algorithm

Gaoshan Hu(✉), Xinyang Ji, and Bin Meng

Shenyang Urban Construction University, Shenyang 110167, Liaoning, China
820094519@qq.com

Abstract. Robots play an important role in our production and life, involving a wide range of fields, such as industry, military, service, medical treatment, agriculture and so on. The traditional mobile robot D_ Star algorithm and manipulator dynamic path planning, the theoretical basis of robot trajectory planning is studied, the body characteristics and robot motion control of construction robot are discussed, and the prediction model of construction robot is studied. The results show that the development trend of trowel height data of construction robot is generally stable, and there are occasional data mutations.

Keywords: D_ Star algorithm · Intelligence · Construction robot · Manipulator dynamic path planning

1 Introduction

With the increase of high-rise buildings and the improvement of people's requirements for building aesthetics, the construction industry puts forward many requirements for building appearance decoration, such as large operation surface, high precision and high quality requirements. At present, the pure artificial exterior wall construction mode is difficult to meet this demand, which has become the bottleneck restricting the improvement of the overall quality and efficiency of the construction industry. In the current environment of rapid development of industrial robot technology, the first choice to solve this problem is to introduce it into building exterior wall construction and realize automation and even intelligence.

With the continuous progress of science and technology, many experts have studied intelligent building robots. For example, Singh R, nagla k s proposed an improved framework - advanced laser and sonar framework (alsf) - by selecting the best distance information corresponding to the selected threshold, the sensory information of laser scanner and sonar is fused to reduce the uncertainty caused by glass in the environment [1]. Tavares P, Costa C M, Rocha l proposed a spatial augmented reality system that projects alignment information into the environment to help the operator locate welded beam attachments. A cooperative welding unit for structural steel manufacturing is proposed, which can automatically coordinate the necessary tasks assigned to human operators and welding robots moving on linear tracks using building information modeling (BIM) standards [2]. Wang Q, Jiao W, Yu r constructed a wireless sensor and

J. H. Abawajy et al. (Eds.): ICATCI 2022, LNDECT 169, pp. 3–10, 2023.
https://doi.org/10.1007/978-3-031-28893-7_1

participant network using ZigBee technology according to the characteristics of ZigBee protocol. In the research of home service robot intelligent space, a highly reliable and easy to build communication network is constructed [3]. Although the research results of intelligent building robot are quite fruitful, it is based on D_ There are still some deficiencies in the research of intelligent building robot modeling and simulation.

This paper is based on D_ The intelligent building robot based on star algorithm is modeled and simulated_ Star algorithm and intelligent building robot are studied, and the vibration mode of suspended high-altitude platform is analyzed. The results show that D_ Star algorithm is conducive to the modeling and Simulation of intelligent building robot.

2 Method

2.1 D_ Star Algorithm

(1) Traditional mobile robot D_ Star algorithm

D of short path planning_ Star algorithm is another important problem to realize autonomous navigation. After fully mastering the obstacle information in the robot operating environment, the map establishment methods include visual graphics method, grid method and so on. Common local path planning methods mainly include artificial potential field method, dynamic window method, etc. [4]. In this case, the planned path cannot be directly used by the robot, but the size and workspace of the construction robot are large. Therefore, the path planning algorithm needs to be improved to meet the requirements of the construction robot. After completing the global path planning, in order to prevent random obstacles on the path, it is necessary to combine the local path planning algorithm for local obstacle avoidance to make the robot reach the working position safely [5].

(2) Dynamic path planning of manipulator.

Path and motion planning technology is an important means to improve the autonomy and intelligence of manipulator, and it is the premise for manipulator to complete various complex tasks. In the future, the working scene of manipulator in intelligent factory must be dynamic cooperation and integration with people. When completing the operation task, the manipulator must be able to perceive the static and dynamic obstacles in its surrounding environment and avoid them in time to ensure the personal safety of surrounding instruments, equipment and personnel. Manipulator path planning is the core technology to meet the above requirements. At the same time, autonomous path planning is also an important problem that must be solved in the process of manipulator intelligence [6]. Among many path planning methods, the fast search random tree algorithm based on random sampling does not map obstacles from task space to configuration space in advance, and the search speed is fast. Compared with other planning methods, it is more suitable for dynamic path planning of high-dimensional spatial manipulator in complex environment. It is a path planning method with complete probability and good scalability.

2.2 Intelligent Building Robot

(1) Theoretical basis of robot trajectory planning

For this high-precision assembly robot, its trajectory needs to be strictly controlled, which is called CP control. In the process of CP control, the end motion trajectory of the robot needs to be given according to the task characteristics [7]. Therefore, it is necessary to obtain a series of sampling points at certain intervals in the motion trajectory, convert the end sampling points into joint space through inverse kinematics, and then fit these sampling points in joint space with appropriate smoothing function. The process of determining the coordinates of each sampling point between two points is called interpolation or interpolation. The point-to-point control process by inserting an intermediate point is called analog CP control. The advantage of using analog CP control method is that it simplifies the task of robot path recognition, saves the storage unit of the controller and improves the operation efficiency of the controller. The path of analog CP control can be a straight line or curve. According to the requirements of robot motion speed, working trajectory position accuracy and computer storage capacity, different interpolation algorithms are selected for calculation. Analog CP control is point-to-point control between multiple interpolation points. The density of interpolation points determines the accuracy of trajectory curve [8]. There are two methods to select interpolation points: timing interpolation and fixed distance interpolation. The basic algorithms of the two interpolation methods are the same, but the former has a fixed time interval and is easy to implement; The latter can ensure the trajectory accuracy, but the time interval changes with the change of working speed, so it is difficult to achieve.

(2) Ontology characteristics of Construction Robot.

Most of the work in building construction has the characteristics of large load, wide operation range and limited working space. Construction robots need to have higher intelligence and wide applicability. In the configuration of construction robot, the series mechanism has the advantages of simple structure, simple control and large workspace. It is mainly used for structural welding, bricklaying, 3D printing and industrialized prefabrication of building components in building construction. Most curtain wall installation and construction robots adopt vehicle mounted form, and their main structure can be divided into two parts, namely, the chassis responsible for moving and the actuator responsible for operation carried by the chassis [9]. From entering the building to completing the operation, two problems can be solved: (1) how to make the robot reach the operation position safely; (2) How to ensure that the robot can complete the operation safely after reaching the operation point. In field operation, due to the limitation of workspace and equipment weight, the collision avoidance of construction robot is particularly important. In the navigation process, it is necessary to combine BIM information to judge the traffic capacity of the robot in the building in real time; The working space of the construction robot is large, and the actual environment space on the site may be narrow during operation. In order to prevent the collision of construction robots, it is necessary to establish a constraint space to further limit the behavior of robots, and carry out trajectory planning in the operation process under the constraint space.

(3) Robot motion control

The motion control system is closely combined with the research object to control the motion of the object through speed control, position control and torque control. The hardware platform is usually composed of controller, driver and motor [10]. Stability, accuracy and robustness are important indexes to evaluate the control system. For the specific research object of robot, from the perspective of control algorithm, on the one hand, the flexibility of system motion should be considered, on the other hand, the anti-interference performancc or robustness of the system should be considered. In the inner loop of the system, the higher the speed control accuracy and robustness of the bottom layer, the easier and more effective the control effect of the upper layer. PID controller, from the perspective of linear system, but not limited to the control of linear system, is still a widely used control mode at present. The principle of PID is to adjust the distribution of the eigenvalues of the characteristic equation of the closed-loop transfer function of the system on the complex plane. The principle is simple, easy to implement and has certain anti-interference effect. As an intelligent control algorithm, fuzzy control is a very active research direction in the field of intelligent control, and has more and more applications in motion control system [11]. By fuzzifying the input variables, establishing the fuzzy rule table and establishing the nonlinear control method, many control problems of complex systems can be solved by using the idea of fuzzification.

2.3 Vibration Modal Analysis of Suspended High-Altitude Platform

Mode is an inherent attribute reflecting the vibration characteristics of mechanical system [12]. Through modal analysis, the easy vibration frequency band of the system can be obtained, and the vibration response of the system in this frequency band can be obtained. Therefore, modal analysis can be used to prevent resonance of the mechanism. The free motion equation of the mechanical system is as follows (1):

$$[M]\{\ddot{u}\} + [C]\{\dot{u}\} + [K]\{u\} = 0 \tag{1}$$

where: [M], [C] and [k] are the mass matrix, damping matrix and stiffness matrix respectively, and u is the node displacement vector of the system.

Before applying quickhull algorithm, the internal voxels of octree graph should be deleted to improve the efficiency of convex hull calculation, because their center point will never be the vertex of convex hull. As mentioned above, the organizational structure of voxels is octomap, so each voxel has a unique index in the octree, as shown in formula (2):

$$Ind_i = [Ind_{ix}, Ind_{iy}, Ind_{iz}]^T \tag{2}$$

where Ind_{ix}, Ind_{iy}, Ind_{iz} is the index value of the ith voxel in the X, y and Z directions, respectively.

Lagrangian mechanical analysis is based on the concept of system energy, that is, the difference between system kinetic energy and potential energy. The Lagrange equation can be obtained from Eq. (3):

$$L(q, \dot{q}) = E_k(q, \dot{q}) - E_p(q) \tag{3}$$

L is a Lagrange function.

3 Experience

3.1 Object Extraction

At present, for high space operation, ropes will be used to meet different operation needs, such as elevators, high-altitude hanging baskets, bridge stay cables, etc. the reason why they are widely used is that ropes have the advantages of strong bearing capacity and light self-weight. However, due to the complex nonlinear dynamic characteristics of the mechanical system composed of ropes, it is difficult to analyze its vibration and establish an accurate mathematical model. There are usually two modeling methods: distributed parameter continuous model and centralized parameter discrete model. The continuous model takes the interception of rope micro unit as the research object, carries out force analysis and establishes the dynamic equation. Then, starting from the dynamic equation of micro unit, the partial differential equation of continuous model is solved through relevant theories to obtain the mathematical model of rope. The discrete model generally discretizes the rope into several segments, each segment is regarded as a rigid block, concentrates the mass at the end point, connects several segments through the motion pair, and adds relevant elastic and damping effects to the motion pair, so as to obtain the discrete model of the rope.

3.2 Experimental Analysis

The constraint space needs to ensure that the construction robot does not interfere with other parts in the current environment, so as to ensure the smooth operation of the robot and the safety of workers. The classical a * algorithm mainly stores nodes in three states around two open seats and closed sets. The specific steps are as follows: Step 1: put the start node into the open table; And prepare to start executing the first layer of loop logic; Step 2: starting from the loop logic of the first layer, take the node with the smallest F value in the open table as the current node, and carry out the following processing. If the current node does not exist, the pathfinding will end in failure. If the current node is end node, the pathfinding is successful. The front node of end node is set to current node, and then the path composed of all nodes from end node to start node is returned.

4 Discussion

4.1 Selection of Time Series Prediction Model

With the in-depth study of time series prediction technology, more and more time series prediction models have been proposed and put into use. When using the prediction model to solve practical problems, the final prediction model will be different due to different demand scenarios and data characteristics. Firstly, in order to narrow the selection range of prediction model, the data characteristics of sensor data of construction robot are analyzed. Taking part of the trowel height data collected by the construction robot as an example, Table 1 shows the visualization results of the data.

It can be seen from the above that when the height of the trowel is 1000 mm, the time is 50 ms; When the height of the trowel is 1500 mm, the time is 100 ms; When the

Table 1. Visualization of timing data of trowel height

Height of trowel (mm)	Time (100 ms)
1000	50
1500	100
2000	200
2500	30

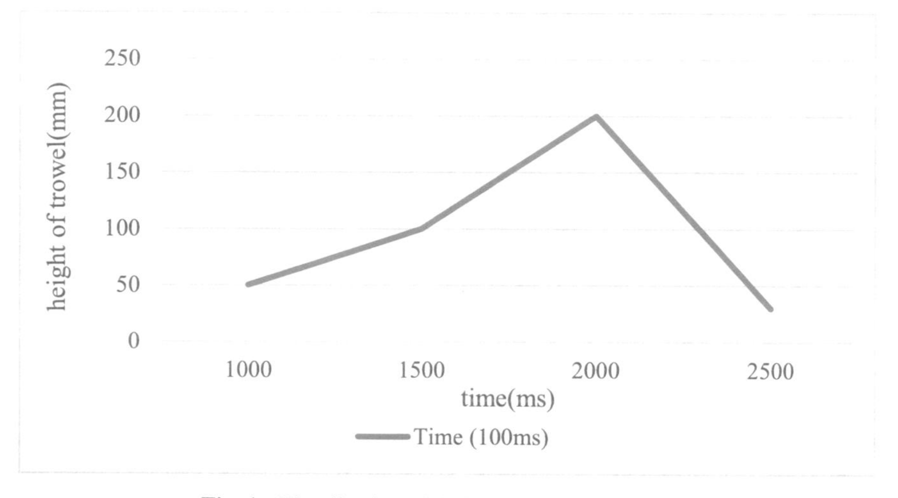

Fig. 1. Visualization of timing data of trowel height

height of the trowel is 2000 mm, the time is 200 ms; When the height of the trowel is 2500 mm, the time is 30 ms. The specific presentation results are shown in Fig. 1.

It can be seen from the curve above that the development trend of trowel height data of construction robot is generally stable, with occasional data mutation. The data fluctuation of the whole cycle is small, and the recent data has a great impact on the future data. Therefore, it is a better choice to predict and improve the missing data of construction robot sensors based on short-term prediction model.

4.2 Static Obstacle Avoidance Path Planning

In Matlab environment, the size of the whole SS is 100 × 900, the obstacle area is a randomly set black rectangular box as the initial position (left blue solid circle) and target position (right red solid circle), and then carry out plane path planning. The smooth srrta is compared with the traditional basic RRT and bidirectional rrta to verify its correctness and SOTA. In order to evaluate and execute Ao, the performance of this algorithm is better than the other two algorithms. Finally, the average search time, average number of sampling nodes and successful Pt in the whole PP are counted and recorded respectively, as shown in Table 2.

Table 2. Curvature variation diagram of s-rrt generated path

10 planning experiments	Average planning time\ms	Average sampling node	Success times
S-RRT	443.5	457.5	8
Basic-RRT	336.7	465.6	6
Bi-RRT	327.6	475.6	4

It can be seen from the above that 8 planning simulations of s-rrt algorithm, 6 Planning simulations of basic RRT and 4 planning simulations of Bi RRT are successful. The specific presentation results are shown in Fig. 2.

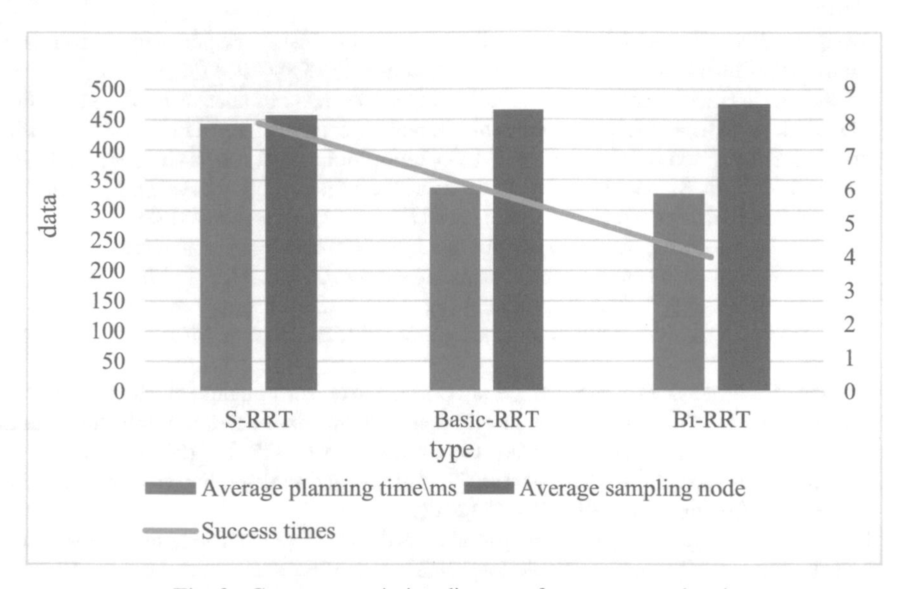

Fig. 2. Curvature variation diagram of s-rrt generated path

From the above relevant data, it can be seen that compared with the other two AB RRT and Bi RRT, the search speed and search efficiency of the basic s-rrt algorithm are significantly improved, and the path optimized by pruning function is smoother. At the same time, the path curvature is also changing, which can better meet the ROSM and no impact of the construction robot in the process of static obstacle avoidance.

5 Conclusion

With the advancement of national urbanization, the past architectural forms and styles can no longer meet people's growing material and cultural needs. The urban population is concentrated, there are many buildings and vast green space, but there are some problems in modern cities all over the world, such as land shortage, high land price and so on. The high-altitude development of buildings is a common phenomenon in urban construction in various countries. This paper studies the static OA path planning problem. The results show that the basic s-rrt can better meet the requirements of stable motion and no collision in the static OA process of engineering robot in terms of search speed and search efficiency.

References

1. Singh, R., Nagla, K.S.: Multi-data sensor fusion framework to detect transparent object for the efficient mobile robot mapping. Int. J. Intell. Unmanned Syst. **7**(1), 2–18 (2019)
2. Tavares, P., Costa, C.M., Rocha, L., et al.: Collaborative welding system using BIM for robotic reprogramming and spatial augmented reality. Autom. Constr. **106**, 102825.1–102825.12 (2019)
3. Wang, Q., Jiao, W., Yu, R., et al.: Modeling of human welders' operations in virtual reality human-robot interaction. IEEE Robot. Autom. Lett. **4**(3), 2958–2964 (2019)
4. Lublasser, E., Hildebrand, L., Vollpracht, A., Brell-Cokcan, S.: Robot assisted deconstruction of multi-layered façade constructions on the example of external thermal insulation composite systems. Constr. Robot. **1**(1–4), 39–47 (2017). https://doi.org/10.1007/s41693-017-0001-7
5. Miyoshi, T., Sato, S., Akashi, T., et al.: Construction of a novel robot system for fixed can-filling works in fishery processing. Jpn. J. Food Eng. **19**(3), 173–184 (2018)
6. Wang, Z., Li, H., Zhang, X.: Construction waste recycling robot for nails and screws: computer vision technology and neural network approach. Autom. Constr. **97**, 220–228 (2019)
7. Mvemba, P.K., Band, S., Lay-Ekuakille, A., et al.: Advanced acoustic sensing system on a mobile robot: design, construction and measurements. IEEE Instrum. Meas. Mag. **21**(2), 4–9 (2018)
8. Dierichs, K., Kyjánek, O, Loučka, M., et al.: Correction to: construction robotics for designed granular materials: in situ construction with designed granular materials at full architectural scale using a cable-driven parallel robot. Constr. Robot. **3**(1), 117–117 (2019)
9. Prabakaran, V., Elara, M.R., Pathmakumar, T., et al.: Floor cleaning robot with reconfigurable mechanism. Autom. Constr. **91**, 155–165 (2018)
10. Zhou, Z., et al.: Pigeon robot for navigation guided by remote control: system construction and functional verification. J. Bionic Eng. **18**(1), 184–196 (2021). https://doi.org/10.1007/s42235-021-0013-3
11. Pldvere, N., Paradis, C.: "What and then a little robot brings it to you?" The reactive what-x construction in spoken dialogue. Engl. Lang. Linguist. **24**(2), 307–332 (2020)
12. Li, Z., Li, Y., Rong, X., et al.: Grid map construction and terrain prediction for quadruped robot based on C-terrain path. IEEE Access **PP**(99), 1 (2020)

Industrial Economic Trade Volume Based on Multi-prediction Model Algorithm

Sonexay Phompida and Donghua Yu(✉)

School of Economics, Shandong University, Jinan 250000, Shandong, China
wanghaobin202204@163.com

Abstract. Industrial economic trade volume is an important indicator of national economic growth, enhance international exchanges, with the development of global economic integration, import and export trade has become an important position in a country's national economy, in the future to promote the rapid development of China's economy and trade, on the international stage, must accurately grasp and predict our import and export trade in a favorable direction. In recent years, with the evolution and maturity of prediction science, at present, nonlinear prediction methods and combined prediction model methods are the hot spots of scholars, and excavated many nonlinear and combined prediction methods, due to different factors due to different directions, in addition, different prediction methods match different use conditions. Therefore, we should choose the prediction method that can improve the prediction accuracy strictly according to the law of things development. The paper aims to study the prediction and analysis of the multi-prediction model algorithm for the industrial economy and trade volume, and build a multi-prediction model, which is of great significance for the development of China's industrial economy and trade. Based on literature research and quantitative analysis, we collected and predicted the data and used the application of industrial economic and trade volume for the accurate prediction of import and export industrial trade volume.

Keywords: Prediction model · Combination prediction · Import and export trade · Neural network algorithm

1 Introduction

Predicting refers to the inference of on the future development trend of things through certain ways and methods, and through reasonable logical reasoning, according to the historical data accumulated in the process of human understanding of things [1]. The rules for the development of things are analyzed qualitatively and quantitatively, and the correctness of such estimates is used as the guiding basis for people's lives or actions. In short, prediction is using past and present information to judge the rules that may happen in the future. In the process of predicting the future development, the unknown knowledge can be speculated according to the knowledge already mastered, and the direction of the future development of things can be expected to some extent, [2].

J. H. Abawajy et al. (Eds.): ICATCI 2022, LNDECT 169, pp. 11–18, 2023.
https://doi.org/10.1007/978-3-031-28893-7_2

As the research continues, the prediction performance of a single prediction model does not reach the user's expectation, because the proposed model presented at this stage cannot perform well in prediction performance in all domains. Therefore, it is very important to combine the advantages of the individual prediction model and establish a better prediction model [3]. The researchers collectively describe this prediction model as combined or mixed prediction models, which are equal to or better than individual models in terms of stability and accuracy by combining the advantages of multiple prediction models. Due to the many advantages of combined prediction models, many research institutions and scholars have done a lot of research work on this. According to the different establishment methods of combined model, such as time complexity, combined framework, dimension reduction effect of data feature space and processing method of data prediction, the combined model is divided into combined model based on weight allocation mode, combined model based on data prepossessing mode, combined model based on model parameters and structure optimization, and combined model based on error correction method [4].

Combination prediction is an important research direction in the field of machine learning. By effectively combining different prediction methods, we avoid the disadvantages of traditional prediction methods and play their own advantages, making the newly generated combined models have a wider range of [5]. For specific application fields, the combined model has better generalization performance and for higher prediction accuracy. In today's world, the trade of each country is the link of the economic relations with other countries, which promotes the growth of the national economy and the global economy [6]. Due to reform and export development, the import and export trade, as it represents the international position on the international stage, introduces a large number of scientific and technological talents, financing scale, development opportunities, and to improve the accuracy of total import and export trade, to correctly predict the trend of domestic economic development and guide the implementation of import and export policy and foreign trade policy.

2 Proposed Method

2.1 Linear Prediction Methods

If the annual import and export data of a city is regarded as a time series, then you can use the time series model to predict the annual import and export data of a city. Generally speaking, the common time series models include auto-regressive model (AR model), mobile average model (MA model) and auto-regressive mobile average model (ARMA model). AR model AR model is an auto-regressive model, this model is a linear prediction model, this model assumes the first N data in the known time series, then the data behind the N point can be derived by fitting the recursion, the mathematical expression of the AR model [7]:

$$X_t = \varphi_1 X_{t-1} + \varphi_2 X_{t-2} + \cdots + \varphi_p X_{t-p} + u_t \tag{1}$$

where, p represents the order of the AR model, φ_i is the fit coefficient of the model, and u_t represents the fit error of the model.

MA model-MA model also known as moving average model, mathematical expression of MA model is shown in (2):

$$u_t = \varepsilon_t - \theta_1\varepsilon_{t-1} - \cdots - \theta_q\varepsilon_{t-q} \tag{2}$$

where, q is the order of the MA model, $\varepsilon_i(= i = t-1, t-2, \cdots, t-q)$ is the random number. The MA model features that the model is stable under any conditions. In the fitting of the MA model, the fitting of the model order q is a key problem. The MA model generally adopts and determines the order of the sample auto-correlation function ρkkA. The order determination process first calculates the partial auto-correlation coefficient sequence of the time series ρkkA, starting from $k = 0$, and viewing the order that significantly is not zero in ρkkA, which can be regarded as the order of the MA model.

2.2 Combination Prediction Ideas

The prediction problem of import and export trade in a city will be transformed into learning a group of time data series to find the law of time development, and predict the future import and export trade in the short term. To this idea, the prediction method based on time series ARIMA and import and export trade of LSTM network is put forward [8]. After the time series ARIMA turns import and export data into stationary data, although the model selects the optimal order, the value through the white noise test, the LSTM network has the ability to learn long-term dependent information, combining the advantages of the two models to achieve better prediction effect, which is also the idea of combined prediction. Combination prediction is finding a way or criterion to extract effective information in each single prediction model and make such effective information effectively combined, such as weighting or in some reasonable way [9]. According to the different ways of building the combination models, they are divided into combined models based on weight allocation, data prepossessing, model parameters and structure optimization, and error correction methods. Because the combined model is to effectively combine different models, and can give full play to the advantages of different models, for example, the weight-based combined model framework, weight allocation to each model by evaluating the relative effect of each model and the product of the weight as the final prediction effect, the original data set after data fusion processing has good prediction effect [10].

3 Experiments

3.1 Experimental Content

In the context of multiple prediction algorithm, the importance and reliability of multiple prediction model algorithm analyzes the development trend of import and export trade and predict import and export trade through big data platform and provide intellectual support for macro-control.

3.2 Experimental Methods

The paper mainly uses the literature research method, Quantitative analysis method and other methods; First, by reading a lot of the literature, master the basic theory of multi-prediction model and the prediction method for China's import and export trade volume, analyze the pros and cons of each method, with a rigorous theoretical foundation, to make knowledge accumulation and preparation for the next experimental analysis; Collect data on 20 years of import and export trade volume in 20 years (20 years, The proportion of domestic import and export volume in the Yangtze River Economic Belt increased from 30.2% to 45.0%). The distribution was analyzed and predicted; Finally, the application of each prediction method in Chinese foreign trade prediction is introduced in detail, Neural network algorithm and auto-regressive mobile averaging method, respectively.

4 Discussion

4.1 Evolution and Status Quo of Foreign Trade Scale – Take the Yangtze River Economic Belt as an Example

4.1.1 The Development Trend of Foreign Trade Scale

The Yangtze River Economic Belt includes Shanghai, Yunnan, Sichuan, Chongqing, Hubei, Hunan, Jiangxi, Anhui, Guizhou, Jiangsu and Zhejiang. Therefore, the development trend of foreign trade scale in these 11 regions determines the production volume of their foreign trade containers. From 1999 to 2018, with the development of the Yangtze River Economic Belt, its foreign trade development level occupied a more and more important position in China. In the past 20 years, the domestic import and export share of the Yangtze River Economic Belt has increased from 30.2% to 45.0%. Combined with the development background of domestic and foreign markets, the development trend of the foreign trade scale of the Yangtze River Economic Belt can be divided into three stages, respectively, for the take-off stage, the development stage of adjustment and the stable development stage.

4.1.2 Regional Distribution of Foreign Trade Scale

From the perspective of the regional distribution of the import and export scale of the Yangtze River Economic Belt, as shown in Fig. 1 and Fig. 2, the export scale of the provinces and cities within the Yangtze River Economic Belt varies greatly. The import and export trade volume in Jiangsu, Zhejiang and Shanghai occupies an absolute advantage in the Yangtze River Economic Belt. In the following, firstly, the distribution of foreign trade scale in the Yangtze River region is analyzed from the export and import aspects, and then the Yangtze River Economic Belt is divided into Jiangsu, Zhejiang, Shanghai, Hubei, Hunan, Sichuan, Guizhou, Yun, Jiangxi, Chongqing and Anhui two parts to further analyze the proportion of import and export trade volume in each region in the Yangtze River Economic Belt.

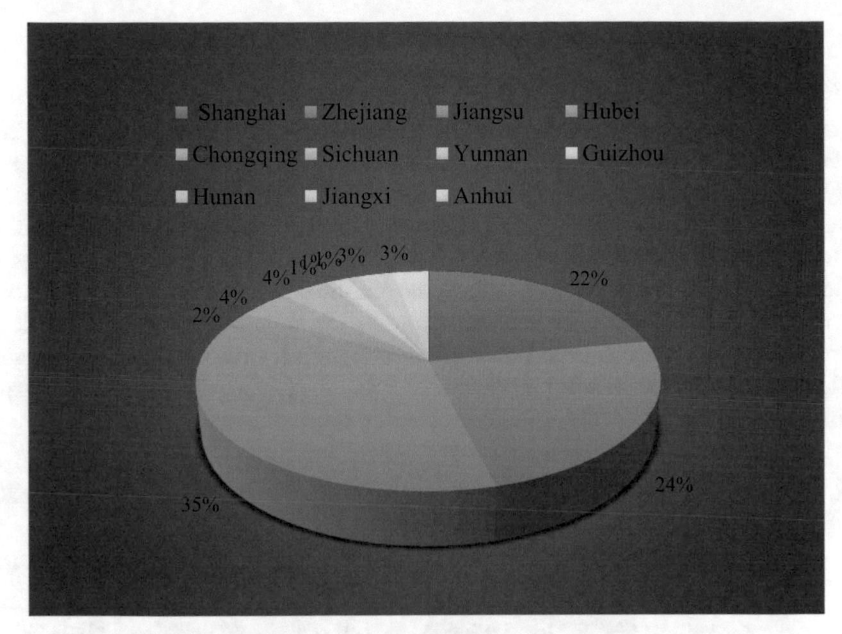

Fig. 1. Exports ratio of 11 provinces and cities along the Yangtze River Economic Belt in 2019

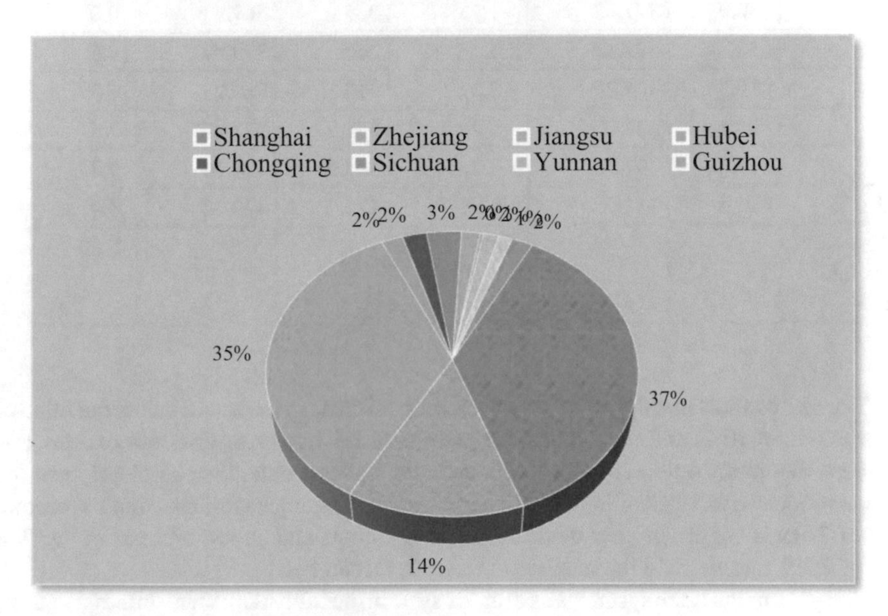

Fig. 2. Import proportion of 11 provinces and cities along the Yangtze River Economic Belt in 2019

4.2 Application of Each Prediction Method in Prediction of Foreign Trade

4.2.1 Artificial Neural Network Algorithm

The data adopts the actual monthly trade of import and export volume from January 2018 to May 2019; as a sample training network, establish TDBPNN model. Then, the above established network was used to predict China's import and export volume from July to December 2019. Because the process of establishing the model is complicated, the following briefly introduces the general idea: removing the seasonal component of the original sequence is completed by checking the non-linearity of the foreign trade system, determine the number of hidden neurons, export trade, simulate the above network and predict China's import and export trade. The following are the results of the six steps predicted from July to December 2019, and the monthly actual data of import and export trade are summarized in Table 1.

Table 1. Monthly forecast and actual value of import and export trade from July–December 2019 (in USD 200 million)

Month	Export trade			Volume of imports		
	Actual value	Predicted value	Relative error (%)	Actual value	Predicted value	Predicted value (%)
7	380.9	377.02	−1	365	367.9	0.8
8	374	382.69	2.3	346	379.89	9.7
9	419.3	405.25	−3.3	416.5	389.23	-6.5
10	409.1	387.62	−5.2	351.9	377.42	7.3
11	417.6	397.62	−4.8	368.9	404.14	9.8
12	480.6	411.95	−14.3	423.4	429.32	1.4
Mean absolute error			5.15			5.89

The results show that the prediction values are all close to the actual value, the relative error is within 10%, and the single-step prediction relative error does not exceed 1%. At the same time, combined with the image and forecast data, we can clearly see the seasonal fluctuation trend of import and export trade. The forecast shows that the second half of 2019 is slightly higher than the monthly import and export amount in the first half of 2019, and reached the maximum value in December.

China's foreign trade system is a complex system, not only with many internal change factors, but also affected by external factors. Therefore, the linear prediction method to predict the trade volume is not in line with the actual situation. Therefore, we use the nonlinear prediction concept to construct a nonlinear artificial neural network to predict the amount of import and export, which is more in line with China's current national conditions. The proposed method requires a moderate number of data samples, and the high accuracy and strong timeliness are very suitable for short-term prediction.

4.2.2 Auto-regressive Moving Averaging Method

This modeling approach is suitable for the prediction of time series that are difficult to judge about the typical features of the time series, and it also does not need to spend much time looking for explanatory variables as in the regression analysis. It only assumes a possible applicable model in advance, and then repeatedly identifies the improvement according to certain procedures to obtain a more satisfactory prediction model. The running software estimated the parameter coefficients passed the statistical test, while the stochastic term of the model also passed the white noise test. Therefore, the above model is used to predict the confidence range of China's export volume forecast value and exports in May and June 2019 in Table 2.

Table 2. Forecast value and forecast range of exports in May and June 2019

Month	predicted value	The lower limit of the 95% prediction interval	Upper bound on the 95% prediction interval
5	340.2	340.2	496.27
6	424.07	345.57	520.39

In 2019, China's foreign trade export volume was actually us $438.23 billion, including us $41.76 billion in May and US $48.07 billion in June. The prediction results were 1.63% and 11.7%, respectively. The average absolute percentage error of the prediction is equal to 6.7%. The following conclusion can be drawn from the forecast results: this model is feasible as a Chinese export volume forecast model.

5 Conclusions

A country's industrial import and export trade volume will face many unpredictable factors, such as the political environment, exchange rate, monetary policy, etc., China's trade system is complex, easy to be affected by the external environment, with great volatility. Our country also mainly adopts a single prediction model to forecast import trade volume, although a single prediction model can intuitive analysis data, interpretation, but there are certain limitations, random volatility, strong by external interference, so the prediction results is not accurate, so based on multiple prediction model algorithm, accurate prediction of import and export industry economic trade volume is of very important significance. If there is a more reliable and effective industrial import and export trade forecast, can accurately grasp the development of a country's economy, international trade, guide decision makers to make scientific and effective policies to deal with possible changes, correctly guide enterprise import and export trade activities, reduce economic fluctuations, promote the development of regulatory foreign trade industry.

Acknowledgements. The National Natural Science Foundation of China (NSFC) project "Research on Appropriate Technology Selection and Dynamic Change of China's Manufacturing

Transformation and Upgrading under the Orientation of High-Quality Development" (Grant No. 71973083); the project of Humanities and Social Sciences Research Planning Fund of the Ministry of Education, "Research on Appropriate Technology Selection, Old and New Dynamic Energy Conversion and Manufacturing Transformation and Upgrading Dynamics Mechanism (Grant No. 19YJA790109); the Natural Science Foundation of Shandong Province, "Research on the Selection of Appropriate Technology and the Transformation of New and Old Kinetic Energy in the Transformation and Upgrading of Manufacturing Industry" (Grant No. ZR2019MG018).

References

1. Al-Najjar, H., Al-Rousan, N., Al-Najjar, D., et al.: Impact of COVID-19 pandemic virus on G8 countries' financial indices based on artificial neural network. J. Chin. Econ. Foreign Trade Stud. **14**(1), 89–103 (2021)
2. Shiraishi, S., Tan, J., Olsen, L.A., et al.: Knowledge-based prediction of plan quality metrics in intracranial stereotactic radiosurgery. Med. Phys. **42**(2), 908–917 (2015)
3. Meza, J., Espitia, H., et al.: Statistical analysis of a multi-objective optimization algorithm based on a model of particles with vorticity behavior. Softuting Fusion Found. Methodol. Appl. **20**(9), 3521–3536 (2016)
4. Khandelwal, M., Marto, A., Fatemi, S.A., et al.: Implementing an ANN model optimized by genetic algorithm for estimating cohesion of limestone samples. Eng. Comput. **34**(2), 307–317 (2018)
5. Ortiz, A.R., Kolinski, A., Skolnick, J.: Tertiary Structure prediction of the KIX domain of CBP using Monte Carlo simulations driven by restraints derived from multiple sequence alignments. Proteins Struct. Funct. Bioinform. **30**(3), 287–294 (2015)
6. Padhy, N., Singh, R.P., Satapathy, S.C.: Cost-effective and fault-resilient reusability prediction model by using adaptive genetic algorithm based neural network for web-of-service applications. Clust. Comput. **22**(6), 14559–14581 (2019)
7. Kota, A., Noriyuki, K., Shinya, S., et al.: Impact of a commercially available model-based dose calculation algorithm on treatment planning of high-dose-rate brachytherapy in patients with cervical cancer. J. Radiat. Res. (2), 2 (2018)
8. Rajabi, M., Shafiei, F.: QSAR models for predicting aquatic toxicity of esters using genetic algorithm-multiple linear regression methods and molecular descriptors. Comb. Chem. High Throughput Screen. **22**(5), 317–325 (2019)
9. Kakarakis, S.D.J., Kapsalis, N.C., Capsalis, C.N.: A semianalytical heuristic approach for prediction of eut's multiple dipole model by reducing the number of heuristics. IEEE Trans. Electromagn. Compat. **57**(1), 87–92 (2015)
10. Scarino, B.R., Minnis, P., Chee, T., et al.: Global clear-sky surface skin temperature from multiple satellites using a single-channel algorithm with angular anisotropy corrections. Atmos. Meas. Techn. **10**(1), 351–371 (2017)

Multi-objective Optimization of FCC Separation System Based on Particle Swarm Optimization

Shanxia Wang(✉)

College of Computer and Information Engineering, Henan Normal University, Xinxiang, Henan, China

wsx_htu@163.com

Abstract. Petroleum is an indispensable and important fuel for the development of modern industry, as well as an indispensable basic source of organic chemical raw materials and transportation fuels. However, the composition of petroleum is extremely complex and can only be fully utilized after proper separation. According to the different processes, the oil processing technology can be divided into primary processing and secondary processing. Among them, secondary processing refers to other processing processes using crude oil distillation products as raw materials, including catalytic cracking (FCC) and hydrocracking (HC). However, in today's society, businesses face major challenges due to issues such as pollution and energy crisis. For example, low energy consumption and excessive pollutant emissions have resulted in a great waste of resources, while high environmental policy costs have caused many economic burdens. Minimizing costs and improving economic benefits have become the goals of modern enterprise management. At the same time, particle swarm optimization, as a calculation method with strong global search ability and fast convergence speed, can provide a new way of thinking for model optimization, and has been widely used in practical production and life. In this paper, the experimental analysis method and data analysis method are used to better understand the results of multi-objective optimization of FCC separation system through experiments. According to the experimental results, the recirculation flow rates in the main fractionation tower were 23100, 38710, 34900, and 42410 kg/h, respectively. It can be seen that the energy consumption is effectively reduced and the yield is improved. The above results can provide an important reference for the design and optimization of the FCC separation system.

Keywords: Particle Swarm Optimization · FCC Separation System · Multi-Objective Optimization · Fractionation Absorption

1 Introduction

In today's industrial production, with the continuous improvement of science and technology and economic level, the demand for energy is also increasing, and the shortage of energy has brought environmental pollution problems. For example, the low energy utilization rate and the excessive pollution emissions have resulted in a large amount of waste of resources, and at the same time, the high cost of environmental governance

J. H. Abawajy et al. (Eds.): ICATCI 2022, LNDECT 169, pp. 19–27, 2023.
https://doi.org/10.1007/978-3-031-28893-7_3

has caused a series of economic burdens. To solve this problem we will seek a new efficient and environmentally friendly, energy efficient and efficient use of resources. Therefore, based on the particle swarm algorithm, this paper studies the multi-objective optimization of the FCC separation system.

At present, many scholars have carried out research on this aspect of catalytic cracking unit, and have obtained quite rich research results. For example, Mamudu A points out that the high cost of fluid catalytic cracking unit (FCCU) catalysts and their growing demand have motivated researchers to develop alternative materials from indigenous sources [1]. According to Thomas A, the closed-loop steady-state performance of a 2 × 2 modulating control structure for a side-by-side FCC unit was evaluated and compared in partial combustion (PC) and complete combustion (CC) modes. In addition, the control structure that controls the temperature of the reactor plenum through the catalyst flow rate has also been found to exhibit good closed-loop steady-state performance [2]. Cheng proposed that the hydrocarbon groups of vegetable oil molecules are very easy to crack, and the hydrocarbon groups of vegetable oil molecules have a strong tendency to aromatize. In the catalytic cracking process of vegetable oils and fatty acids, part of the oxygen is removed in the form of water through hydrogen transfer Olefin saturation ensures proper yields of light olefins and light aromatics [3]. Therefore, this paper starts from a new perspective, combined with particle swarm algorithm, to carry out research on multi-objective optimization of FCC separation system, which has important research significance and reference value to a certain extent.

This article mainly discusses these aspects. Firstly, the particle swarm algorithm and its related research are introduced. Then, the multi-objective optimization of FCC separation system and its related research are expounded. Finally, the experimental research is carried out around the multi-objective optimization problem of FCC separation system, and the corresponding experimental results and analysis conclusions are drawn.

2 Related Theoretical Overview and Research

2.1 Particle Swarm Optimization and Related Research

Particle swarm optimization is a method to find the optimal individual by simulating the competition and movement in nature, and optimize it under different conditions, so as to obtain the advantages of globality and good convergence. It conducts optimization search by simulating natural phenomena, and finally obtains the optimal solution. It is an optimization algorithm based on simulating the predation behavior of birds, and it is also an algorithm based on swarm iteration. The bird swarm corresponds to the population in the particle swarm algorithm, and each bird is a particle in the population.

In this algorithm, individuals dynamically adjust their position and speed according to their own experience and group experience. The update of particle velocity and position can be expressed by formula (1) (2).

$$A_o(l+1) = oA_i(l) + b_1c_1(D_i(l) - X_i(l)) + b_2c_2(D_k(l) - X_il) \tag{1}$$

$$Z_u(r+1) = Z_u(r) + W_u(r+1) \tag{2}$$

Among them, $A_i(l)$ represents the velocity information of the ith particle at time l, $X_i(l)$ represents the position information of the ith particle at time l, $D_i(l)$ represents the individual optimal position, $D_k(l)$ represents the global optimal position, o is the inertia factor, b1 and b2 are acceleration constants, c1 and c2 is a randomly generated value between (0, 1).

Although particle swarm optimization has the advantages of simple model, simple implementation and good robustness, it also has disadvantages such as slow convergence speed and easy to fall into local optimum. In order to give full play to the advantages of particle swarm optimization, many experts and scientists at home and abroad have improved its research [4, 5].

The flying speed of the particles has a great influence on the performance of the algorithm. If the speed is too fast, it is easy to cause particles to fly to other areas through the target value, which reduces the convergence speed of the algorithm. If the particle flying speed is too slow, the local area of the algorithm can be finely searched, but this also reduces the overall detection speed of the algorithm. It can be seen that, in order to control the airspeed of particles, a control parameter, also known as inertia weight, is introduced into the basic algorithm.

Therefore, some researchers propose to dynamically adjust inertial weights when particles are detected. A larger weight value can be used at the beginning of the algorithm and a smaller weight value can be used at the last step of the algorithm. On this basis, they proposed an adjustment strategy that linearly reduces the weight value with the search time [6, 7]. The specific calculation method is shown in formula (3).

$$v = v_1 + (v_t - v_1)\frac{r}{R_{\max}} \tag{3}$$

Among them, v is the inertia weight, v1 is the initial value of the inertia weight, vt is the final value of the inertia weight, r is the current search time, and Rmax is the maximum number of iterations.

The particle swarm optimization model with inertia weight added is called the standard particle swarm optimization algorithm, and its optimization performance has been greatly improved, which can be further used to solve multi-objective optimization problems.

2.2 Multi-objective Optimization of FCC Separation System and Related Research

With the development of the economy, the gasoline and diesel consumption of agricultural machinery and transportation vehicles has increased year by year, and the consumption of liquefied petroleum gas required by industry and residents has also increased year by year. How to formulate an appropriate business plan according to product price and market demand to obtain maximum economic benefits has become an important issue faced by chemical companies in production. In this context, the multi-objective optimization problem of FCC separation system has also attracted more and more attention of chemical engineering scholars and researchers.

At the same time, in today's urgent needs of social and economic development, energy conservation and emission reduction, chemical enterprises not only need to seek production advantages, but also pay attention to various cost control, and also need to pay special attention to the environment related to energy conservation, emission reduction and safe production. This requires the optimization of the operating parameters of the system process in order to achieve maximum economic benefits and minimum energy consumption.

FCC has become an important processing technology for making petroleum products and extracting chemical raw materials. However, there are still some problems in the FCC separation system, which needs to be further optimized to improve resource utilization and reduce economic losses. From the current point of view, the improvement measures include the following aspects [8, 9].

(1) Check the temperature at the bottom of the tower. Among them, the traditional FCC operates around the upper limit and the RFCC operates around the lower limit.
(2) Check the residence time of the oil slurry at the bottom of the tower. It usually takes less than 5 min to maintain the oil suspension in a suitable annular volume, the minimum circulation of the oil suspension is determined by the crude oil and the operation plan.
(3) Check the relative density of the oil suspension, the density should be about 1.0.
(4) The sludge recirculation system maintains a high flow rate, which is generally more than 1.3 times that of oil refining. Removing aromatic-rich components from sludge is also one of the effective ways to reduce coking sales.

In addition to coking in the fractionation tower, salt formation at the top of the tower is also an unavoidable problem. The top temperature of the column is the dew point temperature of the oil vapor below its partial pressure. Only when this value exceeds a certain value of the saturation temperature, that is, corresponding to the partial pressure of the water vapor, the water vapor will flow out of the distillation column and be discharged in a gaseous state. The vapor flows out of the tower and condenses into water, forming salts.

Catalytic cracking is one of the important ways to promote heavy fuel oil, and the FCC separation system usually consists of the following systems.

(1) Reaction regeneration system
The core of the FCC system is the reverse regeneration system, which consists of a reaction unit and a regeneration unit, and the two sub-processes are carried out simultaneously and continuously.
(2) Fractionation system
The composition of the FCC unit fractionation system generally includes a main fractionation unit, a secondary diesel stripping unit, an overhead oil and gas processing unit, a recirculation and reflux unit at all levels and a heat recovery unit. Its function is to fractionate the high-temperature, wide-boiling-range mixed oil and gas at the top of the reaction settler into high-fat gas, raw gasoline, secondary light diesel oil, and bottom oil.
(3) Absorption stabilization system

The system usually consists of compressor unit, absorption tower unit, desorption tower unit, reabsorption tower unit, stabilization tower unit and other heat exchange devices.

(4) Product refining system
The system usually consists of a dry gas refining unit, an LPG refining unit, a gasoline refining unit and a diesel refining unit.

(5) Exhaust gas energy recovery system
The system generally consists of unit devices such as main fan, compressor, gas turbine and waste heat boiler. Its main function is to supply air to burn the catalyst coke, supply compressed air to transport the catalyst, etc., recover the heat in the combustion gas, and reduce the energy consumption of the fuel system.

The function of the FCC separation system is to separate the high temperature and wide boiling range mixed oil and gas from the top of the fractionator of the reaction unit to obtain dry gaseous products, LPG products, gaseous products, etc.

The main task of the catalytic cracking fractionation system is to decompose the oil and the high temperature reaction gas of the reaction regeneration system into fractions such as fatty gas, raw gasoline, light diesel oil, return oil and sludge oil according to different boiling points. Each fraction conforms to the quality index, and the potential high-temperature heat energy of each reflux flowing out of the main fractionation tower can be used as a heat source to stabilize the reboiler of the system after being absorbed [10, 11]. The process of the fractionation system process is shown in Fig. 1.

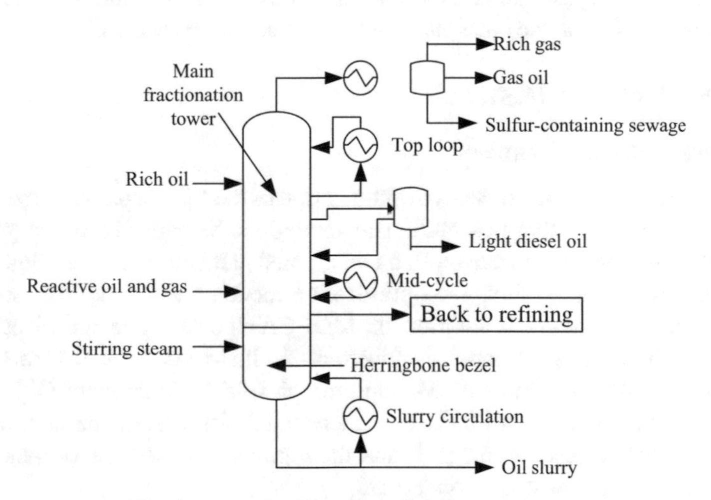

Fig. 1. Process of Fractionation System Process

In the production process of the FCC separation system, the yield of light products (light oil + LPG) has a great impact on the economic benefits. Therefore, efficiency is the main objective for optimizing FCC separation systems. The high energy consumption of the separation system accounts for a large part of the production cost, so energy consumption is the second goal of optimizing the FCC separation system. In terms of optimization, this paper looks at multi-objective optimization of multi-product combination schemes, in any scheme, efficiency and energy consumption are two optimization objectives [12].

In this paper, the following multi-objective optimization models are proposed. The formula expression of the optimization model of the catalytic cracking separation system is shown below.

Model 1: The first optimization objective is to stabilize the yield sum of gasoline and light diesel.

$$\max d_{11}(I) = A_2 + A_3 \tag{4}$$

Model 2: Take the yield sum of LPG, stable gasoline and light diesel as the first optimization objective.

$$\max d_{12}(I) = A_1 + A_2 + A_3 \tag{5}$$

Model 3: Taking the yield sum of LPG and light diesel as the first optimization objective.

$$\max d_{13}(I) = A_1 + A_3 \tag{6}$$

Model 4: Take the yield sum of LPG and stable gasoline as the first optimization objective.

$$\max d_{14}(I) = A_1 + A_2 \tag{7}$$

Among them, d11–d14 are all objective functions, I is the consumption, A1, A2, A3 are the yields of LPG, stable gasoline, and light diesel oil, respectively.

3 Experiment and Research

3.1 Experimental Environment

In this experiment, the multi-objective optimization of the FCC separation system based on particle swarm algorithm was built. The applied mathematics software Matlab and process simulation software were built, the integrated platform of Aspen Plus was used, and MOLCA was used to solve the optimization model, and the optimal solution set Pareto frontier was obtained. In addition, the MOLCA algorithm and optimization model (objective function and constraints) are constructed with MATLAB software. Generally, Aspen Plus can communicate with Matlab through COM (Component Object Model) technology. The parameters of MOLCA are selected as follows: the number of families is 20, the number of descendants is 5, and the number of evolutionary generations is 200. The shrinkage factor was taken as 0.95.

3.2 Experimental Process

In this experiment, the tests were conducted to better understand the results of multi-objective optimization of the FCC separation system. The number of experiments was 4 times in total, and the tests were conducted on the circulating reflux flow, diesel flow rate, supplementary absorbent circulation amount, and top extraction amount of the stabilizing tower in the main fractionation tower of the FCC separation system.

4 Analysis and Discussion

The number of experiments was 4 times in total, and the tests were conducted on the circulating reflux flow, diesel flow rate, supplementary absorbent circulation amount, and top extraction amount of the stabilizing tower in the main fractionation tower of the FCC separation system. The test results are shown in Table 1.

Table 1. Test Results of Multi-Objective Optimization of FCC Separation System

Test times	Circulation return flow (kg/h)	Diesel flow (kg/h)	Absorbent circulation volume (kg/h)	Stable tower top extraction (kg/h)
1	23100	41360	50130	26710
2	38710	43890	53260	29810
3	34900	47658	48370	16980
4	42410	51302	49160	19740

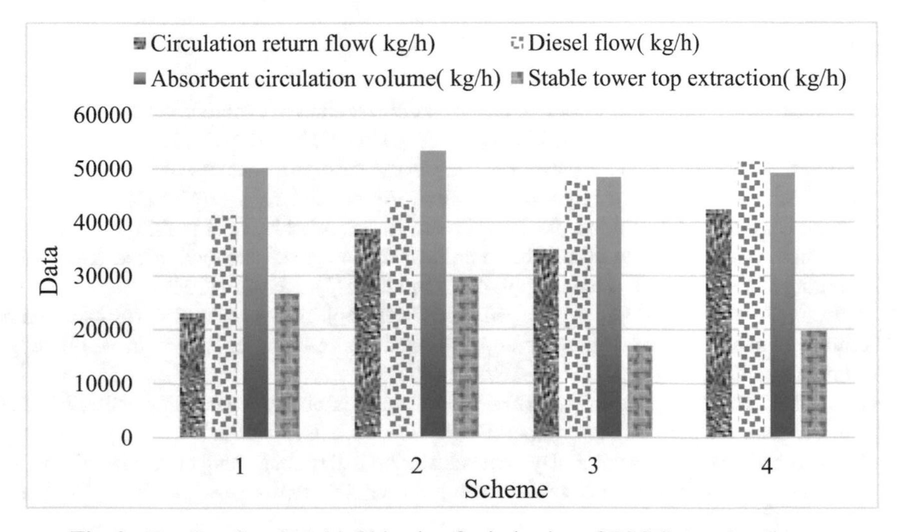

Fig. 2. Test Results of Multi-Objective Optimization of FCC Separation System

It can be seen from Fig. 2 that the recirculation flow rates in the main fractionation tower are 23100, 38710, 34900 and 42410 kg/h respectively. It can be seen that the energy consumption is effectively reduced and the yield is improved. The above results can provide an important reference for the design and optimization of the FCC separation system.

5 Conclusion

With the improvement of petrochemical companies in terms of quality and production capacity, environmental problems also follow. In today's world, with the rapid development of the economy, people's requirements for the quality of life are constantly improving, and technological progress and social productivity are also increasing. Therefore, in order to achieve the goals and tasks of sustainable utilization and recycling of resources, energy conservation and emission reduction, and environmental pollution prevention and control, the country urgently needs to solve the problem of energy shortage. Therefore, under this background, this paper studies the multi-objective optimization of the FCC separation system based on the particle swarm algorithm, hoping to provide an important reference for the design and optimization of the FCC separation system.

Acknowledgement. Project name 2: Henan Normal University 18 Doctor Start-up Project Funding (project code: 5101119170147);

Project name 3: The 2017 Youth Science Foundation Project of Henan Normal University (Project Code: 5101119170305);

Laboratory: Engineering Lab of Intelligence Business & Internet of Things, Henan Province.

References

1. Mamudu, A., Emetere, M., Ishola, F., et al.: The Production of zeolite Y catalyst from palm kernel shell for fluid catalytic cracking unit. Int. J. Chem. Eng. **2021**(7), 1–8 (2021)
2. Thomas, A., Kumar, M.: Comparison of the steady-state performances of 2 × 2 regulatory control structures for fluid catalytic cracking unit. Arab. J. Sci. Eng. **44**(6), 5475–5487 (2019)
3. Fatemi Ghomi, S.M.T., Karimi, B., Behnamian, J., Firoozbakht, J.: A multi-objective particle swarm optimization based on Pareto archive for integrated production and distribution planning in a green supply chain. Appl. Artif. Intell. **35**(2), 133–153 (2021)
4. Alfakih, T., Hassan, M.M., Al-Razgan, M.: Multi-objective accelerated particle swarm optimization with dynamic programing technique for resource allocation in mobile edge computing. IEEE Access **9**, 167503–167520 (2021)
5. Kanwal, S., Younas, I., Bashir, M.: Evolving convolutional autoencoders using multi-objective Particle Swarm Optimization. Comput. Electr. Eng. **91**, 107108 (2021)
6. Jagadeesh, S., Muthulakshmi, I.: Dynamic clustering and routing using multi-objective particle swarm optimization with Levy distribution for wireless sensor networks. Int. J. Commun. Syst. **34**(13) (2021)
7. Einy, S., Oz, C., Navaei, Y.D.: Network intrusion detection system based on the combination of multiobjective particle swarm algorithm-based feature selection and fast-learning network. Wirel. Commun. Mob. Comput. **2021**(10), 1–12 (2021)
8. Ding, W.: Neural network optimized by particle swarm algorithm for prediction of MBR filtering resistance. Comput. Sci. Appl. **11**(5), 1496–1502 (2021)
9. Li, X., Yao, Q., Lu, Z., et al.: Atomic mobilities in liquid and fcc Nd-Fe-B systems and their application in the design of quenching Nd2Fe14B alloys. Metall. Mater. Trans. A **52**(7), 1–11 (2021)
10. Sassykova, L.R., Zhakirova, N.K., Aubakirov, Y.A., et al.: Catalytic cracking using catalysts based on hetero polyacids. Rasayan J. Chem. **13**(3), 1444–1450 (2020)

11. Rao, N.T., Sankar, M.M., Rao, S.P., Rao, B.S.: Comparative study of Pareto optimal multi objective cuckoo search algorithm and multi objective particle swarm optimization for power loss minimization incorporating UPFC. J. Ambient Intell. Hum. Comput. **12**(1), 1069–1080 (2021)
12. Swpu, P.: High efficiency gas-liquid separation system for pumped wells. Petroleum **5**(2), 178–182 (2019)

Design of Prefabricated Building Performance Detection System Based on Optimized Ant Colony Algorithm

Xiaoying Wang, Xianhui Man, Wei Wang(✉), and Yanan Yi

Chongqing College of Architecture and Technology, Chongqing, China
wweiyes@126.com

Abstract. With the advanced development of national construction industrialization, prefabricated buildings are used more and more in our daily life. However, at present, there are some problems in the implementation process of prefabricated buildings that need to be optimized. In this paper, the optimization ant colony algorithm is used to solve the problems faced by the performance detection of prefabricated buildings, and a reasonable optimization scheme is proposed. The prefabricated building performance detection system designed in this paper can well detect the defects existing in the building, such as quality control defects, production structure defects, functional consumption defects and so on. Based on the in-depth analysis of various optimization objectives of prefabricated buildings, the performance detection system designed based on the optimization ant colony algorithm in this paper selects durability and energy conservation and environmental protection as the optimization objectives. The constraints include construction period constraints, rental cost constraints, and production costs. Constraints, decoration area, etc. In the analysis of the deconstruction optimization problem of prefabricated buildings, in the structural optimization of the ant algorithm, the durability and energy saving and environmental protection performance detection are added, and an optimized ant colony algorithm is proposed. The final result of the study shows that the server CPU utilization gradually increased from 10% to 69%, while the memory utilization increased gradually from 30% to 74%. From the experimental data, it can be concluded that as the number of concurrent users increases, the CPU utilization and memory utilization of the server can meet the system performance detection requirements.

Keywords: Prefabricated buildings · Ant colony algorithm · Performance testing · Optimization goals

1 Introduction

Since the 21st century, the country has achieved great success in the field of infrastructure construction. The construction industry has continued to develop and its scale has also increased. Many rural laborers have also stepped into the urban construction industry. The rapid development of the construction industry has The vigorous development of the

J. H. Abawajy et al. (Eds.): ICATCI 2022, LNDECT 169, pp. 28–35, 2023.
https://doi.org/10.1007/978-3-031-28893-7_4

national economy and the expansion of advanced productive forces have contributed a considerable amount of power [1]. However, the production of the construction industry consumes huge natural resources and causes a certain degree of damage to our country's environment. In response to this situation, Chinese researchers and professionals are committed to the optimization and upgrading of the construction industry [2]. Therefore, prefabricated buildings with excellent performance, energy saving and environmental protection have far-reaching significance for the development of my country's construction industry.

In recent years, many researchers have studied the performance detection system design of prefabricated buildings based on optimized ant colony algorithm, and achieved good results. For example, Garay R believes that the development of prefabricated buildings in my country is still in an imperfect stage, and the relevant building performance testing needs to be strengthened. The improvement of the performance of prefabricated buildings will continue to promote the rapid development of prefabricated buildings [3]. Junxia believes that the current proportion of prefabricated buildings is not high, mainly concentrated in large group companies. Due to its fast construction speed and low production cost, it has great potential for application development [4]. At present, domestic and foreign scholars have carried out a lot of research on the performance detection system of prefabricated buildings. These previous theoretical and experimental results provide a theoretical basis for the research in this paper.

Based on the theoretical basis of the optimized ant colony algorithm, combined with the analysis of the construction structure of the prefabricated building, this paper provides a reliable design for the building performance detection system. The optimized ant colony algorithm is an intelligent optimization algorithm because of its good optimization performance, which is widely used in the problem of architectural path planning. By calculating the influencing factors of building structure performance and introducing them into the algorithm, the algorithm can plan the optimal path of building structure under the premise of ensuring the feasibility of prefabricated buildings. Compared with traditional buildings, prefabricated buildings can greatly reduce the generation of construction waste, which is conducive to the promotion of national energy conservation and environmental protection policies.

2 Related Theoretical Overview and Research

2.1 Analysis of the Impact of Insufficient Performance in the Development of Prefabricated Buildings

(1) Lack of standard system for related technologies

The difference between prefabricated buildings and traditional designs is the need to create a relatively complete set of standardized application systems. At this stage, the design of prefabricated buildings still follows the traditional design ideas, not based on prefabricated buildings. Architectural characteristics Good overall design and design, lack of design process, quality of prefabricated components is not up to standard; design process is basically linear, the boundaries of various disciplines are clear, lack of coordination of implementation tasks, clear task allocation, affecting the overall construction performance of the project. There is no single standard for basic prefab building techniques. They are usually native codes, and their application and popularity are poorly accessible [5]. The imperfection of standard specifications seriously affects the quality of prefabricated buildings, and the low degree of standardization also affects the quality of prefabricated buildings. Support technology for integrated components and spare parts. Although various professional technologies (hydropower installation, lifting technology, key node connection technology, etc.) arc relatively mature, there is still a lot of room for improvement in technology integration. In the actual operation of the model house during the assembly, the professional technical safety standards formulated at this stage are combined with the integrated prefabricated construction [6, 7]. There are still many problems, such as safety protection, formwork support, scaffolding, mechanical equipment management, etc.

(2) The public's awareness of prefabricated buildings is not high

At present, the operation of prefabricated buildings is limited to individual cities, and the promotion effect has not yet reached an ideal state. Although each construction site has clarified the work objectives and specific promotion measures in the relevant prefabricated building policy documents, some cities are more "flexible" in implementation, making the project unsatisfactory. Therefore, in the decision-making stage, it is clear which projects implement prefabricated structures, as well as the general ideas and construction standards of prefabricated structures [8]. The development of prefabricated buildings is restricted by the government. State participation limits the development of prefabricated buildings, government regulation is a key factor in the development of prefabricated buildings, the government has taken punitive measures, companies have punished participation in the development of prefabricated buildings, some policies encourage companies to develop prefabricated buildings, and the government provides tax incentives Policies such as policies and financial subsidies will partially subsidize the increased costs.

(3) The prefabricated construction industry chain is immature

Design is the starting point and main link of the entire construction chain of a prefabricated building project. The quality of prefabricated components is more influenced by the design process. However, due to the immature industrial chain of the entire prefabricated building project and the imperfect system, many problems often occur in the design process. Prefabricated elements require multi-disciplinary, multi-stage completion and guided continuation of the workflow [9]. In the process of transforming the traditional operation mode to industrialization, it takes time for designers to change the traditional design concept and accept and improve the design process of prefabricated components. The entire manufacturing process of prefabricated buildings involves many participants, and each participant is involved in the design, production, transportation, and construction of components. The content of work at each stage is different, and the level of professional quality is also uneven, which often leads to poor communication and trust among professionals. Information transmission is not timely, and the quality of information cannot be guaranteed, which affects the overall cooperation process, threatens the construction quality, and delays the duration of the project.

2.2 Prefabricated Building Performance Testing System

As a new type of Web service system, WebService is based on the existing Web technology. With the addition of new protocols and standards, the introduction of new related technologies, platform-independent development platforms, language-independent platform creation of applications Cross-platform connection and integration through limited technical interfaces, and modular, independent, self-describing functions [10, 11]. The underlying Web service platform is XML (extensible) language for describing and exchanging data) + HTTP (Hypertext Internet Protocol), which can use self-describing XML data format and general Internet protocol for accessing services, not limited to specific defined Object Model Protocol. The WebService platform includes an important part of SOAP, an XML-based protocol that defines an extensible message folder format and representation to exchange information between points in a distributed HTTP environment using procedural semantic mechanisms. Therefore, in this paper the prefabricated building performance detection system adopts WebService technology [12].

3 Experiment and Research

3.1 Experimental Method

(1) Optimizing the mathematical model of the ant algorithm

The ant colony system is abstracted from the ant colony food search and only finds connections between two points or a multi-point shortcut optimization problem. The TSP problem is to find the shortest path that passes through all n buildings and passes through each city only once. This paper Take the TSP problem as an example to illustrate the mathematical model of the ant colony algorithm. While the model definition leaves open the problem of the impact of structure, it can be seen from the discussion below that this approach can easily be applied to other solving optimization problems:

$$\mathrm{d}_{ij} = \sqrt{(x_i - x_j)^2 + (y_i - y_j)^2}, 0 \le \mathrm{i} \le \mathrm{n}, 0 \le \mathrm{j} \le \mathrm{n} \tag{1}$$

In the above formula, (xi, yi) are the spatial coordinates of building i, and (xi, yi) are the spatial coordinates of building j. In the ant system, each ant is a simple agent with the following characteristics. The basis for selecting the next one during the movement process is the function of the distance dij between the buildings and the current amount of pheromone contained in the connecting branch.

(2) Pheromone function method

First, m ants are randomly assigned to n different cities, usually $m \le n$. Let $\tau(t)$ be the pheromone intensity of branch (i, j) at time t. Each ant chooses the next city at time t and gets there at time t + 1. The pheromone update formula is as follows:

$$\tau(\mathrm{t}+1) = (1-\rho)\tau_{\mathrm{ij}}(\mathrm{t}) + \Delta\tau_{\mathrm{ij}}(\mathrm{t}) \tag{2}$$

Among them, ρ $(0 < \rho < 1)$ is the pheromone volatilization coefficient. It can be seen from the above formula that the larger the pheromone volatilization coefficient value, the smaller the pheromone retained on the original path.

3.2 Experimental Requirements

This experiment tests the performance of prefabricated buildings based on the optimized ant colony algorithm, which is more similar to the method of leaving pheromones in the path of real ants in the process of searching for food. However, only experiments have found that this model for solving the TSP problem is not ideal. This experiment is mainly aimed at whether the system performance test of prefabricated buildings can meet the needs of use, and the performance indicators of different buildings are tested, and the corresponding results are obtained.

4 Analysis and Discussion

4.1 System Test Performance Analysis

The prefabricated building inspection system must provide access and operation services for a large number of different users, so it is necessary to consider whether the performance of the inspection system can carry a large number of simultaneous services to ensure the normal and stable operation of the system. The purpose of the system performance test is that when there are a large number of concurrent user access functions on the system, if the system server meets the performance indicators such as CPU usage, memory usage, and response delay (Table 1).

Table 1. System performance test table

Concurrent users	Server CPU utilization (%)	Memory usage (%)
100	10	30
200	28	40
500	48	55
1000	69	74

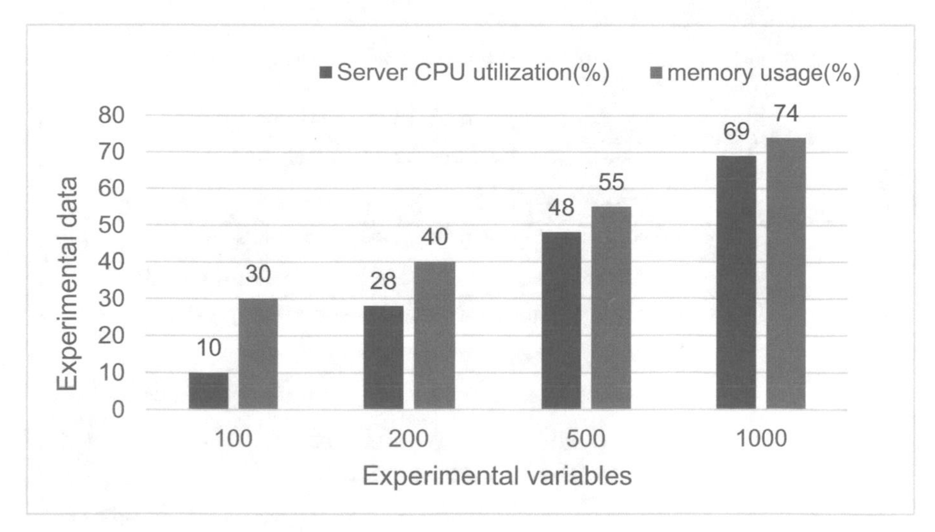

Fig. 1. System performance test figure

As can be seen from Fig. 1, as the number of concurrent users increases, the server CPU utilization gradually increases from 10% to 28%, 48%, and 69%, while the memory utilization gradually increases from 30% to 40%, 55%, and 74%. From the experimental data, it can be concluded that with the increase of concurrent users, the server CPU utilization and memory utilization can meet the system performance detection requirements.

4.2 Analysis of the Detection Accuracy of Different Building Performance Data

In order to detect the prefabricated buildings under the optimized ant colony algorithm, according to the above content, the two performances of durability and energy saving and environmental protection under different building units are counted. The results are shown in the following figure.

As shown in Fig. 2, this experiment uses the prefabricated building performance testing system to test four different buildings. The energy saving rate and durability rate of Building 1 are 85% and 88%, and the energy saving rate and durability rate of Building 2 are 62% and 85%, the energy saving rate and durability rate of Building 3 are 88% and 86%, and the environmental protection rate and durability rate of Building 4 are 71% and 80%. It can be seen from the experimental results that the prefabricated building has achieved good performance in terms of energy saving, environmental protection and durability, and the accuracy of the system is also higher than 80%, indicating that the prefabricated building performance detection system is feasible.

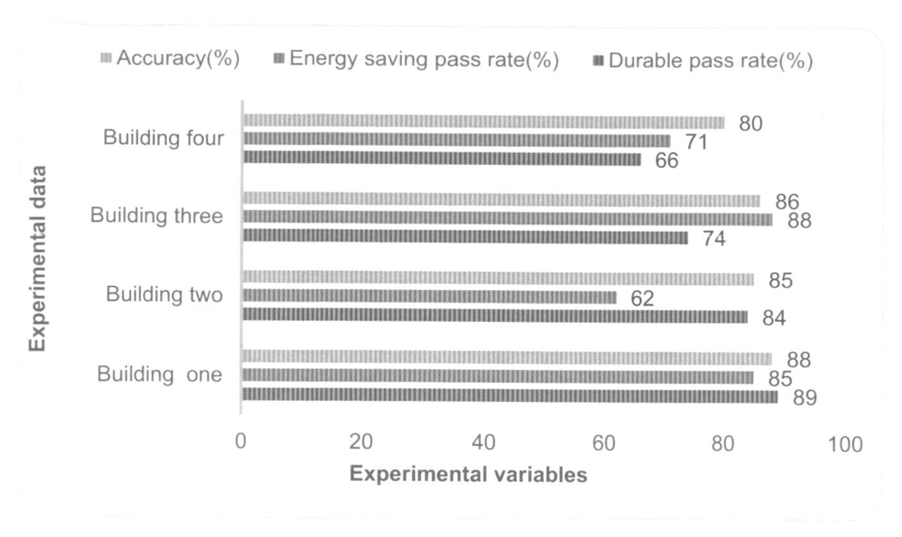

Fig. 2. Analysis of the detection accuracy of different building performance data

5 Conclusions

In this paper, the optimized ant colony algorithm is used to analyze the building performance test, and the relevant experimental simulation analysis is carried out. This shows that it can be rationally researched and developed in the future. At the same time, the performance of the detection system shows a proportional relationship with the increase of the number of concurrent users, which can meet the needs of system detection to a certain extent. After a series of tests, it is proved that the system developed in this paper can meet the actual application requirements. During testing, functionality, security, and ease of use were well documented. In addition, the background of the system. During the test, the power consumption monitoring data can be transmitted to the data center efficiently in real time, and the stability of the underlying system data transmission system is good, which ensures the scientific and rationalization of the statistical analysis of system energy consumption data. Users can query various data of the building through the system. Compared with traditional buildings, prefabricated buildings have the advantages of lower cost and greater environmental protection. Therefore, under the vigorous development of the construction industry in the future, the application of prefabricated buildings in daily life will be more and more, which also means the prefabricated building performance detection system has far-reaching research value.

References

1. Zhang, H., Cong, M., Zhang, R., et al.: The design, development and application of a prefabricated building system: the formation of building-product-model and new type of architecture. New Archit. **5**(2), 4–8 (2017)
2. Griffith, R.: Implementing offsite construction and prefabricated building systems. Des. Cost Data **62**(3), 50 (2018)

3. Garay, R., Arregi, B., Elguezabal, P.: Experimental thermal performance assessment of a prefabricated external insulation system for building retrofitting. In: International Conference on Sustainable Synergies from Buildings to the Urban Scale, no. 1, 1–9 (2017)
4. Sun, J.: The common quality problems and preventive measures of prefabricated building construction, no. (9), pp. 20–26 (2019)
5. Granderson, J., Lin, G., Harding, A., et al.: Building fault detection data to aid diagnostic algorithm creation and performance testing. Sci. Data **7**(1), 65–69 (2020)
6. Krammer, P., Cotte, C., Greber, F., et al.: Cost-efficient battery system with integrated high-performance thermal management. MTZ Worldw. **82**(12), 42–47 (2021)
7. Demirbatır, R.E., Çeliktaş, H., Engür, D.: The effect of the tuning system and instrument variables on modal dictation performance. J. Educ. Train. Stud. **6**(1), 125–130 (2017)
8. Vn, A., Rs, B.: Design of tracking system for prefabricated building components using RFID technology and CAD model. Procedia Manuf. **32**, 928–935 (2019)
9. Ogunde, A.O., Ayodele, R., Joshua, O., et al.: Data on factors influencing the cost, time performance of the Industrialized Building System. Data Brief **18**, 1394–1399 (2018)
10. Jiang, L., Qiao, R., Pan, H.: Study on the application problems and countermeasures of BIM technology in the life cycle of prefabricated buildings. Build. Struct. **12**(8), 2058–2066 (2019)
11. Tumminia, G., Guarino, F., Longo, S., et al.: Life cycle energy performances and environmental impacts of a prefabricated building module. Renew. Sustain. Energy Rev. **92**, 272–283 (2018)
12. Berrada, A., Loudiyi, K.: System performance and testing. Gravity Energy Storage **2**, 105–139 (2019)

Risk Detection System of Bridge Construction Based on Fast ICA Algorithm

Xianhui Man, Wei Wang, Xiaoying Wang(✉), and Yingjia Wang

Chongqing College of Architecture and Technology, Chongqing, China
wxyingvip@126.com

Abstract. With the increase of bridge operation risks and accidents, strengthening the risk management mechanism in the process of bridge operation and putting forward effective measures to solve the risk problem is one of the key problems to be solved at present. This paper discusses the development of fast ICA algorithm, summarizes the fast ICA algorithm and Bridge risk, analyzes the risk identification in the planning and design stage, analyzes the accidents and causes prone to the superstructure of large continuous beam bridge, and studies the high risk of bridge construction. The results show that the bridge accidents caused by construction account for 34%, which is far greater than other reasons.

Keywords: Fast ICA Algorithm · Bridge Construction · Risk Detection · Risk Identification

1 Introduction

In recent years, the construction of large highway and railway bridges has developed rapidly. At the same time, in order to adapt to different engineering environments, the requirements for bridge foundation are much higher than before.

With the continuous progress of science and technology, many experts have studied the bridge construction. For example, huethwohl P, Lu R, Brilakis I proposed a three-stage concrete defect classifier, which can classify potentially unhealthy bridge areas into their specific defect types according to the existing bridge inspection guidelines. The proposed multi classifier helps to develop major or complete inspection modes to achieve more cost-effective and objective bridge inspection [1]. Cha g, park s, oh t proposed a practical shape information model for feasibility study to monitor the deformation or deflection of bridge structure with minimum data loss and maximum calculation efficiency. The three-dimensional position information of the structure is obtained by ground laser scanning, and the octree data structure is used to process the large-scale scanning data efficiently [2]. Lu Z, Wei C, Liu m proposed a risk assessment method for cable system construction of suspension bridge based on cloud model, which can effectively combine the randomness and fuzziness of risk information. By decomposing the construction process of cable system, a multi-level evaluation index system is established. The uncertain analytic hierarchy process (AHP) is used to calculate the index weight [3]. Although the research results of bridge construction risk detection are quite rich,

J. H. Abawajy et al. (Eds.): ICATCI 2022, LNDECT 169, pp. 36–44, 2023.
https://doi.org/10.1007/978-3-031-28893-7_5

the research of bridge construction risk detection system based on Fast ICA algorithm is still insufficient.

In order to study the bridge construction risk detection system based on Fast ICA algorithm, this paper studies the fast ICA algorithm and bridge construction risk detection system, and finds the risk measurement. The results show that the fast ICA algorithm is conducive to the establishment of bridge construction risk detection system.

2 Method

2.1 Fast ICA Algorithm

(1) Development of fast ICA algorithm
The fast ICA algorithm proposed by Oja et al. Is a fast and effective algorithm. It is worth mentioning that non Gaussian plays a very important role in a large class of blind source separation algorithms called independent component analysis. Without non Gaussian distribution, these algorithms cannot be realized at all. Independent component analysis has not been widely concerned until the last two decades. The existing fast ICA algorithms study the maximum non Gaussian characteristics directly from the central limit theorem [4]. Firstly, the fourth-order cumulant (kurtosis) is introduced as the measure of non-Gaussian characteristics, the gradient algorithm based on kurtosis is deduced, and then the fixed-point algorithm with fast convergence speed is introduced. The algorithm processes the whitened data, which belongs to the indirect method. In the form of charts, complex faults are decomposed into simple single small faults, and the accident causes are analyzed and decomposed [5]. The advantage of this method is that it can intuitively and visually analyze the fault causes, but the disadvantage is that it is easy to produce omissions and errors in the analysis of large-scale complex systems. The algorithm uses a large amount of data, reduces the influence of subjective factors on the risk analysis results through various scientific algorithms, and makes the analysis results have the advantages of accuracy, strictness and objectivity. This method is widely used in aviation, nuclear, petrochemical and other industries. The application of computer and the progress of science and technology make accurate risk analysis possible, accelerate the development of quantitative risk analysis, and make it the main direction of risk analysis. The disadvantage of this method is that the analysis process is very complex, occupies a lot of resources and consumes a lot of time [6]. The accuracy of the results is controlled by the accuracy of the calculation model and the integrity of the sample data.

(2) Introduction to fast ICA algorithm
The fast ICA algorithm decomposes the constituent factors of the problem according to the requirements, hierarchizes these factors according to the correlation between the constituent factors, then analyzes them hierarchically, and finally obtains the importance weight of the lowest level factors relative to the highest level (overall goal) [7]. This complex problem can be divided into multiple levels and analyzed one by one. Moreover, the level will be much simpler than the original problem, and people's subjective judgment can be displayed and analyzed in a quantitative way. This method combines quantitative analysis with qualitative analysis, which

is a good choice for analytical research. However, the bridge itself is a complex system, especially for large bridges. During the operation period, there are many and complex factors affecting the risk, so it is impossible to use quantitative function for quantitative evaluation. In addition, many of them rely on the experience of experts for analysis [8]. Sometimes it is difficult to accurately judge complex and cumbersome risk factors. Applying the fast ICA algorithm to bridge operation risk assessment, many complex risk factors can be layered, these risk factors can be classified orderly, and an evaluation system of tomographic analysis can be established for each factor. Then, the fast ICA algorithm is analyzed and studied. Finally, the evaluation results of each risk factor are obtained. Based on the systematic analysis of the project, use the analysis idea of system engineering to find out all possible risk factors in the project, and list them in the table for inspection and review. The checklist can not only be used to judge the existence of risks, but also help to find out the causes of accidents after risks occur [9].

2.2 Risk Detection of Bridge Construction

(1) Development of fast ICA algorithm
The fast ICA algorithm proposed by Oja et al. Is a fast and effective algorithm. It is worth mentioning that non Gaussian plays a very important role in a large class of blind source separation algorithms called independent component analysis. Without non Gaussian distribution, these algorithms cannot be realized at all. Independent component analysis has not been widely concerned until the last two decades. The existing fast ICA algorithms study the maximum non Gaussian characteristics directly from the central limit theorem [4]. Firstly, the fourth-order cumulant (kurtosis) is introduced as the measure of non-Gaussian characteristics, the gradient algorithm based on kurtosis is deduced, and then the fixed-point algorithm with fast convergence speed is introduced. The algorithm processes the whitened data, which belongs to the indirect method. In the form of charts, complex faults are decomposed into simple single small faults, and the accident causes are analyzed and decomposed [5]. The advantage of this method is that it can intuitively and visually analyze the fault causes, but the disadvantage is that it is easy to produce omissions and errors in the analysis of large-scale complex systems. The algorithm uses a large amount of data, reduces the influence of subjective factors on the risk analysis results through various scientific algorithms, and makes the analysis results have the advantages of accuracy, strictness and objectivity. This method is widely used in aviation, nuclear, petrochemical and other industries. The application of computer and the progress of science and technology make accurate risk analysis possible, accelerate the development of quantitative risk analysis, and make it the main direction of risk analysis. The disadvantage of this method is that the analysis process is very complex, occupies a lot of resources and consumes a lot of time [6]. The accuracy of the results is controlled by the accuracy of the calculation model and the integrity of the sample data.

(2) Introduction to fast ICA algorithm
The fast ICA algorithm decomposes the constituent factors of the problem according to the requirements, hierarchizes these factors according to the correlation between

the constituent factors, then analyzes them hierarchically, and finally obtains the importance weight of the lowest level factors relative to the highest level (overall goal) [7]. This complex problem can be divided into multiple levels and analyzed one by one. Moreover, the level will be much simpler than the original problem, and people's subjective judgment can be displayed and analyzed in a quantitative way. This method combines quantitative analysis with qualitative analysis, which is a good choice for analytical research. However, the bridge itself is a complex system, especially for large bridges. During the operation period, there are many and complex factors affecting the risk, so it is impossible to use quantitative function for quantitative evaluation. In addition, many of them rely on the experience of experts for analysis [8]. Sometimes it is difficult to accurately judge complex and cumbersome risk factors. Applying the fast ICA algorithm to bridge operation risk assessment, many complex risk factors can be layered, these risk factors can be classified orderly, and an evaluation system of tomographic analysis can be established for each factor. Then, the fast ICA algorithm is analyzed and studied. Finally, the evaluation results of each risk factor are obtained. Based on the systematic analysis of the project, use the analysis idea of system engineering to find out all possible risk factors in the project, and list them in the table for inspection and review. The checklist can not only be used to judge the existence of risks, but also help to find out the causes of accidents after risks occur [9].

2.3 Risk Measurement

The measurement of risk is shown in formula (1):

$$R = f(p, c) \tag{1}$$

Where: R is the combined value of risk occurrence probability and risk loss level; P is probability; C is the loss caused by risk.

The most commonly used product form is the product of risk probability and risk loss to obtain risk value. As shown in formula (2):

$$R = p \times c \tag{2}$$

Based on the understanding of this formula, we can consider that the risk consequence probability P and the risk consequence caused by loss C are combined into the risk value r in some functional form.

When the continuous density of risk probability and loss can be obtained, the calculation formula of risk value composed of risk probability density and risk loss density is as follows (3):

$$R = \int \int p(x)c(y)dxdy \tag{3}$$

Where: R is the total risk value; P (x) is the density function of risk probability; C (y) is the density function of risk loss.

3 Experience

3.1 Object Extraction

The operation and maintenance of the whole early warning system also needs the participation of personnel. Therefore, it is necessary to establish a perfect and efficient early warning organization system. Therefore, it can be seen that early warning organization should be a part of bridge engineering early warning system. It is an implementation system for risk identification, analysis, evaluation and response. Girder bridge early warning system is an information circulation system, which processes the collected information and provides early warning for the project. In the large-scale bridge early warning system, the early warning organization system is a component system of bridge early warning personnel. From the monitoring of daily risk indicators to the analysis and treatment of alarm conditions, certain personnel must operate. This system is a system to determine the responsibilities and powers of various personnel; Risk monitoring system is a system used for technical monitoring and personnel risk detection. It is a necessary means to understand the alarm situation; The early warning index system is a system that sorts the importance of each risk and determines different risk monitoring intervals; Early warning evaluation system is to process the alarm situation, determine the severity of the alarm situation, and provide help for decision makers; The alarm feedback system is a system that feeds back the alarm and takes measures to control the risk; The response and post-processing system is the final or initial stage of the whole system. People take correct countermeasures through alarm feedback information, adjust the whole system while coping, and maintain the sustainability of the whole system [10].

3.2 Experimental Analysis

Firstly, before the risk analysis of bridge construction, it is necessary to conduct in-depth discussion and communication with the owner on risk events and environment, clarify the object and focus of risk assessment, understand the basic purpose of the owner's assessment, and realize targeted follow-up work. Collect project related data to lay a solid foundation for risk assessment. It can be said that the preparation of data is the basis of follow-up work and an important link of the whole risk assessment, which helps to greatly improve the efficiency and accuracy of later work. Secondly, formulate unified quantitative indicators to measure the scope of various risk losses. In order to find the index with wide adaptability, it is necessary to comprehensively classify the risk losses that may be suffered in the construction stage of Huaihe River Bridge [11]. Considering that various interest groups may have different perspectives on bridge construction risk loss, and the forms of loss are diverse, which is convenient for quantitative analysis and research. Finally, do a good job in quality inspection, strengthen the safety and quality inspection of each process, and avoid rework and related project delay caused by unqualified quality. If the quality is unqualified, it shall be handled in time to make up for the relevant losses in time.

4 Discussion

4.1 There Are High Risks in Bridge Construction

The construction process not only has a lot of uncertainty, but also has the characteristics of high accident risk. In recent years, major damage and collapse accidents often occur in the process of bridge construction in China. According to relevant investigation and analysis, the risk of bridge construction period is much higher than that of service period. According to the statistics, 916 bridge accidents are classified according to the causes of construction accidents, collision accidents, water conservancy accidents, overload accidents, design accidents, etc. See Table 1 for specific statistical results [12].

Table 1. Detailed cause statistics of bridge accidents

Accident type	Percentage
Construction	34%
Collision	21%
Water conservancy	17%
Overload	12%
Design	16%

The construction process not only has a lot of uncertainty, but also has the characteristics of high accident risk. In recent years, major damage and collapse accidents often occur in the process of bridge construction in China. According to relevant investigation and analysis, the risk of bridge construction period is much higher than that of

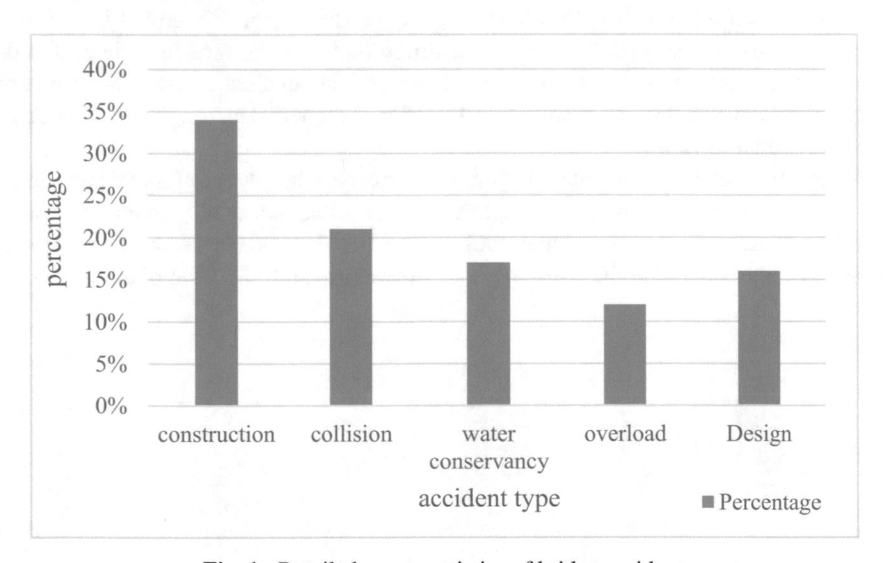

Fig. 1. Detailed cause statistics of bridge accidents

service period. According to the statistics, 916 bridge accidents are classified according to the causes of construction accidents, collision accidents, water conservancy accidents, overload accidents, design accidents, etc. See Table 1 for specific statistical results.

It can be seen from the above Fig. 1 that the number of bridge accidents caused by construction is far greater than that caused by other reasons. Considering that the construction period of bridge is far less than the service life of bridge, it can be seen that the possibility of risk accidents in bridge construction is far greater than that in other stages.

4.2 Comparison and Analysis of Actual Data

Nine experts were invited to rate the impact of risk factors contained in the main tower, steel truss and stay cable on different structural forms. Experts include professors and experts engaged in relevant work in Colleges and universities, construction unit personnel of bridge construction management, bridge designers, etc. the score of influence degree is shown in Table 2.

Table 2. Evaluation of influence degree of main tower on different structural forms

	Single side cable	Double-sided cable	Polyhedral cable	Space cable
Concrete	4	8	8	5
Construction deviation	6	6	4	4
Force majeure	7	5	9	6

Nine experts were invited to rate the impact of risk factors contained in the main tower, steel truss and stay cable on different structural forms. Experts include professors and experts engaged in relevant work in Colleges and universities, construction unit personnel of bridge construction management, bridge designers, etc. the score of influence degree is shown in Table 2.

As can be seen from the above Fig. 2, force majeure has the greatest impact on the four structural forms, followed by construction deviation, and finally concrete. Among the four structural forms, the single cable plane has the highest correlation with the secondary index related to the main tower, followed by multiple cable planes.

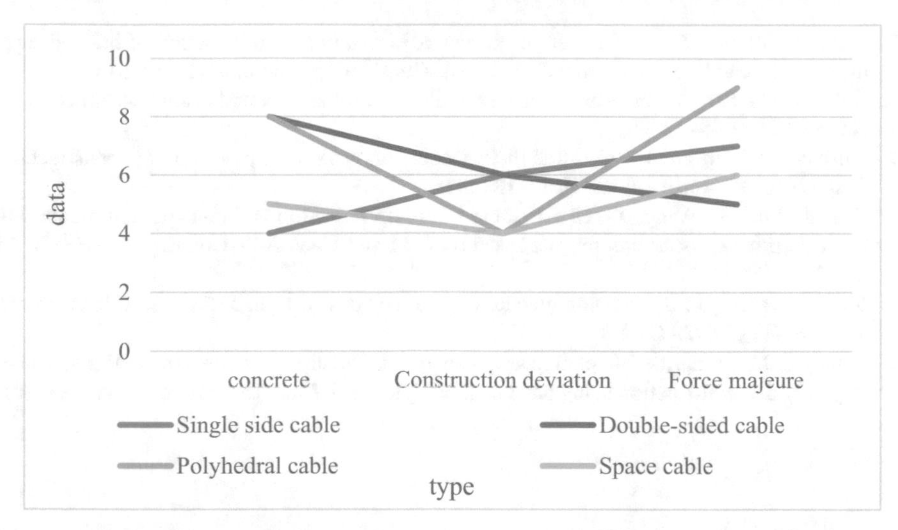

Fig. 2. Evaluation of influence degree of main tower on different structural forms

5 Conclusion

With the continuous development of transportation industry, bridges play a more and more important role in highway transportation. However, large bridge construction accidents occur frequently in China, which not only cause huge economic losses, but also cause a large number of casualties. Risk early warning of large bridges under construction is an important means to ensure construction safety and avoid accidents. This paper studies the influence of the risk factors contained in the main tower, steel truss and stay cable on different structural forms. The results show that force majeure has the greatest impact on the four structural forms.

References

1. Huethwohl, P., Lu, R., Brilakis, I.: Multi-classifier for reinforced concrete bridge defects. Autom. Constr. **105**, 102824.1–102824.15 (2019)
2. Cha, G., Park, S., Oh, T.: A terrestrial LiDAR-based detection of shape deformation for maintenance of bridge structures. J. Constr. Eng. Manag. **145**(12), 04019075.1–04019075.12 (2019)
3. Lu, Z., Wei, C., Liu, M., et al.: Risk assessment method for cable system construction of long-span suspension bridge based on cloud model. Adv. Civ. Eng. **2019**(4), 1–9 (2019)
4. Xu, Y., Kong, F., Gao, K., et al.: The mechanism of mudstone skin friction of large-diameter and long piles based on the pile test of the Longhua Songhua River Bridge in Jilin Province, China. Arab. J. Geosci. **14**(22), 1–12 (2021)
5. Li, X., Guo, X., Sun, G.: Grouting reinforcement mechanism and multimodel simulation analysis of longwall goaf. Geofluids **2021**(5), 1–13 (2021)
6. Ghodoosi, F., Abu-Samra, S., Zeynalian, M., et al.: Maintenance cost optimization for bridge structures using system reliability analysis and genetic algorithms. J. Constr. Eng. Manag. **144**(2), 04017116.1–04017116.10 (2018)

7. Wang, L., Zhang, Z., Han, H., et al.: Recent advances in the construction of bridged rings through cycloadditions and cascade reactions. Chin. J. Org. Chem. **41**(1), 12–51 (2021)
8. Culmo, M.P.: Development of national specifications for accelerated bridge construction. Tr News (320), 49–54 (2019)
9. Tiernan, T.: Fifth circuit overturns district court injunction on bayou bridge construction. Foster Nat. Gas Rep. (TN.3207), 15–18 (2018)
10. Tian, J., Luo, S., Wang, X., et al.: Crane lifting optimization and construction monitoring in steel bridge construction project based on BIM and UAV. Adv. Civ. Eng. **2021**(8), 1–15 (2021)
11. Venkateswaran, B.: Sustainable practices in bridge construction. J. Sustain. Constr. Mater. Technol. **6**(1), 24–28 (2021)
12. Sunaryo, O.L., Magribi, M., et al.: Design analysis of loading structures in the Baruta cable-stayed bridge construction using hand method. Int. J. Sci. Eng. Res. **11**(5), 829–837 (2020)

Research on Optimization of Low-Carbon Logistics Distribution Route Based on Genetic Algorithm

Xiaocui Deng[1(✉)] and Dandan Li[2]

[1] Chongqing College of Architecture and Technology, Chongqing, China
dxcinfo@yeah.net

[2] Sichuan Vocational College of Information Technology, Guangyuan, Sichuan, China

Abstract. Genetic algorithm is a search random algorithm based on the principle of natural evolution. Genetic algorithm is of great significance in terms of location problems, distribution problems, scheduling problems, transportation problems, and layout problems. Based on establishing a mathematical model, a genetic algorithm to solve this problem is constructed. The genetic algorithm uses a commonly used binary code and combines the optimal individual retention strategy and the roulette method in individual selection. Finally, experimental calculations are carried out with this method, and the calculation results show that, using genetic algorithm to optimize the logistics distribution path can easily and effectively find the optimal solution or approximate optimal solution for the problem.

Keywords: Genetic Algorithm · Low Carbon · Logistics · Distribution Path

1 Introduction

Logistics refers to the use of modern information technology and equipment to move items from the place of supply to the place of receipt in an accurate, timely, safe, quality and quantity, door-to-door rationalized service mode and advanced service process [1]. With the laws of physical flow of material materials and applying the basic management principles and scientific methods in the process of social reproduction, so as to achieve the most efficient logistics activities [1]. When studying the logistics distribution path, we often attribute it to combinatorial optimization, which is a NP-complete problem. In the research process, people often use various methods such as scheme evaluation method, dynamic programming method, genetic algorithm and so on for path optimization [2]. However, these algorithms all have a certain degree of shortcomings and are not particularly perfect. In the current research, some algorithms of genetic algorithm in logistics distribution path optimization are improved and perfected, so that the existing shortcomings and deficiencies of the original algorithm can be made up.

J. H. Abawajy et al. (Eds.): ICATCI 2022, LNDECT 169, pp. 45–53, 2023.
https://doi.org/10.1007/978-3-031-28893-7_6

2 Logistics Distribution Routing Problem Description

For the logistics distribution problem, we can generally describe it as follows: In order to meet a certain distribution goal, a vehicles number are sent from the logistics distribution center, and a more suitable transportation route is selected [1]. Under the conditions of mileage, mileage, etc., to complete the distribution goal, it is required to make the formal route to minimize the total transportation cost, and to meet the following constraints:

(1) When the center of distribution sends out a transportation vehicle from the center after completing a distribution task, the transportation vehicle returns to the center of distribution after completing the tasks arranged by the center of distribution.
(2) The maximum transportation volume of transportation vehicles on each route should be greater than or equal to the one-time delivery demand of each station on the route.
(3) The length of each delivery route shall be controlled within the maximum travel distance of a single delivery of the transport vehicle.
(4) There is one and only one transport vehicle to deliver goods to each customer.

3 Mathematical Model of Logistics Distribution Vehicle Scheduling Problem

According to the road traffic conditions of the road section of the logistics distribution, the reasonable logistics distribution path is selected according to the predicted road conditions. The logistics distribution path is optimized. The traffic data is divided into multiple sample groups according to time periods. Among them, the traffic data from 5 to 22 during the day has research value. Therefore, the samples are divided into groups according to time. The road traffic network mainly considers the topological structure of the distribution information, which requires the departure time, the topological structure of the road at the distribution location, and the predicted traffic cost weight of the road section at each moment. It constitutes a time-sharing and weighted traffic network model [3, 4]. At this time, the logistics distribution path optimization problem is transformed into the traveling salesman problem. According to the complexity of the network model. Choose the appropriate algorithm. Finally, find the optimal delivery route.

Prediction of traffic conditions needs to consider weather, road construction, holidays and commuting peaks, etc., considering the same weather conditions. Road traffic on the road sections is affected to varying degrees. For example, when it is raining. Some sections of the road are high. The road is wide. Traffic is less affected. Some road sections have low terrain and narrow roads, which seriously affect road traffic therefore. If a general influence value is set for the weather and meteorology is inappropriate. It is necessary to set the weather influence value by road section. The new version of the meteorological disaster warning signal has changed from five levels to four levels. They are represented by blue, yellow, orange, and red respectively, indicating general, heavy, serious and particularly serious, and are marked in both Chinese and English, which are consistent with all emergency response levels and colors of the country [4]. Therefore, we use the weather impact value according to the warning level into four values. Set the normal weather traffic cost value for working days as the standard value, the warning weather cost value, and the weather impact value as the ratio of the road section cost value during the warning weather period to the normal weather era value. The formula is as follows:

$$T_i = \frac{T_q}{T_0}(q = 1, 2, 3, 4, \ldots\ldots)$$

4 Genetic Algorithm Design of the Model

4.1 Encoding and Decoding

Vehicle routing with time windows is an optimal combination problem based on optimal sorting. To facilitate research and reduce the generation of invalid solutions, natural integer coding is used for chromosomes, that is, ordinal coding [3]. The specific method is as follows: Number the n distribution points in sequence, denoted by 1, 2, ... , n respectively, the vehicles number is m, and the center of distribution is denoted by 0. The chromosome code string after integer coding can be expressed as:

$$(0, i_{11}, i_{12}, \ldots\ldots, i_{1a}, 0, i_{21}, i_{22}, \ldots\ldots, i_{2h}, 0, i_{m1}, i_{m2}, \ldots\ldots, i_{mm})$$

Among them, the chromosome length is n + m + 1. The two adjacent zeros in the chromosome represent a sub-path, that is, the vehicle starts from the center of distribution, completes the customer point distribution service responsible for the vehicle in turn, and then returns to the center of distribution. For example, the chromosome code string for delivery to 9 delivery points and 3 vehicles is (0397012604850), which means that the delivery service to 9 delivery points is completed by 3 vehicles, and there are 3 sub-routes in total. The path arrangement of the corresponding 3 sub-paths is:

Subpath 1: $0 \to 3 \to 9 \to 7 \to 0$
Subpath 2: $0 \to 1 \to 2 \to 6 \to 0$
Subpath 3: $0 \to 4 \to 8 \to 5 \to 0$

The genetic space processed by the genetic algorithm is converted into the actual problem solution space to realize the mapping of the phenotype to the genotype.

4.2 Design of Genetic Operator

(1) Assuming that the population size is N and the fitness of individual i is F i, the probability P i that individual i is selected to be inherited to the next generation population is [4]:

$$P_i = F_i \div \sum_{k=1}^{N} F_i$$

(2) Crossover operator. The genetic algorithm coding of the model uses integer coding; the crossover operator uses Partially Matched Exchange (PMX). PMX differs from traditional crossover operations in that instead of directly swapping two matching regions, the regions to be swapped are placed in front of the first gene of the opposing bit string (such as Fig. 1) [5]. Then, remove the same gene from the original individual as the exchanged gene segment, to obtain the individual after the crossover. The crossover probability is 0.6. The following example illustrates the operation process of PMX:

$$\mathrm{A} = 98|7654|321 \rightarrow \mathrm{A}' = 3456(987654321) \rightarrow \mathrm{A}'' = 3456|98721$$

$$\mathrm{B} = 12|3456|876 \rightarrow \mathrm{B}' = 7654(123421112) \rightarrow \mathrm{B}'' = 7436|97722$$

Randomly generate two chromosomes A and B, generate two crossover points, and place the genes in the crossover segments in front of each other's chromosomes. Then, remove the same genes as the first one, and get new individuals A″ and B″.

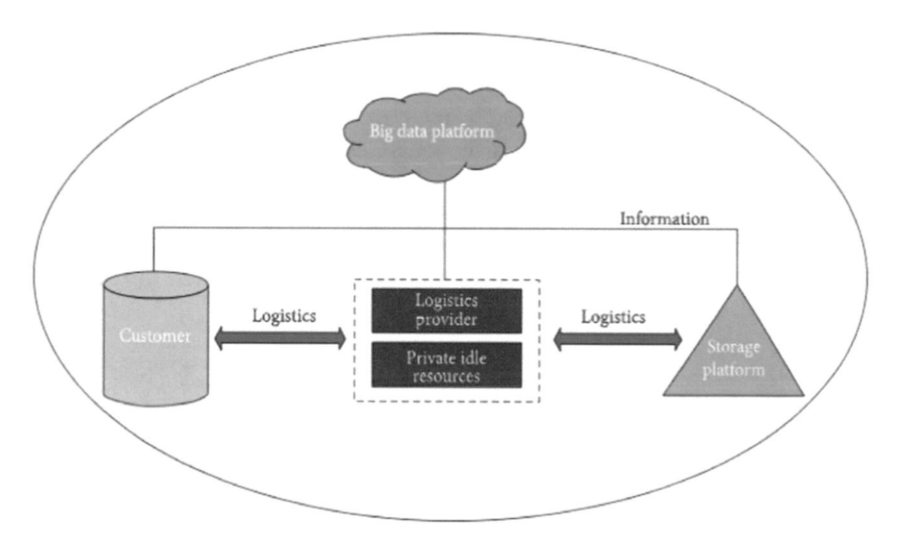

Fig. 1. logistics distribution logical

(3) Mutation operator. Use inverted mutation operator to perform mutation operation. The process is to randomly select two points in a chromosome, perform a complete reverse order operation on the part between the two points, and then realize the idea of mutation in the algorithm and obtain a new individual. Since mutation only plays a supporting role in the evolution of species, the possibility of species mutation is very small [5]. Therefore, the probability of mutation operator in genetic algorithm is smaller. The probability of mutation operator is 0.005. The inverted mutation operator can be illustrated by the following example:

$$A = 98|7654|321 \rightarrow A' = 3456(97|7654|321)$$

4.3 Fitness Function

If the fitness is non-negative, the objective function value needs a certain conversion before it can be used as the fitness function [6]. Therefore, the objective function needs to be converted into fitness. Use the reciprocal method to convert:

$$F_i = \frac{A}{A_i}$$

Among them, F i represents the fitness value of the i-th representative chromosome, and the denominator A i represents the objective function value. The smaller the value of the objective function, the larger the fitness function value, and the more in line with the model optimization goal.

5 The Construction of Logistics Distribution Path Based on Genetic Algorithm

(1) Determination of coding method.
Directly produce L permutations of non-repeating natural numbers between 1 and L, and the permutation constitutes an individual [6, 7]. The elements (customers) in the individual can be divided into each distribution path in turn.
Assuming that the customers in a certain body are arranged as 41235, the corresponding distribution route scheme can be obtained by the following method: first, add customer 4 as the first customer to the distribution route 1, and then judge whether it can meet the constraints of the problem, that is, customer 4 whether the demand of the first vehicle exceeds capacity of the first vehicle, and whether the length of the route 0-4-0 exceeds the maximum travel distance of the first vehicle for one delivery, if it can be satisfied, then customer 1 can be regarded as the third customer add to path 1, and then judge whether the constraints of the problem can be satisfied. If it can still be satisfied, customer 2 can be added to path 1 as the third customer. If the constraints of the problem can still be satisfied, the customer can be added to the path 1. 3 is added to route 1 as the fourth customer. Suppose that the constraints of the problem cannot be met at this time [7], which means that customer 3 cannot be added to route 1, so the first delivery route is feasible: 0-4-1-2- 0.

(2) Determination of the initial group.
Randomly arrangement of L non-repeating natural numbers from 1 to L to form an individual. Assuming that the size of the group is N, the initial group can be formed by randomly generating N such individuals.

(3) Fitness assessment.
In the corresponding distribution route scheme and the delivery vehicles number M (if the distribution routes number ≤ the total delivery vehicles number, take M = 0, indicating the distribution routes number > the total vehicles number, then M > 0), and its objective function value is Z, and M is regarded as the infeasible route of the distribution route scheme corresponding to the individual. And set the as Pw, then the fitness F can be calculated by the formula.

$$F_j = \frac{1}{Z_j + M_j} \times p_w$$

(4) Select operation.
The selection strategy is adopted: arrange the individuals N in each generation group by their fitness, and the individual ranked first has the best behavior, and copy it to a direct enter rank; the other N-1 individuals in the next generation group need to be generated by the wheel selection method through N individuals [8]. The sum of all individual fitness ΣFj, the fitness of each individual Fj/ΣFj ($j = 1, 2, \cdots, N$).
(5) Chromosome reorganization.
In addition to the best individual ranked first, the other N-1 individuals should be paired and recombined through the crossover probability Pc, and the operation method is illustrated with an example:

1) A mating area randomly selection among the parent individuals. E.g. the two parent individuals and the mating area are selected as: A = 47|85631|921, B = 83|46911|257;
2) Add the B mating area to the A front, and add the A mating area to the B front, to get: $A' = 46911|478563921$, $B' = 85631|834691257$;
3) Delete the same natural numbers from the mating area in A' and B' in turn and get the final two bodies: $A'' = 469178532$, $B''= 856349127$.

(6) Mutation operation
Using the continuous multiple swaps mutation technique, the individual has a great change in the arrangement order [9].

6 Optimization of Low-Carbon Logistics Distribution Route Based on Genetic Algorithm

Through the intelligent logistics distribution system including the server, the vehicle-mounted mobile intelligent terminal installed in the cargo compartment of the logistics distribution truck, the sender's, the logistics personnel's mobile intelligent terminal and the recipient's mobile intelligent terminal, the logistics distribution path can be effectively optimized. As shown in Fig. 2, the waste of mailing orders is greatly reduced, thereby greatly reducing the production cost, and the items can be monitored and queried in real time during the transportation process to ensure that the responsibility of the person responsible for the items is clear and prevent the loss of the items; recipients and

The sender can check the location, video, etc. of the item at any time, and it is also convenient for the logistics company to effectively solve the best way to transport the item, such as placement, driving route, etc.

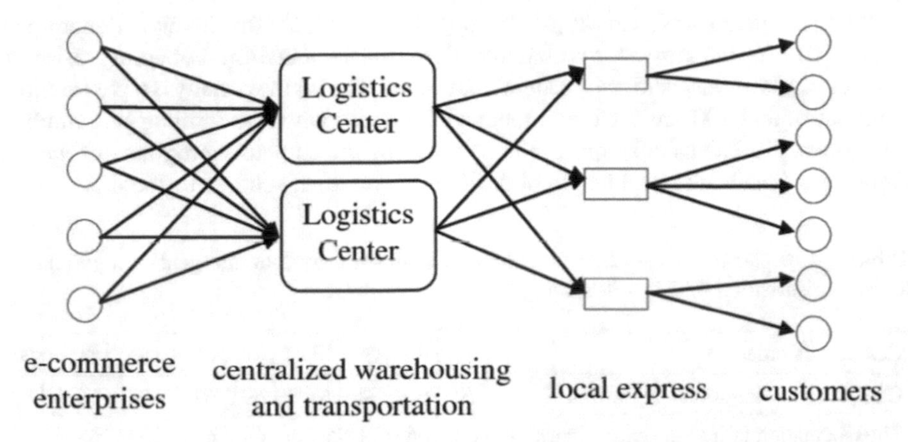

Fig. 2. Low-carbon logistics distribution path optimization logic

6.1 Model Establishment

According to the vehicle routing problem involved in the model, the following hypotheses are proposed:

(1) Distribution center assumption: There is a single distribution center, and several vehicles are required to complete the distribution to each customer point.
(2) Distribution point assumption: The distance between the distribution points and the center of distribution is known, and the demand for goods at each distribution point and the required delivery time window are known.
(3) Vehicle hypothesis: The vehicle departs from the center of distribution and returns to the center of distribution uniformly after completing the goods distribution service to the customer [8].
(4) Assuming that the goods can maintain a constant low temperature during the delivery process, that is, the damage of the goods is only time required for the delivery of the goods.
(5) Road network assumption: The roads in this paper are standard grid-like roads.
(6) Assumption of the temperature inside and outside the refrigerated vehicle: Since the products delivered by the vehicle are of a single category, and the vehicle is uniform, the temperature does not change during transportation.

6.2 Experimental Calculation and Result Analysis

There are 2 distribution vehicles in a logistics center, their load capacity is 8t, the each distribution vehicle maximum driving distance is 50 km, the distance between the center of distribution (its number is 0) and 9 customers and between 9 customers dij and the goods demanded by 9 customers qj (i, j = 1, 2, ···, 8) are shown in Table 1. It is required to arrange the vehicle distribution route reasonably so that the total distribution mileage is the shortest [9].

The parameters are used: 20 for the population size, 25 for the evolutionary generation, 019 for the crossover probability, 0109 for the mutation probability, 5 for the number of gene transpositions during mutation, 100 km for the penalty weight for infeasible paths, and 100 km for the number of iterations when performing hill-climbing operations. 20. The calculation results obtained by the computer programs of the two algorithms (that is, the total length of the delivery route) are shown in Table 1.

Table 1. Comparison of calculation results of traditional algorithm and genetic algorithm for logistics distribution vehicle scheduling problem

Calculation order	①	①	③	④	⑤	⑥	⑦	⑧	⑨
Calculation of traditional algorithm	70	69	74	67	69	67	71	70.5	69
The calculation of the surviving algorithm	68	66	71	67.5	68	67.5	71	68	66

The calculation results of the 9 customers of the genetic algorithm are better than the results of the traditional algorithm [10]. Among the 9 calculations, the genetic algorithm has obtained the optimal distance of 67.5 km for the problem twice and obtained the optimal distance of the problem twice. The suboptimal distance is 66 km. This fully shows that the genetic algorithm has good search efficiency and optimization performance. In addition, an example of a distribution center delivering goods to 9 demand points has been experimentally calculated, and the calculation shows the same effect. The two optimal solutions of the problem can be easily obtained by using the genetic algorithm, and the total length of the distribution path is 66 km. Due to the relatively strong constraints of this problem, using traditional algorithms to solve, sometimes even a feasible solution cannot be obtained [10].

7 Conclusion

Based on the establishment of a mathematical model of the logistics distribution vehicle scheduling problem, this paper constructs a genetic algorithm to solve the logistics distribution selection problem in view of the low efficiency of traditional algorithms in searching for complex problems and the shortcomings of "premature convergence". The experimental calculation results show that the genetic algorithm can overcome the above shortcomings of the traditional algorithm, thereby obtaining better results than the traditional algorithm, fully demonstrating its good optimization performance.

Acknowledgments. Project: Research on the development of the intelligent cold chain logistics system of fresh agricultural products in Guangyuan under the strategy of rural revitalization, and the research on the optimization of logistics packaging unit system under the perspective of supply chain.

References

1. Mohammadi, M., Shahparvari, S., Soleimani, H.: Multi-modal cargo logistics distribution problem: decomposition of the stochastic risk-averse models. Comput. Oper. Res. **131**, 105280 (2021)
2. Stan, M., Borangiu, T., Raileanu, S.: Data- and model-driven digital twins for design and logistics control of product distribution. In: International Conference on Control Systems and Computer Science, pp. 33–40 (2021)
3. Al Theeb, N., Smadi, H., Al-Hwari, T., et al.: Optimization of vehicle routing with inventory allocation problems in cold supply chain logistics. Comput. Ind. Eng. **142**, 106341 (2020)
4. Rodriguez-Melquiades, J., Lujan, E., Segura, F.G.: Sustainable optimization model for routing the process of distribution of products, pickup and transport of waste in the context of urban logistics. In: Gervasi, O., et al. (eds.) ICCSA 2021. LNCS, vol. 12952, pp. 91–106. Springer, Cham (2021). https://doi.org/10.1007/978-3-030-86973-1_7
5. Marchet, G., Melacini, M., Perotti, S., Rasini, M., Tappia, E.: Logistics in omni-channel retailing: modelling and analysis of three distribution configurations. In: International Conference on Service Operations, Logistics and Information Technology, pp. 21–26 (2017)
6. Serna, F., Catalán, C., Blesa, A., et al.: Control software design for a cutting glass machine tool based on the COSME platform. Case study. In: International Conference on Automation Science and Engineering, pp. 501–506 (2011)
7. Cortés, J.A.Z., Serna, M.D.A., Serna, C.A.U.: Multiobjective model to reduce logistics costs and CO2 emissions in goods distribution. Res. Comput. Sci. **147**(3), 75–85 (2018)
8. Alnami, H.M., Mahgoub, I., Al Najada, H.: Enhanced vehicle handling performance for an emergency lane changing controller in highway driving. In: Intelligent Vehicles Symposium, pp. 334–340 (2017)
9. Aijaz, A.: Infrastructure-less wireless connectivity for mobile robotic systems in logistics: why bluetooth mesh networking is important? Comput. Res. Repos. abs/2107.05563 (2021)
10. Abosuliman, S.S., Almagrabi, A.O.: Routing and scheduling of intelligent autonomous vehicles in industrial logistics systems. Soft Comput. **25**(18), 11975–11988 (2021)

Optimization Analysis of Uneven Settlement in Pipe Gallery Based on Ant Colony Algorithm

Haobo Yu[1], Xiaoxi Liu[1], Yuhan Qian[2(✉)], and Yue Liu[3]

[1] University of Science and Technology Beijing, Beijing, China
[2] Aerospace Times FeiHong Technology Company Limited, Beijing, China
yuemailcn@163.com
[3] Changchun University of Technology, Changchun, Jilin, China

Abstract. Nowadays, in many emerging cities in my country, in view of the limitations of economic development at that time, some basic municipal facilities (including pipelines, pipelines, etc.) are laid in the shallow underground soil layer of the road. The expansion of pipeline capacity can not only improve the economical use of urban space, but also an important infrastructure to ensure urban development. However, the uneven settlement of the assembled comprehensive pipe gallery will greatly reduce the performance of the pipe gallery. Therefore, it is an important content to be considered in this paper to study the influence and control of the uneven settlement on the assembled comprehensive pipe gallery. In this paper, after soil survey of a certain piece of land, four monitoring points are selected to detect the influence of the load on its response value and the influence of the underlying weak stratum on the settlement of the pipeline are detected, and the ant colony algorithm is expected to find the optimal method to avoid the settlement of the pipe gallery.

Keywords: Ant Colony Algorithm · Uneven Settlement · Spliced Integrated Pipe Gallery · Load Effect

1 Introduction

Uneven settlement of the foundation will deform the buried pipeline, resulting in pipeline damage or water leakage. However, the underground pipeline structure has a propulsive effect on the development of modern metropolises. Only by eliminating the repeated excavation of the road surface and alleviating traffic congestion can it help to improve the comprehensive capacity of the city and create new economic stimulus for the development of the city.

At present, many scholars have studied the uneven settlement based on ant colony algorithm, and have determined good research results. For example, a scholar used the Winkler elastic beam calculation model to calculate the stress of the buried pipeline, calculated the deformation of the pipeline under the condition of sediment, and obtained the mechanical response characteristics of the pipeline such as longitudinal bending moment, shear force, and angular rotation. However, the Winkler foundation beam model still has

J. H. Abawajy et al. (Eds.): ICATCI 2022, LNDECT 169, pp. 54–62, 2023.
https://doi.org/10.1007/978-3-031-28893-7_7

flaws. First, under conditions of uneven soil deposition, it is difficult to determine the stiffness of the soil around the pipe. Second, the soil around the pipeline is considered to be an independent "soil source", ignoring the forces of interaction between soils [1]. A certain scholar studied the bending moment distribution of buried pipelines under the condition of uneven foundation settlement, and proposed the design method of continuous pipelines and the upper limit solution of pipeline bending moments. A related theoretical study is carried out on the response of the pipeline structure to the layering effect caused by the deformation. The main idea of applying theoretical calculation method to study and analyze the influence of non-uniform settlement of buried pipeline is as follows: apply external force to the buried pipeline, and then calculate the internal force distribution of the flexible deformation buried pipeline. In fact, when the foundation settles unevenly, it is often difficult to determine the external force on the buried pipeline, which makes the above theoretical calculation method directly applicable to practical engineering [2]. Although the research on uneven settlement of pipelines based on ant colony algorithm is very effective, it is necessary to use ant colony algorithm to find the optimal pipeline placement position when facing the problem of pipeline settlement, which provides a new idea for the settlement prediction of underground pipelines in the future.

This paper introduces the concept of ant colony algorithm, and analyzes the reasons for the uneven settlement of the pipe gallery, such as groundwater level changes, vibration loads, water leakage, etc., which can cause the pipeline to settle, and then take the settlement of the spliced pipe gallery as an example to analyze different loads. Under the action of the stress value of the pipe gallery and the influence of different underlying soil qualities on the settlement of the pipe gallery, the ant colony algorithm is used to find the stress point of the pipe gallery to reduce the settlement.

2 Ant Colony Algorithm and Pipe Gallery Settlement

2.1 Ant Colony Algorithm

Ant colony algorithm is a biometric random search algorithm. The movement of ants is chaotic, and it is more difficult to find food alone, but when the whole ant colony works as a group, the results and efficiency of finding food will be different [3]. When the number of ants on the road increases, the more pheromone left behind, other ants can find their way according to the concentration of pheromone. The ant colony algorithm randomly selects the path without prior information, and the research on the path begins to become normal, and gradually tends to the optimal path [4].

2.2 Cause Analysis of Uneven Settlement of Pipe Gallery

When constructing the underground prestressed splicing integrated pipe gallery, the excavation of the trench will cause varying degrees of disturbance to the undisturbed stratum and surrounding soil quality of the underlying layer, which is the fundamental reason for the deformation of the base layer and the surface subsidence [5]. By reviewing the literature summary, the following are several reasons that may cause the uneven settlement of thc pipe gallery.

(1) Uneven distribution of the underlying soil layer

Assuming that the soil quality of the base layer under the pipe gallery is the same, under the same conditions, the overall longitudinal settlement of the assembled pipe gallery will be almost the same, and the longitudinal settlement of the entire pipe gallery will not cause water leakage at the joints of the pipe section. However, the actual engineering situation is that the design length of the underground assembled pipe gallery is long, so the uneven distribution of the underground soil conditions is inevitable. The consolidation characteristics of the layers are different, which leads to the long-term consolidation of the underlying soil layer, the consolidation settlement of different sections may be quite different, so that the time required to achieve the final consolidation and stability will not be long. Therefore, the bottom plate of the integrated pipe gallery may be built on sandy soil with better mechanical properties, or it may be built on the large-porous loess area, silt mud, residual soil or silty clay, resulting in the integrated pipe gallery due to the longitudinal soil quality. Different consolidation characteristics result in uneven settlement. If the pipe gallery is located on the saturated sandy soil with larger compressive modulus, the settlement after disturbance is smaller and the duration is shorter. If the pipe gallery is located on silt soil and soft clay with a small compressive modulus, the settlement will be larger after soil disturbance and the settlement time will be longer. Changes in soil types [6, 7].

(2) Changes in the groundwater level of the stratum where the integrated pipe gallery is located

Because the pipe gallery is below the groundwater level or in a complex stratum such as impermeable, the corresponding changes in the groundwater level over time will also cause the subsidence of the pipe gallery. The rapid decline of the position and the subsidence of the ground cause the structural damage of the assembled pipe gallery [8].

(3) Influence of construction activities adjacent to the underground integrated pipe gallery

The underground comprehensive pipe gallery is generally buried in the downtown area and economically developed areas of the city. The nearby foundation pits and the row upon row of high-rise buildings disturb and squeeze the soil around the pipe gallery, generating new additional loads on the soil layer, causing soil around the pipe gallery.

(4) Settlement of collapsible loess caused by leakage of water from pipelines

In the physical state, the loess wetted by water will settle directly. Therefore, in the area where the pipe gallery passes, special treatment must be taken according to the possible flooding, so as to avoid the collapsible settlement of the substratum loess under the pipe gallery caused by the leakage of the pipe.

2.3 Main Settlement Types of Assembled Pipe Gallery

(1) Saddle settlement of pipe gallery

The saddle-shaped settlement of the assembled pipe gallery is mainly because the pipe gallery encounters two discontinuous areas with poor soil properties when it passes through the longitudinal substratum soil layer, resulting in the undulating posture of the pipe gallery, resulting in assembly The prestressed tendons reach the ultimate tensile strength, which in turn causes the damage of the pipe gallery structure. After the saddle-type settlement of the pipe gallery, the segments dislocate each other. The maximum dislocation amount occurs in the substratum where the soil properties are different, and the dislocation of some pipe gallery segments far from the places with poor soil properties is small. The impact is limited; the upper part of the segment is open, and the lower part is pressed against each other, which will cause the waterproof rubber interface pressure on the upper part of the pipe gallery to fail to meet the standard and reduce the waterproof effect, thereby causing water leakage and other diseases. The waterproofing of the pipe gallery with different physical properties should be strengthened [1, 9].

(2) Concave settlement of pipe gallery

The concave settlement of the assembled pipe gallery is mainly because the pipe gallery encounters an area with poor body properties when it passes through the longitudinal substratum soil layer, causing a certain section of the pipe gallery to sink, resulting in the assembled pipe gallery. The segmental dislocation causes the prestressed tendons to reach the ultimate tensile strength, which in turn causes the damage of the pipe gallery structure [10].

(3) Gradual settlement of pipe gallery

The gradual subsidence of the assembled pipe gallery is mainly because the pipe gallery encounters an area with poor soil properties and a wide range when it passes through the longitudinal substratum soil layer, resulting in the gradual sinking of the pipe gallery, which in turn caused the damage of the pipe gallery structure. After the gradual subsidence of the pipe gallery, diseases such as water leakage will occur, but this kind of subsidence is less damaging than the previous two types of subsidence.

(4) Sudden settlement of pipe gallery

The main reason for the sudden subsidence of the assembled pipe gallery is that the pipe gallery encounters complex geological changes such as ground fissures when it passes through the longitudinal substratum. The structure of the pipe gallery is seriously damaged, which leads to problems such as leakage of the pipe gallery and the effectiveness of the rubber sealing strip. When the sudden settlement of the pipe gallery occurs, the segment of the pipe gallery above the ground fissure is dislocated, which causes the structure of the pipe gallery here to be damaged, and causes excessive prestress tension, which is easy to damage the steel and concrete materials; The pipe gallery segments that are relatively far away from the ground fissure are less affected, and measures such as setting seismic joints in the pipe gallery passing through the cracks and replacing and compacting the underlying soil [11].

2.4 Surface Stress of Prefabricated Pipe Gallery

Underground pipeline loads include layers and transverse and longitudinal pressures. There are two different assumptions about the calculation of the foundation reaction force: one is that the response force is distributed in a straight line, and the other is that the foundation is an elastic semiconductor plane. The stress of the pipe gallery is as follows:

$$w_0 = \sum U_i h_i \tag{1}$$

$$w_1 = w_0 + \frac{2Q_1 + Q_2}{2l} \tag{2}$$

In the formula, w_0 is the constant load earth pressure of the top plate, w_1 is the load on the bottom plate, U_i is the thickness of the soil in the h_i layer, Q1 and Q2 are the masses of the side walls and the middle partition wall, respectively, because the foundation reaction force is assumed to be distributed in a straight line, so it is not necessary to take into account the quality of the bottom plate when seeking w_1.

3 Experimental Research

3.1 Research Purpose

With the continuous deepening of my country's urbanization process and infrastructure construction, the role of underground pipeline systems in social production and life has become increasingly prominent. However, many external unfavorable factors pose challenges to the normal operation of the underground pipeline system. Among them, the uneven settlement of the foundation is particularly serious for the damage to the underground pipeline system. The leakage damage of the underground pipeline caused by it will further erode the soil around the pipeline, and then It will aggravate the uneven settlement of the foundation and cause secondary damage to the underground pipeline system. Therefore, it is very urgent to actively carry out research on the mechanical response characteristics of underground pipelines under the effect of uneven foundation settlement, which is of great significance to my country's economic development and the improvement of people's living standards.

3.2 Research Methods

Field survey method and comparison method: In this paper, soil quality testing is carried out on a certain land, and four monitoring points are selected in the site to compare the stress value of the pipe gallery under different loads at each monitoring point and the longitudinal direction of the pipe gallery under different underlying soil qualities.

4 Finite Element Analysis of Substratum Settlement Under Assembled Pipe Gallery Structure

4.1 Structural Response Test of Assembled Pipe Gallery

As shown in Table 1, the soil quality of a certain piece of land is tested. The soil quality of the land is divided into four categories: one is artificial fill, the other is new loess, the third is saturated soft loess, and the fourth is silty clay. The three mechanical parameters

Table 1. Mechanical parameters of different soil layers

	Compression modulus /MPa	Heavy/(KN/m^3)	Cohesion/KPa
Artificial fill	2.1	15.4	23
New loess	4.7	13.8	37
Bag and soft loess	4.2	19.3	34
Silty clay	6.8	20.5	56

Table 2. Maximum stress of pipe gallery under different soil qualities

	Loading capacity(kg)
Artificial fill	2315
New loess	2068
Bag and soft loess	1573
Silty clay	1629

of compressive modulus, gravity and cohesion of the four types of soil in this land were surveyed, and it was found that the silty clay had the largest value in the three mechanical parameters.

As shown in Table 2, after testing the thickness of various soil materials, it is brought into the surface force algorithm of the prefabricated pipe gallery to calculate the maximum bearing capacity under various soil materials. According to the calculation results, the largest load-bearing capacity is the artificial fill.

Four monitoring points are randomly selected on this piece of land, namely A, B, C, and D. The pipe gallery on this monitoring point is not loaded in the world, and the pressure of the observation point is detected. From the data analysis in Fig. 1, when there is no load on the upper part of the pipe gallery, that is, when the load is 0, the four monitoring points are under pressure, but the pressure is small. When the upper load is applied, the four monitoring points are under pressure. The compression increases rapidly, and with the increase of the upper load, the compression of the observation point also increases. It can be clearly seen from the figure that the two curves at point A and point B are close, and the two curves at point C and point D are close, indicating that the two monitoring points at point A and point B are more sensitive to the upper load, while point C and point D are more sensitive to the upper load. Point pressure growth is more obvious. When different loads are applied above the pipe gallery, it can be seen that the stress at the axillary corner of the pipe gallery is more sensitive than other places, and the stress is concentrated at the junction of the axillary angle of the pipe gallery and the side plate, which is easily damaged.

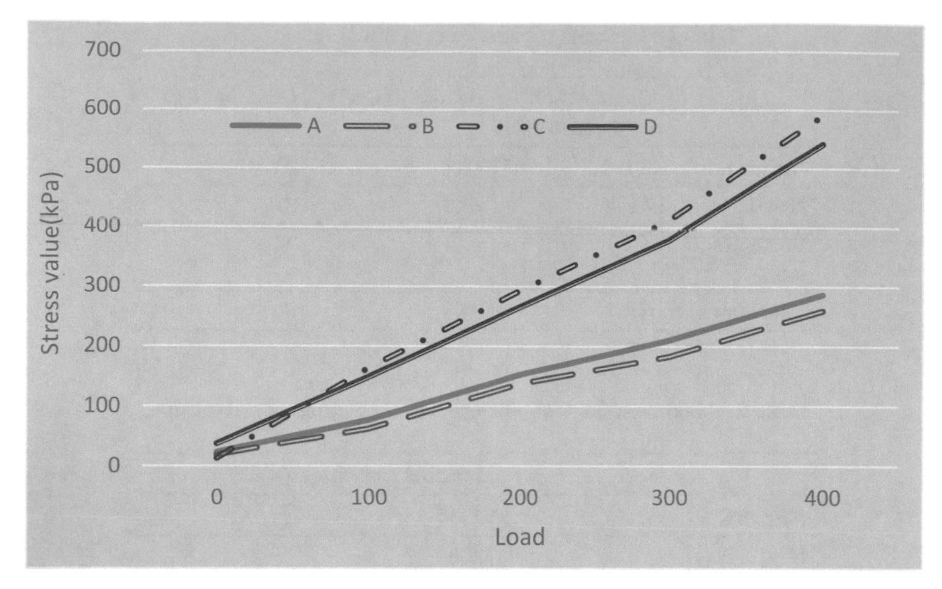

Fig. 1. Stress value (kPa) of the monitoring point of the pipe gallery under different loads

4.2 The Influence of the Soil Quality Difference of the Substratum on the Settlement of the Pipe Gallery

When the soil quality of the substratum is the same, the pipe gallery will sink evenly as a whole, and the probability of water leakage and other damages in the pipe gallery will be greatly reduced. However, in practical engineering applications, it is often found that the geology of the underlying layers is not uniform and invariable, and weak strata often appear, resulting in uneven settlement of the pipe gallery, stress concentration in the pipe gallery, which have adverse effects on engineering applications.

It can be seen from Fig. 2 that when the pipe gallery passes through the substratum with different soil qualities, the longitudinal settlement curve of the pipe gallery changes. It can be seen from this that the elastic modulus of the substratum soil plays a decisive role in the settlement of the pipe gallery. When passing through a certain section of weak stratum, a large settlement occurs in the middle of the pipe gallery, and the smaller the elastic modulus of this weak stratum, the greater the settlement; and it can be seen that due to the reduction of the elastic modulus of this section. As a result, the overall settlement of the substratum increases to varying degrees.

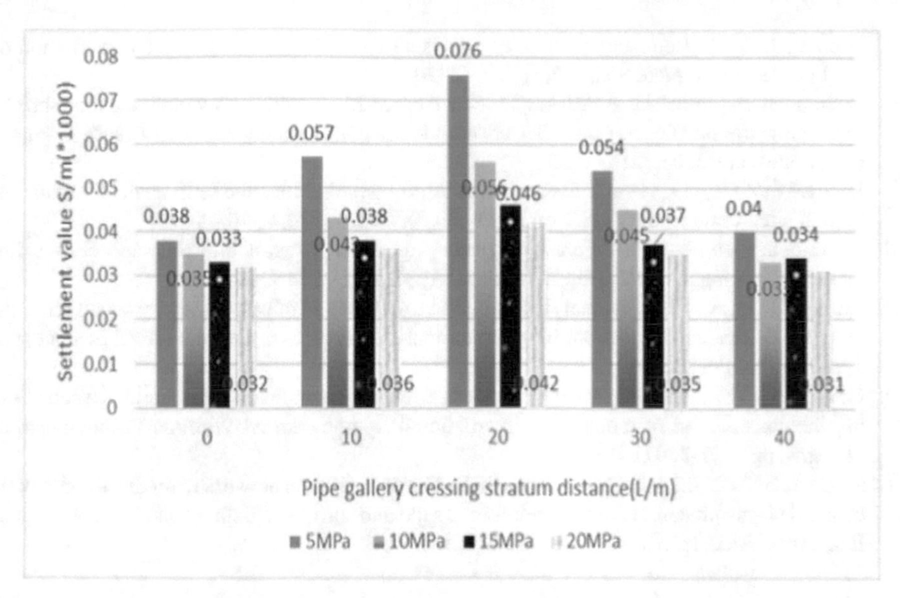

Fig. 2. Longitudinal settlement of the pipe gallery when the elastic modulus of the weak stratum varies

5 Conclusion

In this paper, taking the spliced pipe gallery as an example, it is understood that there is a gap in the bearing capacity of the pipe gallery without soil. By analyzing the stress value of the pipe gallery under different loads, it is found that with the increase of the upper load, the pressure of the pipe gallery increases. Although the settlement phenomenon is difficult to occur when the soil quality of the underlying layers is the same, this phenomenon can only occur in the ideal state. The actual soil quality cannot be the same. Therefore, the influence of different underlying soil materials on the settlement of the pipeline is analyzed. It is found that the smaller the characteristic modulus of weak strata, the greater the subsidence.

References

1. Carrera, E., Jara, O., Dávila, P., Ballesteros, F., Suasnavas, P.: Ergonomics during the construction of the stations and the drilling of the tunnel in the metro of the city of quito. In: Goossens, R., Murata, A. (eds.) AHFE 2019. AISC, vol. 970, pp. 294–301. Springer, Cham (2020). https://doi.org/10.1007/978-3-030-20145-6_29
2. Li, B.: Study on settlement deformation characteristics of reinforced soil subgrade. IOP Conf. Ser. Earth Environ. Sci. **267**(4), 42087–42087 (2019)
3. Anh, T.N.: Analysis of stress-strain behaviours and stability of the ground around the tunnel in Ho Chi Minh city during the construction stages. In: International Conference on Information and Computer Technologies, pp. 356–360 (2020)
4. Ma, L., et al.: Centrifuge modeling of the pile foundation reinforcement on slopes subjected to uneven settlement. Bull. Eng. Geol. Env. **79**(5), 2647–2658 (2020). https://doi.org/10.1007/s10064-020-01723-z

5. Scoular, J., et al.: Retrospective InSAR analysis of East London during the construction of the Lee Tunnel. Remote Sens. **12**(5), 849 (2020)
6. Weinrauch, A., Seidel, H.-P., Mlakar, D., Steinberger, M., Zayer, R.: A variational loop shrinking analogy for handle and tunnel detection and reeb graph construction on surfaces. Comput. Res. Repos. abs/2105.13168 (2021)
7. Ren, L.W., Zhou, G.L., Dun, Z.L., et al.: Case study on suitability and settlement of foundation in goaf site. Yantu Lixue/Rock Soil Mech. **39**(8), 2922–2932, 2940 (2018)
8. Mahmoodzadeh, A., et al.: Forecasting tunnel geology, construction time and costs using machine learning methods. Neural Comput. Appl. **33**(1), 321–348 (2021)
9. Liu, H., Deng, A., Wen, S.: Analytical solution to settlement of cast-in-situ thin-wall concrete pipe pile composite foundation. In: International Conference on Computational Science, no. 3, pp. 1172–1179 (2021)
10. Céspedes, M.M., Armada, A.G.: Characterization of the visible light communications during the construction of tunnels. In: International Symposium on Wireless Communication Systems, pp. 356–360 (2019)
11. Moghaddas Tafreshi, S.N., Mehrjardi, Gh.T.: The use of neural network to predict the behavior of small plastic pipes embedded in reinforced sand and surface settlement under repeated load. Eng. Appl. Artif. Intell. **21**(6), 883–894 (2018)

Multi Objective Optimization of Decoration Engineering Project on Account of Ant Colony Algorithm and BIM

Yuan Zhong, Lu Liu(✉), and Yu Lei

Chongqing College of Architecture and Technology, Chongqing, China
liuluinfo@163.com

Abstract. Project management involves planning, coordinating and controlling the whole project implementation process, so as to successfully achieve the construction objectives under certain resource constraints. The project must strictly control the investment and quality, and implement the objectives under the condition of comprehensive consideration of multiple factors. As long as this mode can be processed and implemented basically according to the requirements, it can basically complete the target task. However, how to optimize the multiple target of the project in an optimal way is a common consideration. The ant colony algorithm and BIM are used to build the data model, optimize the parameters of the input and output of the project data model, and finally complete the task through the multiple target optimization algorithm of the project. This paper studies the knowledge of multiple target optimization of decoration engineering projects based on ant colony algorithm and BIM, and explains a series of viewpoints and theories of multiple target optimization of decoration engineering projects based on ant colony algorithm and BIM. Through the effect analysis of the actual teaching data, the multiple target optimization of decoration engineering project based on ant colony algorithm and BIM is studied. The test results show that the multiple target optimization of decoration engineering project based on ant colony algorithm and BIM achieves 83.52%, 90.11%, 92.95% and 98.60% respectively in the aspects of decoration engineering project coordination, multiple target optimization performance, robustness and obstacle breaking performance.

Keywords: Ant Colony Algorithm · BIM Algorithm · Decoration Engineering · Project Multiple Target Optimization

1 Introduction

The purpose of multiple target optimization of engineering project is to achieve the objectives of shorter construction period, lower cost and higher quality while meeting the requirements of contract construction period, cost and quality. On account of the quantitative relationship between this goal and duration, cost and quality. Ant colony algorithm and BIM are put into multiple target optimization of engineering projects to optimize the model of ant colony algorithm and BIM. After repeated experiments, the

J. H. Abawajy et al. (Eds.): ICATCI 2022, LNDECT 169, pp. 63–70, 2023.
https://doi.org/10.1007/978-3-031-28893-7_8

test shows that ant colony algorithm and BIM have good performance in the planning and control of project duration, cost and quality. Under certain investment, the optimization results of ant colony algorithm and BIM are better, and the effect of data model is very good. The multi-purpose optimization of decoration engineering project on account of ant colony algorithm and BIM enjoys the advantages of curriculum thinking and policy, and provides a solution for the multiple target optimization of decoration engineering project. In terms of communication and learning of multiple target optimization of decoration engineering projects on account of ant colony algorithm and BIM, we must strengthen the application ability of ant colony algorithm and BIM to solve the multiple target optimization problem of decoration engineering projects.

Many scholars at home and abroad have studied the research on account of ant colony algorithm and BIM. In foreign research, a scholar mentioned that the vibration signals of normal transmission and different fault transmission were collected through experiments, and the weak fault features of automobile transmission gears were extracted by combining Teager Energy Operator Demodulation Method with empirical mode decomposition method. on account of the cross validation method, grid search method and particle swarm optimization algorithm are used to search and optimize the important parameters in SVM model, so as to improve the performance of SVM classifier [1]. A scholar proposed a multicast routing algorithm on account of quality of service (QoS), and introduced the search strategy on account of pheromone shortest path in the process of ant colony foraging. Matlab ant colony algorithm is also applied to the local optimization of WSN routing protocol multicast problem. Computer simulation results show that ant colony algorithm has good effect [2]. Eleftheriadis S et al. Combines life cycle assessment (LCA) theory with the capabilities of BIM to investigate the current development of structural system energy efficiency. The engineering dimensions of common decision-making procedures in BIM system are discussed, including optimization methods, constructability, security constraints and regulatory compliance restrictions [3]. However, the multiple target optimization of decoration engineering project on account of ant colony algorithm and BIM is in the primary stage, and there is still a certain gap compared with foreign systems.

The further improvement of ant colony algorithm and BIM in China must start from the following points: first, deepen the research on ant colony algorithm and BIM system; Secondly, the data model on account of ant colony algorithm and BIM is optimized; Finally, strengthen the exchange and cooperation of algorithm theory and improve technical skills.

2 Research on Multiple Target Optimization of Decoration Engineering Project on Account of Ant Colony Algorithm and BIM

2.1 Ant Colony Algorithm and BIM

Ant colony algorithm is a bionic heuristic algorithm, which originates from the simulation of ant foraging process. At present, it has been widely used in many fields. Artificial intelligence ants have been developed according to their powerful behavior characteristics. Artificial intelligence ant simulates the pathfinding characteristics of real ants. It has gradually become an intelligent optimization algorithm. The study found that ants leave an attractive chemical pheromone in the process of foraging. Each ant will be affected by other ant pheromones and release pheromones on the way. When ants choose a path, they are more likely to choose a path with more pheromones. The size of pheromone concentration on the path indicates the length of the path [4–8]. When the pheromone concentration on the path is high, it means that the path is short. Conversely, when the pheromone concentration on a path is low, it means that the path is long. Generally speaking, ants will choose the path with high pheromone concentration and release a certain amount of pheromone. As more and more ants choose this path, the pheromone concentration left in the path increases, forming positive feedback. At the same time, biologists also found that the pheromone on the path will gradually decrease over time [9–11].

The application of BIM starts from three aspects. First, from the perspective of resource sharing in the whole building cycle, BIM technology is applied to realize information sharing and exchange in the whole life cycle through digital expression and information transformation of each physical structure and facilities and equipment of the building [12, 13]. Secondly, BIM technology is applied from the perspective of complete building information model in the whole life cycle. It can realize the visualization of information coordination and sharing throughout the whole life cycle of the building from design, construction to late operation and maintenance [14, 15]. Thirdly, BIM technology is applied from the perspective of the whole process simulation of the construction implementation process. BIM technology can be used for simulation analysis of project design, implementation, operation and maintenance, as shown in Fig. 1.

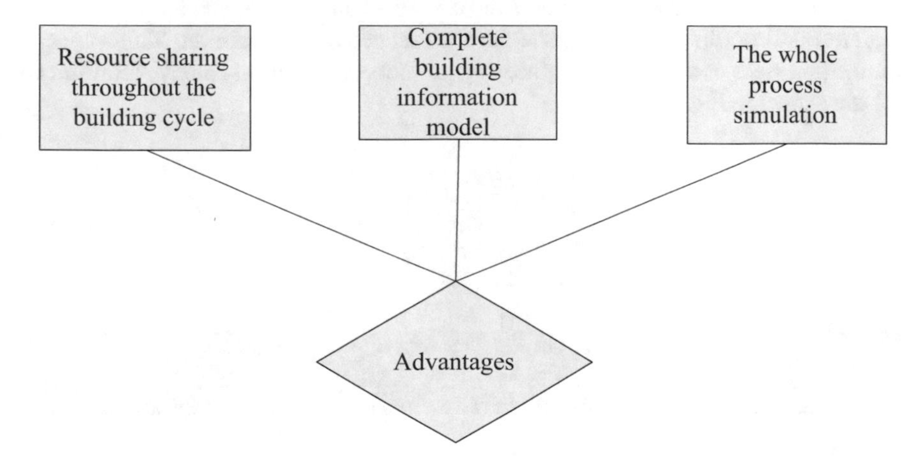

Fig. 1. Advantages of BIM technology

2.2 Multi-objective Optimization of Decoration Engineering Project Based on Ant Colony Algorithm and BIM

Multi-objective optimization of decoration engineering project based on ant colony algorithm and BIM refers to the comprehensive optimization of multiple objectives in decoration engineering project at the same time, that is, the optimal coordination and solution of multiple objectives of time limit, quality, cost and safety.

First, build data model for decoration project.

The ant colony algorithm and BIM are used to establish a comprehensive and multi-dimensional objective function data model for decoration engineering projects. The parameters of the objective function are set and determined respectively. Analyze and deal with all the factors that affect the decoration project.

Secondly, parameters based on ant colony algorithm and BIM data model are initialized. The inertia weight, acceleration constant and speed range of decoration engineering project based on ant colony algorithm and BIM are initialized, and the parameters involved in decoration engineering planning and design, engineering construction and other stages are established.

Thirdly, the parameters contained in the data model are harmonized. The input factors of the model are analyzed. The input items can be transformed in various ways, and the final output values are stored in the form of vectors. The final results are collected and saved, and the data are de-falsified, the unreasonable data are discarded, and the data missing from the data model are filled with scientific means and methods.

Fourth, decoration project system analysis. Multi-objective optimization is a systematic problem. After fully analyzing the components of each sub-objective, the subsystem is comprehensively analyzed, then all sub-objectives are unified, and all objectives are integrated and coordinated to complete the abstract vector optimization. Multi-objective optimization steps of decoration engineering projects based on ant colony algorithm and BIM are shown in Fig. 2.

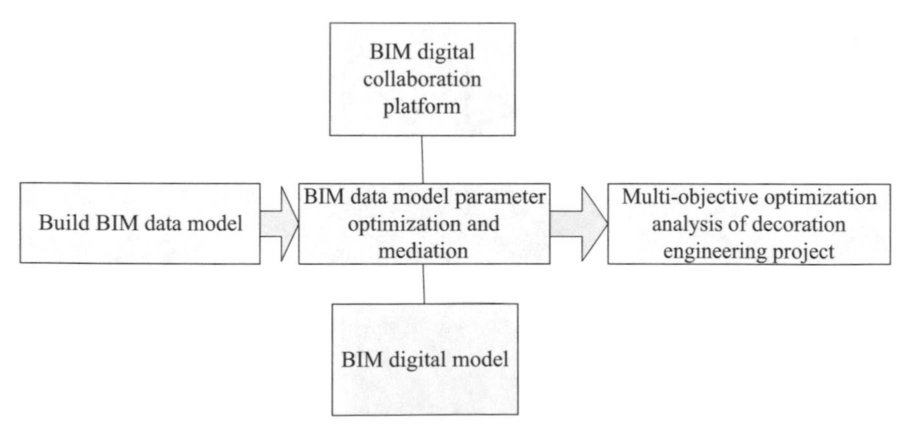

Fig. 2. Multi-objective optimization steps for decoration engineering projects

3 Research on Multiple Target Optimization Effect of Decoration Engineering Project on Account of Ant Colony Algorithm and BIM

3.1 Comparative Analysis

This text uses the comparative analysis method to test the effect of multiple target optimization of decoration engineering project on account of ant colony algorithm and BIM, and analyzes the data model according to the comparative actual effect. The mode of empirical research is to test the process data according to the solution in this text through the multiple target optimization of decoration engineering project on account of ant colony algorithm and BIM, and record and analyze the data of each test point. The method adopted this time is to process data on account of ant colony algorithm, BIM and non algorithm. Stain all sorted data, and use relevant algorithms to add and remove data.

3.2 Brief Description

This text uses the comparative analysis method to record and process the data of the multiple target optimization of decoration engineering project on account of ant colony algorithm and BIM. Analyze the effect of multiple target optimization of decoration engineering project according to the complete process.

3.3 Formula

$$f_3(T) = \frac{(\mathrm{V}_{\max} - 1.2)\left(\frac{T_s - T_{c1}}{T_{c2} - T_{c1}}\right)^2}{v_0} \tag{1}$$

$$\xi = \frac{C}{F_k} \tag{2}$$

It refers to the output item of multiple target optimization of decoration engineering project, the progress of multiple target optimization of decoration engineering project, refers to the temperature of engineering project, indicates the normal temperature of engineering project, refers to another relative temperature of engineering project, refers to the basic progress of multiple target optimization of decoration engineering project, refers to the transformation result of decoration project, and C is the parameter and the number of traversal cycles of the algorithm.

3.4 Calculation Principle

The influencing factors of multiple target optimization of decoration engineering project on account of ant colony algorithm and BIM are analyzed, the effect of multiple target optimization of decoration engineering project on account of ant colony algorithm and BIM is scientifically decomposed, and the parameters of multiple target optimization of decoration engineering project are processed and set.

4 Investigation, Research and Analysis of Multiple Target Optimization of Decoration Engineering Project on Account of Ant Colony Algorithm and BIM

4.1 Test Effect

The test objects are divided into Decoration Engineering Project multiple target optimization and non-algorithm decoration engineering project multiple target optimization on account of ant colony algorithm and BIM. The results are compared and analyzed. The two groups of data are tested respectively to sort out the advantages of ant colony algorithm and BIM multiple target optimization, The multiple target optimization of decoration engineering project on account of ant colony algorithm and BIM records the effect of relevant data: in the comparative analysis method, can the multiple target optimization effect of decoration engineering project on account of ant colony algorithm and BIM improve the goal of decoration engineering project multiple target optimization. The results show that the multiple target optimization effect of Decoration Engineering project on account of ant colony algorithm and BIM has great advantages. The major key problems of multiple target optimization of decoration engineering projects on account of ant colony algorithm and BIM have been solved. The results are shown in Table 1 and Fig. 3.

Table 1. Engineering project data chart

	Coordination	Parameter optimization performance	Vigorous	Barrier breaking performance
Metrics	170	325	485	601
Scale	83.52%	90.11%	92.95%	98.60%
Percentage	90	92	97	99

The multiple target optimization effect of decoration engineering project on account of ant colony algorithm and BIM is better, and the result is better than the traditional method. Achieve high efficiency in the same field in terms of system coordination, parameter optimization performance, robustness and breaking obstacles. Test and analyze the model effect data processing and application. Data processing and analysis can be carried out in many aspects of multiple target optimization performance of decoration engineering projects. Focus on various abilities required in the modeling process, including construction, mathematicization, modeling, model verification and evaluation. Level orientation focuses on the division of modeling ability from the level and law of human cognitive development. Strengthening the research and development of multiple target optimization of decoration engineering projects on account of ant colony algorithm and BIM is conducive to the progress and optimization of decoration engineering projects.

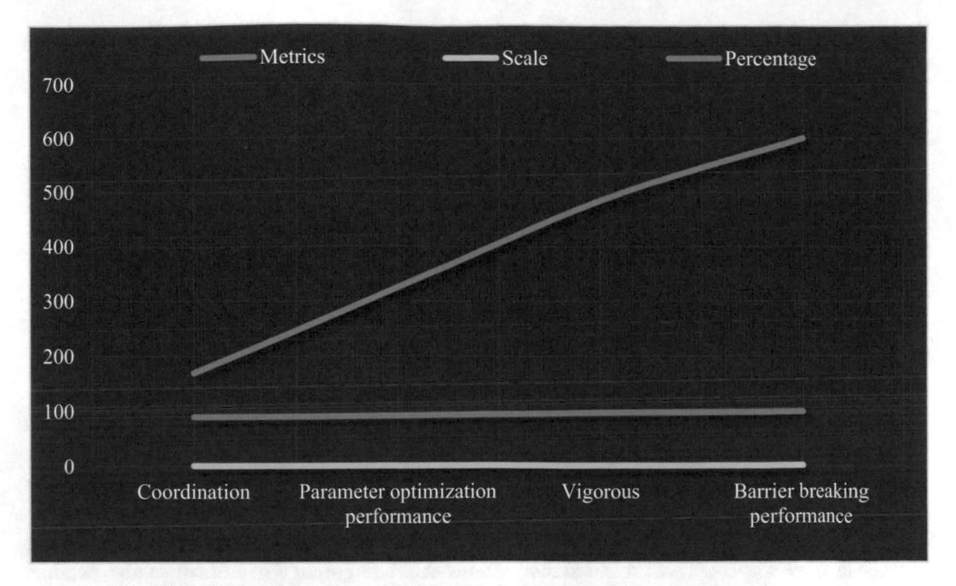

Fig. 3. Engineering project multiple target optimization data chart

5 Conclusions

The combination of BIM Technology and ant colony algorithm is used to solve the design and planning problems of decoration engineering projects. The three-dimensional information model of decoration engineering project is established by BIM Technology as the fire simulation scene to realize the unification of the simulation scene and the actual decoration engineering project. Combined with the optimized ant colony algorithm, the optimal evacuation path is calculated. Finally, the visual display of evacuation path is realized by webgl technology. Multiple target optimization of decoration engineering project on account of ant colony algorithm and BIM, studied the contents and related affairs of decoration engineering project, and made great efforts for multiple target optimization of decoration engineering project. Through all-round analysis and Research on multiple target optimization process of decoration engineering project, it comes into play in improving the quality of decoration engineering project, and improves the innovation of decoration engineering project.

References

1. Yu, C., Zha, W.: Detection and application of breaking of automobile mechanical transmission rod based on ant colony algorithm. Concurr. Comput. Pract. Exp. **31**(10), e4759.1–e4759.7 (2019)
2. Wang, R., He, G., Wu, X., et al.: Multicast optimization and node fast location search method for wireless sensor networks based on ant colony algorithm. J. Digit. Inf. Manag. **15**(6), 303–311 (2017)
3. Eleftheriadis, S., Mumovic, D., Greening, P.: Life cycle energy efficiency in building structures: A review of current developments and future outlooks based on BIM capabilities. Renew. Sustain. Energy Rev. **67**, 811–825 (2017)

4. Moravík, M., Schmid, M., Burch, N., et al.: DeepStack: expert-level artificial intelligence in no-limit poker. Science **356**(6337), 508–508 (2017)
5. Makridakis, S.: The forthcoming artificial intelligence (AI) revolution: its impact on society and firms. Futures **90**, 46–60 (2017)
6. Pask, G.M., Slone, J.D., Millar, J.G., et al.: Specialized odorant receptors in social insects that detect cuticular hydrocarbon cues and candidate pheromones. Nat. Commun. **8**(1), 297 (2017)
7. Tsuji, F., Ishihara, A., Kurata, K., et al.: Geranyl modification on the tryptophan residue of ComX Ro-E-2 pheromone by a cell-free system. FEBS Lett. **586**(2), 174–179 (2019)
8. Gleeson, M.: The digitization of infectious disease: preventing biological threats in the sky. Cut. IT J. **32**(4), 31–36 (2019)
9. Cheung, C.: Panel report: the dark side of the digitization of the individual. Internet Res. **29**(2), 274–288 (2019)
10. Xie, X., Li, Z., Wang, H., Zhu, B.: The combination of three-dimension inverse design and optimization methods for helium circulator's impeller optimization in HTR. In: Rodrigues, H.C., et al. (eds.) EngOpt 2018, pp. 381–393. Springer, Cham (2019). https://doi.org/10.1007/978-3-319-97773-7_35
11. Nithyadevi, N., Gayathri, P., Chamkha, A.J.: Three dimensional MHD stagnation point flow of Al-Cu alloy suspended water based nanofluid with second order slip and convective heating. Int. J. Numer. Meth. Heat Fluid Flow **27**(12), 2879–2901 (2017)
12. Wang, Q.: Full life circle health: a new era. J. Tradit. Chin. Med. Sci. **5**(1), 4–5 (2018)
13. West, J.V.: Circle-to-land life-saving tips. Plane Pilot **55**(1), 22–25 (2019)
14. Gramazio, C.C., Laidlaw, D.H., Schloss, K.B.: Colorgorical: creating discriminable and preferable color palettes for information visualization. IEEE Trans. Vis. Comput. Graph. **23**(1), 521 (2017)
15. Kasabov, N.K., Doborjeh, M.G., Doborjeh, Z.G.: Mapping, learning, visualization, classification, and understanding of fMRI data in the NeuCube evolving spatiotemporal data machine of spiking neural networks. IEEE Trans. Neural Netw. Learn. Syst. **28**(4), 887–899 (2017)

Design of Decoration Energy Saving Integrated System Based on Genetic Algorithm and BIM

Yuan Zhong, Lu Liu(✉), and Yu Lei

Chongqing College of Architecture and Technology, Chongqing, China
liuluinfo@163.com

Abstract. With the improvement of human aesthetic art and living technology, a variety of interior decoration has been produced. While meeting human artistic aesthetics, building interior decoration also poses a major challenge to its energy-saving effect. Taking the low-temperature interior decoration of the experimental adjustable room as an example, this paper analyzes, explores and optimizes the characteristics, structure and working mode of the experimental adjustable room temperature decorative board by using theoretical data analysis, practical detection, genetic algorithm and BIM Technology. Using genetic algorithm, the change law of temperature between the surface of adjustable room temperature decorative board and the indoor temperature of the building with time is obtained. The indoor air temperature and power consumption after the unit works at different times are measured. The results show that the surface air quality can meet the temperature requirements. The adjustable room temperature decorative panel can fully meet the needs of temperature regulation in summer in the tested area, and is evenly distributed in the vertical direction of indoor temperature. The results show that the application of BIM Technology is helpful to the design of decorative energy-saving integrated system, and genetic algorithm also improves the efficiency of accurate data analysis.

Keywords: Genetic Algorithm · BIM System · Decorative Energy-saving Integrated System · System Design

1 Introduction

In this paper, energy-saving decoration is studied by using bequest and BIM Technology. BIM Technology is the latest research achievement of informatization in the field of interior decoration, which promotes the informatization in the field of interior decoration to a higher level. With big data information as the core, it runs through all stages of the whole process and life cycle of design and construction. In terms of collaborative design, it can avoid conflicts and simulate the dynamic process of construction. In addition, collaborative design can prevent conflicts, simulate the dynamic process of construction, and integrate design components with quantity budget.

The design of decoration energy-saving integrated system based on genetic algorithm and BIM has been studied and analyzed by many scholars. TavakkoliMoghaddam r

J. H. Abawajy et al. (Eds.): ICATCI 2022, LNDECT 169, pp. 71–79, 2023.
https://doi.org/10.1007/978-3-031-28893-7_9

proposed a genetic algorithm (GA) to solve the redundancy allocation problem of series parallel system. When a single subsystem can choose the redundancy strategy. Most solutions to the general redundancy allocation problem assume that the redundancy strategy of each subsystem is predetermined and fixed [1]. Dawid h simulated with different coding schemes and explained the amazing differences between the results under different settings by using the mathematical theory of gas with state dependent fitness function. The relationship between coding and gas convergence is explained [2]. It can be seen that the genetic algorithm is very important for the design of decoration energy-saving integrated system in this paper [2].

For the research and analysis of decoration energy-saving integrated system, most of the data analysis methods used in the past cannot be satisfied with the accurate analysis of various data of energy-saving decoration. Based on the previous data analysis methods, this paper introduces genetic algorithm to improve the efficiency of data calculation and analysis and reduce errors. Combined with BIM Technology, a more perfect decoration energy collection system is designed [3, 4].

2 Background Significance

2.1 Principle of Genetic Algorithm

There are obvious differences between genetic evolutionary algorithm and random search algorithm. Genetic operation is directional and purposeful search, while random search is non-directional and purposeless search [11, 12]. The improved genetic algorithm has the characteristics of high search efficiency and easy to find the global optimal solution; The efficiency of genetic algorithm in finding the optimal solution is affected by the selection of coding mode, population size, fitness function and the probability selection of three basic genetic operators.

2.2 Demand Analysis of BIM Technology Applied to Decoration Energy Saving System Design

The use of BIM Technology Combined with other software technologies will help to save manpower and improve cost efficiency: the loss of funds lower than materials can be mastered in time. Collect all material consumption before a specific time point at any time, calculate the corresponding project cost according to the collected material information, and understand the capital consumption at the same time. In order to obtain the requirements for interior decoration materials in the next stage more efficiently and ensure the funds required by the project in time, the requirements for building materials in the next stage are estimated and the corresponding interior decoration materials are planned. Data information affecting decorative materials [5, 6]. With the help of BIM system model, these modified information and data will be filled into the model in time and updated automatically. The clear hierarchy and wide application range should be reflected in the integration, and have certain expansibility [7, 8].

BIM Technology is the digital expression of information. Its application to the decoration energy-saving integrated system represents the new application direction of information technology in China's decoration industry, and realizes the goal of scientific management of the whole decoration project. It provides information function; The function of the drawing can be modified accordingly; It can meet the needs of the development of the construction industry; Integration has been realized; It provides three-dimensional function [9, 10]. In terms of cost control of decoration engineering, its advantages are embodied in that the application of BIM Technology can achieve the goal of communication and collaborative work of all participants. The application of BIM Technology in interior decoration engineering can be seen in Fig. 1.

Fig. 1. Application of IM technology in interior decoration engineering

(1) Visual design: in the early stage of architectural design, BIM can flexibly show the creativity and design concept of architects. The results of each design stage can be fully applied to the next design stage to enhance the continuity of each stage of the design process.
(2) Visual technical disclosure: the application of BIM Technology can make the whole process of decoration engineering expressed in 3D. And through the application of BIM Technology to simulate the complex processes in the project, it can assist the project decision-making, avoid the loss of financial and material resources and save the project duration.

Fig. 2. Application of Bim in decoration projects

(3) BIM based information synchronization: Based on BIM Technology, the physical information of building objects and engineering related information can be synchronized to relevant units, eliminating the time difference and content difference of information transmission between participating units.

(4) Data sharing: in the whole process of decoration engineering, the decoration engineering database of building objects is created based on BIM Technology. Around the design and construction of decoration engineering, all participants can work together to improve the efficiency of data sharing. The application of Bim in decoration projects is shown in Fig. 2.
(5) BIM Technology Assisted scheme selection: BIM Technology can assist scheme selection, whether for design scheme or engineering construction scheme. For the design scheme, the design parameters can be simulated and analyzed through BIM software, and various alternative schemes can be intuitively compared to determine the optimal design scheme. For the project construction scheme, BIM software can simulate the progress, resources and processes of the construction scheme.
(6) Improve the accuracy of material blanking: the BIM model contains the physical information of the building object. In the decoration engineering, for some components that need to be processed and blanking, according to the component information in the BIM model, the decoration engineering components can be mass produced in the factory to avoid the error caused by manual on-site blanking and save materials and costs.
(7) Reverse modeling of existing buildings: the application of BIM Technology in the reconstruction project of existing buildings can reconstruct the BIM model of building objects, integrate the existing data information required by the project into the BIM model, improve the utilization efficiency of existing data, and provide technical and management auxiliary tools for the decoration engineering design and construction of the reconstruction project. The solution is shown in Fig. 3.

Fig. 3. BIM system solution for installation and distribution buildings

BIM can assist the designer to successfully complete the design intention in the design stage, but its application in the construction stage is more inclined to be a technical means of auxiliary construction. Therefore, in order to make BIM Technology give full play to its application value in the construction stage of decoration engineering and achieve the application goal of BIM Technology.

2.3 Energy Saving Design and Decoration of Air Conditioner Based on Genetic Algorithm

(1) Temperature and humidity detection and control. The temperature and humidity sensor can control the opening of the air conditioner according to the detected indoor ambient temperature and humidity data, and adjust the air conditioner temperature within the appropriate temperature range of the human body. If more than one air conditioner needs to be applied in a certain place, each air conditioner can detect the opening and closing separately or jointly;
(2) Human identification control. The pyroelectric sensor installed on the top of the room is connected to the energy-saving controller of the air conditioner through ZigBee wireless sensor to detect the activity of indoor personnel, and the change of micro current is used to detect the static state of the human body to determine whether to turn on and off the air conditioner.
(3) Control strategy. The air-conditioning energy-saving controller supports a variety of air-conditioning control modes, which can be set by the system administrator and is the core function of the air-conditioning energy-saving controller. They are website control, timing control and energy-saving control [11].
(4) Efficiency detection alarm. When the air conditioner continues to operate at full load, but the indoor temperature does not change significantly (the change ratio is set by the manager according to the actual environment), the system will automatically give an alarm, that is, inform the system manager through the information service platform that an air conditioner in one or several rooms is operating at the threshold and has failed;
(5) Monitoring function. This part is mainly realized by the indoor environment information collector module, air conditioning operation information acquisition module and control command execution module of the air conditioning energy-saving controller, including real-time monitoring of indoor environment information, air conditioning operation status, control command execution and outdoor environment information, and regularly reporting the monitored information to the information service platform;
(6) Communication function. At the same time, it has WiFi communication module and wireless ZigBee communication module, and has dual network communication ability. It not only realizes the function of information interaction with the remote information service platform, but also realizes the communication with other air conditioning energy-saving controllers in a specific area and ZigBee wireless sensor information acquisition nodes in the same room;
(7) Air conditioning control. It can send various specific air conditioning control commands through the infrared transmitter to realize various actual control operations of indoor air conditioning;

(8) Handheld devices. Handheld device is mainly a portable device provided for building managers, equipment installers and air conditioning maintenance personnel for on-site management and maintenance. It is composed of ZigBee wireless communication module, LCD module, keyboard operation module and central processing unit module. It can directly operate and maintain the air conditioning energy-saving controller through ZigBee wireless communication, In order to realize the function of direct operation and management of air conditioning in the field. In addition, it also has the function of parameter configuration and simple maintenance of air conditioning energy-saving controller [12].

3 Genetic Algorithm Formula

Assuming that the size of the population is m, where P_I represents the fitness function value of individual I, it represents the probability that individual I is selected, as shown in the following formula (1).

$$Y_i = P_i / \sum\nolimits_{j=1}^{m} P_i \tag{1}$$

Suppose a complex manufacturing task is decomposed into multiple subtasks, and the number of subtasks is much larger than that of manufacturing resources. Now use m to represent the total number of subtasks,

N represents the number of manufacturing resources. The expected time to complete (etc.) matrix in row N and column m represents the time required for each manufacturing resource to complete its task queue, so the load balancing level of the virtual machine can be expressed as the standard deviation of the running time of the virtual machine. For example, formula (2) and formula (3)

$$\text{load} = \sqrt{\frac{\sum\nolimits_{i=1}^{n} (\text{sumTime(i)} - \text{avgTime})^2}{n}} \tag{2}$$

$$\text{avgTime} = \frac{\sum\nolimits_{i=1}^{n} \text{sumTime(i)}}{n} \tag{3}$$

The standard deviation indicates the degree of dispersion of the data. The larger the standard deviation, the greater the degree of dispersion, indicating that the data is more dispersed. For virtual machine operation, when the data is more centralized, each virtual machine runs almost simultaneously and the load is more balanced. Therefore, the smaller the standard deviation, the more balanced the virtual machine load.

4 Research Methods and Data Analysis

In the experimental preparation stage, firstly, the experimental equipment and room are cleaned and checked, and the thermocouple is calibrated with ice water mixture to reduce the error and ensure the reliability of the results. In the experiment, the eight temperature measuring points on the board surface are read out and recorded by the temperature inspection instrument. Due to the rapid change of the board surface temperature after switching on and off, the recording time interval after startup is 1min, the time interval after 20 min is 10 min, and then 30 min. After the operation is stable, it can be recorded once in 1 h; The time interval recorded after shutdown changes from short to long with that after startup, so as to reflect the change trend of temperature as accurately as possible; The indoor temperature is measured by 12 glass thermometers and recorded manually, and the humidity is read manually by hygrometer.

The change trend of indoor and outdoor air parameters is shown in Fig. 4.

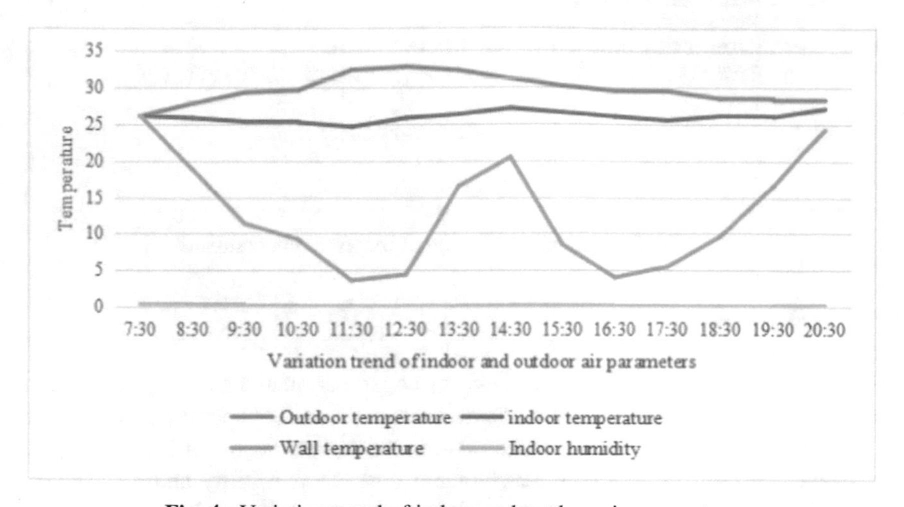

Fig. 4. Variation trend of indoor and outdoor air parameters

It can be seen from the above chart that the relative humidity of indoor air begins to decrease significantly after the unit works normally for 30 min, and continues to drop to the relative humidity level of saturated air. That is, the surface temperature of temperature regulating decorative panel reduces the dew point temperature of indoor air and condensation occurs on its surface, which reduces the relative humidity of indoor air, It gives people a relatively cool and dry feeling.

Next, test the distribution of indoor temperature in the vertical direction with the change of time. The test results are shown in Fig. 5.

The above results show that there is a condensation problem when the temperature regulating decorative board is cooled. The relationship between the medium temperature and the surface temperature of the temperature regulating decorative board is obtained by simulation. In actual use, the air dew point temperature in summer can be obtained through the enthalpy humidity diagram, and the plate surface temperature is selected to be 1–2 °C above its dew point temperature, so as to obtain the temperature of the required refrigerant.

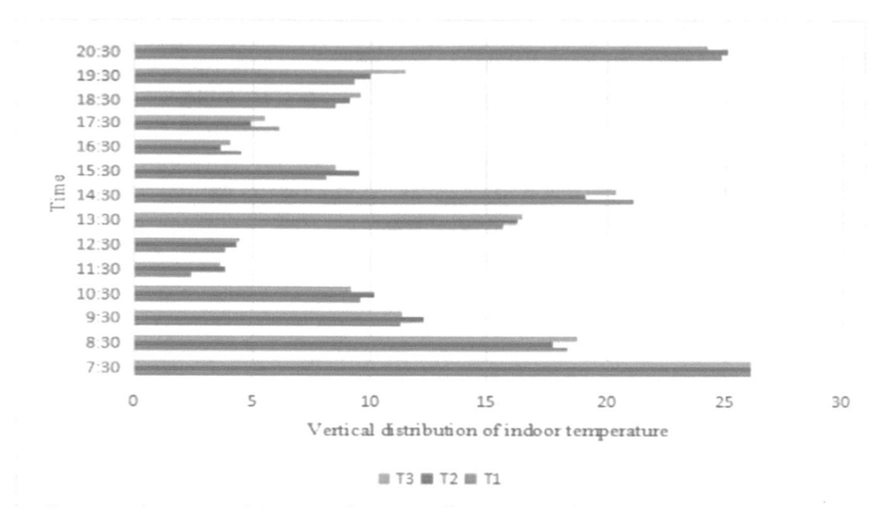

Fig. 5. Vertical distribution of indoor temperature

5 Conclusions

With the increasing shortage of energy, energy conservation and consumption reduction has become the main topic today. In order to achieve the task of energy conservation and emission reduction, with the help of the existing theory and design of BIM Technology, understand the constraints of its development and the feasibility and advantages of BIM Technology in energy-saving decoration management. Study the system operation process of the core functions of BIM Technology into database technology, network, information integration and optimization model data, combined with genetic algorithm, clarify the application process and effect of BIM Technology in decoration energy saving, and provide theoretical support for the development of real-time roaming of data and information. Improve the efficiency of building decoration energy conservation and promote the stable development of BIM Technology Application in decoration energy conservation system. The genetic algorithm and BIM Technology proposed in this paper are applied to decoration energy saving. Although the experimental effect is better than previous research, the measurement of energy-saving decoration requires high accuracy. At present, this research has only played an auxiliary role and there is room for further improvement. The next step is to find problems in practical application and further improve the effect of decoration and energy saving.

References

1. Tavakkoli-Moghaddam, R., Safari, J., Sassani, F.: Reliability optimization of series-parallel systems with a choice of redundancy strategies using a genetic algorithm. Reliab. Eng. Syst. Saf. **93**(4), 550–556 (2017)
2. Ashfaq, M., Minallah, N., ur Rehman, A., Belhaouari, S.B.: Multistage forward path regenerative genetic algorithm for brain magnetic resonant imaging registration. Big Data **10**(1), 65–80 (2022)

3. Dawid, H., Kopel, M.: On economic applications of the genetic algorithm: a model of the cobweb type. J. Evol. Econ. **8**(3), 297–315 (2019)
4. Volkanovski, A., Mavko, B., Boševski, T., et al.: Genetic algorithm optimisation of the maintenance scheduling of generating units in a power system. Reliab. Eng. Syst. Saf. **93**(6), 779–789 (2017)
5. Shegay, M.V., Svedas, V.K., Voevodin, V.V., Suplatov, D.A., Popova, N.N.: Guide tree optimization with genetic algorithm to improve multiple protein 3D-structure alignment. Bioinform. **38**(4), 985–989 (2022)
6. Yoshitomi, Y., Ikenoue, H., Takeba, T., et al.: Genetic algorithm in uncertain environments for solving stochastic programming problem. J. Oper. Res. Soc. Jpn. **43**(2), 266–290 (2017)
7. Arabasadi, Z., Alizadehsani, R., Roshanzamir, M., Moosaei, H., Yarifard, A.A.: Computer aided decision making for heart disease detection using hybrid neural network-Genetic algorithm. Comput. Methods Programs Biomed. **141**(C), 19–26 (2017)
8. Paes, F.G., Pessoa, A.A., Vidal, T.: A hybrid genetic algorithm with decomposition phases for the unequal area facility layout problem. Eur. J. Oper. Res. **256**(3), 742–756 (2017)
9. Fetouh, T., Zaky, M.S.: New approach to design SVC-based stabilizer using genetic algorithm and rough set theory. IET Gener. Transm. Distrib. **11**(2), 372–382 (2017)
10. Rajeswari, K., Neduncheliyan, S.: Genetic algorithm based fault tolerant clustering in wireless sensor network. IET Commun. **11**(12), 1927–1932 (2017)
11. Sokhangoee, Z.F., Rezapour, A.: Energy saving of course keeping for ships using CGSA and nonlinear decoration. IEEE Access **8**, 141622–141631 (2020)
12. Krishnamoorthy, R., Krishnan, K.: Optimal operation of integrated energy system based on exergy analysis and adaptive genetic algorithm. IEEE Access **8**, 158752–158764 (2020)

Electronic Logistics Distribution System with Improved Entity Recognition Algorithm

Yu Chen(✉)

Chongqing College of Architecture and Technology, Chongqing, China
chywin@yeah.net

Abstract. As an effective means of logistics enterprise management, logistics distribution system can improve the efficiency of logistics distribution, shorten transportation distance, reduce transportation cost and time. Considering the path optimization problem in the logistics distribution system, this paper constructs a mathematical model of the logistics distribution path optimization problem based on the k-means algorithm, and designs an electronic logistics distribution system. By studying the efficiency of sporting goods in the distribution management module of the logistics distribution system based on the improved entity recognition and the traditional entity recognition algorithm, it is proved that the improved algorithm can improve the distribution efficiency, and the k-means algorithm and the particle swarm algorithm are applied to the route. And the vehicle scheduling optimization, got the best distribution plan.

Keywords: Improved entity recognition algorithm · Logistics distribution · K-means algorithm · Route optimization

1 Introduction

The introduction of various algorithms in the logistics distribution system can reduce the difficulty of solving the problem and meet the requirements of planning agility. On the other hand, the logistics distribution planning problem can be decomposed according to the planning period and strategic importance, so that different decision support methods can be adopted for different problems to meet the needs of the systematic and integrated development of the distribution process.

Many scholars at home and abroad have conducted in-depth research on the optimization of electronic logistics distribution system with improved entity recognition algorithm. For example, when a scholar analyzed the logistics distribution system, he found that the benefits of the logistics distribution subsystem often do not always depend on the overall benefits of the logistics distribution system. Therefore, when planning the enterprise logistics distribution network, through the integration and integration of each subsystem, the improved entity recognition algorithm is used to coordinate the relationship between the distribution subsystem and the entire system, so that the benefits of the distribution subsystem are subject to the overall distribution system. Benefit, so as to optimize the efficiency of the logistics distribution system [1]. In order to solve the

J. H. Abawajy et al. (Eds.): ICATCI 2022, LNDECT 169, pp. 80–87, 2023.
https://doi.org/10.1007/978-3-031-28893-7_10

problem of long logistics distribution lines, a scholar has studied related algorithms for optimizing routes, such as insertion algorithm, scanning algorithm and saving algorithm. to the specified line until all customer points are inserted into the line [2]. The main key point of the insertion algorithm is the choice of customer location and the choice of customer path location insertion. The saving algorithm has been applied in the vehicle route optimization problem, which mainly refers to the selection of the vehicle driving route according to the distance between the customer locations and the shortest distance principle, including the customer points within the vehicle driving route range [3]. Although the research results on the optimization of the electronic logistics distribution system by improving the entity recognition algorithm are good, it is necessary to further study the performance optimization of the algorithm to make the distribution system play a role in practical applications.

This paper firstly introduces the RFID technology and entity recognition technology used in logistics distribution. Both technologies can identify the order barcode information during order sorting. Then, three kinds of distribution and distribution route optimization algorithms and vehicle scheduling optimization algorithms are listed respectively. Finally, Design the electronic logistics distribution system and test whether the application of each algorithm to the system can improve the distribution efficiency.

2 Related Technologies and Algorithms

2.1 Technology Used in Logistics and Distribution

(1) RFID technology

RFID is a technology that can receive non-contact information through radio frequency and automatically identify it. Its widespread use makes the logistics and distribution process more technical. Using RFID technology in the process of entering and leaving the warehouse, the staff can obtain the logistics information of the goods through the card reader, and the logistics system can also automatically count and analyze the data to understand the whole process of logistics and distribution [4].

(2) Entity Recognition Technology

The goal of entity recognition is to identify tuples in a dataset that refer to the same real entity. The traditional entity recognition method recognizes the similarity between objects, but the improved entity recognition method proposes a new class of rules that can describe the complex matching conditions of tuples and entities [5]. Based on this kind of rules and the application in logistics distribution, a rule-based entity recognition problem is proposed and an online entity recognition algorithm is designed. In this algorithm, rules can be used for each tuple to determine the entity it refers to.

2.2 Distribution and Transportation Route Optimization Algorithm

Optimizing the logistics distribution path is the key to improving the economic benefits of enterprises. It is the magic weapon for logistics enterprises to be invincible in an increasingly competitive environment. There are many algorithms that can optimize the logistics distribution path, such as dynamic programming algorithm, segmentation algorithm, branching. However, these algorithms also have disadvantages such as large amount of calculation and long time to find the optimal solution [6].

(1) mileage saving method

The milcagc saving method is also called the saving method, which usually solves the transportation problem that takes the shortest route as the goal when there are many vehicle route choices. In recent years, with the development of the logistics industry and customer demand, the share of small-batch, multi-level distribution requirements has gradually increased. However, due to its particularity, this mode of transportation not only wastes resources, but also significantly increases transportation costs and brings financial burdens to enterprises. To this end, choosing the cheapest transportation route becomes the key to competition among enterprises. Considering that the goods are delivered to the customer with guaranteed quality within the specified time, aiming at the shortest route and the shortest time, choosing the route with the lowest cost in the delivery process is the basis for making the request [7].

(2) Genetic algorithm

Genetic algorithm, also known as genetic evolution method, is an adaptive search algorithm based on the genetic evolution mechanism of the biological world [8]. Genetic algorithm is one of the main solving algorithms in the research of logistics distribution route optimization problem. It has the advantages of simple mathematical model solution, strong robustness and parallelism, and is of great significance for optimizing logistics distribution routes.

(3) k-means clustering algorithm

The k-means clustering algorithm is a hard clustering algorithm, which can improve the efficiency of logistics distribution and save the cost of logistics enterprises in the distribution link. As a tool of data analysis, clustering algorithm means to divide a large amount of data into sets according to certain rules, so that similar data are in one set, and dissimilar data are in a set center [9].

The genetic algorithm has the shortcomings of insufficient local search ability and precociousness. The k-means algorithm is used to divide the data that the customer's location coordinates most need to be divided, and then the calculation steps of the k-means algorithm are used to divide the data. Finally, the local distribution center and the customer location are classified as similar. One group, after the customer location is divided, the path is optimized for the customer location. The k-means algorithm looks at n data at the location of n customers, and then divides the data into k sets according to the parameter k, and randomly specifies the initial cluster center to divide the data and the cluster center into a group, and the dissimilar ones into one group. The group definition guidelines are shown in the following formulas.

$$R = \sum_{i=1}^{k} \sum_{x \in C_i} |x - x_i|^2 \tag{1}$$

R is the squared error and x is the mean. Among all customers, k customers are selected as the initial cluster centers for dividing the data. Divide the remaining customers into the set with the closest distance to the cluster center according to the formula:

$$D_S(C_a, C_b) = \min\{d(x, y)|x \in C_a, y \in C_b\} \tag{2}$$

Among them, C_a and C_b represent the two combinations where the customer is located, min represents the number of elements in the set, and $D(C_a, C_b)$ represents the distance between the customer location and the cluster center.

$$x_i = \frac{1}{n_i} \sum_{x \in C_i} x \tag{3}$$

Among them, C is the set where a customer is located, x is a customer in C_i, that is, $x \in C_i$, and n is the customer in the kth set. Repeat the above operations until the customers in the collection have not changed.

2.3 The Method of Distribution Vehicle Scheduling Optimization

(1) Tabu search algorithm

In order to solve the problem that it is difficult to obtain the global and global optimal solution during local search, the tabu search algorithm uses the amnesty criterion to release some excellent solutions that have been tabooed, so that a variety of solutions can be provided to obtain the overall optimal solution. The tabu search algorithm was first used to solve the delivery vehicle routing problem [10].

(2) Particle swarm algorithm

Particle swarm optimization is similar to genetic algorithm. The particle swarm algorithm regards each bird in the flock as a particle, and uses mathematical methods to quantify the position of each bird in the search area as the particle position of the current cycle. The position of the food in space corresponds to the optimal solution of the objective function to be optimized, and the distance between the bird and the food is used to evaluate the advantages and disadvantages of each particle solution. Therefore, the historical optimal position of each particle currently existing in the population is regarded as the individual optimal position, and the historical optimal position currently found by all particles in the population is regarded as the total optimal position. In the full optimization process, through the continuous competition and cooperation of each particle, the group is constantly approaching the optimal solution [11].

(3) Simulated annealing algorithm

The simulated annealing algorithm is a random search algorithm that simulates the principle of solid welding. The process of the algorithm is controlled by the boiling temperature of the solid material. In vehicle scheduling optimization, the optimal number of vehicle usage is found by analyzing the number of permutations of vehicle usage scenarios [12].

3 Design of Electronic Logistics Distribution System

3.1 Overall System Design

The system considers the transportation distance, cost and time of logistics distribution, which can increase the stability, scalability and compatibility of the logistics distribution system so that more systems can be applied. The logistics distribution system adopts Java EE-based Spring, SpringMVC, MyBatis (SSM) three-tier framework, relational database SQLServer, Tomcat server, JDK and MyEclipse software to implement. The logistics distribution system is divided into five modules, as shown in Fig. 1.

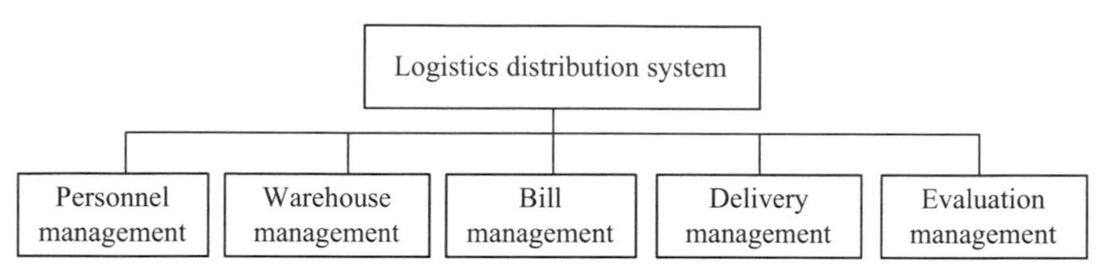

Fig. 1. Architecture of logistics distribution system

In the logistics distribution system, the main function of personnel management is to add, edit and modify information for managers and customers; the warehouse management module, the commodity in-out function, can view information, add and delete information for the commodities purchased by customers, and other operations; bills Management includes bill query, addition and deletion, and can be specific to a day's bill settlement and other functions; delivery management includes viewing the order information of products, the information of vehicle-delivered products, and the driving route of vehicle-delivered products; evaluation management is the customer's response to this. Evaluation of the delivery service of the product.

3.2 Database Design

As shown in Table 1, the electronic logistics distribution system database includes the recipient's name, address, telephone number, cargo name, cargo quantity, order date, logistics status and other information. To keep these information in the database, it is necessary to set the reserved fields length. The lengths of the above information fields are 5, 20, 15, 30, 5, 10, and 30, respectively.

Table 1. Database related shipping information

	Field length
Recipient's name	5
Receiver's address	20
Recipient's phone	15
Item name	30
The quantity of goods	5
Order date	10
Logistics status	30

4 System Application

4.1 Comparison of Sorting Efficiency

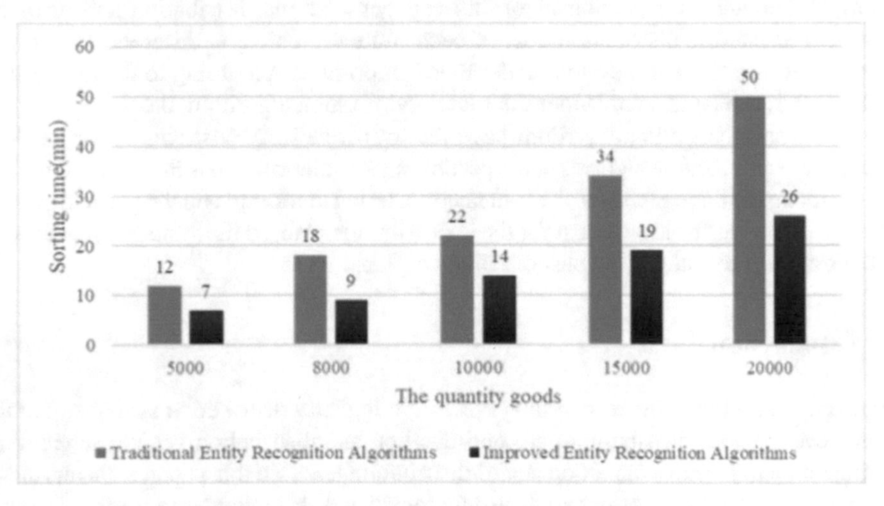

Fig. 2. Comparison of system sorting efficiency under two algorithms

Introduce the traditional entity recognition algorithm and the improved entity recognition algorithm into the logistics distribution system, use the system to sort the goods, and compare the speed of the two algorithms to sort the goods. The results are shown in Fig. 2. When sorting 5,000 pieces of goods, the system sorting under the traditional algorithm takes 12 min, and the system sorting under the improved algorithm takes 7 min. The sorting efficiency of the improved algorithm is nearly twice that of the traditional algorithm. This is because the improved algorithm integrates RFID technology has improved the speed of sorting and scanning goods, thereby improving sorting efficiency.

Table 2. Distribution cost comparison (yuan)

	Plan A	Plan B	Plan C
Saving algorithm	527	662	540
Genetic Algorithm	564	598	536
K-means clustering algorithm	413	476	395
Tabu search algorithm	622	581	554
Particle swarm algorithm	506	497	428
Simulated Annealing Algorithm	575	632	519

4.2 Comparison of Logistics and Distribution Costs

To transport 10,000 pieces of goods from place M to place N, the routes of different delivery schemes are different, and the number of vehicles to be called is also different. Because the saving algorithm, genetic algorithm, and k-means clustering algorithm can optimize the distribution route, while the tabu search algorithm, particle swarm algorithm, and simulated annealing algorithm can optimize the distribution and dispatch vehicles, so that the distribution cost of each route or vehicle is different. Hence the scheduling cost of the test system under these algorithms. According to the test results in Table 2, it can be known. Under the route optimization algorithm, the three schemes of the k-means clustering algorithm have the lowest delivery cost. Under the vehicle scheduling optimization algorithm, the particle swarm algorithm has the lowest delivery cost. Among the three schemes, the k-means clustering algorithm and the particle swarm The algorithm has the lowest cost for the C distribution plan, so the system can use these two algorithms to analyze the best distribution plan.

5 Conclusion

Many companies are now researching electronic logistics distribution systems, mainly to achieve logistics distribution path optimization and distribution vehicle scheduling optimization to improve the efficiency of distribution tasks. In this paper, k-means algorithm is added to the constructed electronic logistics distribution system for path optimization, particle swarm algorithm is added for vehicle scheduling optimization, and the best distribution scheme is found, which reduces distribution costs and saves vehicle resources, thereby improving the economy of society and enterprises. In addition, adding an improved entity recognition algorithm to the distribution system can speed up the sorting of goods. Therefore, the application of the optimization algorithm can be considered when designing the distribution system to improve the distribution efficiency.

References

1. Pegado, R., Rodriguez, Y.: Distribution network reconfiguration with the OpenDSS using improved binary particle swarm optimization. Latin Am Trans. **16**(6), 1677–1683 (2018)

2. Patil, N., Patil, A., Pawar, B.V.: Named entity recognition using conditional random fields. Procedia Comput. Sci. **167**(3), 1181–1188 (2020)
3. Bazi, I.E., Laachfoubi, N.: Arabic named entity recognition using word representations. Int. J. Comput. Sci. Inf. Secur. **14**(8), 956–965 (2018)
4. Os, A., Kj, A., Ms, A.: Using the operations research methods to address distribution tasks at a city logistics scale. Transp. Res. Procedia **44**(8), 348–355 (2020)
5. Mejjaouli, S., Babiceanu, R.F.: Cold supply chain logistics: system optimization for real-time rerouting transportation solutions. Comput. Ind. **95**(12), 68–80 (2018)
6. Hurtado, P.A., Dorneles, C., Frazzon, E.: Big Data application for E-commerce's logistics: a research assessment and conceptual model. IFAC-PapersOnLine **52**(13), 838–843 (2019)
7. Ron, C.: Blockchain in logistics, transportation and distribution. Apics Perform. Adv. **50**(3), 313–322 (2020)
8. Tasbirul, I.M., Nazmul, H.: Reverse logistics and closed-loop supply chain of waste electrical and electronic equipment (WEEE)/E-waste: a comprehensive literature review. Resour. Conserv. Recycl. **137**(5), 48–75 (2018)
9. D'Ascenzo, M.: GST and electronic distribution platforms. Int. VAT Monit. **31**(1), 14–18 (2019)
10. Koppers, M., Farías, G.G.. Organelle distribution in neurons: logistics behind polarized transport. Curr. Opin. Cell Biol. **71**(05), 46–54 (2021)
11. Mairizal, A.Q., Sembada, A.Y., Tse, K.M., et al.: Electronic waste generation, economic values, distribution map, and possible recycling system in Indonesia. J. Clean. Prod. **71**(6), 126096–126098 (2021)
12. Macías, R.G.: Logistics optimization through a social approach for food distribution. Socio-Econ. Plan. Sci. **2020**(1), 1–3 (2020)

Application Analysis of Electric Vehicle Intelligent Charging Based on Internet of Things Technology

Wenqiang Wang[1(✉)], Shanpeng Xia[1], Yongtao Nie[1], and Ali Rizwan Jalali[2]

[1] College of Automotive Engineering, Weifang Engineering Vocational College, Weifang 262500, Shandong, China
wwqzmr@163.com

[2] American University of Afghanistan, Kabul, Afghanistan

Abstract. Electric vehicle is a new type of mobile intelligent power equipment and energy storage terminal. Electric vehicle energy service infrastructure network is an important part of smart grid. In order to solve the automation and intelligence problems of wide area electric vehicle charging and changing business, this paper first analyzes the charging and changing business scenarios of electric vehicles and the technical support requirements of operation monitoring business. Therefore, this paper proposes the information perception application and communication networking mode of Internet of things technology in electric vehicle charging and swapping network. On this basis, a wide area electric vehicle charging and switching network operation monitoring and management platform architecture based on GIS and SOA is designed. The results show that: the total number of public charging piles in China increases with the growth of years, and the largest increase is 593000 in 2020.

Keywords: Internet of Things · Electric Vehicle · Intelligent Charging · Charging Pile

1 Introduction

New energy vehicles have been recognized by the market, many experts devote themselves to the research of new energy vehicles. The necessity of using compensation technology and mathematical model to determine circuit parameters to achieve effective transmission is proposed. This paper discusses the concept of intelligent charging station, which uses the technology of Internet of things to realize the automation of charging process. A special IPT system is provided, which uses the Internet of things to connect the charging system with the off grid solar panel to power the inductive coupler of the transmitter. A prototype detector that can handle up to 100W power transmission has been built and combined with existing IOT solutions. These solutions give evidence of the automatic charging process, and deduce and suggest the design parameters of Renault twizy. According to the information flow distribution of electric vehicle integrated station, the detailed parameters of station yard and vehicle terminal can be

J. H. Abawajy et al. (Eds.): ICATCI 2022, LNDECT 169, pp. 88–95, 2023.
https://doi.org/10.1007/978-3-031-28893-7_11

obtained through monitoring system [1]. Some experts have studied the information collection of smart grid, using RFID technology to collect the data of electric vehicle battery system, combined with the Internet of things and GPS, can quickly and accurately obtain the battery information. In addition, it can diagnose the state of the battery and deal with the fault in time. The research method of Internet of things terminal data integration is proposed. Based on middleware technology, the physical architecture of EV charging is analyzed. The heterogeneous hierarchical structure of EV charging network and the middleware of EV charging station network are designed. The structure and topology are adjusted. The main controller S3C2440 module is designed, the interface circuit of the system is designed, and the interface circuit which can communicate with 485 network and Ethernet is designed. The data resource integration algorithm of multi station electric vehicle charging terminal is improved. This paper introduces a vehicle two-way electric vehicle battery charger, describes the topology of the vehicle two-way battery charger and the control algorithm of three working modes. In order to verify the topology, a laboratory prototype is developed, and the experimental results of three working modes are obtained [2]. Parallel load forecasting algorithm is implemented by using parallel computing framework, and parallel local weighted linear regression model is proposed. Through the study of energy efficiency grade model, the energy consumption level of power users can be comprehensively evaluated. Using the Internet of things technology, It provides universal perception of the physical world and real-time interactive view through a variety of sensors and wireless devices, and provides an intelligent and effective charging management system (CMS) for electric vehicles (EV). Based on the distributed optimization capability of AC directional multiplier (ADMM), an effective decentralized charging scheme is proposed in order to provide more thoughtful and efficient charging service for electric vehicles in residential areas, A charging load management scheme for electric vehicles based on IEEE wave standard and IEC61850 standard is proposed. Based on IEEE wave and IEC 61850 standards, the communication and service model of electric vehicle based on RSU and CSS is established. An electric vehicle charging scheduling algorithm based on the communication between electric vehicle, RSU and CSS is proposed [3]. In order to study the intelligent charging of electric vehicles under the Internet of things, this dissertation studies the intelligent charging of the Internet of things and electric vehicles, and finds the GPS technology. It is indicated that the Internet of things is significant to the intelligent charging of electric vehicles.

2 Method

2.1 Internet of Things

(1) Concept of Internet of things

The Internet of things is to use multi-sensor data integration to improve the measurement accuracy of equipment and instruments and the correctness and integrity of data [4]. The traditional architecture of Internet of things includes perception layer, network layer and application layer [5]. The network layer realizes the safe and reliable transmission of information through various heterogeneous networks, including various wired and wireless networks, switches, gateways, etc.; the application layer provides users with information cooperation, data sharing and display functions, including data storage, cloud computing, big data processing, platform services, information display, etc. [6].

(2) The relationship between Internet of things and electric vehicles

The application of Internet of Things technology in electric vehicles will become a trend and also the mainstream development direction of electric vehicles [7]. On the one hand, the Internet of things collects the working state of the electric vehicle through the wireless sensor device, and transmits it to the background operation support management system through the wireless network, and then uses the powerful background management program to complete the real-time monitoring of the driving state of the electric vehicle, so as to ensure the safe driving and even unmanned driving. [8]. On the other hand, it can realize the intelligent perception of the charging station information, process the charging station data in real time, shorten the charging service time, and then reduce the number of infrastructure of the charging station [9].

(3) Application of Internet of things in electric vehicle charging

The charging network assistant management subsystem can provide charging equipment account management, operation monitoring, metering and billing, operation and maintenance management and other functions for managers, and can automatically and intelligently manage the resources and operation status of the whole charging network; China electric vehicle network can provide information services related to electric vehicles and charging network for the public and member users, including charging network search and management news consulting search provides members with monitoring and management of electric vehicle status information [10].

2.2 Intelligent Charging of Electric Vehicles

(1) Intelligent charging of electric vehicle

Electric vehicle is a new type of mobile intelligent power equipment and energy storage terminal [11]. Electric vehicle energy service infrastructure network is an important part of smart grid [12]. The traditional charging point network is divided into wired and wireless. Wired network generally uses Ethernet, which needs to lay network lines, and the cost is high. This method is only suitable for the construction of centralized charging station. For the construction of distributed charging pile, this method is not practical, so it is abandoned. Charging pile is responsible for providing personalized charging service for the owner. The car owner can start the charging pile by swiping the card or scanning the code on wechat to provide various ways of recharging, such as amount, time, automatic recharging, appointment recharging, etc. At the same time, the charging point has its own function of state detection and fault diagnosis. In the process of charging, the data of the vehicle battery pack and the running state of the battery stack are collected and reported to the cloud platform management system. In case of any abnormality, it will be shut down immediately to ensure the personal and property safety of users.

(2) Integrated mode of charging point Internet of things

In the operation and management of charging piles, the different battery types of electric vehicles in the vehicle network determine the diversity of users' charging needs; the use of charging parking spaces in different regions determines the diversity of charging options of electric vehicle users; the diversity of vehicle load characteristics; the power grid in different periods in the same region determines the diversity of charging power control modes; at the same time, different charging piles have different charging characteristics. The diversity of charging equipment information is determined by the power difference and different states of the same charging point. Therefore, the multi network integration mode of charging pile is to build an information bridge between charging network and vehicle network, parking network and power supply network.

(3) Operation law of electric vehicle

The operation law of electric vehicles is highly closed to the usage habits of drivers, involving driving distance, charging time and charging location between electric vehicles and charging facilities and so on. There is no big difference between electric vehicles and ordinary fuel vehicles, so the charging habits of electric vehicle users can be analyzed and summarized according to the driving data of fuel vehicles. If the user's charging behavior is inconsistent, a large number of users will choose to charge after work or on the rest day, and stop charging before going to work or when the power battery is fully charged, which is the disordered charging mode of electric vehicles.

2.3 GPS Technology

User equipment, also known as GPS receiving equipment, is mainly responsible for receiving satellite signals, and necessary conversion, amplification and processing, so as to calculate the coordinates, speed, heading, time and other data of the measured position. If the travel time data of single vehicle segment meets the condition of Eq. (1), it is the delay calculation data.

$$t_p - t_r \leq t < t_p \tag{1}$$

The system uses 3.3V and 5V voltage output value to supply power for each module. The calculation formula of variable voltage is shown in Eq. (2):

$$V_{out} = V_{REF}(1 + \frac{R_2}{R_1}) \tag{2}$$

The daily mileage probability distribution function of electric vehicles is shown in formula (3):

$$R_2 = R_1(\frac{V_{OUT}}{V_{REF}} - 1) \tag{3}$$

After setting the expected cost, the social insurance company. E when the EV starts, the user can calculate thc initial state of charge soci. According to the daily mileage D and power battery parameters of the electric vehicle, the charging time ti of the electric vehicle is calculated. It is shown in Eq. (4):

$$f_m(d) = \frac{1}{\sqrt{2\psi_m d}} \exp \tag{4}$$

3 Experience

3.1 Extraction of Experimental Objects

The communication module is responsible for the network construction and data transmission of the charging pile information management system. According to the overall design requirements of the system, the gateway is composed of coordination node, processor and GPRS module. ZigBee protocol is used to communicate between gateway and terminal node. The coordination node is responsible for the establishment of the network and receives the data and instructions sent by the terminal node; the GPRS module uses the mobile data network to send the data to the information management center, and the RS232 communication is used between the coordination node and the GPRS module.

3.2 Experimental Analysis

After the system resource is initialized, self-check is performed. If it fails, it will jump to the corresponding failure interface and report back to the background. Jump to the main interface after self-check. The main interface mainly integrates charging and system setting functions. One is the charging function. Users can get the permission to enable the charging point by swiping the card or scanning the code. After obtaining the license, the user can select the corresponding charging gun and connect it to the electric vehicle. Then, select the charging mode, start charging formally, calculate the charging cost in real time, and monitor the status of equipment and vehicle battery pack. If the charging stop condition is reached, first judge whether the charging is terminated due to abnormal conditions. If not, analyze the cause of the exception and report to the cloud platform. If not, settle the amount according to the normal ending process and return to the initial interface. First, the sensor on the terminal node collects data, and then sends it to the coordinator. According to the original path return information, the terminal makes the corresponding action to the received feedback information, so as to realize the remote monitoring of the terminal.

4 Discussion

4.1 Public Charging Infrastructure

Since entering the 21st century, great changes have taken place in the world energy system. The traditional energy system with fossil fuels as the main body has gradually changed into a new energy system with renewable energy as the main body. At the same time, transportation occupies an important proportion in various energy consumption modes. In such an environment, the new energy vehicle industry has a good development prospect, and China's charging infrastructure is also developing. As demonstrated in Table 1.

Table 1. General situation of public fee infrastructure in 2018–2020

Particular year	Total power supply(10000)	Year on year
2018	52.1	67%
2019	57.4	71%
2020	59.3	87%

It can be seen from the above that in 2018, there were 521000 public charging posts nationwide in China was, up 67% year on year; in 2019, the total quantity of public charging posts nationwide was 574000, up 71% year-on-year; in 2020, the total number of public charging posts nationwide was 593000, year-on-year growth of 87%. The results are exhibited in Fig. 1.

According to the table, we can see that the total number of public charging piles in China increases with the growth of the year, and the maximum increase of the total number of public charging piles in 2020 is 593000.

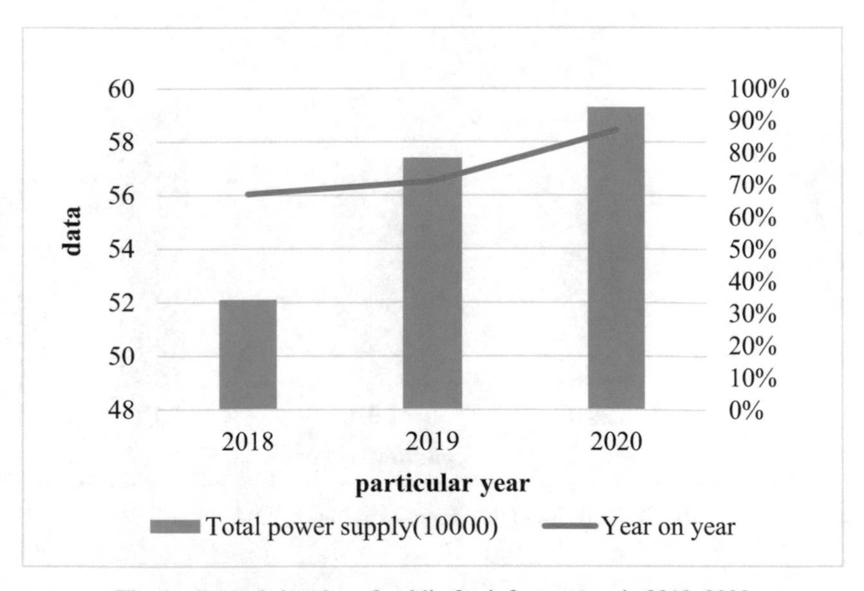

Fig. 1. General situation of public fee infrastructure in 2018–2020

4.2 Statistics of New Energy Vehicles

As a low-carbon green vehicle, electric vehicle meets the requirements of sustainable development of smart grid. The flexible interactive communication network provided by smart grid provides platform and technology for realizing intelligent operation and charging of electric vehicles. In the development plan of China's energy-saving and new energy vehicle industry, it is expected that by 2020, the number of electric vehicles in China will reach 6–12 million, and the construction of charging stations, charging piles and other facilities in major cities is in full swing. See Table 2 for the growth trend of new energy electric vehicles.

Table 2. Statistics of new energy vehicles in 2018–2020

Particular year	Total number of vehicles(vehicle)
2018	12000
2019	20000
2020	25000

It can be seen from the above that China will produce 12000 new energy vehicles in 2018, 20000 in 2019 and 25000 in 2020. The results are shown in Fig. 2.

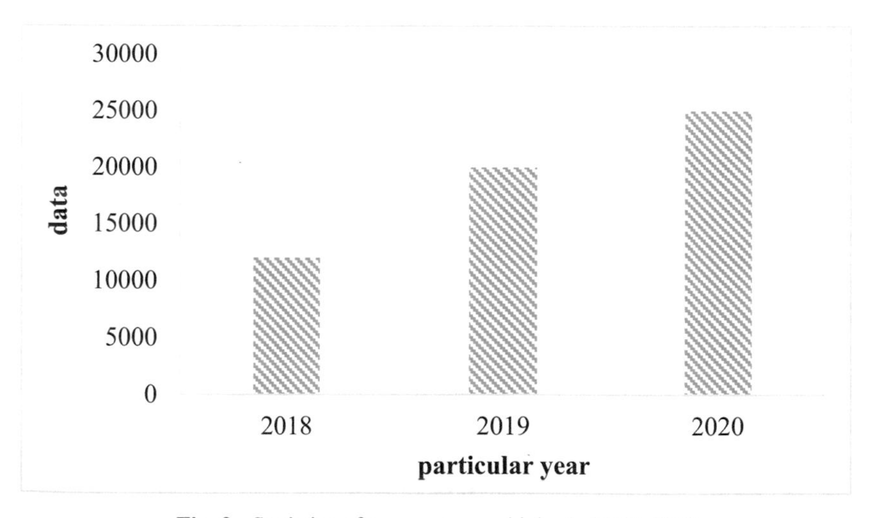

Fig. 2. Statistics of new energy vehicles in 2018–2020

It can be seen from the above that China will produce at least 12000 new energy vehicles in 2018 and at most 25000 new energy vehicles in 2020.

5 Conclusion

With the continuous increase of electric vehicles and the widespread application of new energy technologies, effective road network modeling plays a significant role in the development of the electric vehicles. This dissertation introduces a road network model which can better reflect the actual road conditions. In addition, the waiting time before charging is also considered. Compared with the minimum distance standard, the proposed time minimum standard and Dijkstra algorithm are used for electric vehicle path tracking. Finally, the speed limit of electric vehicle and other influencing factors are considered in the simulation to reflect the actual running state.

References

1. Li, H., Han, D., Tang, M.: A privacy-preserving charging scheme for electric vehicles using blockchain and fog computing. IEEE Systems Journal **15**(3), 3189–3200 (2020)
2. Metwly, M.Y., Abdel-Majeed, M.S., Abdel-Khalik, A.S., et al.: IoT-based supervisory control of an asymmetrical nine-phase integrated on-board EV battery charger. IEEE Access **8**, 62619–62631 (2020)
3. Cecil, J., Albuhamood, S., Ramanathan, P., et al.: An Internet-of-Things (IoT) based cyber manufacturing framework for the assembly of microdevices. Int. J. Comput. Integr. Manuf. **4**, 1–11 (2019)
4. Kaur, K., Garg, S., Kaddoum, G., et al.: Demand-response management using a fleet of electric vehicles: an opportunistic-SDN-based edge-cloud framework for smart grids. IEEE Network **33**(5), 46–53 (2019)
5. Meng, W., Cai, L., Yang, W., et al.: Mining subsidence prediction method based on geomagic. Metal Mine (1), 52–96 (2017)
6. Tang, Q., Wang, K., Song, Y., et al.: Waiting time minimized charging and discharging strategy based on mobile edge computing supported by software-defined network. IEEE Internet Things J. **7**(7), 6088–6101 (2020)
7. Liang, Y., Tao, J., Zou, Y.: Distributed online energy management for data centers and electric vehicles in smart grid. IEEE Internet Things J. **3**(6), 1373–1384 (2017)
8. Lin, C., Pan, J., Lian, Z., et al.: Networked electric vehicles for green intelligent transportation. IEEE Communications Standards Magazine **1**(2), 77–83 (2017)
9. Chaudhary, R., Jindal, A., Aujla, G.S., et al.: BEST: blockchain-based secure energy trading in SDN-enabled intelligent transportation system. Computers & Security **85**(AUG.), 288–299 (2019)
10. Hong, C., Li, G.: Optimization design of network structure based on genetic algorithm. In: 2018 International Conference on Virtual Reality and Intelligent Systems (ICVRIS), 206–209 (2018)
11. Korkas, C.D., Baldi, S., Shuai, Y., et al.: An adaptive learning-based approach for nearly optimal dynamic charging of electric vehicle fleets. IEEE Transactions on Intelligent Transportation Syst. **19**(7), 2066–2075 (2017)
12. Dabbaghjamanesh, M., Kavousi-Fard, A., Zhang, J.: Stochastic modeling and integration of plug-in hybrid electric vehicles in reconfigurable microgrids with deep learning-based forecasting. IEEE Trans. Intell. Transp. Syst. **99**, 1–10 (2020)

Urban Ecotourism Evaluation System Based on Ant Colony Algorithm

Ni Cheng[1](✉) and Anli Teekaraman[2]

[1] Wuhan Railway Vocational College of Technology, Wuhan 430205, Hubei, China
342603058@qq.com
[2] Vrije Universiteit Brussel, Brussels, Belgium

Abstract. Under the background of today's world economic development, tourism, as a new type of industry, is developing very fast. In order to promote the development of urban economy, it is necessary to construct its tourism resources and improve its service system. The purpose of this paper to study the evaluation system of urban ecotourism is to attract more tourists to the city and promote economic development through the design of ecotourism. This paper mainly uses the questionnaire survey method and the example method to conduct a related in-depth research on the urban ecotourism evaluation system. The survey data shows that 52.1% of the people believe that the urban ecotourism evaluation system needs to be improved from its natural environment, social history, human environment and service system.

Keywords: Ant Colony Algorithm · Urban Ecology · Tourism Development · Evaluation System

1 Introduction

The evaluation of urban ecotourism refers to the scientific and reasonable analysis and prediction of the number of tourists and the consumption capacity of tourists in a certain area by using certain methods and means. It can reflect the past development status, future trends and development of tourism resources of the region or a scenic spot.

There are many practical theoretical achievements in the research of urban ecotourism evaluation system based on ant colony algorithm. For example, Tobias Brandt et al. showed the potential value that spatial and semantic analysis of social media messages can bring to a smart tourism ecosystem [1]. Pedro Fernandes Anunciação et al. analyzed the degree of integration of public and private economic activities by assessing the location of the main players in the tourist area, comparing two important cities [2]. F. Dahan proposed that service composition has attracted increasing attention as a promising paradigm for optimizing data accessibility, integrity, and interoperability in cloud computing. He proposed an efficient agent-based ant colony optimization (ACO) algorithm to solve the cloud service composition (CSC) problem [3]. Therefore, this paper intends to start with the ant colony algorithm, and conduct related research on the urban ecotourism evaluation system.

J. H. Abawajy et al. (Eds.): ICATCI 2022, LNDECT 169, pp. 96–103, 2023.
https://doi.org/10.1007/978-3-031-28893-7_12

This paper firstly studies the ant colony algorithm and analyzes it in detail. Secondly, the basic theory and health indicators of the tourism system are described. Then the urban ecotourism system is expounded. Afterwards, a method for evaluating the ecological security of tourist cities is proposed. Finally, the research is carried out by means of a questionnaire survey, and a conclusion is drawn.

2 Urban Ecotourism Evaluation System Based on Ant Colony Algorithm

2.1 Ant Colony Algorithm

In the absence of available means of communication and timely information, the ant colony can always find the shortest path between the nest and food, and it can adjust its food-seeking behavior and make new decisions even if the environment changes. This intelligent swarm behavior arises because individual ants leave a special substance called a pheromone on the path they traverse and use this substance to guide them in choosing a route. Other ant individuals can also perceive the presence and intensity of this information, and switch to a direction with greater information intensity when choosing a path. Since there were no ants on either route and no pheromone to refer to, they chose randomly first. After a period of time, the remaining pheromones will appear in the path of individual ants, which other individual ants can refer to when choosing a route. Pheromones are substances that can dissipate at a rate and slowly disappear over time. Generally speaking, the shorter the path, the more pheromone accumulated, thus attracting more ants to choose, and the ant colony can finally find the optimal path through this self-organizing behavior [4, 5].

The total number of ants is t, and the variable b is the number of cities. When artificial ants construct an exploitable path for the TSP problem, the city nodes b in the path selection should first be numbered from 1 to b, and then the selection probability of each node is determined according to the selection formula of the given probability, and according to some rules the selection has a transition value node. Ant l calculates the probability that the next node k visits the current node i according to formula (1) [6, 7].

$$P_{ik}^{1}(s) = \begin{cases} 0, k \in allowed \\ \frac{\varsigma_{ik}^{\mu}(s)\Psi_{ik}^{v}(s)}{\sum_{s \in allowed}\varsigma_{ik}^{\mu}(s)\Psi_{ik}^{v}(s)}, k \notin allowed \end{cases} \quad (1)$$

where $\varsigma_{ik}(s)$ is the concentration of pheromone on the line composed of nodes i and k, usually when s = 0, take $\varsigma_i(0) = 0$ (constant).

After the ant colony builds the route, it will update the pheromone, thereby preventing the accumulation of pheromone on the route and making the heuristic information ineffective. It is updated as follows:

$$\varsigma_{\mathrm{ik}}(\mathrm{s}+\mathrm{S}) = (1+\rho)\cdot\varsigma_{\mathrm{ik}}(\mathrm{s}) + \sum_{\mathrm{l}=1}^{\mathrm{b}}\Delta\varsigma_{\mathrm{ik}}^{\mathrm{l}}(\mathrm{s},\mathrm{s}+\mathrm{S}) \quad (2)$$

where $\Delta\varsigma_{\mathrm{ik}}^{\mathrm{l}}$ is the pheromone released by ant l on the path (i, k).

(1) Advantages of Ant Colony Algorithm

Self-organization: Self-organization refers to the process in which the state of the entire system gradually changes from disorder to order under the influence of the internal interaction of the system. In the initial stage of the ant colony algorithm, each ant conducts a random unordered search, and then, as pheromones play a prominent role in this process, each ant progresses toward the optimal solution. The search process of ant colony algorithm reflects its own self-assembly characteristics [8, 9].

Robustness: The robustness of the ant colony algorithm is mainly reflected in that it does not rely heavily on the selection of the initial solution when searching for the optimal solution of the problem and the solution space of the problem, and does not depend on the constant adjustment of algorithm parameters.

Positive feedback: When ants are foraging, taking a shorter route will release more pheromones, and an increase in the amount of information will cause more ants to choose the shorter route. Relatively speaking, pheromones become shorter and shorter, slowly disappearing over time. This process reflects the positive feedback characteristic of the algorithm [10, 11].

Distributed computing: The process of each ant looking for a solution to the problem is relatively independent, and they communicate indirectly through pheromone, and the failure of one ant does not affect the final optimal solution of the problem.

2.2 Tourism System

Tourism system is a collection of different tourism activities, and it is an organic whole formed by the contact and interaction of tourists' tourism activities. It has the overall function of realizing tourism value. Modern tourism is a large industry with great influence, wide participation, many factors, great changes, high status and globalization [12].

Urban ecotourism mainly focuses on the product of urban tourism development, the purpose of urban tourism, the role of architects in urban tourism, the form of urban ecotourism, and the combination of urban tourism and culture. These activities provide tourists and local residents with more opportunities to appreciate the city's natural and cultural resources, respect and protect urban resources and cultural diversity, protect local cultural relics and arts, promote urban ecological health, and encourage people to actively participate in sports, social activities and support the growth of communities and economy.

First, urban ecotourism is a sustainable tourism concept based on the theory of eco-city and urban ecosystem. Secondly, from a geographical point of view, urban ecotourism is a tourist behavior in cities and suburbs. Third, it includes two streams of natural ecotourism and cultural ecotourism. Finally, urban ecotourism is a healthy form of tourism that combines economic, social and environmental benefits.

Ecosystem health refers to the stability and sustainability of ecosystems. That is have ability to maintain organizational structure, self-regulation, and ability to cope with stress over time. For the most man-made urban ecosystems. Health means not only the health and integrity of natural and man-made environmental ecosystems that provide ecological services to people, but also the health and social health of urban residents.

The Urban Ecosystem Health Index system is used to assess changes in population health, social health and well-being, and natural ecosystem integrity caused by social, economic, cultural, environmental, and political interactions in urban areas.

2.3 Urban Ecotourism System

A definition of an urban ecosystem was discussed and developed. The urban ecosystem is a human-centered ecosystem, which has been artificially transformed in structure, transformed in material circulation, and partially transformed in energy conversion, and has been affected by human activities for a long time. The urban environment subsystem includes the overall natural, economic and social environment of urban ecotourism activities.

The natural ecological environment mainly includes atmospheric environment, water environment, acoustic environment, green space environment and biological species environment. The economic environment mainly refers to the "big" external environment of urban economic development and the "small" internal environment of tourism economic development. The social environment mainly refers to the environment of the participants in tourism activities, such as the population density of local residents, the ratio of tourists to men and women, age structure, education level and departments, social management policies, etc. The details of urban ecotourism are shown in Fig. 1:

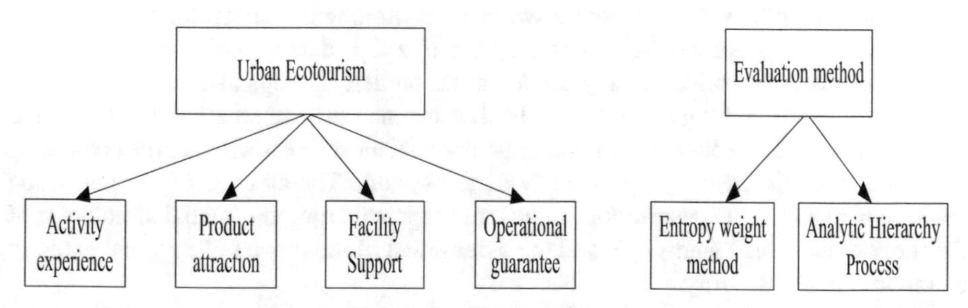

Fig. 1. Urban Eco-Tourism System

The experience subsystem of urban ecotourism activities mainly includes the perception of tourists' experience and the comfort of local residents in tourism activities. Tourist perception mainly refers to tourists' satisfaction with hotels and attractions, as well as their satisfaction with entertainment facilities, tourist safety, reception services, etc. The comfort level of residents indicates whether tourism activities affect the lives of local residents. Some bring comfort to local residents, provide more employment opportunities, and increase economic income, while others bring inconvenience to life, causing traffic congestion and pollution.

The attractive subsystem of urban ecotourism products, including urban ecotourism products and their combinations, travel route design, urban ecotourism resources, landscape value and other components. Urban ecotourism products are the core of the urban ecotourism system and have higher requirements for tourism types. With the constant changes of market competition and market demand, the combination of tourism products is also constantly changing. The combination of quantity, quality, advantages and disadvantages determines the quality of the tourism system to a large extent.

The supporting subsystem of urban ecotourism equipment mainly includes three categories: tourism reception facilities, tourism public service facilities and tourism infrastructure. It mainly includes accommodation facilities, cultural and entertainment facilities, catering facilities, shops, public leisure facilities, tourism security facilities and other tourism reception facilities, ecological interpretation systems and public health facilities such as transportation, water, electricity and lighting. These facilities are designed to meet the needs of ecotourists, and the quantity and quality of supporting systems that ensure the normal operation of ecotourism spaces will also affect the quality of urban ecotourism systems.

The urban ecotourism operation guarantee subsystem, the urban ecotourism system is vital to tourists and residents, and is an important social public product. In order to protect the public's ecotourism rights and help some enterprises and tourists fulfill their obligations to regulate tourism behavior, it is necessary to improve the quality of urban ecotourism system and provide a solid guarantee for its operation. These assurances include political and regulatory assurances, management assurances and technical assurances.

2.4 Evaluation Methods of Ecological Security in Tourist Cities

When using the rating index system to evaluate environmental safety, the first step is to determine the weight value of the index. At present, weight determination methods can be divided into four categories, namely mathematical models, ecological models, ecological landscape models and digital soil models. The essence of mathematical models is the quantification of indicator data. The ecological model must assess whether the ecological load state exceeds the carrying capacity of the system. The essence of the landscape ecological model is the description of the landscape structure, the spatial simulation of the three-dimensional landscape, and the assessment of the impact of regional ecology on environmental security.

The entropy weight method can be completely based on objective quantitative data, and can be evaluated by scientific calculation methods, which excludes many uncertain factors and the one-sided influence of personal subjective consciousness and experience, and has great scientific and reliability. Gaining abstraction means losing information. The higher the degree of order, the lower the entropy, and the more information it contains; on the contrary, the higher the degree of disorder, the greater the entropy, and the lower the amount of information.

Since the indicators of the evaluation index system cannot be compared due to the inconsistency of the dimensions between the coefficients, the original data of the indicators must be unified before determining the weights to eliminate the influence of different units and different measurement standards. For a scoring index system with multiple hierarchical structures, according to the additivity of the concept of entropy, the weight value corresponding to the upper structure can be determined proportionally by using the index calculation value of the lower structure.

AHP is an in-depth analysis of complex decision-making problems, combining qualitative and quantitative analysis methods to examine the nature or influencing factors of the problem, transforming analytical thinking into an objective model, and using a small amount of information to prioritize human resources. Thought process, mathematics and effective analysis. AHP is flexible, easy to understand, and easy to use. It is well integrated with the existing theoretical system, accepted by practical decision makers, and can solve their problems. Attribute Hierarchy Model is a new unstructured decision-making method that adapts to the correlated index system. In addition, AHM can avoid many tedious calculations and facilitate decision-making.

Construction of ecological security index model According to the structural characteristics of the ecological security index evaluation index system of tourist cities established in this paper, this paper adopts the weighted average method to comprehensively evaluate the ecological security status. Among them, the evaluation criteria include driving force, pressure, condition, influence and response.

3 Investigation on the Design of Ecotourism Evaluation System

3.1 Background of Urban Ecotourism System Design

Ecotourism refers to the realization of harmonious development between people and the environment through the optimization of the natural environment based on the natural ecological environment. In this process, it is necessary to take into account that there is a certain degree of difference between different tourists. The evaluation system of urban ecotourism mainly includes natural factors, social and historical conditions, humanistic characteristics, and resource utilization value. Ecotourism is based on the concept of environmental protection, through the protection of the natural environment, so that tourists can experience physical and mental pleasure and spiritual satisfaction in the process of experience.

3.2 Questionnaire Survey of Urban Ecotourism Evaluation System

In order to design an excellent urban ecotourism evaluation system, it is necessary to master its basic requirements and people's needs for the ecotourism evaluation system. Therefore, this paper specially invited local residents to conduct a questionnaire survey on urban ecotourism. The questionnaire survey of this paper is mainly carried out around the ecotourism evaluation system. These include residents' suggestions on the evaluation system, pointing out existing ecotourism problems, etc.

3.3 Questionnaire Survey Process

This questionnaire firstly designed relevant questions, and used the entropy weight method to judge the reliability and index factors of the questionnaire. Secondly, 200 enthusiastic citizens were invited to answer the questions. The questionnaire lasted for a week. 190 questionnaires were effectively recovered, and the data were sorted.

4 Analysis of Questionnaire Results

4.1 Indicator Attitude of Urban Ecotourism Evaluation System

According to the results of the questionnaire survey in this paper, this paper analyzes the natural environment, social history, human environment and service system of urban tourist attractions. The data are shown in Table 1.

Table 1. The Indicator Attitude of Urban Ecotourism Evaluation System

	Agree	Disagree	General
Natural environment	29	8	11
Social history	22	10	16
Cultural environment	25	6	20
Service system	23	6	14

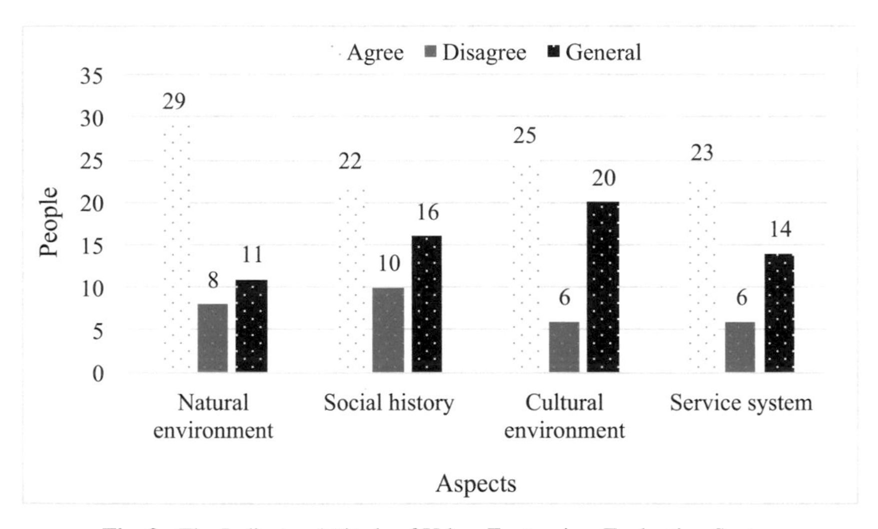

Fig. 2. The Indicator Attitude of Urban Ecotourism Evaluation System

As shown in Fig. 2, we can find that in terms of natural environment, 29 of the surveyed citizens agree with it. In terms of social history, 22 of the citizens surveyed agreed with it. In terms of humanistic environment, 25 people in the surveyed citizens agreed with it. In terms of service system, 25 of the citizens surveyed agree with it.

5 Conclusion

Ecotourism refers to the scientific management of various artificial systems based on the natural landscape and human environment, so that tourists can feel the physical and mental pleasure and cultivate their sentiments when viewing the natural scenery. The development of urban ecotourism requires a deep understanding of its evaluation indicators. In this paper, an urban ecotourism system model is constructed from the natural environment factors, and the optimal solution is obtained by using the ant colony algorithm. In the end, this paper analyzes the evaluation index criteria in people's minds by means of a questionnaire survey, and believes that the construction of the natural environment of the city is the most important.

References

1. Brandt, T., Bendler, J., Neumann, D.: Social media analytics and value creation in urban smart tourism ecosystems. Inf. Manag. **54**(6), 703–713 (2017)
2. Anunciação, P.F., Peñalver, A.J.B.: Information urbanistic perspective in the context of blue economy: analysis of setúbal and cartagena tourism offer. Int. J. Sociotechnology Knowl. Dev. **9**(3), 65-83 (2017)
3. Dahan, F.: An effective multi-agent ant colony optimization algorithm for QoS-aware cloud service composition. IEEE Access **9**, 17196–17207 (2021). https://doi.org/10.1109/ACCESS.2021.3052907
4. Dahan, F., El Hindi, K.M., Ghoneim, A., Alsalman, H.: An enhanced ant colony optimization based algorithm to solve QoS-aware web service composition. IEEE Access **9**, 34098–34111 (2021)
5. Pohan, M.A.R., Trilaksono, B.R., Santosa, S.P., Rohman, A.S.: Path planning algorithm using the hybridization of the rapidly-exploring random tree and ant colony systems. IEEE Access **9**, 153599–153615 (2021). https://doi.org/10.1109/ACCESS.2021.3127635
6. Muteeh, A., Sardaraz, M., Tahir, M.: MrLBA: multi-resource load balancing algorithm for cloud computing using ant colony optimization. Clust. Comput. **24**(4), 3135–3145 (2021)
7. Hamim, M., El Moudden, I., Pant, M.D., Moutachaouik, H., Hain, M.: A hybrid gene selection strategy based on fisher and ant colony optimization algorithm for breast cancer classification. Int. J. Online Biomed. Eng. **17**(2), 148–163 (2021)
8. De Melo Menezes, B.A., Herrmann, N., Kuchen, H., de Lima Neto, F.B.: High-level parallel ant colony optimization with algorithmic skeletons. Int. J. Parallel Program **49**(6), 776-801 (2021)
9. Jelokhani-Niaraki, M., Samany, N.N., Mohammadi, M., Toomanian, A.: A hybrid ridesharing algorithm based on GIS and ant colony optimization through geosocial networks. J. Ambient Intell. Humaniz. Comput. **12**(2), 2387-2407 (2021)
10. Kanso, B., Kansou, A., Yassine, A.: Open capacitated ARC routing problem by hybridized ant colony algorithm. RAIRO Oper. Res. **55**(2), 639–652 (2021)
11. Rathee, M., Kumar, S., Gandomi, A.H., Dilip, K., Balusamy, B., Patan, R.: Ant colony optimization based quality of service aware energy balancing secure routing algorithm for wireless sensor networks. IEEE Trans. Eng. Manage. **68**(1), 170–182 (2021)
12. Vafaei, M., Khademzadeh, A., Pourmina, M.A.: A new QoS adaptive multi-path routing for video streaming in urban VANETs integrating ant colony optimization algorithm and fuzzy logic. Wirel. Pers. Commun. **118**(4), 2539-2572 (2021)

License Plate Recognition System Based on Neural Network

Jinhai Zhang(✉)

Shandong Jiaotong University, Jinan, Shandong, China
zhangjinhai76@163.com

Abstract. Artificial neural network has excellent data fitting ability and is widely used to predict the development trend of complex nonlinear systems, However, the research on the application of artificial neural network in intelligent traffic regulation has the problems of difficult access to traffic sample data caused by insufficient development in the field of traffic monitoring in China, and it is impossible to verify the effectiveness of artificial neural network in the regulation of regional traffic system only by predicting the traffic flow of a single road. Intelligent transportation system applies advanced information technology, computer technology, sensor technology and image processing technology to realize the automatic management of vehicles. As an important part of intelligent transportation system, license plate recognition plays a vital role in traffic management. Therefore, the research on license plate recognition technology has very important practical significance.

Keywords: OpenCV · SVM · Intelligent Transportation System · Artificial Neural Network

1 Introduction

Nowadays, with the rapid development of society and the increasing number of social vehicles and private vehicles, the society is always faced with traffic congestion, frequent accidents, energy shortage and atmospheric environment deterioration caused by exhaust emissions. With the development of the times and the progress of science, people not only have a strong dependence on transportation, but also put forward faster, safer and more environmental requirements for the urban transportation system. How to meet this requirement, for many cities, the main ways are: expanding service capacity, increasing the construction of transportation facilities, vigorously developing public transportation, limiting people's traffic demand and so on.

The concept of intelligent transportation system has entered people's vision. Compared with the past traffic control system, the intelligent transportation system has great superiority [1]. One of its core technologies is the artificial neural network, which can reliably predict short-term traffic statistics and adjust traffic flow in advance. This not only releases a lot of human and material resources, but also can efficiently tap the potential of the transportation system itself and alleviate the traffic pressure.

License plate recognition technology has been studied in western developed countries around the 1980s. At the same time, some recognition technologies with practical value

J. H. Abawajy et al. (Eds.): ICATCI 2022, LNDECT 169, pp. 104–110, 2023.
https://doi.org/10.1007/978-3-031-28893-7_13

have been gradually extended to the scenes of access monitoring and electronic charging. British alpha technology company developed an automatic license plate recognition system called Argus around 1985. The system can track and process black-and-white images. The recognition time of vehicles passing through the system is within 0.1 s, and the speed of vehicles can reach 100 miles per hour; The vehicle license plate recognition system in Asia includes the VLPRS system developed by Singapore Optasia group [2]. This system can well complete the license plate recognition in Singapore and the VECON system developed by Hong Kong ATV group. This system can be effectively applied to the recognition of Hong Kong license plates. At the same time, most developed countries and regions have developed relevant systems that can be used for vehicle license plate recognition in this region. In the aspect of license plate recognition, at present, it is mainly through template matching, support vector machine classifier, artificial neural network classifier and a series of methods related to cluster analysis.

Automatic license plate recognition system has developed rapidly in various fields and tends to be mature. At present and in the future, the rapid development of intelligent transportation system in the whole country will make the development of automatic license plate recognition technology develop in the direction of high definition, intelligence and integration, and play its unique role in different fields [3].

2 Overview of License Plate Recognition System

License plate recognition system has the advantages of small volume, low cost, convenient movement and so on. It is widely used in intelligent parking lot, highway toll station, safety monitoring bayonet and so on. With the rapid development of electronic technology and computer technology, the processing speed of embedded license plate recognition system will be faster and faster, so its development prospect will be better and better.These features are described below [4]:

(1) Bionic characteristics

Bionic characteristic is one of the important characteristics of artificial neural network algorithm. Bionic (imitating Biology) characteristic not only affirms the effectiveness and enforceability of artificial neural network algorithm in reality, but also helps to improve the artificial neural network algorithm because the artificial neural network algorithm comes from the bionics of human brain.

(2) Characteristics of massively integrated parallelism

Artificial neural network is composed of large-scale artificial neurons connected with each other. The calculation and information storage of each neuron are real-time, that is, parallel. This characteristic of artificial neural network brings many benefits, such as fast response speed, strong computing power and so on.

(3) Nonlinear characteristics

Artificial neural network itself is composed of nonlinear neurons. This nonlinear characteristic will make artificial neural network have advantages in dealing with nonlinear information.

(4) Unity

Although today's artificial neural network algorithms are different, their representation method is unified, that is, using unified representation symbols, which is conducive to the dissemination and development of artificial neural network algorithms.

(5) Generalization properties

The generalization characteristic means that the artificial neural network will have a reasonable output for the untrained input information. This characteristic helps to expand the work field of the artificial neural network[5]. For example, the artificial neural network used for the traffic flow prediction training of a certain area can also be used for the traffic flow prediction of another area directly or simply through training.

(6) Adaptive characteristics

The external environment of neural network generating input data may change for some reasons, resulting in the change of input data characteristics. When the change range is small, the neural network can be reused by simple training, or directly set as a neural network that can adjust itself in real time.

(7) Global properties

The global characteristic of artificial neural network is that the information of each neuron will receive the potential influence of all neurons.

(8) Fault tolerance characteristics

The fault tolerance characteristics of neural networks come from the integration and parallelism of large-scale neurons. The more complex the structure and the more neurons, the higher the fault tolerance[6]. Whether the implementation form of artificial neural network is hardware or software simulation, it has good fault tolerance, but the fault tolerance of hardware implementation is better than that of software simulation, because the artificial neural network implemented by hardware has more computing nodes and storage nodes.

Artificial neural network BP model is an algorithm based on artificial neural network (ANN). It was studied and designed by Rumelhart, McCelland and their research team in 1986. BP algorithm has become the most widely used neural network learning algorithm at present. Structurally speaking, BP neural network is a typical multi-layer network, which is divided into input layer, hidden layer and output layer, as shown in the Fig. 1. It is characterized by: layers adopt full interconnection mode, there is no interconnection between units in the same layer, and there is no feedback connection between neurons in each layer. The connection between elements is realized by a specific excitation function. It is nonlinear, non limiting, very qualitative and non convex, and has a wide range of adaptability.

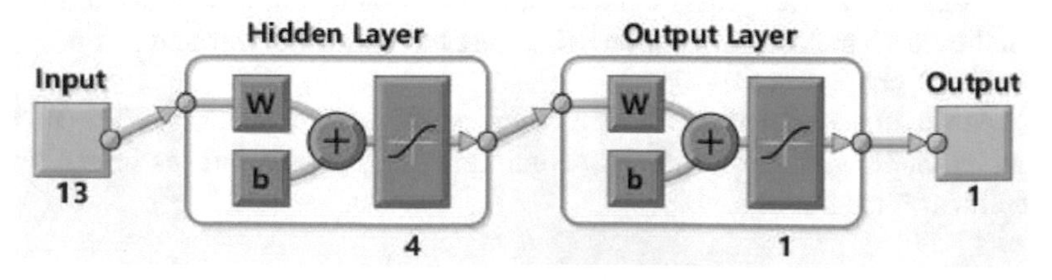

Fig. 1. BP model structure

It has excellent adaptability to BP model. Firstly, the variable structure of this study conforms to the characteristics of ANN stratification. The input layer is composed of 13 secondary variables such as main business income, and the output layer is composed of enterprise market value [7]. The number of hidden layers does not need to be calculated according to the empirical formula. The previous factor analysis concluded that the best number of hidden layer units is 4. Secondly, for the nonlinear relationship from primary variable to dependent variable, ANN (BP) can solve the defects of multiple linear regression equation and improve the accuracy of the model.

The fitting ability of artificial neural network is not only affected by the algorithm, but also the structure is very important A link in the. Due to the ultra-high degree of freedom of neural network, different hidden layers and different numbers of hidden layers can be determined, and the best fitting scheme can be given after the training process. Too few hidden layers or too few hidden layers will lead to the network under fitting, while too many hidden layer neural nodes or too many hidden layers may lead to the network over fitting. Therefore, when using BP network modeling and calculation, it is necessary to determine the optimal number of hidden layers and the number of hidden layer neurons (Fig. 2).

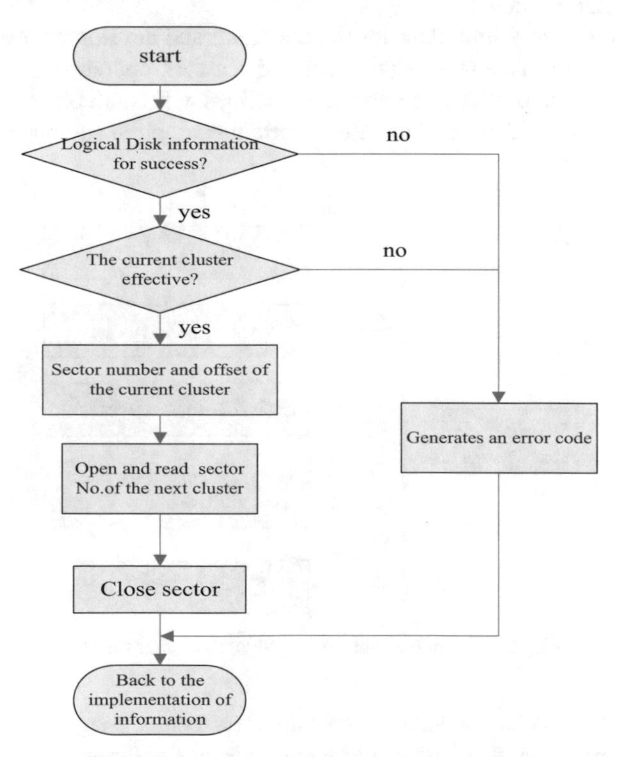

Fig. 2. Flow chart for the next cluster number

3 Design of License Plate Location Platform

At present, the prediction models applied to the traffic regulation methods of intelligent transportation system include multiple linear regression model, historical trend model, time series model and Kalman filter model in addition to artificial neural network [8].

(1) Historical trend model

The historical trend model assumes that the traffic conditions occur periodically, that is, each section of the road has the same traffic flow in the same period in a day with the same historical trend. The key to establishing the model is to classify the working days with similar historical trend. Although the historical trend model can solve the problem of traffic flow change in different time and different periods to a certain extent, its dynamic prediction is insufficient because it cannot solve the problem of traffic flow prediction in unconventional and sudden traffic conditions, such as traffic flow prediction in traffic accidents.

(2) Time series model

The theoretical basis of time series analysis is mainly mathematical statistics and random process theory. [9].

(3) Kalman filtering model

Security is very important for the operation and development of the network, such as the use of network equipment and network operating system with good network security, the use of communication lines with small bit error rate and good anti-interference ability, and the use of certain encryption equipment (Fig. 3).

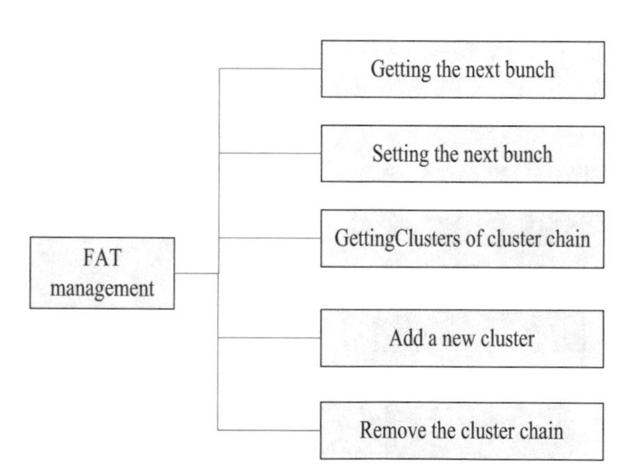

Fig. 3. Kalman structure for the next cluster number

Generally, the image obtained from the camera is a color image, which is composed of R (red), G (green) and B (blue). For each color in the three primary colors RGB, 256 levels can be used as a gray image, that is, each color can be represented by 8-bit binary data. Therefore, the three primary colors need a total of 24 bit binary numbers, so the number of color types that can be represented is 256 $\times$ two hundred and fifty-six $\times$ 256 = 224, about 16 million kinds. Therefore, there will be a huge amount of data for each license plate image, which is very unfavorable to the later calculation and affects the speed of license plate recognition [10]. Because license plate recognition only processes the brightness of license plate measurement, and there is no requirement for color. Therefore, we convert the color image into gray image in order to improve the speed of license plate recognition.

Due to the randomness and diversity of noise, there are many methods to eliminate noise, including mean filter, median filter, wavelet transform filter, Wiener filter and so on. When the linear smoothing filter contains noise points in the processed pixel field, the existence of noise will always affect the calculation of the pixel value of the point more or less (for Gaussian smoothing, the degree of influence is directly proportional to the distance of the noise points), but in the median filter, the noise is often directly ignored; Moreover, compared with the linear smoothing filter, the median filter has lower blur effect while reducing noise [11].

Binarization is to convert a gray image into a binary image with only black and white pixels. Binarization plays a very important role in image segmentation and character segmentation, because the binarized image has only black and white pixels. We can set the region of interest as white (or black). For segmentation, we only need to segment the white (or black) pixels, so as to complete the task of image segmentation

(1) Clustering connected domain character segmentation method [12].

    Its basic idea is that the pixels of each character form a separate connected domain, and the characters can be segmented as long as they reach the starting and ending positions of the rows and columns of each character. However, the first character is a Chinese character, which often has connectivity. This requires a priori data such as the width and height of a single character in the license plate and the spacing between characters, so as to decide whether to re cluster and segment. For example, if the word "Xiang" in the license plate of Hunan is segmented into the right part of the word "Xiang" and the three points of water on the left are lost, it is necessary to re cluster and segment the characters according to some a priori data.

(2) License plate character segmentation based on projection

    The extended SVC handler is the application program interface of the subsystem, which is responsible for receiving the request of the upper program. It can be called like an ordinary system call. The request is transmitted through a parameter data packet and a function code. All the entries of the subsystem call are registered in this handler.

Function code is a 32-bit value. Its lower 8 bits are the subsystem ID. The remaining high bits can be used by the subsystem for any purpose. Usually, they are used as function codes within the subsystem. The function code of the subsystem must be a positive number, that is, its highest bit must be 0. If it is a negative number, it means that the function code is an SVC call, not an extended SVC call.

In the HFS file subsystem, in order to make the coding of the function code clearer, the following rules are specified: for a 32-bit subsystem function code, 0–7 bits are the subsystem ID, 8–15 bits are the number of parameters of the function, 16–23 bits are the module sub serial number of the function in the HFS subsystem, 24–27 bits are the module serial number in the HFS subsystem, 28–31 bits are reserved, and the highest bit must be 0.

Vehicle number plate is the only identity mark of the vehicle. Because of its particularity and importance, the most important part of intelligent traffic management system is license plate recognition system. The English abbreviation of license plate recognition system is LPR. The technical basis of LPR is image processing, pattern recognition and other technologies. The main recognition method is to recognize the license plate photos taken by the camera on the road.

The rapid development of urbanization in China has brought the increasing pressure of transportation, which has also created many opportunities for the development of intelligent traffic management. Experts believe that in the future, the development of license plate recognition technology can be rapidly expanded in many fields, and the whole industry of license plate recognition system is bound to undergo great changes. Among them, companies that master the core technology can occupy a certain market position. Of course, this is an important way for the rapid development of license plate recognition technology.

References

1. Degond, P., Delitala, M.: Modelling and simulation of vehicular traffic jam formation. Kinetic & Related Models **21**(4), 96–107 (2019)
2. Souza, A.M.D., Yokoyama, R.S., Maia, G., et al.: Real-time path planning to prevent traffic jam through an intelligent transportation system. Computers and Commun. IEEE **32**(2), 196-206 (2019)
3. Njord, J., Peters, J., Freitas, M., et al.: Safety applications of intelligent transportation systems in Europe and Japan. Eur. J. Neurosci. **45**(4), 456–463 (2020)
4. Walsh, G.C., Ye, H.: Scheduling of networked control systems. IEEE Control Systems Magazine **32**(3), 98-111 (2019)
5. Zhu, X., et al.: Traffic flow prediction based on artificial life and RBF neural network. Energy Procedia **56**(3), 81–88 (2019)
6. Huang, S., Sadek, A.W.: A novel forecasting approach inspired by human memory: the example of short-term traffic volume forecasting. Transp. Res. Part C **43**(1), 142–153 (2019)
7. Fallah-Tafti, M.: The application of artificial neural networks to anticipate the average journey time of traffic in the vicinity of merges. Knowl.-Based Syst. **33**(5), 62–70 (2020)
8. Wan, L., Zeiler, M., Zhang, S., et al.: Regularization of neural networks using dropconnect. International Conference on Machine Learning **59**(4), 216-224 (2020)
9. Chollet, F.: Xception: deep learning with depthwise separable convolutions. IEEE Conference on Computer Vision and Pattern Recognition **46**(2), 155–166 (2019)
10. Rastegari, M., Ordonez, V., Redmon, J., et al.: Xnor-net: Imagenet classification using binary convolutional neural networks. European Conference on Computer Vision. Springer **38**(5), 94-101 (2019). https://doi.org/10.1007/978-3-319-46493-0_32
11. Bottou, L., Curtis, F.E., Nocedal, J.: Optimization methods for large-scale machine learning. SIAM Rev. **60**(2), 223–311 (2019)
12. Potena, C., Nardi, D., Pretto, A.: Fast and accurate crop and weed identification with summarized train sets for precision agriculture. International Conference on Intelligent Autonomous Syst. **40**(3), 105–121 (2020)

Design of Mechanical Equipment Fault Detection Robot Based on Electronic Tracing Algorithm

Yizun Yao[1(✉)] and Jumshaid Ullah Khan[2]

[1] Guilin University of Electronic Science and Technology, Guilin, Guangxi, China
yaoyizun@163.com
[2] Commune d'Akanda, Akanda, Gabon

Abstract. With the rapid development of modern industrial technology, robot fault detection technology has been widely used in many technical sectors and fields. Due to the multiple limiting factors in the mechanical structure of traditional robotics, it can only complete some relatively monotonous and simple tasks, and the process of parameter design is also very complicated. The robot designed based on the electronic tracing algorithm in this paper is not only capable of adapting to the environment, but also has a relatively low cost and can be used in more complex working environments. The robot design model based on the electron tracing algorithm proposed in this paper is constructed by the transfer of photoelectrons to the cathode, through the combination of the electron differential and the robot model structure, and obtains different electron distribution algorithms, and applies the electron tracing algorithm to the robot. It is very important for the detection of mechanical equipment failure in the design of the machine. This paper mainly studies the design of mechanical equipment fault detection robots, proposes a series of data model analysis, and analyzes the main theories and characteristics of the electronic tracing algorithm. The final results of the research show that the accuracy of the MMLT algorithm under four faults is 0.89 and 0.91, 0.96 and 0.88, the highest accuracy is 0.96, in contrast, the MMLT algorithm is better than the other two algorithms.

Keywords: Equipment Failure · Electronic Tracing · Robot Design · Electronic Transportation

1 Introduction

In the current production process of enterprises, machinery and equipment play a vital role in product production and improving work efficiency. Machinery and equipment greatly improve the utilization rate of workers, avoid complicated and repetitive labor services, and also ensure their work safety. Compared with traditional industrial production, it also greatly reduces industrial energy consumption [1]. Nowadays, the development of mechanical equipment is developing in the direction of high speed, advanced and heavy load. Therefore, the safety problem of mechanical equipment needs to be solved

J. H. Abawajy et al. (Eds.): ICATCI 2022, LNDECT 169, pp. 111–117, 2023.
https://doi.org/10.1007/978-3-031-28893-7_14

urgently, which is mainly manifested in monitoring the working status of mechanical equipment [2]. The risk brought by mechanical equipment failure cannot be ignored. Robot fault detection technology needs to be perfected for the application of mechanical processing workshop. The research on mechanical equipment fault detection robot has far-reaching significance for improving equipment stability and safety.

In recent years, many researchers have studied the robot design of mechanical equipment fault detection based on electronic tracing algorithm, and achieved good results. For example, Deb A believes that the mechanical equipment fault detection robot should mainly study the detection effect presented in the simulation process, analyze the overall structure of the robot to detect the fault, and detect and use the detected fault information [3]. Kindt P H believes that the robot can accurately detect the failure of mechanical equipment during the action process, and when the robot system is fully designed, it can complete the task autonomously without human intervention [4]. At present, domestic and foreign scholars have carried out a lot of research on mechanical equipment failure robots. These previous theoretical and experimental results provide a theoretical basis for the research in this paper.

Based on the theoretical basis of the electronic tracing algorithm, combined with the analysis of mechanical equipment failure machine testing technology, this paper explores the method of mechanical equipment failure robot detection. Combining the PSMLT algorithm and the improved PSSMLT algorithm and the MMLT algorithm, it has a strong role in the field of fault detection robots. Influence. The normal operation of machinery and equipment can improve the efficiency of industrial production and greatly help people's daily life.

2 Related Theoretical Overview and Research

2.1 Classification of Mechanical Equipment and Robots

(1) Vacuum adsorption robot

The vacuum adsorption detection robot mainly uses the method of air compression to set the internal device to a vacuum state, so that the detection robot can be well adsorbed on its surface. Vacuum suction robots are mainly divided into two types: single-disc suction robots and multi-disc suction robots. The suction force of the vacuum suction detection robot comes from the different pressures on both sides of the air pressure, and is composed of a large suction cup that can carry 40 kg [5, 6]. This adsorption component is relatively simple, light in weight, and the method of controlling adsorption is relatively simple, and is often used in small detection robots [7]. However, this single suction cup structure also has major shortcomings: First, the flatness of the suction wall of the robot is required to be high, and poor wall flatness may cause problems in the airtightness between the suction cup and the wall, causing the detection robot to overturn or fall off; Secondly, the ability of detecting robots to cross obstacles is poor, and the scope of use is relatively narrow.

(2) Permanent magnet adsorption robot

The vacuum adsorption inspection robot has the disadvantage of weak adsorption force, while the permanent magnet adsorption inspection robot has the advantage of greater adsorption force, and has lower requirements for the adsorption environment, and has better adaptability to the uneven and rugged wall environment.

Because of the continuous growth of the steel industry, the detection technology of permanent magnet adsorption robots has been widely used [8]. There is an inclination angle between the permanent magnet adsorption device of the detection robot and the wall surface, and each adsorption device is driven by a separate motor, which makes the detection machine not only has a large adsorption force, but also moves flexibly and freely. The robot adopts permanent magnet adsorption technology, and adopts the method of attaching the permanent magnet of the arc surface to the tube wall, which greatly improves the magnetic adsorption force and the adsorption effect between the robot and the contact surface [9]. The use of the robot not only greatly shortens the operation time of cleaning and testing of mechanical equipment, but also reduces the work intensity of staff and improves the detection accuracy of damage to mechanical equipment.

(3) Bionic adsorption robot

Biomimicry is a discipline that obtains inspiration and assumptions by learning biological phenomena, and realizes and effectively applies biological functions in engineering [10]. The bionic adsorption detection robot is mainly based on the composition of the simulated organism, and the mutual detection system is designed according to its basic structure. However, the current research on the bionic robot is not very mature, and most simulation products simply imitate the body surface of the organism. The structure is still in the stage of "similarity but dissimilarity", and it is still unable to inherit a certain special function of biology. The bionic adsorption detection robot is a special detection robot developed by imitating the adsorption principle of biology. The morphological structure of protists [11].

2.2 Research Difficulties of Fault Detection Robots

There is a constraint relationship between the two key points of the detection robot's adsorption force and movement flexibility. Adsorption function and moving function are two key functions of the detection robot. In order to ensure that the adsorption stability of the detection robot reaches a certain strength, and to ensure that there will be no external interference factors during detection, generally the larger the adsorption force, the better. The mobile flexibility of the detection robot is reflected in the robot's ability to climb over obstacles and move, but the strong adsorption capacity at this time means that the flexibility will be affected to a certain extent [12]. At the same time, the detection robot requires less magnetic leakage of the adsorption unit, and preferably has a unilateral magnetic field to avoid the adsorption force on the stator and rotor on both sides of the air gap, resulting in a large running resistance, which will affect the detection robot. Movement has a hindering effect.

3 Experiment and Research

3.1 Experimental Method

(1) Electron transport method in cathode

The electron transport in the primer is mainly expressed by the electron transport model, and the electron transport model is mainly affected by the electric field strength, scattering and strength and reflection strength, where the scattering type is:

$$\tan\left(\frac{\varphi}{2}\right) = \frac{0.1093}{E_0 b_e} \quad (1)$$

In the above formula, E is the energy before photoelectron scattering, in eV, and b is the vertical distance from the ion Be to the extension line of the photoelectron path, in nm. Simulation Model of Electron Movement in Cathode.

(2) Proximity system electronic transport method

The proximity system is suitable for between the cathode and the incident end of MCP and the exit end of MCP and the phosphor screen. When the electrons are emitted at the angle θ between the initial velocity v and the normal line and are longitudinally displaced by a distance of L, their speed and displacement in the Z and R directions can be determined. Expressed as:

$$\mathrm{L(x)} = \int_{-00}^{00} \mathrm{p(x, y)dy} \quad (2)$$

In the above formula, P(x, y) is the light intensity on the image plane of the phosphor screen, and L(x) is the line spread function. The discrete Fourier transform is performed on the line spread function, and the modulation transfer function is obtained by calculation.

3.2 Experimental Requirements

In this experiment, the robot body was developed according to the characteristics and structure of mechanical equipment robots to detect robot materials. The four corners of the body were extended to the arms close to the body and connected to the wheel structure. Two engines installed at the front of the body can drive the wheels to rotate, and the rear wheels are connected to the axle to drive the drive wheels to rotate. Under the action of the power supply and the generator, the driving wheel moves forward. When the mechanical equipment fault detection robot starts to work, it can monitor the working state of the mechanical equipment; the data collected through the fault detection monitoring can be fed back in time, and Real-time analysis based on feedback results. After discovering the failure of the mechanical equipment, use the wheel to walk with the support of the interactive equipment, and conduct a comprehensive analysis and inspection of the mechanical equipment failure.

4 Analysis and Discussion

4.1 Scene Average Per Pixel Sample Analysis

The parameters in the scene rendering process in this paper need to be set, and some other parameters are also set to default values. The experimental data is obtained by analyzing the average per-pixel sample value of the scene under three different algorithms: PSMLT, PSSMLT, and MMLT, as show in Table 1.

Table 1. Scene Average Sample Per Pixel Values

Scenes	PSMLT	PSSMLT	MMLT
Cornell	384	256	512
Room	384	256	512
Veach	256	128	256

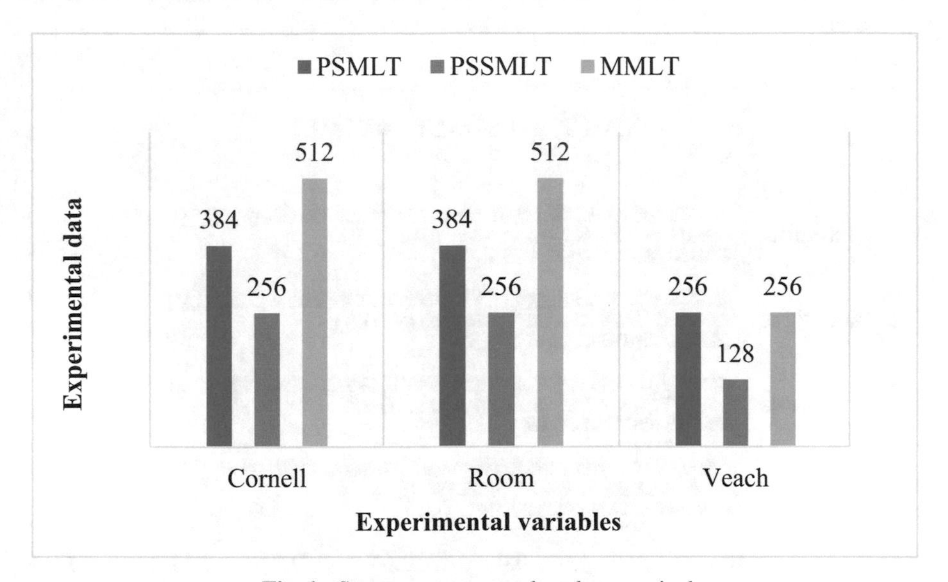

Fig. 1. Scene average sample value per pixel

It can be seen from Fig. 1 that the average sample values per pixel of the PSMLT algorithm are 384, 256 and 512, which are better than 256, 256 and 128 of the PSSMLT algorithm. The MMLT algorithm is better than the PSMLT algorithm and PSSMLT in the direct illumination scene, because its mutation strategy can cover the path in the scene better. The addition of the three functions on the basis of MMLT does not have a great impact on the generated image. By comparison, it can be seen that the image generated by the improved algorithm based on reception probability has less noise than the other two improved algorithms. The image is more realistic in the shaded part than the improved MLT algorithm based on reception probability.

4.2 Comparative Analysis of Robot Fault Detection Accuracy Under Three Algorithms

In order to check the comparison of the robot fault detection accuracy under the three algorithms, according to the above content, the mechanical normal and four kinds of fault data under different numbers of data are counted, and the results are shown in the following figure.

As shown in Fig. 2, this experiment uses the electronic tracing algorithm to establish a robot fault detection system and conducts a series of related simulation experiments to detect, and compares and analyzes the blue fault detection accuracy of the mechanical equipment fault detection robot based on the electronic tracing algorithm. It can be seen from the experimental data that the accuracy of the MMLT algorithm under four faults is 0.89, 0.91, 0.96 and 0.88, the highest accuracy is 0.96, and the accuracy of the PSMLT algorithm is 0.33, 0.39, 0.38 and 0.48. And the accuracies of the PSSMLT algorithm are 0.59, 0.61, 0.58 and 0.54. In contrast, the MMLT algorithm outperforms the other two algorithms.

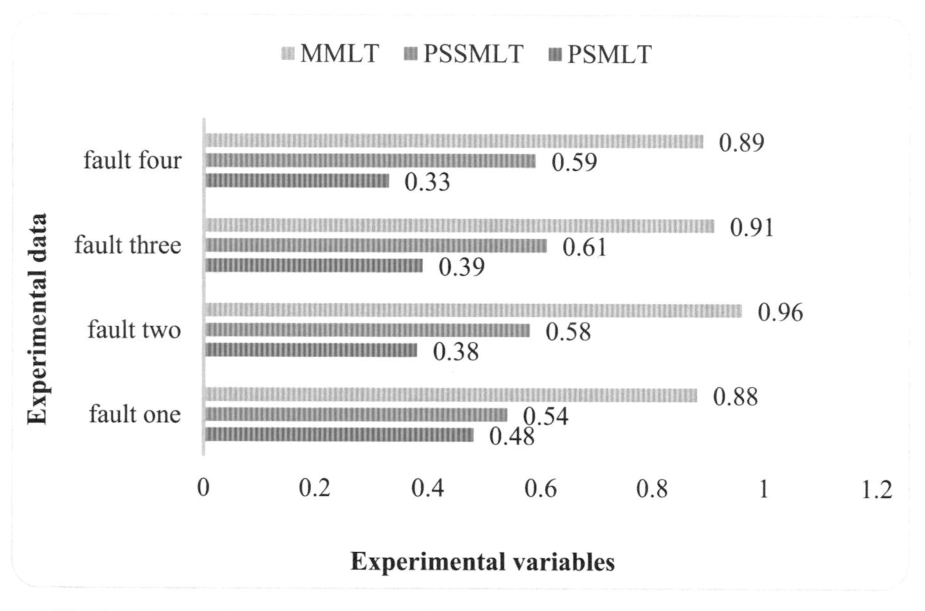

Fig. 2. Comparative analysis of robot fault detection accuracy under three algorithms

5 Conclusions

In this paper, the electronic tracing algorithm is used to conduct experimental simulation analysis on the mechanical equipment fault detection robot. Under the MMLT algorithm, the robot has the highest fault detection accuracy. The results show that the fault detection robot can reflect the average sample value per pixel of the mechanical equipment scene. Knowing the operating status of the equipment and testing data, there is no time and space limitation, which improves the work efficiency. The researchers can obtain the first-hand information and data of the mechanical equipment site through the robot, so as to realize further scientific research. The fault detection robot greatly improves the efficiency of mechanical equipment, so that the source of the fault can be found and found at the first time. The fault detection robot can give feedback and remind the staff in time. The accurate fault detection rate is the basis for ensuring the good operation of the equipment. The current high development of industrialization and industrialization has made machinery and equipment an important factor in the productivity of enterprises. Therefore, the research on fault detection robots plays a crucial role in improving the competitiveness of enterprises, accelerating the development of industrialization, and enhancing the productivity of enterprises. In view of the instability of mechanical equipment failures in the current industrial development, flexible fault detection robots play a very critical role, providing a certain guarantee for the safety of workers' working environment. The mechanical equipment fault detection robot has the advantages of simplicity and speed in human-computer interaction detection, which provides a feasible way to analyze the fault detection results.

References

1. Mart Nezcastel, J.N., Villarrealcervantes, M.G.: Integrated Structure-Control Design of a Bipedal Robot Based on Passive Dynamic Walking **9**(8), 1482 (2021)
2. Nguyen, A.T., Vu, C.T.: A Study and Design of Localization System for Mobile Robot Based on ROS. **186**(5), 106-110 (2021)
3. Deb, A., Wypych, Z., Lonner, J., et al.: Design and control of an autonomous robot for mobility-impaired patients. J. Medical Robotics Res. **5**(1), 1–9 (2022)
4. Kindt, P.H., Chakraborty, T., Chakraborty, S.: How Reliable is Smartphone-based Electronic Contact Tracing for COVID-19? **5**(9), 20–26 (2020)
5. Karakaya, M., Celik, E.T.: Effect of pupil dilation on off-angle iris recognition. J. Electron. Imaging **28**(3), 1–15 (2019)
6. Rivera-Alvarado, E., Torres-Rojas, F.J.: APU performance evaluation for accelerating computationally expensive workloads. Electronic Notes in Theoretical Computer Sci. **3**(9), 103–118 (2020)
7. Jorfi, M., Luo, N.M., Hazra, A., et al.: Diagnostic technology for COVID-19: comparative evaluation of antigen and serology-based SARS-CoV-2 immunoassays, and contact tracing solutions for potential use as at-home products **133**(4), 23–43 (2020)
8. Chaikovsky, I., Lebedev, E., et al.: Inter-relationships of different electrocardiographic indicators of left ventricular hypertrophy in 25,000 Chinese adults. Eur. Heart J. **5**(99), 1 (2021)
9. Gaurav, K., Singh, R., Kumar, A.: Modified simple chemical plume tracing algorithm. Journal of Physics: Conference Series **1455**(1), 012006 (2020)
10. Tarapore, D., Christensen, A.L., Timmis, J.: Generic, scalable and decentralized fault detection for robot swarms. PLoS ONE **12**(8), 2058–2066 (2017)
11. Mirzaei, M., Hosseini, I., Ghaffari, V.: MEMS gyroscope fault detection and elimination for an underwater robot using the combination of smooth switching and dynamic redundancy method. Microelectron. Reliab. **109**, 37–40 (2020)
12. Filaretov, V., Zuev, A., Procenko, A., et al.: Fault detection of actuators of robot manipulator by vision system. Appl. Mech. Mater. **865**, 457–462 (2017)

Application of Mobile Learning App in Smart English Classroom Under the Background of Information Technology

Qiang Cui(✉)

Foreign Languages School, Dalian University of Science and Technology, Dalian, Liaoning, China
769300086@qq.com

Abstract. The era of mobile Internet has spawned the emergence of a new learning method called mobile terminal learning, and countless mobile education APPs have swept in, breaking the time and space limitations of learning and communication. The purpose of this paper is to study the application of mobile learning apps in smart English classrooms under the background of information technology. of students as the research object, one of the classes was given a three-month mobile learning APP-assisted English teaching, and the other class adopted the traditional mode of English teaching. Finally, the students of the two classes were investigated by means of a questionnaire. The survey results show that the smart English teaching model based on mobile learning APP has a very large audience among students.

Keywords: Information Technology · Mobile Learning App · Smart Classroom · English Teaching

1 Introduction

In modern society, we are paying attention to the rapid development of science and technology. At present, the work of the digital office is basically successful, and the school education has introduced information-based educational equipment such as intelligent multimedia voice systems. The emergence of the concept of mobile learning has broken the essential concept of education [1, 2]. One of the advantages of mobile teaching is that it not only adapts well to the integration with traditional teaching, but also makes use of its own advantages and cutting-edge educational technology to make up for the traditional teaching mode. Mobile learning breaks people's traditional learning methods in the past, and the focus on learners' innate learning thinking and learning methods is also increasing. Mobile learning has become a hot spot in the field of education [3, 4].

The emergence of mobile learning APP has promoted the development of mobile learning and better supported teacher education and student learning. However, at the same time, mobile learning APP lacks rationality, lacks formalism, and there are many problems worthy of in-depth study. After analyzing mobile learning applications, some

J. H. Abawajy et al. (Eds.): ICATCI 2022, LNDECT 169, pp. 118–126, 2023.
https://doi.org/10.1007/978-3-031-28893-7_15

scholars believe that the use of mobile learning applications can support students' informal learning, create different learning scenarios, and solve problems in the process of mobile learning. In the future, we should try to integrate mobile learning applications into different learning scenarios [5, 6]. Other scholars believe that mobile learning applications need to be enhanced with the goal of student motivation, participation, constructiveness and sociality. We will provide appropriate support for student learning [7, 8]. Some researchers have analyzed the recently released literature on mobile learning applications, arguing that in order to better integrate theory with practice, there is a need for researchers to increase their learning principles and principles in a mobile learning environment [9, 10]. Establish a clearer connection between application design features, strengthen the connection between theoretical foundations and educational tools, and how mobile learning applications can support users of more disciplines and abilities to explore. To analyze the development of mobile learning applications from the four perspectives of learners, teachers, programmers and distributors, the development of mobile learning applications should focus on improving the essential motivation of students [11, 12]. As such, gamification is and will be mobile learning for a long time to come, and application development needs to focus on what motivates students. In conclusion, play is not the only factor that fosters student motivation. In the future, relevant research needs to be strengthened to avoid overemphasizing games and ignoring the educational nature of APPs.

On the basis of consulting a large number of relevant references, combined with the characteristics of mobile learning, the concept of English smart classroom and the problems existing in English teaching, this paper selects students from two classes in a university in this province as the research object, and conducts a study on them for a period of three months of experimental teaching. Finally, through a questionnaire survey, we can understand the attitudes of students' mobile learning APP.

2 Application of Mobile Learning App in Smart English Classroom under the Background of Information Technology

2.1 Features of Mobile Learning

(1) Mobility
Neither the traditional classroom nor the current distance learning can truly achieve learning anytime, anywhere. Even in distance learning, students still have to sit in front of a computer to study. The portability, mobility, and wireless communication advantages of portable learning devices enable students to study anywhere, anytime. Students can enjoy learning anytime, anywhere, whether on the bus, on the subway, in the waiting room, or outside the classroom.

(2) Interactivity and timeliness
The current teaching method is that after leaving the classroom, it is difficult to communicate with teachers, let alone interact with information. With the advent of mobile teaching, this problem has been well alleviated. The high portability of mobile electronic devices and the ubiquity of wireless networks enable both students to use email, QQ, WeChat, Fetion and other software to achieve better communication. And this communication method is more easily accepted by teachers and

schools. For students who are usually shy and introverted, this method of communication is more helpful in stimulating emotional expression and enthusiasm for learning.

(3) Popularity
The advent of mobile devices and their superior features have given learners a wealth of course options. The decreasing prices of some mobile devices and the increasing penetration of smartphones have created a strong mass base for mobile learning.

(4) Networking and Personalization
At present, the vigorous development of software and hardware technologies such as computers and mobile devices, wireless network information technology, and mobile communication technology provides an important scientific and technological foundation for students to learn through mobile devices, and is a powerful source for students to learn independently and individually and guarantees are provided. In the process of traditional teaching, the total amount of knowledge that teachers master is often limited. When teachers face problems, the effectiveness of immediate teaching is often higher than that of traditional teaching. Through mobile learning, students can choose an appropriate time period and place, or teach in a normal mode according to their actual needs, the course content they are interested in, and the speed of mastering the progress. Mobile courses best fit the learner's personality and meet their individual needs, so that learners can get the most personalized educational services.

(5) Situational
Students will be exposed to new knowledge in new situations, mobile learning will be able to provide more situational information, and learning English will be more relaxed, vivid and contextualized.

2.2 The Concept of English Smart Classroom

Smart classrooms use new-generation information technologies such as "cloud computing and big data", and build an online English education environment and lesson preparation environment based on new-generation information technologies such as big data and artificial intelligence. The educational process is integrated to create an intelligent and effective new classroom environment. Focusing on intelligent infrastructure, advanced education management, and personalized English education, dynamic real-time learning data analysis, evaluation and feedback, 3D communication and interaction, intelligent resource boosting, and English classroom teaching will completely change the form and content of classrooms. The function of teaching English in the classroom in the era of big data. Effectively integrating information technology in the English teaching process and creating an information-based English teaching environment not only plays a leading role for teachers, but also implements the information-based English teaching environment, which is also the initiative, enthusiasm and creativity of students, emphasizing the main body of students. The characteristics of status, education and learning have changed the traditional teaching structure of the English classroom from "teacher-centered", from educational structure to "combination of dominant themes" in educational structure. With the continuous development of information technology and its application to English education in school education, information technology has gradually evolved from early assistance to deep integration with English education, and traditional classrooms have gradually evolved into smart classrooms.

2.3 Problems Existing in English Teaching

(1) The classroom teaching mode is single
In the classroom, teachers often indoctrinate students with little attention and attention to their feedback activities. This method of teaching one-way communication often makes educational content boring. It does not support the diversified development of talents, nor does it support students' learning interest and motivation. Teachers have a certain level of education, but reaching higher levels requires teaching skills and good teaching skills. The teaching method is relatively simple, and students' evaluation of advanced teaching methods and strategies such as ability education and theoretical practice is low, reflecting the difficulty of meeting students' needs.
(2) The implementation effect of teaching activities is not good
The implementation of university education activities is dominated by classroom education. Due to the short class time, teachers rarely ask students questions to prevent them from learning or progressing. In addition, when psychological teaching to students, teachers cannot effectively organize the lessons. In addition, the educational content of universities is highly specialized, broad and profound, and classroom discipline is often difficult to manage. Students are busy playing with their phones from time to time, which can lead to a vicious cycle and even a tired attitude towards vocational training. In general, there are some problems that need to be further improved in the university's classroom education activities.
(3) Teaching content lags behind
Due to the highly academic and scientific nature of university education, it is difficult to have a clear understanding of education. Universities tend to place less emphasis on the living image of education, and neither do teachers. When teachers teach, most of the content is scripted. In this sense, there is still a gap between college English teaching and social needs.
(4) Lack of emotional communication between teachers and students
Traditional classroom teaching methods limit teaching activities to the classroom, limit teaching time, and enable students to absorb and internalize what they have learned. In terms of classroom communication, the teacher-student relationship is relatively loose. Character teaching lacks good interaction, and students who learn less are not mentored. Lack of communication between teachers and students leads to apathy between teachers and students.

3 Experiment

3.1 Questionnaire Design

This paper selects students from a certain college in this province as the research object, and selects two classes from the school, which are recorded as Class A and Class B respectively. Among them, there are 48 people in class A and 45 people in class B, a total of 93 people are the subjects of this experiment. During the experiment, class A was used as an experimental class with mobile learning APP assisted English teaching, and class B was used as a control class taught in traditional mode. After a three-month teaching experiment, the students of the two classes were investigated by means of questionnaires. This teaching experiment emphasizes the operation control of independent variables, some control of irrelevant variables, and no random sampling of subjects. It is an educational quasi-experiment.

3.2 Reliability Test of the Questionnaire

The half-reliability method divides the survey elements into two parts, calculates the correlation coefficient between the two halves, and then evaluates the reliability of the entire scale. Divide reliability is an internal consistency factor that measures agreement between the two halves of the score. This approach is generally not suitable for fact-based questionnaires, but for credibility analysis on issues such as attitudes and perceptions. The Spieljan Brown formula expresses the relationship between the change in test length and the value of test reliability, namely:

$$r_{kk} = \frac{kr_{xx}}{1 + (k - 1)r_{xx}} \tag{1}$$

where k is the ratio of the increased test length to the original test length. When calculating the split-half reliability. Tests are only half as long, and reliability is reduced. And the length of the whole test is twice the length of the half test, substituting 2 into the above formula, that is

$$r_{kk} = \frac{2r_{xx}}{1 + r_{xx}} \tag{2}$$

4 Discussion

4.1 Attitudes Towards the English Teaching Mode of Mobile Learning Apps

Table 1. Attitudes towards the English teaching model of mobile learning apps

	Class A	Class B
strongly agree	33.7%	34.1%
relatively agree	32.5%	35.8%
remain neutral	28.2%	25.6%
disagree	3.4%	3.1%
strongly disagre	2.2%	1.4%

As can be seen from Table 1 and Fig. 1, when answering whether they agree that the smart English teaching model based on mobile learning APP can effectively improve their English learning ability, the students on the whole expressed a positive attitude, and 33.7% and 33.7% agreed very much. 34.1%, 32.5% and 35.8% agree, and 28.2% and 25.6% are neutral. The proportion of students who disagreed and disagreed was 3.4%, 3.1% and 2.2%, 1.4% respectively. This shows that, on the whole, the smart English teaching model based on mobile learning APP has a very large audience among students.

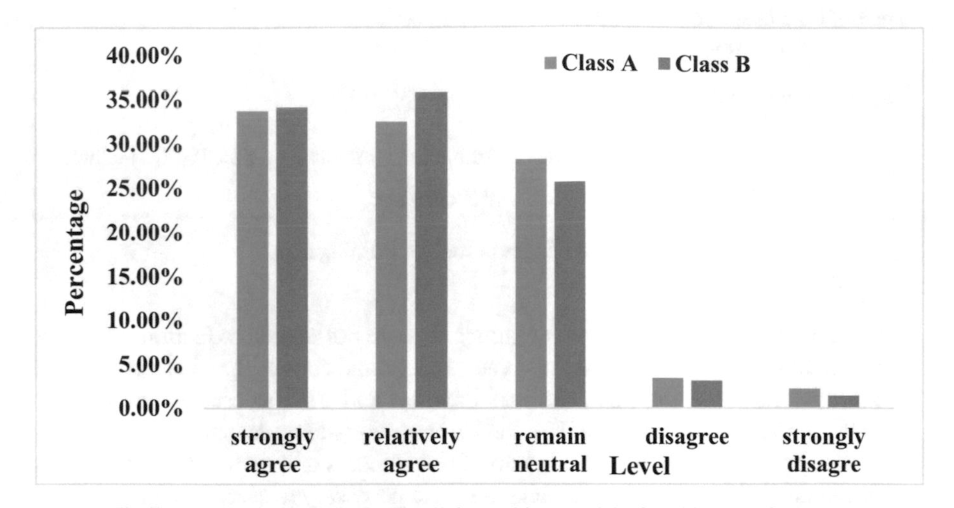

Fig. 1. Attitudes towards the English teaching model of mobile learning apps

4.2 Advantages of Mobile Learning Apps

As can be seen from Fig. 2, in the two classes, most students believe that the content of the mobile learning APP is more authoritative and can provide detailed guidance for their learning; the second is that the content is rich and updated quickly. It can be seen that the use of APP for students still pays attention to the use of functions and experience in their learning, while the consideration of appearance and interface is relatively small.

4.3 Strategies to Improve the Effectiveness of Smart English Teaching Mode Based on Mobile Learning App

(1) Selected online teaching resources
The Internet environment has put forward new requirements for traditional English education. Teachers need to digitize and update educational content, enrich English teaching resources, and expand English learning channels. The uneven quality of network resources is the main obstacle for students to obtain useful information resources. Therefore, teachers need to help students carefully monitor and select appropriate digital educational resources based on their cognitive development.

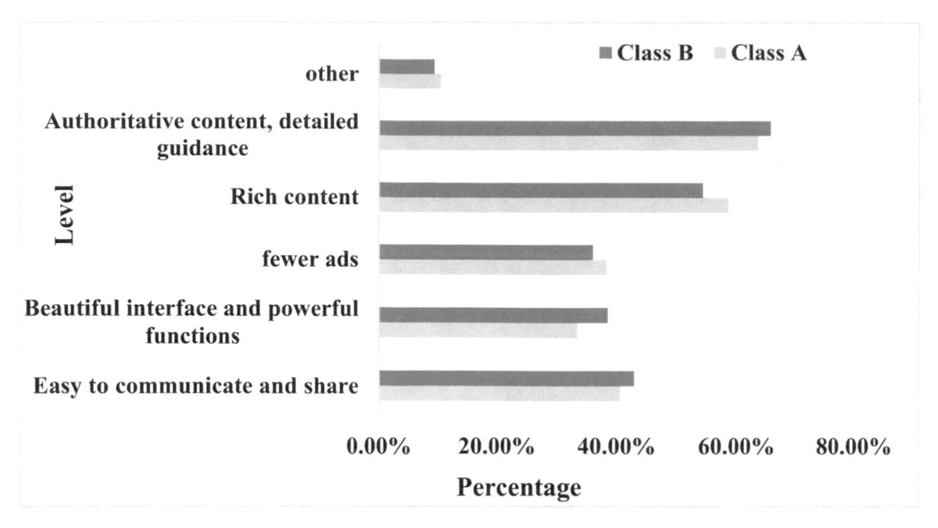

Fig. 2. Advantages of mobile learning apps

For the purpose of learning, we do things that are not related to learning. Teachers should have the courage to innovate and explore, and constantly update educational concepts. At the same time, teachers gave up and tried to create conditions for students to learn on the move. Use the various educational resources of the mobile APP to fully mobilize the enthusiasm of the students and strengthen the students' dominant position. At the same time, it is also necessary to give more guidance and encouragement to students with weak foundations, so that mobile applications can become real learning tools in their hands.

(2) Improve self-learning ability
Whether it is traditional classroom paper education or APP-based portable learning, students arc always the main body of learning. The way to improve the comprehensive ability and basic quality of a language and improve self-learning ability is the beginning and end of all students' learning activities. With the beginning of Internet education, students have entered a new era of learning based on mobile applications. Improving autonomous learning ability is an important reference. It is necessary to prepare for learning before learning activities, adapt learning methods to self-feedback in learning activities, and conduct self-evaluation and self-reflection. Learning outcomes after learning activities.

(3) Using multiple evaluation methods
No matter what teaching method or learning tool is used, the ultimate goal of English teaching is to improve students' basic literacy and language skills. To achieve this goal, the two-way interaction between teachers and students is inseparable. If the establishment of a harmonious teacher-student relationship is a prerequisite, then it is important to ensure that the education plan for mobile devices is optimized, the selection of online educational resources, and a variety of evaluation methods are very important. These four aspects are interrelated and complement each other,

shaping the overall process of students' use of APP for teacher-led learning. Different from traditional teacher evaluation methods, APP-based teaching methods have higher requirements on teachers. In short, teachers should not only achieve multi-channel and differentiated assessment, but also adopt different assessment methods according to the knowledge base of students. At the same time, ensure that the evaluation is reasonable, comprehensive and scientific.

5 Conclusions

With the advent of mobile learning apps, students can overcome time and space constraints, continue to truly appreciate learning everywhere, and spend less time continually improving their English. The birth of these new learning methods will significantly change the traditional teaching mode, further optimize the way of teaching on mobile devices, further enrich the learning resources of mobile device learning, and provide technical support for mobile teaching, so that the mobile device learning APP It has more and more practical value in English teaching.

References

1. Fatoni, P., Rosalina, M.: Efektifitas penggunaan games edukasi untuk meningkatkan kemampuan dan hasil belajar siswa dengan aplikasi mobile learning pada mata kuliah computer programming. Jurnal Informatika dan Sistem Informasi **13**(1), 80–96 (2021)
2. Oz, C.T., Uzunboylu, H., Ozcinar, Z.: The effect of visual design self-efficacy of language teachers on mobile learning attitudes during the pandemic period. J. Univ. Comput. Sci. **27**(5), 524–542 (2021)
3. Rochmah, N., Cahyana, U., Purwanto, A.: Development of mobile learning: basis of ethnopedagogy of baduy community, Banten province. IOP Conference Series: Materials Science and Eng. **1098**(2), 022092 (2021)
4. Akour, I., Alshurideh, M., Barween, et al.: Using machine learning algorithms to predict people's intention to use mobile learning platforms during the COVID-19 pandemic: machine learning approach. JMIR Medical Education **7**(1), 1–17 (2021)
5. Oktariyana, A.M., Sulaiman, I., et al.: Design of mobile learning rhythmic gymnastics materials for high school / vocational high school levels as a distance learning media during the Covid-19 pandemic. Sports Engineering **9**(3), 394–402 (2021)
6. Sungur-Gül, K., Ate, H.: Understanding pre-service teachers' mobile learning readiness using theory of planned behavior. Educational Technology & Society **24**(2), 44–57 (2021)
7. Kellam, H.: A conceptual framework and evaluation tool for mobile learning experiences. International Journal of Virtual Personal Learning Environ. **11**(1), 1–22 (2021)
8. Guerrero, C.H., Domínguez, E.L., Velazquez, Y.H., et al.: Kaanbal: a mobile learning platform focused on monitoring and customization of learning. International Journal of Emerging Technologies in Learning **16**(1), 1–26 (2021)
9. Widiastika, M.A., Hendracipta, N., Syachruroji, A.: Pengembangan media pembelajaran mobile learning berbasis android pada konsep sistem peredaran darah di sekolah dasar. Jurnal Basicedu **5**(1), 47–64 (2020)
10. Kurniasih, S., Darwan, D., Muchyidin, A.: Menumbuhkan kemandirian belajar matematika siswa melalui mobile learning berbasis android. Jurnal Edukasi Matematika dan Sains **8**(2), 140 (2020)

11. Nurdiansyah, E., Waluyati, S.A., Dianti, P.: Pelatihan Pembuatan Mobile Learning Berbasis APP Inventory bagi Guru-Guru PPKn SMP di Kota Palembang. Jurnal Anugerah **2**(2), 67–72 (2020)
12. Silva, K.: Object of study of literacy: a learning object based on mobile learning to aid in the process of child literacy. Int. J. Innovation Educ. Res. **8**(11), 32–40 (2020)

Technical Application and Design of Power Intelligent Platform Based on Winter Olympics

Zhidong Yang, Fengshi Luan, Jinxin Liu, Di Yang, and Haobo Xu(✉)

State Grid Beijing Electric Power Company, Beijing 100031, China
15506050203@163.com

Abstract. In order to solve the factors such as the complex environment of power support for the Winter Olympics, the scattered and complex support data, and the unintelligent support command. Based on the structured analysis method, with the power data as the core, the internal and external systems have full access, full coverage and full integration of Olympic-related data. The design includes monitoring function, command function and display function. It is decomposed layer by layer from top to bottom from the integration of power protection elements, and data traceability is carried out. Combined with the use of digital twin, artificial intelligence, knowledge graph, intelligent speech recognition and other technologies, it can realize comprehensive perception and intelligent analysis of video surveillance.

Keywords: Smart Platform · Artificial Intelligence · Platform Design · Power Guarantee

1 Introduction

During the construction of the Winter Olympics intelligent platform [1], it is necessary to monitor all elements of power supply security, including: main grid and operation data, security stadiums and related data, work orders, weather, traffic, race schedule, video information, etc. Therefore, it is necessary to integrate the models and data information of the existing scheduling, distribution network management and control, marketing management and control, urban operation big data center, stadium data, green energy, video and other systems, and the splicing workload is huge; at the same time, different equipment has different Private coding, in order to realize the correlation of all-element information, it is necessary to compare and correlate different system codes, and there are many technical challenges [2, 3].

2 Platform Design and Technology Application

The data architecture of the platform adopts the design mode of vertical and horizontal sub-theme, combined with the business architecture for data architecture design. The functional modules mainly include monitoring functions, command functions and display functions (Fig. 1).

J. H. Abawajy et al. (Eds.): ICATCI 2022, LNDECT 169, pp. 127–135, 2023.
https://doi.org/10.1007/978-3-031-28893-7_16

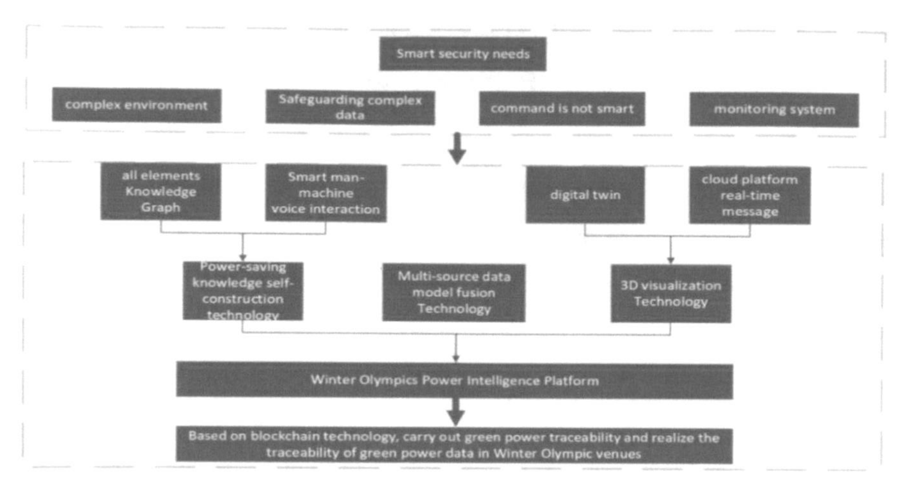

Fig. 1. Main modules of the platform

2.1 Monitoring Function Design

The Winter Olympics power command and support system takes the strong and smart grid as the platform and multi-energy complement as the main line to comprehensively display the monitoring of external grid transmission, power generation monitoring of power plants, and monitoring of load and storage of power grids. Run the overall information to realize the monitoring of the power transmission of the external network. This overall monitoring mode improves the emergency response capability of power grid accidents and reduces the time for the commander to understand different information.

1. On the venue side, power security personnel can view the overview information of the current operation of the Winter Olympic venues, and realize the traceability analysis of the power supply of the venues. On the 3D view of the stadium, you can view the operation of the stadium, the follow-up arrangement of the game, the access status of energy equipment and energy-consuming equipment, access real-time video, monitor the real-time status of the power distribution room, help the monitoring personnel understand the structure information of the transmission line, improve the efficiency The speed and efficiency of information transfer.
2. On the transmission side, the power protection personnel can use the system to have an overview of the operation and maintenance of power transmission and transformation, and comprehensively monitor the operation status of transmission lines, cables, and substations related to power protection. Compared with previous competitions, the inspection of transmission lines mainly relies on periodic inspections by operation and maintenance personnel. Although equipment problems can also be found, but the lack of monitoring of special environment and climate, it is very easy for line accidents to occur before the next inspection cycle. The use of the Winter Olympics power intelligent platform can timely monitor and correct various hidden dangers in the power transmission process.

2.2 Command Function Design

In terms of management methods, power supply guarantee work relies too much on assault and sports management, and power supply command decisions mainly rely on manual experience, lacking the intelligent auxiliary support and information linkage system of the information system. The Winter Olympics power intelligent platform can overview the distribution of power protection elements in spatial information, and grasp the geographical distribution of power protection-related stations, lines, venues, and equipment. Panoramic monitoring and visual command.

1. In terms of guarantee flow line tracking, commanders can use the system to have an overview of the power supply process plan and the maintenance of the guarantee flow line, and realize the functions of import, maintenance and confirmation of the guarantee process; to show the relationship between the competition schedule and the guarantee flow line, as well as to show the various competition areas, The completion status and details of the power protection tasks of each venue provide data service support for ensuring simulation deductions. In terms of security risk assessment, commanders can use the system to have an overview of security risk levels and regional distribution, master the zoning, hierarchical distribution and specific risk content of line security risks, and divide risk zoning according to venue lines, etc. After selecting specific security objects, they can View risk details. Commanders can also use the system to combine meteorological information to assess the risk of the safeguarded objects within the scope of meteorological warnings, and to carry out safeguards risk warnings in a timely manner, so as to realize active warnings and responses to emergencies.
2. In terms of fault handling, commanders can use the system to have an overview of the fault handling situation, grasp the briefing information on fault research and judgment, the scope of existing faults and the content of emergency plans. When a fault occurs, according to the fault information and related characteristics, the fault handling information is intelligently pushed and displayed in a panoramic view, and emergency plans for similar faults are actively pushed. In the past, there was a lack of interaction and linkage between business systems, and data collection and fusion capabilities were poor. Now, through the application of this platform, the problem of data and information islands between the company's business systems has been broken, and real-time multi-source data sharing and full data coverage have been realized. It saves the time that the commander needed to find the optimal solution by himself, and the commander can directly understand the type and severity of the fault based on the relevant information pushed, which reduces the judgment time and improves the fault response speed.

2.3 Demonstrate Functional Design

In terms of power supply display, the platform provides an intelligent power supply display function, showing that during the Winter Olympics guarantee period, green energy will be transported into the venue from the Zhangbei Rou direct project, energy storage station, etc. through the Beijing power grid to provide the power supply guarantee for each terminal of the venue, and the venue will also be sent to the venue. The primary wiring diagram related to the station room is bound to its carrier such as the venue power distribution room, substation, etc., and the linkage of the power supply system diagram is realized. Through this linkage, the time spent calling to view different screens is effectively reduced. Not only can the problem be grasped in time, but also the support team and task execution status can be accurately grasped, and can even be traced back to the individual, which improves the speed and efficiency of information transmission and feedback.

1. In terms of power protection display, the platform provides a large-screen display page of the system, which can perform viewing angle operations, layer-by-layer drilling and other functions on the GIS map, and conduct command, monitoring, and display related to power protection. Through the theme scene on the large screen, the theme scenes in the process of power protection are divided according to the business, and they are organized into different theme functions, and relevant themes are presented by adjusting map elements, power grid elements, and power protection elements. The reason why the large-screen display is adopted is that the large-screen interactive system is very stable, energy-saving and environmentally friendly, easy to maintain, and has a longer service life than other display devices. It can not only display more clearly, but also greatly improve efficiency.
2. In terms of venue display, through the establishment of the Winter Olympic venue model, the three-dimensional venue view is presented, and the visualization of the surrounding and interior of the venue is realized. Through the modeling of the external facade and internal structure of the venue, a highly restored digital model of the venue can be presented. The related browsing and presentation functions can be performed in the model, and the related pipeline networks and connections in the venue and equipment can be modeled. Presenting its connection relationship is more clear, intuitive and three-dimensional, which is convenient for the next step of supervision and management.

3 Platform Technical Case

The platform integrates the data fusion between different types of business systems based on the multi-dimensional calculation method of the matching degree of the all-element model of power preservation based on the minimum edit distance algorithm; integrates the power conservation knowledge of artificial intelligence technologies such as knowledge graph, deep learning and natural language processing. Self-construction method to form a power-protection knowledge brain with intelligent command function; innovative fusion architecture of real-time monitoring service bus and artificial intelligence engine platform base technology, using digital twin method to create a panoramic three-dimensional real-time information model of power-protection objects, to achieve The power grid operation monitoring at full voltage level and the penetrating command of multi-level security resources; the green power traceability method based on the block chain certificate storage technology has realized the whole process of the green power supply and consumption data of the Winter Olympic venues. Retrospectively.

3.1 Apply Artificial Intelligence Technology

In order to meet the needs of the Winter Olympics operation and power supply, the platform applies a number of new technologies such as artificial intelligence [4, 5], which can not only realize the simultaneous display of the three-level command interface of the city, the competition area, and the venue(Fig. 2). It can also carry out multi-scenario applications of intelligent support and command to meet the needs of power grid operation monitoring at the full voltage level of 500 kV to 380 kV, thus ensuring the effective integration between various levels, improving the convenience of interface integration, and ensuring the efficient operation of the command system.

Fig. 2. Artificial intelligence penetration

3.2 Apply Dynamic Knowledge Graph Technology

Apply the dynamic knowledge graph technology to form the knowledge brain of the Winter Olympics. The platform organically integrates multi-source heterogeneous models and power protection data, and aggregates fragmented power protection plans into a large amount of "knowledge", forming a knowledge brain in the field of power protection, and then mining and discovering complex relationships between data. It has functions such as automatic generation of decision-making plans, automatic distribution of fault handling task lists, and timely early warning of emergencies, which improves the intelligent and scientific decision-making level of fault handling, and effectively improves the reliability of power supply and distribution networks [6, 7].

3.3 Apply Intelligent Speech Recognition Technology

Apply intelligent voice recognition technology to realize voice recognition query in the field of power protection (Fig. 3). The platform relies on advanced speech recognition, semantic understanding, speech synthesis and other voice full-link interaction technologies to build a proprietary model of vocabulary training in the field of power protection, and realize instruction recognition in the field of power protection. Combined with the data of the power protection business, the commander can quickly and directly query the target information through voice commands [8, 9].

Fig. 3. Human-Computer Interaction & Speech Recognition

3.4 Apply Block Chain Technology

The platform adopts the green power traceability method. Confirmation, uploading, sharing and life cycle management of green electricity data have realized the whole process of reliability and traceability of green electricity supply and consumption data in all Winter Olympic venues, and provided solid and reliable technical support for promoting green Winter Olympics.

4 Economic Benefit Analysis

The task of ensuring power supply for the 2022 Winter Olympics will comprehensively improve the effectiveness of the power supply support command, and fully coordinate with the Winter Olympics Organizing Committee and the Hebei Company to efficiently deploy support elements; it will be linked with the operation of the event in real time to achieve layer-by-level penetrating support for the guaranteed objects. Command; Apply massive data analysis results to actively summarize power protection information. The traditional power supply guarantee work management requires on-site personnel to patrol continuously. After problems are found, they are reported to the headquarters for decision-making and processing, and the information is transmitted from the bottom to the top. According to the information managed by the Panorama Intelligent Command

System for Power Supply Guarantee of the Winter Olympics, the Power Supply Guarantee General Headquarters, the sub-headquarters and the on-site headquarters will grasp the panoramic real-time information in an all-round way during the Winter Olympics, and actively discover problems through the system at the first time., issues instructions in a timely manner, and information is conveyed from top to bottom. Compared with the traditional power supply guarantee work management mode, it can transmit information quickly and accurately, reduce labor consumption, simplify the management process, and greatly reduce the response time.

The platform realizes the application of massive data analysis, actively summarizes power protection information, realizes active fault warning, real-time positioning, traceability, and automatic research and judgment. The platform has introduced 3D models of all Winter Olympic venues, connected to all power supply interfaces in the venues, and can give early warning of possible risks and failures, effectively ensuring power supply stability, minimizing power outage losses, and avoiding power outages. International reputation issues. The command platform also has an expanded level penetration preview function of the power port, which can quickly locate the source of the fault in seconds for the power failure, and sink to the terminal equipment through animation navigation, which greatly reduces the fault traceability workload.

The main source of revenue of the platform is the saving of personnel and information costs, the avoidance of power outage losses, the benefits of energy digitalization, and future sustainable benefits. By using the comparative analysis method, compared with previous large-scale events and activities, the Winter Olympics power protection command platform takes advantage of its highly integrated all-round monitoring and visibility of the entire venue to effectively reduce the number of operators required. The economic benefits transformed from this are mainly the daily subsidy fees paid. It is estimated that the proportion of personnel saving during the Winter Olympics is about 30%, and the labor cost saved during the period is shown in the Table 1 below.

Table 1. Operator Cost Savings Statistics Checklist

The total manpower of the venue	reduce the number of people	Labor cost (10,000 yuan/person per day)	Duration period (days)	Save labor cost (ten thousand yuan)
3000	1000	0.05	120	6000

In order to measure the economic benefits brought by the shortening of time, the unit value of the operator is set as the quantitative basis, and there are n types of operators in the test competition, let S (salary) be the daily subsidy amount of the operator, in yuan; T (time) is the average daily working time of the operator, in hours. Then Si represents the daily subsidy amount of the type of operator, and Ti represents the average daily working time of the type of operator, then the average unit value V (value) of the operators in the test competition is:

$$V = \frac{1}{n}\sum_{i=1}^{n}\frac{S_i}{T_i} \quad (1)$$

The average unit value can be calculated according to the type of workers actually dispatched and the working time in the test competition. According to the past experience of ensuring electricity, the information collection and transmission process in the past was collected manually by the venue operators and then submitted to the headquarters. After the Winter Olympics operation guarantee platform was put into operation, the main console could remotely gather the front lines of each venue and terminal equipment. Information. According to the calculation, the average time for collection and transmission of single-item electricity protection information has been shortened from 300 s in the past to about 5 s. If a fault is traced to the source, it will take a long time and many people. Even if the maintenance personnel can quickly reach the place where the fault occurs, it is necessary to manually check the power line. According to the calculation, in the traditional power protection mode, the time span of traceability is relatively large, and the average value is about 20 min. The fault tracing only needs to perform the tracing and sinking operation in the command platform, which takes about 30 s, and the total emergency effect efficiency is increased by more than 95%. According to the above parameters, the information cost calculation model saved during the Winter Olympics is constructed:

$$C = \sum_{i=1}^{n}\left[(M_{trans} - M_{trans}{}^{*}) \times V \times N_{i,trans} + (M_{Trace} - M_{Trace}{}^{*}) \times V \times N_{i,Trace}\right] \quad (2)$$

In the formula, C is the information cost saved (yuan); M(trans) is the information transmission time in the traditional mode (hours); M(trans*)is the information transmission time of the Winter Olympics command platform (hours); V is the average unit value of operators (yuan/hour);Ni is the information transmission amount on the day; M(trace) is the fault traceability time in the traditional mode (hours); M(trace*) is the fault traceability time of the Winter Olympics command platform (hours);Ni is the day Fault traceability.

5 Concluding Remarks

The Winter Olympics intelligent platform integrates advantageous resources such as power grid information, Winter Olympics events, venue energy, and urban operation, and uses technologies such as digital twins, artificial intelligence, dynamic knowledge maps, and intelligent voice recognition to meet the real-time perception of Winter Olympics security elements and make overall use of energy interconnection. Application value, realize the penetrating command of the Winter Olympics service guarantee, through data sharing and open platform, support the communication and interaction between the Winter Olympics Organizing Committee and the Hebei Company, and create a unified monitoring and command display platform for the power supply service guarantee of the Beijing 2022 Winter Olympics.

References

1. Gulijiazi, Y., Chunci, C.: Public perceived effects of 2022 winter olympics on host city sustainability. Sustainability **13**(7), 3787 (2021)
2. Manickavasagam, K., Hariharan, R.: Assessment of power system security using Security Information Index. Generation Transmission & Distribution **13**(04), 3040–3047 (2019)
3. Ramtin, M., Javad, L., Ross, B.: Constraint screening for security analysis of power networks. Trans. Power Syst. **32**(3), 1828–1838 (2017)
4. Erickson, B.J.: Basic artificial intelligence techniques: machine learning and deep learning. Radiol. Clin. North Am. **59**(6), 933–940 (2021)
5. Inkyung, S., Matthias, B., Peter, N.: Unpacking the role of artificial intelligence for a multimodal service system design. Electronics **11**(4), 549 (2021)
6. Su, H.: Construction model and evaluation of dynamic knowledge map for deep learning. Conference Series **1915**(4), 292-309 (2021)
7. Woo, J.-H., et al.: Dynamic knowledge map: reusing experts' tacit knowledge in the AEC industry. Autom. Constr. **13**(2), 203–207 (2003)
8. Abdel-Hamid, O., Mohamed, A.R., Jiang, H., et al.: Convolutional neural networks for speech recognition. IEEE/ACM Trans. Audio Speech & Language Processing **22**(10), 1533–1545 (2014)
9. Miramont, J.M., Restrepo, J.F., Codino, J., et al.: Voice signal typing using a pattern recognition approach. J. Voice **36**(1), 34–42 (2020)

Water Ecological Health Evaluation System Based on Data Fusion Technology

Chengjie Liu(✉)

Weifang Xiashan Reservoir Management Service Centre, Weifang City 261325, Shandong, China
jsyz167@163.com

Abstract. In order to ensure that ecosystems can perform their service functions normally, serve human beings, and ensure sustainable social development, the health of ecosystems has gradually attracted more and more attention from human beings, and the assessment of ecosystem health has become a research hotspot. The purpose of this paper is to study the water ecological health evaluation system based on data fusion technology. The vector data collected in this study can be converted into raster data by means of interpolation. After the image is spatially registered, the attributes of each pixel are calculated and fused to obtain the pixel value of the new image. Through this data fusion method, the comprehensive calculation of multi-source and multi-form data is realized. On the basis of comprehensive research and analysis of the natural environment characteristics, water ecological factors and social and economic activities of the M watershed, ArcGIS software is used as a platform, and data fusion technology is used. The M watershed monitoring point attributes of the M watershed are evaluated by the importance level, and the water chemistry index of section A bridge is 3.2514.

Keywords: Data fusion · Water ecological health · Evaluation system · Evaluation index

1 Introduction

River is a complex water ecosystem, which has the characteristics of openness, complexity, nonlinearity and non-equilibrium, and changes with the requirements of the times [1]. Thus, to date, the definition of "healthy" river ecosystems has remained vague. There is no unified and clear statement, and the understanding is not comprehensive enough [2]. However, using the "healthy" state to reflect the status of the river water ecosystem can more vividly reflect the impact of human life and production activities on the river water ecosystem. This appropriate metaphor is enough to cause the impact of human activities on the river water ecosystem [3].

With the continuous development of river health evaluation methods, a variety of evaluation methods with different characteristics have appeared in the world [4]. Vinod found that the average values of Cu, Co, Zn, Pb, As and Cr in Indian sediments were higher than the Australian Temporary Sediment Quality Guidelines, the world averages

J. H. Abawajy et al. (Eds.): ICATCI 2022, LNDECT 169, pp. 136–143, 2023.
https://doi.org/10.1007/978-3-031-28893-7_17

for surface rocks and threshold effect levels for freshwater ecosystems. The main factors associated with HM contamination and sand intrusion were observed by cluster analysis and principal component analysis. Pollution index results indicated that HM pollution in sediments ranged from moderate to high [5]. Larbi L studied the biological composition of macroinvertebrates and aquatic macrophytes in coastal Ada, Ghana, to determine the status of biodiversity. The results showed that 70% of the sampled aquatic ecosystems had water parameter concentrations within the range of natural background levels. However, the concentrations of nitrate and phosphate were significantly higher than the World Health Organization (WHO) recommended standards for healthy aquatic ecosystems [6]. Establishing a system to comprehensively evaluate rivers can provide scientific knowledge for decision makers in river planning and management, and create a sustainable development environment for river water environment [7].

On the basis of the health evaluation analysis report, combined with the theoretical basis and practical operation experience of water ecological health evaluation at home and abroad, this paper uses the principal component analysis method in the statistical method to screen the monitoring sections and indicators, and optimizes them through operational operation., which can greatly reduce the workload of sample collection and data processing and increase the accuracy of the work without affecting the accuracy and representativeness of the evaluation results. Finally, using the optimized monitoring section and index system of the M watershed to conduct sample collection, follow-up testing and data analysis, and evaluate the water ecological health of the M watershed.

2 Research on Water Ecological Health Evaluation System Based on Data Fusion Technology

2.1 Data Fusion Technology

Data fusion is a new research guideline for dealing with problem-specific data using multiple sensors and multiple types in systems. The most accurate definition of data fusion can be summarized as: using computer technology, under certain conditions, automatically analyze and compile the observation data of multiple sensors obtained in a series of times, and complete decision-making and needs assessment [8]. In summary, data integration is the complete processing of data from multiple sensors or multiple sources to obtain accurate and more reliable conclusions [9].

There are many ways to achieve fusion, and the method of fusion varies by application and application needs. Mainly based on data group technology, computing technology and identification technology [10]. Computational analysis is mainly to verify, analyze, complete and monitor the most relevant sensor observation data, and analyze and compile the recently discovered non-essential observation data to form a perfect situation. Combined "global review" analysis is done in real-time based on observations from multiple sensors.

2.2 Basin Ecological Health Evaluation Indicators

(1) Water quality.

The water quality of the monitoring section in the basin is good, the water environment quality index is high, and the ecological health level of the water system is high. Use the statistical data of the water quality status of the water quality monitoring section of the river basin to calculate [11].

(2) The proportion of natural vegetation reaches.

Constructing a buffer zone of the river edge 50 m from the water ecosystem to the periphery will have a greater impact on the health of the water area due to its close proximity to the water area. The larger the proportion of natural vegetation reaches, the better the health of the waters [12].

(3) Industrial wastewater discharge index.

The watershed is located in an old industrial base of an industrial city in Shandong Province. The discharge of a large amount of industrial waste water will seriously damage the health of the water ecosystem, thereby affecting the ecological health of the entire watershed. The discharge volume of industrial wastewater per unit area is used to indicate the discharge index of industrial wastewater in the basin. The greater the discharge, the worse the health of the waters.

2.3 Water Ecological Health Assessment System Technology

(1) CUP module.

In the hardware design of the water environment data acquisition system, the CUP module selects STM32103ZE. The ARM processor is designed with 32-bit Cortex-M3 architecture. In the working state, the STM32103ZE has a maximum frequency of only 72 MHz, but each Hz can process a program of 12.5 million instructions per second. Multiplication operation and hardware division operation can be realized simultaneously in one cycle. The chip integrates two kinds of memory, flash memory and static random access, and these two kinds of memories are used as containers for temporarily storing data. Their memories are 512KB and 64KB respectively;

(2) LCD module.

In the structure of the water environment data acquisition terminal, human-computer interaction is an important part of it. It is naturally essential to choose a reliable human-computer interaction mode. In this paper, the human-computer interaction mode will use the graphics support in embedded applications. Therefore, in the hardware design idea of human-computer interaction, a complex programmable logic device is used to drive the 4.3-in. thin-film transistor liquid crystal display. The display pixels meet the requirements of the overall design, the height ratio is moderate, the contrast ratio is adjustable, and the maximum Up to level 9, the viewing angle is in line with people's viewing angle whether in the horizontal or vertical direction.

(3) GPRS module.

The GPRS module is a module for data transmission, and it is a module that determines whether the entire water environment data acquisition system has practical significance. The design of this module considers the actual cost and specific application environment, and will use the SIM900A module as the data transmission guarantee in this paper.

3 Investigation and Research on Water Ecological Health Evaluation System Based on Data Fusion Technology

3.1 The Overall Framework of the System

According to the work content and process of water ecological health assessment, the logical structure of the system can be divided into three layers: application layer, logic layer and data layer. The application layer includes functional modules such as data management, map management, water ecological health assessment, surface water pollution simulation, 3D scene and system maintenance. Various information needs and analysis functions. The logic layer processes the spatial data through the API interface, and is responsible for the realization of the business logic of the spatial database, such as the access, performance and operation of the spatial data. The data layer adopts SQL Server 2008, and realizes unified storage and operation of spatial data and attribute data through ArcSDE.

3.2 Overview of the Study Area

The Weihe River Basin has a warm temperate semi-humid continental monsoon climate, with cold winters and hot summers and four distinct seasons. The annual average temperature is 12.2 °C, and the annual average precipitation is 646 mm. Due to the influence of the monsoon, the inter-annual variation of precipitation is large and the distribution is uneven within the year. The precipitation in winter and spring is scarce, and the precipitation in summer and autumn is concentrated, and the rainfall is mostly concentrated in June-September, accounting for 73.6% of the annual rainfall. Under normal circumstances, the ice begins to freeze from the end of December to the following February, and the freezing period lasts for 1 to 2 months.

3.3 Algorithm for Water Ecological Health Assessment

The comprehensive evaluation of water ecological environment quality adopts the comprehensive index method, and the WQI index is used to represent the water environment quality of each evaluation unit:

$$WQI = \sum_{i=1}^{n} x_i w_i \tag{1}$$

In the formula: xi is the evaluation index score, wi is the evaluation index weight.

Select the water-retention capacity, soil texture and topography of vegetation to establish an evaluation index system for water-retention function of the watershed. The water conservation function index S can be calculated by the following formula:

$$S_j = \sqrt[3]{\prod_{i=1}^{3} C_i} \tag{2}$$

In the formula: Sj is the water conservation function index of the jth region, and Ci is the grade of factor i.

4 Analysis and Research of Water Ecological Health Evaluation System Based on Data Fusion Technology

4.1 Water Ecological Health Evaluation Module

The application sets the default rating system. Without modifying the unit index system, the system evaluates the ecological health of the river water body according to the default rating system and its default parameters. In order to make the evaluation more flexible, the system provides the function of dynamically modifying the evaluation system and weights. Indicators in the rating system can only be selected from the existing system data structure framework, and cannot be added or deleted at will. The corresponding index is stored in the two-dimensional database table in the background. After the user selects the tags and weights as required, the corresponding XML files are created in the directory associated with the application. After XML will evaluate the tagging system, different tagging rating systems can be selected as needed to evaluate the ecological health of the river.

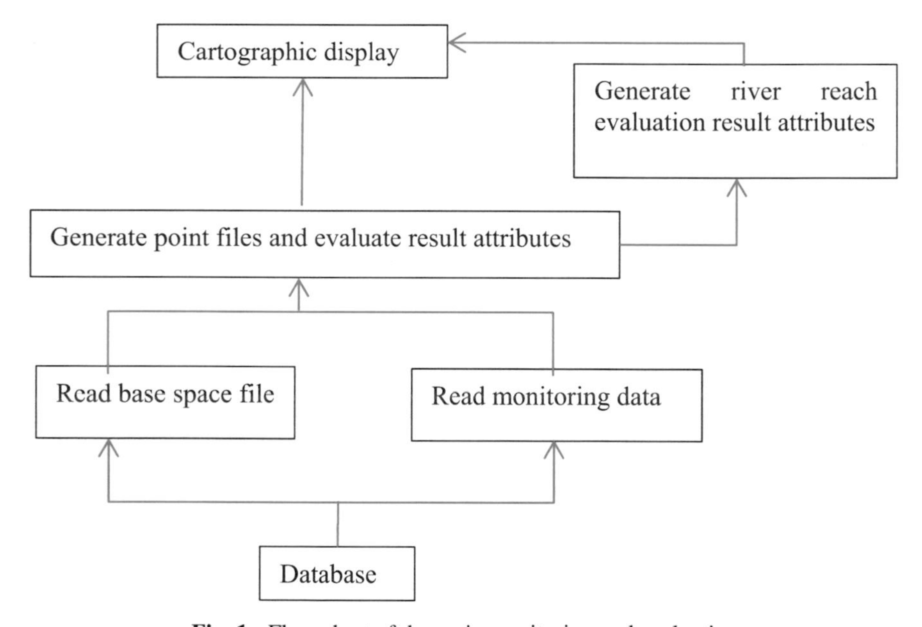

Fig. 1. Flow chart of dynamic monitoring and evaluation

Since the monitoring points are fixed, after selecting the corresponding evaluation index system and setting the corresponding weights, the system will require the user to input the scores of all secondary indicators of each monitoring point according to the monitoring information table, and calculate the score of the monitoring point. Score, so as to obtain the evaluation of the river reach by weighted average. Its internal principle is that the system first reads the attribute table of the monitoring point file, adds a field for storing the score of each secondary indicator, a field for the primary indicator score, and a field for the final evaluation score to the attribute table, and then passes the score entered by the user. Based on the data and the index system and its weights that have been set up, the score of the comprehensive evaluation of the monitoring points is calculated. After the evaluation results are obtained from the monitoring points, the river reach document obtains the water ecological health evaluation results of the river reach according to the weighted evaluation of the monitoring points it owns, and renders it into an evaluation grade map. For each evaluation, a new copy of monitoring point and its evaluation result data, river reach and its evaluation result data will be stored in the system. The file name is identified by time plus index system. This is not only convenient for users to view, but also convenient for computers to obtain evaluation results at different times according to time, forming a historical retrospective demonstration of river water ecological health, as shown in Fig. 1.

4.2 Vector Data Fusion

Table 1. Attributes of monitoring points in the M watershed

OBJBCT ID	Section	Aquatic organisms	Habitat	Water chemical index
1	A bridge	3.5214	4	3.2514
2	River B	3.5540	2	1.6852
3	C River	2.8862	3	3.2651
4	D Garden	3.1547	3	2.6662

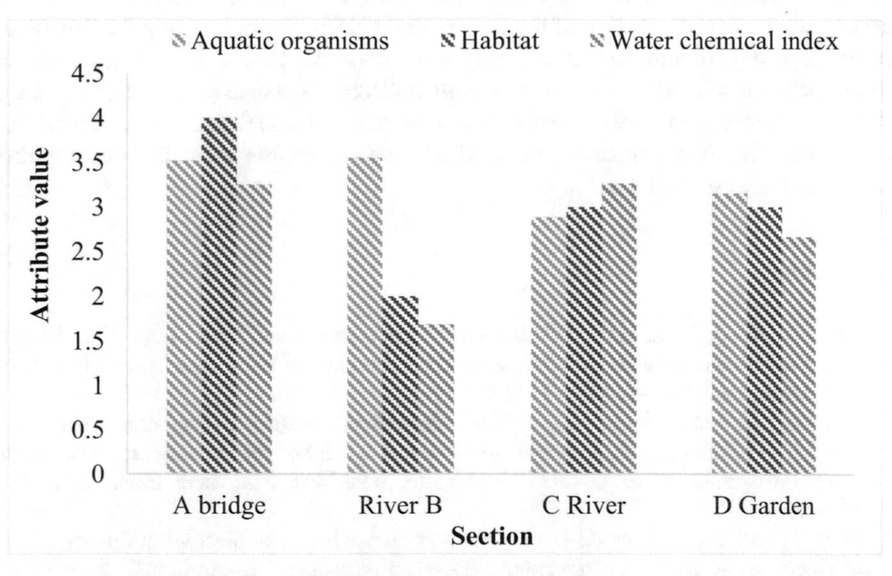

Fig. 2. Properties of monitoring points in the M watershed

Vector data includes fields, lines, and regions. Point data should be processed when evaluating residential care and various maintenance services. The results of the comprehensive analysis are computerized, i.e. the data aggregation process shown in Fig. 2 and Table 1. The entire liquid is then analyzed using Thiessen polygon analysis or interpolation analysis. The characteristic of the Thiessen polygon is that any position in the polygon is closest to the sampling point in the polygon, and each polygon has only one sampling point, so different points can be used as representatives of the Thiessen polygon area, and a different point can be used as a local Thiessen Polygon representation. Interpolation analysis predicts the value of other cells in a raster based on known data at multiple sample points. The relative distance dial and the natural neighborhood method are the two most common interpolation methods. The inverse distance method is to determine different densities according to the distance from the interpolation point to the sampling point, and then measure the average value to obtain the value of the interpolation point. Natural Neighbor Correlation uses an algorithm that finds sample sites that are closest to scale based on interpolated points, and analyzes the area to combine these samples with density. The natural areas of all sites are associated with Thiessen polygons.

5 Conclusions

In recent years, with the rapid growth of my country's GDP, the gradual advancement of industrialization and urbanization, the large-scale discharge of industrial and domestic wastewater has led to many problems such as runoff, biodiversity and the degradation of various river ecological environments. In accordance with the principle of index system selection, this paper draws on the development experience of domestic and foreign watershed health evaluation systems, and selects evaluation indicators in combination with the actual situation of the Huahe River Basin. By optimizing the functional operation of departments and monitoring indicators, on the premise of accurate and representative results, the workload of sample collection and data processing can be significantly reduced, subsequent operations can be reduced, and the accuracy of work can be improved. Watershed management methods and sustainable watershed development provide a clearer scientific basis.

References

1. Shen, Y., Kong, D., et al.: An assessment of the presence and health risks of endocrine-disrupting chemicals in the drinking water treatment plant of Wu Chang China. Hum. Ecol. Risk Assessment **24**(3–4), 1127–1137 (2018)
2. Rehman, I.U., Ishaq, M., Ali, L., et al.: Enrichment, spatial distribution of potential ecological and human health risk assessment via toxic metals in soil and surface water ingestion in the vicinity of Sewakht mines, district Chitral Northern Pakistan. Ecotoxicol. Environ. Saf. **154**, 127–136 (2018)
3. Chen, J., Fan, B., Li, J., et al.: Development of human health ambient water quality criteria of 12 polycyclic aromatic hydrocarbons (PAH) and risk assessment in China. Chemosphere **252**(4), 126590 (2020)

4. Kumari, P., Chowdhury, A., Maiti, S.K.: Assessment of heavy metal in the water, sediment, and two edible fish species of Jamshedpur Urban Agglomeration, India with special emphasis on human health risk. Hum. Ecol. Risk Assess. **24**(5–6), 1477–1500 (2018)
5. Vinod, K.A., Sharma, S., et al.: A review of ecological risk assessment and associated health risks with heavy metals in sediment from India. Int. J. Sediment Res. **35**(05), 90–100 (2020)
6. Larbi, L., Nukpezah, D., Mensah, A., et al.: An integrated assessment of the ecological health status of coastal aquatic ecosystems of Ada in Ghana. West Afr. J. Appl. Ecol. **26**(1), 89–107 (2018)
7. Aazami, J., KianiMehr, N., Zamani, A.: Ecological water health assessment using benthic macroinvertebrate communities (case study: the Ghezel Ozan River in Zanjan Province, Iran). Environ. Monit. Assess. **191**(11), 1–9 (2019). https://doi.org/10.1007/s10661-019-7894-1
8. Muratoglu, A.: Water footprint assessment within a catchment: a case study for Upper Tigris River Basin. Ecol. Indic. **106**, 105467.1–105467.13 (2019)
9. Tasian, G.E., Ross, M., Song, L., et al.: Ecological momentary assessment of factors associated with water intake among adolescents with kidney stone disease. J. Urol. **201**(3), 606–614 (2018)
10. Joloda, R.N.R., Karimi, S., Bouteh, E., et al.: Human health and ecological risk assessment of pesticides from rice production in the Babol Roud River in Northern Iran. Sci. Total Environ. **772**(4), 144729 (2021)
11. Porowska, D.: Hydrogeological assessment of the mineral composition of the bottled waters available for sale in Poland. J. Ecol. Eng. **21**(4), 103–111 (2020)
12. Guarda, P.M., et al.: Assessment of ecological risk and environmental behavior of pesticides in environmental compartments of the Formoso River in Tocantins, Brazil. Arch. Environ. Contam. Toxicol. **79**(4), 524–536 (2020). https://doi.org/10.1007/s00244-020-00770-7

Identification and Analysis of Wind Turbine Blade Cracks Based on Multi-scale Fusion of Mobile Information Systems

Yongjun Qi and Hailin Tang(✉)

Faculty of Megadata and Computing, Guangdong Baiyun University, Guangzhou 510450, Guangdong, China
linht88@163.com

Abstract. With the development of industry and the consumption of resources, environmental and energy issues have gradually become factors restricting the progress of human civilization. In recent years, people have been continuously researching and trying in the field of exploring and developing new energy. With the gradual consumption of coal, oil and other energy sources, wind energy, which is rich in reserves and has a long history of development, has gradually been developed and utilized by various countries. Wind energy can convert mechanical energy into electrical energy, contributing to the development of large-scale design of wind power blades in my country, and for the design of wind power blades. This paper mainly studies the crack detection problem of wind turbine blades based on the multi-scale fusion technology of mobile information system. By improving the detection efficiency of wind turbine blades, the hidden risks caused by the loss of generator blades can be reduced, and the safety performance of wind power generation can be improved. Multiscale fusion techniques for mobile information systems facilitate crack detection and analysis by analyzing relevant mathematical framework models and concepts such as invariant scale and invariant displacement based on statistical significance. The final result of the study shows that the total number of transverse cracks is 98, the number of identification is 95, the accuracy rate is 96.94%, the total number of longitudinal cracks is 86, the number of identification is 81, the accuracy rate is 94.17%, the total number of transverse cracks is 94.17%, and the total number of longitudinal cracks is 86. It is 84, the number of identifications is 79, and the accuracy rate is 94.05%. It can be seen from the experimental data that the identification accuracy rate of transverse cracks is the highest.

Keywords: Wind power generation · Crack identification · Multi-scale fusion · Information system

1 Introduction

Since the 21st century, the economies of all countries in the world have been growing rapidly, and the corresponding energy consumption is also increasing. Traditional

J. H. Abawajy et al. (Eds.): ICATCI 2022, LNDECT 169, pp. 144–151, 2023.
https://doi.org/10.1007/978-3-031-28893-7_18

non-renewable energy sources are facing depletion, and the deep-seated energy crisis is approaching step by step. Forcing us to look for a new type of energy to replace conventional energy, the first option is to develop and utilize wind energy, and wind turbines can efficiently convert wind energy into electricity. The use of wind energy reduces the emissions, protects the environment, maintains the ecological balance, and improves the energy structure [1]. Wind energy is not only "green energy", but also the most promising renewable energy with the lowest cost, and has very important development and utilization value.

In recent years, many researchers have studied the crack identification and analysis of wind turbine blades based on multi-scale fusion of mobile information systems, and achieved good results. For example, Dameshghi A believes that European countries started early in the stage of wind power generation technology and have sufficient wind energy reserves, and the relevant technical theory has been very mature [2]. Mangalraj P believes that the operating cost and operating efficiency in the establishment and operation of wind turbines are very critical, and improving operating efficiency and reducing operating costs are the first things to consider [3]. At present, scholars at home and abroad have carried out a lot of research on the crack identification and analysis of wind turbine blades.

In this paper, based on the theoretical basis of multi-scale fusion technology, combined with the analysis of wind turbine blade crack identification, from a theoretical perspective, the theoretical study of the vibration behavior of cracked blades is the basis for non-destructive testing of fatigue cracks. In the early analysis of the dynamic characteristics of the cracked beam, it is generally believed that the crack can be treated as a crack. In practical engineering, such simplification may lead to wrong estimation of natural frequencies and mode shapes, etc.

2 Related Theoretical Overview and Research

2.1 Modeling Analysis of Generator Blade Crack Identification

CAD utilizes computer data processing power and image acquisition capabilities, and then assists designers with manually entered commands to design and analyze 2D graphics or 3D product images to create the ideal project. The development of a technology that people want to achieve, such as CAD, has undergone many technological changes, the first caused by the CATIA model system that appeared in the 1970s [4]. The extensive application of solid modeling technology has become the basis of the second revolution of CAD technology. In terms of technology, parametric modeling techniques were widely used in the 1990s [5]. Therefore, what drives the third technological revolution is the large-scale use of parametric modeling methods; CAD technology has developed so far, which can be used for various methods such as free-form surface modeling, assembly modeling, and intelligent features. The use of technology (SWIFT), the complete automation of the design process, the design of molds, and industrial design with the help of computers, etc., can be said to be very powerful.

Solidworks has been well received by market participants for its simple operation, simple commands, excellent performance, and fast update speed. Solidworks is also a highly used 3D modeling software in business, research institutions and many research universities. SolidWorks provides a variety of panels, such as component models, component related mounting panels, all transistors, power panels, machine designers, and other related panels [6, 7]. SolidWorks is based on a Windows environment. Solidworks software interface is based on the ease of use and visualization of Windows software, and has strong visualization ability. Solidworks is the first of many 3D modeling programs. The Function Manager is located on the left side of the design interface. The most important function of the Function Manager is to automatically record all work phases corresponding to the design phase.

Solidworks component design provides design guidance from multiple perspectives, including front reference level, top reference level, right side reference plane, and reference levels that operators can create themselves [8, 9]. For data exchange, Solidworks parts can be stored in a variety of formats, including STEP format files (a common format for 3D components), dxf format files, and dwg format files for 2D exchange. It can also be saved as a picture format such as a jpg file. At the same time, Solidworks also provides secondary development, and the corresponding interface is related design blocks such as Application Programming Interface. SolidWorks 3D modeling software, together with CATIA, PRO/E and UGNX software, is a large-scale 3D CAD design software with simple operation and easy-to-use features [10]. Solidworks software has full functionality and many types of components. Entire models created with SolidWorks can be redesigned and edited, and component designs, assemblies, and 2D mechanical drawings can be integrated.

2.2 Analysis of Multi-scale Information Fusion Technology

Multi-scale information fusion can reflect the different characteristic information of generator cracks from various aspects, and realize the complexity and uncertainty of the description sequence of multiple scales, thereby improving the shortcomings of single-scale entropy description and reducing the deviation [11, 12]. However, the feature information set with high dimensionality will also introduce additional insensitive features, which may cause redundancy and conflict of some features, which will seriously affect the accuracy and reliability of fault diagnosis. In order to more effectively identify feature information and reduce interference components.

(1) Study the cantilever beam with transverse penetrating cracks, and simplify the first-order bending vibration as a cantilever beam with time-varying stiffness and a single-degree-of-freedom system with two degrees of freedom and damping; build an experimental bench, using Hilbert The transformation analyzes the experimentally measured response information. The time-varying stiffness and damping of the signal and the excitation signal are obtained and compared with the cosine stiffness model in the literature; the identification results further modify the cosine stiffness number model and study the parameters in the new model within a certain range. Varies with external conditions. The system responses before and after the model improvement are obtained by numerical calculation, and the results are compared with the experimental results. The results show that the response characteristics of the system calculated by the improved cosine stiffness model are consistent with the actual situation.

(2) According to the improved cosine stiffness model obtained from the experiment, the direct disturbance method is used to determine the vibration stability limit, the influence of each parameter on the stability is analyzed, and the multi-scale method is used to obtain the periodic solution of the steady-state response of the parameter system to explain the phenomenon in the experiment, to verify the correctness of the improved cosine stiffness model. The nonlinear response characteristics are obtained under different system parameters.
(3) Check the blades with semi-elliptical cracks, discretize through 3D solid elements, and have no friction contact function. This model simulates the crack breathing effect during the vibration process. Under the action of simple harmonic loads, the standard finite element model is used, and the cantilever beam is nonlinear. The analysis of the relationship between dynamic characteristics and external excitation frequency, damage degree and crack position lays a theoretical foundation for the identification of blade damage.

3 Experiment and Research

3.1 Experimental Method

On the section of the blade root bolted connection, the effect of all bolts is equivalent to the external bending moment, but each bolt cannot share the load equally. Therefore, calculations must be made to determine the maximum bolt axial force:

$$F = \frac{M}{\sum_{i=1}^{N} r_i^2} r_1 = \frac{M}{R^2 \sum_{i=1}^{N} (\sin\theta_i)^2} r_i \tag{1}$$

$$F_{\mathrm{i}} = \frac{2M}{R^2 N} r_i + \frac{F_{ax}}{N} \tag{2}$$

In the above expression: F represents the equivalent axial force (kN) of the bolt under the action of bending moment, Fi represents the equivalent axial force (kN) of the bolt under the action of bending moment and axial force, and R represents the bolt circle radius (m).

3.2 Experimental Requirements

In this experiment, a wind turbine blade crack identification analysis based on multi-scale fusion of mobile information system is proposed. Due to the limitations of some traditional time-domain signal analysis methods in application, wavelet transform is developed from the initial application in signal processing. It has become a powerful time and frequency domain analysis tool. Wavelet transform is a research hotspot in the field of signal processing, and has important applications in signal filtering, feature extraction, image compression and data fusion. The multi-scale theory is based on wavelet transform and multi-scale representation theory, and has achieved rapid development and application because of its perfect multi-scale structure, solid theoretical foundation and flexible structure.

4 Analysis and Discussion

4.1 Accuracy Analysis of Wind Turbine Crack Identification

The types of cracks include transverse cracks, longitudinal cracks, and network cracks. Statistical analysis of crack classification results is carried out for different types of cracks. The experimental results are shown in the following Table 1.

Table 1. Analysis table of wind turbine crack identification accuracy

Crack category	Total number of cracks	Number of identifications	Accuracy(%)
Lateral cracks	98	95	96.94
Longitudinal cracks	86	81	94.17
Mesh cracks	84	79	94.05

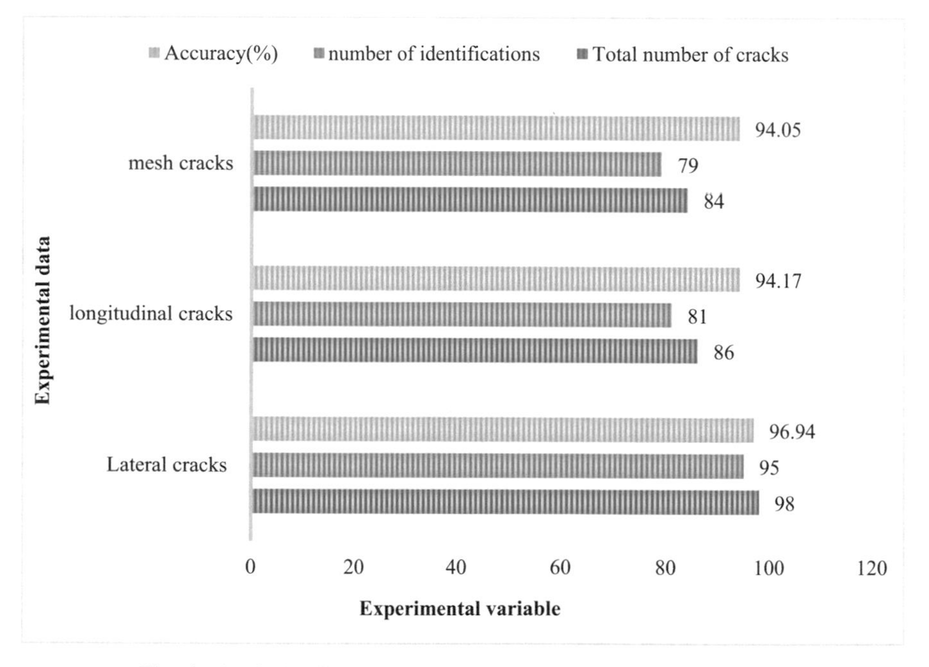

Fig. 1. Analysis of the accuracy of wind turbine crack identification

It can be seen from Fig. 1 that according to the analysis of the detection accuracy of wind induced cracks, the total number of transverse cracks is 98, the identification number is 95, the accuracy degree is 96.94%, the total number of longitudinal cracks is 86, the identification number is 81, the accuracy degree is 94.17%, and the total number of transverse cracks is 84.79 cases were identified, the accuracy was 94,05%. The experimental data show that the identification accuracy of transverse cracks is the highest.

4.2 Quantitative Analysis of Transverse Crack Detection

It can be seen from the above experiments that in the detection of crack types, the accuracy of lateral detection is the highest. Next, the accuracy of lateral crack detection is further analyzed quantitatively through lateral crack detection. The results are shown in the following figure.

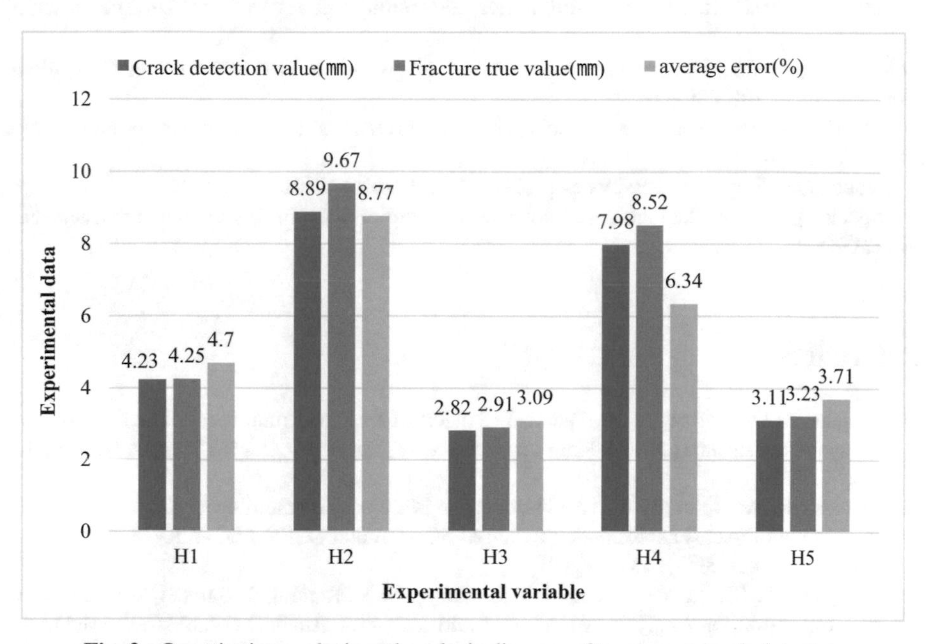

Fig. 2. Quantitative analysis and analysis diagram of transverse crack detection

As shown in Fig. 2, in the analysis of the actual and detected values of transverse cracks, as well as the average error rate, it can be seen from the data results that when the detected value is 4.23 mm and the true value is 4.25 mm, the average error is 4.7%, and the detected value is 4.7%. 8.89 mm, when the true value is 9.67 mm, the average error is 8.77%, the detection value is 2.82 mm, when the true value is 2.91 mm, the average error is 3.09%, it can be seen that the average error is basically kept within 10%, with a certain feasibility.

5 Conclusions

In this paper, the multi-scale fusion technology of mobile information system is used to identify the cracks of wind turbine blades to analyze the problem of crack identification of wind turbine blades. Among them, the identification rate of transverse cracks is the highest, but the overall identification accuracy rate is more than 90%. The quantitative detection and analysis of transverse cracks shows that the average error between the actual value and the detection value is kept within 10%, indicating that many The scale fusion technology has a certain feasibility for the detection of wind turbine blade crack identification, but due to the existence of errors, more experimental data are needed for analysis. Crack identification has done some research and exploration, but there are still many areas to be improved. It is believed that with the advancement of technology, related algorithms will become more and more perfect, and many technical problems will also be solved. Automatic detection technology for cracks in generator blades Will be more deeply applied to the maintenance and management of wind power generation.

Acknowledgements. Feature innovation project of colleges and universities in Guangdong province, No. 2020KTSCX163.

Feature innovation project of colleges and universities in Guangdong province, No. 2018KTSCX256.

Guangdong Baiyun university key project, No. 2019BYKYZ02.

Special project in key fields of colleges and universities in Guangdong province, No. 2020ZDZX3009.

References

1. Sghaier, M.O., Hadzagic, M., Patera, J.: Fusion of SAR and multispectral satellite images using multiscale analysis and dempster-shafer theory for flood extent extraction. IEEE **6**(10), 6–16 (2020)
2. Dameshghi, A., Refan, M.H.: Wind turbine gearbox condition monitoring and fault diagnosis based on multi-sensor information fusion of SCADA and DSER-PSO-WRVM method. Int. J. Model. Simul. **39**(1), 48–72 (2019)
3. Mangalraj, P., Sivakumar, V., Karthick, S., Haribaabu, V., Ramraj, S., Samuel, D.J.: A Review of Multi-resolution Analysis (MRA) and Multi-geometric Analysis (MGA) tools used in the fusion of remote sensing images. Circuits Systems Signal Process. **39**(6), 3145–3172 (2019). https://doi.org/10.1007/s00034-019-01316-6
4. Sedrette, S., Rebai, N.: A GIS approach using morphometric data analysis for the identification of subsurface recent tectonic activity. Case Study Quaternary Outcrops North West of Tunisia. **14**(1), 19–20 (2022)
5. Pribadi, C.B., Hariyanto, T., Hadi, Z.: Advisability analysis of gold exploration based on remote sensing and geographic information systems (case study: Banyuwangi regency). In: IOP Conference Series: Earth and Environmental Science, vol. 731, no., 1, 012017 (7pp) (2021)
6. Sakamoto, L., Fukui, T., Morinishi, K., et al.: Blade dimension optimization and performance analysis of the 2-D Ugrinsky Wind Turbine. Energies **15**(5), 6–8 (2022)
7. Sghaier, M.O., Hadzagic, M., Patera, J.: Fusion of SAR and multispectral satellite images using multiscale analysis and dempster-shafer theory for flood extent extraction. In: Fusion of SAR and Multispectral Satellite Images Using Multiscale Analysis and Dempster-Shafer Theory for Flood Extent Extraction, vol. 5, no. 7, 37-42 (2019)
8. Nweke, H.F., Teh, Y.W., Mujtaba, G., Alo, U.R., Algaradi, M.A.: Multi-sensor fusion based on multiple classifier systems for human activity identification. HCIS **9**(1), 1–44 (2019). https://doi.org/10.1186/s13673-019-0194-5
9. Sitaula, C., Shahi, T.B., Aryal, S., et al.: Fusion of multi-scale bag of deep visual words features of chest X-ray images to detect COVID-19 infection. Sci. Rep. **11**(1), 22–36 (2021)

10. Wahid, F.F., Sugandhi, K., Raju, G.: A fusion based approach for blood vessel segmentation from fundus images by separating brighter optic disc. Pattern Recogn. Image Anal. **31**(4), 811–820 (2021). https://doi.org/10.1134/S105466182104026X
11. Roman, J., Legal-Ayala, H., Noguera, J.: Applications of multiscale mathematical morphology to contrast enhancement and images fusion. In: 2020 15th Iberian Conference on Information Systems and Technologies (CISTI), vol. 16, no. 1, pp. 25–37 (2020)
12. Jin, L.S., Xu, Y.Q., Chen, Z.S., et al.: Relative basic uncertain information in preference and uncertain involved information fusion. Int. J. Comput. Intell. Syst. **15**(1), 655–682 (2022)

Study on Intelligent Terrain Monitoring System and Key Rescue Technology of Construction Machinery

Xiaowei Jiang[1(✉)], Yiming Zhao[1], and Craig Pearsall[2]

[1] Changchun University, Changchun, Jilin, China
hyqq901@163.com

[2] BA Components Ltd, Kirk Sandall Industrial Estate, Sandall Stones Rd, Doncaster DN3 1QR, UK

Abstract. This paper mainly studies the design of structure, drive system, control system and other parts of crawler all-terrain rescue engineering machinery. Based on the understanding of the functions to be realized and the actual environmental background, the system composition and function design of crawler all-terrain rescue construction machinery are studied. The crawler all-terrain rescue engineering machinery is suitable for the rescue of the affected people after the disaster, especially for any complex terrain after the disaster, such as snow, swamp, pothole, water and so on. Strive to reach the rescue scene as soon as possible, so that the environment will no longer become an obstacle to life. The design is similar to other vehicles using crawler walking mechanisms except for the drive system. Car body structure has a certain unique design. The appearance is compact and light, easy to operate.

Keywords: Rescue Engineering Machinery · Construction Machinery · Key Rescue Technology · Structural Design · Drive System

1 Introduction

Disaster is a time scene that human beings never want to see. However, with the rapid development of society, highway, fire, power and communication, aviation and other departments of emergency rescue demand is also growing. Natural disasters and accidents emerge one after another. In recent years, even more intense, the market demand for rescue and rescue machinery has risen sharply, and put forward higher requirements for the performance of vehicles, showing a trend of diversified functions. In ancient times, there was "Dayu flood control, three through the house and not into", and now there is "earthquake relief, unity", the selfless spirit of rescue in the everlasting inheritance, countless rescue equipment in the constant upgrading. Awe life, urgent, people for the rescue machinery research and development toward high-tech, efficient direction is the general trend, which is also the expression of the "first time rescue" consciousness. Earthquakes, mudslides and other natural disasters after the devastation. Ordinary vehicles could not move forward, and traditional rescue and rescue vehicles are increasingly

J. H. Abawajy et al. (Eds.): ICATCI 2022, LNDECT 169, pp. 152–159, 2023.
https://doi.org/10.1007/978-3-031-28893-7_19

unable to meet the needs of the market. Most rescue vehicles are limited to size, weight and other factors, and could not reach the disaster site, thus delaying the best time to rescue the affected people. Therefore, it is urgent to develop a new and suitable crawler all-terrain rescue engineering machinery.

In short, there are many deficiencies in both speed and safety when rescue teams use rescue machinery for rescue [1]. It is likely to lead to the failure of the rescue operation, and may injure the personal safety of the rescuers, and may also lead to the life safety of the rescuers threatened [2]. Combined with the shortage of rescue engineering machinery in emergency rescue, it is very important to analyze how to improve the application of rescue engineering machinery in emergency rescue, how to develop more advanced rescue engineering machinery, and how to improve the safety and reliability of rescue engineering machinery in rescue.

2 Technical Status Quo and Existing Problems of Rescue Construction Machinery

2.1 Technical Status of Rescue Engineering Machinery

At present, there are many rescue and rescue machinery manufacturers, but the rescue machinery produced by each manufacturer is single, lacks serialization, and the mechanical function is less, and the degree of integration is low [3]. The technical characteristics of the existing rescue machinery are as follows.

(1) Based on the chassis of large and medium-sized trucks or buses, structural modification of the carriages is carried out to facilitate the layout and loading of various rescue and rescue equipment. For example, an earthquake rescue machinery is modified based on the German MAN car chassis, the compartment is divided into a variety of different sizes of boxes, in order to facilitate the loading of all kinds of equipment.
(2) Emergency rescue machinery with power supply, lifting, traction and other basic rescue capabilities. Emergency machinery generally loaded alternator, truck crane and traction winch. For example, a rescue machinery is equipped with a crane with a maximum lifting mass of 5.4t, a traction winch with a maximum traction force of 53kN and a vehicle-mounted generator with a maximum traction force of 22kVA.
(3) The main task of machinery is to load the main body of rescue equipment. Rescue machinery generally needs to load a large number of rescue equipment. For example, a medium-sized rescue vehicle is loaded with more than 100 sets of life-saving equipment such as hand-held thermal imagers, protective equipment such as breathing apparatus and chemical protection suits, demolition equipment such as hydraulic clamps and electric sparks, and first aid kits.
(4) A large number of on-board devices adopt independent function mode. A large number of demolition and rescue equipment loaded on the vehicle, such as hydraulic pliers, hydraulic expander and other hydraulic pump stations are generally driven by small engines for energy supply, air hammers, picks and other demolition tools are generally driven by small engines for pneumatic machine function [4]. All kinds

of other investigation and inspection and first aid equipment are also mutually integrated in energy supply.

2.2 Problems Existing in Rescue Construction Machinery

It can be seen from the technical status quo of the above rescue and rescue machinery that there are many technical defects in the existing rescue and rescue vehicles, mainly including the following points.

(1) Function separation between rescue and rescue equipment and vehicle, vehicle only becomes the loading tool of rescue and rescue equipment, and the degree of integration is low.
(2) In order to load all kinds of equipment as much as possible, truck or bus chassis is generally modified. The vehicle's large size, poor passability and poor cross-country performance make it difficult to reach the rescue site in the first time.
(3) The rescue equipment has a wide variety of energy resources, large volume of support equipment, and logistical supply is difficult. For example, if a rescue vehicle needs to have the four functional requirements of hammer, hydraulic shear, welding and lighting at the same time, two small engines are needed to provide power for the pneumatic machine and hydraulic pump, and a large power internal combustion engine power station is also needed to provide power for welding and lighting equipment. Different fuel types of various power sources cause difficulties in logistics supply.
(4) Small power and low efficiency of rescue equipment. Since most rescue equipment is supplied by independent power source, the power of power source is severely limited by volume and weight. Small internal combustion engine is generally adopted to achieve the rescue equipment, which has a small power and is difficult to complete the demolition task required by high power and has low working efficiency.

3 Research and Development of Crawler All-Terrain Rescue Engineering Machinery

3.1 Structural Design

This design is a crawler type all terrain rescue engineering machinery. It mainly consists of three parts: chassis, suspension and drive system. The material used in the suspension is aluminum alloy steel, and the connection mode is six points independent connection. Also equipped with a continuously variable transmission [5]. Frame, are included on the rescue mechanical loading on the frame of dynamic driving system, connected to the power drive device of shift, front at the bottom of the handle and direction control component, at the bottom of the front wheel, axle, after the upper shaft, shock absorbing device, track wheel, two pairs of rubber tracks, mounted on a frame at the top of the pedal, it also includes mounted on the frame in front of the frame position of the motor, brake and crawler meshing track driving wheel [6].

The rescue machinery is installed with rubber caterpillar to increase the bearing area to reduce the pressure, and will not sink when walking in the swamp or mud. Not only that, the track is mainly cross-country ability than the wheeled strong, can cross the ditch, over the low wall, climbing ability is also very good.

The rescue machinery has the advantages of easy steering, strong grip, good stability, easy climbing, simple structure, and high transmission efficiency. It is a crawler all-terrain rescue engineering machinery that can be carried out on mountain slopes and soft soil surfaces even in cold weather conditions.

3.2 Selection of Driving System

Compared with electric motors, oil-fired engines have the following advantages [7].

(1) fuel saving: excellent combustion efficiency produces extremely high economic benefits.
(2) Quiet: a low noise engine set that could be used at anytime and anywhere.
(3) Reliable: stable automatic voltage regulation system and oil warning system, safe to use.

Honda has been a leader in internal combustion engine technology for nearly half a century, and its superb technology and rich experience are reflected in every engine, ensuring stable performance and reliable operation in the long run of each engine, thus greatly saving time in maintenance. To sum up, the design of the rescue machinery uses Honda four-stroke 14 horsepower oil-fired generator [8]. Here are some of the engine's advantages over others.

(1) Purpose and performance: easy to start; The body inside and outside the zinc block; Overheat and low oil pressure alarm protection system; Engine overspeed protection [9];
(2) fuel saving: four-stroke engine adopts advanced half ball combustion chamber, so that the gasoline completely burns, can turn every drop of gasoline into power. Overall, the four-stroke generator saves up to 50% fuel.
(3) Strong horsepower: the four-stroke engine is larger, so the horsepower comes earlier. On the same horsepower label, the horsepower is more abundant than that of the two-stroke engine, and it is more obvious at medium and low speeds.
(4) Easy to use: the four-stroke engine design is perfect, equipped with a variety of different models such as: fuselage length, front (rear) control, electric lift, etc., to meet the needs of different models and operation, there is no need to add oil in the gasoline when refueling.
(5) Environmental protection: the clean pure gasoline fuel and advanced hemisphere combustion chamber make the emission index of four-stroke engine meet the standards of California "CARB 2008" and Europe "Bordensee", making a contribution to the promotion of environmental protection.
(6) durable: four-stroke engine lubrication system and advanced auto lubrication system, adopting pressurized circulating oil, make each bearing and need to get the direct supply lubrication oil lubrication components, unlike two-stroke engine needs to adopt mixed oil lubrication indirectly, in addition to the component to accurate

(too much cause carbon deposition [10], too little cylinder), more will cause pollution and high emissions. A piston in a four-stroke engine cools down twice as long after each burn as a two-stroke engine. A four-stroke engine can last four times longer than a two-stroke engine with superior lubrication and cooling systems.

(7) smooth and quiet operation: the four-stroke engine adopts precision supporting bearing bush and direct cycle oil supply and lubrication system to minimize the sound of the machine operation. In addition, the unique design of three-chamber muffler, propeller center bottom jet exhaust and high efficiency acceleration pump makes the machine run smoothly and quietly [11].

3.3 Technical Parameters

(1) Appearance size (mm): 1194 × 686 × 1270 (length × width × height);
(2) Left and right tilt Angle of pedal: 15 o;
(3) Vertical height of handle: 1270 mm;
(4) Vehicle weight: 145 kg;
(5) Maximum driving speed: 40 km/h.

3.4 Overcoming Vertical Walls

The process of crawler all-terrain rescue construction machinery to overcome the vertical wall can be divided into three stages:

The first stage (Fig. 1) -- the front wheel rises to the wall edge;

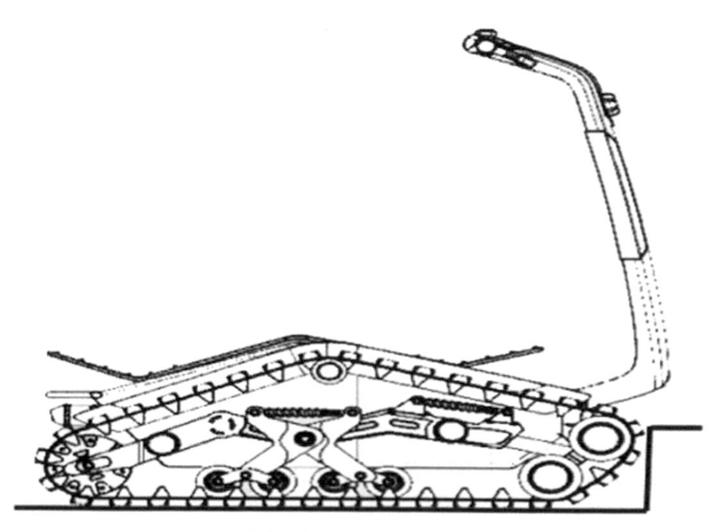

Fig. 1. The first stage

The second stage (Fig. 2) -- the center of gravity of crawler-type all-terrain rescue construction machinery advances to the position coincides with the vertical line of the vertical wall;

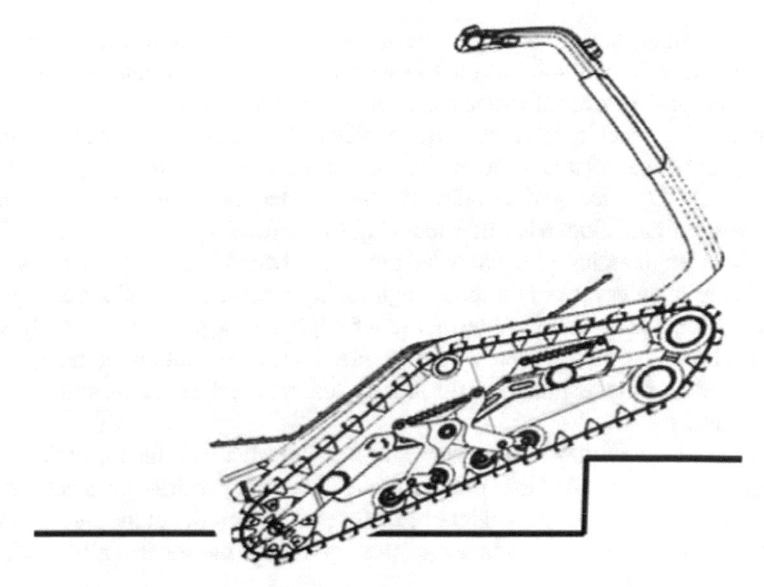

Fig. 2. The second stage

The third stage (Fig. 3) -- the sliding crawler all-terrain rescue construction machinery gently contacts the top surface of the vertical wall.

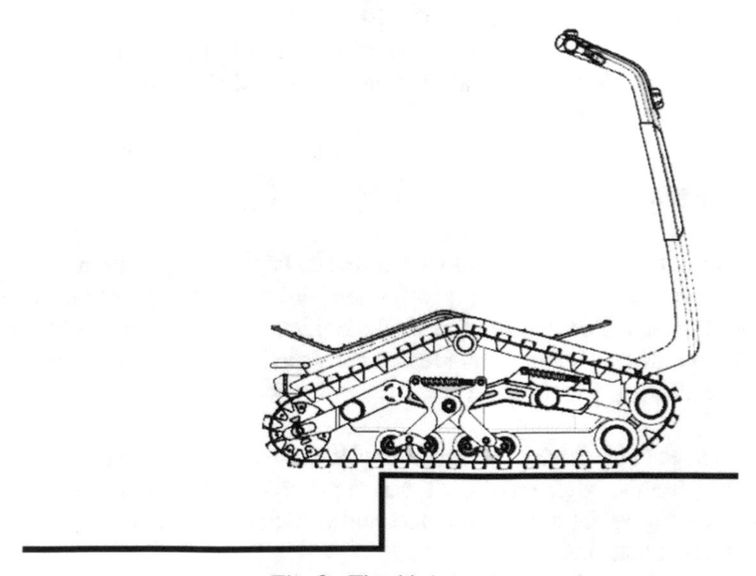

Fig. 3. The third stage

The following is to analyze the motion process, dynamic conditions and influencing factors of each stage one by one.

In the first stage: at this stage, the caterpillar all-terrain aid construction machinery with low speed in D first point of contact with the vertical wall, and then tracked all-terrain rescue project overall mechanical rotary motion, make the crawler all-terrain relief engineering machinery front along the vertical wall up gradually, and slide type as determined by the point D all-terrain rescue vehicle center of gravity position of g.

When point D is lower than point G, the crawler all-terrain rescue engineering machinery will rotate clockwise after touching the vertical wall. Make the crawler all-terrain rescue engineering machinery vertical wall become impossible to overcome, in the existing crawler all-terrain relief engineering machinery g point above point D, therefore, in the caterpillar of all-terrain relief engineering machinery, nearly vertical wall should be fierce, so that increase the counterclockwise rotation moment, make the crawler all-terrain relief engineering machinery sink rear and front rise, will rise to more than g D o 'clock.

When the balance elbow of the rear weight-bearing wheel hits the travel limiter, there are two kinds of movement of the crawler all-terrain rescue construction machinery [12]. The translational motion of the center of gravity and the rotational motion of the body.

At this point, there are the following forces acting on the tracked all-terrain rescue construction machinery.

(1) Gravity G of crawler all-terrain rescue construction machinery;
(2) Ground normal reaction force F;
(3) The difference between the ground tangential reaction force $F(\varphi\text{-}f)$, the adhesion force $F\varphi$, and the ground deformation resistance Ff;
(4) The reaction force of the vertical wall F1.
(5) Tangential reaction force $F1(\varphi'\text{-}f')$ along the vertical wall, where$\varphi'$and $f'$are the adhesion coefficient of vertical wall and track and the resistance coefficient of ground deformation.

4 Conclusion

In the design process of crawler all-terrain rescue engineering machinery, the whole design is divided into three parts: drive system design, steering design, throttle and brake device design. The design of these three parts is carried out one by one. After the basic design of crawler all-terrain rescue construction machinery is completed, it is modified according to its rationality.

(1) In the design of the driving system of crawler all-terrain rescue construction machinery, the gasoline engine and CVT can more efficiently complete the purpose of use according to the use environment and use mode of crawler all-terrain rescue construction machinery.
(2) In the steering design of crawler all-terrain rescue construction machinery, the steering is realized through the connection of sensors and pedals.
(3) In the throttle and brake device design of crawler all-terrain rescue engineering machinery, electronic throttle and hydraulic brake are adopted according to the design principle of this part of motorcycle.

Acknowledgments. This research is supported by the Art Project of 2019 National Social Science Foundation under the grant No. 19BH148.

This research is also supported by 2021 Social Science Research Planning Project of Education Department of Jilin Province under the grant No. JJKH20210662SK. The name of this project is "Research on the Visual Transformation Mechanism and Realization Path of Folklore in Jilin Province".

This research is also supported by the Teaching Research Subject of Changchun University under the grant No. XJYB19-04.

This research is also the achievement of the Teaching Reform Research General Subject of Vocational Education and Adult Education of Education Department of Jilin Province under the grant No. 2019ZCY376. This subject is approved in 2019 and its name is "research on the construction of project school-based teaching material of visual identity in higher vocational college".

References

1. Chang, H., et al.: Melatonin successfully rescues the hippocampal molecular machinery and enhances anti-oxidative activity following early-life sleep deprivation injury. Antioxidants (Basel, Switzerland) **10**(5), 774 (2021)
2. Zhang, Y., Gao, P., Ahamed, T.: Development of a rescue system for agricultural machinery operators using machine vision. Biosys. Eng. **169**, 149–164 (2018)
3. Ramalho, S.S., Silva, I.A.L., Amaral, M.D., Farinha, C.M.: Rare trafficking CFTR mutations involve distinct cellular retention machineries and require different rescuing strategies. Int. J. Mol. Sci. **23**(24), 24 (2022)
4. Khan, U., James, B., Late, D.: ESCRT machinery mediates the recycling and Rescue of Invariant Surface Glycoprotein 65 in Trypanosoma brucei. Cell Microbiol **22**(11), 32–44 (2020)
5. Andrew, B.: You got your hand stuck in what? machinery rescue tool kit essentials. Fire Eng. **172**(2), 45–50 (2019)
6. Formosinho, M., Sousa Reis, C., Jesus, P.: Rescuing the ghost from the machine: towards responsive education and beyond explanatory machinery systems. Int. J. Lifelong Educ. Lead. **1**(1), 32–40 (2015)
7. S. Madhuri, S. Chetna. Nano-agrotechnology: nanoscale machinery for the rescue of farmers. Agricult. Res. J. **51**(1), 1–7 (2014)
8. Heydari, P., Varmazyar, S., Fallahi, A., Hashemi, F., Jafarvand, M.: Investigating the awareness of rescue groups rather than the warning signs (safety-health) installed on the heavy road machinery of carrying hazardous materials. Res. J. Med. Sci. **10**(7), 725–728 (2016)
9. Nadler, F., Lavdovskaia, E., Richter-Dennerlein, R.: Maintaining mitochondrial ribosome function: the role of ribosome rescue and recycling factors. RNA Biology, 1–15
10. Goffin, P., et al.: Lactate racemization as a rescue pathway for supplying D-lactate to the cell wall biosynthesis machinery in Lactobacillus plantarum. J. Bacteriol. **187**(19), 6750–6761 (2005)
11. Shahid, S.: The rules of attachment: REC8 cohesin connects chromatin architecture and recombination machinery in meiosis. Plant Cell **32**(4), 808–809 (2020)
12. Sbastien, M., Alan, U., Arnaud, H., Fabrice, L., Christiane, B., Yuri, M.: Deficiency of the tRNATyr:Ψ35-synthase aPus7 in Archaea of the Sulfolobales order might be rescued by the H/ACA sRNA-guided machinery. Nucleic Acids Res. **37**(4), 1308–1322 (2008)

Construction Management Mode of Building Engineering Based on BIM Technology

Chun Ding(✉)

Shandong Transport Vocational College, Weifang, Shandong, China
xingjiliehu@163.com

Abstract. With the development of society and economy, the construction industry is constantly improving, and the construction management mode of construction projects is also changing. The construction of the construction management mode of a construction project is a very complicated and long process, which requires relevant staff to consider from many aspects. At the same time, BIM technology has been gradually applied to engineering construction, it can effectively improve the quality of engineering construction, and promote the maximum economic and social benefits of enterprises. Therefore, this article is based on BIM technology to carry out research on the construction management mode of construction engineering and its related content. This article adopts experimental analysis method and data analysis method to better understand the effect of BIM model in construction cost management of construction project through experiment. According to the research results, the model can reflect the dynamic changes of the construction process progress and cost in a timely manner, and has a certain effect on controlling the construction progress and cost, which has a certain positive significance for improving the level of construction management.

Keywords: BIM technology · Construction engineering · Construction management · Dynamic simulation

1 Introduction

Construction management of construction projects refers to the process of planning, organizing and controlling the entire project. In project construction, many factors need to be considered, such as material cost and construction period. These influencing factors interact with each other to form an organic overall structure. At the same time, BIM technology can make full use of the information data model to deal with complex, changeable, cumbersome and diverse problems. It can also allocate the time required for each construction phase in a reasonable manner, allowing various departments to coordinate and optimize the allocation of resources. The level of quality, reducing the number of rework and reducing costs are of great significance.

At present, many scholars have conducted corresponding research on BIM technology and its application, and have obtained very rich research results. For example, Jiang Zhihao pointed out that BIM technology is based on three-dimensional digital technology and uses modeling methods to expand building information, penetrates the entire

J. H. Abawajy et al. (Eds.): ICATCI 2022, LNDECT 169, pp. 160–167, 2023.
https://doi.org/10.1007/978-3-031-28893-7_20

life cycle of buildings, and is conducive to the sustainable development of construction projects [1]. Duan Zhichao believes that construction management is an important work to ensure construction quality, and BIM technology is applied in construction management of construction projects, which is conducive to promoting the normal operation of construction management [2]. Wang Haijun pointed out that the project cost occupies a large proportion in the whole process of the construction project, and the rational use of BIM can significantly improve the efficiency of project cost management [3]. Therefore, this article combines BIM technology to study the construction management mode and related content of construction engineering, which to a certain extent has a certain positive research significance and reference value for improving the level of construction management.

This articlc mainly discusscs thcsc aspects. First of all, it elaborates on BIM technology and related research. Then, it discusses the construction management mode of construction engineering and its related research. In addition, the application of BIM technology in construction management of construction projects is introduced. In addition, this article also carried out experimental research around the BIM model, and drew the corresponding experimental results and analysis conclusions.

2 Related Theoretical Overview and Research

2.1 BIM Technology and Related Research

BIM is the process of creating and using digital models during the planning, construction, and operation management stages of a construction project. BIM technology can integrate information on time, cost, quality, safety, building energy efficiency, performance analysis, etc. into the BIM model.

Some researchers also define BIM as a building information model. It is a new methodology for the overall design, construction, operation, and maintenance management of a building project. It runs through the entire construction process and the human activities of the project and contains many uncertain factors. BIM technology has the following characteristics.

One is visualization. BIM technology can simulate the rendering after completion of the project, intuitive and descriptive three-dimensional modeling, and provide comprehensive construction information, so that the entire life cycle of project decision-making is in a visual state.

The second is coordination. The personnel involved in the construction phase of the project coordinate and collaborate with each other to discuss construction issues and related issues to ensure that the information is correct for follow-up inspections.

The third is optimization. Some engineering projects have large scale, tight schedules, and complex processes, leading to some problems. The BIM model can be adjusted and optimized, and the content before and after the change can be compared intuitively, simplifying the work content and optimizing the project.

Fourth, the output efficiency is high. At present, most of the project results are still presented in the form of two-dimensional drawings. BIM output drawings are different from traditional two-dimensional CAD drawings. BIM drawing is the use of BIM technology to display, simulate, collide and visualize the output of the project [4, 5].

BIM technology has certain application value. After BIM technology was introduced as a new concept, it has aroused widespread attention in China.

It uses the Internet and computers to improve project management, which can reduce rework, save time, reduce costs and add more value. The acceleration of project progress, the reduction of costs, the improvement of quality and safety management quality, the reduction of rework and significant economic advantages are the manifestations of specific application values.

So far, BIM technology has been applied in construction projects one after another. It has played an important role in all stages of the entire project implementation process. On the one hand, the BIM model database is used to retrieve historical engineering data similar to the proposed project for cost estimation; on the other hand, the automatic calculation function of the BIM model can quickly extract engineering data and apply the company's internal quota. Using BIM model to simulate construction can improve the competitiveness of technical solutions, automatically and quickly calculate the quantity, and facilitate the selection of construction strategies [6, 7].

In addition, comprehensive consideration of customers' design wishes and cost control. Dynamic control of each stage of the construction process, management of on-site monitoring and progress warning, so as to realize the reasonable planning of the construction progress and the reasonable allocation of capital investment.

2.2 Construction Management Methods and Related Research of Construction Projects

The construction management of a construction project is a complicated process. The project implementation consumes a lot of financial and material resources, and invests a lot of manpower and equipment. In the construction management of the project, all parties must consider schedule, cost, quality, safety, etc. Multi-dimensional information collection and release make full use of information.

At present, some construction projects have different degrees of problems in construction management.

First, there are violations of laws and regulations. Some construction companies have illegal subcontracting, which has laid a lot of hidden dangers to the actual management of process construction.

Second, the qualification management and monitoring system is not sound. In some construction departments, the supervision status is relatively backward.

Third, some construction companies have not done well in terms of quality control, cost, and construction period. In this case, on the one hand, the construction cost remains high, and on the other hand, there are serious problems with quality. At the same time, it is necessary to strengthen the management of the existing site to effectively control the scope of the construction project appraisal, thereby ensuring the quality of the construction project. However, under the condition of extensive construction management, some construction projects have problems during the construction year [8, 9].

To carry out project management, we must first understand the construction progress. After confirming the progress of signing the contract, formulate a project implementation plan. The supporting work plan is formulated according to the technical implementation plan. The project planning method is mainly horizontal road map method and key road method, and the formulation of the schedule is too dependent on the manager's experience. When formulating a schedule, due to factors such as project implementation costs, engineering technical feasibility, and material use, the schedule needs to be optimized and adjusted again and again. When adjusting the lower-level plan, it is necessary to gradually adjust the upper-level progress plan, which is difficult and time-consuming to prepare.

Second, the construction cost needs to be controlled. Budget and plan the project cost and material consumption reserve; at the end of the month, determine the actual investment cost and material usage based on the overall situation of the construction, and summarize them according to the ledger. Once budgeted expenses are compiled by budgeting software, Excel spreadsheets are usually used for statistical management. This method of cost control is now widely used by most construction companies.

In addition, the construction quality must be checked. Among them, the construction part pays special attention to project quality. The quality control process basically includes three stages before construction, during construction and after construction, which requires specific adjustments in quality control. The preliminary inspection shall determine the project quality target and formulate a corresponding construction plan. In terms of quality control, after the completion of the sub-project, the quality of the project is checked and accepted through acceptance and testing. The quality control data in the construction process is mainly recorded and managed in the form of tables and lists [10, 11]. The process of project quality control is shown in Fig. 1.

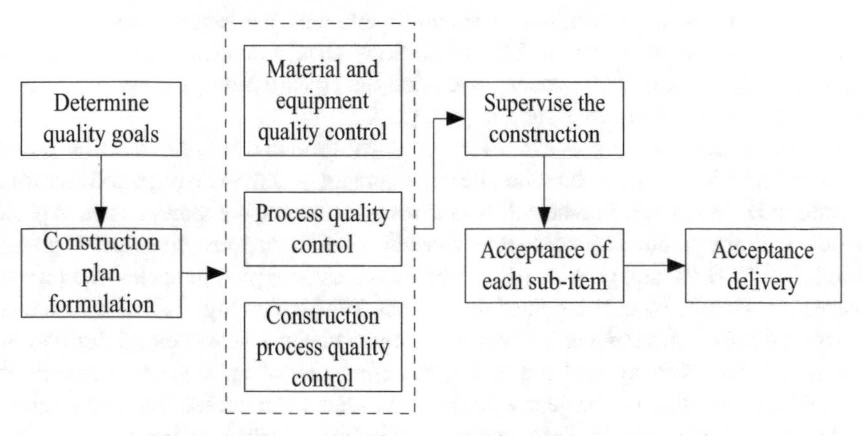

Fig. 1. The Process of Engineering Project Quality Control

2.3 Application of BIM Technology in Construction Management of Building Engineering

Construction companies use BIM technology to strengthen construction management, integrate construction resources, safety quality, layout, cost changes and other information to create dynamic comprehensive management and visual simulation of the construction process based on BIM. In addition, BIM technology can also be used to control construction costs. The application of BIM in structural engineering management, production management and enterprise management, as well as multidisciplinary integrated management, has effectively improved the quality and efficiency of project management, and improved the overall construction management level of construction companies.

Introduce BIM technology in the construction preparation stage. According to the two-dimensional CAD drawings provided by the bidder, the bidder converts the sample into a three-dimensional model, automatically calculates the quantity and accurately extracts the quantity data, and combines the project attributes to create a more accurate quantity list to reduce the occurrence of calculation errors and thereby reduce economic disputes, and improve work efficiency and give the components enough time to develop a bidding strategy. Apply BIM technology to simulate the construction of various construction plans, choose the best plan, and help design the site plan, optimize the construction plan for collision detection, and avoid unnecessary rework. In addition, the exchange of information and data enables the two parties to establish a transparent connection when signing the contract, simulate the issue of the exemption clause and propose countermeasures to avoid customer complaints.

Apply BIM technology to construction technology management, through professional BIM collision control software, perform virtual calculations, check for errors, omissions, collisions and gaps in construction drawings, optimize before construction, reduce construction rework, save costs, and shorten construction period, and optimize the construction plan through construction simulation. Before the project starts construction, it is necessary to formulate a scientific and reasonable construction plan. The intuitiveness and analyzability of BIM technology clearly control the time nodes and processes of the construction process, optimize the construction plan and improve the scientific rationality of the construction plan [12].

During the construction process, a 4D model was formed by linking the BIM components to the schedule. Use the 4D schedule management system to manage and control the schedule and resources during the entire construction process, compare the actual project schedule with the planned schedule under visual conditions, promptly warn of possible future schedule risks, and provide an optimized adjustment plan to ensure the project's construction Time reduces the risk of the project. BIM technology is based on a three-dimensional information model, conveying design and construction plan information in a visual and intuitive way, and improving the efficiency of information transmission. In addition, connecting the project with the BIM model can realize the quality control of the project construction production process and improve the construction production management level.

3 Experiment and Research

3.1 Experimental Environment

In this experiment, a BIM model is constructed based on BIM technology. The model is implemented using software systems such as Revit, Navisworks and Glodon. The BIM model is based on importing 2D CAD engineering drawing information into Revit software to generate a 3D BIM model, which contains the display size, material, and reinforcement configuration of each component. Import the 3DBIM modeled by Revit into the NavisworkManange software to predict any collision nodes that may occur during the construction process in advance, which is beneficial to shorten the construction time and reduce the cost.The calculation method of CAD scan data is shown in formula (1) (2).

$$mp + nq + uo - f = 0 \tag{1}$$

$$f_r = |mp_r + nq_r + uo_r - f| \tag{2}$$

Among them, m, n and u are all coefficients, (pr, qr, or) are the coordinates of the scanning point, and f is the distance between the tangent plane and the origin of the coordinates.

3.2 Experimental Process

In this experiment, in order to better understand the effect of the BIM model in the construction cost management of construction projects, the relevant data of a certain project was selected as the experimental sample, and the dynamic simulation analysis of the model on the project schedule and cost control was tested. The test items Including cost deviation, schedule deviation and cost performance index. Among them, CV stands for cost deviation, SV stands for schedule deviation, and CPI stands for cost performance index. The test results are shown below.

4 Analysis and Discussion

In this experiment, the dynamic simulation analysis of the project schedule and cost control of the model is tested. The test items include cost deviation, schedule deviation and cost performance index. The test results are shown in Table 1.

Table 1. Analysis of Project Schedule and Cost Control

Stage	CV (cumulative)	SV (cumulative)	CPI (cumulative)
1	0.14	1.78	1
2	0.21	1.19	0.99
3	0.18	0.94	1
4	0.36	1.23	0.98
5	0.29	1.03	0.97

It can be seen from Fig. 2 that in these five construction stages, the cost deviations calculated by the model are 0.14, 0.21, 0.18, 0.36, and 0.29, respectively, and the cost performance indexes are 1, 0.99, 1, 0.98, and 0.97, respectively. It can be obtained that the model can reflect the dynamic changes of the construction process progress and cost in a timely manner, and has a certain effect on controlling the construction progress and controlling the cost, which has a certain positive significance for improving the level of construction management.

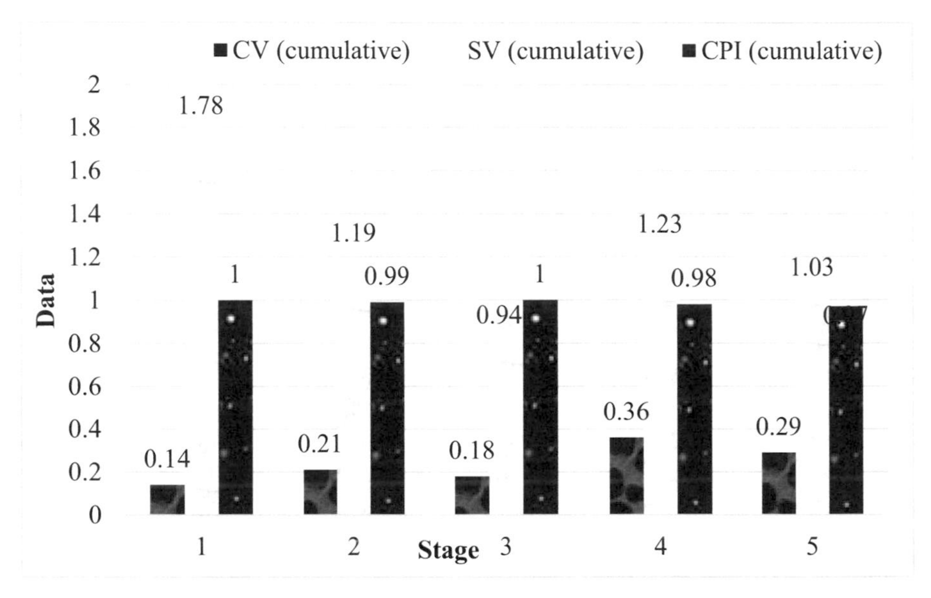

Fig. 2. Analysis of Project Schedule and Cost Control

5 Conclusion

The process of construction management of building engineering is relatively complicated, and it will be affected by many factors during the entire project implementation process. At the same time, BIM technology is the most important content in construction engineering. It can not only plan construction, but also provide effective means for construction management. It plays a very important role in the entire process of construction projects. It can combine traditional building information modeling with digitization and informatization. Therefore, based on the BIM technology, this article conducts research on the construction management mode and related content of construction projects, which is of great significance for improving the level of project construction management, reducing the number of rework and reducing costs.

References

1. Singh, M.M., Sawhney, A., Borrmann, A.: Integrating rules of modular coordination to improve model authoring in Bim. Int. J. Constr. Manag. **19**(1), 15–31 (2019)
2. Kim, D.G.: Development of building information modeling (Bim) and mobile linking technology based on construction it convergence technology. Asia Life Sciences **4**, 2245–2258 (2018)
3. Keyvanfar, A., Shafaghat, A., Ya'Acob, N., et al.: Sustainable post-disaster settlement (SPS) assessment model for evaluating performance of construction management in post-flood risk-reduction and recoVERY. J. Sustain. Sci. Manage. **16**(5), 174–199 (2021)
4. Yussof, F., Hasbi, H.A., Zawawi, E.: Employability forecast among construction management from the employer's perspective in Malaysia construction industry. Built Environ. J. **18**(1), 79 (2021)
5. Parolise, A.: Engineered smoke control in a metro tunnel renovation using construction management project delivery. Eng. Syst. **36**(1), 29 (2019)
6. Netto, J.T., Santos, J., Filho, W.P.: Proposal of improvements in the management of construction companies: an international case study. Interações (Campo Grande) **21**(3), 499–512 (2020)
7. Al, S.: Improve a human resource allocation guide in construction management based on case study. Turkish J. Comput. Math. Educ. (TURCOMAT) **12**(6), 626–637 (2021)
8. Kadaei, S., Sadeghian, S., Majidi, M., et al.: Hotel construction management considering sustainability architecture and environmental issues. Shock. Vib. **2021**(1), 1–13 (2021)
9. Tam, N.V., Toan, N.Q.: Research trends on machine learning in construction management: a scientometric analysis. J. Appl. Sci. Technol. Trends **2**(3), 96–104 (2021)
10. Hathiwala, A.M., Pitroda, J.R.: Application of 5d building information modeling for construction management. Solid State Technol. **64**(2), 2156–2163 (2021)
11. Petrescu, T.C., Voordijk, H., Toma, I.O.: Then and now: Construction Management Practices in Romania and the Netherlands. Int. J. Technol. Policy Manage. **21**(2), 91 (2021)
12. Shaikh, A., Sankhe, O., Ansari, S., et al.: Construction management of a high rise structure using msp software. Int. J. Innovat. Eng. Sci. **6**(5), 01–04 (2021)

Simulation of Electronic Equipment Control Method Based on Improved Neural Network Algorithm

Zhenghong Jiang[1(✉)] and Chunrong Zhou[2]

[1] School of Big Data, Chongqing Vocational College of Transportation, Chongqing 402247, China
jiangzhenghong@cqjy.edu.cn

[2] Science and Technology Department, Chongqing Vocational College of Transportation, Chongqing 402247, China

Abstract. In recent years, with the continuous improvement of the degree of industrial automation in our country, the intelligent requirements for electronic equipment are also rising. For this reason, some scholars propose to apply artificial neural network algorithm to electronic equipment to control the equipment. However, the current artificial neural network algorithm is generally implemented by FPGA. This scheme has the disadvantages of high cost and lack of applicability. Therefore, the purpose of this paper is to study the simulation of electronic equipment control method based on improved neural network algorithm. This paper uses PSpice circuit simulation software to model, simulate and optimize the design of the circuit. And the nonlinear function generator circuit, adder circuit and analog multiplier circuit are modeled, simulated and optimized by PSpice simulation software, and finally the adder circuit and the multiplier circuit are used to form a neural network overall circuit that meets the requirements to achieve Algorithmic function. After training and simulating the curve obtained by the boiling water experiment with the method in this paper, the experiment shows that the curve fitted by the polynomial is basically consistent with the curve obtained by the experiment, and the maximum error obtained by calculation is 1.35%. Through the research of this paper, it fills the blank of our country's neural algorithm in the control simulation of electronic equipment, and promotes the economic development of our country.

Keywords: Matlab · Neural Network Algorithm · Pspice · Adder Circuit · Multiplier Circuit

1 Introduction

Based on the characteristics and capabilities of intelligent information processing, the neural network algorithm is widely used, and its role is becoming more and more obvious. Many problems that cannot be solved by traditional information systems have achieved good results after using neural network algorithms. Neural network algorithms can run

J. H. Abawajy et al. (Eds.): ICATCI 2022, LNDECT 169, pp. 168–175, 2023.
https://doi.org/10.1007/978-3-031-28893-7_21

through all aspects of information retrieval, transmission, retrieval, processing and utilization. It can be seen that neural network algorithms have a large number of application channels in artificial intelligence, process control methods, state space modeling, and fault analysis and processing.

In the research on the simulation of electronic equipment control method based on improved neural network algorithm, many experts and scholars have conducted in-depth research on it, and achieved good results, such as: Jae-Hong L published a different from the perceptron. Bramslw L, a professor at Boston University, proposed several nonlinear dynamic system structures based on the research on biology and psychology, which played an important role in promoting the research of neural network [1, 2]. Physicist Jahn T studied the dynamic characteristics of neural networks, introduced the concept of energy function, gave the stability criterion of the network, and proposed a new approach for associative memory and optimized computation [3]. It can be seen that the characteristics and capabilities of brain-based intelligent information processing of neural networks make its application fields increasingly expanded, and its potential is becoming more and more obvious.

This paper studies the neural network algorithm and introduces the neural network theory in detail. On the basis of the theory, the PSpice software is used to simulate each part of the circuit of the ANN, and then the self-programming method is used to train the ANN. Finally, the function of the whole circuit is analyzed.

2 Functional Analysis and Improvement Method of Neural Network Algorithm in Electronic Equipment

2.1 Fault Diagnosis and Analysis of Electronic Equipment Based on Neural Network

(1) Selection of the number of layers of the neural network

In order to apply the neural network to fault diagnosis through programming, the first problem to be solved is how to select the number of layers of the network. The theory shows that three layers can satisfy any uninterrupted function of any accuracy on the basis of setting the number of nodes in the hidden layer of the network by oneself. Under normal circumstances, increasing the number of networks is used to improve accuracy and reduce bias, but the result is that the network becomes more complicated, and the number and time of network iterations are increased [4, 5].

(2) Selection of initial weights

In a nonlinear system, if the initial weight value is too large, the network cannot be adjusted in time, so the initial weight value is usually a random number between -1 and 1.

(3) Selection of expected error

When designing the program, we can set two different target error parameters. After running the program, we can determine which expected error parameter to use by comparing the effect of the final training curve [6].

(4) Self-learning stage of neural network

Arbitrarily assign an initial value to the neural network's ownership value and threshold, set the desired error accuracy parameter, and select a suitable learning rate and maximum training times. Determine the input sample set P and the target output sample set T. Assuming that there are N samples in total, by conducting the samples from the input to the output step by step and finally obtaining the output result, then calculate the error E according to the error expression. If the total error is less than E < e or the maximum number of training times is reached, output their computation results and stop learning. Otherwise, the learning rate and momentum coefficients are modified according to the network learning expression. Starting from the output layer, the bias signal is sent back to the input layer, and the weight threshold is modified according to the network weight learning step. Re-enter until requirements are met [7, 8].

2.2 Improvement of Neural Network Algorithm

Due to some defects in the neural network, when the neural network corrects the connection weights, it mainly uses the steepest descent method for the quadratic function of the deviation to obtain the minimum value of the deviation. This method searches in the direction of negative gradient, which is easily considered to be the best search method. For some systems with multiple local minima problems, traditional neural networks cannot easily solve them. At the same time, the core of the neural network algorithm is the correction of connection weights. The correction of connection weights is obtained by correcting in the direction of negative gradient, which is easy to fall into a local minimum value during the search process, which is often used in practical engineering mathematical models. There are multiple local minima. When searching in all directions near the local minimum value, the deviation will increase, so that the search in the direction of negative gradient cannot escape the local minimum value. In view of the defects and deficiencies of conventional neural networks, this paper improves to a certain extent. Usually the second-order Newton method can be used to greatly speed up the convergence speed, but the second-order derivative of the weight (wk), the quadratic function Ep of the deviation, has a large amount of computation, so this paper adopts the variable scale method [8, 9].

The basic principles of the variable scaling method are briefly introduced below:

For an unconstrained objective function f(X) with a second-order continuous partial derivative, take X(k) as the approximate minimum point of the objective function. Perform a second-order Taylor approximation of the objective function f(X) around this point:

$$f(X) = f(X^{(k)}) + \nabla f(X^{(k)})^T \Delta X + \frac{1}{2} \Delta X^T H(X^{(k)}) \Delta X \quad (1)$$

If f(X) is a quadratic function, its Hessian matrix is a constant matrix, and the difference between the gradients at any two points X(k) and X(k + 1) can be calculated.

In the process of using the variable scaling method to find the optimal solution to the objective function in the neural network algorithm, the connection weights at the optimal point are often obtained. Based on the variable scaling method, the connection weight correction process of the BP neural network is obtained as follows:

$$\Delta w(k+1) = \eta((1-\beta)H(k)D(k) + \beta H(k-1)D(k-1)) + \alpha \Delta w(k) \quad (2)$$

Another important reason for the slow convergence speed of the steepest descent method is that the learning rate is not easy to choose. In order to reduce the error, the learning rate of the network is usually set to be relatively small, but it greatly reduces the speed of the learning process; secondly, a large learning factor Not only can it speed up learning, but it can also lead to overcorrection, causing oscillations. It can be seen from the above formula that in the process of correcting the connection weights twice in a row, if the search direction is the same, it means that the descending speed is too slow, and the search step size can be increased; long [10].

3 Self-programming to Learn and Train Neural Network Algorithms

3.1 Neural Network Learning Algorithm Learning Rules

There are three commonly used learning algorithms:

(1) Unsupervised Hebb learning algorithm:

$$\Delta\omega ij = \eta oi(k)oj(k) \tag{3}$$

(2) Supervised delta learning algorithm: The teacher signal is introduced on the basis of the Hebb learning algorithm, and the main purpose is to change the excitation of one of the neurons into the teacher signal.
(3) Supervised Hebb learning algorithm: By combining the above two algorithms and taking the advantages of the two, a supervised Hebb learning algorithm is finally obtained.

3.2 The Process of Neural Network Learning Algorithm

(1) Set the corresponding synaptic weights, bias values and parameters of the activation function for each layer of the neural network. Generally speaking, the synaptic weight and bias value are selected as Gaussian distribution or random quantity with uniform distribution [11].
(2) The training samples in the training sample space are input into the neural network one by one for forward calculation. The calculation process is:
1) Calculate the output of neurons in each layer;
2) After calculating the output of each layer, use the expected response of the training sample space to calculate the error signal of the neural network;
3) Reverse calculation to generate corrections for the synaptic weights of neurons in each layer. The process is to calculate the respective local gradients separately;

(3) Finally, continuously using the samples in the training library to train the neural network can make the neural network converge to the optimal weight, which indicates that the neural network has already recognized the knowledge of the sample space.

3.3 Training the Neural Network Algorithm

This paper adopts the training method based on power series. First, let X0 be a normalized sample group composed of M samples, that is, [fi(xi)] = 1, where i = 1, 2......M.

The output of the corresponding initialized neural network model is set to:

$$S(xi) = \sum_{j=1}^{N} ajsj(xi) \tag{4}$$

Among them, N is the size of the finite orthogonal basis function Sj, so the neural network algorithm training can be simplified as a mathematical problem. That is to find a set of coefficients A = (a1, a2, a3,...aN) so that it satisfies the requirements [12].

$$Q = \sum_{i}^{M} (fi(xi) - \sum_{j}^{N} ajSj(xi))^2 \tag{5}$$

After finding this coefficient, take the minimum value. In this paper, the least squares method is used to obtain the value.

3.4 Neural Network Algorithm Recognition Sample Method

The most commonly used neural network algorithm is the learning of control knowledge for identifying samples. This paper will simulate a learning of three sample spaces to achieve the purpose of identifying the target. Set three categories as A, B, C. Each target contains a set of training data, and the neural network has three outputs, each output representing a target. When the output is the training sample of the target, it is expected that the output of the neural network is greater than the threshold at the output, and if the output is not satisfied, the training and learning of the neural network is performed. At the same time, if the training samples of the target activate the output of other targets, then the corresponding learning is performed to correct the error. In the recognition process, the feature vector to be recognized is input into the neural network. If the output of the activated neural network meets the threshold requirements, the recognition result will be output, and if it cannot be satisfied, the output will be rejected.

4 Experimental Simulation

4.1 Simulation Process of Nonlinear Function Generator

Firstly, establish the circuit schematic diagram in PSpice simulation software, set the value of each component, select DC sweep, thc sweep voltage range is −9 V–+9 V, and the sweep voltage interval is set to 0.1. The simulation results obtained are shown in the Table 1.

The experimental results are shown in the Fig. 1. When the scanning voltage is −9 V, the theoretical calculation voltage is 6.0 V, and the simulation results are in good agreement with the theoretical calculation results.

Table 1. Simulation results

Scan voltage(V)	−9	−8	−7	−6	−5	−4	−3	−2	−1	0	1	2	3	4	5	6	7	8	9
Simulation voltage(V)	6.0	5.8	5.6	5.5	5.3	5.1	4.8	3.9	3.2	0	−2.1	−4.2	−4.7	−5.2	−5.3	−5.8	−6.0	−5.8	−6.1

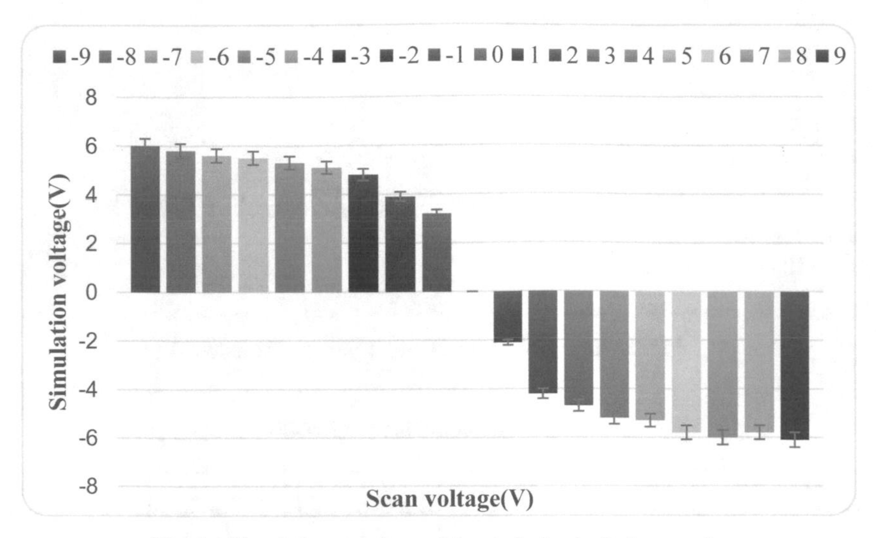

Fig. 1. Simulation results and theoretical calculation results

4.2 Simulation of the Adder Circuit

First, establish the circuit schematic model of the adder in PSpice, set the value of each component, select the DC voltage sweep, the voltage sweep range is −20 V–+40 V, and the sweep interval is 0.01. The simulation results obtained are shown in the Table 2.

As can be seen from the Fig. 2, by taking the output voltage as the ordinate and the input voltage as the abscissa, through multiple verifications, after substituting the experimental data into the formula for calculation, it is basically consistent with our expected goal. The simulation curve is consistent with the theoretical calculation, and the experimental results are satisfactory.

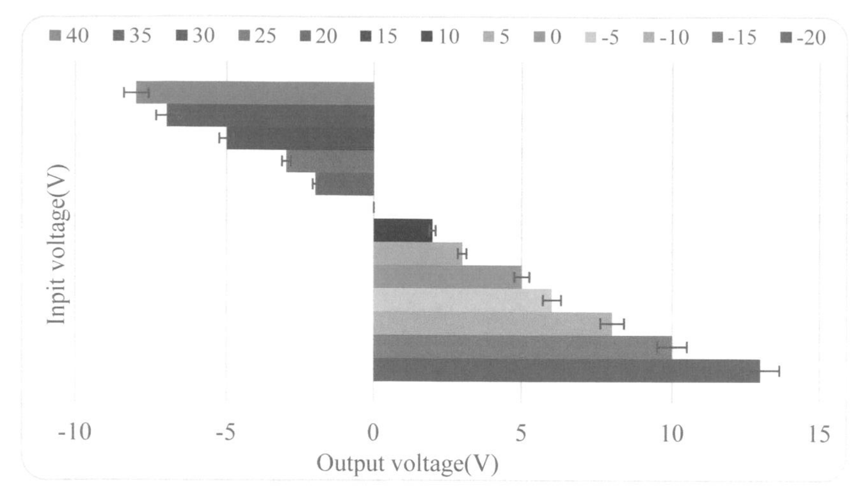

Fig. 2. Input Voltage and Output Voltage Variation

Table 2. Circuit Simulation Results

Input voltage(V)	−20	−15	−10	−5	0	5	10	15	20	25	30	35	40
Output voltage(V)	13	10	8	6	5	3	2	0	−2	−3	−5	−7	−8

5 Conclusions

Intelligence is a major trend in the development of household appliances in the future. By applying this method to industrial production, it can provide a faster and more favorable production environment for industrial production, and has a significant effect on improving the quality of products. Through, because the fault diagnosis function of faulty electronic equipment is also added in this paper, the When the method is applied to electronic equipment, the maintenance difficulty of the electronic equipment can be effectively reduced, the fault diagnosis time is greatly reduced, and the correct rate of fault identification is significantly improved. It provides convenience for maintenance personnel and improves the satisfaction of users when using the product. In this paper, a simple function neural network algorithm is realized by using the analog circuit. By training the neural network by self-programming and normalizing the variables, the objective function can be expanded into a finite-term polynomial under the premise of a certain precision, so as to realize the simulation of the electronic equipment control method of the neural network algorithm.

References

1. Jae-Hong, L., Do-Hyung, K., Seong-Nyum, J., et al.: Diagnosis and prediction of periodontally compromised teeth using a deep learning-based convolutional neural network algorithm. J. Periodontal Implant Sci. **48**(2), 114–123 (2018)

2. Bramslw, L., Naithani, G., Hafez, A., et al.: Improving competing voices segregation for hearing impaired listeners using a low-latency deep neural network algorithm. J. Acoust. Soc. Am. **144**(1), 172–185 (2018)
3. Jahn, T., Ziaukas, Z., Kobler, J.P., et al.: Neural observer for nonlinear state and input estimation in a truck-semitrailer combination. IFAC-PapersOnLine **53**(2), 14306–14311 (2020)
4. Nugroho, D.C., Mayaratri, Y., Syai'In, M., et al.: Household electricity network monitoring based on IoT with of automatic power factors improvement using neural network method. IOP Conf. Ser. Mat. Sci. Eng. **1010**(1), 012045 (7pp) (2021)
5. Logvin, V., Karlova, T.: Automated system for quality control of tool processing in glow discharge based on neural network monitoring. Bull. Bryansk State Tech. Univ. **2021**(3), 16–24 (2021)
6. Beskostyi, D.F., Borovikov, S.G., Yastrebov, Y.V., et al.: Use of aposteriori information in the implementation of radar recognition systems using neural network technologies. J. Russ. Univ. Radioelectronics **22**(5), 52–60 (2019)
7. Fapi, C.B.N., Wira, P., Kamta, M., et al.: Simulation and dSPACE hardware implementation of an improved fractional short-circuit current MPPT algorithm for photovoltaic system. Appl. Solar Energy **57**(2), 93–106 (2021)
8. Tantciura, S., Qiao, Y., Andersen, P.Ø.: Simulation of counter-current spontaneous imbibition based on momentum equations with viscous coupling, brinkman terms and compressible fluids. Transp. Porous Media **141**(1), 49–85 (2021). https://doi.org/10.1007/s11242-021-01709-9
9. Chodey, M.D., Shariff, C.N.: Neural network-based pest detection with k-means segmentation: impact of improved dragonfly algorithm. J. Inf. Knowl. Manag. **20**(3), 2150040 (2021)
10. Kwon, H.Y., Kim, N.J., Lee, C.K., et al.: Searching magnetic states using an unsupervised machine learning algorithm with the Heisenberg model. Phys. Rev. B, Condensed Matter Mat. Phys. **99**(2), 024423.1–024423.7 (2019)
11. Alsaade, F.W., Aldhyani, T., Al-Adhaile, H.M.H., et al.: Developing a recognition system for classifying COVID-19 using a convolutional neural network algorithm. Cmc **68**(1), 805–819 (2021)
12. Zvarevashe, K., Olugbara, O.O.: Recognition of speech emotion using custom 2D-convolution neural network deep learning algorithm. Intelligent Data Analysis **24**(5), 1065–1086 (2020)

Size Detection System of Building Rebar Based on LightGBM Algorithm

Manli Tian(✉)

School of Road Bridge and Architecture, Chongqing Vocational College of Transportation, Chongqing 402247, China
tianml0129@163.com

Abstract. With the development of my country's economy and the continuous development of real estate, steel bars are an indispensable material in the construction industry, and the stability of their quality is self-evident. However, to understand their quality, we must test them. Traditional testing There are generally the following five methods: non-damage detection method, magnetic induction method, radar method, chisel in-situ detection, sampling detection, and the above five methods basically lack applicability and inaccurate measurement. Therefore, the purpose of this paper is It is based on the LightGBM algorithm to study the construction steel bar size detection system to solve the drawbacks of the traditional detection method. This paper first compares and evaluates the LightGBM algorithm with other algorithms to establish the final algorithm of this paper, and then uses CART as the machine learning to iterate K rounds to establish the building steel rebar detection system model, and uses the grid search algorithm to optimize and improve the LightGBM algorithm., thereby greatly shortening the data search time. Finally, the key parameters of the final algorithm are determined through experiments. Finally, the detection accuracy of the detection model is improved to 95.38%. By applying the lightGBM algorithm-based building steel bar size detection system proposed in this paper in the actual detection process, the time spent on detection can be greatly reduced, thereby improving work efficiency and greatly helping our country's infrastructure construction.

Keywords: Lightgbm algorithm · Building steel bar size detection · CART · Machine learning

1 Introduction

Under the background of the national "14th Five-Year Plan", my country's scientific and technological level has improved rapidly, infrastructure construction across the country has developed rapidly, and the demand for steel bars is also increasing day by day. For this reason, the variety of steel bars has become very important, and traditional The method not only has the problem of low detection accuracy and low detection efficiency, but with the wide application of machine algorithms, in order to solve the above problems, a building steel bar size detection system based on the LightGBM algorithm emerges as the times require.

J. H. Abawajy et al. (Eds.): ICATCI 2022, LNDECT 169, pp. 176–183, 2023.
https://doi.org/10.1007/978-3-031-28893-7_22

In the research of building steel bar size detection system based on LightGBM algorithm, many experts and scholars have studied it and achieved good results. For example, Mousa Y A studied the relationship between cancer and ion channels in the literature, and proposed A LightGBM-based model to predict the number of ion channels [1]. To help researchers find out the relationship between electrical signals and ion exchange, Majhi S et al. studied the energy system of photovoltaic power generation in the literature, and proposed a fusion model based on LightGBM and LSTM to predict short-term photovoltaic power generation [2]. To obtain the best forecast results, MLP is applied to dynamic weight assignment and combination to generate forecast results. It can be seen from this that LightGBM is widely used.

In this paper, the LightGBM algorithm is analyzed and introduced in detail by looking for a large number of documents, and the improved grid optimization algorithm is used to find the optimal parameters, and other models in ensemble learning, XGBoost, CatBoost are selected for comparative experiments, and it is found that based on LightGBM The model effect is higher than several other models. Then the model based on LightGBM algorithm is analyzed experimentally. Finally, a complete set of construction steel bar size detection system is proposed, which meets the needs of social development, and has good detection effect, small error and high precision.

2 Establishment of Building Steel Bar Size Detection System Based on Lightgbm Algorithm

2.1 Evaluation and Selection of Detection Models

In the field of machine learning, there are various evaluation criteria for evaluating the generalization performance of mathematical models. For classificationThe combined form of the predicted and true classes of the mathematical model includes four types: TP, FP, TN, and FN. This paper evaluates model performance by employing the "receiver operating characteristic" curve, a powerful tool. The ROC curve is drawn from both the "true rate" on the ordinate and the "false positive rate" on the abscissa. Combine the preprocessing method with the model at full wavelength to determine the optimal combination. The training set has 1848 average spectra for training the model, and the most suitable prediction model is selected. The test set has 792 spectra to verify the accuracy of the model, and the total accuracy is obtained, and then the sub-test set (264 each) is used to verify the model pair. Accurate recognition rates for the three time periods. This study uses five preprocessing methods combined with four classification models of SVM, RF, LR, and LightGBM to detect the accurate recognition rate of the surface area of rebar. In order to prevent the default parameters of the model from reducing the classification accuracy of the model, the grid search method is used to adjust the main parameters affecting the classification accuracy of the model, and the appropriate parameter value range is set. The grid search method only changes one parameter value per iteration. After several iterations, the parameters with the highest classification accuracy are recorded [3, 4].

2.2 Lightgbm Detection Model Establishment

In this paper, LightGBM is used to extract the time-varying delay feature of the time series data of the time series input window. The selected data set is set as:

$$A = \{ (m_i, n_i), i = 1, 2, \ldots, k\} \tag{1}$$

$$\hat{n}_i = \sum_{j=1}^{K} f_j(m_i), f_j \in \varphi \tag{2}$$

$$D(f_j) = \sum_{i=1}^{k} g(n_i, \hat{n}_i) + \sum_{j=1}^{K} \Omega(f_j) \tag{3}$$

Among them, the first item on the right side of the formula is the loss function, and the second item is the regularization part.

2.3 Lightgbm Algorithm Optimization Design

(1) Establishment of the LightGBM algorithm flow in this paper

First, input the processed data set, use the histogram algorithm to find the optimal segmentation point of the feature, and use the Leaf-wise leaf growth strategy with depth limit to generate the CART regression tree; then, calculate the residual of the first CART regression tree. The residuals of the previous round are used as the training samples of the next CART regression tree, and the residuals are continuously fitted and trained repeatedly; finally, the weighted summation of the CART regression trees generated by each round of training is used to obtain the final building steel bar size detection model [5].

(2) Optimization of hyperparameters of LightGBM algorithm

The LightGBM algorithm itself is an improved algorithm of gradient boosting decision tree. The weak classifier in essence is still a decision tree model. Therefore, when setting hyperparameters, in addition to some hyperparameters of the algorithm itself, the parameters of the decision tree algorithm should also be considered, such as Among the many hyperparameters, such as max_depth in the decision tree algorithm, there are the following commonly used hyperparameters, namely learning_rate, objective, num_leaves, max_depth, bagging_fraction, feature_fraction, verbosity, boosting, these hyperparameters determine the use of LightGBM prediction thickness model The algorithm, learning rate, etc., are of great significance to the effect of the model. Therefore, during parameter optimization, if only one parameter is adjusted at a time, it will affect the training time and accuracy of the model. In order to improve this process, a grid search algorithm is introduced to optimize hyperparameter selection and obtain the optimal hyperparameters [6, 7].

3 Processing of Key Parameters in the Size Detection if Building Steel Bars

3.1 Data Preprocessing

The steel bar size detection process is a process operation. The process operation environment is complex. The environment where the sensors in each link are located is extreme, and the data collection and storage are easily affected by noise, which makes the data very prone to data missing and abnormal data. For this reason, this paper preprocesses the data to improve the quality of the data, strengthen the useful information in the data, and provide the basis for the subsequent model establishment and feature extraction.

(1) Missing value processing and outlier removal

The missing data in production data usually faces many situations, which can be roughly divided into transient missing and long-term missing. Since the time series input window is designed in this paper, for the case of long-term missing, this paper adopts data truncation and restarts the sliding of the time series input window in the next continuous normal data. For the instantaneous missing data, the linear interpolation method is used to fill in the instantaneous value, and for the instantaneous missing data, the linear interpolation method is used to calculate the change slope [7, 8].

(2) Data normalization

Through the analysis of the data mechanism and the method of mutual information correlation analysis, the key variables of the key parameters of the steel bar size detection process are selected. Data for model training is likely to cause the weights of some variables to be too small, reducing the effectiveness of some features. For this reason, this paper uses variable data normalization to map all variable values between 0 and 1, so as to eliminate the interference caused by non-uniform dimensions to model feature extraction.

(3) Selection of performance evaluation index of detection model

The performance evaluation index of the prediction model is the key to model selection. For regression prediction models, the mean square error (MSE) is the most commonly used evaluation parameter, which represents the square of the difference between the predicted value and the actual value. Mean value of. Although this indicator is easy to interpret, it is more sensitive to outliers. In addition, in the industrial application process, the accuracy required by different application scenarios is also different. Some production processes require an accuracy of not less than 90%, but in some cases, not less than 80% It is considered acceptable, and the mean square error does not fully explain whether the mode under the current parameters is suitable for the current task. Therefore, this paper uses multiple criteria to judge the training situation of the model, namely RMSE, MRE, MAE [9, 10].

3.2 Improvement of Lightgbm Algorithm

In machine learning algorithms, the selection of hyperparameters directly affects the performance of the entire model, so the selection and optimization of hyperparameters is an optimization method for model performance. The grid search algorithm is a method of automatically finding optimal hyperparameters, originally used to optimize SVMs, and later introduced by various machine learning models. The essence of the algorithm is to apply the mathematical exhaustive method to the model according to mathematical ideas. All combinations in the grid are used for model training. Cross-validation evaluates the effect of the model, and the parameter combination with the best effect is selected as the optimal parameter [11, 12]. In the experiment, it is found that there are many parameters that need to be optimized, the range is also relatively large, and the search time is long, so we improve the speed of the traditional grid search algorithm, and convert the direct search for the best parameters to search for the best interval first, and then The precise search is carried out inside the optimal interval, which turns the global fine search into a coarse search interval first, and then the interval fine search, which greatly reduces the search time.

4 Experiment Analysis of Rebar Dimension Detection System

4.1 Lightgbm Model Experimental Results

Optimize the important parameters of the LightGBM prediction model based on the grid search algorithm. First, set the learning rate to 0.1, set the algorithm to gbdt, and then adjust the number of iterations. It is found that when the number of iterations is 1200, the model effect tends to be stable, so the The number of model iterations this time is set to 1200. Determine the number of iterations, and use the grid search algorithm to optimize parameters such as max_depth, num_leaves, bagging_fraction, and feature_fraction to obtain the optimized LightGBM thickness detection model. The experimental results are shown in the Table 1:

Table 1. Experimental results

Learning rate	Number of iterations	Root mean square error
0.1	100	0.25
0.1	200	0.15
0.1	300	0.08
0.1	400	0.05
0.1	500	0.025
0.1	600	0.02
0.1	700	0.019
0.1	800	0.017
0.1	900	0.015
0.1	1000	0.013
0.1	1100	0.012

(*continued*)

Table 1. (*continued*)

Learning rate	Number of iterations	Root mean square error
0.1	1200	0.01
0.1	1300	0.01
0.1	1400	0.01
0.1	2000	0.01

It can be seen from the Fig. 1 that the thickness detection model of LightGBM tends to be stable when the number of iterations reaches 1200, and the root mean square error is 0.01 at this time, so we can draw the following conclusions, the thickness model of LightGBM in this paper is on the test set. The detection effect is very good, the error is small, and the accuracy is high.

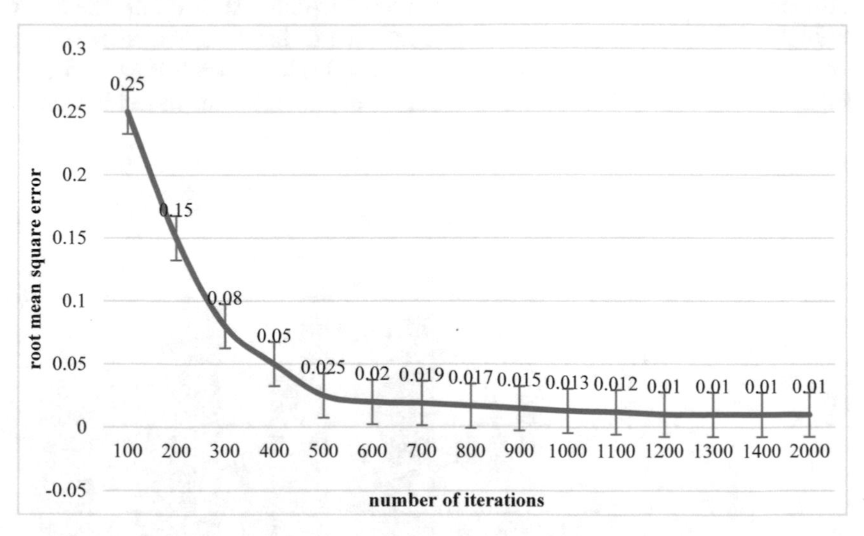

Fig. 1. Variation trend of root mean square error with the number of iterations

4.2 Lightgbm Model Depth Adjustment Experiment

After determining the number of iterations, adjust the experiment for the depth of the LightGBM tree model. Similarly, the learning rate is 0.1 and the number of iterations is 1200. Under the condition that the number of iterations and learning rate remain unchanged, by observing RSME, MAE and MRE The error value varies with the model depth to determine the appropriate model depth.The experimental results are shown in Table 2.

It can be seen from the Fig. 2 that when the depth of the LightGBM model is 2, the error is large, which means that when the depth of the tree model is too shallow, the

Table 2. Data from LightGBM Model Depth Tuning Experiments

Model depth	RSME	MAE	MRE
2	0.38	0.32	0.215
3	0.37	0.31	0.205
4	0.34	0.275	0.18
5	0.33	0.27	0.17
6	0.32	0.26	0.16
7	0.34	0.28	0.19
8	0.345	0.29	0.20

nonlinear ability of the model is too low, and it is difficult to converge to a lower accuracy. When the model depth increases from 3 to 6, the error value decreases continuously with the model depth, and finally reaches the lowest value. At this time, the accuracy is the highest, but when the model depth continues to increase, the error value also increases, indicating the complexity of the model at this time. If it is improved, there is a phenomenon of over-fitting. For this reason, the depth of the tree model is selected to be 6.

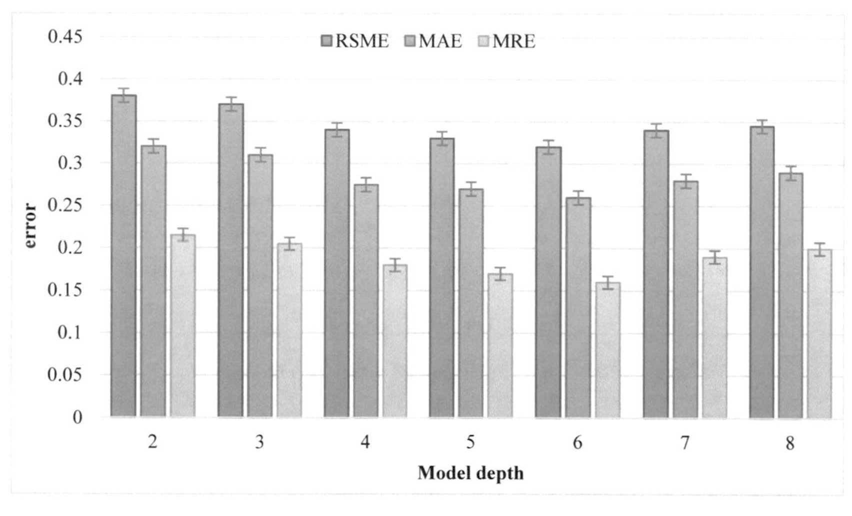

Fig. 2. Variation of error values of RSME, MAE and MRE with model depth

5 Conclusions

This paper proposes a set of solutions to the problems of poor accuracy in traditional building steel bar size detection methods. By applying this solution to the actual detection process, it not only meets the actual detection requirements, but also improves the detection efficiency. It provides an idea for the rapid detection of steel bar size. Helped the development of our country's construction industry.

References

1. Mousa, Y.A., Helmholz, P., Belton, D., et al.: Building detection and regularisation using DSM and imagery information. Photogram. Rec. **34**(165), 85–107 (2019)
2. Majhi, S., Mukherjee, A., George, N.V., et al.: Corrosion detection in steel bar: A time-frequency approach. NDT & E international **107**, 102150.1-102150.16 (2019)
3. Maisuradze, M., Yudin, Y., Kuklina, A.A., et al.: Formation of microstructure and properties during isothermal treatment of aircraft building steel. Metallurgist **65**(9–10), 1008–1019 (2022)
4. Iyama, J.: Detection of fracture in steel members of building structures by microstrain measurement. Int. J. Steel Struct. **20**(5), 1720–1729 (2020). https://doi.org/10.1007/s13296-020-00408-3
5. Ferreiro-Cabello, J., Fraile-Garcia, E., Lara-Santillán, P.M., et al.: Assessment and optimization of a clean and sustainable welding procedure for rebar in building structures. Appl. Sci. **10**(7045), 1–15 (2020)
6. Lee, K.-H., Lim, S., Cho, D.-H., Kim, H.-D.: Development of fault detection and identification algorithm using deep learning for nanosatellite attitude control system. Int. J. Aeronaut. Space Sci. **21**(2), 576–585 (2020). https://doi.org/10.1007/s42405-019-00235-9
7. Khafajeh, H.: An efficient intrusion detection approach using light gradient boosting. J. Theor. Appl. Inf. Technol. **98**(5), 825–835 (2020)
8. Chidambaram, S.: The application of clash-detection processes in building information modelling for rebar. In: Proceedings of the Institution of Civil Engineers - Smart Infrastructure and Construction, vol. 172, no. 2, pp. 1–17 (2020)
9. Alsharkawi, A., AlFetyani, M., Dawas, M., et al.: Poverty classification using machine learning: the case of Jordan. Sustainability **13**(3), 1412 (2021)
10. Shin, Y., Kim, S., Chung, J.M., et al.: Emergency department return prediction system using blood samples with LightGBM for smart health care services. IEEE Consumer Electron. Mag. **PP**(99), 1–1 (2020)
11. Ma, S.: Predicting the SP500 index trend based on GBDT and LightGBM methods. E3S Web Conf. **214**(5), 02019 (2020)
12. Shahini, M.: Machine learning to predict the likelihood of a personal computer to be infected with malware. SMU Data Sci. Rev. **2**(2), 9 (2019)

Integrated Model of Building Energy Consumption Prediction Based on Different Algorithms

Honghong Wang(✉)

College of Road, Bridge and Architecture, Chongqing Vocational College of Transportation, Chongqing 402247, China
w18375829963@163.com

Abstract. Energy is very important in social development and scientific and technological progress, which guarantees the transformation of the human renewal era. However, the energy on the earth is fixed and will be exhausted one day. Therefore, the research on energy conservation and related technologies in the field of architecture has become the focus of domestic and foreign scholars. Affected by factors such as personnel distribution, weather conditions and equipment operation time, BE consumption data is highly uncertain and random, and it is difficult to accurately predict BE consumption. Based on this, the purpose of this paper is to study different algorithms based on the comprehensive prediction model of building energy (BE) consumption. Firstly, this paper summarizes the energy consumption in the building field by consulting a large number of documents, and simulates the hourly and monthly energy consumption of the building based on EnergyPlus software. The monthly energy consumption is characterized by periodic oscillation, and a combined forecasting model of cumulative TGM-RBF is proposed. The experiment shows that the accuracy of the model in predicting the monthly BE consumption is 1.81% and 3.30% higher than that of the cumulative TGM (1,1) model (T-M) and GM (1,2) model respectively. Compared with G-M and cumulative TGM (1,2) model, the prediction accuracy of energy consumption is increased by 1.11% and 1.53% respectively.

Keywords: BE consumption prediction · Energyplus · Tgm-Rbf · Cumulative Tgm Model · Gm Model

1 Introduction

According to data from the World Watch Research Organization, as the world's largest energy-consuming sector, buildings account for 40% of global annual energy consumption and 36% of total carbon emissions. Buildings have high energy consumption and great energy-saving potential. Key to the "peak" and "carbon neutrality" strategies, it is extremely important to design and implement energy-efficient building technologies to reduce carbon emissions in an efficient and sustainable way. Therefore, this paper

J. H. Abawajy et al. (Eds.): ICATCI 2022, LNDECT 169, pp. 184–192, 2023.
https://doi.org/10.1007/978-3-031-28893-7_23

proposes a solution of BE consumption prediction integration model based on different algorithms to solve the above problems.

In the research on the integrated model of BE consumption prediction based on different algorithms, many scholars have conducted in-depth research on it, and achieved incomparable results, such as: Wang L et al. combined stack autoencoders (SAEs) and limit Learning Machine (ELM) proposes an extreme deep learning model, which is compared with some popular machine learning methods. From the experimental results, the model has the best prediction accuracy under different BE consumption conditions [1]; Nam et al. proposed a deep fully connected neural network for the monthly BE consumption prediction problem under the real data of 1 million customers. Models and convolutional neural network models [2]. Bedi G et al. developed the ability of deep learning techniques to automatically extract informative features to estimate BE [3]. Their results show that deep learning-based feature engineering can significantly improve BE consumption prediction results.

In this paper, the correlation analysis method is used to analyze the influencing factors of BE consumption, and the correlation between BE consumption and influencing factors is reflected by the correlation coefficient. Then, according to the analysis results, an appropriate algorithm is selected to establish a prediction model, and the experimental policy is compared with the actual value to verify the accuracy of the algorithm.

2 BE Consumption Analysis and Establishment of Other Models

2.1 Research on the Relativity of Factors Influencing BE Consumption

Through the relevant literature survey, the influencing factors of BE consumption include equipment power, lighting density, heating and air-conditioning energy consumption, air-conditioning on time, personnel density, enclosure structure, equipment operation mode, window-to-wall ratio, and meteorological factors. This paper mainly selects 10 factors from the number of building air conditioners on time and meteorological factors, namely weather temperature, relative humidity, outdoor wind speed, outdoor wind direction, direct radiation, scattered radiation, indoor humidity, indoor temperature, dew point temperature and air conditioner on time number. However, there is a nonlinear and random relationship between the influencing factors of BE consumption and BE consumption. First, the correlation analysis method is used to explore the correlation between BE consumption and influencing factors, and the correlation coefficient is used to reflect the degree of correlation [4, 5].

The scatter plots of BE consumption and direct radiation, scattered radiation, weather temperature, weather humidity, dew point temperature, outdoor wind speed, outdoor wind direction, indoor humidity, indoor temperature and the number of hours when the air conditioner is turned on are respectively drawn., it can be preliminarily inferred that BE consumption has a high correlation with direct radiation, scattered radiation, indoor temperature, indoor humidity, weather temperature, weather humidity, outdoor wind speed and outdoor wind direction, but is not significantly correlated with the number of hours when air conditioners are turned on and dew point temperature. The BE consumption levels corresponding to different levels of meteorological factors other than the number of air-conditioning on-time and dew point temperature are quite different.

SPSS software was further used to calculate the Pearson correlation coefficient between 10 influencing factors and BE consumption. From the correlation analysis, it can be seen that BE consumption is significantly correlated with direct radiation, scattered radiation, indoor temperature, weather temperature, outdoor wind speed, and weather humidity. It is weakly correlated with outdoor wind direction and indoor humidity, and has no significant correlation with the number of hours when air conditioners are turned on and dew point temperature. From the correlation direction, BE consumption is positively correlated with indoor temperature, weather temperature, outdoor wind direction, outdoor wind speed and indoor temperature, and negatively correlated with indoor humidity and weather humidity, which confirms the inference of the scatter [6].

2.2 T-M Establishment

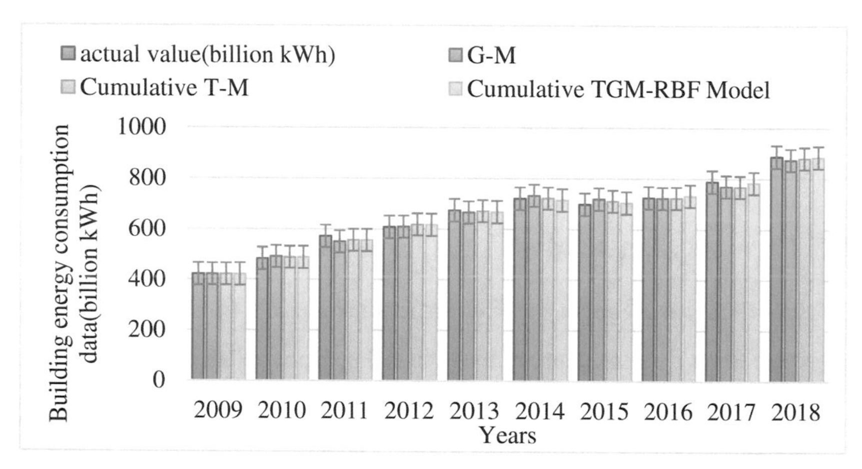

Fig. 1. Comparison of actual and predicted energy consumption in China's construction industry

Aiming at the characteristic of periodic oscillation of monthly energy consumption of public buildings, this paper designs a combined transformation method of inverse trigonometric function and power function to convert the original data sequence into a data sequence with an increasing trend. On this basis, in order to ensure the reliability of the model parameters, firstly, the scale test is carried out, and then a gray G-M (G-M) based on functional transformation is established, that is, the T-M. The specific modeling steps are as follows:

(1) Input the original sequence, and perform the level test, record Q(0)

$$Q^{(0)} = \left\{ q^{(0)}(1), q^{(0)}(2), \cdots, q^{(0)}(n) \right\} \tag{1}$$

In the formula, k = 1,2,…,n.

(2) Preprocess the original sequence to control the range of all data within the interval [0, 1], marked as Y(0)

$$y^{(0)}(k) = \frac{q^{(0)}(k)}{M}(k = 1, 2, \cdots n) \tag{2}$$

In the formula, $q^{(0)}(k)$ is the k-th value in the initial sequence, M is all the numbers in the sequence greater than $Q^{(0)}$, which can be selected according to the actual situation, and $y^{(0)}(k)$ is the preprocessed value. The kth value of the sequence.

(3) Perform Y(0) functional transformation on the pair, as shown in the formula:

$$p^{(0)}(k) = \arcsin\left(y^{(0)}\right)^{a} + N(K) \times 2\pi \tag{3}$$

In the formula, $N(k) \times 2\pi$ is the period value of a sine function that P(k) becomes a positive monotonically increasing sequence, and N(k) is 0, 1, 2,… In order to reasonably determine the value of α, it is traversed from 0.1 to 0.9 in 0.1 steps, and the GM(1, 1) model is established respectively and the average relative error is calculated. Finally, the α value corresponding to the model with the smallest error is selected, that is, 0.5.

2.3 Establishment of Cumulative T-M

As an improved curve fitting method, the accumulation method ignores the error assumption of curve fitting by accumulating data, and requires less calculation, so it has been widely used in the field of economics and engineering. The GM(1,1) grey prediction model has low prediction accuracy for data series with large random volatility, and the parameters of the model are solved by the least square method, which is complicated in calculation process and large in calculation amount [7, 8].

(1) First, perform the row-level comparison test on the original sequence and then perform logarithmic transformation to obtain a new sequence. Finally, use the T-M to model the new sequence from steps 1 to 3, and finally obtain a new sequence.
(2) Calculate the cumulative generation sequence and the background value sequence once, and then obtain the order cumulative operator of the sequence according to the definition of the cumulative operator.

2.4 Performance Evaluation Indicators of Prediction Models

The commonly used metrics to evaluate the performance of machine learning models are: mean square error (MSE), mean absolute percentage error (MAPE) and relative coefficient. In this paper, MSE and MAPE are used to reflect the prediction error, the ratio of the error to the actual data, and the overall fitting level [9, 10].

3 Construction of Cumulative TGM-RBF Prediction Model

3.1 Modeling Rationality Analysis

Grey theory can weaken the randomness of time series and explore the ability of system evolution, and can be well combined with other models. Therefore, it is integrated with other models, so that each model can fully combine its own advantages, so that the prediction accuracy of the combined model is greatly improved. Studying the output process of the neural network shows that the output result is close to a certain fixed value because it is accompanied by a certain precision. Thus, with constant error, the output will fluctuate around this fixed value. According to the grey system theory, the output value of the neural network model is actually a grey number. In short, the two are closely related. Therefore, it is feasible to establish a GM-RBF residual correction series combination model. The core of its thought is: not only can use the neural network to examine the gray system, but also use the gray system to study the neural network [11].

3.2 The Need for Modeling

In deformation monitoring, when predicting time-observed sequence samples, although a non-equidistant model of dimensional new information such as GM(1,1) can be established, which can greatly improve the prediction accuracy of the gray system, due to the gray G-M itself lacks self-adaptive ability, the result obtained is the performance of the change trend of the observation sequence, and the information of the random part is seriously lost, so that the prediction result cannot truly reflect the real situation of the deformed body. Therefore, it is necessary to establish a residual correction model to correct the trend term. However, the nonlinearity and randomness of the residual sequence determines that the selected model must have good nonlinearity and randomness fitting ability. The RBF neural network has this feature. Its self-learning, nonlinear, and self-organizing capabilities enable the model to predict the residual sequence well and obtain effective information of the residual sequence. Therefore, it is necessary to combine the trend prediction of non-equidistant weighted GM (1,1) with the residual correction obtained by the RBF neural network prediction to establish a cumulative series combination model [12].

3.3 Establishment of Cumulative Tgm-Rbf Model

Since both the cumulative T-M and the RBF neural network can be regarded as function prediction algorithms of numerical, non-mathematical models. The cumulative T-M is used to solve the problem, which has a small amount of calculation and can meet the prediction requirements in the case of a small number of samples. The residual sequence of the RBF neural network prediction model is used for prediction, and the prediction accuracy is high and the error is controllable. Therefore, combining the two with each other can learn from each other and effectively improve the prediction accuracy of the model. The specific modeling steps are as follows:

(1) The original data sequence of BE consumption is processed by the cumulative T-M, and the processed data is used as the original input data of the cumulative T-M.

(2) Use the cumulative T-M to predict the original sequence and obtain the corresponding residual sequence.

4 Experimental Analysis of BE Consumption Prediction Integrated Model

4.1 Forecast of Energy Consumption in China's Construction Industry

Table 1. Prediction results and precision analysis of total electricity consumption of three models

Years	Actual value (billion kWh)	G-M	Cumulative T-M	Cumulative TGM-RBF Model
2009	421.9	421.9	421.9	421.4
2010	483.2	491.51	488.7	488.3
2011	571.8	550.28	556.8	556.4
2012	608.4	609.7	618.7	616.9
2013	675.1	666.3	671.9	668.3
2014	721.7	733.2	722.8	715.1
2015	698.7	719.9	711.4	704.8
2016	725.6	723.2	723.9	732.7
2017	789.2	770.1	768.3	784.1
2018	888.1	874.3	881.2	885.1

In this paper, a cumulative TGM-RBF model is established for the energy consumption data of the building industry in a certain country, and the results are compared with the G-M and the cumulative T-M. The original BE consumption data are shown in Table 1.

As can be seen from the Fig. 1, the RMSE of the G-M is much larger than that of the cumulative TGM-RBF model and the cumulative T-M, and there is an overfitting problem; the cumulative TGM-RBF model is similar to the GM(1, 1) model and the cumulative T-M, the prediction accuracy of this model is improved by 1.11% and 1.53%, respectively. It is further illustrated that the cumulative TGM-RBF model is superior to the G-M and the cumulative T-M in predicting BE consumption.

4.2 Monthly Energy Consumption Forecast of Buildings

In this paper, the numerical simulation analysis of an office building in a city is carried out, and the office building model is established by SketchUp, as shown in Figure, with a total of five floors, the indoor height is 3.5m, the thickness of the outer wall is 0.24m, and the south side of the building is made of glass curtain wall and steel Frame composition;

Table 2. BE consumption data

Month	Actual data	G-M	Cumulative T-M	Cumulative TGM-RBF Model
1	9.64	8.30	9.35	9.73
2	8.81	7.61	8.41	8.65
3	7.36	7.12	7.17	7.31
4	6.61	7.01	6.99	6.88
5	6.32	7.11	7.03	6.45
6	7.68	8.49	8.37	7.55
7	8.73	7.61	7.82	8.61
8	10.87	10.45	10.67	10.89
9	9.48	10.25	8.26	9.17
10	8.13	9.25	7.75	8.35
11	6.98	6.94	7.37	6.87
12	9.16	8.01	8.38	9.03

the air conditioner uses a two-pipe fan coil unit plus a fresh air system. Combined with the building design parameter standards, it can be known that the heat transfer coefficient of the building envelope is. Roof 0.55W/(m2·K), exterior wall 0.6W/(m2·K), exterior window 2.5W/(m2·K), exterior window 2.5W/(m2·K), The shading coefficient is 0.6; other parameters choose the default value. The main energy consumption data of the building are shown in the Table 2.

As can be seen from the Fig. 2, the accuracy of GM(1,1) is 94.05%, the accuracy of the cumulative T-M is 96.54%, and the accuracy of the cumulative TGM-RBF model is 98.35%. In contrast, the prediction accuracy of the cumulative T-M is improved by 2.49% compared with the G-M; the prediction accuracy of the cumulative TGM-RBF combination model is better than that of the cumulative T-M and GM The (1, 1) models improved by 1.81% and 3.30%, respectively. It is further demonstrated that the cumulative TGM-RBF combined model is superior to the G-M and the cumulative T-M in predicting BE consumption.

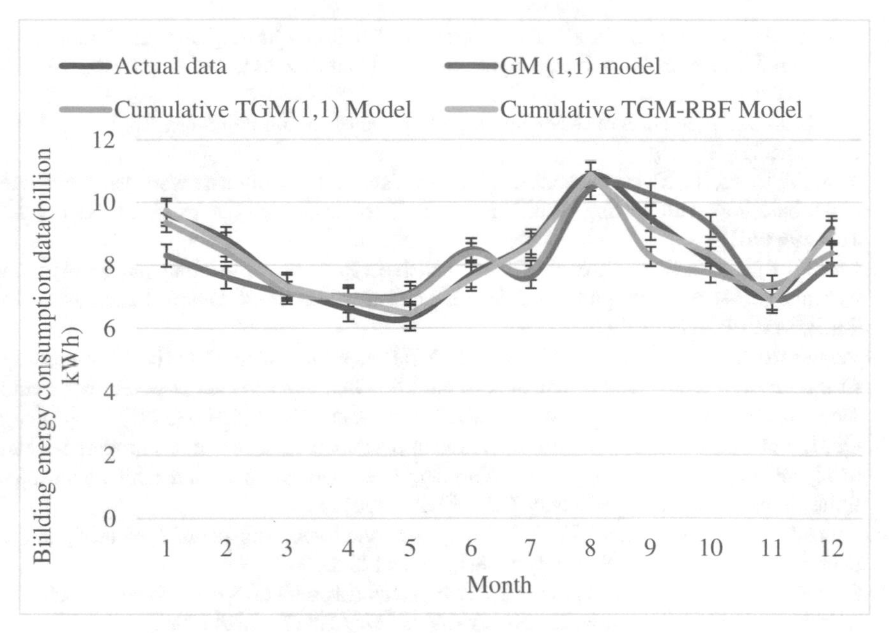

Fig. 2. Comparison curve between simulated and predicted BE consumption

5 Conclusions

The results of this paper are of great value for current BE consumption forecasts. By applying this prediction model in the field of BE consumption prediction, it can point out the direction for BE conservation and provide great help to the sustainable development of energy conservation in our country. In addition, the method used to construct the composite model and the general laws found in this paper have reference and reference significance for the extensive application of artificial intelligence in the energy field, and also have important significance for promoting the intelligent development of energy systems.

References

1. Wang, L., Liu, X., Brown, H.: Prediction of the impacts of climate change on energy consumption for a medium-size office building with two climate models. Energy Build. **157**, 218–226 (2017)
2. Nam, Chul, Seong, et al.: Development of optimization algorithms for be model using artificial neural networks. J. Korean Soc. Living Environ. Syst. **24**(1), 29–36 (2017)
3. Bedi, G., Venayagamoorthy, G.K., Singh, R.: Development of an IoT driven building environment for prediction of electric energy consumption. IEEE Internet Things J. **99**, 1 (2020)
4. Jarošová, P., Vala, J., et al.: Computational prediction and control of energy consumption for heating in building structures. AIP Conf. Proc. **1863**(1), 1–4 (2017)

5. Sun, L., Wei, Q., He, L., et al.: The prediction of building heating and ventilation energy consumption base on Adaboost-bp algorithm. IOP Conf. Ser. Mat. Sci. Eng. **782**(3), 032008 (5pp) (2020)
6. Alshibani, A.: Prediction of the energy consumption of school buildings. Appl. Sci. **10**(17), 5885 (2020)
7. Jiang, T., Li, Z., Jin, X., et al.: Flexible operation of active distribution network using integrated smart buildings with heating, Ventilation and air-conditioning systems. Appl. Energy **226**, 181–196 (2018)
8. Liu, K., Liu, T.Z., Fang, P., et al.: Comprehensive approach to modeling and simulation of dynamic soft-sensing design for real-time BE consumption. Int. J. Distrib. Sensor Networks **13**(5), 155014771770493 (2017)
9. Javanmard, M.E., Ghaderi, S.F., Hoseinzade, M.: Data mining with 12 machine learning algorithms for predict costs and carbon dioxide emission in integrated energy-water optimization model in buildings. Energy Conversion and Management **238**(114153) (2021)
10. Oh, J., Koo, C., Hong, T., et al.: An economic impact analysis of residential progressive electricity tariffs in implementing the building-integrated photovoltaic blind using an advanced finite element model. Appl. Energy **202**, 259–274 (2017)
11. Ahmad, J., Tahir, A., Larijani, H., et al.: Energy demand forecasting of buildings using random neural networks. J. Intell. Fuzzy Syst. **38**(1), 1–13 (2020)
12. Soleymani, S.A., Goudarzi, S., Kama, N., et al.: A hybrid prediction model for energy-efficient data collection in wireless sensor networks. Symmetry **12**(12), 2024 (2020)

Cost Optimization Control System of Prefabricated Building Based on BIM Technology

Jie Wang(✉)

School of Road Bridge and Architecture, Chongqing Vocational College of Transportation, Chongqing 402247, China
jiujiwoniu@163.com

Abstract. At present, the cost of prefabricated buildings in my country is higher than that of traditional buildings. Cost is the blood and lifeline of enterprise development. Cost is an important factor restricting the rapid and large-scale development of prefabricated buildings at this stage. How to provide the owners with the best quality residences and services at the lowest cost is the core competitiveness of the development of prefabricated buildings. The purpose of this paper is to study the cost optimization control system of prefabricated buildings based on BIM technology. Based on cost management and BIM technology, this paper studies the project cost management system based on BIM technology, and analyzes the relevant business process and system process of the system; through the analysis of the existing cost management system, find out the problems existing in the existing system, and then it is concluded that the application of BIM technology to the project cost management system can directly solve many problems faced by the traditional system. Using BIM technology to reduce the cost of prefabricated construction is the core content. The cost of prefabricated buildings is analyzed from the perspective of the whole life cycle and the whole industry chain. The experimental study shows that after the calculation of Matlab software, it can be seen from this functional coefficient that the prefabricated building with a prefabrication rate of 50% has the highest cost optimization index for prefabricated components.

Keywords: BIM technology · Prefabricated buildings · Cost optimization · Cost control system

1 Introduction

With the in-depth development of the modern construction industry, the advantages of BIM technology and prefabricated buildings are becoming more and more obvious, and they have shown significant development advantages in many aspects such as building construction, cost control, personnel use and mechanical investment [1, 2]. However, in the domestic construction field, the application of BIM technology and prefabricated buildings is in its infancy, and many aspects such as technology research, practical

J. H. Abawajy et al. (Eds.): ICATCI 2022, LNDECT 169, pp. 193–201, 2023.
https://doi.org/10.1007/978-3-031-28893-7_24

application and cost control are still in the exploratory stage, and have not yet been applied and promoted in the entire construction industry chain, and thus have relatively Broad application space.

In the research of prefabricated building cost optimization control system based on BIM technology, many scholars have studied it and achieved good results. For example, Li T integrated BIM5D technology with cost and construction management software, combined with cost system control process and cost Control model, use BIM cloud data management to achieve integrated and refined cost control [3]. Lu S applied BIM technology to the whole life cycle cost control of prefabricated buildings, and realized the coordination, visualization and process of cost control [4]. It can be seen that the research on the cost optimization control system of prefabricated buildings based on BIM technology is of great significance.

This paper establishes the next-generation construction cost budgeting workflow based on BIM technology, and clarifies the functional positioning of the next-generation construction cost budgeting software. This workflow clearly depicts the process of using BIM-based architectural design software and construction cost budgeting software to complete construction cost budgeting, thereby clarifying the position of construction cost budgeting software in the entire construction cost budgeting process.

2 Design and Research on Cost Optimization Control System of Prefabricated Building Based on BIM Technology

2.1 Model Design of Building Cost Budgeting Software System Based on BIM

(1) BIM data management platform

The platform user interface includes two sub-modules: the BIM graphical interaction module is used to display the geometric topology information of the building in the form of a 3D model, and provides users with a 3D graphical interface for interactively operating BIM data; the BIM resource manager module uses It is used to describe the architectural logic hierarchy expressed by BIM data, and provide users with a tree interface for interactively operating BIM data [5, 6]. They all realize the access to the BIM data container by calling the BIM data application interface.

(2) Functional module design of construction cost budget

In the cost item setting, the cost item can be generated and set according to the quota or bill of quantities specification, and at the same time, the association between the cost item and the component can be realized intelligently by judging the construction information of the component. The inventory library and the quota library are used as data sources. In the IFC standard data import, IFC standard files can be exported. The files include component geometry information, construction process information, cost item information, etc., which is convenient for downstream software such as construction management system and information reuse system.

2.2 Research on Requirements and System Design of Prefabricated Construction Project Cost Management System Based on BIM

(1) Business process

When the company obtains the design drawings of the project, on the one hand, it can model the content of the drawings according to the information already grasped, and simulate the building information model. On the other hand, if conditions permit, electronic drawings can be directly converted into building information models. After the budget is completed, the amount of information obtained from the simulation model of the building information data of the relevant project can be used when bidding for a plan; before the project is constructed, the relevant target cost can be calculated based on the building information of the simulation model; The realistic simulation model establishes the information cost to adjust the project department in time, such a reasonable process management [7, 8]. Finally, the financial department will calculate the cost of the entire project and complete the cost management task of the entire project.

(2) System process

Compared with the traditional cost management, in addition to the whole process of the traditional cost related to the role of participation in management, it is also the role of information management that we need to establish a simulation model. Cost management can extract all the data and information for the construction and operation of the simulation model, so for the system operator, the role of building the information simulation model is very important. Therefore, each role can extract the information they want according to their needs, while traditional cost management, to a certain extent, is not functionally related to each other, information is not shared and exchanged, and information is not related to business processes. Computer application system problems where applications are disconnected from each other.

(3) Information management system

Attendee rights management is system administrator, main management, management roles and management user information. For example, it involves costs associated with new management of relevant personnel, system administrators need to give people the home environment, including: staff affiliations, belonging roles.

(4) Bid management cost

Regarding the cost of the bidding management information officer, responsible for planning and acting as three bidder planners, the main person in charge of the bidding project, and completing the registration of bidding planning, thus forming a bidding cost. After the registration is completed, the responsible planner designates the affiliated branch or representative office under the project, before the bidding of the entire project cost plan is completed [9, 10]. After the project evaluation, the information officer needs to register the evaluation and release information of the relevant project, and the entire bidding cost project management is completed.

(5) Target cost management

Target cost management involves accountants, project department and production management leaders. The accountants completed the unit cost of the project at this stage, the registration of the rules, the contract price registration, the cost of development and change, and finally completed the task cost, the leader of the production management department knew the cost of this rule at the beginning of the project, related business audits Task.

(6) Actual cost management

The actual cost of the three user roles related to the role of the project management unit, the representative office or branch office and the head office role. The main role of the representative office is to generate the actual monthly expenses from the accounting expenses of the head office expenses generated by the report. The main role of this company is to complete the audit of the cost information report and to form the actual cost data at the beginning of this month [9, 11]. For example, during the construction process of this month, the cost of people flow, logistics, machinery and other expenses incurred by some projects, and the cost of project personnel registered every day.

2.3 Application Strategy of BIM Technology in Cost Control of Assembly Building

(1) Cost data sharing

1) Appropriate project value personnel can input data into the system, and BIM technology can support the exchange of company information. The program administrator can analyze the data and information contained in the system at any time, whether in the system server or in the BIM system, there is a place where data and information can be shared, and the corresponding staff can request data. Anytime there is a need to improve employee efficiency and improve project efficiency.
2) BIM technology allows operators to retrieve the data they need as soon as possible, and the content downloaded to the database may change from time to time. Therefore, the application of BIM technology is an important way to improve the company's data processing level. The relevant billing management technical information can also play an effective role in improving the data processing capability of the relevant data.

(2) Fine control

Data can be recovered faster and with high accuracy using BIM technology. The model created on this basis can take into account the exchange of information, and the cost of building materials and applications can also be clearly seen in the software. Additionally, companies can use this technology to calculate and process information faster. In the BIM technology system, there are many data algorithms related to project cost control and management. Employees can jump right into fields and then let the program perform complex tasks. The processing is fast and the accuracy of the results is very high.

(3) Scientific control of engineering quantity

The software solves the data connection problem and provides accurate and in-depth analysis of the computer performance results in the data. In this system, only the employee needs to put a project value file in it, and the world can be modeled using BIM technology to collect a lot of data, the software can analyze the results of the program and develop the results. To help the construction team better understand the relevant data. As a result, technical research and mapping work did not require much human input, and the accuracy of the work was improved, overcoming many of the previous weaknesses.

2.4 Order Analysis of Organizational Structure

In information theory, entropy can be used as a measure of the chaos or disorder of a state, with uncertainty. The order degree of the system is related to entropy, and there is the following relationship between them:

$$R = 1 - \frac{H}{H_m} \tag{1}$$

Among them, H is the entropy of the system structure, which represents the maximum entropy of the system. When the R is larger, the order of the organizational structure is higher, and the organizational structure is more reasonable. According to the definition of entropy, the order degree R of organizational structure, the order degree of information transmission speed 1 and the order degree of information transmission quality 2 have the following relationship:

$$R = R_1 + R_2 \tag{2}$$

In: $R_1 = 1 - \frac{H_1}{H_{1m}}$, $R_2 = 1 - \frac{H_2}{H_{2m}}$.

3 Experiment Research on Cost Optimization Control of Prefabricated Buildings Based on BIM Technology

3.1 Experimental Subjects

This paper selects the product production of company A in this city as the case analysis object, and uses the data from 2014 to 2020 as the analysis basis.

3.2 Basic Data for Operation Effect Evaluation

The data obtained in this paper are mainly obtained in the following two ways:

(1) Annual financial report of the company

The annual financial report of Industrial Science Electronic Technology Co., Ltd. Refers to the comprehensive financial report of the company's current operating conditions provided at the end of each year, which contains relevant data on quality costs and is an important basis for the company to formulate strategic planning. Annual financial accounting reports usually include cash flow reports, operating results reports and year-end financial status reports.

(2) Company accounting vouchers

It is mainly the accounting voucher of the electronic products produced by Industrial Science Electronic Technology Co., Ltd., which involves the original data of quality cost. The quality cost accounting voucher is the second-level subject of prevention, appraisal, internal loss and external loss cost. Product quality cost accounting voucher, which can provide reference and comparison for input and output indicators.

4 Experimental Research and Analysis on Cost Optimization Control of Prefabricated Buildings Based on BIM Technology

4.1 Application of Value Engineering Optimization Model

Table 1. Judgment matrix of building prefabrication rate

Building prefabrication rate	Prefabrication rate 10%A1	Prefabrication rate 30%A2	Prefabrication rate 50% A3
Prefabrication rate 10%A1	0.93	0.16	0.14
Prefabrication rate 30%A2	4.26	0.96	0.45
Prefabrication rate 50% A3	6.63	2.06	0.96

Through on-site visits to some prefabricated component factories, the basic situation and research purpose of the project were introduced to the prefabricated factory managers, skilled workers and researchers related to prefabricated buildings, and questionnaires were issued. Finally, according to the survey results of the questionnaires, a judgment matrix was constructed. The specific data are shown in Table 1.

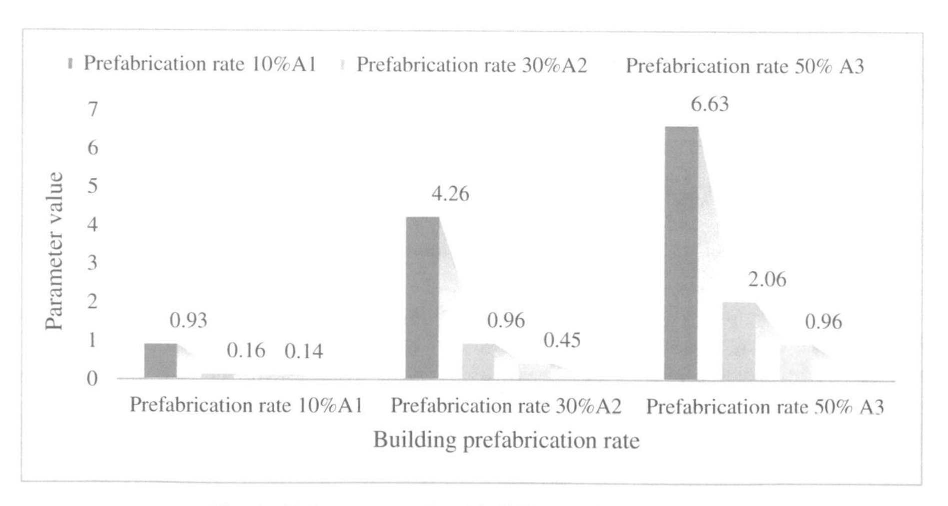

Fig. 1. Judgment matrix of building prefabrication rate

As shown in Fig. 1, after the operation of Matlab software, $\lambda max = 3.135$, $CI = 0.0049$, $CR = 0.0264$ are obtained, which pass the consistency test; and use this as the function coefficient. From this function coefficient, it can be seen that the prefabricated building with a prefabrication rate of 50% has the highest cost optimization index for prefabricated components, followed by a prefabrication rate of 30%, and a prefabrication rate of 10% is the worst.

4.2 Analysis of Average Efficiency of Cost Control

Add the 2014–2020 input and output data of the investigated company A in this city respectively, calculate the arithmetic mean, and use the obtained mean value as the new input and output of each year. Using DEAP software, you can get the 2014 ~ 2020 company Annual average technical efficiency, pure technical efficiency and scale efficiency. The descriptive statistics of the average efficiency of quality cost control of company A in this city from 2014 to 2020 are shown in Table 2.

Table 2. Annual Average Efficiency of Quality Cost Control of Industrial Science Companies from 2014 to 2020

	Minimum (%)	Maximum value	Mean (%)	Standard deviation (%)
Technical efficiency	26.4	1	59.3	15.3
Pure technical efficiency	38.4	1	66.8	14.7
Scale efficiency	28.5	1	84.1	11.6

As shown in Fig. 2, we first look at the minimum and maximum values of technical efficiency, pure technical efficiency and scale efficiency. The minimum values of each efficiency parameter are relatively small, 26.4%, 38.4% and 28.5% respectively, indicating that the quality and cost control process of company A in this city has serious extensiveness. The maximum value of each efficiency parameter is 1. According to statistics, the year in which the technical efficiency is equal to 1, the year in which the pure technical efficiency is equal to 1, and the year in which the scale efficiency is equal to 1 account for 5.5%, 10.1% and 8.09% of the total number of samples under investigation, respectively. That is to say, the number of years in which the quality cost control of company A in this city is in an effective state is very small, and a considerable number of years are in an ineffective state of quality cost control. Looking at the average value of each efficiency parameter, the pure technical efficiency is 66.8%, the scale efficiency is 84.1%, and the average technical efficiency only reaches 59.3% of the effectiveness. This result is consistent with the previous analysis of time changes. Company A's quality and cost control economic benefits are generally unsatisfactory. In terms of the degree of dispersion, the standard deviation of each efficiency parameter is not small, which shows that there are significant differences in the efficiency level of quality cost control of company A in this city.

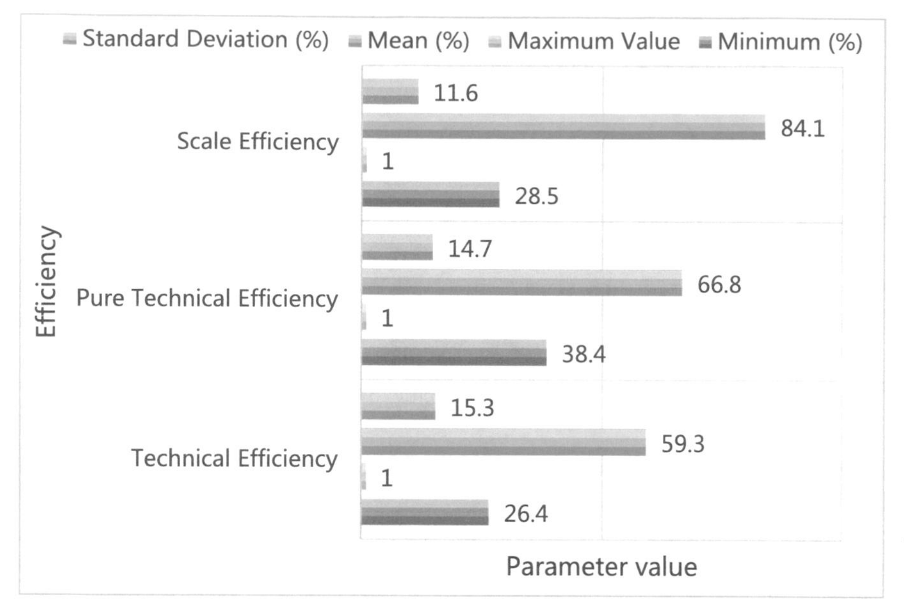

Fig. 2. Average annual efficiency of quality cost control of industrial science companies from 2014 to 2020

5 Conclusions

Combine BIM technology with prefabricated building construction cost control, use BIM technology visualization, digitization and other functions to assist process project design, control the construction cost of prefabricated building projects in advance and adjust them in time to avoid bringing problems to the later stage The stage brings cost waste to the project implementation, makes up for the shortcomings of traditional cost control such as untimely data update and lag in message transmission, and improves the control efficiency of prefabricated building construction costs.

References

1. Abbasianjahromi, H., Hosseini, S.: A risk-cost optimization model for selecting human resources in construction projects. SN Appl. Sci. **1**(11), 1–13 (2019). https://doi.org/10.1007/s42452-019-1570-5
2. Khalesi, M.H., Salarieh, H., Foumani, M.S.: Dynamic modeling, control system design and MIL–HIL tests of an unmanned rotorcraft using novel low-cost flight control system. Iranian J. Sci. Technol. Trans. Mech. Eng. **44**(3), 707–726 (2019). https://doi.org/10.1007/s40997-019-00288-x
3. Li, T., Liu, H., Wang, H., et al.: Hierarchical predictive control-based economic energy management for fuel cell hybrid construction vehicles. Energy **198**, 117327.1-117327.10 (2020)
4. Lu, S., Li, Y., Xia, H.: Study on the configuration and operation optimization of CCHP coupling multiple energy system. Energy Conv. Manage. **177**, 773–791 (2018)

5. Zhang, Q.: Research on the construction schedule and cost optimization of grid structure based on BIM and genetic algorithm. J. Phys. Conf. Ser. **1744**(2), 022065 (2021). (5pp)
6. Sun, X., Hu, Z., Li, M., et al.: Optimization of pollutant reduction system for controlling agricultural non-point-source pollution based on grey relational analysis combined with analytic hierarchy process. J. Environ. Manage. **243**, 370–380 (2019)
7. Vitiello, U., Ciotta, V., Salzano, A., et al.: BIM-based approach for the cost-optimization of seismic retrofit strategies on existing buildings. Autom. Const. **98**, 90–101 (2019)
8. Spinnraeker, E., Pauen, N., Schnitzler, A., et al.: Webtool zur risikobasierten-probabilistischen Lebenszyklus-kostenanalyse auf Basis digitaler Gebäudemodelle - BIM_(2P) LCC. Bauingenieur **94**, 37–44 (2019)
9. Sandberg, M., Mukkavaara, J., Shadram, F., et al.: Multidisciplinary optimization of life-cycle energy and cost using a BIM-based master model. Sustainability **11**(1), 286 (2019)
10. Zeng, T., Ren, X.M., Zhang, Y., et al.: An integrated optimal design for guaranteed cost control of motor driving system with uncertainty. IEEE/ASME Trans. Mechatron. **PP** (99), 1–12019
11. Tsumura, R., Hardin, J., Bimbraw, K., et al.: Tele-operative low-cost robotic lung ultrasound scanning platform for triage of covid-19 patients. IEEE Robot. Autom. Lett. **6**(3), 4664–4671 (2021)

Three Models of Digital Transformation of High-End Equipment Manufacturing Industry in the Internet Era

Chun Cui(✉)

School of Economics, Shenyang University, Shenyang 110031, Liaoning, China
shaonianxing12@163.com

Abstract. Under the background of the rapid development of the Industrial Internet, the digital transformation of the high-end equipment manufacturing industry has become an inevitable trend. This article puts forward the content of the digital transformation of the high-end equipment manufacturing industry, including the digital transformation of production factors, the digital transformation of the production process, and the digital transformation of production products. According to the content of the digital transformation, three models of the digital transformation of the high-end equipment manufacturing industry are proposed, namely "digital transformation". The production model of "twin factory+physical factory", the business application model of "reform and update+interconnection", and the product cycle model of "internal incubation+external output".

Keywords: High-end equipment manufacturing · Industrial internet · Digital transformation

Entering the "Industry 4.0" era, the direction of high-end equipment manufacturing transformation is to achieve digitization. According to the definition of the Development Research Center of the State Council, digital transformation refers to the use of a new generation of information technology to build a closed loop of data collection, transmission, storage, processing and feedback, to break through data barriers between different levels and different industries, and to improve the overall operating efficiency of the industry, to build a new digital economy system. Developed countries attach great importance to digital transformation. According to McKinley's forecast, with existing technology, the digital transformation of the government can generate more than $1 trillion in value globally each year [1]. Subject to the technical conditions and organizational conditions for obtaining information, the information structure of the industrialized system has the characteristics of time lag and incomplete content. Although it has gone through hundreds of years of industrialization, the information is not timely, continuous and refined [2].

The Industrial Internet provides important infrastructure for the digital transformation of the high-end equipment manufacturing industry and is an important engine for the digital transformation of the high-end equipment manufacturing industry. The concept of "Industrial Internet" was first proposed by GE in the United States in 2012. By

J. H. Abawajy et al. (Eds.): ICATCI 2022, LNDECT 169, pp. 202–210, 2023.
https://doi.org/10.1007/978-3-031-28893-7_25

making full use of data resources and carrying out cross-border cooperation in various ways, enterprises can create new profit sources other than products and services, form a multi-dimensional profit model, and expand the space for value realization [3]. Verhoef et al. (2019) divide digital transformation into three stages: Digitization, Digitalization, Digital transformation [4]. Although the first two stages both translate to digitization, their meanings are not the same. The first stage of Digitization is the process of converting analog information into digital information. [5] Figures for the second stage of digitalization is the use of digital technologies to change existing business processes, including communications, distribution, business relationships management, etc. [6]. In short, the second stage is defined as the exploitation of digital opportunities, and the first stage is the basic framework of the second stage [7]. The third stage of digital transformation involves strategic changes in business models [8]. Regarding the digital transformation of the manufacturing industry, existing research has conducted preliminary discussions from the following perspectives: First, from the perspective of the effectiveness of digital transformation, the study found that the application of information technology has a positive effect on the improvement of enterprise performance in terms of cost saving, production and management efficiency improvement, etc. effect [9]. The second is the process perspective of digital transformation, that is to explore the path of modern information technology-driven manufacturing transformation represented by "Internet+", and believes that it is necessary to break the path dependence based on latecomers, imitation, catch-up and low cost, and rebuild the basic capabilities of the industry. Enhance the technological innovation capability of the manufacturing industry, or use the cluster development and integrated development of the information technology industry as a carrier to achieve value chain climbing through industrial innovation [10].

Since 2017, the digital transformation of China's manufacturing industry has accelerated, and its market share has continued to increase, as shown in the Fig. 1 below:

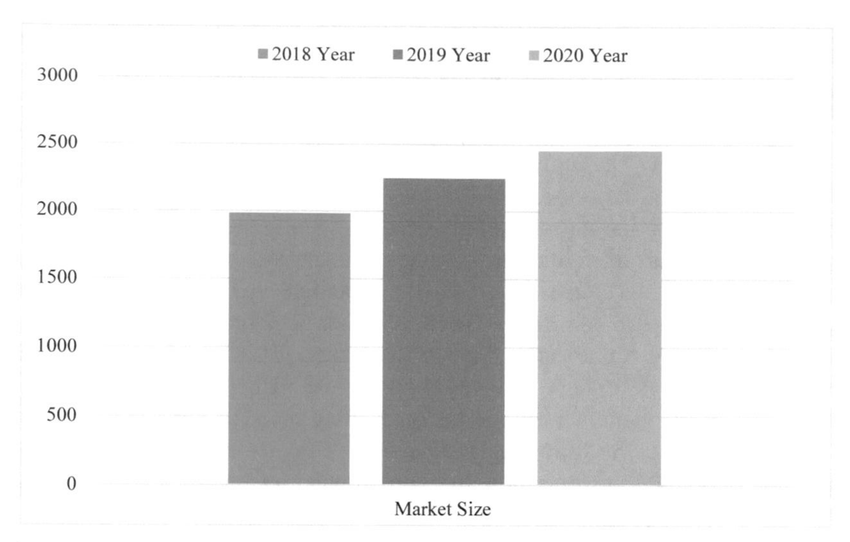

Fig. 1. Market Size of Digital Transformation of China's Manufacturing Industry (Unit: 100 million yuan)

The content of the digital transformation of the high-end equipment manufacturing industry mainly includes: the digital transformation of production factors, the digital transformation of the production process, and the digital transformation of production products. According to the content of the digital transformation of the high-end equipment manufacturing industry, this article proposes three models for the transformation of the high-end equipment manufacturing industry:

1 High-End Equipment Manufacturing Industry Establishes a Production Model of "Digital Twin Factory+Physical Factory"

In the high-end equipment manufacturing industry, data has become a new production factor. On the one hand, the use of intelligent equipment will generate data; on the other hand, data participates in product design, process production, and equipment maintenance. The development of the high-end equipment manufacturing industry needs to transform the past production-driven model based on the technology of human resources as the core to digitization as the driving model of the core production factors, and use the knowledge in the digital model to assist or replace the human production model.

The high-end equipment manufacturing industry can transform data into production factors through the establishment of a "digital twin factory" production model. On the basis of making full use of data-driven features, the digital twin factory tries to simplify the mechanical interaction between people and equipment as much as possible. Through the digital twin factory, it is possible to comprehensively apply digital and network technologies, and use the inheritance of numerical control equipment and information systems to provide control and execution of production plans, thereby improving the flexibility of the production system, reducing production costs, controlling risks, and improving production efficiency.

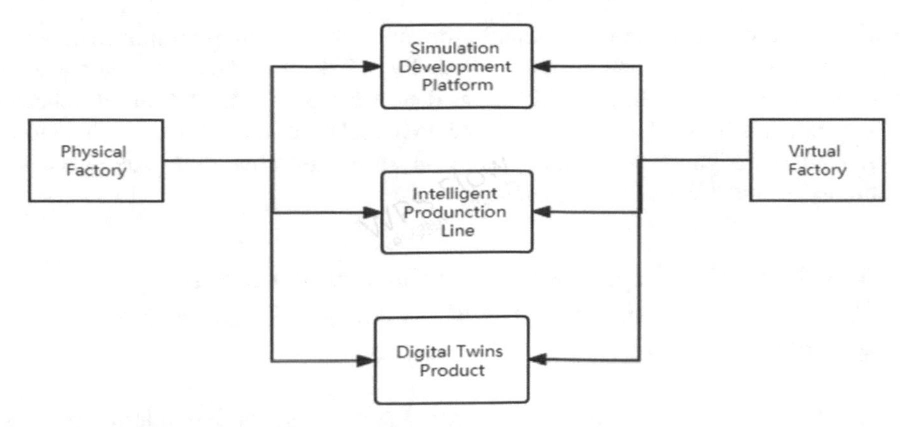

Fig. 2. "Digital Twin Factory+Physical Factory"

Therefore, through the integration of the virtual factory and the physical factory, the production and operation of the physical factory are comprehensively monitored, analyzed and predicted in the virtual factory, and the construction, transformation, and production process optimization of the physical factory is guided through the virtual factory. How to combine "virtual factory and physical factory"? As shown in the Fig. 2.

1.1 Digitization of R&D and Design

Build a design and simulation research and development platform, and realize product design through the establishment of a unified digital model. In the entire production process, the product design and process design are simulated, so that the production control program and the program design are coordinated and consistent; in the entire industry chain, the product design at all levels of the system, sub-systems, stand-alone machines, components, and parts form coordination and consistency; Form coordination and consistency in the design of different majors in structure, electrical, and software.

1.2 Digitization of the Production Process

Establish intelligent processing production lines, intelligent inspection production lines, and intelligent assembly production lines to realize flexible production lines featuring multi-variety mixed lines and man-machine integration to meet the downstream needs of multi-variety and small batches. By opening up the MES and other software systems and the information interaction and control of production equipment, the boundary between man and machine is opened up. Through the simulated production of the data model, the data management of the production process and the quality of data support are realized.

1.3 Product Digitization

Build digital twins of products and improve product quality. Use full-level supplier management and product data packages as tools to build digital twins of model products, make full use of the value of data, and promote product design optimization, product quality improvement, and guarantee capabilities. Take the aerospace industry as an example, provide design data, manufacturing data, acceptance data, test data and operating data through physical rockets. Use digital twin rockets as tools to realize simulation operation and design optimization, quality analysis and improvement, fault query and health management Wait for the goal.

2 The High-End Equipment Manufacturing Industry Establishes a Business Application Model of "Renovation and Interconnection+Interconnection"

The digital transformation of the high-end equipment manufacturing industry not only includes the conversion of data into production factors into production, but also the application of digital technology to the business of the production process. The digital transformation of the production process can adopt the iterative application model of "reform and update + interconnection" with production as the center to other links. In the manufacturing of high-end equipment manufacturing, the use of machine-based production of precision and high-quality products has laid a foundation for digital transformation. Digital transformation starts from the manufacturing process, and other departments such as R&D, supply chain, sales and other production businesses perform digital applications in production as needed, and are connected with the digital manufacturing system to realize the digital transformation of the entire production process (see Fig. 3).

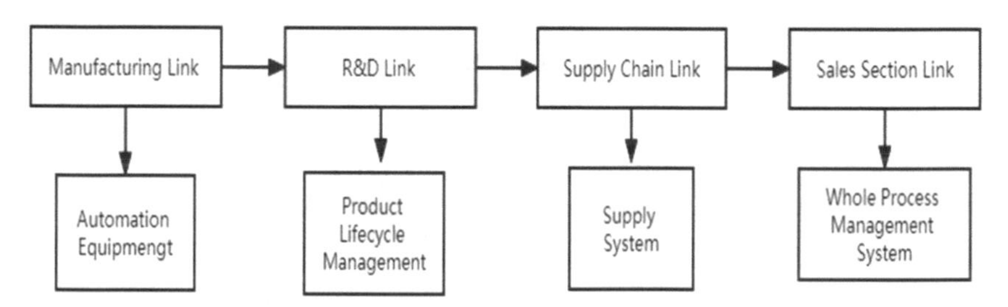

Fig. 3. Digitization of the production process

2.1 Digital Transformation of Manufacturing Links

The digital transformation of production and manufacturing is mainly reflected in "reconstruction", such as the use of data acquisition cards, data communication cards, and the installation of RF (radio frequency) collectors for old equipment to transform equipment networking capabilities to achieve on-site production Real-time perception of critical equipment operating status. By transforming the existing automation equipment, optimizing the production process, and using digital means, the accumulated technical experience over the years is transformed into a substantial system management tool, which improves the utilization rate of production capacity and increases the profit margin of the enterprise. On the basis of continuous optimization and improvement, we have successively increased the introduction of advanced automation equipment such as AGV (Automatic Guided Vehicle) and robots with large investments.

2.2 Digital Transformation of Other Links

In the R&D link: establish a digital product design process based on CAX, and gradually establish a R&D knowledge base to achieve data accumulation in the R&D link, and then to PLM (product life cycle management). Establish product research and development and data management systems, product project management systems, product modular systems, quality management systems, EDE data systems, standard inter-standard management systems, case sharing systems, etc., to realize the full production cycle management of products, and to improve the speed of R&D product design And quality; in the supply chain, establish a production-sales coordination system, production planning management system, industrial planning system, supplier management system, material management system, etc., and integrate and optimize information in the supply chain by improving the relationship between the upstream and downstream supply chains flow, logistics, and capital flow, effectively coordinate the internal and external resources of the enterprise to jointly meet the needs of the sales side, reduce inventory and costs, and improve the reliability and flexibility of the plan; in the sales link, build support for omnibus-channel, multi-category, online and offline integration The whole-process management system of sales business realizes customer-related data sharing, scene connection, and online and offline integration. Digital marketing applications are seamlessly integrated with front-end systems to achieve agile development and quickly respond to the flexibility of front-end business models. It can be highly integrated with back-end ERP (Enterprise Resource Planning) systems to automatically convert sales operations into financial processing vouchers.

2.3 The Digital Interconnection Between the Production Link and Other Links

By connecting the IMS (Workshop Execution System) and the PLM system, product design data can be directly and automatically synchronized to the IMS system, and then the maneuverability analysis can be synchronized during the product design process, and the parallel business model of the product from the design end to the manufacturing end can be realized. Change the serial mode between the past business, shorten the product from research and development to market-oriented listing cycle; docking IMS system and ERP system, the basic information in the ERP system and automatic interaction on-site production information are automatically synchronized to the IMS system, which can realize the factory Production transparency, combined with IMS planning and scheduling, all-in-one analysis and other functions to achieve real-time dynamic production resource scheduling management, improve enterprise capacity application rate; dock IMS system and SCM (supply chain management system), and realize materials with suppliers State sharing, combined with IMS to provide material identification analysis services, purchase order delivery management and other functions to achieve cross-enterprise resource collaboration, greatly reducing inventory costs.

3 High-End Equipment Manufacturing Industry Establishes a Product Circulation Model of "Internal Incubation + External Output"

3.1 Internal Incubation Mode

Digital technology companies "internally incubated" in the high-end equipment manufacturing industry have strong competitiveness. Digital technology companies use data as production factors to produce knowledge products. Knowledge products are non-exclusive and non-competitive. There will be "black boxes" in the development process. During the research process, not only a large amount of capital investment is required, but it is very likely that the product will not be recognized in the market. Therefore, digital technology The risk of corporate R&D is relatively high. On the one hand, digital technology companies incubated by high-end equipment manufacturing companies have a large amount of financial support, and at the same time they have accumulated experience in digital transformation, reducing the uncertainty of the product market.

3.2 Advantages of Internal Incubation

Digital technology companies that are "internally incubated" in the high-end equipment manufacturing industry master specialized digital service technologies, which can reduce the transaction costs between companies caused by information asymmetry. For high-end equipment manufacturing enterprises, they have higher requirements for providing scientific and technological service products, which are mainly reflected in: high requirements for the professionalism and accuracy of digital technology service products; high-end equipment manufacturing enterprises need to apply to digital technology companies Provide the corresponding data materials, and the safety requirements for the use of data materials are relatively high. Digital technology companies that are "internally incubated" are more professional, and at thc same time provide higher protection for the security of data usage.

3.3 The Role of Internal Incubation on the Manufacturing Industry Chain

Digital technology companies that are "internally incubated" by high-end equipment manufacturing companies can increase the value of the entire equipment manufacturing industry chain. When the high-end equipment manufacturing industry in the upstream of the industrial chain undergoes digital transformation, it will inevitably drive the transformation of downstream equipment manufacturing enterprises. The downstream equipment manufacturing companies' demand for digital service products has the characteristics of universality and standardization. Professional digital service companies provide products for downstream equipment manufacturing companies, which not only saves the cost of repeated R&D by unit companies, but also reduces The risk of R&D by individual companies.

3.4 Transformation from Internal Incubation to External Output

The products provided by digital technology companies "internally incubated" by high-end equipment manufacturing companies are exported to the "outside". Here, "external" refers to other high-end equipment manufacturing companies that have failed to complete digital transformation, but also refers to digital transformation. Other industries in demand (see Fig. 4).

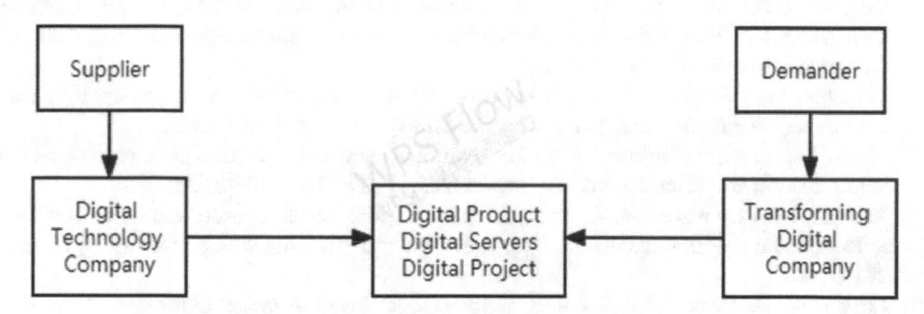

Fig. 4. Transformation from internal incubation to external output

The "external output" products of digital technology companies can be divided into three types: digital products, digital services and digital solutions. Digital products are the transformation of traditional products, mainly to realize the internalization of products; digital services are to provide value-added services on the basis of digital products, such as personnel training, product system transformation, etc., to achieve a product+service profit model; digital solutions The plan is a tailor-made digital transformation plan for the enterprise to help the enterprise achieve the goal of digital transformation. The challenges faced by the digital transformation of the high-end equipment manufacturing industry mainly lie in the difficulty of applying digital elements, the slow promotion of digital technology, and the existence of data islands, which make high-end equipment manufacturing companies think that digital transformation is too complicated and dare not try. Professional digital technology companies provide personalized and customized digital transformation solutions, which can not only solve the problem that most high-end equipment manufacturing companies cannot start digital transformation, but also bring new profitability for digital technology companies. Realize the digital transformation of high-end equipment manufacturing industry.

4 Conclusion

This paper introduces three modes of digital transformation of China's equipment manufacturing industry, which are digital twin factory+physical factory and renovation and interconnection+interconnection and internal incubation+external output. A variety of models will inevitably appear in the transformation of the high-end equipment manufacturing industry. We need to promote the digital transformation model to promote the smooth digital transformation of the high-end equipment manufacturing industry.

Acknowledgements. Fund Project: Shenyang Social Science Federation Project "Research on Promoting the Integrated Development of Shenyang's Advanced Manufacturing Industry and Modern Service Industry (E-commerce) (Project Number: SYSK2021–01-144)."

References

1. Corydon, B., Ganesan, V., Lundqvist, M.: Digital by default: a guide to transforming government 01 March 2020. https://www.mckinsey.com/industries/publicsector/our-insights/transforming-government-through-digitization
2. Thatcher, J.B., Wright, R.T., Su, N.H., et al.: Mindfulness in information technology use: definitions, distinctions, and a new measure. MIS Q. **42**(3), 831–847 (2018)
3. Dyer, J.H., Singh, H., Hesterly, W.S.: The relational view revisited: a dynamic perspective on value creation and value capture. Strategic Manag. J. **39**(12), 3140–3162 (2018)
4. Verhoef, P.C., Broekhuizen, T., Bart, Y., et al.: Digital transformation: a multidisciplinary reflection and research agenda. J. Bus. Res. Bus. (2019). https://doi.org/10.1016/j.jbusres.2019.09.022
5. Loebbecke, C., Picot, A.: Reflections on Societal and business model transformation arising from digitization and big data analytics: a research agenda. J. Strat. Inf. Syst. **24**(3), 149–157 (2015)
6. Schallmo, D., Williams, C.: Digital Transformation Now! Guiding the Successful Digitalization of Your Business Model. Springer, Cham (2019). https://doi.org/10.1007/978-3-319-72844-5
7. Michael, R., Romana, R., Christiana, M., et al.: Digitalization and its influence on business model innovation. J. Manuf. Technol. Manag., vation. J. Manag. (2018). https://doi.org/10.1108/JMTM-01-2018-0020
8. Sebastian, I.M., Ross, J.W., Beath, C., et al.: How big old companies navigate digital transformation. MIS Quart. Execut **16**(3), 197–213 (2017)
9. Comin, D.A., Lashkari, D., Mestieri, M.: Structural Change with Long-Run Income and Price Effects. NBER Working Papers No. 21595 (2015)
10. Huang, Y.Y., Handfield, R.B.: Measuring the benefits of ERP on supply management maturity model: a "big data" method. Int. J. Oper. Prod. Manag. **35**(01), 2–25 (2015)

Distribution Route Planning of Fresh Food E-commerce Based on Ant Colony Algorithm

Xiaoxiao Wang[1](✉) and Rasha Almajed[2]

[1] Dalian Vocational and Technical College, Dalian 116035, Liaoning, China
xxwang12345678@126.com

[2] American University in the Emirates, Dubai, UAE

Abstract. With the development of e-commerce, the e-commerce industry has ushered in new opportunities in our country. As a new industry, fresh agricultural products have great potential. However, problems such as high logistics costs and slow delivery time restrict its further expansion. Therefore, in order to save costs, it is necessary to plan the distribution route. Based on the ant colony algorithm, this paper conducts a research on the distribution route planning of fresh food e-commerce. This article mainly uses the survey method and interview method to understand the citizens' attitudes and views, and uses the data method to study the fresh food distribution route planning. The survey results show that 27% of people feel that the ability to respond to emergencies is very important. Therefore, for the route planning of fresh food e-commerce distribution, the probability of emergencies on the route needs to be considered, and the cold chain logistics and distribution capabilities should also be improved.

Keywords: Ant colony algorithm · Fresh food e-commerce · Distribution route · Route planning

1 Introduction

When people shop online, they have higher and higher requirements for the quality of fresh products. However, due to my country's logistics technology, transportation methods and costs, there are many problems in the distribution of fresh food e-commerce. Fresh agricultural products have the characteristics of high value and perishable. Therefore, a fast delivery method is required to ensure the quality of fresh produce when it reaches customers.

There are many researches on fresh food e-commerce distribution route planning based on ant colony algorithm. For example, Lazarowska said that in practice, the choice of the optimal logistics distribution route is one of the most important issues in the logistics vehicle configuration system. The ant colony algorithm has strong overall search ability and robustness, which can better solve the problem of logistics distribution path involved, thereby improving the quality of logistics service [1]. Liu Changshi said that, taking into account economic and ecological costs, research the time-varying vehicle

J. H. Abawajy et al. (Eds.): ICATCI 2022, LNDECT 169, pp. 211–219, 2023.
https://doi.org/10.1007/978-3-031-28893-7_26

routing problem of online distribution of fresh products under the background of time-varying networks, and comprehensively examine the time-varying speed of vehicles, vehicle fuel consumption, and CO2 emissions. The perishability of fresh agricultural products, such as availability. Client time window and minimum update limit, etc. [2]. Zhang Qian believes that in today's economically developed, e-commerce platform is the center of people's daily life, and fresh food e-commerce has also emerged. How to better ensure the safety of fresh food on e-commerce platforms and how to enable operators of fresh e-commerce platforms to reduce cold chain logistics costs and improve cold chain logistics efficiency has become a hot issue in recent years [3]. Therefore, this article is from the perspective of ant colony algorithm, the distribution path of fresh e-commerce graduate students. This research theme is of epochal significance.

This article first studies the relevant theories of the logistics and distribution of fresh agricultural products, and secondly studies the logistics and distribution of fresh food e-commerce. The final research is the vehicle path planning, and the path planning is carried out from the ant colony algorithm, and finally a questionnaire survey is conducted to draw conclusions.

2 Fresh Food e-Commerce Distribution Path Planning Based on Ant Colony Algorithm

2.1 Logistics and Distribution of Fresh Agricultural Products

Fruits, vegetables, poultry eggs, meat products and other fresh foods are rich in nutrients and can provide people with important vitamins, protein and fiber that have a direct impact on human health. As an important source of human nutrition, fresh food has become an indispensable part of people's nutritional life. It is becoming more and more important on people's dining table, and its share is also increasing [4, 5].

(1) Importance of fresh produce.

People are accustomed to collectively refer to fresh fruits and vegetables, fresh meat and freshwater products as "three fresh". In fact, in addition to the above three types of fresh products, fresh agricultural products also include other primary forms of agricultural products such as flowers, eggs, and milk. Most of these types of fresh produce are produced through planting or direct selection [6, 7].

(2) Circulation characteristics of fresh agricultural products

1) Fresh agricultural products are perishable. This feature of fresh agricultural products requires that its circulation and logistics are different from ordinary products, and the circulation can only run smoothly under a certain environment [8, 9].
2) The logistics and distribution of fresh agricultural products must be fresh. Fresh agricultural products are characterized by long production cycles, strict storage conditions, and large losses in circulation. This has created a situation where the added value of fresh agricultural products is large, and freshness is an important indicator to measure the value of fresh agricultural products. This requires companies that provide fresh agricultural products sales services to ensure the quality, safety and freshness of fresh agricultural products [10, 11].

3) The logistics and distribution of fresh agricultural products are subject to time constraints. The distribution of fresh agricultural products is mainly for chain supermarkets, commercial supermarkets, fruit and vegetable stores, school canteens, hotels and other places [12].
4) The logistics distribution network of fresh agricultural products is scattered. There are many distribution networks for fresh agricultural products, which are distributed in every corner of the city and surrounding areas, which sometimes brings difficulties to distribution. Scientific planning of distribution routes and reasonable planning of distribution points are difficult problems faced by enterprises specializing in distribution services.

2.2 The Logistics and Distribution of Fresh Food E-Commerce

The fresh food e-commerce distribution model refers to the organization of transportation, sorting, loading and unloading, and transshipment, and under certain time and conditions, through logistics nodes, fresh products are delivered to consumers. The manufacturer (or supplier) delivers the goods directly to the supermarket, and then is sorted by the distribution center, that is, the last link is obtained and sent to the end user to complete the entire process. The e-commerce distribution model of fresh groceries usually has three methods: internal operations, third-party logistics and outsourcing. Internal construction, that is, the company builds one or more warehouses by itself. Build or lease these warehouses to companies that need to transport or deliver goods. They are responsible for the production and processing of products. At present, most consumers in China have a positive attitude towards online shopping, but there are also problems such as lack of supply and high damage to goods. Express companies need to collect payment or deliver goods to designated locations.

2.3 Vehicle Path Planning

(1) Path problem

The vehicle routing problem is also called the vehicle scheduling problem. The problem of vehicle routing with time windows stems from the strict requirements of customers on arrival time in actual logistics and distribution. The problem of the difficult time window is related to the strict time window set by the customer for the maintenance of the vehicle. No advance or delay is allowed. The problem of soft time window introduces a penalty function, so if no time window is provided, additional logistics costs will be incurred. The time window can increase the delivery speed, thereby increasing the customer's satisfaction with the logistics service.

(2) Classification of vehicle routing problems

According to the type of restriction, it can be divided into capacity restriction, time period restriction and mileage restriction. According to the characteristics of distribution tasks, it is divided into one-time delivery, one-time delivery and integration of delivery. According to the charging situation of the vehicle, there are full problems and incomplete charging problems. According to the number of parking spaces involved, it is divided into a single parking problem and multiple parking problems. According to the type of transportation vehicle, the problem is divided into single model and multiple model. According to the number of optimization goals, it is divided into single-objective problems and multi-objective problems. According to whether you want to return after completing the distribution task, it is divided into closed questions and open questions. According to whether the preconditions and boundary conditions are specified, it is divided into static problems and dynamic problems.

(3) Problem description of fresh food distribution channels

The lack of delivery time constraints for delivery vehicles and the lack of standardization of delivery route planning are two major delivery problems.

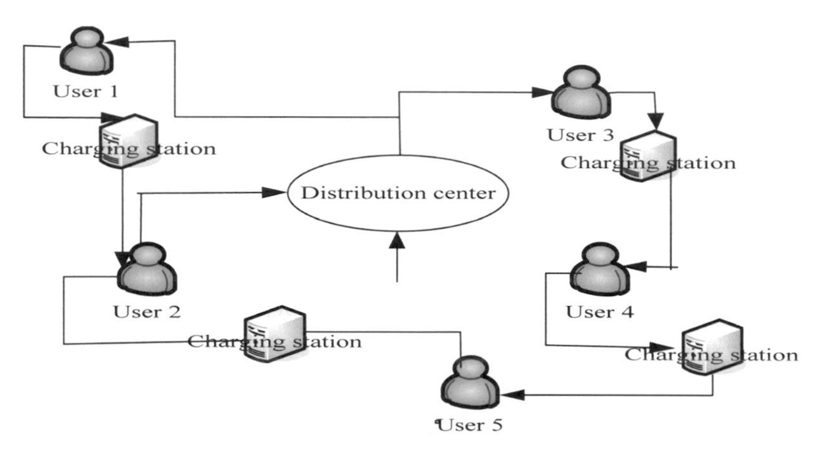

Fig. 1. Fresh food distribution electric refrigerated truck path

In a distribution network composed of a cold chain distribution center and multiple customer demand points, the number of customers to be served, the distribution location of the customer point, the specific needs and time of the customer, the arrival of the goods, the arrangement of electric refrigerated vehicles, and the customer Provide fresh groceries. As part of the vehicle carrying capacity, electric refrigerated trucks are driven to each customer point one by one, and each customer point is only served once. The flexible time window limit includes the customer-specified receiving satisfaction time and the final acceptable receiving time. Its existence can ensure that the time when the vehicle arrives at the customer point is not delayed, and can improve customer satisfaction. If the remaining vehicle power is not enough to reach the next customer point, the electric car must drive into a nearby charging station to charge, and then fully charge. Charging time affects the time required for the vehicle to reach the remaining customer points. Fresh food distribution route planning must not only cover all customer needs, but also consider practical factors such as battery capacity, charging time, air-conditioning, and time windows, and reasonably optimize the number of electric refrigerated vehicles, distribution routes and delivery orders. Obtain a plan plan for the minimum sum of fixed transportation costs, travel costs, cooling costs, window fines, and freight damage costs. The details are shown in Fig. 1.

Fresh food distribution has special conditions such as traffic, vehicles, weather, etc., but this is a typical NP problem, because the timetable problem of electric refrigerated trucks is extremely difficult to solve. To make the solution simpler, this article makes the following assumptions when planning the route of electric refrigerated trucks to deliver fresh produce:

Assume that the customer's location, demand, and the location of the charging station are known. Assume that the customer's demand does not exceed the maximum load capacity of the electric refrigerated truck and cannot be split. Assume that the power consumption of the electric refrigerated truck is proportional to the distance traveled. Assume that electric of the refrigerated truck will be delivered after leaving or visiting the central charging station, and it will have the maximum output. Assuming that the charging capacity of the electric refrigerated truck is proportional to the charging time.

(4) Cost

1) Freight and travel expenses. The cost of transporting standard vehicles is a fixed constant and has no direct relationship with the number of customers and the number of kilometers traveled. They usually include buying or renting vehicles, routine maintenance, and staff costs.
2) Cooling costs. In order to keep the transportation of fresh food at low temperature, electric refrigerated trucks have to continue to operate the refrigeration unit, which consumes the electric energy of the electric refrigerated trucks, resulting in refrigeration costs. Fresh food is perishable, and refrigerant must be used to keep the delivery temperature low.
3) Cost of time period. If the refrigerated truck arrives too early, you must wait until the customer is satisfied with the time window. Customers, thus define a smaller penalty coefficient. However, if the electric refrigerated truck arrives in the last hour after the expiration of the customer satisfaction window, the customer can accept the delivery and start work immediately; wait or deliver, reducing customer satisfaction or increase the new cost of sales, that is, set a higher penalty coefficient.
4) The cost of damage to the goods. In the process of distribution and transportation of fresh food, fresh food deteriorates slowly in the refrigerated truck compartment

(5) Path planning algorithm

1) The basic model of ant colony algorithm

When implementing the ant colony algorithm, the following variables and constants are used: the number of ants, reflecting the total amount of pheromone released by the ants, the force of the edge arc trajectory, the probability of ants transition, and the total path and length of the ants. The trajectory intensity update equation of the ant colony algorithm is:

$$\ell_{mn}(s+x) = (1-l)\cdot \ell_{mn}(s) + \sum_{i=1}^{a} \Delta\ell_{mn}^{i}(s) \tag{1}$$

$$\Delta\ell_{mn}(s) = \sum_{i=1}^{a} \Delta\ell_{mn}(s) \tag{2}$$

Among them, ℓ is the track intensity, and l is the pheromone volatilization coefficient. Ant colony algorithm can form different types of ant colony algorithm models according to the different update methods of pheromone traces, which are ant week model, ant density model and ant population model.

(6) The basic process of ant colony algorithm

1) Initialize parameters and set the maximum number of iterations
2) Place the ants at the starting point
3) Let the ant taboo table index number L = 1
4) Calculate the transition probability
5) Choose the maximum distance and move the ant to that place
6) If $l < x$, then $l = l + 1$; otherwise, go to step 7
7) Update the pheromone on the path according to formula (1)
8) Determine whether the maximum number of iterations has been reached.

(7) Path planning based on ant colony algorithm

In order to increase the visibility of the nodes to be selected, improve the predictability of the algorithm, and speed up the convergence speed of the algorithm, it is essential to introduce heuristic factors into the ant colony algorithm. The state transition rule of ant colony algorithm applied to path planning problem is similar to the algorithm transition rule applied to TSP problem. The possibility of ants from one node to another depends mainly on the information heuristic and expectation heuristic. The method of space planning is the grid method. After the space is divided by the grid method, the number of each grid is calculated according to the grid size, start point, end point, and coordinates. If all the ants start from the starting point and reach the end point, after completing a path search, the pheromone of the route taken by the ants needs to be updated.

3 Questionnaire Survey

3.1 Investigation Background

With the rapid development of the economy and the Internet, consumers have higher and higher expectations for the logistics speed, food quality and nutrition of fresh food, but the current development of fresh food distribution cannot meet the demands of consumers. Fresh products are not easy to store, easy to damage, and relatively high delivery requirements: some of the characteristics of fresh products themselves lead to higher requirements for storage, transportation, and distribution. In addition, e-commerce orders are small and regionally distributed. Nowadays, fast-paced modern life has put forward higher requirements on the quality of fresh products and delivery time. How to effectively ensure the quality of fresh products and meet the time needs of customers, and improve customer satisfaction has become the logistics and distribution service of fresh food e-commerce. One of the key considerations.

3.2 Questionnaire Design

This questionnaire survey was mainly launched around the path planning of fresh products e-commerce distribution. The content of the questionnaire includes the following aspects:

(1) The residents' familiarity with the fresh food market
(2) Residents' understanding of fresh food distribution
(3) Residents' use of fresh food e-commerce
(4) Residents' suggestions on the delivery method of fresh food
(5) Residents' choice of the path of fresh products

3.3 Questionnaire Process

A total of 200 copies of the questionnaire were distributed and 50 people were interviewed on the spot. 180 questionnaires were effectively collected, and the effective response rate reached 92%. The questionnaire is distributed locally, and the questionnaire is filled out offline. The questionnaire survey lasted for a week. After collecting the questionnaire, the computer was used to sort out the data and analyze it.

4 Investigation and Analysis

4.1 Fresh Product Distribution Countermeasures

According to the results of the questionnaire survey, this paper selects fresh food distribution strategies for analysis. In the questionnaire, citizens have four attitudes toward fresh food delivery: familiar, unfamiliar, general, and unaware, and their support for specific countermeasures is shown in Table 1:

Table 1. Residents' opinions on the countermeasures for the distribution of fresh products

	Strengthen distribution cost control	Enhance the ability to respond to emergencies	Strengthen the informatization construction	Improve the cold chain logistics	Improve relevant regulations and supervision
Familiar	12	15	13	14	9
Generally	11	16	10	12	11
Unfamiliar	13	11	9	13	8
Don't know	9	10	11	12	11

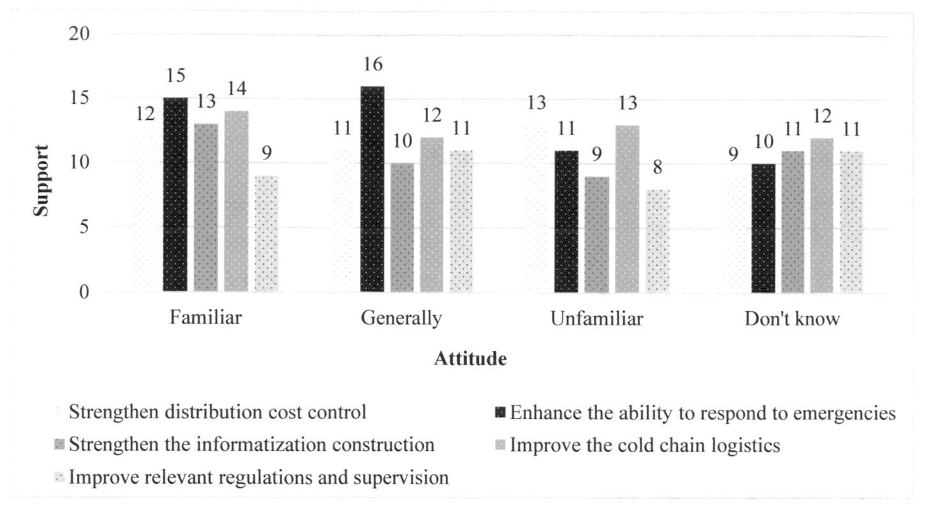

Fig. 2. Residents' opinions on the countermeasures for the distribution of fresh products

As shown in Fig. 2, we can see that in the countermeasures, people's support rates for the five countermeasures are similar. Among them, 45 people think it is important to strengthen the control of distribution costs, 52 people think it is important to strengthen the ability to respond to emergencies, and 43 people think it is important to strengthen the information construction of the cold chain industry. In addition, 51 people paid attention to improving the cold chain logistics industry chain, and 39 people paid attention to improving the relevant laws and regulations and supervision of the cold chain industry.

5 Conclusion

Fresh food e-commerce uses the Internet as a platform to transform the traditional food sales model into an e-commerce model. The distribution model of fresh food e-commerce platform is to transport fruits and vegetables from the place of production to consumers. Fresh food e-commerce has broad development prospects, but there are also some problems in its distribution path. Because fresh products have the characteristics of perishable, strong seasonality and so on. Makes cold chain transportation an indispensable link. For the planning of fresh food routes, various factors need to be taken into consideration, such as cost, coping ability, informationized data analysis, etc., all need to be improved. Route planning is inseparable from aspects such as economy, time and effort.

References

1. Lazarowska, A.: Discrete artificial potential field approach to mobile robot path planning – science direct. IFAC-PapersOnLine **52**(8), 277–282 (2019)
2. Changshi, L., Xiancheng, Z., Huyi, S., et al.: Research on TDVRPTW of fresh food e-commerce distribution: based on the perspective of both economic cost and environmental cost. Control Dec. **5**, 1273–1280 (2020)

3. Qian, Z., Wuyi, Z., Rattapon, I.: Research on the Optimization of Cold Chain Logistics Path in E-commerce Environment. Special Economic Zone, **000**(011), 103–105 (2018)
4. Santos, A.M., Lima, R.D., Pereira, C.S., et al.: Optimizing routing and tower spotting of electricity transmission lines: an integration of geographical data and engineering aspects into decision-making. Electric Power Syst. Res. **176**, 105953.1–105953.12 (2019)
5. Mutar, M.L., Aboobaider, B.M., Hameed, A.S., et al.: Enhancing solutions of capacity vehicle routing problem based on an improvement ant colony system algorithm. J. Adv. Res. Dyn. Control Syst. **11**(1), 1362–1374 (2019)
6. Amina, I.K., Bayda, A.K.: A new algorithm to estimate the parameters of log-logistic distribution based on the survival functions. J. Phys. Conf. Ser. **1879**(3), 032037 (2021). (8pp)
7. Izah, R.N., Kurnia, A., Sartono, B.: Classification of paddy growth phase based on landsat-8 image with convolutional neural network algorithm. J. Phys. Conf. Ser. **1863**(1), 012074 (2021) (6pp)
8. Sadiq, A.T., Raheem, F.A., Abbas, N.: Ant colony algorithm improvement for robot arm path planning optimization based on D* strategy. Int. J. Mech. Mechatron. Eng. **21**(1), 96–111 (2021)
9. Raheem, F.A., Abdulkareem, M.I.: Development of path planning algorithm using probabilistic roadmap based on modified ant colony optimization. World J. Eng. Technol. **07**(4), 583–597 (2019)
10. Deolia, V.K., Sharma, A.: Optimal path planning approach for unmanned vehicles using modified ant colony algorithm. J. Adv. Res. Dyn. Control Syst. **11**(11-SPECIAL ISSUE), 266–270 (2019)
11. Saad, M., Salameh, A.I., Abdallah, S., et al.: A Composite metric routing approach for energy-efficient shortest path planning on natural terrains. Appl. Sci. **11**(15), 6939 (2021)
12. Amar, L.B., Jasim, W.M.: Hybrid metaheuristic approach for robot path planning in dynamic environment. Bull. Electr. Eng. Inform. **10**(4), 2152–2162 (2021)

SCA-ECPE: Emotion-Cause Pair Extraction Based on Sentiment Clustering Analysis

Xin Xu[1,2], Houyue Wu[1,2], and Guangli Zhu[1,2](✉)

[1] School of Computer Science and Engineering, Anhui University of Science and Technology, Huainan 232001, China
glzhu@aust.edu.cn

[2] Institute of Artificial Intelligence, Hefei Comprehensive National Science Center, Hefei 230088, China

Abstract. As a branch task of cause-effect extraction, ECPE (Emotion-Cause Pair Extraction) focuses on analyzing the causes of emotional effects. The existing models are usually directly extracted by neural networks, resulting in low recognition accuracy of emotion clauses. To further improve the extraction accuracy, we proposed an ECPE extraction model named **SCA-ECPE**. Firstly, the k-means clustering method is used to execute a separate cluster analysis of emotion clauses to obtain prior knowledge. Then, the clauses of the original document are classified and matched, and then filtered based on prior knowledge. Finally, the probability is calculated by a multi-layer neural network to extract the emotion-cause pairs. Experimental results show that prior sentimental knowledge can effectively enhance extraction performance.

Keywords: ECPE · Sentiment analysis · Clustering algorithm · Prior knowledge

1 Introduction

As a narrow cause-effect extraction task, ECPE [1] has important research value and is widely used in case reasoning, risk prediction, and other fields. Due to the diversity of emotional information contained in documents, which makes implicit emotional semantic information difficult to identify. Implicit emotional semantic information can further support the accuracy improvement of ECPE, but the existing models have few studies in this respect.

In the field of emotional analysis, some of our research has been published [2, 3]. Similar to the BERT [4] providing prior knowledge for downstream NLP tasks, we use the emotion analysis model to provide technical support for ECPE. To obtain prior knowledge of emotion, two issues need to be considered:(1) How to divide emotion clauses in the dataset and determine the standard of division; (2) How to quantify the divided emotion clauses for subsequent ECPE extraction process.

Based on the above discussion, we propose an ECPE model based on sentiment clustering analysis, named **SCA-ECPE**. The final EC-Pairs are obtained from three

J. H. Abawajy et al. (Eds.): ICATCI 2022, LNDECT 169, pp. 220–228, 2023.
https://doi.org/10.1007/978-3-031-28893-7_27

steps: list raw EC-Pairs, clean EC-Pairs through prior knowledge, and rank cleaned EC-Pairs to extract the ground-truth EC-Pairs. The workflow of SCA-ECPE is shown in Fig. 1.

1) **List Raw EC-Pairs**. At the initial step, each clause in the document may be an emotion clause or a cause clause. In order to avoid missing information, we need to list all possible EC-Pairs.
2) **Obtain Cleaned EC-Pairs Through Prior Knowledge**. For the raw EC-Pairs listed in the previous step, the prior knowledge obtained by SCA is used to clean up the combination with low possibility.
3) **Rank and Extract the Ground-Truth EC-Pairs**. The cleaned EC-Pairs are ranked by a neural network classifier. The highest probability EC-Pair will be regarded as the final extraction result of the document.

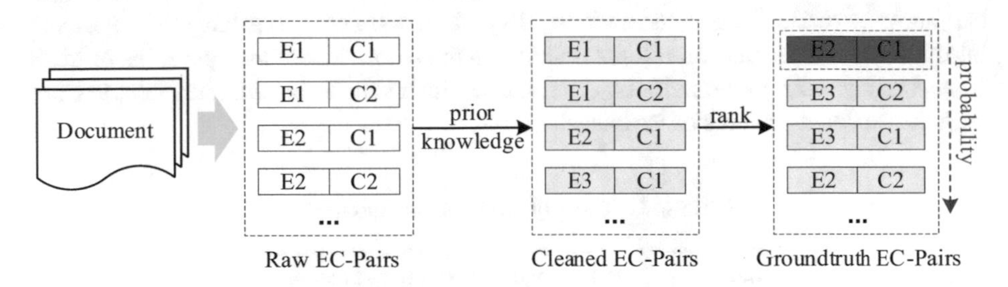

Fig. 1. The workflow of SCA-ECPE

Compared with the classical ECPE model, the SCA-ECPE model obtains prior knowledge through sentiment clustering analysis, which improves the accuracy of subsequent extraction. The advantage of SCA is that it can provide emotional evaluation criteria for EC-Pair and enhance the learning ability of the model in emotional information.

The following is organized as follows: Sect. 2 introduces the relevant research of ECPE. Section 3 illustrates the details of SCA-ECPE. The experimental discussion is in Sect. 4, and the summary and prospect are carried out in Sect. 5.

2 Related Work

As the pre-task of ECPE, ECE was first proposed by Lee [5]. ECE extracts cause clauses from the document based on the words provided that contain emotional attributes. Study [6] identified the different clause relations in the document and based on this, proposed a recurrent neural network extraction model RTHN. Different from ECE, ECPE extracts cause through emotion clauses, which makes ECPE application more feasible in the real scene.

Xia et al. proposed the basic two-step extraction model after improving the ECE task [1]. In order to learn the internal relationship between emotion clauses and cause clauses, the author divides the two-step model into two seed models, Inter-CE and Inter-EC. Inter-CE uses the information of emotion clauses for the classification of cause clauses, which is opposite to Inter-EC. Study [7] proposed an end-to-end ECPE network, which can avoid error propagation in a two-step network. In [8], Tang et al. combined the possible EC-Pairs and then used the semantic information between pairs learned from the attention network to filter.

Considering the task characteristics of ECPE, it is helpful to improve the accuracy of ECPE by learning emotional prior knowledge. Therefore, we propose the SCA-ECPE model to learn the sentiment distribution of the ECPE dataset.

3 Method

3.1 Sentiment Clustering Analysis

The emotional attributes of the ECPE dataset are divided into six categories, including happiness, surprise, fear, anger, sadness, disgust. The six emotions have differences in intuition and have a clear emotional polarity. In the word embedding space, the original dataset contains six emotional subspaces, and sentiment clustering analysis helps to learn high dimensional features of various emotions.

Table 1. The proportion of six emotions

Emotion	Percentage	Number of clauses
Sadness	26.94%	567
Happiness	25.83%	544
Fear	18.00%	379
Anger	14.35%	302
Disgust	10.69%	225
Surprise	4.18%	88

The process of sentiment clustering analysis can be divided into three steps:

(1) Build a new sub-dataset based on the original benchmark dataset, which contains only annotated emotion clauses. The proportion of six emotions is shown in Table 1.
(2) Introducing the sentiment analysis model in [3], each emotion clause is mapped to the emotional semantic space to prepare for the subsequent clustering analysis.
(3) The emotion vectors mapped in the previous step are obtained by the text clustering method (k-means, k = 6) to obtain six emotional subspaces, as shown in Fig. 2. The radius and spatial center of the six emotional subspaces will be used for the subsequent EC-Pairs cleaning process.

Figures 2 and 3 more intuitively describe the distribution of six emotional words in the benchmark dataset. It can be seen from Figs. 2 and 3 that sadness and happiness

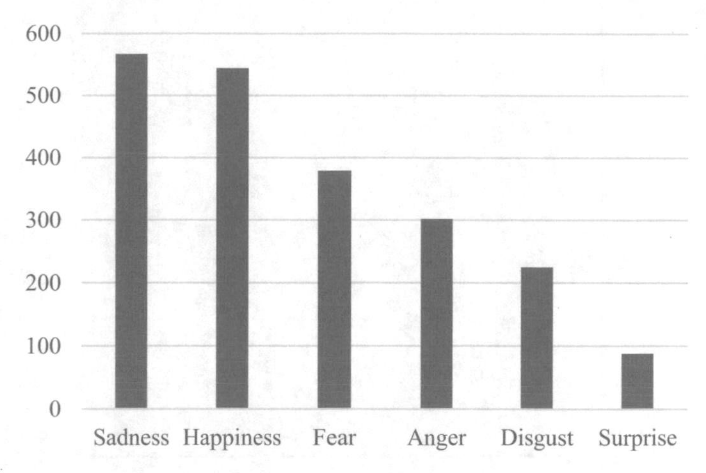

Fig. 2. The number of six emotions

appear most frequently, and these two emotional words have obvious emotional polarity differences.

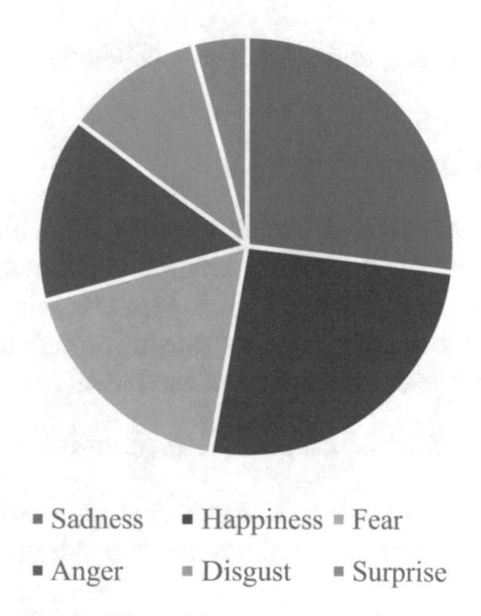

Fig. 3. The percentage of six emotions

In formula 1, $\overline{D}$ represents the sub-dataset constructed in step 1 that contains only emotion clauses. Through k-means based sentiment clustering analysis, the model will generate a series of emotional subspaces, denoted as $Espace_k$ ($k = 1, 2, \ldots, 6$). In formula 2, χ represents the analytic function of emotional subspace, and the spatial radius rad_k, central coordinates $coord_k$ and embedding representation of each subspace e_k are calculated by this function. Figure 4 abstractly shows the distribution characteristics of emotional subspace.

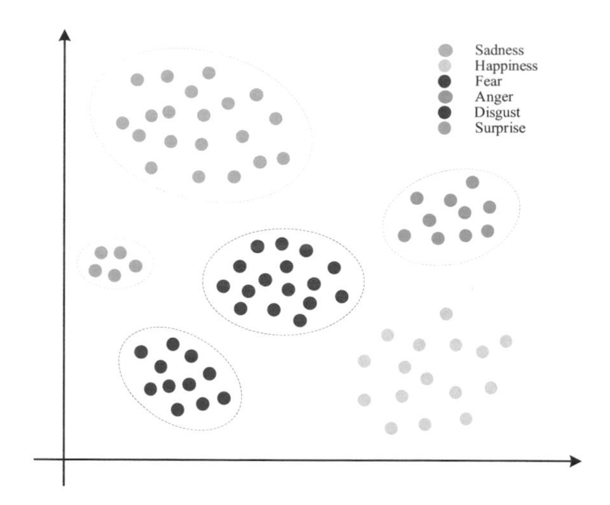

Fig. 4. The emotional subspaces in the dataset

$$Espace_k = SCA\left(k - means(\overline{D})\right) \tag{1}$$

$$\{rad_k, coord_k, e_k\} = \chi(Espace_k) \tag{2}$$

3.2 Clean and Rank EC-Pairs

To ensure the integrity of the original semantic information, all possible EC-Pairs need to be constructed. Assuming the number of clauses contained in a document is *m*, the number of all possible EC-Pairs should be m^2. However, the time complexity of this approach is high. Using the emotional subspace information obtained by SCA as a priori knowledge can improve the efficiency of the model.

$$Econfidence_i = Dis(e_i, coord_k) \tag{3}$$

In formula 3, $Econfidence_i$ (i = 1, 2, ..., m^2) is obtained by *Dis* (Euclidean distance calculation function), where e_i represents the embedding representation of emotion clause in the *i-th* EC-Pair.

In the cleaning process of EC-Pairs, $Econfidence_i$ is an important evaluation index. The network will calculate the distance between e_i and six emotional subspaces in turn. If all *Econfidence* are smaller than the corresponding rad_k, the EC-Pair where e_i is located will be eliminated.

The cleaned EC-Pairs have a high probability of containing emotion clauses. We embed two clauses in each EC-Pair into a full connection layer, and finally extract the highest ranking EC-Pair as the extraction result. The loss of the model is the classification loss for each cleaned EC-Pair:

$$Loss = \frac{1}{N}\sum_{j=1}^{N} CrossEntropy(y_j, \hat{y}_j) \tag{4}$$

where y_j is the ground-truth result, $\hat{y}_j$ is the predicted result, N is the total number of all cleaned EC-Pairs.

4 Experiments and Analysis

4.1 Dataset of ECPE

The ECE dataset [9] is the basis of the ECPE dataset, which consists of more than 20,000 news documents from SINA NEWS website[1]. Study [1] extends the dataset from word-level to clause-level, increasing the proportion of emotional information in extraction. The information of the benchmark ECPE dataset is listed in Table 2.

It can be seen from Table 2 that most documents contain only one EC-Pair, and about 10% of documents contain multiple EC-Pairs. Since the causal situation in the network news documents is relatively simple, it reduces the difficulty of extraction.

Table 2. Dataset of ECPE

	Percentage	Number
Number of EC-Pair = 1	89.77%	1746
Number of EC-Pair = 2	9.10%	177
Number of EC-Pair in Document >2	1.13%	22
Total	100%	1945

4.2 Experimental Discussion

In order to verify the effectiveness of SCA-ECPE, we conducted multiple sets of comparative experiments on the benchmark dataset. The experimental results are shown in Table 3.

The comparison models are **Baselines** [1], **RANKCP** [10], **E2EECPE** [7], **DQAN** [11], **Trans-ECPE** [12]. The comparison model has not only two-step model, but also end-to-end model.

The following conclusion can be summarized through the data in Table 3: Compared with other models, **SCA-ECPE** achieves **the highest accuracy and F1 score**. This proves that prior sentimental knowledge is helpful to the extraction performance. In the process of constructing emotional subspace by clustering, the higher the dimension of word embedding, the more obvious the discrimination between subspaces. The pre-training mode of the original dataset can provide more semantic information, which is helpful in the NLP tasks.

Figure 5 shows the comparison result of **SCA-ECPE** and pipeline model. It can be seen that the model has improved in all aspects, especially in F1 value. The **Baselines**

[1] https://news.sina.com.cn/.

Table 3. The statistics of extraction results on the benchmark dataset

Model	EC-Pair Extraction		
	P (%)	R (%)	F1 (%)
Baselines (Indep)	68.32	50.82	58.18
Baselines (Inter-CE)	69.02	51.35	59.01
Baselines (Inter-EC)	67.21	57.05	61.28
RANKCP	66.10	**66.98**	65.46
E2EECPE	64.78	61.05	62.80
DQAN	67.33	60.40	63.62
Trans-ECPE (LSTM)	65.15	63.54	64.34
SCA-ECPE	**70.19**	64.15	**67.03**

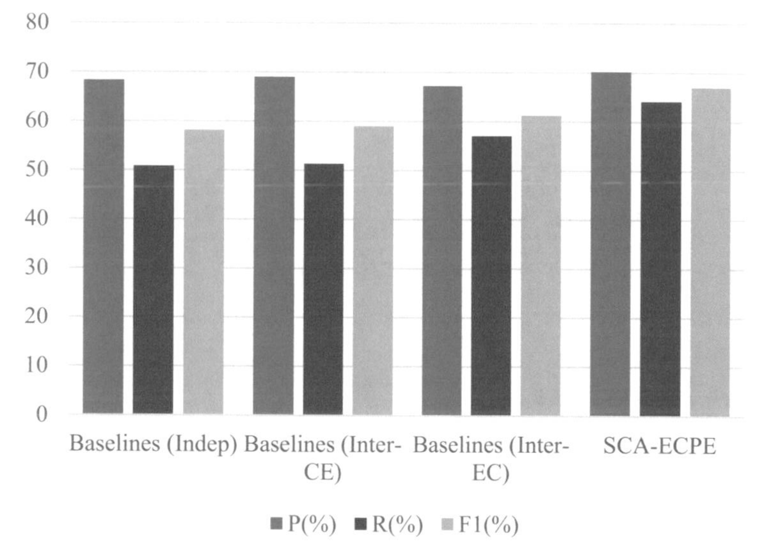

Fig. 5. Comparison results with pipeline model

model is limited by the depth of semantic understanding and cannot break through the bottleneck of accuracy, so the focus of comparison is the end-to-end model.

Figure 6 shows the comparison results between **SCA-ECPE** and end-to-end models. It can be seen that the model has improved in all aspects, especially in precision. In terms of recall rate, RANKCP is superior to the **SCA-ECPE** model due to its unique ranking scoring mechanism. In addition, our model is superior to the comparison model.

In general, **SCA-ECPE** has achieved good results in both pipeline model and end-to-end comparison models. This shows that the analysis of ECPE dataset is correct and effective.

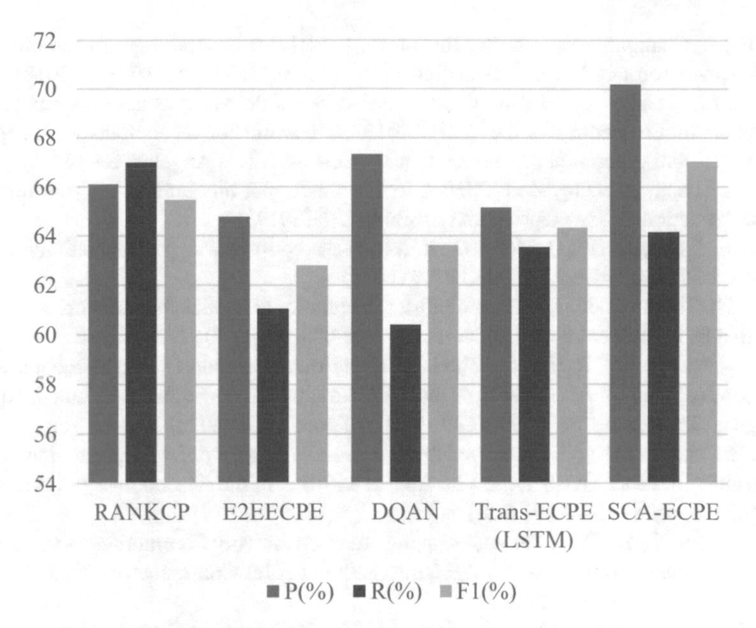

Fig. 6. Comparison results with end-to-end model

5 Conclusion

This paper proposes the **SCA-ECPE** model based on prior knowledge for the ECPE task, which **improves the extraction accuracy**. The ECPE task can provide technical support for emotional reasoning applications. As a kind of pre-learning information, prior knowledge can be used to improve performance bottlenecks in special fields. Taking practical application as a starting point, ECPE can mine more information at the data level. In the future, our research will involve the expansion and further learning of ECPE dataset.

Acknowledgements. This work was supported by the National Natural Science Foundation of China (Grant NO. 62076006), the University Synergy Innovation Program of Anhui Province (GXXT-2021–008), and the Anhui Provincial Key R&D Program (202004b11020029).

References

1. Xia, R., Ding, Z.: Emotion-cause pair extraction: a new task to emotion analysis in texts. In: Proceedings of the 57th Annual Meeting of the Association for Computational Linguistics, pp. 1003–1012. Florence, Italy (2019)
2. Zhang, S., Hu, Z., Zhu, G., Jin, M., Li, K.-C.: Sentiment classification model for Chinese micro-blog comments based on key sentences extraction. Soft. Comput. **25**(1), 463–476 (2020). https://doi.org/10.1007/s00500-020-05160-8
3. Zhang, S., Yu, H., Zhu, G.: An emotional classification method of Chinese short comment text based on ELECTRA. Connect. Sci. **34**(1), 254–273 (2022)

4. Devlin, J., Chang, M.W., Lee, K., Toutanova, K.: BERT: pre-training of deep bidirectional transformers for language understanding. arXiv preprint arXiv:1810.04805 (2018)
5. Lee, S.Y.M., Chen, Y., Huang, C.R.: A text-driven rule-based system for emotion cause detection. In: Proceedings of the NAACL HLT 2010 workshop on computational approaches to analysis and generation of emotion in text, pp. 45–53. Los Angeles, CA (2010)
6. Xia, R., Zhang, M., Ding, Z.: RTHN: A RNN-transformer hierarchical network for emotion cause extraction. arXiv preprint arXiv:1906.01236 (2019)
7. Song, H., Zhang, C., Li, Q., Song, D.: End-to-end emotion-cause pair extraction via learning to link. arXiv preprint arXiv:2002.10710 (2020)
8. Tang, H., Ji, D., Zhou, Q.: Joint multi-level attentional model for emotion detection and emotion-cause pair extraction. Neurocomputing **409**, 329–340 (2020)
9. Gui, L., Wu, D., Xu, R., Lu, Q., Zhou, Y.: Event-driven emotion cause extraction with corpus construction. In: Proceedings of the 2016 Conference on Empirical Methods in Natural Language Processing, pp. 1639–1649. Austin, Texas, USA (2016)
10. Wei, P., Zhao, J., Mao, W.: Effective inter-clause modeling for end-to-end emotion-cause pair extraction. In: Proceedings of the 58th Annual Meeting of the Association for Computational Linguistics, pp. 3171–3181 (2020)
11. Sun, Q., Yin, Y., Yu, H.: A dual-questioning attention network for emotion-cause pair extraction with context awareness. In: 2021 International Joint Conference on Neural Networks (IJCNN), pp. 1–8. IEEE (2021)
12. Fan, C., Yuan, C., Du, J., Gui, L., Yang, M., Xu, R.: Transition-based directed graph construction for emotion-cause pair extraction. In: Proceedings of the 58th Annual Meeting of the Association for Computational Linguistics, pp. 3707–3717 (2020)

CCRFs-NER: Named Entity Recognition Method Based on Cascaded Conditional Random Fields Oriented Chinese EMR

Xiaoqing Li[1,2], Zhengyan Sun[1,2], and Guangli Zhu[1,2](✉)

[1] School of Computer Science and Engineering, Anhui University of Science & Technology, Huainan 232001, China
glzhu@aust.edu.cn

[2] Institute of Artificial Intelligence, Hefei Comprehensive National Science Center, Hefei 230000, China

Abstract. Named Entity Recognition(NER) for Chinese Electronic Medical Record (EMR) is to identify medical entities and entity boundaries. Nested entities exist in Chinese electronic medical records, which makes NER low accuracy. To improve the recognition accuracy, this paper proposes a NER method, named CCRFs-NER. Firstly, the feature set is constructed by word features, part-of-speech features, and entity-identifier features. Then, Bi-LSTM and attention mechanism are used to process the feature set to extract the global and local features. Finally, the features are input into the CCRFs for probability prediction, and the recognition results are obtained. The experimental results show that the CCRFs-NER can improve the accuracy of named entity recognition for Chinese electronic medical records.

Keywords: Chinese EMR · Named Entity Recognition · Nested entities · Attention mechanism · CCRFs

1 Introduction

Named Entity Recognition of Chinese EMR is of great significance to the construction of medical knowledge graph and medical question answering system, etc.[1]. There are nested entities in Chinese EMR, mostly composed of nouns and basic disease words, and often accompanied by entity identifiers. However, the current NER research methods do not deeply capture the potential characteristics of nested entities, resulting in low recognition accuracy.

Regarding the research on nested entities, previous work proposed a span-based model and improved runtime performance [2–4]. The NER task can also be seen as a neural multi-classification task and the accuracy of nested entities can be improved by the sequence tags [5]. In [6, 7], our research group proposed that the recognition accuracy in product reviews is improved by concatenating multiple features to capture domain information and dependence.

J. H. Abawajy et al. (Eds.): ICATCI 2022, LNDECT 169, pp. 229–237, 2023.
https://doi.org/10.1007/978-3-031-28893-7_28

For nested entities in Chinese EMR, an effective NER method needs to consider the following two points: (1) How to construct the feature set by using the part-of-speech and entity-identifier features. (2) How to improve the accuracy of NER for Chinese EMR.

Based on the above considerations, this paper proposes a NER method for Chinese EMR, named CCRFs-NER. The motivation is to enrich the feature set by using features of part-of-speech and entity-identifiers. And the results are obtained by using CCRFs, to improve the recognition accuracy. The specific process is shown in Fig. 1.

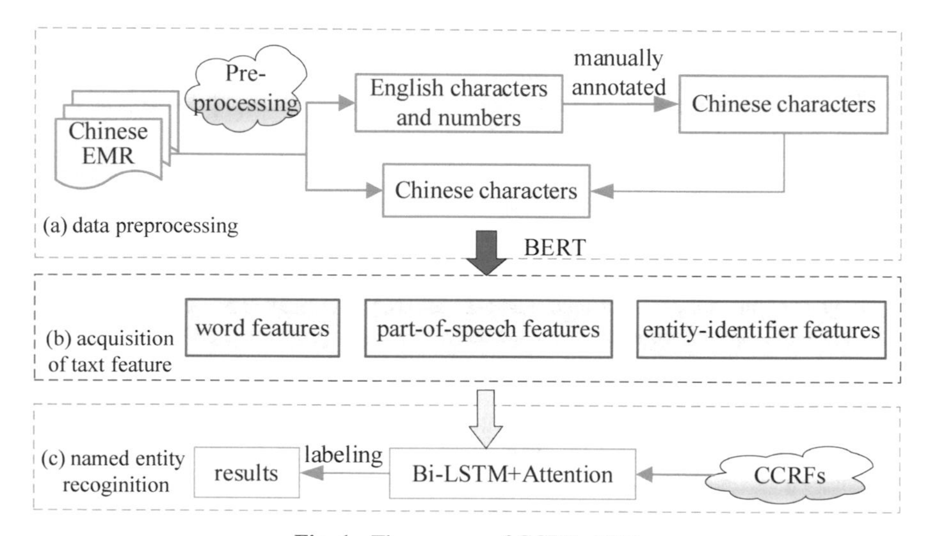

Fig. 1. The process of CCRFs-NER

Compared with the classical NER model, the CCRFs-NER model constructs the feature set by the pre-training model, which enriches the input information of the NER model. Moreover, the CCRFs model can fully consider the interdependence between adjacent labels, thus the optimal label sequence can be obtained.

This paper is organized as follows: Sect. 2 introduces the related work of NER. Section 3 illustrates the details of CCRFs-NER. The experimental discussion is in Sect. 4, and the summary and prospect are carried out in Sect. 5.

2 Related Work

As early as 1994, Carol Friedman et al. developed a universal NLP machine, which can identify clinical medical information and express it in structured language[8]. Xu et al. developed MedEx to extract drug information from clinical notes and other texts[9]. With the deepening of sequence labeling research, the NER task can be regarded as a sequence labeling problem. The commonly used methods include the Hidden Markov Model[10] and the Conditional Random Field Model [11, 12].

With the hot research of neural networks, the application of neural network models in NER tasks can improve the accuracy. Study [13] matched the disease entities with the disease dictionary and proposed a method based on the Bi-LSTM and the CRF model. To identify clinical information, the Bi-LSTM model and the CRF model are used to make more extensive use of context information[14]. Tao et al. combined CRF model with word embedding method to extract drug names and prescription information from medical prescriptions[15]. Aras et al. used BiLSTM network and Transformer network to identify named entities and achieved good results[16].

Considering the data characteristics of Chinese EMR, it is helpful to improve the recognition accuracy by the features of part-of-speech and entity-identifiers. Therefore, a NER method based on CCRFs Oriented Chinese EMR is proposed.

3 Method

Chinese EMR text contains numbers, letters, and other characters, such as '80/min'. And there are nested entities, such as '先天性糖尿病(congenital diabetes)'. The text semantic representation can enrich the input information of the NER model and improve the accuracy, as shown in Tables 1 , 2. Before entity recognition, it is necessary to preprocess the text.

Table 1. Examples of nested entities

disease name	modifier	basic disease term
congenital heart disease	congenital	heart disease
congenital diabetes	congenital	diabetes
chronic bronchitis	chronic	bronchitis
(aplastic anemia	aplastic	anemia
Allergic Rhinitis	Allergic	Rhinitis

Desensitization treatment. EMR text contains a large number of private information related to patients, doctors, and hospitals. Desensitization processing is needed to protect patient privacy information.
Translation. The English words and digital symbols in Chinese EMR text are translated into Chinese, and the number is converted into Chinese characters in a semi-artificial way.
Word segmentation. Jieba word segmentation tool can obtain the corresponding part-of-speech while obtaining the text segmentation results.

4 Embedding Layer

For a given Chinese EMR text $S = (S1, S2...Si...Sn)$, $1 \leq i \leq n$, Si represents the ith sentence. $Si = (x1, x2...xj...xm)$, $1 \leq j \leq m$, xj denotes the word that consistutes of a sentence. $xj = (t1, t2...tk...tc)$, $1 \leq k \leq c$, where tk represents the characters that compose the words. For the input text, the sentence feature, word feature, and character feature can be obtained by formula (1–3).

$$sentence\_embeddings = BERT(S) \tag{1}$$

$$word\_embeddings = BERT(sentence) \tag{2}$$

$$character\_embeddings = BERT(word) \tag{3}$$

Part-of-speech and entity-identifiers can be regarded as words, so the feature representation can be obtained by formula (4–5).

$$part - of - speech\_embeddings = BERT(part - of - speech) \tag{4}$$

$$entity - identifier\_embeddings = BERT(identity - identifier) \tag{5}$$

The BERT model can be used to obtain the features of words, part-of-speech, and entity-identifiers. Thus, the feature set can be constructed.

4.1 Representation Layer

Bi-LSTM is composed of forward LSTM and backward LSTM, which can simultaneously obtain data features of future and past time. It can effectively solve the problem of gradient explosion and gradient disappearance. From the formula (6–8), the word, part-of-speech, and entity-identifiers can be expressed.

$$word\_hi = [\overrightarrow{h}i; \overleftarrow{h}i] \tag{6}$$

$$part - of - speech\_hi = [\overrightarrow{h}i; \overleftarrow{h}i] \tag{7}$$

$$entity - identifier\_hi = [\overrightarrow{h}i; \overleftarrow{h}i] \tag{8}$$

In the text data of EMR, different words have different importance for identifying the correct entity. The introduction of attention mechanism in NER task of Chinese EMR can give different degrees of attention to words. It can strengthen the feature weight related to entities and weaken the feature weight unrelated to entities, which is conducive to improving recognition accuracy.

4.2 Prediction Layer

Assuming that there are L state characteristic functions $E = (e1, e2...el)$, the weights are $u1, u2...ul$. And V state transition characteristic function $F = (f1, f2...fv)$, whose weight is $\lambda 1, \lambda 2 ... \lambda v$. For the input sequence $xj = (t1, t2...tk...tc)$, the output probability y can be calculated as formula(12).

$$P(A|S) = \frac{1}{Z(s)} exp(\sum_{i,v} \lambda vfv(A_{i-1}, Ai, S, i) + \sum_{i,l} ulsl(A_i, S, i)) \tag{9}$$

$$Z(S) = \sum_{A} exp(\sum_{i,v} \lambda vf_v(A_{i-1}, Ai, S, i) + \sum_{i,l} ulsl(A_i, S, i)) \tag{10}$$

$$P(Y|X) = \frac{1}{Z(x)} exp(\sum_{i,v} \lambda vfv(y_{i-1}, yi, X, i) + \sum_{i,l} ulsl(y_i, X, i)) \tag{11}$$

$$Z(X) = \sum_{x} exp(\sum_{i,v} \lambda vf_v(y_{i-1}, yi, X, i) + \sum_{i,l} ulsl(y_i, X, i)) \tag{12}$$

where, X can be obtained by formula(9–11).

The core idea of CCRFs-NER is to enrich the feature set by using part-of-speech and entity-identifiers. It is introduced into the CCRFs model to enrich the input information, and high accuracy recognition results can be obtained. Based on the recognition of basic disease terms, the nested entities can be identified to improve the recognition accuracy.

5 Experiments and Analysis

5.1 Experimental Method

To verify the effectiveness of the proposed method, this paper takes the data in Task 2 released by CCKS 2019 as the research object. The specific experimental operations are as follows.

Step1: Data preprocessing. Firstly, the data are densensitized. Then, all English words and numbers in the data set are translated into Chinese. Finally, jiaba tool is used to process the text to obtain the word segmentation results and the corresponding part-of-speech.

Step2: Feature acquisition. The EBRT model is used to obtain the features of word, part-of-speech, and entity-identifier respectively. Then, the feature set of the Chinese EMR is constructed.

Step3: The obtained feature set is introduced into the first layer of the Bi-LSTM + Attention + CRF framework to obtain entities. It includes basic disease, treatment, symptoms, medical check-ups, and body parts.

Step4: The basic disease words are combined with part-of-speech and entity-identifier to construct a new feature set. The new feature set is introduced into the second layer Bi-LSTM-Attention-CRF framework to obtain the final recognition results.

5.2 Annotation Method

The BIO sequence labeling method annotates the recognition results, as shown in table 2. Some part of speech tagging is shown in Table 2.

Table 2. The BIO sequence labeling method

sign	meaning	sign	meaning
D	Independent disease name	B-T	The beginning of the treatment name
B-D	The beginning of the disease name	I-T	The middle or end of the treatment name
I-D	The middle or end of the disease name	B-C	The beginning of the check name
BD	Basic disease words	I-C	The middle/end of the check name
B-BD	The beginning of the basic disease name	B-P	The beginning of the part name
I-BD	The middle or end of the basic disease name	I-P	The middle or end of the part name
B-S	The beginning of the symptom name	B-E	The beginning of the entity identifier
I-S	The middle or end of the symptom name	I-E	The middle or end of an entity identifier

Table 3. The experimental results of each model

Model	NER of Chinese EMR		
	P(%)	R(%)	F1(%)
Method 1 Baseline (Bi-LSTM + CRF)	84.28	85.78	85.02
Method 2 BERT + Bi-LSTM + CRF	85.20	85.99	**85.60**
Method 3 BERT + Bi-LSTM + CCRFs	85.39	86.49	85.94
Method 4 BERT + Attention + CRF	85.68	86.01	**85.84**
Method 5 BERT + Attention + CCRFs	85.98	86.67	86.32
Method 6 Bi-LSTM + Attention + CRF	86.69	87.39	**87.04**
Method 7 Bi-LSTM + Attention + CCRFs	86.90	87.89	87.40
Method 8 BERT + Bi-LSTM + Attention + CRF	87.67	88.21	**87.94**
Method 9 BERT + Bi-LSTM + Attention + CCRFs	88.90	88.52	**88.71**

5.3 Experimental Discussion

In order to verify the effectiveness of CCRFs-NER, several groups of comparative experiments are set and analyzed. The experimental results of each model are shown in Table 3.

It can be seen from Table 3 and Fig. 2–4 that the F1 value of the CCRFs-NER model reaches 88.71, which is about 3.69 percentage points higher than the baseline. An important reason for the improvement of accuracy is that the proposed model **enriches the feature set by using part-of-speech and entity-identifier**. Furthermore, the fusion of basic

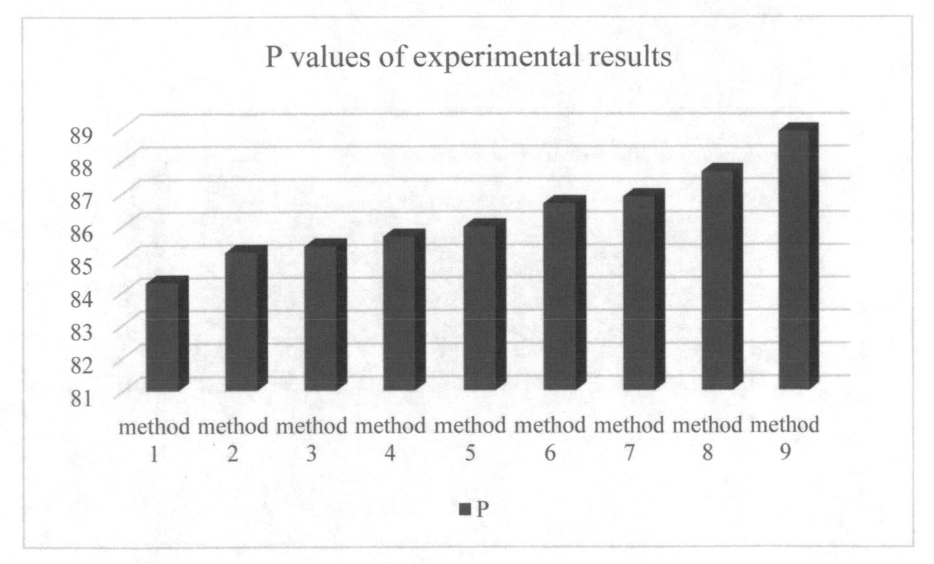

Fig. 2. The P values of experimental results

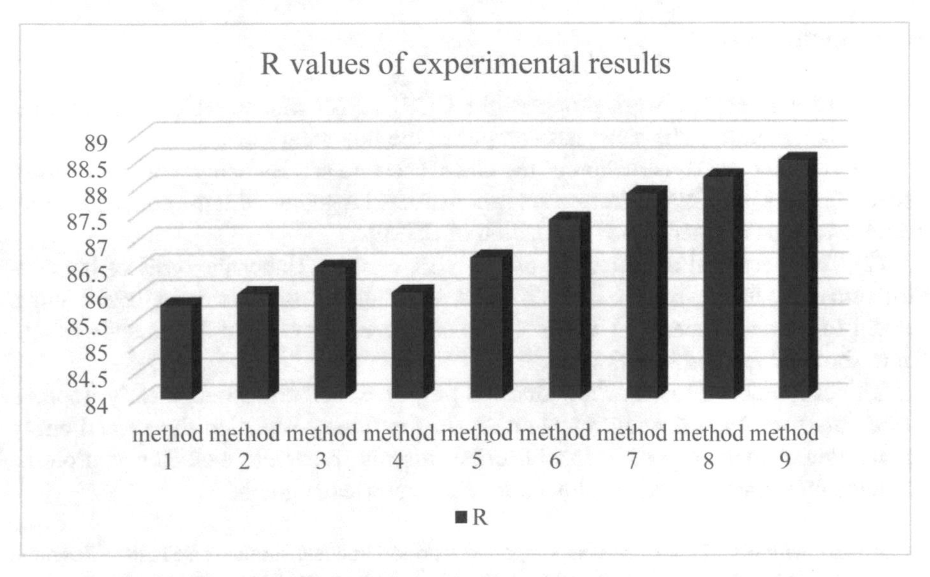

Fig. 3. The R values of experimental results

disease words with part-of-speech and entity-identifier enriches the input information of the CCRFs model.

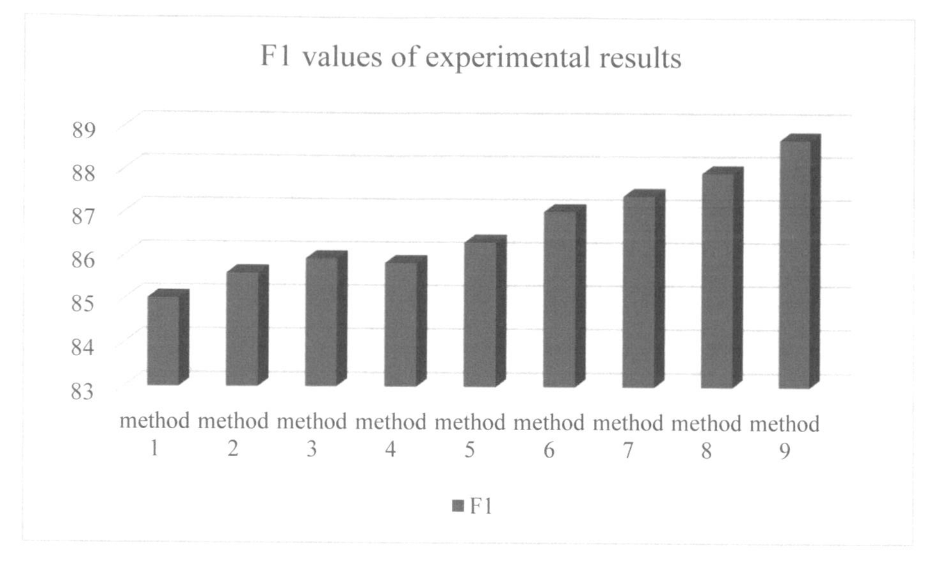

Fig. 4. The F1 values of experimental results

6 Conclusions

For nested entities, this work proposes the CCRFs-NER model, which improves the recognition accuracy. The paper has completed the following aspects:

(1) The effective construction of the CEMR feature set. By using word and part of speech features, the CEMR feature set is constructed together with the entity identifier features to mine rich semantic information of CEMR.

(2) The successful optimization of the NER method. Under the basis of the construction of the feature set, the Bi-LSTM and Attention are used to extract local features related to entities. Then, the CCRFs model can be used to obtain the named entities. Thus, the NER method is optimized.

The experimental results show that the proposed method can effectively improve the accuracy of Named entity recognition. In the future, based on the named entity recognition method proposed, we will further improve recognition efficiency, promote relationship extraction, and construct a medical knowledge graph.

Acknowledgements. This work was supported by the Graduate Students Scientific Research Program of Anhui Province(YJS20210402), the National Natural Science Foundation of China (Grant NO.62076006), the University Synergy Innovation Program of Anhui Province (GXXT-2021–008), and the Anhui Provincial Key R&D Program(202004b11020029).

References

1. Zhang, R., Zhao, P., Guo, W., Wang, R., Lu, W.: Medical Named Entity Recognition based on Dilated Convolutional Neural Network. Cognitive Robotics **2**, 13–20 (2022)
2. Li, F., Wang, Z., Hui, S., Liao, L., Zhu, X., Huang, H.: A segment enhanced span-based model for nested named entity recognition. Neurocomputing **465**, 26–37 (2021)
3. Li, D., Yan, L., Yang, J., Ma, Z.: Dependency Syntax Guided BERT-BILSTM-GAM-CRF for Chinese NER. Expert Syst. Appl. **196**(15), 116682 (2022)
4. Yu, G., et al.: Adversarial Active Learning for the Identification of Medical Concepts and Annotation Inconsistency. J. Biomed. Inform. **108**, 103481 (2021)
5. Jiang, D., Ren, H., Cai, Y., Xu, J., Liu, Y., Leung, H.: Candidate region aware nested named entity recognition. Neural Netw. **142**, 340–350 (2021)
6. Zhang, S., Zhu, H., Xu, H., Zhu, G.: A named entity recognition method towards product reviews based on BiLSTM-Attention-CRF 2021, Int. J. Comput. Sci. Eng. https://www.inderscience.com/info/ingeneral/forthcoming.php?jcode=ijcse (Accessed 4 March 2022)
7. Zhu, G., Liu, W., Zhang, S., Chen, X., Yi, C.: The method for extracting new login sentiment words from Chinese micro-blog based on improved mutual information. Int. J. Comput Syst Sci Eng **35**(3), 223–232 (2020)
8. Carol, F., Alderson, P., Austin, J., Cimino, J., Johnson, S.: A General Natural-language Text Processor for Clinical Radiology. J Am Med Inf. Assoc. **1**(2), 161–174 (1994)
9. Hua, X., Stenner, S., Doan, S., Johnson, K., Waitman, L., Denny, J.: MedEx: a Medication Information Extraction System for Clinical Narratives. J Am Med Inf Assoc **17**(1), 19–24 (2010)
10. Yan, X., Xiong, X., Cheng, X., Huang, Y., Zhu, H., Hu, F.: HMM-BiMM: Hidden Markov Model-based Word Segmentation Via Improved Bi-directional Maximal Matching Algorithm. Comput. Electr. Eng. **94**, 107354 (2021)
11. Sharma, R., Morwal, S., Agarwal, B.: Named Entity Recognition using Neural Language Model and CRF for Hindi Language. Comput. Speech Lang. **74**, 101356 (2021)
12. Gandhi, H., Attar, H.: Extracting Aspect Terms using CRF and Bi-LSTM Models. Proc Comput Sci **167**, 2486–2495 (2021)
13. Xu, K., Yang, Z., Kang, P., Wang, Q., Liu, W.: Document-level Attention-based BiLSTM-CRF Incorporating Disease Dictionary for Disease Named Entity Recognition. Comput. Biol. Med. **108**, 122–132 (2019)
14. Catelli, R., Casola, V., Pietro, G., Fujita, H., Esposito, M.: Combining Contextualized Word Representation and Sub-document Level Analysis Through Bi-LSTM+CRF Architecture for Clinical De-identification. Knowl.-Based Syst. **213**, 106649 (2021)
15. Tao, C., Filannino, M., Uzun, N.: Prescription Extraction using CRFs and Word Embeddings. J. Biomed. Inform. **72**, 60–65 (2017)
16. Aras, G., Makaroglu, D., Demir, S., Cakir, A.: An Evaluation of Recent Neural Sequence Tagging Models in Turkish Named Entity Recognition. Expert Syst. Appl. **182**, 115049 (2021)

AS-GSI: Aspect-Level Sentiment Analysis Integrating Global Semantic Information

Mingxing Su[1,2], Subo Wei[1,2], and Shunxiang Zhang[1,2](✉)

[1] School of Computer Science and Engineering, Anhui University of Science and Technology, Huainan 232001, China
sxzhang@aust.edu.cn

[2] Institute of Artificial Intelligence, Hefei Comprehensive National Science Center, Hefei 230088, China

Abstract. The purpose of aspect-level sentiment analysis is to predict the sentiment polarity of a given aspect in context. At present, most of the existing models use the attention mechanism to capture sentiment information with a given aspect. That did not fully consider the global semantic information, resulting in low accuracy of sentiment polarity recognition. To address this problem, this paper proposes an aspect-level sentiment analysis model that integrates global semantic information(AS-GSI). Firstly, AS-GSI captures the semantic information of context by using BiLSTM. The self-attention mechanism and aspect-based attention mechanism are introduced into AS-GSI, which to obtain global semantic information and sentiment information. Then, a simple and effective fusion mechanism is designed to fuse the two pieces of information fully. Finally, the fused information is input into the softmax-based aspect level sentiment analyzer to calculate the sentiment score. The validity of AS-GSI is verified on the Restaurant14, Restaurant15, and Restaurant16 of SemEval. Experimental results show that the accuracy of AS-GSI can reach 78.27%, 78.14%, and 80.33% on three data sets.

Keywords: Aspect-level sentiment analysis · BiLSTM · Attention mechanism · Self-attention mechanism

1 Introduction

Aspect-level sentiment analysis (ABSA) [1] is a fundamental task in sentiment analysis. The purpose of ABSA is to predict the sentiment polarity(e.g. positive, neutral, and negative) of the given opinion target in the review sentence. Currently, the mainstream research methods employ an attention mechanism to capture sentiment information related to the given aspect. Then aggregate them to predict the sentiment polarity. Despite the effectiveness of aspect-based attention mechanism, we argue that it fail to pay enough attention to global semantic information. In consequence, a meaningful but challenging research issue is how to fuse global semantic information into ABSA. To achieve the goal, this article proposes the AS-GSI model. AS-GSI obtains global semantic information by introducing self-attention mechanism. The self-attention mechanism

J. H. Abawajy et al. (Eds.): ICATCI 2022, LNDECT 169, pp. 238–246, 2023.
https://doi.org/10.1007/978-3-031-28893-7_29

can fully mine the global semantic information by calculating the attention scores of each word and other words in the input. The same as the mainstream method, AS-GSI capture sentiment information by aspect-based attention mechanism. The attention mechanism based on a given aspect can capture the sentiment information by calculating the attention scores of the aspect and each word. Then, the global semantic information and sentiment information are fused, and the fused information is input into the sentiment classifier based on softmax to calculate sentiment score.

The innovations of AS-GSI are (1) the self-attention mechanism is introduced to fully mine the global semantic information in the review sentence, (2) the aspect-based attention mechanism is introduced to obtain the sentiment information with a given aspect. (3) a simple and effective fusion mechanism is designed to make the fusion of global semantic information and sentiment information more complete.

2 Related Work

Early works usually adopt traditional machine learning methond based on lexical and syntactic features. Turney et al.[2] complete the classification task by calculating the probability that words and seed words appear simultaneously. Kiritchenko et al.[3] extracted sentiment information by sentiment dictionary and feature engineering. They established sentiment classification model by support vector machine (SVM). Nasukawa et al.[4] first proposed dependency parsing of sentences, and then added predefined rules to judge an aspect of sentiment. Jiang et al. [5] proposed goal-dependent sentiment analysis. That paper established goal-related features based on syntactic structure of sentences to achieve the emotional polarity judgment of specific goals.

The performance of early methods is highly dependent on the quality of manually labeled sentiment dictionaries. Therefore, in recent years, a large number of sentiment classification methods based on deep learning have been proposed. The LSTM model based on attention mechanism(ATAE-LSTM) is proposed by Wang et al.[6]. ATAE-LSTM combines the input sentence representation with aspect word, which uses LSTM to encode to obtain implicit vector representation. Ma et al. [7] considered the interaction between a given scheme and its content words and proposed Interactive AttentionNetwork (IAN). IAN uses two LSTM based on attention mechanism to interactively capture the connection between aspect terms and context. Chen et al. [8]. Proposed a RAM model based on the memory network and the cyclic attention mechanism. RAM adopt GRU network to extract attention from multiple layers. Liu[9] proposed a new global semantic memory network. That realizes global semantic information sharing in all aspects and generates semantic representations in specific fields.

Recently, most of the models that have performed well in ABSA use the corresponding attention mechanism. However, to capture the sentiment informaton,these models often ignore the global semantic information of sentence. Taking this issue into account, this article applies the self-attention mechanism to self-learn the global semantic information of sentences in the AS-GSI. Finally, sentiment information and global semantic information are combined to guide sentiment prediction.

3 Model

Figure 1 shows the overall architecture of AS-GSI. It mainly consists of six parts: Embedding layer, Feature extraction layer, Global semantic extraction layer, Aspect sentiment perception layer, Feature fusion layer, and Output layer In this part, we will first give the task formalization of ABSA, then introduce Embedding layer and Feature extraction layer. Finally, we present the details of the Global semantic extraction layer, Aspect sentiment perception layer, and Feature fusion layer.

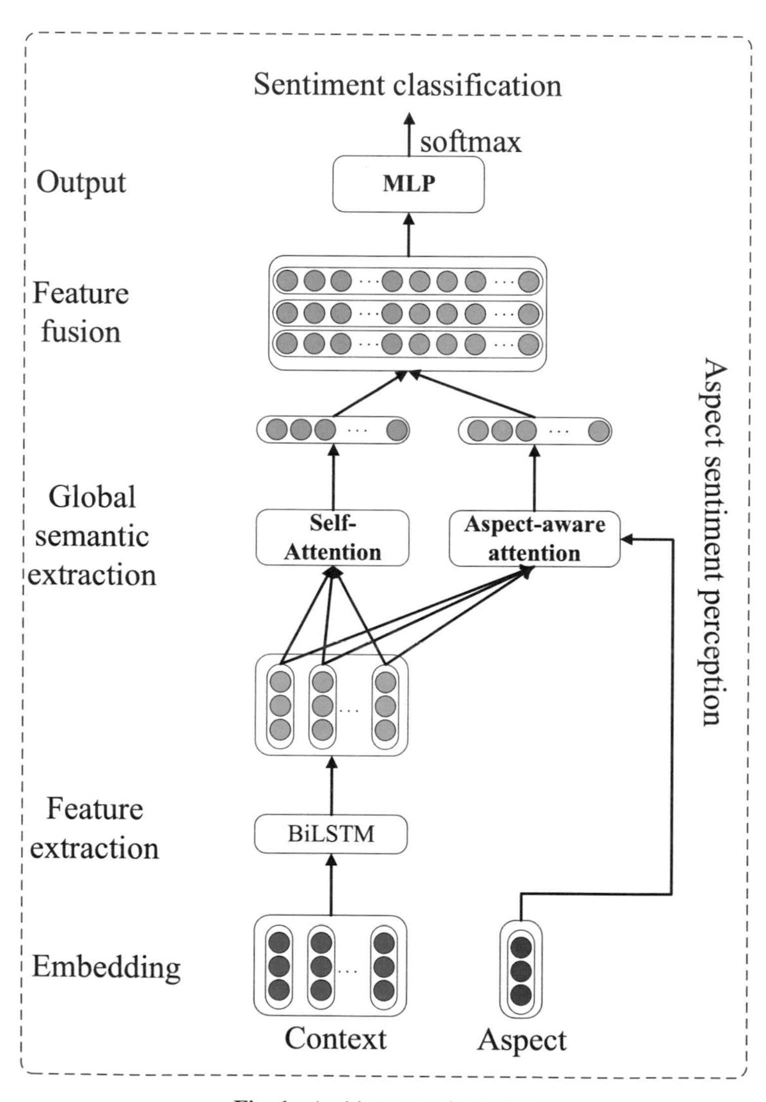

Fig. 1. Architecture of AS-GSI

3.1 Task Formalization

ABSA Formalization: Formally, $s = \{w_1, w_2, ..., w_n\}$ is a review sentence consisting of n words. $t = \{t_1, t_2, ..., t_a\} = \{w_l, w_{l+1}, ..., w_r\}$ is a given aspect containing $a = |r - l|$ words. The purpose of ABSA is to speculate the sentiment polarity (i.e., positive, neutral, and negative) of the aspect word t in the sentence s.

3.2 Embedding Layer and Feature Extraction Layer

As shown in Fig. 1, the word embedding layer maps each word w_i in the review sentence to a low-dimensional dense vector space. This work uses a pre-trained language model $\mathbf{BERT_{base}}$[10] to obtain a fixed word embedding for each context and given aspect word. For given the sentence and aspect, we obtain the corresponding word embedding vector-matrix:$S = BERT(s)$,$T = BERT(t)$. Considering that aspect words may contain more than one word, we simplify the aspect vector by using the traditional additive averaging method. Input the context vector from word embedding into BiLSTM to extract the text feature vector $H = \{h_1, h_2, ..., h_n\} = BiLSTM(S)$.

3.3 Global Semantic Extraction Layer

Self-attention mechanism is a special form of attention mechanism by calculating the attention scores of each word and other words in a sentence, the dependence between each pair of words are obtained. AS-GSI uses the self-attention mechanism to obtain the word dependence in the sentence, so to obtain the global semantic information. Equations (1) and (2) from its internal specific principles:

$$r = \sum\nolimits_{i=1}^{n} \alpha_i h_i \tag{1}$$

$$\alpha_i = \frac{\exp(h_i W^Q h_i + b)}{\sum_{j=1}^{n} \exp(h_j W^Q h_i + b)} \tag{2}$$

where W^Q is the matrix transformation parameter, b is the bias parameter of the network, α_i is the attention calculation score.

3.4 Aspect Sentiment Perception Layer

For capturing the sentiment information of a given aspect, AS-GSI uses the attention mechanism based on aspect perception. Different from the self-attention mechanism, the attention mechanism based on aspect perception takes the given aspect vector as the query vector, to calculate the correlation between each word in the sentence and the given aspect. The calculation formula of attention score is similar to the formula (1)–(2). The only change is that h_i in the formula is replaced by a vector representation of a given aspect.

3.5 Feature Fusion Layer

To fully integrate the global semantic information into the ABSA task, a simple but effective fusion mechanism is proposed: the global semantic information and the sentiment information are calculated in the following three ways, which are concatenation, element-wise product, and their difference.

Concatenation is a simple and effective combination method, which can maximize the retention of information of two sets of vectors. The element-wise product can measure the similarity of the two vectors, and the difference can capture the distribution containing the degree of each dimension.

Then, AS-GSI input the fusion vector into the multi-layer perceptron for deep fusion.

4 Experiment and Analysis of Results

This section will give experimental methods, results, and experimental analysis.

4.1 Experimental Data

Three benchmark datasets (Rest14, Rest15, Rest16) from SemEval were used for experiments. Table 1 shows the details of the experimental data set.

Table 1. Experimental datasets

Dataset	Type	Positive	Neutral	Negative
REST14	Train	2164	637	807
	Test	728	196	196
REST15	Train	912	36	256
	Test	326	34	182
REST16	Train	1240	69	439
	Test	469	30	117

4.2 Evaluation Indicators

The experimental evaluation standard uses two indexes of accuracy (Acc) and F1 as the evaluation method.

4.3 Experimental Methods

To verify the effectiveness and generalization of the AS-GSI, two-dimensional experiments are carried out.

Performance evaluation of the model compared with existing work. The specific experimental models are as follows:

(1) **ATT-CNN.** A Convolutional Neural Network Based on Attention Mechanism.
(2) **TD-LSTM.** It uses two LSTM networks to model a given aspect's previous and subsequent sections.
(3) **ATAE-LSTM.** It is a kind of modeling context through the LSTM network and then embedding the hidden state and the given aspect jointly to supervise the generation of the attention vector.
(4) **IAN.** It uses two LSTM networks to model sentences and the given aspect respectively and interactively produces two parts of attention vectors for sentiment classification.

At the same time, to explore the importance of global semantic information for ABSA, a set of ablation experiments was designed. The specific experimental models are as follows:

(1) **AS-GSI(-GSEL).** Remove the global sentiment extraction layer based on our model.
(2) **AS-GSI(-FFL).** Based on the model in this paper, the fusion layer is removed, and the global semantic information and aspect emotional information is predicted, respectively. Finally, the expected scores are weighted and averaged.
(3) **AS-GSI:** Our model

4.4 Experimental Results and Analysis

According to the above comparison model and ablation model, the test data was tested, the specific experimental data is shown in Table 2.

Table 2. Experimental result

Model		Rest14		Rest15		Rest16	
		Acc	F1	Acc	F1	Acc	F1
Comparison	ATT-CNN	75.13	66.24	76.45	63.42	77.53	64.08
	TD-LSTM	76.52	65.75	77.53	63.37	78.32	64.21
	ATAE-LSTM	77.36	64.37	78.05	62.82	79.64	64.23
	IAN	77.46	63.21	**78.26**	63.12	79.96	63.21
Ablation	AS-GSI(-GSEL)	77.07	62.47	77.72	63.21	78.43	65.00
	AS-GSI(-FFL)	77.29	66.13	77.58	63.59	79.02	64.53
Ours	AS-GSI	**78.27**	**67.12**	78.14	**64.90**	**80.33**	**66.62**

Through the analysis of the experimental results in Table 3, we can obtain:

(1) The ATT-CNN model performed the worst at accuracy in the comparison experiment model. In natural language processing, the adjacent words in the sentence

are not necessarily related. There may be some modifier words between the related terms, which leads to the incomplete information obtained by the CNN convolution operation.

(2) Considering global semantic information in ABSA can effectively improve prediction accuracy. The experimental results show that the best performance is achieved on the data set Rest14 and Rest16, and the accuracy rates are 78. 27% and 80. 33%. Although Rest15 does not reach the optimal performance, it is almost equivalent to the IAN model with the optimal performance, and the accuracy is only 0.12% lower.

(3) The feature fusion layer of the model in AS-GSI can effectively and simply integrate global semantic information and aspect sentiment information. The experimental results show that the accuracy of AS-GSI on the three data sets increased by 0.98%, 0.56%, and 1.31%, respectively, compared with AS-GSI(-FFL).

In order to enhance the comparison of experimental results, this paper presents the resulting data with bar graphs, as shown in Fig. 2 A, B and C respectively represent the experimental results under three different data sets.

5 Conclusion

Focusing on that global semantic information is not fully considered in some aspect-level sentiment analysis models, this article proposes an AS-GSI model.AS-GSI mainly obtains the global semantic information by self-attention mechanism and captures the sentiment information by attention mechanism based on given aspects. Then a simple and effective fusion mechanism is designed to retain all information, similar information, and difference information of the two to the greatest extent. Finally, the obtained fusion information is input into the aspect sentiment classifier based on Softmax. The main contributions of AS-GSI are: (1) In the ABSA task, the global semantic information of the review sentence is used to improve the accuracy of the prediction. (2) A simple and effective fusion mechanism is designed to provide the basis for fully fusing global semantic information and sentiment information.

The AS-GSI model is trained based on the dataset, in which the implicit sentiment expression has been removed before training. The future task is to explore the implicit sentiment of aspect words.

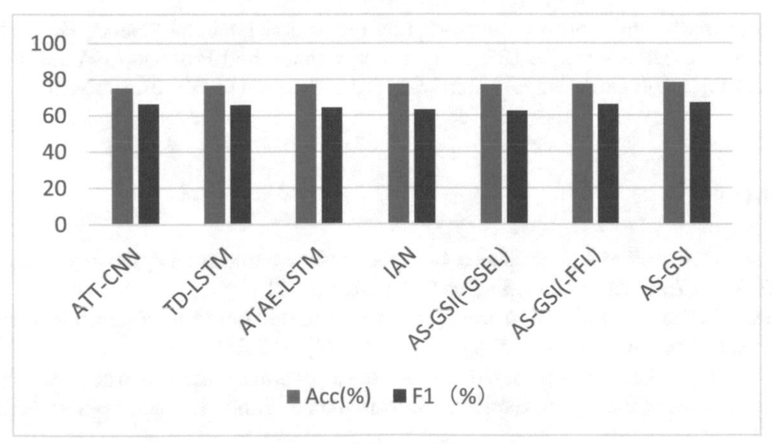

(a) Rest14

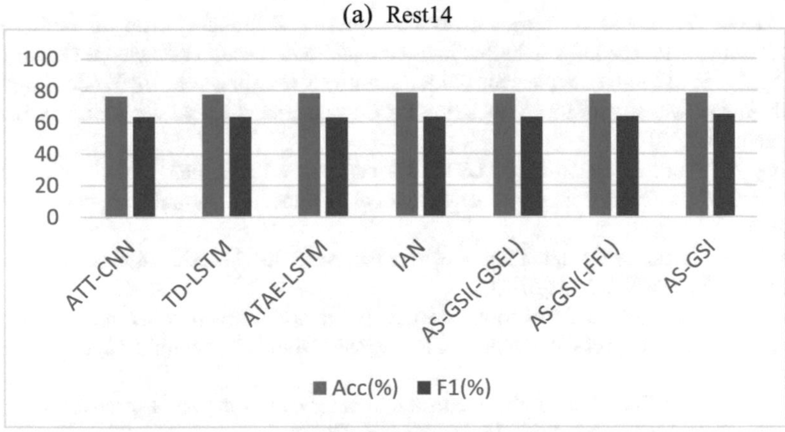

(b) Rest15

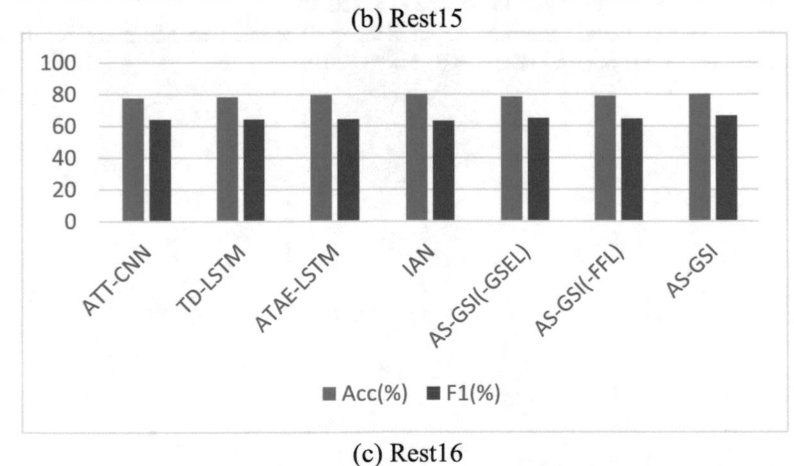

(c) Rest16

Fig. 2. Comparison of experimental results

Acknowledgement. This work was supported by the National Natural Science Foundation of China (Grant NO. 62076006), the University Synergy Innovation Program of Anhui Province (GXXT-2021-008), and the Anhui Provincial Key R&D Program (202004b11020029).

References

1. Nazir, A., et al.: Issues and challenges of aspect-based sentiment analysis: a comprehensive survey. IEEE Transactions on Affective Computing (2020)
2. Turney, P.D., Littman, M.L.: Measuring praise and criticism: Inference of semantic orientation from association. ACM Trans. Inf. Syst. (TOIS) **21**(4), 315–346 (2003)
3. Kiritchenko, S., et al.: Nrc-canada-2014: detecting aspects and sentiment in customer reviews. In: Proceedings of the 8th International Workshop on Semantic Evaluation (SemEval 2014) (2014)
4. Nasukawa, T., Yi, J.: Sentiment analysis: capturing favorability using natural language processing. In: Proceedings of the 2nd International Conference on Knowledge Capture (2003)
5. Jiang, L., et al.: Target-dependent twitter sentiment classification. In: Proceedings of the 49th Annual Meeting of the Association for Computational Linguistics: Human Language Technologies (2011)
6. Wang, Y., et al.: Attention-based LSTM for aspect-level sentiment classification. In: Proceedings of the 2016 Conference on Empirical Methods in Natural Language Processing (2016)
7. Ma, D., et al.: Interactive attention networks for aspect-level sentiment classification. arXiv preprint arXiv:1709.00893 (2017)
8. Chen, P., et al.: Recurrent attention network on memory for aspect sentiment analysis. In: Proceedings of the 2017 Conference on Empirical Methods in Natural Language Processing (2017)
9. Liu, Z., et al.: GSMNet: global semantic memory network for aspect-level sentiment classification. IEEE Intell. Syst. **36**(5), 122–130 (2020)
10. Devlin, J., et al.: BERT: pre-training of deep bidirectional transformers for language understanding. arXiv preprint arXiv:1810.04805 (2018)

LeakGAN-Based Causality Extraction in the Financial Field

Zhengyan Sun, Xiaoqing Li, and Guangli Zhu(✉)

School of Computer Science and Engineering, Anhui University of Science and Technology, Huainan 232001, China
glzhu@aust.edu.cn

Abstract. Causality extraction model can quickly extract causality in text. It can be applied to event prediction, question-answering systems, and scenario generation. The traditional causality extraction pays more attention to the extraction of entities and ignores the deep semantic representation between cause entities and effect entities of the text. So the accuracy of causality extraction is low in the end. A causal relationship extraction model based on LeakGAN is proposed to solve this problem. The core task of the model is to analyze the causality existing in the review text (this paper focuses on explicit one-cause and one-effect analysis). It also realizes deep extraction under semantic enhancement. Four levels of causality extraction model based on LeakGAN are proposed: data preprocessing, pre-training, Bi-LSTM + Attention, and LeakGAN learning. Experimental results show that the model can improve the accuracy of causality extraction.

Keywords: LeakGAN · Bi-LSTM + Attention · Causal relation extraction

1 Introduction

In recent years, causality technology has been widely used in various aspects of natural language processing tasks. Due to the uniqueness of causality patterns, the complexity of the semantic structure, the diversity of expression, and other factors, it is inevitable to increase the difficulty of causality extraction in different fields.

In the financial field of event prediction, question-answering systems, and scenario generation, causality extraction technology has a high application value. At the same time, the information redundancy in the financial field is considerable, which requires machine learning to extract valuable information quickly. Therefore, causality extraction is an essential task for the financial field. Because of the characteristics of complex proper nouns, low-value density, and incomplete information, the text has the problem of semantic ambiguity, which leads to the low accuracy of causality.

Given the above problems in the task of causality, the main considerations are as follows: (1) Construction of causality candidate library. Building a causality candidate library in the financial field can better learn the characteristics of proper nouns. (2) Multi-feature fusion. In addition to the location features, also increases the weight of the word, the emotional intensity of the word, and the correlation degree of the word, making the learning characteristics more comprehensive. (3) Feature extraction.

J. H. Abawajy et al. (Eds.): ICATCI 2022, LNDECT 169, pp. 247–255, 2023.
https://doi.org/10.1007/978-3-031-28893-7_30

We propose a method BALG(Bi-LSTM + Attention + LeakGAN, BALG), Bi-LSTM learning context global features, join the Attention mechanism, learning word-level features. Through LeakGAN to filter the characteristics of learning, learning higher discrimination features.

Based on the above considerations, this paper combines BERT pre-training technology and LeakGAN confrontation neural network model, to extract causality phrases. The specific process is shown in Fig. 1. The advantage of this model is to learn more complicated semantic information for features of high discrimination.

This paper is organized as follows: Sect. 2 gives the construction method of the causality extraction model. Section 3 provides the experimental analysis of the model. Section 4 summarizes this article.

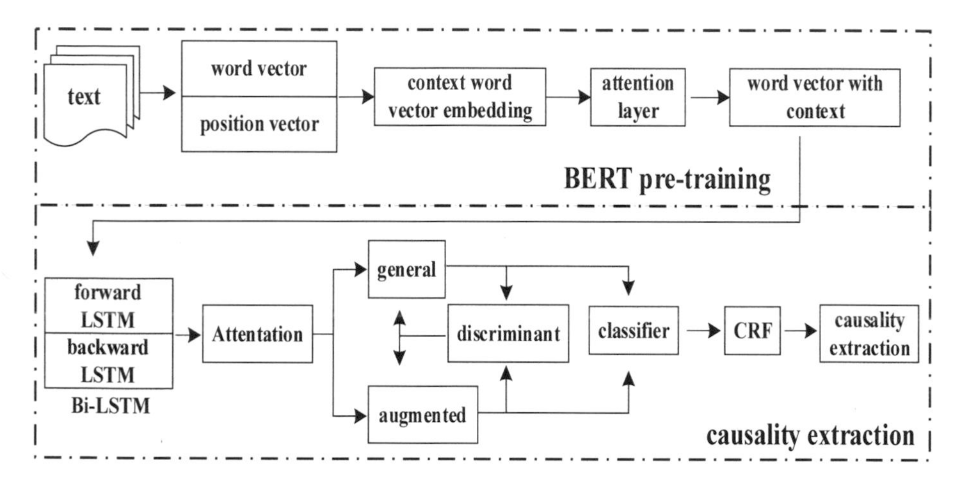

Fig. 1. The framework of the causality extraction model

2 Construction of Causality Extraction Model

2.1 Data Preprocessing

The main tasks of data preprocessing include two aspects. One is to preliminarily screen the content of the text, delete the sentence components of error or unify the sentence format. The second is to label the selected sentences. For example, causality pairs use cause phrases or sentences to represent a cause, and result phrases or sentences mean result. Since this article involves sequence annotations, punctuation is also annotated as words (labeled 'O'). The annotations are shown in Fig. 2

Due to the absence of causal connectives, the causal extraction in this paper is not limited to explicit causals with labels. In this paper, C-B denotes the beginning of the cause, C-I denotes the middle part of the cause, C-E denotes the end of the cause, O denotes the other, E-B denotes the beginning of the result, and E-I denotes the middle of the result. E-E indicates the end of the result.

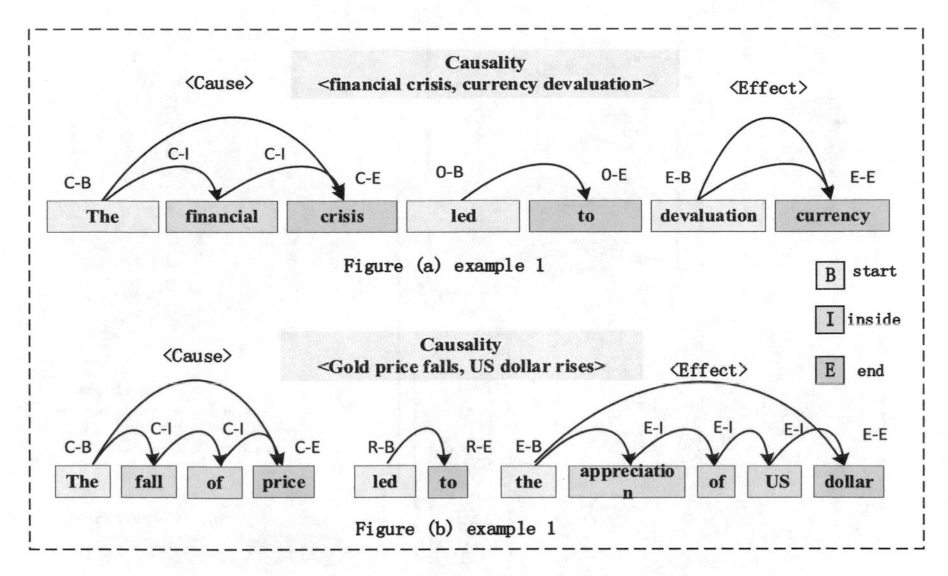

Fig. 2. Sequence labeling example graph

2.2 Model Architecture

For the causal relationship extraction task, the GAN model can make the two data distribution close to the characteristics, so the semantic enhancement method is adopted, especially for the sentence with fuzzy text semantics, the causal relationship in the sentence is extracted, and the causal relationship text is added to the basic model, and this model is called the enhanced model. This enables the relation classification model to learn highly discriminative classification features in adversarial learning, thereby improving the accuracy of relation extraction.

Since the selected comment text is in the financial field, this paper uses the data set FNP 2021 Shared Task 2. Therefore, this paper makes the following assumptions:

Hypothesis 1: All causality sentences in this paper contain a pair of causality;

Hypothesis 2: Both the data set of the basic model and the data set of the enhanced model satisfy that the vector is a simple independent unit Gaussian variable;

Hypothesis 3: Co-occurrence words in the same sentence are semantically related;

Hypothesis 4: This article uses the same Softmax classifier for the basic model and the enhanced model, and optimizes with the same parameters.

For the preprocessed data set, the word segmentation is performed by 'jieba', followed by word embedding, and the features of the context are predicted by the Bi-LSTM double loop network learning context. In the adversarial neural network model proposed in this paper, both data are pre-trained by BERT to obtain semantic word vectors. The adversarial learning process is mainly driven by the generator and the discriminator. Finally, it is continuously optimized in adversarial learning (Fig. 3).

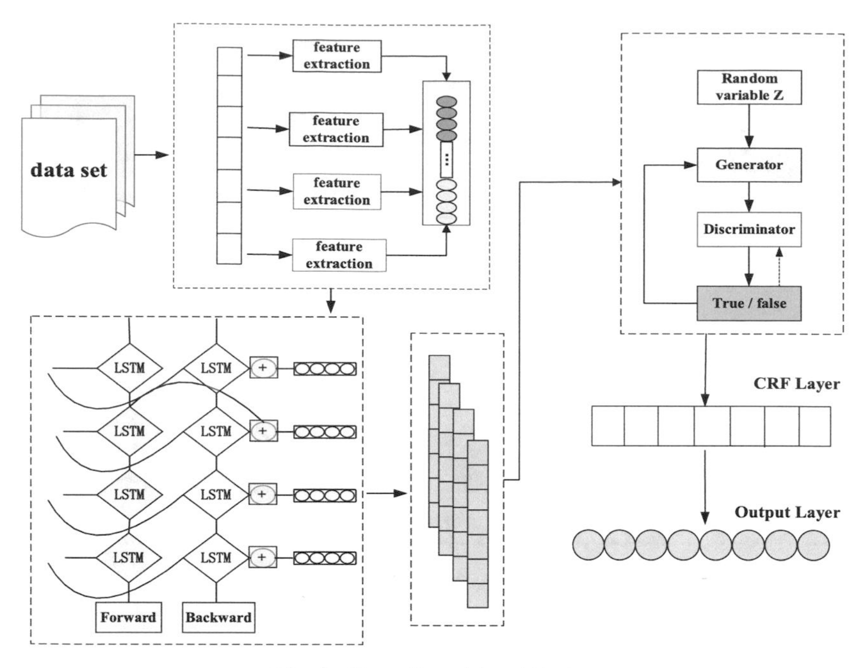

Fig. 3. Overall model architecture

2.3 Feature Fusion Based on Causality

Although the traditional causality extraction uses the syntactic structure and semantic information of sentences, it often only uses a single location feature to learn, which makes the feature dimension learned insufficient. Therefore, this paper adopts multi-feature fusion technology to learn with more multi-dimensional features. In addition, except for the basic location features, to select more accurately, the correlation of words is considered to measure the correlation of causal words and context words. Mainly by calculating the similarity to calculate the correlation, as shown in the formula (1).

$$C_i = \frac{\sum (Ru, i - Ru)(Ru, j - Ru)}{\sqrt{\sum (Ru, i - Ru)}^2 \sqrt{\sum (Ru, j - Ru)^2}} \times \left(\frac{Ri \times Rj}{Ru^2}\right) \tag{1}$$

R_u denotes the average of causal words in the whole review corpus. $R_{u,\ i}$ denotes the correlation between i and R_u. $R_{u,\ j}$ denotes the correlation between j and $R_{u.}$ The feature fusion process is shown in Fig. 4.

2.4 Feature Learning of Causality

BALG (Bi-LSTM + Attention + LeakGAN, BALG) refers to learning the semantic features of the complete sentence through the Bi-LSTM bidirectional loop network. We concentrate the attention of the feature vector on the words that affect the causality by adding the word level attention. And LeakGAN realizes text confrontation learning to learn high discrimination features. The specific algorithm is as follows.

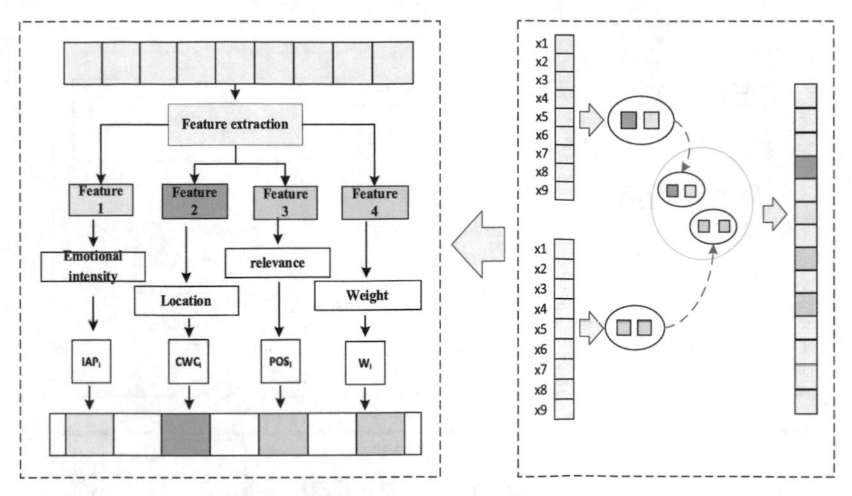

Fig. 4. Feature fusion based on causality

Algorithm 1: BALG construction algorithm

Input: Multi-feature fusion word vector $w=\{w_0,w_1,\ldots\ldots,w_{n-1}\}$

Output: True / False (YES or NO)

1. simultaneous pretraining of BERT parameters
2. for i in w,$w=\{w_0,w_1,\ldots\ldots,w_{n-1}\}$;
3. Bi-LSTM context semantic $\{h_0,h_1,\ldots\ldots,h_{n-1}\}$;
4. Add word-level attention a_i;
5. if the discriminator is YES :
6. train feature $\{f_1,f_2\cdots\cdots f_n\}$;
7. end if;
8. else:
9. Discriminator feeds back to the generator;
10. Generator update parameters θ_g ;
11. Discriminator update parameter θ_d
12. end for

Algorithm 1: Step 1–3 is Bi-LSTM for feature learning. Step 4 is the attention layer that joins the word level. And Step 5–12 identifies the true and false data generated by the generator through the LeakGAN discriminator. If the result is False, the parameter is improved to train the generator, and if it is True, the training ends.

2.5 LeakGAN Generative Adversarial Network

The generative adversarial network is mainly composed of a basic model, causality enhancement model and softmax classifier. When using the enhancement model, we add the extracted causality based on the original sentence to improve semantics.

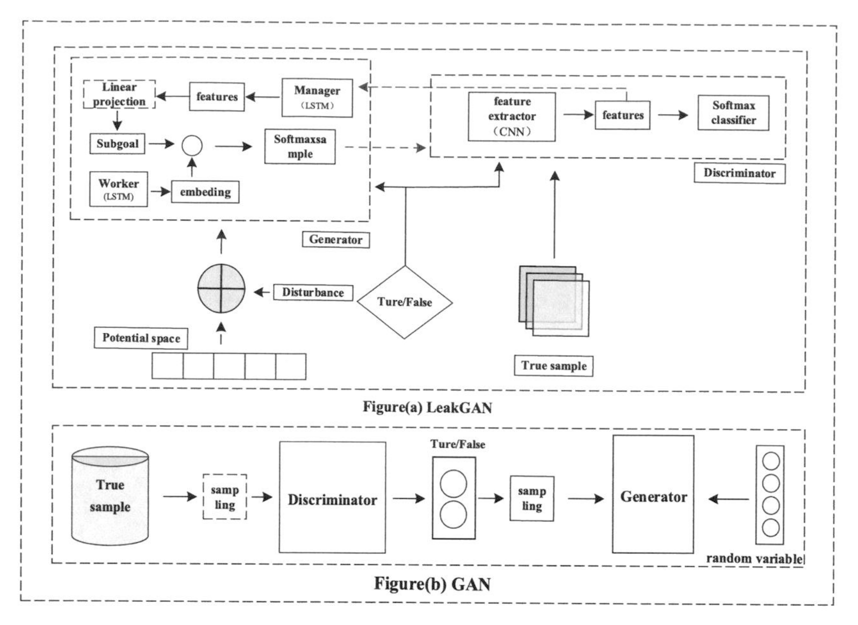

Fig. 5. LeakGAN (a) and GAN (b).

Figure 5 (a) is the model of the LeakGAN generative adversarial network. The Manager module in the generator receives the feature vector from the discriminator and generates the target embedding and the Worker module. The module extracts the features of the current generated word and outputs a potential vector to guide the Worker module to generate the next word, so such a policy gradient algorithm can be directly end-to-end used to train the generator. Figure 5 (b) is an ordinary GAN adversarial neural network, in which the discriminant network calculates the loss function after the discrimination is completed, but the ordinary GAN is prone to the disappearance of the gradient, and the generator and the discriminator may not be matched when learning, resulting in the inability to converge. Therefore, LeakGAN is superior to ordinary GAN in processing text.

GAN mainly learns through confrontation and uses the confrontation of the basic model and discriminant model as supervised learning. For the representation learning of GAN, a common problem is the instability of the training process. Due to the addition of a random disturbance, it is likely to cause the generator to produce meaningless output. G and D are a process of confrontation, and in this confrontation process, G is constantly learning, D is also constantly learning, and it is necessary to ensure that the learning rate of the two is the same, that is, the generator G is improved by the result of the discriminator feedback. Similarly, discriminator D can also continuously learn from each other to improve its accuracy. The whole extraction network is composed of the word vector layer, Bi-LSTM layer, LeakGAN layer and softmax output classification layer after BERT pre-training. The last layer of Softmax is shared by two models, also known as the classifier.

The BERT pre-training process in the model is enhanced, added to the original text by using the selected causality, and then pre-trained. The training motivation of the adversarial network is to make the data of the basic model and the enhanced model as close as possible. So that the loss function is minimized, to optimize the model parameters and improve the model accuracy.

3 Experiments

3.1 Experimental Data

Because of the selection of comments in the financial field, this paper uses the datasets FNP 2021 Shared Task 2 and FNP 2020 Shared Task 2. Based on this, we make the following assumptions.

Hypothesis 1: All causality sentences in this article contain a pair of causality.

Hypothesis 2: Both the dataset of the basic model and the enhanced model satisfy that the vector is a simple independent unit Gaussian variable.

Hypothesis 3: Co-occurrence words in the same sentence are semantically related.

Hypothesis 4: This paper uses the same Softmax classifier with the same parameters for the basic and enhanced models.

The experimental evaluation standard uses three indexes of precision, recall and F as the evaluation method.

$$precision = \frac{analyze\ the\ correct\ number\ of\ film\ reviews}{Analysis\ of\ the\ number\ of\ film\ reviews} \tag{2}$$

$$recall = \frac{analyze\ the\ correct\ number\ of\ film\ reviews}{Total\ correct\ number\ of\ film\ reviews} \tag{3}$$

$$F = \frac{2 \times recall \times precision}{recall + precision} \tag{4}$$

The dataset used in this paper is from the open-source database. The causal dataset in the financial field contains a total of 1753 valid data. Among them, there are 1420 causal data in the filtered data used for training. The dataset is mainly divided into a training set, test set, and verification set, which are divided according to 8:1:1. The construction of the dataset is shown in Table 1.

Table 1. Dataset construction description

Dataset	Total	Filtered
FNP 2021 Shared Task	2395	2200
FNP 2020 Shared Task	1111	1000
SemEval data set	2000	896

3.2 Experimental Analysis

In this model, Bi-LSTM can be long-distance learning for relational reasoning. BALG, CCN, Bi-LSTM, Rule-based, and GCN are used to extract causality from FNP 2021 shared and FNP 2020 shared task datasets, and the extraction F is calculated. Finally, the experimental results are compared, and the specific experimental results are shown in Table 2.

Table 2. Experimental of the financial dataset

Model	SemEval			FinCausal		
	R	P	F	R	P	F
BALG	0.856	0.801	0.828	0.907	0.860	0.883
CCN	0.816	0.795	0.805	0.828	0.799	0.813
Bi-LSTM	0.784	0.726	0.754	0.822	0.777	0.799
Rule-based	0.739	0.711	0.725	0.797	0.744	0.770
GCN	0.838	0.812	0.825	0.890	0.813	0.821

The results in Table 2 show that the precision, recall, and F score of the BALG method on the dataset is significantly higher than those of other methods. The above experiments show that this method has higher accuracy. It further proves the proposed method's effectiveness in causality in the financial field.

4 Conclusion

To effectively extract causality, this paper proposes a causality model based on LeakGAN semantic enhancement. This model analyzes the causality of one cause and one effect in the review text. To realize the deep extraction under semantic enhancement and improve the accuracy of causality.

The method of the BERT pre-training model is proposed, and BERT pre-training is used to obtain the word vector with semantics. This paper combines BERT pre-training with the LeakGAN network, by the LeakGAN generative adversarial network, high discrimination features are obtained for causality. It further improves the accuracy of cause and effect extraction.

In the future, the method in this paper can be considered to be applied to multi-cause and multi-effect text or to extract implicit causality text. At the same time, the causality model based on LeakGAN semantic enhancement can also provide methods for causality in other fields, such as the medical field.

Acknowledgement. This work was supported by the National Natural Science Foundation of China (Grant NO.62076006), the University Synergy Innovation Program of Anhui Province (GXXT-2021-008), and the Anhui Provincial Key R&D Program (202004b11020029).

References

1. Xu, J.H., Zuo, W.L., Liang, S.N., Wang, Y.: Causality extraction based on graph attention network. Comput. Res. Develop. **57**(1), 159–174 (2020)
2. Huan, L., Naveed, A., Ajmal, M.: Spherical kernel for efficient graph convolution on 3D point clouds. IEEE Trans. Pattern Anal. Mach. Intell. **43**(10), 3664–3680 (2021)
3. Pechsiri, C., Piriyakul, R.: Causal pathway extraction from web-board documents. Appl. Sci.-Basel **11**(21), 10342 (2021)
4. Duc-T, V., Feras, A.O., Ebrahim, B.: Extracting emporal and causal relations based on event networks. Inf. Process. Manage. **57**(6), 102319 (2020)
5. De, S.T.N., Xiao, Z.B., Zhao, R., Mao, K.Z.: Causal relation identification using convolutional neural networks and knowledge based features. World Acad. Sci. **11**(6), 697–702 (2017)
6. Zhao, H.C., Li, Y.H., Wang, J.X.: A convolutional neural network and graph convolutional network-based method for predicting the classification of anatomical therapeutic chemicals. Bioinformatics **37**(18), 2841–2847 (2021)
7. Li, P.F., Mao, K.Z.: Knowledge-oriented convolutional neural network for causal relation extraction from natural language texts. Expert Syst. Appl. **115**, 12–523 (2018)
8. Hu, J.J., Wang, Z.Q., Chen, J.Q., Dai, Y.H.: A community partitioning algorithm based on network enhancement. Connect. Sci. **33**(1), 42–61 (2021)
9. Xu, K., Wang, P., Chen, X., Luo, X.F., Gao, J.Q.: Causal event extraction using causal event element-oriented neural network. Int. J. Comput. Sci. Eng. **24**(6), 621–628 (2021)
10. Gao, J., Liu, X., Chen, Y., Xiong, F.: MHGCN: multiview highway graph convolutional network for cross-lingual entity alignment. Tsinghua Sci. Technol. **27**(4), 719–728 (2022)

Research on Aspect Extraction for Chinese Commodities Reviews

Subo Wei[1,2], Mingxing Su[1,2], and Shunxiang Zhang[1,2](✉)

[1] School of Computer Science and Engineering, Anhui University of Science and Technology, Huainan 232001, China
sxzhang@aust.edu.cn

[2] Artificial Intelligence Research Institute of Hefei Comprehensive National Science Center, Hefei, China

Abstract. In the field of aspect extraction of Chinese reviews on commodities, the static word vector obtained by traditional word embedding cannot represent polysemy, resulting in low accuracy of aspect extraction. To solve this problem, this paper proposes an aspect extraction model of Chinese commodity reviews based on BERT. Firstly, the BERT pre-training language model is used to embed the words in the commodity review text. Secondly, the BiGRU network is used to extract the word vector obtained by word embedding to obtain the text features. Then different attention degrees are given to each word through the attention mechanism so that the text features and attention degrees are fully integrated to obtain new text features. Finally, the network output is input into the CRF layer to extract the relevant aspects of the evaluation object. The model can fully learn the semantics of words in the encoding stage. The experimental results show that the proposed method improves the extraction accuracy.

Keywords: BERT · Aspect extraction · Commodity reviews in Chinese · BiGRU-CRF

1 Introduction

Aspect extraction is one of the core subtasks of text-level sentiment analysis tasks, which determines whether subsequent sentiment analysis tasks can be carried out smoothly [1]. The methods of aspect extraction can be divided into methods based on traditional machine learning and methods based on deep learning. (1) Manually formulated rules and statistical knowledge methods extract the corresponding research objectives from the complex review information texts [2, 3]. In the research task of attribute extraction, the corresponding association rules are introduced into the attribute extraction task [4, 5], which improves the accuracy of extraction. On this basis, an extraction method integrating a dependency relationship is proposed to use the dependency relationship between comment objects and comments [6, 7]. (2) Deep learning method applies word embedding of deep learning to domain entity attribute extraction tasks[8]. A deep neural network is used to encode the input text and obtain the corresponding target object

J. H. Abawajy et al. (Eds.): ICATCI 2022, LNDECT 169, pp. 256–263, 2023.
https://doi.org/10.1007/978-3-031-28893-7_31

from the comment text [9]. The combination of Bi-LSTM and attention mechanism can realize the feature sharing of the process of attribute extraction and the process of emotion analysis [10–13].

BERT + BiGRU ensures that the model can fully learn the long-term dependence [11] between the context information of the text sequence and the word vector. The attention mechanism considers the importance of different words, which is conducive to reducing the impact of non-target words on the extraction results and further improving the accuracy of extraction. The major contributions of our work include two aspects: (1) A new aspect extraction method for Chinese commodity reviews is proposed, and its effectiveness is proved. (2) The BERT model is applied to the aspect extraction task, which provides a basis for aspect-level sentiment analysis of Chinese commodity reviews.

The rest of this paper is organized as follows: Sect. 2 explains the model introduction. We give the experimental results and analysis in Sect. 3. Finally, Sect. 4 provides the conclusion.

2 Model Introduction

The extraction model proposed in this paper consists of the following four parts: coding layer, text feature extraction layer, attention layer, and CRF layer. The specific structure of the model is shown in Fig. 1.

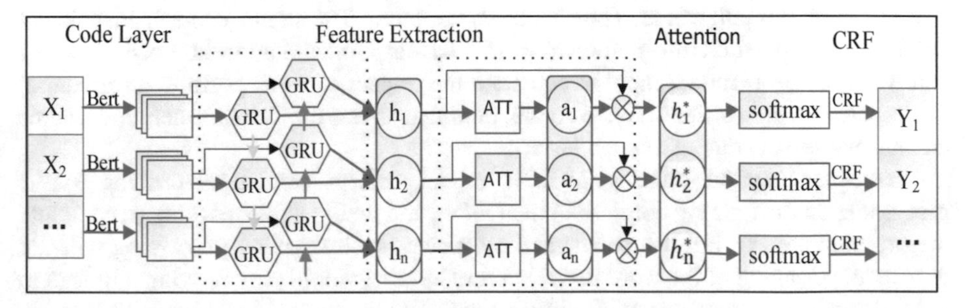

Fig. 1. Bert-based extraction model of Commodity Chinese reviews

2.1 Coding Layer and Text Feature Extraction Layer

In the field of commodity Chinese reviews extraction, the traditional word embedding technology can only obtain static word vectors and cannot dynamically characterize the polysemy of words according to the current context. In recent years, the BERT pre-training language model has been widely used. In recent years, with the widespread use of BERT pre-trained language model, it has been proved on many English NLP tasks that the model can effectively solve the problem that traditional word vectors cannot represent polysemy. Therefore, this paper uses BERT pre-trained language model for

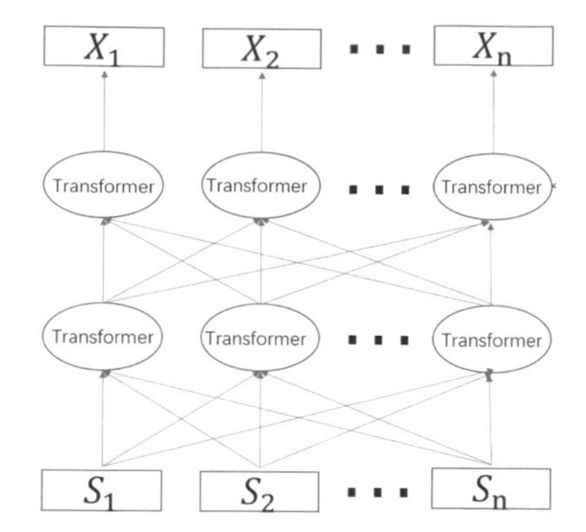

Fig. 2. Structure diagram of BERT pre-training language pattern

character embedding. The specific framework of the BERT pre-training language model is shown in Fig. 2.

The BERT model is mainly based on the bidirectional Transformer network architecture. The Transformer model is a new network architecture proposed by Google that abandons the RNN-dependent self-Attention mechanism. The existence of the Self-Attention mechanism allows any unit of the input text sequence to interact without length constraints, and efficiently captures long-distance context semantic features. Coupled with the bidirectional Transformer encoder structure, the left and right context information of the current unit can also be effectively captured. The attention mechanism is not sensitive to the position information of words. To this point, Google adds position vector and word vector as model input to constrain the impact of word position information. At the same time, in order to speed up the training process, the Transformer model sums and normalizes the output of each layer.

In the process of pre-training, the BERT model performs word-level masking (Mask) processing on the training corpus and then infers and predicts this masking to complete the pre-training. For English segmented by spaces, such masking processing will not destroy the meaning of the word itself, but for Chinese, word-level masking will lead to some deviations in vocabulary. Aiming at how BERT is applied to Chinese NLP tasks, HIT and IFLYTEK jointly released the Chinese BERT model with full-word coverage. The masking processing is not based on words, but on words. The performance on multiple Chinese datasets surpasses the original BERT model.

After preprocessing the original Chinese comment data, the sequence text $S = \{s_1, s_2, s_3, \ldots, s_n\}$, n is the number of input text words, which is input into the BERT model through one-hot encoding. The word vector of the output text sequence of the model is expressed as $X = \{x_1, x_2, x_3, \ldots, x_n\}$.

Considering the length of the commodity review text and the defects of the original RNN network, this paper adopts the gating loop network GRU. GRU is a variant of RNN. To solve the gradient problem in long-term memory and backward propagation, the input and output structure of GRU is the same as that of ordinary RNN: Time t, Input x_t, and the hidden state passed down by the previous node $h_{t\text{-}1}$. This hidden state contains the historical information of the sequential test. The formalization is as follows:

$$r_t = \sigma\left(W_r \cdot \left[h_{t-1}\ x_t\right]\right),\ z_t = \sigma\left(W_z \cdot \left[h_{t-1}\ x_t\right]\right) \tag{1}$$

$$h'_t = \tanh\left(W_{h'} \cdot \left[r_t * h_{t-1}\ x_t\right]\right),\ h_t = (1 - z_t) * h_{t-1} + z_t * h'_t \tag{2}$$

$$y_t = \sigma(W_o \cdot h_t) \tag{3}$$

Formula 1 represents reset door and update door. Reset gate controls information about the previous state and writes to the current candidate set h'_t. Formula 2–3 represents the hidden state passed to the next unit and the node's output.

To fully explore the long dependency relationship of context, the bidirectional GRU model is adopted. The hidden layer state output by forwarding GRU and backward GRU is expressed. The calculation formula is shown in Formula 4.

$$\overrightarrow{h}_t = GRU\left(\overrightarrow{h}_{t-1}\ x_t\right),\ \overleftarrow{h}_t = GRU\left(\overleftarrow{h}_{t-1}\ x_t\right) \tag{4}$$

Stitch the hidden states of forwarding GRU and backward GRU together to form a new state vector denoted h_t as t, which is the output of the BiGRU hidden layer.

2.2 Attention Layer and CRF Layer

When dealing with sequential text problems, the text feature vector output from the text feature extraction layer is input into the attention calculation layer to calculate the importance of each word vector. The calculation formula is shown in Formula 8.

$$\alpha_i = soft\max(S(K, Q)),\ S(K, Q) = \frac{K^T Q}{\sqrt{D}} \tag{5}$$

$S(K, Q)$ is the attention scoring mechanism, and the attention distribution $W(X, Y) = \sum_{\mathrm{i}=1}^{n} S_{y_{i-1},y_i} + \sum_{i=1}^{n} Z_{i,y_i}$ can be interpreted as the degree of attention to the i information when the context query Q.

Although the BiGRU layer can capture the long-term context information in the text sequence, there may be a strong dependency between the output words (a strong dependency tag word). Therefore, after the output of the above network model, the classical linear CRF model is used to label the commodity terms in the review text. CRF model can model the transition probability between states (the dependency relationship between labels) and obtain a global optimal annotation sequence by considering the transition probability relationship between labels. The specific formula of the CRF model is shown in Formula 6:

$$W(X, Y) = \sum\nolimits_{\mathrm{i}=1}^{n} S_{y_{i-1},y_i} + \sum\nolimits_{i=1}^{n} Z_{i,y_i} \tag{6}$$

$W(X, Y)$ represents the predicted score of the given input sequence X and its corresponding label sequence Y. where S denotes the probability transfer matrix between tags, that is, S_{y_{i-1},y_i} is the probability value of the transfer from tag y_{i-1} to tag y_i. Z_{i,y_i} represents the unnormalized probability of the i mapping to label y_i. In the model prediction stage, a set of globally optimal label sequences is obtained by Formula 7.

$$Y^* = \underset{y' \in Y_X}{\arg\max}\, W(X, y') \tag{7}$$

3 Experiment and Analysis of Results

This section will give experimental methods, results, and experimental analysis.

3.1 Experimental Data

Crawling the product reviews from Taobao and Jingdong to build the experimental data set, removing the sentences without emotional tendencies, removing the reviews containing implicit terms, and classifying the data set. The final product reviews contain only 'Mobile phone', 'Camera' and 'Clothes of men' without neutral statements. The preprocessed commodity review texts were manually labeled, and 7/10 of the labeled text data were used as the training data of this experiment. The remaining labeled data were used as the test data. Table 1 gives the experimental data details.

Table 1. The experimental data

Commodity categories	Total number of comments	Number of comments on annotations
Mobile phone	2178	2013
Camera	1782	1518
Clothes of men	2956	2874

3.2 Experimental Parameters and Evaluation Indicators

The coding layer of this model adopts the Chinese BERT pre-training language model jointly issued by the Harbin Institute of Technology and IFLYTEK. The transformer has 12 layers, the hidden layer dimension is 768, and 12 attention heads, all of which have 110 M parameters. To prevent overfitting, the batch is set to 64 and dropout is set to 0.5 during model training. At the same time, the dimensions of forwarding and backward hidden states of BiGRU in the text feature extraction layer are set to 128, and the optimizer selects Adam to minimize the training loss of the model.

3.3 Experimental Methods

To verify the effectiveness of the model proposed in this paper in the extraction task of Chinese commodity reviews, the specific experimental operations are as follows:

(1) **Get the Dataset**. Crawling goods online review text, and the crawled text to remove implicit aspects of the text, classification, removal of neutral comments and other data preprocessing operations, data processing results are shown in Table 1 experimental data.
(2) **Label Terms of Goods.** The processed data set is manually annotated. To ensure the accuracy of annotation, multiple cross-validations are used.
(3) **Extract Commodity Terms.** The proposed BERT-based commodity Chinese comment extraction model is used for the aspect word extraction task of comment text, and finally, the evaluation index of the experiment is calculated. Two traditional extraction methods are selected as comparative experiments.

3.4 Experimental Results and Analysis

According to the above experimental steps, the test data was tested. The specific experimental data is shown in Fig. 3.

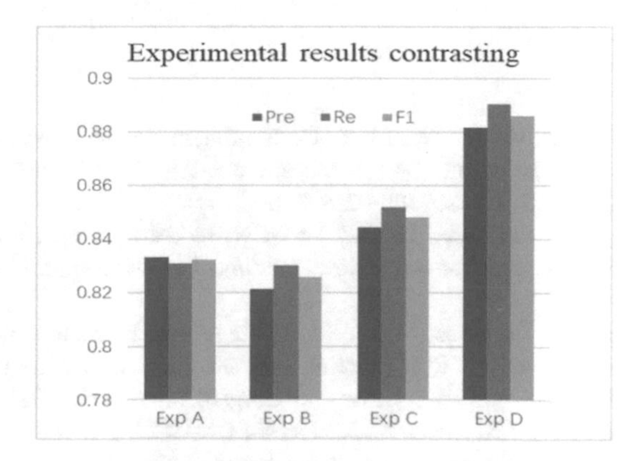

Fig. 3. The experimental results are compared with the bar chart

From the comparative data in Fig. 2, the word extraction model based on BERT proposed in this paper is superior to the traditional word embedding extraction method in accuracy, recall rate and F1. The comparative experimental results of Experiment C and D show that the attention mechanism is conducive to improving the accuracy of the extraction task. It not only reduces the impact of irrelevant words on the CRF layer but also makes the model pay more attention to the corresponding task objectives to reduce the impact of non-target words on the extraction results.

In addition, we found that different sequence length model parameters had different effects on the experimental results, which we believed was caused by the BERT layer and BiGRU layer. The internal structure of the BERT layer and BiGRU layer took into account the context information, and the context information brought by different sentence lengths was also different. In this paper, it is determined by experiments that good results can be obtained when the sequence length is set to 512.

4 Conclusion

In the research field of commodity Chinese comment extraction, aiming at the problem that the word embedding of traditional extraction methods cannot represent the polysemy of Chinese words in specific contexts, this paper proposes a commodity comment extraction model based on BERT. The BERT pre-training language model is applied to the field of commodity Chinese comment extraction, and it is proved that it improves the accuracy of extraction, which provides a research basis for the subsequent commodity Chinese aspect-level sentiment analysis task. The experimental results show that compared with the traditional extraction methods, the proposed method has achieved higher accuracy in the extraction task.

Acknowledgements. This work was supported by the National Natural Science Foundation of China (Grant NO.62076006), the University Synergy Innovation Program of Anhui Province (GXXT-2021–008), and the Anhui Provincial Key R&D Program (202004b11020029).

References

1. Zhang, S., Hu, Z., Zhu, G., Jin, M., Li, K.-C.: Sentiment classification model for Chinese micro-blog comments based on key sentences extraction. Soft. Comput. **25**(1), 463–476 (2020). https://doi.org/10.1007/s00500-020-05160-8
2. Zhou, J., Huang, J.X., Hu, Q.V., He, L.: Sk-GCN: modeling syntax and knowledge via graph convolutional network for aspect-level sentiment classification. Knowl.-Based Syst. **205**,(2020)
3. Zhang, S., Xu, H., Zhu, G., Chen, X., Li, K.: A data processing method based on sequence labeling and syntactic analysis for extracting new sentiment words from product reviews. Soft. Comput. **26**(2), 853–866 (2021). https://doi.org/10.1007/s00500-021-06228-9
4. Zhu, X.F., Zhu, L., Guo, J.F., Liang, S., Dietze, S.: GL-GCN: global and local dependency guided graph convolutional networks for aspect-based sentiment classification. Expert Syst. Appl. **186**,(2021)
5. He, Y.X., Sun, S.T., Niu, F.F.: A sentimental semantic enhanced deep learning model for weibo sentimental analysis. J. Comput. **10**(4), 773–790 (2017)
6. Zuo, E.G., Zhao, H., Chen, B., Chen, Q.: Context-specific heterogeneous graph convolutional network for implicit sentiment analysis. IEEE Access **8**, 37967–37975 (2020)
7. Xu, Z., et al.: Hierarchy-cutting model based association semantic for analyzing domain topic on the web. IEEE Trans. Industr. Inf. **13**(4), 1941–1950 (2017)
8. Xu, H. Q., Zhang, S.X., Zhu, G.L.: ALSEE: a framework for attribute-level sentiment element extraction towards product reviews. Connection Sci. **34**, 1–19 (2021)
9. Zargari, H., Zahedi, M., Rahimi, M.: GINS: a global intensifier-based N-gram sentiment dictionary. J. Intell. Fuzzy Syst. (Preprint) **40**, 1–14 (2021)
10. Wei, J.Y., Liao, J., Yang, Z.F., Wang, S.G.: Bi-LSTM with multi-polarity orthogonal attention for implicit sentiment analysis. Neurocomputing **383**, 165–173 (2020)

11. Li, X., Lam, W.: Deep multi-task learning for aspect term extraction with memory interaction. In: Proceedings of the 2017 Conference on Empirical Methods in Natural Language Processing, pp. 2886–2892 (2017)
12. Wang, B., Wang, H.: Bootstrapping both product features and opinion words from Chinese customer reviews with cross-inducing. In: Proceedings of the Third International Joint Conference on Natural Language Processing, vol.-I (2008)

LDM: A Location Detection Method of Emotional Adversarial Samples Attack for Emotion-Cause Pair Extraction

Houyue Wu[1,2], Xin Xu[1,2], and Shunxiang Zhang[1,2](✉)

[1] School of Computer Science and Engineering, Anhui University of Science and Technology, Huainan 232001, China
sxzhang@aust.edu.cn

[2] Institute of Artificial Intelligence, Hefei Comprehensive National Science Center, Hefei 230088, China

Abstract. ECPE as the task of extracting emotion-cause pairs provides technical support for some fields, such as public opinion analysis. The existing ECPE methods are attacked by emotional adversarial samples, the precision of extracted EC pairs will decrease. The current adversarial defense methods cannot accurately detect the attacked location that affects the extraction results. To address this problem, this paper proposes LDM (A Location Detection Method of Emotional Adversarial Samples Attack for Emotion-Cause Pair Extraction), to location attack. Firstly, according to the results of ECPE dataset extraction, the Emotion-Cause scores are calculated. The EC domain (Emotional Cause Domain) is constructed by the upper and lower limits of the Emotion-Cause scores. Then, EC pairs are extracted from ECPE data attacked by emotional adversarial samples, and their Emotion-Cause scores are calculated. The Emotion-Cause scores distribution in the EC domain is analyzed to obtain the outlier data information. According to the location of the outlier data, capture the attack location. Finally, remove outlier data, and the ECPE secondary extraction is carried out to obtain the final extraction results. The experimental results show that LDM can accurately capture the attack location and help the model prominent improve the extraction precision.

Keywords: ECPE · Emotional adversarial samples attack · Location detection · Emotion-Cause scores

1 Introduction

The ECPE (Emotion-Cause Pair Extraction) [1] task is derived from the ECE task [2]. ECPE is one of the important research tasks in the field of natural language processing, which provides technical support for many fields such as psychotherapy and public opinion analysis. The core of the ECPE task is to extract emotional clauses and cause clauses from data. When the core emotion is attacked by emotional adversarial samples, the extraction results will deteriorate.

J. H. Abawajy et al. (Eds.): ICATCI 2022, LNDECT 169, pp. 264–273, 2023.
https://doi.org/10.1007/978-3-031-28893-7_32

The existing research on the ECPE task only considers changing the extraction method to improve the extraction precision [1, 3, 4]. They ignore the data changes will reduce the extraction precision. When ECPE tasks were attacked by emotional adversarial samples, the existing adversarial defense technology [5, 6] cannot accurately locate the attack location.

In our previous research, some achievements have been made in the research on sentiment analysis [7, 8], which provides a research basis for this work. Different from previous work, this paper integrates sentiment analysis into the detection of attack location. To better apply to ECPE tasks, the following two points need to be considered: (1) The scope of the EC domain needs to adapt to the emotion-cause pair extraction in most task areas. (2) Emotion-Cause scores calculation needs to fully combine the causal relationship between emotional clause and cause clause. The motivation of this paper is to propose an adversarial location detection method to help detect the location of modified data in ECPE tasks attacked by emotional adversarial samples. The process of LDM is shown in Fig. 1.

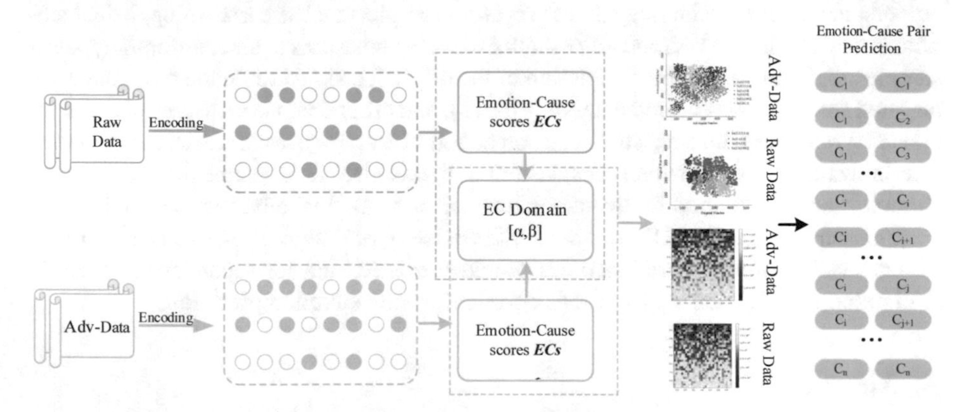

Fig. 1. The process of LDM

LDM is a simple and effective method. When LDM determines the EC domain, calculate the Emotion-Cause scores according to the extraction results of the ECPE benchmark dataset. The Emotion-Cause scores make full use of the original information of the data itself while considering the context semantic information. Which reduces the spurious correlations between different emotional clauses and cause clauses. LDM can more intuitively find outlier data according to the distribution of Emotion-Cause scores. These outlier data are data attacked by emotional adversarial samples.

The main contribution of this paper is to propose a detection method LDM. Which can detect the data attacked by emotional adversarial samples in ECPE data and visually display the data location.

2 Related Work

2.1 ECE and ECPE

The task of extracting emotional cause was proposed by Lee et al. [2] in 2010. Scholars found that the time of forming emotional Cause is generally composed of multiple words, Gui et al. proposed the ECE task in clause [9, 10]. Xia et al. [1] pointed out that there was a correlation between emotion and cause, and cause clauses should not be identified only. Therefore, the ECPE task was proposed to match emotion clauses and cause clauses into Emotion-Cause Pair (EC-Pair). Later scholars derived different ECPE extraction methods [4, 11-13]. The existing ECPE tasks have a strong correlation with emotional words, so there are great hidden dangers. When the core emotional words change, the extraction results will change.

2.2 Adversarial Defense

Adversarial defense technology and adversarial sample attack are a set of opposite technologies. The purpose is to actively establish defense measures against different types of sample attacks to ensure the completion of the original task. At the same time, there are targeted fonts [5, 14] replacement words [6, 15], and other defense measures of different attacks. From these different studies, it can be found that these tasks are defenses against a certain attack method, which is consistent with this idea. At the same time, there is no suitable detection method for the attack location of emotional adversarial samples.

This paper proposes LDM to improve the existing problems of ECPE. Detecting the attack location of adversarial samples provides trusted data for secondary extraction. And reduces the influence of spurious correlations on extraction precision.

3 Method

3.1 EC Domain

EC domain is constructed by calculating the emotional cause scores of each clause in the document. The emotional Cause score is mainly obtained by weighting the emotional score and Cause score. Suppose there are M documents, the i $(1 \leq i \leq M)$ document is d_i, the j $(1 \leq j \leq N)$ clause of d_i is $c_{i,j}$. For ECPE tasks, the extracted emotional clause and the corresponding cause clause need to be annotated in the original document data. Each document performs text preprocessing and word embedding in terms of clauses to obtain the word vector $c_{i,j}$.

The cause score is measured by the cause information contained in the clause. The emotional score is measured by the emotional information contained in the clause. Cause score is expressed by Cause information in Cause clause C^C. The emotional score is represented by the information contained in the emotional clause (result) E^C. Cause scores and Emotion scores are calculated by the formula (1–2).

$$Causescores\left(c_{i,j}\right) = DKL\left(\left(c_{i,j}|c_i\right), C^C\right) \tag{1}$$

$$Emotionscores(c_{i,j}) = DKL\left((c_{i,j}|c_i), E^C\right) \tag{2}$$

where DKL (Kullback-Leibler divergence) represents the difference between all clauses and cause clauses or effective clauses.

Then the final score *ECs* of emotional causes calculate by:

$$\begin{aligned} ECs(c) &= \lambda i\langle Causescores(c_{i,j})\rangle + \lambda j\langle Emotionscores(c_{i,j})\rangle \\ &= \lambda_i \sum_{ci,j \in di} p(c_{i,j}) DKL\left((c_{i,j}|c_i), C^C\right) + \lambda_j \frac{1}{N} \sum_{ci,j \in di} DKL\left((c_{i,j}|c_i), E^C\right) \end{aligned} \tag{3}$$

where λ_i and λ_j are calculation parameters.

The final numerical range is obtained by calculating the upper and lower limits of *ECs*, named EC Domain.

3.2 EC Strength Distribution

The intensity distribution of emotional causes highlights the position of opponent attacks. When scoring *ECs* for emotional Cause, the *ECs* results are numerical. To increase contrast, scatter plot distribution and thermal diagram distribution are used for data distribution. The scatter plot and thermal diagram can intuitively observe the obvious outlier data information.

The core idea of LDM is to construct EC Domain and study the intensity distribution of emotional causes based on the score of emotional causes. There is a causal relationship between the cause clause and emotion clause, which is an intuitive representation of ECPE extraction results. When analyzing in the context of document events, there are several clauses with a weak causal relationship with emotional clauses before the cause clause. These words are the deep-seated reasons for this emotion. Therefore, we should also calculate the scores of emotional causes between deep-level cause clauses and emotional clauses. When the sample encounters an emotional adversarial sample attack, the core emotional words change, but deep-seated reasons will not change.

4 Experiments and Analysis

4.1 ECPE Dataset

The experiment uses the Chinese benchmark data set by [1], and the data come from Sina City News. There are 2105 documents in the dataset, including news events in multiple fields, covering multiple emotional words. Multiple types of emotional words make the impact on the final results greater when attacked by emotional adversarial samples. Use the synonym replacement attack method proposed by Zhou et al. [6] to replace emotional words. The attack uses synonym substitution and antonym substitution. The original data and the classification of the changed emotional types are shown in Table 1.

In order to more intuitively and accurately show the proportion of emotional words under adversarial example attacks, Fig. 2 below shows the data in Table 1. It can be

Table 1. Distribution of emotion types

Emotion	Original Percentage	Adversarial Percentage
Surprise	4.18%	18.32%
Disgust	10.69%	20.12%
Anger	14.35%	18.90%
Fear	18.00%	19.25%
Happiness	25.83%	12.36%
Sadness	26.95%	11.05%

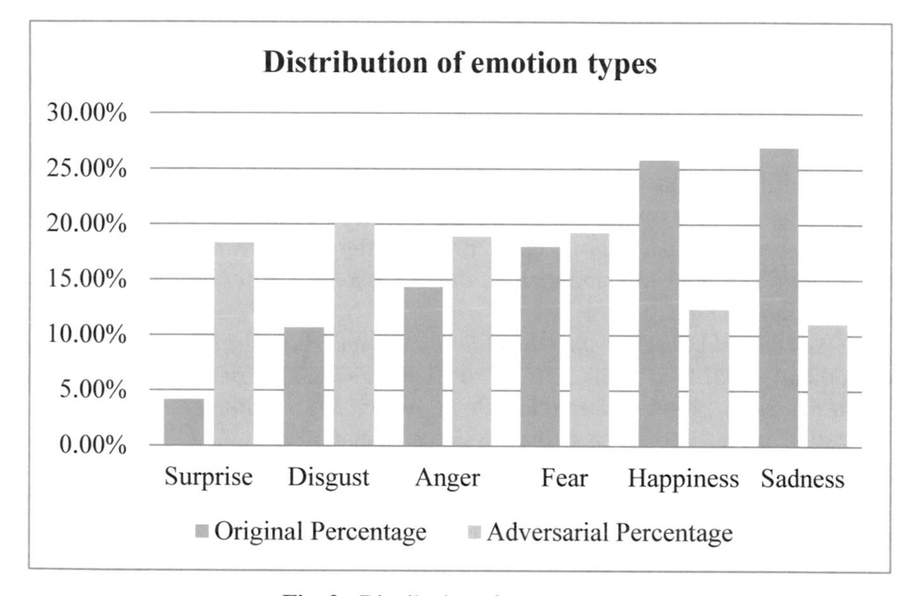

Fig. 2. Distribution of emotion types

clearly found that happy and sad words with strong emotional polarity change more greatly when they are attacked by adversarial samples attack.

Figure 2 and Table 1 show the changes of emotional words in the original data when encountering adversarial sample attacks. When the data encounters emotional adversarial sample attacks, the data has undergone tremendous changes. Among them, happiness and sadness decline most, because these two emotions are often expressed and easily attacked.

4.2 Experiments on Adversarial Dataset

The experimental results are shown in Table 2.

Table 2. Distribution of Extraction Results in EC domain(Negative number indicates the difference between the number of attack locations identified by LDM and the number of actual attack locations)

Emotional Causes Strength S	Original Number	Adversarial Number	LDM Identifies Outlier data
S ∈ (0.215,0.3]	23	268	225 (**−20**)
S ∈ (0.3,0.4]	125	410	273 (**−12**)
S ∈ (0.4,0.5]	254	370	105 (**−9**)
S ∈ (0.5,0.6]	365	139	216 (**−10**)
S ∈ (0.6,0.7]	412	224	169 (**−19**)
S ∈ (0.7,0.8]	560	198	347 (**−15**)
S ∈ (0.8,0.9]	236	166	64 (**−6**)
S ∈ (0.9,0.961]	130	330	191 (**−9**)

The experiment is carried out on the Chinese benchmark dataset, and the emotional cause score is calculated to construct the EC Domain. An adversarial dataset is constructed based on the emotional adversarial attack of all clauses in the baseline dataset. According to statistics, the score range of emotional causes calculated by Formulas (1–3) is 0.215–0.961(data are normalized). Therefore, it is concluded that EC Domain is [0.215,0.961]. According to EC Domain, the intensity of emotional causes is analyzed on the emotional adversarial dataset.

From Table 2, it can be found that the LDM Identified Outlier data is a comparison result display of the data of the original data sample and the adversarial sample. Even if the number is marked, the display of the results is not very convenient.

Here, the data in Table 2 is presented in the bar chart in Fig. 3. The number of adversarial samples and the number of original samples are displayed in the same column, which can compare the difference in price and quantity. This also better shows the differential value of the LDM Identifies Outlier data. Based on Fig. 3 and Table 2, the following conclusions can be easily drawn:

(1) **The adversarial sample attack can change the score distribution of the original EC domain.** Emotion-cause pair extraction will change the extraction results when using adversarial example attack. Different adversarial sample attack location changes have different extraction results.
(2) **LDM can successfully identify most of the attack locations.** When the core emotional words change, the corresponding emotional reasons will change. LDM is calculated by deep causality and emotional words, and the final results are obtained. Using LDM to judge the attack location of adversarial examples can improve the accuracy of extraction.

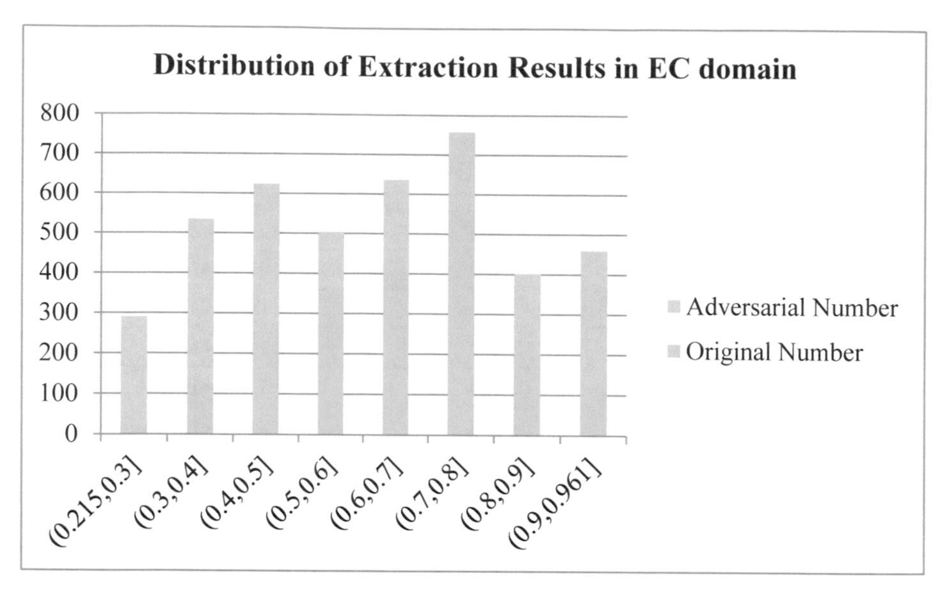

Fig. 3. Distribution of extraction results in EC domain

4.3 Experiments on Test Dataset

3/10 of the original data are selected to replace random sentiment words, and the generated adversarial samples are doped into the original data as a test set to ensure that the original data is unchanged. Use LDM to remove the attacked clauses found, and the results are as follows:

Table 3. Comparison of Several Different Models

Model	Benchmark ECPE Corpus		
	Original Text Set	Adversarial Text Set	LDM Removes Attacked Clauses
ECPE-2Steps[1]	64.10%	60.17%	72.36%
ECPE-MLL[11]	65.82%	62.01%	69.23%
Trans-ECPE[12]	70.14%	64.57%	76.95%
ECPE-2D[4]	65.32%	60.33%	69.56%
E2E-PExt$_E$[13]	60.92%	56.25%	67.32%

In order to compare the test results on the original test set and the test results on the adversarial example test set in Table 3 with the LDM method more conveniently, the results in Table 3 are split into different histograms for comparison, as shown in Fig. 4 and Fig. 5. The data results in Table 3 and Fig. 4 show the effectiveness of the proposed method LDM in detecting the attack location of adversarial examples to improve the ECPE task.

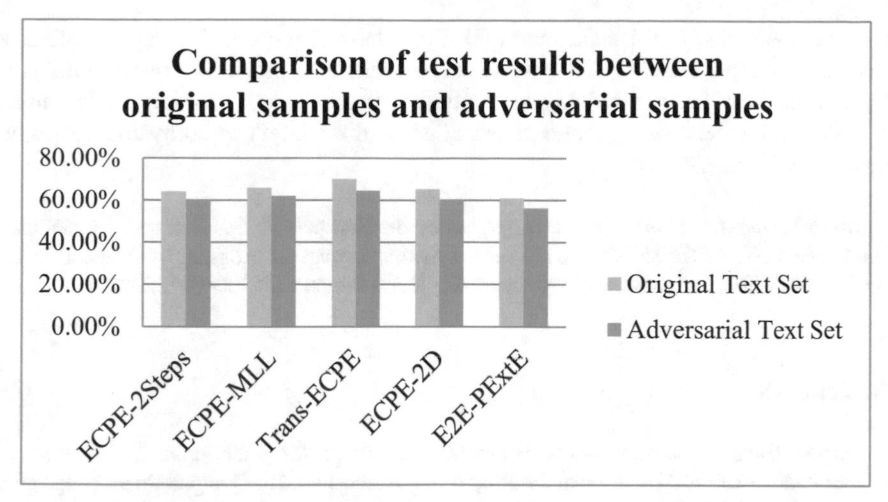

Fig. 4. Comparison of test results between original samples and adversarial samples

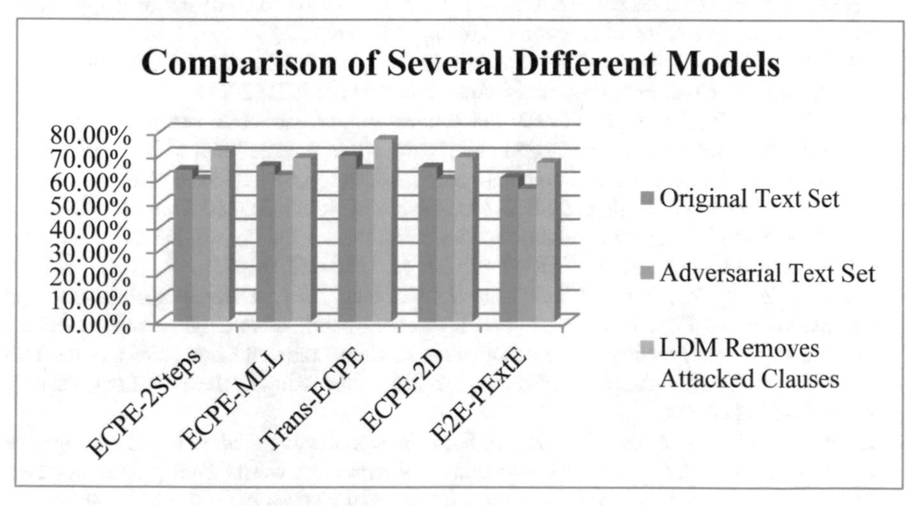

Fig. 5. Comparison of several different models

LDM helps existing models improve extraction precision. When the data received by the attack are removed, the small overall data volume will lead to a large amount of accurate data, making the extraction precision accurate.

5 Conclusion

Propose LDM in order that locate the position of ECPE task attacked by emotional adversarial samples. LDM combines emotions and causes to ensure that the model fully considers the relationship between them. By calculating the emotional cause score, the deep causal is integrated into the cause extraction process to ensure that the data is highly sensitive to attacks. LDM makes the ECPE task have defense ability against affective adversarial sample attacks. The experimental results also prove that our method is an effective method that can quickly locate the emotional attack location. In the future, we will combine the background information of the context to study the deep false correlation of emotional causes.

Acknowledgements. This work was supported by the National Natural Science Foundation of China (Grant NO.62076006), the University Synergy Innovation Program of Anhui Province (GXXT-2021–008), and the Anhui Provincial Key R&D Program(202004b11020029).

References

1. Xia, R., Ding, Z.: Emotion-Cause pair extraction: a new task to emotion analysis in texts. In: Proceedings of the 57th Annual Meeting of the Association for Computational Linguistics, pp. 1003–1012 (2019)
2. Lee, S.Y.M., Chen, Y., Huang, C.R.: A text-driven rule-based system for emotion cause detection. In: Proceedings of the NAACL HLT 2010 workshop on computational approaches to analysis and generation of emotion in text (pp. 45–53) (2010)
3. Yu, J., Liu, W., He, Y., Zhang, C.: A mutually auxiliary multitask model with self-distillation for emotion-cause pair extraction. IEEE Access **9**, 26811–26821 (2021)
4. Ding, Z., Xia, R., Yu, J.: ECPE-2D: emotion-cause pair extraction based on joint two-dimensional representation, interaction and prediction. In: Proceedings of the 58th Annual Meeting of the Association for Computational Linguistics, pp. 3161–3170 (2020)
5. Keller, Y., Mackensen, J., Eger, S.: BERT-Defense: a probabilistic model based on BERT to combat cognitively inspired orthographic adversarial attacks. Findings of the Association for Computational Linguistics: ACL-IJCNLP 2021, pp. 1616–1629 (2021)
6. Zhou, Y., Zheng, X., Hsieh, C.J., Chang, K.W., Huang, X.J.: Defense against synonym substitution-based adversarial attacks via dirichlet neighborhood ensemble. In: Proceedings of the 59th Annual Meeting of the Association for Computational Linguistics and the 11th International Joint Conference on Natural Language Processing (Volume 1: Long Papers), pp. 5482–5492 (2021)
7. Zhang, S., Xu, H., Zhu, G., Chen, X., Li, K.: A data processing method based on sequence labeling and syntactic analysis for extracting new sentiment words from product reviews. Soft. Comput. **26**(2), 853–866 (2021). https://doi.org/10.1007/s00500-021-06228-9
8. Xu, H., Zhang, S., Zhu, G., Zhu, H.: ALSEE: a framework for attribute-level sentiment element extraction towards product reviews. Connect. Sci., 1–19 (2021)
9. Gui, L., Xu, R., Wu, D., Lu, Q., Zhou, Y.: Event-driven emotion cause extraction with corpus construction. In: Social Media Content Analysis: Natural Language Processing and Beyond, pp. 145–160 (2018)
10. Gui, L., Xu, R., Lu, Q., Wu, D., Zhou, Y.: Emotion cause extraction, a challenging task with corpus construction. In: Li, Y., Xiang, G., Lin, H., Wang, M. (eds.) SMP 2016. CCIS, vol. 669, pp. 98–109. Springer, Singapore (2016). https://doi.org/10.1007/978-981-10-2993-6_8
11. Ding, Z., Xia, R., Yu, J.: End-to-end emotion-cause pair extraction based on sliding window multi-label learning. In: Proceedings of the 2020 Conference on Empirical Methods in Natural Language Processing (EMNLP), pp. 3574–3583 (2020)
12. Fan, C., Yuan, C., Du, J., Gui, L., Yang, M., Xu, R.: Transition-based directed graph construction for emotion-cause pair extraction. In: Proceedings of the 58th Annual Meeting of the Association for Computational Linguistics, pp. 3707–3717 (2020)

13. Singh, A., Hingane, S., Wani, S., Modi, A.: An end-to-end network for emotion-cause pair extraction. In: Proceedings of the Eleventh Workshop on Computational Approaches to Subjectivity, Sentiment and Social Media Analysis, pp. 84–91 (2021)
14. Li, J., et al.: TextShield: robust text classification based on multimodal embedding and neural machine translation. In: Proceedings of the 29th USENIX Conference on Security Symposium, pp.1381–1398 (2020)
15. Bao, R., Wang, J., Zhao, H.: Defending pre-trained language models from adversarial word substitution without performance sacrifice. In: Findings of the Association for Computational Linguistics: ACL-IJCNLP 2021, pp. 3248–3258 (2021)

Cyber Intelligence for CV Process and Data Mining

The Application of Computer Vision Synthesis Technology in the Editing of Film and TV Shooting

Fangfang Chen[1(✉)] and Malik Alassery[2]

[1] Jiangxi University of Software Professional Technology, Nanchang, Jiangxi, China
jxchenfangfang@163.com
[2] Amman Arab University, Amman, Jordan

Abstract. Post-compositing technology is an important final stage in the film or TV production process. The post-construction technology is to first compose the film according to the camera table, and then combine the previously produced video and image elements through the software. This paper studies the application of computer vision synthesis technology in the post-production editing of FAT (FAT). Based on the literature, we understand the post-production editing of FAT, and then investigate the application of computer vision synthesis technology in the post-production editing of FAT. The result is that computer vision synthesis technology in FAT can enrich the level of the FAT and the form of visual presentation is more colorful.

Keyword: Computer vision · Synthesis technology · Post-editing film · Television shooting

1 Inductions

In recent years, post-synthesis technology has fully penetrated into the FAT production process, and plays an important role in all stages of production. Many original equipment is gradually being replaced by computers [1, 2]. In the past, FAT production required very expensive professional hardware and software. Production teams with low budgets are generally difficult to reach [3, 4]. With the maturity of computer technology, PC performance has been significantly improved, and FAT production has gradually shifted to the PC platform [5, 6]. With the development of the FAT industry, more and more industries have crossed into the FAT industry. Cross-industry creators can use professional FAT production systems to create their own films, recreate more creative ways of thinking, and bring them to the industry rich works [7, 8].

For the research of post-composite technology, some researchers have proposed an automatic segmentation algorithm for moving video objects based on dynamic background construction. The algorithm divides the foreground by dynamic background modeling technology, that is, uses the background information of multiple image frames next to the current frame to establish a moving background, and then uses the removal of

J. H. Abawajy et al. (Eds.): ICATCI 2022, LNDECT 169, pp. 277–284, 2023.
https://doi.org/10.1007/978-3-031-28893-7_33

the background information to segment the moving foreground, while detecting similar foreground areas, such as zero frame difference motion the area around the object, etc., are merged into the newly segmented moving foreground area to make it more tidy and consistent. The algorithm finally takes the foreground area as the starting partition, and divides the area based on the color gradient, and optimizes the color boundary of the foreground area through active contour snake calculation to achieve a more accurate contour of the moving object [9, 10]. Some developers have also extended the powerful keys technology to 3D applications. By dealing with the problem of the front and rear video flickering when the front and rear frames are separated, the video data is transformed into a space-time cube. This method is also suitable for processing the temporal continuity of the background and the video scene segmentation. At the same time, the effect of the algorithm changes with the number of each foreground segmentation. The higher the color contrast between the foreground and the back boundary, the higher the foreground segmentation performance. However, the segmentation effect reduces the color flicker of the front boundary between adjacent frames, which will cause the problem of blurring the segmentation boundary, thus losing the segmentation characteristics [11]. Therefore, some researchers provide an image-based algorithm based on the environment key. It regards the environment as using a set of textures to describe the three kinds of light effects in the object. Although this method can obtain a very good combination result, the algorithm cannot accurately describe the irregular diffuse reflection phenomenon. In addition, because the environment-based image synthesis process requires large-scale acquisition and is costly, it is more suitable for the synthesis of small scenes [12]. In summary, with the increasing market value of FAT, people are studying more and more professional facilities for FAT. With the maturity of computer technology, the post-production work of FAT has been replaced by computing software, so computer vision there are more and more researches on technology.

This paper studies the application of computer vision synthesis technology in FAT shooting post-editing. Based on the literature, the necessity of computer vision technology and the collection of FAT shooting and post-editing as well as the application of computer vision technology in FAT shooting and post-editing are carried out, and then use the method of questionnaire survey to show the impact of the application of computer vision technology in FAT shooting and post-editing.

2 Research on Editing of FAT Shooting

2.1 The Necessity of Combining Computer Vision Synthesis Technology with FAT Shooting

2.1.1 Enrich the Image Hierarchy

Through the "real" shooting of real scenes, the "virtual" production of animations produced by digital technology, and the combination of "real" and "virtual" artistic expression techniques, it is a conflict between different levels of FAT and artistic expression. Let the audience appreciate the dazzling sparks produced by the collision of art.

2.1.2 Meet the Emotional Needs of the Audience

Video short films need to show all the content of the script within a limited time. The actual shooting details are added to the short film using computer technology. Such as live-action movies, TV shorts, etc., this method can stimulate the audience's strong sense of substitution, feel the emotions and thoughts conveyed by the FAT works in a limited time, and have a deeper understanding of the meaning of the works.

2.1.3 Show Different Visual Experience

Combining digital technology with real shots can not only make real shots look real, but also enrich the visual experience of FAT. Various script settings show the use of digital technology software to produce creative ideas and ideas that cannot be realized in movies and TV shows. FAT are not limited to the same one-dimensional foreground and background, but a combination of multi-dimensional visual performance. This new form of expression brings a brand new visual enjoyment to the audience.

In short, the combination of digital technology and live shooting has changed the traditional way of FAT production and brought new vitality to FAT. In order to combine digital technology with real shots and express different forms of art in FAT, to show the audience different visual experiences that meet the emotional needs of the audience. The combination of different images will reflect the dimensions of the new art of FAT, allowing the audience to experience the real and wonderful the world of FAT.

2.2 Application of Computer Vision Synthesis Technology

(1) There are many digital 2D drawing editing software, but these drawing editing software are a set of programs that use digital computer drawing. According to its functions, digital computer technology can be divided into two types: painting and image design. That is, bitmap technology and appropriate technology. Typical software such as Painter, Illustrator, Corel Draw, Photoshop. For most image editing work, FAT creators prefer to use Photoshop for editing and Painter for compositing. Photoshop is the most popular and powerful 2D drawing editing software. Although there are many softwares, such as Painter and Illustrator, Photoshop has powerful layer editing functions, which make it more efficient in the digital synthesis process, and the operation is relatively simple. It is currently the leader in the field of digital synthesis software applications.

(2) Digital synthesis technology is an important technology for synthesizing special effects in modern film production and occupies an important position. For 2D plane production, Photoshop is a 2D editing technology that can effectively combine and combine, while FAT need to combine and synthesize dynamic effects, which is a more complex and advanced synthesis technology. The application of digital synthesis technology is not only the stacking of materials and simple and beautiful processing, but more importantly the processing of dynamic images. This processing combines elements such as light and shadow, and can be adjusted and processed with special technology. This kind of synthesis includes the synthesis of various elements of FAT, and if a certain element is ignored, it will have a significant impact on the realism of the synthesized image, thereby significantly reducing the effectiveness of the image. This is also one of the main problems that the digital post processor must solve in the working process.

(3) Due to the abundance of special effects technology, the core technology of digital models is also very powerful, and its advantages go beyond the scope of digital synthesis software. After Effects is a typical special effects technology software. After Effects allows to create simple 3D objects, customize the texture of 3D objects, define 3D cameras, and so on. In addition to the boring weakness of node synthesis, popular 3D software such as Maya and Houdini usually supports multiple types of materials. If the encoding significantly reduces the resolution of the material, you can use 3D software such as Maya to render the material. Therefore, After Effects completely completed the last step of the later stage, in which the integration of resources made up for the lack of 3D software.

3 The Application Investigation of Computer Vision Synthesis Technology in the Post-editing of FAT Shooting

3.1 Research Questions

The problems studied in this paper are in two aspects: the first is the impact of computer vision synthesis technology on FAT, and the second is the development direction of computer vision synthesis technology in the editing of FAT shooting.

3.2 Research Process

3.2.1 Survey Object

This article investigates the application of computer vision synthesis technology in the post-editing of FAT shooting. The subject of investigation is FAT post-producers. In order to ensure the accuracy of the subject, the investigation activity was carried out in the TV station center of this city.

3.2.2 Survey Sample

Determination of the number of survey samples According to the relevant information consulted and the actual conditions of the survey, the number of samples for this survey was determined to be 138. After distribution, the number of questionnaires returned was 130, and the number of valid questionnaires was 125.

3.3 Data Processing

The standard error is the standard deviation of the sampling distribution of a specific statistic (average sample value, average sample difference, sample ratio, correlation coefficient, etc.). The standard error is used to measure the degree of variance of a statistical sample. Parameter estimation and hypothesis testing are important measures used to measure the gap between statistical samples and commonly used parameters. In practical applications, the standard error is usually calculated based on data samples.

Common standard errors include the mean error of the standard sample and the ratio error of the standard sample. The calculation formula is shown in (1) (2):

$$\sigma = \sqrt{\frac{\sum (X_i - \mu)^2}{M}} \tag{1}$$

$$s = \sqrt{\frac{\sum (x_i - \overline{x})^2}{m - 1}} \tag{2}$$

Among them, μ is the overall mean, M is the number of overall data, and $\overline{x}$ is the sample mean.

4 Analysis of Survey Results

4.1 The Impact of Computer Vision Synthesis Technology on FAT

This paper investigates the application of computer vision synthesis technology in the post-production editing of FAT through a questionnaire survey. The results of the impact of computer vision synthesis technology on post-production editing of FAT are obtained by sorting out the questionnaire data, as shown in Table 1:

Table 1. The influence of computer vision synthesis technology on FAT

	FAT Center 1	FAT Center 2	FAT Center 3
Rich picture hierarchy	44%	46%	45%
Meet the emotional needs of the audience	24%	22%	21%
Show different visual experience	32%	32%	33%

As can be seen from Fig. 1, in the impact of computer vision synthesis technology on FAT, the rich picture levels accounted for more than 44%, and then it showed different visual feelings, accounting for more than 32%.

4.2 The Development Direction of Computer Vision Synthesis Technology in Post-production Editing of FAT

This paper investigates the application of computer vision synthesis technology in the post-production editing of FAT through the questionnaire survey method, and obtains the development direction of computer vision synthesis technology in the post-production editing of FAT by sorting out the questionnaire data. The results are shown in Table 2:

It can be seen from Fig. 2 that in the development direction of computer vision synthesis technology in the post-editing of FAT shooting, it is mainly the re-interpretation of the visualization of virtual products, accounting for more than 44%, and then the transformation of expression methods, accounting for 33% about.

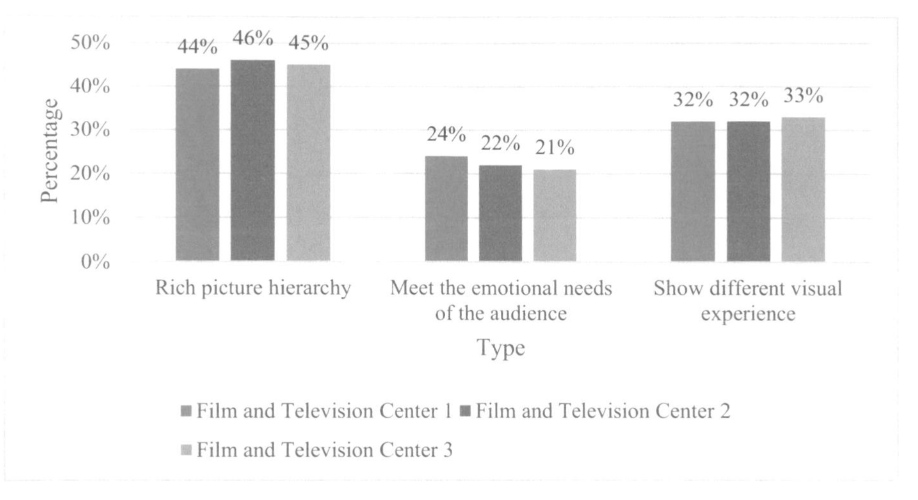

Fig. 1. The influence of computer vision synthesis technology on FAT

Table 2. The development direction of computer vision synthesis technology in the post-production editing of FAT

	FAT Center 1	FAT Center 2	FAT Center 3
Reinterpretation of the visualization of virtual products	46%	44%	45%
Change in performance	32%	32%	34%
Promote the artistic process of FAT	22%	24%	21%

4.3 Outlook

(1) The evolution of the FAT industry is coordinated with science and technology in the fields of computers. For example, the application of new technologies such as virtual reality technology and 3D holographic technology in many fields. To this end, film creators need to constantly update their knowledge, look for the development direction of new technologies, fully adapt to the mainstream design concepts of the times, and expand more ways of expression. There is reason to believe that in the near future, the diversification of technology will accelerate the rapid development of the FAT industry, expand the form of communication between film and the public, and achieve the best effects of FAT. Specifically, creating virtual product images through computer technology will be better and more vivid, which can ensure a brand-new visual experience.

(2) In the digital age, the combination of digital and real film photography technology and FAT production has made important discoveries in many directions and fields, and has had a subtle influence in the field of aesthetic art. The combination of digital and real-world shooting technology challenges the traditional aesthetic theory of art, completely breaks the traditional FAT shooting methods in real movies, and creates creative

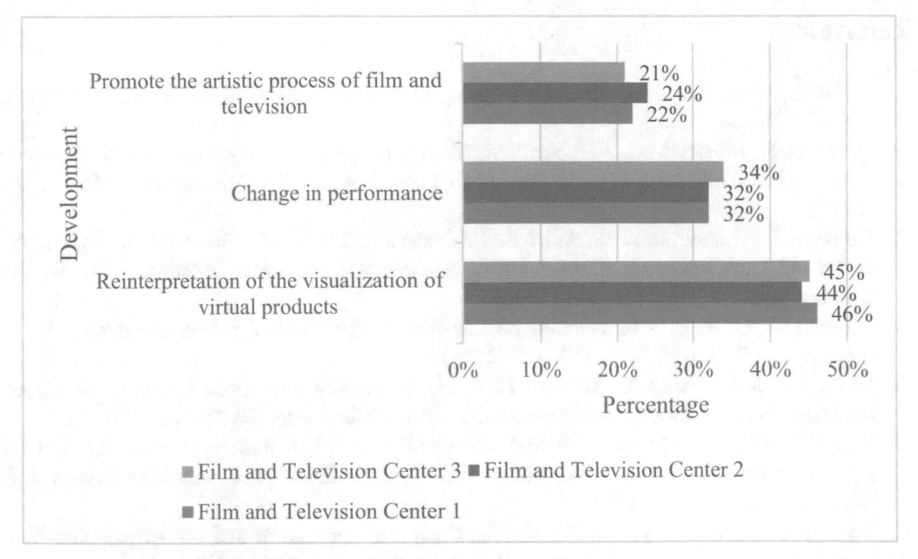

Fig. 2. The development direction of computer vision synthesis technology in the post-production editing of FAT

expression channels in FAT. The differentiation of styles presents the public with a sense of artistic beauty of fiction and reality.

(3) The magic of the combination of digital and real shots is a combination of illusion and reality. The animated characters in the 3D animation software and the virtual and real motion scenes shot by real and virtual cameras are shocking, but at the same time it is difficult to distinguish between true and false. The technology of combining digital animation and real shooting has freed FAT from the objective connection between reality and virtuality, turned "fake reality" into reality, and created an unprecedented super-realistic visual experience. Therefore, the development of digital technology will surely subvert the traditional FAT Aesthetics.

5 Conclusion

This article focuses on the application of computer vision synthesis technology in the editing of FAT shooting. After understanding the relevant theories, it investigates the application of computer vision synthesis technology in the editing of FAT shooting. The main impact of technology on FAT is the richness of content levels and the richness of visual expressions.

References

1. Williams, M.B., Davis, S.H.: Nonlinear theory of film rupture. J. Colloid Interface Sci. **90**(1), 220–228 (2016)
2. Perreault, F., Tousley, M.E., Elimelech, M.: Thin-film composite polyamide membranes functionalized with biocidal graphene oxide nanosheets. Environ. Sci. Technol. Lett. **1**(1), 71–76 (2016)
3. Hultman, L., Bareño, J., Flink, A., et al.: Interface structure in Superhard TiN-SiN nanolaminates and nanocomposites: film growth experiments and ab initio calculations. Phys. Rev. B Conden. Matt. **75**(15), 1418–1428 (2016)
4. Altman, C., Cory, H.: The generalized thin-film optical method in electromagnetic wave propagation. Radio Sci. **4**(5), 459–470 (2016)
5. Khan, M., Rahman, M.A., Yasmin, P., et al.: Formation and characterization of copper nanocube-decorated reduced graphene oxide film. J. Nanomater. **2017**(4), 1–6 (2017)
6. Ketmaneechairat, H.: Probabilistic and biologically inspired feature representations michael pelsberg synthesis lectures on computer vision morgan & claypool publishers 2018. J. Inf. Secur. Res. **10**(1), 29 (2019)
7. Venkatesh, V., Thong, J., Xu, X.: Unified theory of acceptance and use of technology: a synthesis and the road ahead. J. Assoc. Inf. Syst. **17**(5), 328–376 (2016)
8. Kokaram, A., Singh, D., Robinson, S., et al.: Motion-based frame interpolation for film and television effects. IET Comput. Vision **14**(6), 323–338 (2020)
9. Pumarola, A., Agudo, A., Martinez, A.M., Sanfeliu, A., Moreno-Noguer, F.: GANimation: one-shot anatomically consistent facial animation. Int. J. Comput. Vision **128**(3), 698–713 (2019). https://doi.org/10.1007/s11263-019-01210-3
10. Khadzhiev, S.N., Kolesnichenko, N.V., Ezhova, N.N.: Slurry technology in methanol synthesis (Review). Pet. Chem. **56**(2), 77–95 (2016). https://doi.org/10.1134/S0965544116020079
11. Pei, Z., Wyman, I., Hu, J., et al.: Silver nanowires: Synthesis technologies, growth mechanism and multifunctional applications. Mater. Sci. Eng. B **223**, 1–23 (2017)
12. Tolendiuly, S., Alipbayev, K., Fomenko, S. et al.: Properties of high-temperature superconductors (HTS) and synthesis technology. Metalurgija -Sisak then Zagreb- **60**(1-2), 137–140 (2020)

Application of 3S Technology in Land Use Landscape Ecology

Ziwen Qu(✉)

Nanjing Normal University School of Geographical Sciences, Nanjing, Jiangsu, China
qzwchampionv@163.com

Abstract. As human intervention in the land ownership system continues to increase, the contradiction between land use and human activities has become increasingly prominent. Therefore, it is of great significance to apply 3S technology and based on data information to establish a model of land use and landscape design changes, and to study land use to maintain ecological health. This article uses the investigation method and the data method to understand the 3S technology accordingly, and conducts an in-depth study on the land use situation in this area. The survey results found that the non-farm agricultural land in this area accounted for 50.1%, and the green area accounted for 36% of it. This shows that the local land use release conforms to landscape ecology.

Keywords: 3S technology · Land use · Landscape ecology · Applied research

1 Introduction

"3S" technology is an important means for dynamic land use monitoring and finding areas of land change. Because of the differences in soil structure types in different regions. In addition, because people use various methods of pesticides and fertilizers to cause serious water pollution problems, land degradation has become increasingly severe. Therefore, regional land use planning needs to use 3S technology for a reasonable layout.

There are many theoretical results for the application of 3S technology in land use landscape ecology. For example, Some scholars said that land resources are an important part of the urban construction process. With the continuous development of the land market, various illegal phenomena of land resources have become more and more serious. In order to protect arable land resources and effectively crack down on various violations of laws and regulations, it is necessary to strengthen land construction in the new era. And 3S technology is the main technical means of land use inspection work [1]. Some scholars said that as a high-tech, 3S technology is very effective in the application of information management. Therefore, it is applied to the management of land resources and has a very broad prospect [2]. Some scholars believes that the application of 3S technology in the scientific exploration of land resources provides new ways and technical support for the research and exploration of land resources [3]. Therefore, this article explores the application of 3S technology from land use and landscape ecology, aiming to make full use of the existing methods and technologics for land exploration, planning and design.

J. H. Abawajy et al. (Eds.): ICATCI 2022, LNDECT 169, pp. 285–293, 2023.
https://doi.org/10.1007/978-3-031-28893-7_34

This article first studies 3S technology and dynamic remote sensing monitoring. Secondly, it analyzes the relationship between landscape ecology and land use. Finally, an in-depth analysis of the application of 3S technology, and the use of technology to understand the situation of local land use and the proportion of the area, and finally draw a conclusion.

2 Application of 3S Technology in Land Use Landscape Ecology

2.1 3S Technology and Dynamic Remote Sensing Monitoring

The high and new technology represented by "3S" includes the use of remote sensing technology (RS) to obtain land use change information, and the use of global positioning system (GPS) technology to quickly and accurately obtain spatial coordinate information changes to receive the use of geographic information systems (GIS). A new operating mechanism for monitoring and managing ground dynamic remote sensing [4, 5].

2.1.1 Geographic Information System Technology

Geographical Information System is an information subject in which computer applications are applied to geography. It covers many fields of computer science, geography and related fields, and it also derives many research fields. GIS is still based on computer maps. Many applications have been developed around the visualization of geographic information. The expression, processing, storage and analysis of geographic information are all related to graphics and images. Therefore, computer graphics and digital image processing are the most direct subjects of computer application in geography [6, 7].

2.1.2 Global Positioning System Technology

The Global Positioning System is a space-based satellite navigation and positioning system jointly developed by the Army, Navy and Air Force.

2.1.3 RS Technology

Remote sensing refers to the technology of telemetry and detection of objects. Generally, it does not directly contact the object itself, but collects and receives information from the target object through the instrument.

2.1.4 Dynamic Remote Sensing Monitoring

The so-called dynamic land use monitoring includes comparing land use data at different time periods. Dynamic land use monitoring can provide dynamic information on the quantity, quality and spatial distribution of various types of land use [8, 9].

2.1.5 The Role and Advantages of "3S" Technology in Land Use

"3S" integration technology is the three major supporting technologies for geographic information acquisition. Apply remote sensing, geographic information system and global positioning system to conduct comprehensive research on land resource environment and environmental changes [10].

The global positioning satellite system has the characteristics of high precision, user-friendliness and wide application. The ability to use GPS to perform high-precision positioning and testing of land use/cover change results is a reliable guarantee for the spatial accuracy of remote sensing information, various ecological environment and regional research resources and thematic data [11, 12].

2.2 Landscape Ecology

2.2.1 Landscape Ecological Planning and Land Use Planning

The principles of landscape ecology provide new possibilities for land use planning, ecological protection, landscaping and integrated land management. They are extremely important for various scales of land in the region. With the change of land use, each change process will be affected by ecological or human disturbance, forming different landscape spatial patterns, and then determining the function and change of the land. The analysis of regional landscape design and ecological process is the basis of land use planning. Successful planning lies in understanding the planned landscapes, and ecological processes help us use them as prerequisites for spatial planning. The important significance of ecological planning is reflected in the adaptation of land use patterns. Therefore, the analysis of landscape structure and its functions and the dynamic evaluation of landscape models are of great significance to planning [13].

2.2.2 Evaluation Method of Landscape Ecological Quality

Landscape ecological quality is an index to measure the ecological conditions of landscape ecosystems, reflecting the ability of landscape ecosystems to maintain ecological functions. The ability of a landscape ecosystem to perform its ecological functions is determined by the structure of the landscape ecosystem and is also affected by the open environment. In landscape ecology, landscape structure not only refers to the composition of the landscape, but also emphasizes the importance of landscape space design. The selection principles of landscape ecological quality evaluation indicators include systemicity and scientificity, dominant factors and differences, measurability and feasibility, and emphasize the reflection of natural ecological characteristics. The evaluation index system can be divided into three levels: the first level is the target level. The second layer is the standard layer, including the composition, model and disturbance of the ecological state of the landscape ecosystem. The third layer is the performance index, and the specific index layer is the specific index that characterizes each standard.

2.2.3 Ecological Components of the Landscape

The ecological components of the landscape are the basic components of the ecological environment and the realization of ecological functions. The ecological functions of the ecosystem must be supplemented by a certain amount and quality of ecological components.

(1) High-functional Ecological Components

Its ingredients mainly involve natural ecological landscape. Ecosystem service level is based on ecosystem function, and the evaluation of ecosystem service value indirectly reflects the overall capacity of the regional ecosystem. In order to eliminate the impact of the difference in the area of the evaluation unit on the evaluation results, the equivalent of the ecosystem service function (EVE) value per unit area of the evaluation unit is used as the level result to evaluate the ecological function of the landscape.

$$EVE = \sum\nolimits_{a=1}^{m} t_a \times r_a, r = s_a/s_g \quad (1)$$

In the formula, EVE represents the value equivalent of ecosystem services in a regional unit area, t_a represents the total value equivalent of ecosystem services in each ecosystem, and r_a is the ratio of ecological land to the total area of the region.

(2) Ecological Landscape Aggregation Index

The agglomeration index is an index established based on the distribution relationship between the boundaries of various landscape elements in the landscape and is not affected by the ecological differences between adjacent heterogeneous landscapes. It can be expressed as:

$$\text{aggregation index} = \left[\sum\nolimits_{a=1}^{n} (f_{ab}\max \to f_{ab})\right] \times 100 \quad (2)$$

In the formula, f_{ab} is the number of adjacent pixels of type patch; max $\to f_{ab}$ is the maximum number of adjacent pixels of type a patch.

(3) Degree of Separation from Ecological Landscape

The degree of separation describes the degree of dispersion of individual distribution of different elements in a certain type of landscape. The higher the value, the wider the geographic distribution of the landscape type, and the lower the landscape connectivity.

(4) Fragmentation of Ecological Landscape

Landscape fragmentation describes the degree of fragmentation of the land use landscape. The fragmentation process is mainly manifested as an increase in the number of parcels and a decrease in area.

(5) Edge Density of Ecological Landscape

Boundary density refers to the total length of the boundaries of all plots in the landscape divided by the total area of the landscape. The higher the boundary density, the more complex the boundary.

2.3 3S Technology Application

2.3.1 GPS Control Point Data

The collection and correction of remote sensing images require control points, and the number of control points is determined by the correction model. The general calibration model has at least 12 control points per scene, and the number of control points needs to be increased in areas that are difficult to identify. The control points should be evenly distributed, clearly marked, and able to control the entire screen of the scene. Field surveys use GPS to conduct surveys, locate accurate location data, obtain accurate location data, and collect images to evaluate the accuracy of monitoring DOM.

2.3.2 Image Preprocessing Based on RS

Due to changes in the position of SPOT satellites, the interaction between the atmosphere and electromagnetic waves, the uneven detection medium, the height of the sun, the curvature and rotation of the earth, and the shape of the earth's surface, remote sensing images have some distortions during the imaging process. Therefore, it is necessary to restore the true information of the picture as much as possible. The specific corrective actions are: Import the image and view the entire image. Start the geometric correction model, and then select the keyboard to input control points. Define image mapping parameters and set height projection parameters. After the setting is completed, multiple control points will be collected. Finally, the image is resampled.

2.3.3 Detection of Change Information Based on RS

Computer automatic detection methods can quickly find the location and quantity of changed information, reduce labor and cost, and prevent misjudgment and omission of subjective factors. Image difference method is one of the most widely used methods in dynamic monitoring. This involves subtracting the remote sensing images of two time periods pixel by pixel according to the band to generate a new difference image representing the spectral development between the two time periods. The comparison process after classification includes classification of remote sensing images during alignment and subsequent comparative analysis of classification results to discover and extract change information.

2.3.4 Extraction of Change Information Based on GIS

Use the image texture change information found by the image difference method and the post-classification comparison method to assist human visual interpretation, help identify specific change areas, and determine the type of change. The basic elements of interpretation are hue/color, size, shape, texture, structure, height, tone, combination configuration and geographic location.

2.3.5 GPS Monitoring Information Polling

On-site inspection According to the requirements of the order, after making preparations such as the measurement base map and the on-site record sheet, go to the monitoring area for verification. The changed model after visual inspection is not completely accurate, and there are some suspicious models, so the field industry should review each model separately.

3 Accuracy Assessment of Remote Sensing Monitoring Results of Land Use Dynamics

3.1 Evaluation Method

There are generally two methods for DOM accuracy rating: absolute accuracy rating and relative accuracy rating. In order to ensure the continuity of remote sensing dynamic land use monitoring, the method of relative accuracy assessment is more objective and practical. The calculation formula is as follows:

$$rms = \sqrt{\frac{\sum_{q} (a_q - A_q)^2 + (b_q - B_q)^2}{x}} \tag{3}$$

In the formula: rms is the error in the position of the point, x is the number of checkpoints, a_q - A_q represents the a, b coordinates of the checkpoint on the front phase image, and b_q - B_q is the a, b coordinate of the checkpoint on the later phase image.

3.2 Evaluation Process

Using the ERDAS platform, import two phases of remote sensing orthophotos, select 20 features with the same name, and calculate the error in the position. Through statistical calculations, there are a total of 20 control points in the A direction and B direction of the local land use, and their errors are all within one pixel. And there is no ghosting phenomenon on the orthorectified map.

3.3 Sources of Error

The attribute error of the land use category directly reflects the correctness of the interpretation of the revised land category. The area error of the map change point is the difference between the detected area of the extracted map change point and its actual area, which is affected by the resolution of the data source used, correction, recording, merging quality, change information detection and extraction accuracy. The complete performance of multiple connection errors such as mixed pixels and card restrictions.

4 Analysis of Monitoring Results

4.1 Occupied Land Situation of Newly-Added Construction Land

The newly-added construction land in this area occupies 10 arable maps with an area of 30 hectares; 15 non-cultivated agricultural land with an area of 43 hectares; 6 unused land with an area of 12 hectares. The specific situation is shown in Table 1:

As shown in Fig. 1, we can see that according to GIS and RS technology, this paper monitors the local land use. There are three types of land occupation. Among them, the largest area is non-cultivated agricultural land, accounting for 50.1%, followed by cultivated land area, and the next is unused land.

Table 1. Newly-added construction land occupation land situation

Occupied land	Number	Proportion	Area	Proportion
Cultivated land	10	32.2%	30	35.3%
Non-cultivated agricultural land	15	48.4%	43	50.1%
Unused land	6	19.4%	12	14.6%

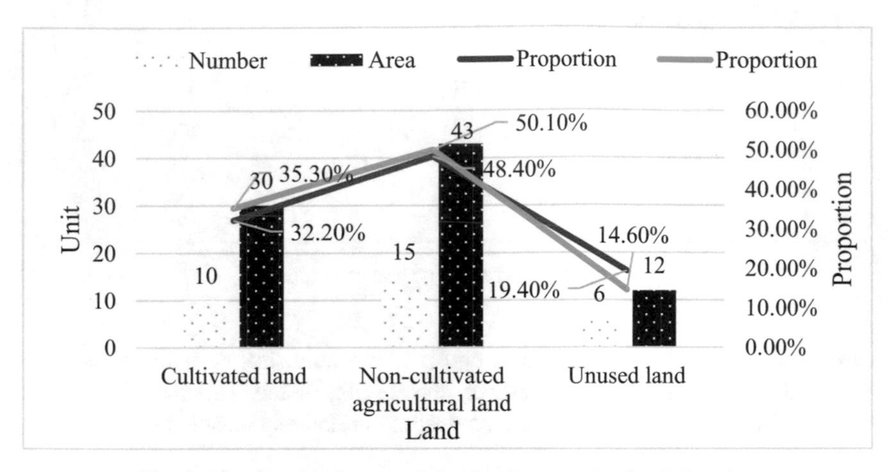

Fig. 1. Newly-added construction land occupation land situation

4.2 Quantity Structure of Land Use

After analyzing the land use situation according to the 3S technology, combined with the analysis method of landscape ecology, the non-cultivated agricultural land is laid out. The specific situation is shown in Table 2:

Table 2. Land use type and area ratio

	Area	Proportion
Garden	5.16	12%
Woodland	3.87	9%
Grassland	6.45	15%
Land for transportation	4.73	11%
Construction land	20.21	47%
Other land	2.58	6%

As shown in Fig. 2, we can see that non-cultivated agricultural land is mainly divided into garden land, grassland, woodland, construction land and other land. Among them,

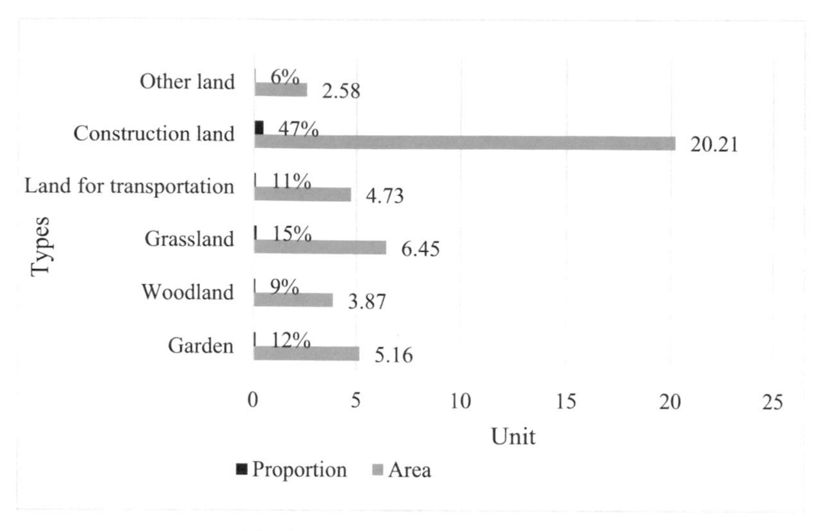

Fig. 2. Land use type and area ratio

the area with the largest proportion is construction land, followed by grassland and garden land, and the green area is relatively evenly distributed. This shows that the area covered by the local green land landscape design is relatively reasonable.

5 Conclusion

Under the conditions of declining land resources, decreasing biodiversity, deteriorating ecological environment, shortage of urban construction land and traffic congestion, necessary measures need to be taken to maintain the balance of the ecosystem. To this end, this article uses 3S technology to gain an in-depth understanding of the new land use in this area. According to the investigation and analysis of this article, it is found that the area of non-cultivated agricultural land is the largest, and among them, the area of construction land is the first. The landscape land ranks second among non-farm farms. It shows that the local land use situation is relatively reasonable.

References

1. Harker, K.J., Arnold, L., Sutherland, I.J., et al.: Perspectives from landscape ecology can improve environmental impact assessment. FACETS **6**(1), 358–378 (2021)
2. Jsdsa, C., Pd, B., Jefo, C., et al.: Landscape ecology in the Anthropocene: an overview for integrating agroecosystems and biodiversity conservation. Perspectives Ecol. Conserv. **19**(1), 21–32 (2021)
3. Tavakoli, M., Monavari, M., Farsad, F., Robati, M.: Ecotourism spatial-time planning model using ecosystem approaches and landscape ecology. Environ. Monit. Assess. **194**(2), 1–14 (2022). https://doi.org/10.1007/s10661-021-09558 1
4. Mansourian, S.: From landscape ecology to forest landscape restoration. Landscape Ecol. **36**(8), 2443–2452 (2021). https://doi.org/10.1007/s10980-020-01175-6

5. Milovanović, A., Milovanović Rodić, D., Maruna, M.: Eighty-year review of the evolution of landscape ecology: from a spatial planning perspective. Landscape Ecol. **35**(10), 2141–2161 (2020). https://doi.org/10.1007/s10980-020-01102-9
6. Cumming, G.S., Epstein, G.: Landscape sustainability and the landscape ecology of institutions. Landscape Ecol. **35**(11), 2613–2628 (2020). https://doi.org/10.1007/s10980-020-00989-8
7. Fedyń, I., Figarski, T., Kajtoch, Ł: Overview of the impact of forest habitats quality and landscape disturbances on the ecology and conservation of dormice species. Eur. J. Forest Res. **140**(3), 511–526 (2021). https://doi.org/10.1007/s10342-021-01362-3
8. Schroder, W., Murtha, T., Golden, C., et al.: The lowland Maya settlement landscape: environmental LiDAR and ecology. J. Archaeol. Sci. Rep. **33**(1/2), 102543 (2020)
9. Mahmoudzadeh, H., Masoudi, H.: The analysis of structural landscape changes in Tabriz city using landscape ecology principles with an emphasis on the connectivity concept. Town Country Plann. Quart. Rev. Town Country Plann. Assoc. **11**(2), 179–204 (2020)
10. Showler, A.T., Adalberto, P.: Landscape ecology of Rhipicephalus (Boophilus) microplus (Ixodida: Ixodidae) outbreaks in the south Texas coastal plain wildlife corridor including man-made barriers. Environ. Entomol. **3**, 3 (2020)
11. Delaney, L., Di Stefano, J., Sitters, H.: Mammal responses to spatial pattern in fire history depend on landscape context. Landscape Ecol. **36**(3), 897–914 (2021). https://doi.org/10.1007/s10980-020-01186-3
12. Fusco, N.A., Carlen, E.J., Munshi-South, J.: Urban landscape genetics: are biologists keeping up with the pace of urbanization? Current Landscape Ecol. Rep. **6**(2), 35–45 (2021). https://doi.org/10.1007/s40823-021-00062-3
13. González-Saucedo, Z.Y., González-Bernal, A., Martínez-Meyer, E.: Identifying priority areas for landscape connectivity for three large carnivores in northwestern Mexico and southwestern United States. Landscape Ecol. **36**(3), 877–896 (2021). https://doi.org/10.1007/s10980-020-01185-4

Application of Computer BIM Technology in Architectural Design

Zhijun Bao(✉)

College of Urban Construction and Design, Urban Vocational College of Sichuan, Chengdu 610110, Sichuan, China
2000312472@scuvc.edu.cn

Abstract. Since the Industrial Revolution, environmental destruction and energy depletion have seriously threatened our own survival and development. As one of the largest resource-consuming industries of mankind, the construction industry consumes a large proportion of the total energy in society. BIM theory is gradually being accepted by industry insiders. This emerging technology in my country is still in its infancy. Compared with the countries that began to apply BIM technology in the early days, there is still a big gap in the research and application of BIM in my country. Therefore, the application of this technology in AD (architectural design) is the focus of this article. Judging from the analysis of the current status of the domestic construction industry, the total output value is showing an upward trend. From 1995 to 2020, the total output value has increased by nearly 480 billion, but the gap with the manufacturing industry has widened again, and the design and production mode has been changed. It is particularly important. It analyzes the effect of computer BIMT (BIM technology)in AD.The results show that 46% of people believe that computer BIMT can improve design efficiency, and 22% believe that computer BIMT can enhance the accuracy of project decision-making. The remaining 19 people and 13 people think that this technology can reduce later rework and improve the quality of the project.

Keywords: BIM technology · Architectural design · Designing process · Application

1 Introduction

The design methods and processes of traditional technology platforms are difficult to meet all relevant needs of AD, but from the current development situation, the development of BIMT in the field of structural design is extremely slow. Compared with foreign research, the popularization and application of BIM technology in China is relatively late. BIM theory reading and research textbooks are few, and the scope of application is narrow. The application of BIM technology can more directly reflect the commercial value of the construction phase. The application of IM technology in the design stage can not only transmit information and complete manner, but also optimize the AD steps. The most concerned problem for structural engineers is how to better apply BIMT to structural design.

J. H. Abawajy et al. (Eds.): ICATCI 2022, LNDECT 169, pp. 294–301, 2023.
https://doi.org/10.1007/978-3-031-28893-7_35

Park believes that while the emergence of the Internet has promoted the process of globalization, it has also improved the ability of information exchange and processing [1]. Alsafouri explored the application of BIMT in landscape design to help the development of BIMT applications in the future. The application development of BIM in the current landscape design field is very slow. The main reason for this phenomenon is that the main body of the architectural landscape construction and the construction party have lower requirements for BIM technology during the landscape construction process [2]. Ferreira pointed out that BIM technology has been used as a symbol of the development of information technology in the construction industry since it was first proposed. BIM technology has advantages that the traditional working mode does not have, and has changed the disadvantages of the traditional mode [3].

This article carried out a specific analysis of the current situation of the domestic construction industry, and it is understood that the total output value of the construction industry is showing an upward trend, but the gap with the manufacturing industry has widened again; secondly, after analyzing the application effect of BIMT in architectural design, I learned that this new technology can not only improve efficiency, but also reduce later rework and improve project quality.

2 Research on the Application of Computer-Based BIM Technology in Architectural Design

2.1 BIM Technology

BIM is the abbreviation of Building Information Modeling. The building here is not just a chivalrously understood house, it can be a part of a building or a house or a construction project. The information is mainly divided into geometric information and non-geometric information. Geometric information is information that can be measured in a building. Non-geometric information includes Time, space, physics and other related information [4]. With the continuous development of information technology in the construction industry, the concept of BIM has also been given more different meanings and functions. Since the concept of BIM was first proposed to the present, professionals have conducted endless researches on BIM related theories for more than 30 years. Up to now, international organizations have not given a very accurate definition of the concept of BIM [5, 6]. Regardless of the definition of BIM, the word "information" is particularly emphasized, which is also the root of BIM. BIM is a powerful but complex, slow, and expensive tool. It is not equivalent to some software, but a kind of work flow [7].

Since the beginning of this century, the rapid development of computer technology and the Internet of Things has made the application of BIM technology more and more [2].With the gradual application of BIM in the industry in recent years, people continue to discover its advantages. This technology fundamentally promotes the forward development, and construction professionals can better improve architectural planning and design by establishing and using three-dimensional information models [8]. Among them, the building information model plays a very important role. In the BIM building information model, this model can not only be used as architectural rendering display, but also can export useful building data reports. A considerable number of technicians believe that BIM can reduce the probability of rework. All participants in the project can view possible conflicts through the shared BIM model, prompting them to resolve these issues in advance. In fact, BIM can not only solve various problems that may occur in the early stage of project construction, but also design coordination and optimization to reduce data errors and information loss on the project. Due to the lack of real-time communication between different majors, information "incompatibility" is prone to appear. Through the sharing platform of BIM, all parties in the project can easily change the saved and updated information [9]. Finally, BIM can perform building performance simulations for building projects, and provide performance simulations such as building energy efficiency and sunlight analysis in the form of 3D images. The problems in the construction organization plan formulated through simulation are found to guide the entire actual construction process [10].

2.2 Architectural Design

Architectural plan design refers to the design of a plan that meets the architectural functions according to the needs of the project and the design requirements of the project, and in the process, proposes the concept of architectural space, expresses the design process, and provides guidance. AD is mainly divided into four stages, including conceptual design, schematic design, preliminary design and construction drawing design [11]. Building design is easier to control and manage the design cycle. The models built by previous engineers in structural design software are all analytical models. The data under this model is often simplified and processed data. Compared with the traditional design process, the data transmission method and work process have also changed. For example, the current design cycle is usually compressed very short, usually requiring designers to complete the design overnight. The emergence of new technologies can shorten the entire construction project cycle within a reasonable range. In addition, the work tasks of designers in each design stage will be advanced, they no longer need to spend a lot of time to draw horizontal and vertical sections, and they can focus on the design itself. In the BIM mode, designers can also solve problems such as collisions, errors, and omissions more quickly and effectively, thereby greatly improving design quality.

2.3 The Specific Application of BIM in the AD Process

The specific application of BIM in the AD process includes two main parts: the planning stage and the specific application. First of all, the site and data must be analyzed in the planning stage. The BIM GIS system can be used to analyze the site of the building. We only need to add some site data, such as climate. Then input these characteristics and the surrounding environment into the GIS system, and finally through the calculation of the GIS system, the desired analysis results can be easily obtained [12]. After the site analysis is completed, the architect can proceed to the next step of a more in-depth design based on the specific results of the site analysis. Therefore, the BIM technology combined with GIS software can make the AD more accurate and efficient. Perform data analysis on the functional space according to relevant standards. The first step is to import relevant policies and regulations into the database, and then use spatial analysis and simulation to consider complex spaces [13]. Doing so can spend more time and energy on project innovation and optimization, making the plan more scientific and reasonable.

The second is the specific application. The specific application is divided into two stages: visual design and effect analysis. The first stage is the visual design stage. In this stage, computer graphics and image processing technologies are mainly used to convert data into graphics, and conduct interactive processing. This technology not only brings more intuitive feelings to the project in terms of visual expression, but also enables architects to design their own works from a multi-dimensional perspective. Throughout the design process, the visual advantage of the architectural model allows the various participants of the project (designer, owner, government, construction party) to communicate more effectively. Owners can see the intuitive design results, and the government can clearly know that the design is sufficient to meet the approval requirements. Second, analyze the impact of decisions made in the process of AD on the entire building cannot be ignored. Any small mistakes or mistakes can cause fatal problems in the subsequent stages of the design.The digital analysis using Ecotact software mainly includes the following aspects: visibility analysis, sunlight analysis, thermal radiation analysis, etc. Applying Ecotact software to the design stage of the BIM model scheme can effectively optimize the building layout. The following formulas are used in the process of digital analysis and analysis of BIM models:

$$x_j^c(k) = y_j(k-1) + x_i(k-1) \tag{1}$$

$$S^2 = \frac{(A - X_1) + (A - X_2) + \cdots (A - X_N)^2}{N} \tag{2}$$

3 Research on the Application of Computer BIMT in AD

3.1 Experimental Background

The shortage of resources and global environmental problems have aroused people's attention. BIM, as a kind of emerging computer-aided AD technology, has exerted its own advantages in the field of AD. However, according to the current development situation, the development of BIM technology in AD is very slow.

3.2 Experimental Process Steps

This article describes the process of BIM technology in architectural design. In order to better study the application of computer BIMT in AD, the investigation of this experiment is mainly carried out with the help of the questionnaire software and the printed paper questionnaire. These architects are from different parts of the country, 50 men and women each. A total of 100 questionnaires, 100 questionnaires were returned. The status quo of the construction industry and the effect of BIM technology on AD. Finally, the results of the questionnaire statistics were analyzed.

4 Experimental Analysis of the Application of Computer BIMT in AD

A specific analysis was carried out on the current situation of the domestic construction industry and the effect of BIM technology on AD.

4.1 Analysis of the Status Quo of the Construction Industry

In order to understand the current situation of the construction industry, the production efficiency growth rate of the construction industry and other non-agricultural industries were compared. The data results are shown in the following Fig. 1:

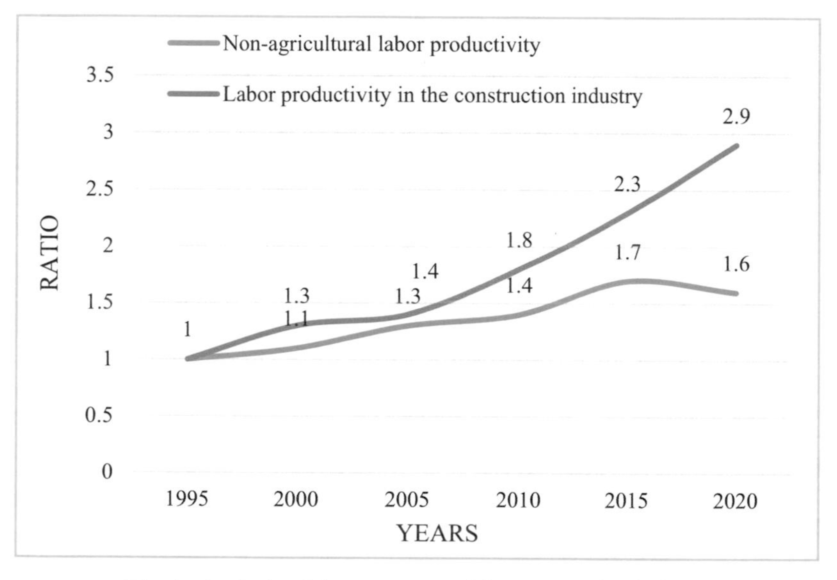

Fig. 1. Analysis of the status quo of the construction industry

It can be seen from Fig. 1 that in recent decades, the labor productivity of the construction industry has been rising year by year. From 1 in 1995 to 2.9 in 2020. Among them, the labor productivity growth effect of the construction industry after 2000 is particularly significant. In 2020, labor productivity has nearly doubled compared with the beginning of the 21st century. The labor productivity of the non-agricultural industry has also been improved, but compared with the construction industry, its growth is still relatively slow. From 1995 to 2020, the labor productivity of the non-agricultural industry only increased by 0.6.

As can be seen from Fig. 2, the total output value of the construction industry has shown an overall upward trend during the 25 years from 1995 to 2020. The total output value of the CI (construction industry) in 1995 was only 778.9 billion, and by 2020, the total output value of the CI has risen to 5,579 billion, an increase of nearly 480 billion. After 2010, the growth rate of the total output value of the CI was significantly faster than that of previous years, and the average growth output value in the five years reached 1.5 trillion yuan. On the whole, the domestic construction industry has maintained a momentum of growth, but the gap with the manufacturing industry has widened again, and it is particularly important to change the design and production mode.

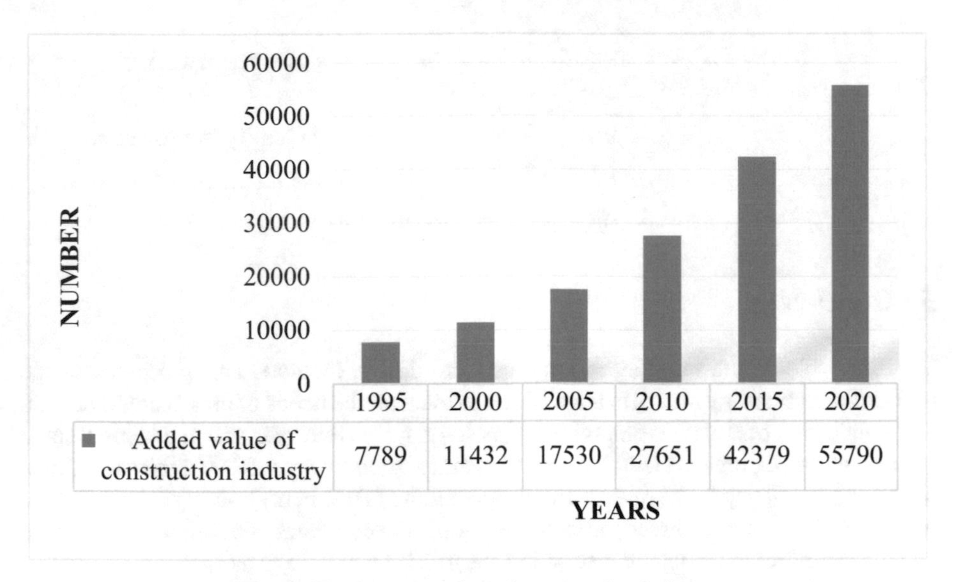

Fig. 2. Total output value of construction industry

4.2 Analysis of the Effect of the Application of Computer BIMT in AD

Table 1. The application effect of BIMT in AD

	Improve efficiency	Strengthen accuracy	Reduce rework	Quality improvement
Number	46	22	19	13
Proportion	46%	22%	19%	13%

The effective application of computer BIMT plays a very important role in promoting the vigorous development of the construction industry. It can be seen from Table 1 and Fig. 3 that 46 people think that computer BIM can improve design efficiency. 22% of people believe that computer BIM can enhance the accuracy of program decision-making. For example, the coordination between civil air defense zoning design and pipeline design and layout, and the coordination between room clearance and other design layouts are inseparable from BIM technology. There are 19 people who think that this technology can reduce later rework and avoid a lot of unnecessary troubles. The remaining 13 people believe that computer BIM technology can improve the quality of the project.

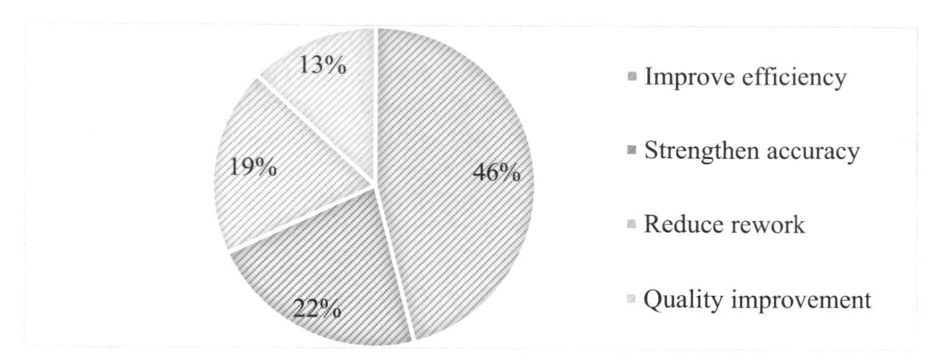

Fig. 3. The application effect of BIMT in AD

5 Conclusion

The life cycle of the building is long, and the design methods and processes of the traditional technology platform are difficult to meet all the needs of architectural design. The application of BIM this emerging technology in my country is still in the initial stage of exploration. Compared with the countries that started to apply BIM in the early stage, there is still a big gap in the research and application of BIM in my country. The AD stage mentioned in this paper directly affects the smooth development of building construction and the use effect of the later construction stage. BIM technology has great development potential. The use of BIM technology in the design phase is critical to the successful implementation of the entire project. Therefore, this paper focuses on the characteristics of BIM technology and its application in the construction industry. It is hoped that the in-depth research on the application of BIM technology in the architectural design stage will encourage more architectural experts to learn and apply BIM technology in the future. In turn, it will help promote the computerization process of China's mechanical engineering industry.

References

1. Jun, P.K.: A study on improvement of BIM (Building Information Modeling) working environment in architectural design area using API (Application Program Interface). Korea Sci. Art Forum **30**, 107–117 (2017)
2. Alsafouri, S., Ayer, S.K.: Mobile augmented reality to influence design and constructability review sessions. J. Archit. Eng. **25**(3), 04019016 (2019)
3. Ferreira, I.B.: Documentao Arquitetnica do Casaro de Zé Pereira na Cidade de Princesa Isabel-PB. Revista Principia - Divulgação Científica e Tecnológica do IFPB **1**(40), 94 (2018)
4. Alqlami, T., Al-Alwan, H.: The application of BIM tools to explore the dynamic characteristics of smart materials in a contemporary Shanashil building design element. Int. J. Sustain. Dev. Plan. **15**(2), 193–199 (2020)
5. Gokuc, Y.T., Arditi, D.: Adoption of BIM in architectural design firms. Archit. Sci. Rev. **60**(6), 483–492 (2017)
6. Lee, K., Choo, S.: A hierarchy of architectural design elements for energy saving of tower buildings in Korea using green BIM simulation. Adv. Civil Eng. **2018**, 1–13 (2018)

7. Diao, P.H., Shih, N.J.: Trends and research issues of augmented reality studies in architectural and civil engineering education—a review of academic journal publications. Appl. Sci. **9**(9), 1840 (2019)
8. Choi, J., Lee, S., Kim, I.: Development of quality control requirements for improving the quality of architectural design based on BIM. Appl. Sci. **10**(20), 7074 (2020)
9. Malewczyk, M.: The usage of the openBIM idea in architectural design on the example of blender and blenderBIM add-on. Architectus **2**(66), 99–104 (2021)
10. Schroeder, R.: Comprehensive BIM integration for architectural education using computational design visual programming environments. Build. Technol. Educat. Soc. **2019**(1), 46 (2019)
11. Akinade, O.O., Oyedele, L.O., Ajayi, S.O. et al.: Designing out construction waste using BIM technology: stakeholders' expectations for industry deployment. J. Clean. Prod. **180**, 375–385 (2018)
12. Miller, D.: Putting BIM at the heart of a small practice. Archit. Des. **87**(3), 42–47 (2017)
13. Pavlovskis, M., Antucheviien, J., Migilinskas, D.: Conversion of industrial buildings and areas in terms of sustainable development by using BIM technology: analysis and further developments. Mokslas - Lietuvos ateitis **7**(5), 505–513 (2016)

Creative Design of Digital Platform Under Information Technology

Ming Lv, Xue Feng(✉), and Cen Guo

Shenyang Jianzhu University, Shenyang, Liaoning, China
1445452064@qq.com

Abstract. Now the emergence of the Internet for information technology (it) is not affected by time and space constraints, modern society rapidly into the digital era, information at very fast speed, people's lives, learning, entertainment, get a huge change, the fusion of digital technology and the cultural creative industry, not only provides the technical support, and expand the development way, to show the cultural characteristics, Passing on history in a new medium. In the digital era, every field is changing the economic model. Digital economy drives Internet consumption, and consumers can enjoy convenient services on the digital platform at any time. Digital cultural and creative industry is also showing a new ecological scene in the development of the Internet.

Keywords: Digitization · Cultural and creative industries · Innovation · Interaction design

1 Introduction

Artificial intelligence to promote the diversity of cultural creative industry development, promote the transformation and upgrade of the cultural creative industry platform, computer and digital media technology break the traditional mode of barriers, deep influence and ecological industrial development, pay attention to more excellent novel creative content, promote cultural soft power, build a multi-level cultural value system, wen gen industry does not stick to traditional form, Combine with modern science and technology, enrich and complete the construction of cultural creative industry ecological system.

2 Digitization

In the digital era, with the emergence of the Internet, anyone can look for their favorite things on the Internet and express their views. The distance between the public and art is no longer far away, and interactive thinking is also generated. In the course of human society, it has gone through a long process from the wave of agriculture to the wave of industrialization. Finally, the thinking mode of big data brought by Internet technology and the interconnection of resources are changing the development of major industrial

J. H. Abawajy et al. (Eds.): ICATCI 2022, LNDECT 169, pp. 302–310, 2023.
https://doi.org/10.1007/978-3-031-28893-7_36

ecology in human society. The information revolution began to break out, so the human society has quickly entered the digital era, the Internet makes information technology not constrained by time and space, spread at a very fast speed, people's life, learning, entertainment has been greatly changed. Digital sharing economy is mastered by Internet platforms and can share various material and non-material resources [1]. Digital economy brings opportunities and new market benefits to developing countries [2]. Digital cultural industry has become an emerging industry, integrating with people's life. Various innovative and creative activities and services have surrounded us. Electronic devices are changing people's consumption patterns.

3 The Article

Cultural and creative industry is an emerging industry with creativity as its core. With culture as its core, it integrates with diversified cultures through different carriers to construct new cultural phenomena. Cultural and creative industry is receiving attention and comprehensively promoting the prosperity and progress of emerging industries. Under the influence of the Internet, the economic form of product creation needs to develop in the direction of innovation as the core, and the forms of expression brought by digitalization are gradually diversified.The design of Miyi County brown sugar designed in Fig. 1 is promoted through the cultural and creative products designed by the outer packaging and font patterns, and combined with the local culture.

Fig. 1. Mi Yi brown sugar visual design

3.1 Traditional Culture

Culture is a spiritual word, but it is also created for the human mind and hands. Culture is embodied in all material and immaterial products. Chinese traditional culture is extensive and profound, and is usually expressed in immaterial form. Ancient books and cultural relics are easy to be damaged, and the colors and handwriting are unclear. It takes manpower and time to repair them. Gen from the Palace Museum in Beijing to build the brand, you can see that the palace is not limited to the traditional ideas, in a new era of change and have a certain innovation concept, to create their own brand image and cultural symbols, have developed many Chinese style elements of product, also received a lot of young people's favorite, thereby giving impetus to the spread of Chinese excellent traditional culture. Immersive interactive technology can tell ancient stories in different forms, such as museums and galleries [3]. Virtual environment can be created through different devices, head-mounted display devices can be wearable like helmets, and panoramic display can be holographic projection with 3D glasses [4]. The Palace Museum in Beijing not only develops cultural and creative products, but also creates a digital museum. As long as you enter the website, you can enjoy the cultural relics displayed in an all-round way and let people understand them in more detail. This is a more three-dimensional and diversified Palace Museum presented by digital technology, and also makes the cultural relics come alive. In addition, the Palace Museum has also established VR4D immersive experience hall, which realizes immersive experience interaction design through virtual reality technology, allowing people to feel as if they are really in the real historical environment. VR equipment, lighting and atmosphere control system and other modern technical facilities, so that traditional cultural treasures can also retain traces of ancient civilization through digital technology, "Along the River at Qingming Festival" can also be active, through the combination of immersive interactive technology and digital images to show the colorful life of people in the Song Dynasty.

3.2 Education

More and more countries began to pay attention to the cultural creative industry, China is also continuously upgrade the cultural power in the reform and development of positioning, expand emerging development areas, at the same time pay attention to the training creative talent, determine the cultural creative industry in 2009 as a national backbone industry, break through the traditional form, fusion AI technology culture connotation. Zhang Weimin, dean of the College of Cultural and Creative Arts, once proposed to solicit ideas of cultural and creative innovation from all over the world by holding international conferences and founding academic journals to build an international platform for cultural and creative exchange. Due to the integration of various forms of expression with advanced technological means, classroom education encourages the participation of multimedia education theories and practices [5].

The development of cultural and creative industry needs innovative talents. Human beings can build social relations in different dimensions, and interpersonal relations can only be developed through interaction [6]. Nowadays, creative people are an important part of the development of cultural and creative industry, and more fresh blood is needed to join in. Digital technology and social network content also need people to carry out creative productivity and promote the development of digital economy [7]. Digital cultural and creative fields need to keep pace with The Times, take innovative ideas as the pillar, expand multiple digital resource channels, expand the space for communication and interaction, create brand image characteristics of cultural and creative fields, and sublimate the value of cultural and creative fields. Cultural and creative industry needs talent cultivation, innovative talents to strengthen creative thinking, explore the way ahead, cultivate the spirit of practice, applied to practical operation, young and energetic thinking to strengthen the practical ability in the field of cultural and creative. The development of information and communication technology has accelerated the establishment of a global information civilization, so modern communication technology makes closer social relations and promotes the development of globalization [8].

3.3 The Tourist

The combination of culture and tourism is the way of cultural travel. Local cultural characteristics can be effectively publicized through the Internet platform, which not only promotes the development of innovative economy, but also achieves the purpose of spreading and carrying forward local culture. Figure 2 depicts a festival illustration of the Lisu ethnic minority, and the form of digital illustration can also spread the characteristics of local festivals.Actively expand areas and scope of modernization construction, create good ecological civilization construction, cultural tourism project development space and gradual increase, digital text brigade industry promote the development of high quality, combining with multimedia devices, introduce advanced technology, break through the traditional tourism mode, further reveals the cultural connotation, digital brigade motivate industry, has great development space, Digital technology and the cultural tourism depth fusion, let the cultural connotation is widely spread, digital technology have penetrated into cultural tourism industry of online and offline services, experience, not only online, a short video to promote the public, immersive experience respectively scenic spot design and development, using the holographic projection, VR, AI technology, such as promoting cultural tourism will be a new height.

4 Digital Innovation

The integration of digital technology and cultural creative industry not only provides technical support, but also innovates development approaches, improves product benefits, and provides a broader development platform, showing unique cultural charm. The combination of artificial intelligence technology and design thinking, exists in every corner of our life, efficient information communication, break through the traditional form, like digital movies, games, music, through the multimedia transmission, from production to the user experience a variety of forms, fusion path of virtual reality to promote healthy development of the digital economy, the emergence of artificial intelligence resource cost, reduce the time Driven by digital technology, cultural and creative industries have entered a new growth space.

The virtual network world under the control of big data can accurately control any hot topic of current events, and also inspire cultural creativity. Culture creative industry for the center with artificial intelligence, the gradual shift from technology as the center for creative centered, widely distributed AI technology, cultural creativity and design the extended more diverse cultural forms, security mechanisms such as data mining function

Fig. 2. Lisu harvest festival illustration

of artificial intelligence can bring more diversified forms of product innovation, promotes the human efficient implementation of creative inspiration.

4.1 Smart Era

Fig. 3. CCTV News AI sign language anchor

Artificial intelligence changes and innovates multi-level fields in the society, and improves production efficiency. As the core driving force, AI technology drives the new development of the intelligent era. It can be seen that the generation and review of intelligent content and other key links are undergoing profound changes, and the cultural and creative production system is also being reshaped. CCTV news is the first AI sign language anchor with speech recognition function and natural language understanding. This technology can see the character image of virtual AI sign language anchor on the display screen and translate text or speech into sign language movements, which is convenient for deaf and dumb people to watch and understand. Figure 3 is an overview diagram of the AI sign language anchor from Baidu Encyclopedia.Digital images and videos are based on computers and networks and have richer performance characteristics. Mann Norwich had thought culture in the 21st century by the interface definition, can see clearly now the digital era is changing the way people live and the way of thinking, the emergence of human-computer interaction design and produce interactive thinking, computers as auxiliary tool to help people find a large number of data, avoid people neglect, computing, storage, modification, also deep analysis of the content, To build a new interactive mode, modern society enters into intelligent transformation and upgrading, people start intelligent life, smooth use experience, artificial intelligence creates value of The Times, and digital development potential is infinite [9].

AI can free people's hands and replace mechanical work. Data analysis tools of ARTIFICIAL intelligence can perfectly analyze work content by collecting all aspects of data, while human work increasingly attaches importance to creative ability. Virtual reality technology has brought the field of cultural and creative activity, and the development of scientific and technological concepts has entered a new stage in technological progress. Now the occurrence of all kinds of intelligent products, which gives the function of information communication, only need to upload the information on the Internet platform, products of the usage of information users can understand at any time, using technologies such as intelligent induction of artificial intelligence, data calculate understand the user's habits, the user can through the interactive instruction, to achieve effective communication in the life. AI participates in human activities, has complex computing power, and becomes a tool for producing data. Artificial intelligence enables wearable devices to realize interactive mode and help people provide reliable suggestions. The emergence of virtual voice assistants that can carry out interactive dialogues and other functions makes human-computer interaction deeper into human life and provides people with in-depth interactive experience. Data along with the rapid development of science and technology, the development of new technology applied in many areas, human society both transportation and medical treatment, etc., through artificial intelligence speech recognition technology, design of electronic medical records can quickly record the change of the condition, can help doctors to save time, fast accurate access to information, improve the efficiency of treatment, artificial intelligence can even unmanned operation, The operation is performed by a computer controlled robot on the operating table. No supermarkets such as convolution of artificial intelligence neural network, biometric frontier of artificial intelligence technology to realize face recognition, and code can store shopping, and face the payment function, biological recognition technology can record every consumer behavior, determine the spending habits, not to cooperate with facial recognition technology can accurately capture facial and body features of consumers. Unmanned driving technology also uses external equipment such as radar sensors and cameras to achieve safe driving. Artificial intelligence algorithms can also help the distribution of logistics through accurate identification technology, which is more accurate and fast than human distribution, and accelerate the development of Internet digital economy. In the era of big data, people are constantly receiving new information, followed by more and more material and spiritual needs, and people's lifestyle is richer with the progress of science and technology. Artificial intelligence is mainly based on people's needs, as a tool to solve people's problems in work and life, and better adapt to the social environment.

The multi-faceted cooperation of cultural and creative industries promotes the development of the industry, greatly improves the production efficiency through AI technology, and promotes the excellent traditional culture accepted by the public. In the era of intelligence, the coverage of digital technology promotes social development and reform, digital creative industry leads the cultural industry to flourish, artificial intelligence improves the production efficiency of traditional forms, and artificial intelligence can accurately match user needs and play an important role. In order to cultivate the vitality of culture, expand the influence of cultural and creative brand image, in-depth research causes cultural value, innovates and develops the driving force of cultural and creative production, digitization stimulates innovation power, the progress of emerging technology, and changes human life. Science and technology will expand people's ability infinitely, creative thinking is more important, digital cultural and creative needs to have characteristics, promote the upgrading of cultural and creative production mode. The interactive display presents the design advantages to bring spiritual pleasure and the dissemination of cultural core. Internet platform to promote a variety of channels and resources, digital product carries human emotions and cultural connotation, more can get social identity, by raising the consumption demand, driven by digital economic consumption system, balance the market rules and the macroeconomic regulation and control, create and ecological thriving achievements, by building a complete cultural symbol, improve the image of the local culture, Traditional culture is spread through digital transformation, and the Internet + cultural mode is gradually integrated and deepened, promoting the upgrading of the business form of digital creative industry. Application of virtual reality technology, the digital content as the core, the multimedia experience device can be integrated with the projection technology, to the user immersive experience, create brand image by accumulating IP popularity resources, to carry out the relevant business activities, to the user to emotional experience, and the product promotion, product quality is guaranteed, is rich in cultural spirit, cultural transmission characteristics, There is plenty of creative space to explore.Data sharing in the Internet platform, building an intelligent ecological system together with various fields, utilizing cultural advantages, giving play to greater benefits, enhancing the attraction to consumers, promoting cross-border integration, establishing innovative thinking, and mining the value of cultural and creative industries through big data technology.

4.2 Emotional Design

Young people are easier to understand and accept new technologies, and they also provide the impetus for economic development [10]. User experience true feelings is the important basis of emotional design and emotional design experience on the spirit of the user as the center, the deep research and the product of the emotional interaction with people, the core content of mining people-oriented, service for the user experience, through the analysis of psychological state and wen gen product environment systemic convey the cultural connotation of temperature and moved, This also makes cultural and creative industries more valuable. Although the big data computing speed of computers is fast, the existence of designers makes up for the humanization element that ARTIFICIAL intelligence lacks. The people-oriented design principle makes artificial intelligence no longer a cold machine with emotional design elements, but with cultural quality and humanistic care. McLuhan predicted the future of the electronic media of speech of the global village, through artificial intelligence of the virtual environment, all people can at any time in the Internet connection, reduce the distance the social interaction, the electronic media to two-dimensional plane into three-dimensional stereo image, even thinking will become a solid culture from paper text jumps out can see interaction, Humanization with temperature, emotional design is an important element in the future development trend of cultural and creative field. Interactive technology realizes interactive mode, so that users can carry out visual thinking communication activities [11].

5 Conclusion

With the progress of The Times, artificial intelligence has become an indispensable factor in contemporary human society, changing the way humans perceive the world, promoting social progress and broadening people's horizons. The cooperation between cultural and creative industries and science and technology drives the development of digital cultural and creative industries. Science and technology coordinate culture and culture complement science and technology. In the database logic principle, the control is in the hands of the user, and it is user-oriented, focusing more on the user's needs and experience. The future world is highly intelligent experience can bring more innovation, positive response national policy, increase the spread of influence of the cultural creativity, promote cultural creative industries more stereo diversification, create a development road of scientific and technological innovation, build to attract the public's cultural characteristics, shorten the distance between the public and wen gen, consumption structure, promote the contemporary culture.

References

1. Pouri, M.J., Hilty, L.M.: The digital sharing economy: a confluence of technical and social sharing. Environ. Innov. Societal Trans. **38**, 127–139 (2021)
2. Williams, L.D.: Concepts of digital economy and industry 4.0 in Intelligent and information systems. Int. J. Intell. Netw. **2**, 122–129 (2021)
3. Verhulst, I., Woods, A., Whittaker, L., Bennett, J., Dalton, P.: Do VR and AR versions of an immersive cultural experience engender different user experiences? Comput. Human Behav. **125**, 106951 (2021)
4. Paes, D., Irizarry, J., Pujoni, D.: An evidence of cognitive benefits from immersive design review: comparing three-dimensional perception and presence between immersive and non-immersive virtual environments. Autom. Constr. **130**, 103849 (2021)

5. Bateman, J.A.: What are digital media? Discourse Context Media **41**, 100502 (2021)
6. Chuluunbaatara, E., Ottavia, Luh, D.-B., Kung, S.-F.: The role of cluster and social capital in cultural and creative industries development. Procedia - Social Behav. Sci. **109**, 552–557 (2014)
7. Boccella, N., Salerno, I.: creative economy, cultural industries and local development. Procedia - Social Behav. Sci. **223**, 291–96 (2016)
8. Horoshko, O.-I., Horoshko, A., Bilyuga, S., Horoshko, V.: Theoretical and methodological bases of the study of the impact of digital economy on world policy in 21 century. Technol. Forecast. Soc. Change **166**, 120640 (2021)
9. Kucker, S.C.: Processes and pathways in development via digital media: examples from word learning. Infant Behav. Develop. **63**, 101559 (2021)
10. Youssef, A.B., Boubakerb, S., Dedajc, B., Carabregu-Vokshi, M.: Digitalization of the economy and entrepreneurship intention. Technol. Forecast. Social Change **164**, 120043 (2021)
11. Oti, A., Crilly, N.: Immersive 3D sketching tools: implications for visual thinking and communication. Comput. Graph. **94**, 111–123 (2021)

Application of Database Testing Techniques in Web Development

Yiding Sun[1](✉) and Romany Viju[2]

[1] Xi'an Technological University, Xi'an, Shaanxi, China
2468128237@qq.com
[2] New Valley University, Kharga, Egypt

Abstract. In recent years, the saying of the information age has been deeply rooted in people's hearts. In the context of the rapid development of Internet technology, big data technology, and network information technology in China, the security of Web development and application have been paid more attention to. In the process of Web development and application, database testing technology is one of the indispensable links, the results of which will provide technical suggestions for Web development and improve the efficiency of the application. Based on it, this paper discusses the application of database testing techniques in web development, aiming to provide some help for related systems and platforms.

Keywords: Database test · Web development · Application

1 Introduction

Based on the rapid development of big data technology, database testing technology in the process of web application development also needs to improve the technical mode and optimize the technical means according to the actual situation to ensure that the system can carry out data processing in a stable and orderly manner. At the same time, technical personnel need to choose the appropriate method to continuously adjust the database testing work, to better play the value of the application of database technology in computer web development.

2 Characteristics of Database Testing

Database technology from its birth to the present, formed a solid theoretical foundation, mature commercial products, and a wide range of applications, becoming the infrastructure of enterprises, departments, and even individuals in their daily work, production, and life everywhere in less than half a century. The database is like an abstract warehouse, where goods are stored in the warehouse, and data are stored in the database. Many characteristics of the DBS (Database System) explain its wide application. First, the data in the DBS is structured, not only in the sense of data internal but also in the sense that the whole is structured and correlated. Second, the data in the DBS can be

J. H. Abawajy et al. (Eds.): ICATCI 2022, LNDECT 169, pp. 311–318, 2023.
https://doi.org/10.1007/978-3-031-28893-7_37

shared and used by multiple applications and users, which also reduces data redundancy and saves storage space. Finally, the user's application and the data in the database stored on the disk are independent of each other. In other words, when the data's physical storage or logical structure is changed, the application will not change, which is ensured by the DBMS (Database Management System) secondary image function. Thus, database testing technology provides the necessary environment and support for data collection and utilization. Then, it is especially important to standardize the database operation methods and improve the accuracy of testing when conducting Web development and applications. And the Table 1 shows the mainstream categories of DBMS today.

Table 1. Database management system categories

Category	Definition	Applicable scene
Relational DBMS	Support table data management mode database	Complex SQL Query
Document DBMS	Support document storage data optimization	Shopping Cart
Time series DBMS	Support time series data optimization	Sensor
Graph DBMS	Support for graphically structured efficient databases	Social Network
Hybrid transaction DBMS	Support for working with both relational and non-relational databases	Heterogenous Data

3 Reasons for Using Database Technology for Web Development

3.1 Subjective Reasons

At the beginning of common project development, having a database server, technicians create the tables and stored procedures they need on the server with administrator status to generate data is all that is needed. However, because many software development technicians now focus their research on software function setting and system programming, the lack of targeted testing of database technology directly leads to the maintenance and management of the database being more and more cumbersome and more difficult to release. At this time, some good enterprises will have a full-time DBA (Database Administrator) to unify the database-related development, but the change script is also very difficult to manage and generate [1]. But the general enterprise technical staff will have difficulty in encountering various problems, such as tracking the changes of database objects or identifying the abandoned database objects in the development process. It is even difficult to deploy changes to different copies of the database due to the presence of test data. As part of the project deliverables, the database is dynamic and changing. To ensure that all databases use the same objects and that scripts can be managed through source code control, we need to use database testing techniques for web development so that researchers have a clear understanding of the role of the database and use more scientific techniques to drive the software development process.

3.2 Objective Reasons

3.2.1 Reasons for Conceptual Structure Design

Conceptual structure design techniques occupy a strong position in database testing. In specific tests, as the number of tests increases, the technical personnel are required to delineate the overall boundaries to ensure that technical boundaries can be created to support the needs of Web development and applications so that the development views can be used to describe the static organization of the Web development environment and the relationships between static entities.

3.2.2 Reasons for Physical Structure Design

In essence, physical structure design technology is a means of using software programs or other tools to perform targeted checks in the database [2]. The stability of the physical structure design reduces the duplication of extended system functionality and the total cost of the implementation process during version updates. A good physical structure will make the test design more efficient in Web development and minimize the workload caused by iterations.

3.2.3 Reasons for Logical Structure Design

Before the actual testing of the data in the database using logical structure design techniques, the technician has to conduct a global search for the information that needs to be used in the database such as data types, character segment names, and column name reports. The logical structure design determines the nature of changes that occur in the system, and in Web development, the impact of changes in functional requirements on the system will be limited, thus avoiding a huge amount of regression testing and providing security for other applications.

From Fig. 1, we can find that the conceptual, logical, and physical structure design parts are the most critical among the above six stages, and these three parts are described separately above.

4 Database in Web Development Applications

4.1 JDBC Technology

4.1.1 Overview of the JDBC Technology

JDBC refers to JAVA Database Connectivity, a standard JAVA application programming interface, used to connect the JAVA programming language and database. The API library has some common uses related to databases, such as making database connections, creating or executing SQL statements, and being able to view their records. Compared to the ODBC approach that uses C language as an interface to implement the API, the JDBC method allows portable access to the underlying database, writes different types of executables, and admits JAVA programs to contain code that is not related to the database [3].

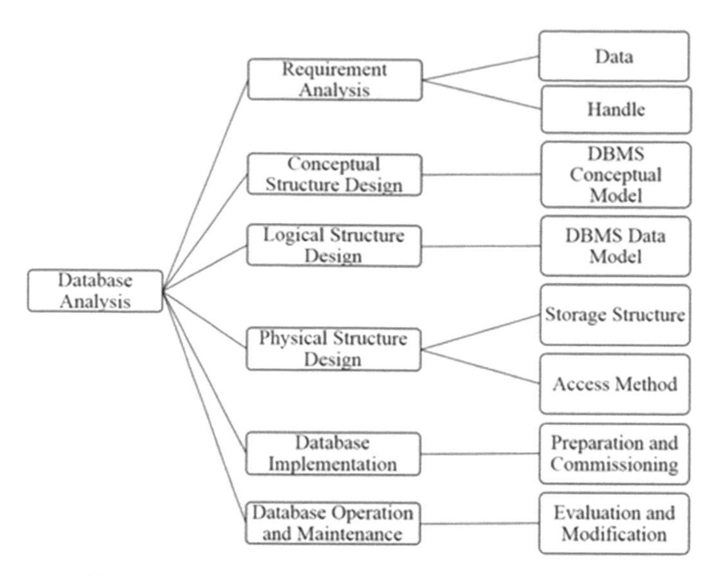

Fig. 1. Database analysis composition structure diagram

4.1.2 JDBC Technology in Web Development

The JDBC method is widely used in Java Web development. Usually, Web programs operate the database through JDBC, and the database framework is built by relying on the JDBC API to carry out. From the perspective of the development model, the use of JDBC in Java Web development should follow the MVC design philosophy, so that the Web program has a certain degree of scalability [4]. The MVC design philosophy refers to the division of software into three layers: The Model layer (M), the View layer (V), and the Control layer (C). The model layer refers to the business logic of the overall program; the view layer refers to the interactive interface of the program when users use it, and the control layer refers to the distribution of various requests from users through the controller. The following Fig. 2 clearly shows the operation flow of MVC design philosophy.

In the case of a very large amount of data, it is difficult to display all the data on one page, so it is necessary to use paging query technology. There are many ways to achieve paging query through JDBC, here mainly gives two typical paging methods. The first is paging through the cursor of the ResultSet, an object encapsulated in the JAVA API, which has the concept of a "cursor" that records the start and end position of the object and can be used on a variety of databases. The second type of paging is done through the database mechanism. Now widely used database software, itself provides a paging mechanism, such as SQL Server provides the "top" keyword, MySQL database provides the "limit" keyword, and so on [5]. Compared with the first method, its mechanism can significantly reduce the overhead of database resources and improve the performance of the program, but the application is relatively single, which can only be used in the corresponding database.

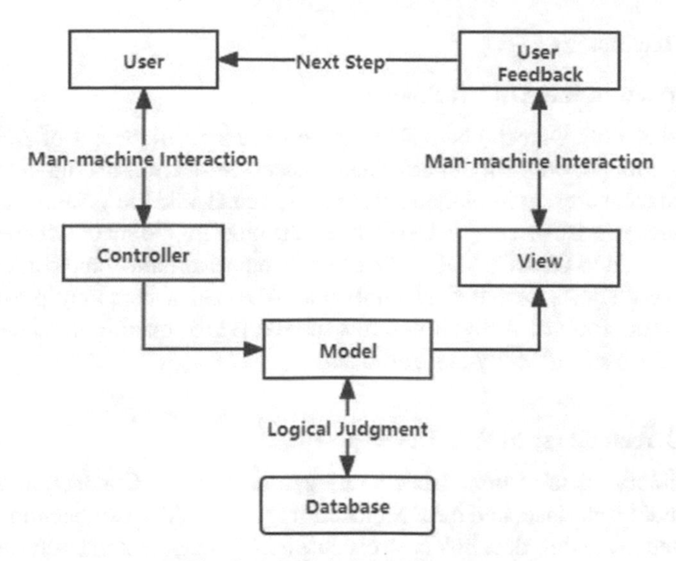

Fig. 2. MVC design philosophy

4.2 CGI Technology

4.2.1 Overview of the CGI Technology

CGI (Common Gateway Interface) is a standard interface for external extension applications to interact with the Web server, and there are generally two types of CGI: standard CGI and buffered CGI. Standard CGI is through standard I/O, while buffered CGI allows a wider choice of development tools, but because of this, only a few kinds of WWW servers support buffered CGI, making its compatibility negatively affected. In addition, programs written according to standard CGI are not related to specific WWW servers but programs written according to cached CGI are related to them, so, client users can query data through the server.

4.2.2 CGI Technology in Web Development

As a standard extension technology of the WWW server, CGI is closely related to the actual development and design of Web applications, especially in the HTTP protocol. CGI communicates between the client and the server through the HTTP protocol, and the CGI request submitted by the client user is an HTTP request, mainly including "GET" and "POST" methods [5]. After the extension is processed, the standard output stream passes the reply information to the server, and then the reply returned to the client is the HTTP reply message, which mainly consists of the reply header and reply data. In Web development, the use of a CGI database can enhance the convenience of user operation, but there are also some problems. For example, when the number of applications submitted during operation is too high, it will take up many resources in the computer system and the operation efficiency cannot be guaranteed. Therefore, technical personnel should selectively choose CGI technology in combination with database characteristics to ensure the smooth progress of Web development.

4.3 ADO Technology

4.3.1 Overview of the ADO Technology

ADO (ActiveX Data Objects) belongs to the core technical content of ASP access to database systems, providing high-performance access to all kinds of data sources, including relational databases, non-relational databases, e-mail, and file systems [6]. There are several features of ADO access to data sources, the most important of which is its convenience. Compared to ODBC, ADO technology is object-oriented and can directly construct record set objects if needed. Not only that, ADO can access multiple data sources and comes in the form of ActiveX controls like OLE DB, making the development of Web applications substantially more efficient.

4.3.2 ADO Technology in Web Development

The ADO library contains three basic interfaces, namely the ConnectionPtr interface, the CommandPtr interface, and the RecordsetPtr interface. The ConnectionPtr interface is usually used to create a data link or to execute a SQL statement that returns no results, i.e., a record set or a null pointer. CommandPtr provides a simple way to execute stored procedures and SQL statements that return a recordset, and when frequent access to the database is required and a significant number of recordsets are to be returned, the use of CommandPtr interface is an efficient means of doing so when frequent access to the database is required and a significant number of recordsets are to be returned. Compared to the first two, the RecordsetPtr interface provides more control over recordsets, and when multiple recordsets need to be used, stored procedures can be executed through record locking, cursor control [7]. Web development can write a base model for scripting database access through the setup of ADO technology, and use ODBC with the help of data sources from native devices to the database system. After specifying the object, it is easy to use ADO to call the object directly or give SQL commands to achieve the effect of hierarchical data transfer.

4.4 IDC Technology

4.4.1 Overview of the IDC Technology

IDC belongs to the Internet Data Interface, is no need to program multiple times to complete the task of database access technology. Compared with the traditional CGI technology, IDC to a certain degree to solve the CGI script's slower access speed and the use of complex maintenance problems, by calling the appropriate ODBC driver to access the database, running on the server IIS (Internet Information Services)/IDC [8].

4.4.2 IDC Technology in Web Development

IDC uses the Internet database connector file to access the database, using HTML language construction output Web interface. Specify the ODBC data source to be connected through the database connector, call the parameters in the request, data operations on the data source, and then the results will be passed back to the client through the HTX file specified in the IDC file [9]. After that, the server will return to the browser's part and complete all the database access tasks. With the continuous maturity of e-commerce, future IDC technology will gain further development in CDS (Content Delivery Service) technology, especially after the reasonable use of intelligent allocation technology, according to the principle of proximity access to different nodes, and thus save a large amount of money to build mirror sites.

5 Database Technology Development Prospects

Database testing technology has an important guiding role and significance to the development of programming language databases, through the development and in-depth analysis of database testing technology, it can improve the efficiency of data processing in the database and obtain more accurate test results [10]. Combined with the characteristics and situation of web development application system, to deal with all kinds of development technology, and thus enhance the security, stability, convenience, advice of database access, highlight the value of all kinds of technology advantages, to provide efficient data support for smooth software development work, to avoid hidden dangers, so that its role is given full play, to further improve the stability and reliability of computer software systems [11]. To achieve the need to promote the sustainable development of software, to better meet people's needs when facing users, to have a more positive impact on human lifestyles, such as the development of cloud database and hybrid data, more focused and extensive solution to the problem of sharing resources in different places, or the content of data integration and data warehouse, to find more valuable information for enterprise development to provide decision support, or in the AI In the direction of achieving greater technical breakthroughs to promote the development of human civilization. After gaining more people's recognition in this field, so that the development and application of computer software are more contemporary and technological [12]. Therefore, this paper provides a forward-looking perspective on three aspects: Hardware, Deployment, and Data Model, and explains the possible future representative projects and their theoretical foundations in these three aspects, which are summarized in Table2.

Table 2. Possible directions of database development

Main theoretical basis	Represent item	Future direction of development
Classical control theory	Random access memory	Hardware
Computational intelligence	Artificial intelligence	Deployment
Multi-model	Time series/Graph	Data model

6 Conclusion

To sum up, this paper discusses the application of database testing technology in web development. Firstly, the characteristics of database testing are analyzed, and the subjective and objective reasons are pointed out respectively, especially the concept, logic, and physical structure design of the database in the objective reasons. Then it briefly describes the database technology in the mainstream network development and application, and emphatically compares the advantages and disadvantages of various methods. Finally, it explores the development prospect of database technology and provides the development direction for how to meet human needs and achieve greater technological breakthroughs in the future.

References

1. Riva, A., Bellazzi, R., Lanzola, G., et al.: A development environment for knowledge-based medical applications on the world-wide web. Artif. Intell. Med. **14**(3), 279–293 (1998)
2. Jones, H.D.: Wheat transformation: current technology and applications to grain development and composition. J. Cereal Sci. **41**(2), 137–147 (2004)
3. Guerrero, L.A., Fuller, D.A.: A pattern system for the development of collaborative applications. Inf. Softw. Technol. **43**(7), 457–467 (2001)
4. David S. Carlson Craniofacial embryogenetics and development. Am. J. Orthod. Dentofacial Orthop. **155**(6) (2019)
5. Čeke, D., Milašinović, B.: Early effort estimation in web application development. J. Syst. Softw. **103**, 219–237 (2015)
6. Salas-Zárate, M.d.P., Alor-Hernández, G., Valencia-García, R., et al.: Analyzing best practices on Web development frameworks: the lift approach. Sci. Comput. Program. **102**, 1–19 (2015)
7. Gupta, J., Bavinck, M.: Inclusive development and coastal adaptiveness. Ocean Coastal Manag. **136**, 29–37 (2017)
8. Santek, B.: Current development of biotechnology in Croatia. J. Biotechnol. **256** (2017)
9. Mouna, D., Ahmed, T., Fathi, E., et al.: Effects of gender and personality differences on students' perception of game design elements in educational gamification International. J. Human-Comput. Stud. **154** (2021, prepublish)
10. Li, F.Y., Jen, H.G., Ying, C.P.: et al.: Effects of a concept mapping-based two-tier test strategy on students' digital game-based learning performances and behavioral patterns Comput. Educ. 2021(prepublish)
11. Élise, L., Qinjie, J., Stuart, H., et al.: Analyzing the relationships between learners' motivation and observable engaged behaviors in a gamified learning environment. Int. J. Hum. Comput. Stud. **154** (2021)
12. Souha, B., Ahmed, M., Henda, B.G.: Adaptive gamification in E-learning: a literature review and future challenges. Comput. Appl. Eng. Educ. **30**(2), 628–642(2021)

A Model Prediction Algorithm-Based Driving Path Planning Model for Autonomous Vehicles

Yangyong Liu(✉)

Intelligent Manufacturing and Automobile School, Chongqing Vocational College of Transportation, Chongqing 402247, China
lyy457796148@163.com

Abstract. Unmanned driving technology is an important development direction of the future automobile industry. It has received attention and attention from enterprises and has become a current research hotspot. Path planning plays an important role in the system of unmanned vehicle driving technology, and its performance will directly determine the success of unmanned vehicle technology. Therefore, this paper conducts in-depth research on path planning, conducts a simulation experiment of unmanned vehicles based on model prediction algorithm, and analyzes the success rate of path planning for three roads: straight road, intersection and U-shaped road when there are barrier-free vehicles, to examine the effect of model prediction algorithms in path planning.

Keywords: Model prediction algorithm · Unmanned vehicle driving · Path planning · Simulation experiment

1 Introduction

With the advancement of technology, the research and development of driverless cars has gradually become the focus of public attention. The implementation of driverless cars can not only facilitate people's travel, but also save the driver's time, but its safety issues cannot be ignored. This paper will take the self-driving car as the research object, and study the path dynamic planning problem of the model prediction algorithm to realize the safety of the self-driving car.

Many scholars at home and abroad have studied the path planning model of unmanned vehicle driving based on model prediction algorithm, and achieved good research results. For example, a researcher designed a model predictive controller. Considering the dynamic constraints based on the predictive model, a MPC predictive control method was proposed, which can not only avoid static obstacles, but also control dynamic obstacles such as pedestrians and vehicles on the road. Avoidance, to achieve a full range of path optimization [1]. A scholar proposed an improved algorithm for manually designing the route potential to prevent the car route planning process from falling to a local minimum value, ensuring that the car can accurately follow different reference routes without a driver, and has better applicability [2]. There have been some achievements in the research on the driving path planning model of unmanned vehicles based on model

J. H. Abawajy et al. (Eds.): ICATCI 2022, LNDECT 169, pp. 319–326, 2023.
https://doi.org/10.1007/978-3-031-28893-7_38

prediction algorithms, but if the safety of unmanned vehicles is maximized, relevant algorithms should be used to plan the optimal path to avoid driving collisions.

This paper expounds the related concepts of path planning, puts forward the factors that affect path planning, and focuses on analyzing the impact of road problems in environmental factors on path planning. According to the model prediction algorithm, the success rate of path planning under different road conditions is analyzed under the condition of barrier-free vehicles, which confirms the effectiveness of the algorithm in path planning.

2 Model Prediction Algorithms and Path Planning Theory

2.1 Overview of Path Planning

Route planning functions play a key role in the field of autonomous vehicle driving. The most basic planning requirement is to avoid obstacles in the driving path of the car to ensure driving safety. The path planning function is extremely important in unmanned vehicle driving technology. In many cases, the car is required to follow a specific route to the destination to successfully complete the driving work, and it is required to avoid a collision resulting in a car accident [3, 4].

2.2 Analysis of Factors Affecting Path Planning

Path planning elements mainly include vehicle factors and environmental factors [5]. The vehicle element refers to the unmanned vehicle as a kind of wheeled mobile robot, which is a kind of incompletely constrained robot. A non-integrity constraint is a constraint that contains the generalized coordinate derivative of the system and is not integrable. Therefore, the unmanned vehicle is a typical non-integrity restraint system. For example, for unmanned vehicles, because they cannot move laterally, not any position on the road can be directly reached, which is limited by the steering ability and speed of the body. Therefore, the paths generally planned by robot planning methods often cannot meet the requirements of local path planning of unmanned vehicles. In addition, if it is a manned unmanned vehicle, although there is no driver, there are passengers riding, so the comfort of the ride must be considered. Therefore, the unmanned vehicle path planning system eventually becomes a complex nonlinear system under the nonlinear constraints such as road constraints, geometric constraints of the environment, and non-holonomic constraints of the system [6]. Environmental factors refer to external environmental factors. As the main information input of unmanned vehicles, they have a very important impact on the environment perception, system decision-making, path planning and system control of the entire unmanned vehicle [7]. External environmental factors themselves are also a huge and complex system. Broadly speaking, external environmental factors mainly include static environmental factors and dynamic environmental factors. The static environmental factors include different types of roads, traffic signs, scenes on and on both sides of the road, and static obstacles; the dynamic environmental factors include vehicles, pedestrians, traffic lights, and dynamic obstacles [8]. This paper mainly focuses on the road conditions of environmental factors.

(1) Structured road

The main static environment elements that affect unmanned vehicles are roads and static obstacles. Scenes such as buildings have little impact on the path planning of vehicles, so they will not be discussed. Among them, roads are divided into structured roads and unstructured roads. In structured roads, there are straight roads, curves, intersections, Y-shaped intersections, roundabouts, exits and entrances of arterial roads, and combination roads of these elements [9].

As the simplest component of structured roads, straight roads are the main form of existence of structured roads. Most roads are built on the principle of being as straight as possible. Straights are further divided into single-lane and multi-lane. The boundaries of straights are straight lines. The main parameters are the length, width and boundary position of the road. Vehicles drive in a fixed way on straight roads, generally straight, lane changing and some obstacle avoidance behaviors [10].

Curves are usually used as a connection of roads, and curves are also typical elements in typical road conditions. Due to the particularity of the curve, the condition of the curve is a frequent place for traffic accidents. Therefore, when planning the path of the unmanned vehicle, it is necessary to focus on the driving ability of the vehicle under the curvature value of the curve. The type of vehicle driving on a curve is generally to steer along the curve [11].

Incoming and outgoing roads are formed by the merging of two or more roads. A single road is divided into multiple roads to form an outgoing road. Vehicles generally change lanes and avoid obstacles on the inbound and outbound roads.

Crossroads are formed by the intersection of two or more roads, and the path planning of unmanned vehicles mainly considers vehicles and pedestrians. Vehicles generally have steering, U-turn and straight ahead driving modes at intersections.

The U-shaped road is usually a curve with relatively large curvature, and it may appear in the position where a U-turn is required on a two-way straight road. Similar to the curve, the vehicle on the U-shaped road generally makes a U-turn along the U-shaped road. Different from the curve, the steering angle is larger, and the target point is generally behind [12].

(2) Traffic signs

The role of traffic signs is to remind and instruct vehicles to drive properly. They convey rich road information and are designed to be simple and clear at a glance. The impact of traffic signs on unmanned vehicles is more reflected in the decision-making level of unmanned vehicles, but traffic signs also have an important impact on the path planning results of unmanned vehicles. Whether the planned path conforms to traffic signs and whether it is within the scope permitted by regulations is also an issue that should be considered in the path planning of unmanned vehicles [13].

(3) Pedestrians

Pedestrians generally walk on the road with great randomness, which is an uncertain factor in road traffic. At the same time, due to the lack of corresponding protection for pedestrians, they often suffer great injuries in the event of a traffic accident. Pedestrians walking on the road can generally be divided into two categories, one is walking along the road, and the other is walking across the road. Due to the randomness of pedestrians, detecting and avoiding pedestrians is one of the key and difficult points of research for unmanned vehicles.

2.3 Model Prediction Algorithms

When planning the collision avoidance path based on the model prediction algorithm, first, the maximum lateral avoidance distance is determined according to the safe position to be achieved; secondly, the midpoint slope A of the collision avoidance curve is determined by the maximum lateral acceleration and jerk constraints; finally, Determine the longitudinal avoidance distance to ensure that the lane change is complete. When the parameter A is determined, both the maximum lateral acceleration and the acceleration constraint need to be considered [14]. The corresponding inequality expression is as follows:

$$\left|a_p(X)\right| \leq a_{p\max} \tag{1}$$

$$|j(X)| \leq j_{\max} \tag{2}$$

Among them, $a_{p\max}$ is the maximum lateral acceleration, which can characterize the lateral stability of the vehicle; $j_{\max}$ is the maximum lateral acceleration, which can reflect the driver's comfort.

The design principles of the model prediction algorithm are as follows:

(1) Security

Safety is the number one concern for driverless vehicles. Due to the high speed of the vehicle, the danger of a traffic accident is greater, so there are higher requirements for the safety of the vehicle system. At the same time, because the vehicle is subject to non-integrity constraints, it cannot move freely like a general mobile robot, so the path searched by the algorithm must meet the constraints of the vehicle, and at the same time, it is necessary to consider whether the performance of the vehicle can complete the tracking of the planned path, otherwise Unmanned vehicles not only cannot complete the tracking of the planned path, but also cause mechanical damage to the vehicle's actuators, which cannot meet the requirements of safe driving.

(2) Real-time

As a transportation vehicle, the ultimate goal of unmanned vehicles is to efficiently complete the driving task of manned or loaded. Therefore, whether the driving path can be efficiently searched and the driving task can be completed efficiently is an important requirement of the algorithm. The real-time performance of the algorithm is reflected in the success rate of the path search results.

(3) Adaptability

The external traffic environment is complex and diverse, and the driving scene changes rapidly. Therefore, for different traffic environments and changing driving scenarios, the algorithm must be able to search for an appropriate path. The adaptability of the algorithm is mainly reflected in the search ability in different scenarios.

(4) Comfort

The role of the vehicle is mainly to carry people or goods. If it is a vehicle for passengers, then when the algorithm searches for the route, the feelings of the passengers must be taken into account to make the ride more comfortable. The comfort of the searched path is mainly reflected in the parameters such as the number of turns and the curvature of the searched path. The searched path should meet the non-completion constraints of the vehicle as much as possible.

3 Experimental Research

3.1 Research Content

Unmanned vehicle driving provides passengers with a good riding experience while ensuring driving safety and optimal paths. Therefore, this paper constructs three road plane models, conducts simulation experiments for the model prediction algorithm, and analyzes the paths of the algorithm under different road conditions. Planning success rate.

3.2 Experimental Method

This paper mainly adopts the comparison method, which refers to using the model prediction algorithm to compare the success rate of each path planning in the case of obstacles and obstacle-free vehicles on three roads: straight road, intersection, and U-shaped road.

4 Experimental Results

4.1 Simulation Experiment Design

The straight road, intersection and U-shaped road driving scenes are built respectively, and the plane modeling is shown in Fig. 1:

In the simulation environment, the unmanned vehicle is simplified to one type, and the vehicle parameters are shown in Table 1. In the experiment, it is assumed that the length of the driverless vehicle is 4200 mm, the width of the vehicle is 1310 mm, the height of the vehicle is 1090 mm, and the wheelbase is 2270 mm.

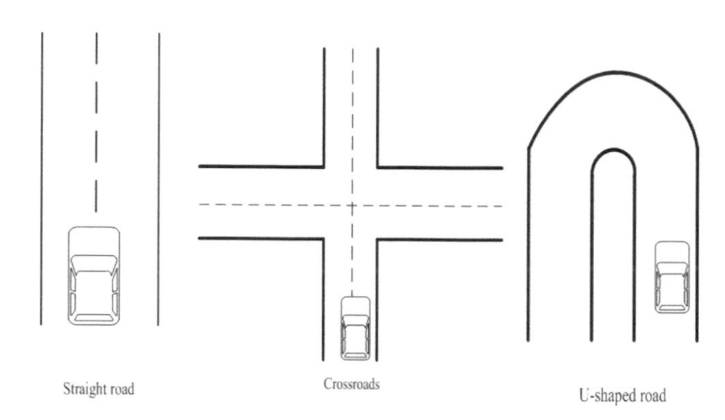

Fig. 1. Typical road type classification

Table 1. Unmanned vehicle parameters

Parameter	Unit (mm)
Car length	4200
Car width	1310
Car height	1090
Wheelbase	2270

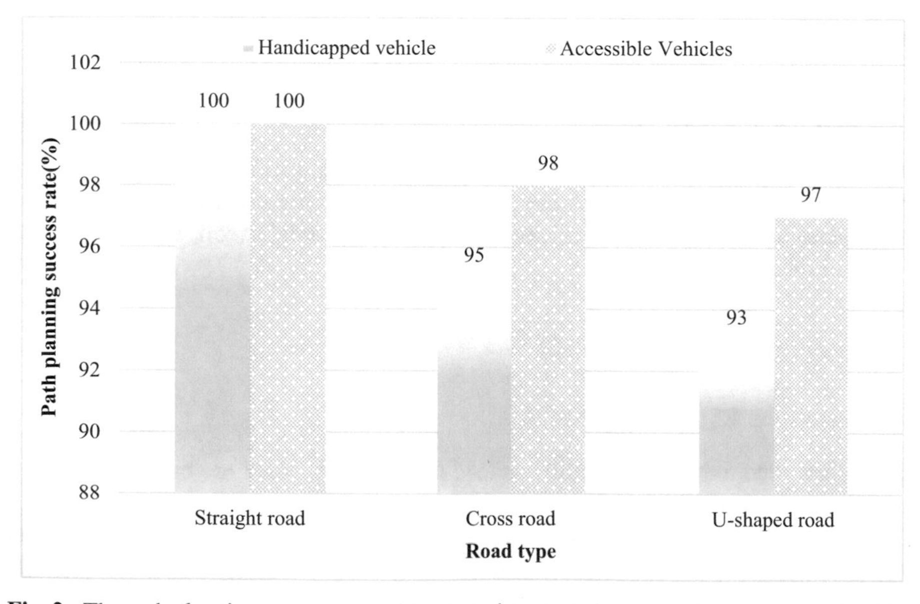

Fig. 2. The path planning success rate of three road types under the model prediction algorithm

4.2 Simulation Results of Path Planning Based on Model Prediction Algorithm

As shown in Fig. 2, the path planning success rate of the model prediction algorithm for three road types is tested in the case of obstacle-free vehicles and obstacle-free vehicles. It can be seen from the data in the figure that the path planning success rate of the algorithm reaches 100% regardless of whether there are barrier-free vehicles on the straight road, while the path planning success rate for barrier-free vehicles at intersections and U-shaped roads is higher than that of vehicles with barrier-free vehicles. The success rate at the time of obstacles, and the success rate of the intersection path planning is higher than that of the U-shaped road, indicating that among the three road types, the path planning of the straight road is the easiest, while the planning of the intersection and U-shaped road is slightly more difficult. Increased, and the planning difficulty is relatively greater when there are obstacle vehicles. The path planning success rate of each road is more than 90%, which also confirms the feasibility of the model prediction algorithm in the path planning problem.

5 Conclusion

In this paper, the simulation environment of straight road, cross road and U-shaped road is built, and the simulation experiment of path planning based on model prediction algorithm is carried out. The path planning success rate of the prediction algorithm on three roads with barrier-free vehicles. The experimental results show that the model prediction algorithm has the highest path planning success rate for the straight road under three typical road conditions, followed by the cross road, and then the U-shaped road. The path planning success rate is lower when there are obstacle vehicles.

References

1. Oh, K.S., Park, S.Y., Yi, K.S.: Model prediction based fault detection algorithm of sensor for longitudinal autonomous driving using multi-sliding mode observer. Trans. Korean Soc. Mech. Eng. A **43**(3), 161–168 (2019)
2. Sungyoul, Park, Kwangseok, O., et al.: Model predictive control-based fault detection and reconstruction algorithm for longitudinal control of autonomous driving vehicle using multi-sliding mode observer. Microsyst. Technol. **26**(1), 239–264 (2020)
3. Lefkopoulos, V., Menner, M., Domahidi, A., et al.: Interaction-aware motion prediction for autonomous driving: a multiple model Kalman filtering scheme. IEEE Rob. Autom. Lett. **6**(1), 80–87 (2020)
4. Biswas, S., Anavatti, S.G., Garratt, M.A.: Multiobjective mission route planning problem: a neural network-based forecasting model for mission planning. IEEE Trans. Intell. Transp. Syst. PP(99), 1–13 (2019)
5. Chakraborty, R., Pramanik, A.: DCNN-based prediction model for detection of age-related macular degeneration from color fundus images. Med. Biol. Eng. Compu. **60**(5), 1431–1448 (2022). https://doi.org/10.1007/s11517-022-02542-y
6. Sandzimier, R.J., Asada, H.H.: A data-driven approach to prediction and optimal bucket-filling control for autonomous excavators. IEEE Rob. Autom. Lett. PP(99), 1–1 (2020)
7. Jeong, Y., Kim, S., Jo, B.R., et al.: Sampling based vehicle motion planning for autonomous valet parking with moving obstacles. Int. J. Autom. Eng. **9**(4), 215–222 (2018)

8. Kim, T., Heo, S., Yi, K., et al.: Robust autonomous emergency braking algorithm for vulnerable road users. Trans. Korean Soc. Mech. Eng. A **42**(7), 611–619 (2018)
9. Pyun, B., Seo, M., Kim, S., Choi, H.: Development of an autonomous driving controller for articulated bus using model predictive control algorithm with inner model. Int. J. Automot. Technol. **23**(2), 357–366 (2022). https://doi.org/10.1007/s12239-022-0033-y
10. Makantasis, K., Kontorinaki, M., Nikolos, I.: Deep reinforcement-learning-based driving policy for autonomous road vehicles. IET Intel. Transp. Syst. **14**(1), 13–24 (2020)
11. Park, C., Jeong, N.-T., Yu, D., Hwang, S.-H.: Path generation algorithm based on crash point prediction for lane changing of autonomous vehicles. Int. J. Automot. Technol. **20**(3), 507–519 (2019). https://doi.org/10.1007/s12239-019-0048-1
12. Hoel, C.J., Driggs-Campbell, K., Wolff, K., et al.: Combining planning and deep reinforcement learning in tactical decision making for autonomous driving. IEEE Trans. Intell. Veh. PP(99), 1–1 (2019)
13. Kuru, K., Khan, W.: A framework for the synergistic integration of fully autonomous ground vehicles with smart city. IEEE Access PP(99), 1–1 (2020)
14. Okamoto, K., Itti, L., Tsiotras, P.: Vision-based autonomous path following using a human driver control model with reliable input-feature value estimation. IEEE Trans. Intell. Veh. **4**(3), 497–506 (2019)

Design of Control System of Automated Production Line Based on PLC and Robot

Min Tan[1](✉), Jingfang Chen[1], and Ramya Radhakrishnan[2]

[1] Chongqing Aerospace Polytechnic, Chongqing 400021, China
tanmin@cqht.edu.cn
[2] Jimma University, Jimma, Ethiopia

Abstract. With the development of my country's industrial technology, automated production lines have been applied in various industries. PLC is composed of a variety of components. It has the advantages of good versatility, high precision and programmable. The purpose of this paper to study the production line control system is to design and optimize the automated production line control system through a deep understanding of PLC and robots, and improve the efficiency of the system. This article mainly conducts performance testing and analysis by means of experimental and comparative methods. Experimental results show that the coverage rate of the control system in this paper can reach more than 90%, the accuracy rate can reach 85%, and the time-consuming is about 2 s.

Keywords: Programmable logic controller · Automated production · Control system · Design research

1 Introduction

The birth of the automated production line is a product of the combination of the high concentration of productivity and the mechanized production mode in the development of human society to a certain stage. Industrial robots originated in steam engines, internal combustion engine vehicles and other fields. With the accelerating process of industrialization and urbanization and the increasing labor cost in our country, automatic control technology has become indispensable in the product design and manufacturing process.

There are countless research theories on the design of automated production line control systems based on PLC and robots. For example, Zhang Linfang has developed a PLC-based automatic production line control system for environmentally friendly energy-saving bricks in order to improve the operating capacity of the production line [1]. In order to study the construction method and mutual communication method of each link of the automatic control system, Han Guirong designed an automatic transportation system for automobile wheels [2]. Chang Qingqi introduced the composition and working process of the elevator hall door robot packaging system. System control is realized through upper and lower control methods. The lower control system is realized on the basis of PLC, and mainly controls various industrial actuators and robots [3]. So

J. H. Abawajy et al. (Eds.): ICATCI 2022, LNDECT 169, pp. 327–334, 2023.
https://doi.org/10.1007/978-3-031-28893-7_39

this article intends to study the automatic production line control system, and conduct an in-depth study on it.

This article first studies the programmable controller, and analyzes its characteristics and categories. Secondly, related researches on the packaging and palletizing automatic production line are carried out. Then design and analyze the system. Afterwards, the control system design is explained. Finally, the system was tested by experimental methods, and the data results were obtained.

2 Design of Automatic Production Line Control System Based on PLC and Robot

2.1 Programmable Controller

PLC is a new generation of industrial controller developed on the basis of microprocessor, integrated computer and automation technology. Specifically, it is a CNC electronic system mainly used in industrial environments [4, 5].

It is used in its internal memory program to execute user-oriented instructions, such as sequence control, logic operations, counting, time and arithmetic operations, and to control various types of machines or production processes through input/digital or analog output. Its high reliability, high temperature resistance, impact resistance and vibration resistance have become the most effective tool for solving automatic control problems [6, 7].

The PLC has high reliability, convenient control, small size, and low price. The new generation PLC also has PID setting function. Its application has been expanded from switch control to analog control, and it has achieved success in the air-and is used in aerospace, metallurgy, light industry and other industries. The complexity and difficulty of the control system make PLC develop towards integration, and now it is mainly the integration of PLC and PC, the integration of PLC and DCS, the integration of PLC and PID, etc. Strengthened networking and communication skills. Due to the rapid development of data communication technology in recent years, users have put forward high requirements for openness, and fieldbus technology and Ethernet technology are also developing simultaneously [8, 9].

Siemens PLC occupies a large share in the domestic market with its extremely high cost performance, and is widely used in various fields of life in our country. PLC provides almost perfect modern automatic control device for industrial automation [10, 11].

The API adopts sequential scanning and continuous loop mode, as shown in Fig. 1, the steps are as follows: Compile the program according to user control requirements and save it in the memory. When the controller is in execution, the CPU will execute the command and follow the command sequence number (or address number) sequentially Perform periodic recurring requests. If there is no jump command, start from the first command defined and run the user-defined program in sequence until the program ends. Then launch in some way to control the external load [12].

The basic principle of PLC control is classified according to the role of different objects in the production process, which can be divided into: industrial belt conveyors, packaging robots and punching machines. Industrial tape production line control system

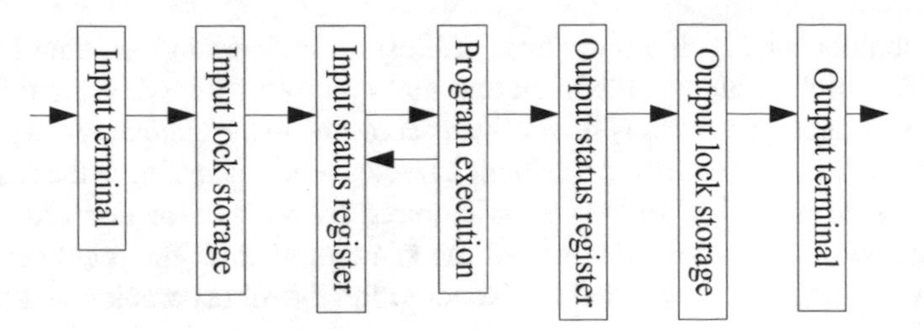

Fig. 1. Scanning process of PLC

is mainly composed of main program module, sensor/actuator/drive circuit. The main control unit uses relays to transport and transfer parts to complete the automatic operation; the control signal of the sensor or PLC is realized by the analog conversion device. In the entire process of controlling the automated production line, it is mainly composed of the main controller, the sensor detection unit, the drive system and the controlled object.

This article mainly studies an automated production line control system based on the combination of PLC and robot. The system uses a relay logic controller to realize the coordination of the action requirements between the workpiece handling equipment and the controlled device. As well as the electrical and other functional modules between the intermediate actuators, complete automatic control, detection and protection of the working state. Use industrial manipulators instead of manual operations. In order to achieve real-time monitoring of the processing procedures during the production process of the factory.

2.2 Packaging and Palletizing Automatic Production Line

In industrial production, a metering control system composed of metering devices is widely used to accurately measure and control the flow of small particles or powdery bulk materials. At present, it has applications in various industries such as cement building materials, metallurgical power, food, mining, chemical industries and other industries. In the building materials industry, in the cement manufacturing process, it is used as components, fillers and other technological compounds in the production process of pulverized coal raw materials, raw powder clinker, etc. In the metallurgical industry, the example of an automatic sintering feed system is very typical. Quantitative feeding reflects its importance in industrial production. Therefore, it is particularly important if the automation and accuracy of the feeding system can be continuously improved. In recent years, the integration of traditional mechanical systems and computer control has continued to increase, which has obvious advantages in industrial production efficiency and corporate economic benefits. Therefore, the use of electromechanical integration technology and computer control technology to complete quantitative analysis is the future development direction.

The production line system consists of machines that can perform certain functions automatically, and requires independent or unreliable controlled equipment. It includes the design of the product processing process and represents a complete process cycle and drive components. At the same time, it must have a certain weighing relationship and control ability to support the operation and control of the whole station. In PLC control, the production process is streamlined according to the technological process and equipment to realize an automated production line system. According to the requirements of the processed parts, the various tools required by the workstation are transported to the machine tool or automatic line. At the same time, after the robot sends out the manipulator, it should perform a series of actions after the robot completes the processing operation and return to the driving machine to continue working until the assembly process Finish. This is a complete machining process.

A digital filter is a discrete-time system, algorithm, or device. Digital signal processing technology. Its application can make the automatic production control line of packaging and palletizing more efficient. Suppose the order of the filters is:

$$P(g) = 1/(g^3 + 2g^2 + 2g + 1) \quad (1)$$

Perform frequency conversion from low pass to high pass to find the analog high pass Q(t):

$$Q(\mathrm{t}) = P(g)\Big|g\frac{\Omega}{t} = \frac{T^2}{T^3 + 2\Omega_z T^2 + 2\Omega_z T^2 + \Omega_z^3} \quad (2)$$

In the design process of Butterworth low-pass filter, as the digital frequency gradually increases, the amplitude of the system drops rapidly. With a larger filter structure, the attenuation of the system amplitude is also greater.

2.3 System Design

The main task of human-computer interaction is to quickly save parameters and data in the application system, process them correctly, monitor the entire running state of the control in real time, and display it on the touch screen as accurately as possible. Different methods will appear in the design and planning of the human-computer interaction interface, and finally different ideas, methods and methods, different interfaces and control models can be designed. Human-computer interaction interface Graphical user interface is also often used for PLC control. When designing the menu interface, users usually follow the following design principles:

The user must reasonably conceive and design the different levels and overall structure of the interface. Users should use accurate titles and titles when selecting and using each menu.

Users should use English or use frequently used levels in alphabetical order. Users should sort the menu items in the menu system appropriately and wisely, and organize them into groups.

The structure of the automatic bag feeder includes several parts such as suction bag machine, bag feeder, bag feeder, bag receiving machine, and bag receiving machine. The automatic bagging machine is composed of a transfer hopper, bag taking, bag opening clamping device, revolving door and tightening device, frame, etc. Indexing conveyor is a combination of indexing conveyor and bagging machine. Its main function is to transport and index the delivered bags according to the designated maneuver plan.

The bag indexing adopts 90° indexing, and the indexing is controlled by photoelectric signal. The structure of the palletizer is composed of an end guide plate, a sliding plate mechanism and a forming device. In the packaging process, it is necessary to automatically perform weighing, bag filling, bag filling, and bag filling.

2.4 Control System Design

Common pneumatic and electrical control adopt a relatively complete pneumatic control system, which is resistant to external interference, relatively easy to control, and very flexible in program control to cope with very complex working conditions and relatively strict control requirements. The cylinder drive and output power are supplemented by a pneumatic circuit. The processing of input/output signals and the change of logic state can be performed by the servo loop.

The main control valve is divided into two categories: the first category is an electromagnetic directional valve with a coil. The control circuit inside the valve is relatively intuitive and simple. However, if the power is cut off, it may automatically return to its original state, and some failures may cause the device to loosen and over tighten. The second is the dual-coil direction control solenoid valve, which controls the air intake of the cylinder, and the solenoid valve control circuit is slightly complicated Some. But the main reason is that it has a memory function, which can still maintain the original working state after power failure, so it has better safety performance. The programmed PLC program executes the corresponding designated electromagnetic switch. The preset pneumatic control program can only be executed and completed when the electromagnet is energized.

3 Debug and Run

3.1 Debug and Run Tasks

Generally speaking, debugging and running can accomplish the following tasks:

Check and correct errors in the PC user program through debugging and execution to meet the requirements of the production machine;

Check and correct PC I/O wiring and other wiring errors through commissioning and operation.

Use the debugging process to adjust the timer setting value. Better adapt it to the needs of the process.

Through debugging and execution, master the working mode of the automatic line in each working mode.

3.2 Implementation of Commissioning Operation

Commissioning and work-in-progress are implemented step by step within the prerequisite framework in a targeted and planned manner. Therefore, before debugging and running, please create a debugging and running step.

3.3 Simulation Software

Before installing the entire system on site, use simulation software to debug and verify the compiled PLC ladder program. The simulation software for debugging this system is TRiLOGI software.

3.4 System Test Indicators

During the operation of the system, the PLC control program is modified to enable it to work as expected. Because this design uses an industrial robot control system. Therefore, when the workpiece is moved from the tote to the designated position, the corresponding operation can be completed by simply placing the workpiece in the designated position.

In the process of experimental operation, the performance, function, time and other aspects of the data are recorded, and then the automatic control system is analyzed according to the data results.

4 Analysis of Data Results

4.1 Analysis of Performance and Time

According to the test experiment mentioned above, this article selects 4 sets of test results for comparative analysis. The content of the analysis includes the control coverage accuracy rate, coverage rate and time-consuming of the control system. The details are shown in Table 1:

Table 1. Analysis of performance and time

	Coverage accuracy	Coverage	Time
1	89%	91%	2.1
2	88%	90%	2
3	85%	92%	1.6
4	87%	91%	1.8

As shown in Fig. 2, we can find that when the coverage rate reaches 90%, the control accuracy rate is 88%, and it takes 2 s at this time. In addition, when the coverage rate reaches 92%, its control accuracy rate reaches 85%, which takes 1.6 s. This shows that the longer the time-consuming, the higher the control accuracy.

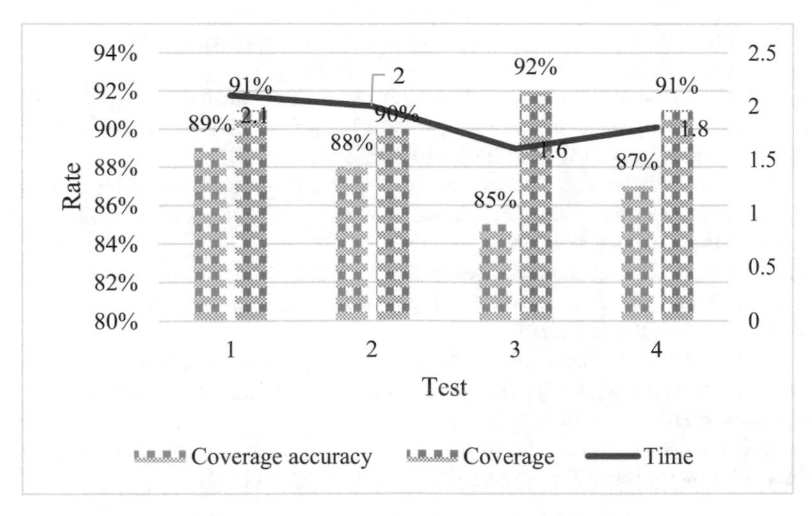

Fig. 2. Analysis of performance and time

5 Conclusion

Under the control of PLC, the workpiece is automatically transported according to the set program, and the entire production line is a production process. Through the calculation of the mathematical model parameters (proportional relations) of the controlled object, the actual state or running-time value of the controlled object can be obtained over time. Therefore, the corresponding functional modules can be designed according to the system requirements and the interconnection and coordination between various parts can be realized. This article introduces the PLC control technology, and designs an automated production line control system based on PLC and robots. There is a difference between the time required to handle and move the workpiece in the processing process and the actual demand.

References

1. Zhang, L.: Design of automatic production line control system for environmentally friendly energy-saving bricks. J. Guangdong Inst. Petrochem. Technol. **028**(006):46–49,53 (2018)
2. Guirong, H., Xubing, C.: Design of a fully automatic car wheel handling system based on PLC, robot and vision system. Mod. Manuf. Eng. **000**(009), 52–58 (2017)
3. Qingqi, C., Haiwen, Z., Bin, H., et al.: Design of elevator hall door robot packing control system based on PLC. Mach. Tool Hydraul. **046**(003), 75–78 (2018)
4. Murali, P.K., Darvish, K., Mastrogiovanni, F.: Deployment and evaluation of a flexible human-robot collaboration model based on AND/OR graphs in a manufacturing environment. Intell. Serv. Rob. **2020**(C):1–19 (2010)
5. Mrugalska, B., Stetter, R.: Health-aware model-predictive control of a cooperative AGV-BASED PRODUCTION SYSTEM. Sensors **19**(3) (2019)
6. Franek, J., Nalepa, B.: Study of the vision system's impact on increasing the reliability of the production system. New Trends Prod. Eng. **2**(2), 46–56 (2019)

7. Vocetka, M., Huczala, D.: The use of the two-handed collaborative robot in non-collaborative Application. Acta Polytechnica **60**(2), 151–157 (2020)
8. Abbood, W.T., Abdullah, O.I., Khalid, E.A.: A real-time automated sorting of robotic vision system based on the interactive design approach. Int. J. Interact. Des. Manuf. **14**(1), 201–209 (2019). https://doi.org/10.1007/s12008-019-00628-w
9. dit Eynaud, B., Amélie, K.N., Roucoules, L., et al.: Framework for the design and evaluation of a reconfigurable production system based on movable robot integration. Int. J. Adv. Manuf. Technol. **118**(7), 2373–2389 (2022)
10. Vijayakumar, S., Dhasarathan, N., Devabalan, P., et al.: Advancement and design of robotic manipulator control structures on cyber physical production system. J. Comput. Theor. Nanosci. **16**(2), 659–663 (2019)
11. Idrissi, M., Salami, M., Annaz, F.: A review of quadrotor unmanned aerial vehicles: applications, architectural design and control algorithms. J. Intell. Rob. Syst. **104**(2), 1–33 (2022). https://doi.org/10.1007/s10846-021-01527-7
12. dos Santos, P.J.S., Vargas J D S., Vogel Cé, A., et al.: Graphical analysis of the movements of a line-following robot. Phys. Educ. **57**(2), 023004 (8pp) (2022)

Research of Multi Sensor Fusion Positioning Algorithm Based on Kalman Algorithm

Faying Li, Juan Xiao(✉), Wenyi Huang, and Sijie Cai

School of Computer and Artificial Intelligence, XiangNan University, Chenzhou, Hunan, China
keyanxj@163.com

Abstract. This paper proposes a multi-sensor fusion algorithm, combined with the traditional Kalman algorithm, to achieve UAV precise positioning. At the same time, consider more external constraints affecting the stability and safety of UAV flight, and build the extended Kalman multi-sensor filter model. The application of this model to the Intelligent forest inspection system will improve the positioning accuracy of UAV positioning, conducive to shorten the UAV cruise time and improve the reliability and stability of UAV.

Keywords: UAV · Kalman Algorithm · Multi Sensor Fusion · Precision

1 Introduction

With the rapid development of sensor technology and automatic recognition technology, the wide application of 5G network communication, all kinds of intelligent devices more and more appear in all walks of life, various application fields, to meet various application needs. Among them, UAV is a widely used, convenient, flexible, stable, safe and reliable intelligent device [1,2].Since the 1970s, drones have been widely seen in the public eye, used in emergency search and rescue, aerial photography, military inspection, forest inspection and other fields in harsh environments. Because UAV is unmanned tools, need to rely on all kinds of sensors, lidar and other independent obstacle perception, and according to the perceived obstacle type and position change real-time, therefore, how to realize the accurate positioning of obstacles, the comprehensive identification of various types of obstacles, finally achieve high precision obstacle positioning and recognition, improve the UAV in the process of mission reliability [3,4], safety and stability, become a key problem of UAV research.

This paper relies on the traditional Kalman filtering algorithm, combines the multi-sensors carried by the UAV itself to realize the precise positioning of the UAV itself and the obstacles, design and realize the autonomous positioning system of the UAV, and improve the autonomous ability of the UAV.UAV autonomous system usually use sensors to perceive their state behavior and surrounding environment, collect all kinds of information during flight, such as location, speed, acceleration, air pressure, using a

J. H. Abawajy et al. (Eds.): ICATCI 2022, LNDECT 169, pp. 335–343, 2023.
https://doi.org/10.1007/978-3-031-28893-7_40

single sensor to obtain data noise components, cause inaccurate data, limited accuracy of sensor measurement, therefore, scholars at home and abroad are committed to study multiple sensor fusion, using appropriate fusion algorithm in induction or cognitive fusion, expect to obtain the most appropriate estimate of the current state.Multi-sensor fusion can fully collect UAV information, and the estimated state is more accurate and reliable.

2 Intelligent Inspection of the UAV Positioning Strategy

(1)Basic principle of UAV multi-sensor fusion.

Multi-sensor fusion refers to the "merging" of data from multiple sensors based on known fusion algorithms [5], with the aim of generating relatively accurate information.In fact, data fusion in a broad sense is an efficient and accurate processing method of massive data in the current big data and cloud computer environment. According to the application situation and accuracy requirements, data fusion can be divided into pixel-level fusion, feature-level fusion and decision-level fusion.UAV multi-sensor fusion is pixel-level data fusion, embedded in the UAV speed sensor, displacement sensor, acceleration sensor, pressure sensor and other various types of sensors for multi-level, multi-directional information complementary and optimization combination processing [6,7], eventually produce the consistency of the observation environment, accurate access to the UAV current status, behavior, surrounding environment information.There are two ways of multi-sensor data integration: one is to integrate the data obtained by different types of sensors to improve the accuracy of each sensor; the other is to integrate multiple sensor data, combination deduction and collaboration, improve the intelligence and autonomy of the whole sensor system, effectively realize the obstacle avoidance function, and improve the overall safety and reliability of the UAV.

(2)UAV multi-sensor fusion model.

UAV are all kinds of data collected by multi-sensors, diversified data sources and different data fusion processing methods [8]. According to the generation time of sensor target data, whether various sensors have their own independent perception modules, multi-sensor fusion algorithm has post-fusion algorithm and front fusion algorithm. Figure 1 is the structure diagram of fusion algorithm.

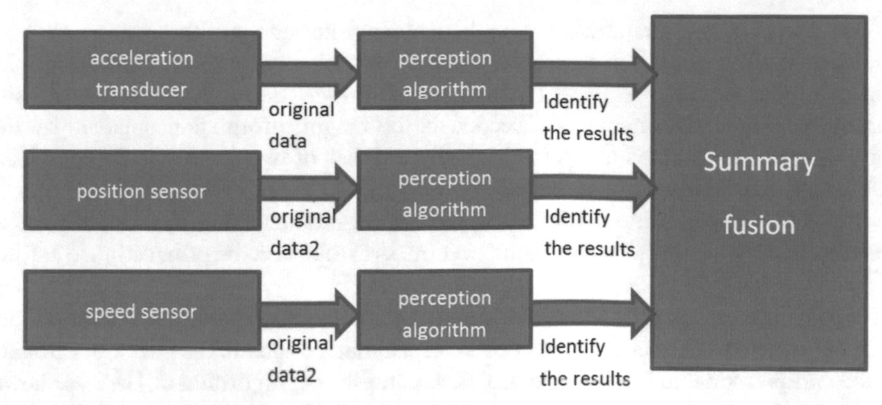

Fig. 1. Structure diagram of the UAV multisensor fusion model

The UAV multi-sensor fusion algorithm first generates the target data needed by the autonomous system independently by each sensor of the UAV [9,10]. Since each sensor has its own independent perception, all the UAV sensors complete the target data generation, and then the data fusion is unified by the main processor.Will UAV built-in sensors such as position sensor (collect UAV current location information), speed sensor (collect UAV speed information), acceleration sensor (get UAV acceleration, ensure the UAV fly in a smooth way), air pressure sensor (sensing UAV working environment air pressure, ensure the bad weather pressure impact on the impact of UAV obstacle avoidance function).

3 Multi-sensor Fusion UAV Positioning is Realized Based on the Kalman Algorithm

Combined with the UAV's location sensor, speed sensor, attitude sensor, attitude sensor, ultrasonic sensor, infrared sensor, satellite positioning and GPS parameters and other multisource data, Based on the basis of the traditional Kalman filtering algorithm, Data parameters affecting the position of the UAV are fully introduced to the equation of state in the Kalman filtering algorithm [11].In particular, the impact of severe weather and terrain complexity are important factors affecting the location of the UAV, The own status of the UAV (position, speed, acceleration) and the surrounding environment status information have the best estimated, To accurately estimate the distance and location relationship between the UAV and the obstacles, Provide favorable conditions for drones to avoid obstacles in the shortest possible time.To achieve safe and reliable flight of UAV, improve the quality of UAV inspection, reduce the loss caused by UAV collision with obstacles, and extend the service life of UAV.

The sensing system used for UAV sensing and obstacle avoidance function mainly includes ultrasonic, infrared TOF, laser, millimeter wave radar and depth sensing camera [12,13]. Among them, the volume of ultrasonic and infrared TOF modules is relatively small, but the measurement distance is close, the anti-interference is weak, the volume of millimeter wave radar is large, but the measurement distance is long and strong interference resistance.

(1) A localization strategy based on the Kalman filtering algorithm.

Kalman filter algorithm is used in a single sensor environment, considering the multiple input signals, we focus on two factors: one is the tilt Angle sensor for the current tilt Angle information, the second is the height information obtained by the height sensor.The Kalman algorithm basically consists of two stages: prediction phase and update phase [14].

Prediction stage: estimate the state of the current moment (time k) according to the posterior estimate of the previous moment (time k-1) to obtain the prior estimate at time k.

Update stage: Based on the recursive algorithm, use the measurement value of the current moment to correct the prediction stage estimate, to obtain the posterior estimate of the current moment.The flow chart of Kahman filtering algorithm of UAV as shown in Fig. 2:

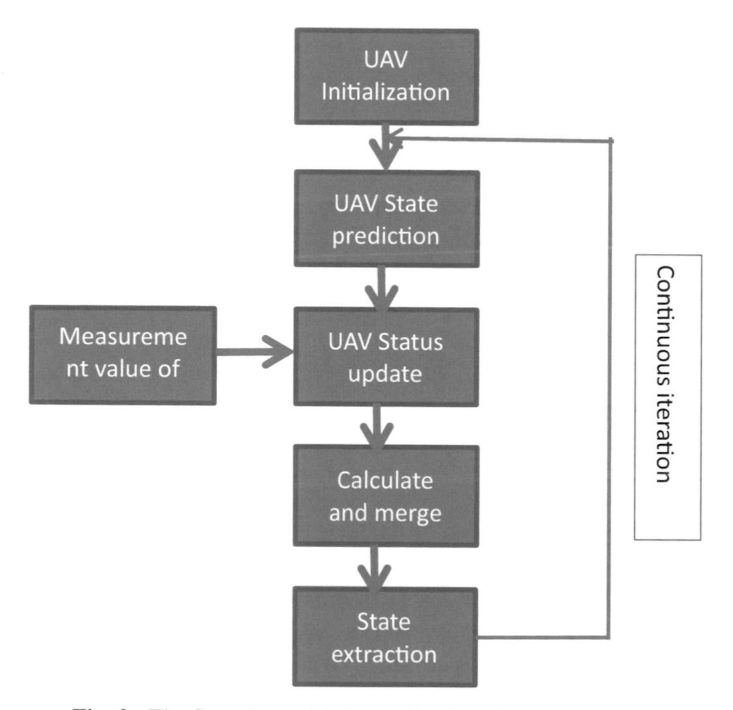

Fig. 2. The flow chart of Kahman filtering algorithm of UAV

(2) A multi-data fusion model based on the Kalman filtering algorithm was constructed.

When the UAV flies in complex terrain environment or in severe convective weather, the state prediction value of the Kalman filtering algorithm has great error, which affects the estimate value, resulting in the inaccurate positioning position of the UAV, and the reduced accuracy, and directly affects the execution efficiency of the current mission of the UAV [15].

If the target state prediction value produces a large error value at time K, the target estimate is largely not at time K-1, which needs to be corrected.If the target speed estimates and target position were at time K-1, the information at time K was compensated.Meanwhile, the filtered variance matrix at the K-1 moment at the K-1 moment.

For convenience, assuming that the position coordinates of the UAV in space are (x, y, z) and the rate is (Vx, Vy, Vz), the position of the UAV is k by the fusion of GPS, position sensor and speed sensor. In order to obtain the optimal estimate, the weather and terrain factors are considered as binding conditions and the measurement noise affecting the Kalman filter.

The Kalman filter algorithm is suitable for dynamic systems with more uncertainty factors, by making based predictions about the next step of the system, and then optimizing the estimates through the update stage, as in the five steps.

Assuming to estimate the current position information of the UAV, the distance in the horizontal direction is obtained by the ultrasonic sensor, and the vertical distance is obtained by the distance sensor, here, to simplify the construction and analysis of the model, temporarily ignore the GPS positioning device.We focus on two state variables: the position information obtained by ultrasonic sensor is Si and by distance sensor is Di.

1. Build the equation of state expression:x(t) = Ax(t-1) + Bu(t) + w(t)

among:

x(t):The state vector, the estimated physical quantity, here is the current position of the drone.

A:State transfer matrix, how the state at the previous moment impacts on the state at the current moment.

B:Controls the input matrix, indicating how the external physical quantities affect the state quantity.

u(t):Control the quantity, the physical quantity externally acting on the current state quantity at the current moment.

w(t):Process noise, to obey the overall distribution of the process noise.

2. Build the UAV observation equation expression:z(t) = Hx(t) + v(t)

among:

z(t):measurement result;measuring result.

H:Observation matrix, mapping the real state space into the observation space.

v(t):Observed noise following a normal distribution.

3. Forecast updates

Predicted status:

$$\hat{x}(t \mid \mathrm{t}-1)=A\hat{x}(t-1)+Bu(t) \tag{1}$$

Predictive error covariance matrix:

$$P(t|t-1) = AP(t-1)AT + Q \tag{2}$$

4. Real-time estimation of the target status

1) Optimum estimated state quantity:

$$\hat{x}(t) = \hat{x}(t \mid t-1) + K(t)[z(t) - H\hat{x}(t \mid t-1)] \tag{3}$$

2) Calculating error gain

$$k(t) = \frac{P(t \mid t-1)H^T}{[R + HP(t \mid t-1)H^T]} \tag{4}$$

3) Error covariance matrix:

$$P(t) = [I - k(t)H]P(t|t-1) \tag{5}$$

k(t)The above k (t) is the measurement gain matrix. The larger the k value is, the closer the optimal estimate is to the measurement value, and the smaller the k value is, and the more close the optimal estimate is to the predicted value.

Extended Kalman filtering is the addition of a computational transfer matrix module according to the conventional Kalman filter, as shown in Fig. 3:

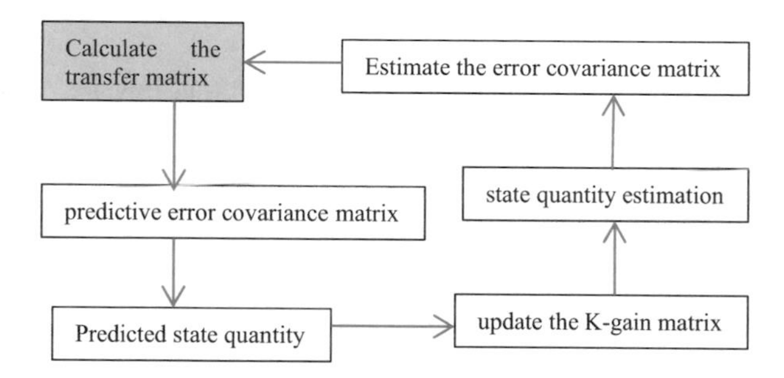

Fig. 3. Extended Kalman model structure diagram

The computational transfer matrix module in Fig. 3, recomputed at each iteration, is suitable for the dynamic characteristics in the UAV spatial environment and achieves the optimal estimation of the location of the UAV.

Considering the dynamics during UAV flight, an extended Kalman discrete system model was used:

$$X_k = \phi_{k,k-1}X_{k-1} + \Gamma_{k-1}W_{k-1} \tag{6}$$

$$Z_k = H_k X_k + V_k \tag{7}$$

In the above formula:

X_k represents ote the state vector at time K, estimator.

Z_k represents ote the measurement vector at time K.

$\phi_{k,k-1}$: One-step state transfer matrix of the system at times k-1 to k.

W_{k-1} is system noise at the time of k-1.

Γ_{k-1} is the system noise matrix.

H_k is the system measured the matrix at time k.

V_k is the system noise is measured at time k.

The state vector, measurement vector, system noise, system measurement matrix, and system measurement noise are given below:

The UAV flies in 3 D space, let its position coordinates are (x, y, z), and the rate in 3 D space is (Vx, Vy, Vz) state vector, since the position coordinate x has the same name as the state vector, the state vector is represented by Ψ.

$$\dot{\psi} = \left[\dot{x}, \dot{y}, \dot{z}, \dot{V}_x, \dot{V}_y, \dot{V}_z\right] + W \tag{8}$$

where w is the system noise

Let $u = [Vx, Vy, Vz]$ be the system input signal.

The measurement vector (since the position coordinates share the same name as z) is:

$$\gamma = \left[x_p, y_p, z_p, x_d, y_d, z_d\right] + \mathrm{V} \tag{9}$$

where V is the system measurement noise

The xp, yp, zp are the coordinate values of data collected in 3 D by location sens

The xd, yd, zd are the coordinate values of the horizontal and vertical UAV collected by the attitude sensor, respectively.

4 Multi-sensor fusion of UAV positioning software design

```
predict(P):
  P=F*P*F.transpose()
  Return x,P
  def  update(x,P,z):
  Z=matrix([z])
  y=Z.transpose()-(H*x)
  S=H*P*H.transpose()+R
  K=P*H.transpose()*S.inverse()
  x=x+(K*y)
  P=(I-(k*H))*P
  return  x,P
  plot_position_variance(x,P,edgecolor='r')
  for z in measurements:
    x,P=predict(x,P)
    x,P=update(x,P,z)
    plot_position_variance(x,P,edgecolor='b')
    print(x)
    print(P)
```

5 Conclusions

It is the key to UAV intelligent perception and obstacle avoidance control system to quickly and accurately identify air obstacles and choose appropriate obstacle avoidance strategies to ensure the safety and stability of UAV in the mission execution process.It is of great significance to realize the precise positioning of the UAV before the intelligent perception and identification of obstacles, and to carry out reasonable and reliable path planning after the subsequent identification of the UAV of obstacles.Based on the traditional Kalman filtering algorithm, using the multi-sensor data carried by the UAV itself, and considering the external constraints such as terrain and weather, the extended Kalman filter model is constructed, and the obstacle positioning accuracy of the UAV is improved. The next step will study the path planning strategy and algorithm after obstacle identification.

Acknowledgments. Scientific Research Project of Hunan Provincial Department of Education Xiangtong No.264 (20C1694); Open project of Hunan Key Laboratory of embedded and network computing (NO.20210103, NO.20210104).

References

1. Boutayeb, M., Aubry, D.: A strong tracking extend-ed Kalman observer for nonlinear discrete-timesystems. IEEE Trans. Autom. Control **44**(8), 1550–1556 (2019)
2. Joseph, J., LaViola, J.R.A.: comparison of unscentedand extended Kalman filtering for estimating qua-ternion motion. In: Proceedings of the 2006 American Control Conference, IEEE Press, 2435–2440, (2020)
3. Farrell, J.A.:High-speed sensor-assisted navigation. Beijing: Electronic Industry Press (2019)
4. Groves, P.D.: Principles of GNSS and multi-sensor navigation system. Beijing: National Defense Industry Press (2018)
5. United States Department of Defense (Do D). Version 2.0,Unmanned Aircraft System Airspace Integration Plan
6. Rosen, P.A., Hensley, S., Wheeler, K.: UAVSAR: new NASA airborne SAR system for research. IEEE Aerosp. Electron. Syst. Mag. **22** (11), 21–28 (2017)
7. Osborne, R.W., Bar-Shalom, Y., Willett, P.: Design of an adaptive passive collision warning system for UAVs. IEEE Trans. Aerosp. Electron. Syst. **47**(3), 2169–2189 (2021)
8. Miah, S., Milonidis, E.: Adaptive multi-sensor data fusion positioning algorithm based on long short term memory neural networks. In: ICCT, pp. 566–570 (2020)
9. Gross, J., Gu, Y., Rhudy, M., et al.: Flight Test evaluation of sensor fusion algorithms for attitude estimation. IEEE Trans- actions on Aerospace and Electronic Systems **48**(3), 2128–2139 (2021)
10. Simon, Z., Davide, S., Stephan, W., et al.: MAV navigation through indoor corridors using optical flow. IEEE International Conference on Robotics and Automation. Anchorage: IEEE, 3361–3368 (2020)
11. Bethke, B., How, J.P., Vian, J.: Group health management of UAV teams with applications to persistent surveillance. In: Proceedings of the American Control Conference.Seattle, USA (2018)
12. Kaparias, I., Karcanias, N.: An Innovative Multi-Sensor Fusion Algorithm to Enhance Positioning Accuracy of an Instrumented Bicycle. IEEE Trans. Intell. Transp. Syst. **21**(3), 1145–1153 (2020)

13. Bertuccelli, L.F., Wu, A., How, J.P.: Robust adaptive Markov decision processes:planning with model uncertainty. IEEE Control Syst. Mag. **32**(5) 96–109 (2017)
14. Lentilhac, S.: UAV flight plan optimized for sensor requirements. IEEE Aerosp. Electron. Syst. Mag. **25**(1) 11–14 (2020)
15. Popescu, D.C.: Ioan Dumitrache, Simona Iuliana Caramihai, Mihail Octavian Cernaianu: high Precision Positioning with Multi-Camera Setups: Adaptive Kalman Fusion Algorithm for Fiducial Markers. Sensors **20**(9), 2746 (2020)

Business Personalized Automatic Recommendation Algorithm Based on AI Technology

Yi Zhou(✉)

Chongqing College of Architecture and Technology, Chongqing, China
zhouyicq@yummail.cn

Abstract. With the advent of the era of AI, the continuous development of the Internet and the rapid development of e-commerce, the personalized e-commerce recommendation system as a part of e-commerce has also attracted more and more attention from enterprises. The personalized e-commerce recommendation system focuses on recommendation algorithms, which will enable many researchers at home and abroad to concentrate on the study of recommendation algorithms. This paper aims to study the business personalized automatic recommendation algorithm based on AI technology. On the basis of analyzing the personalized recommendation technology, the collaborative filtering algorithm is optimized for the data sparsity, scalability, cold start and other problems of the collaborative filtering algorithm. And the performance of the improved algorithm was tested experimentally. The experimental results show that the algorithm improved in the paper produces the best recommendation effect, and the error value is always smaller than the others, which also verifies that the optimized algorithm in this paper is reasonable.

Keywords: AI Technology · E-commerce · Personalized Automatic Recommendation Algorithm · Collaborative Filtering Algorithm

1 Introduction

With the rapid development of the Internet, e-commerce has emerged as the times require. Because of its own advantages such as price advantages and purchasing convenience, people are more inclined to online shopping and e-commerce purchases. However, e-commerce also has its shortcomings. The explosive increase of information resources reduces the efficient use of information [1, 2]. In the era of e-commerce, users want to get rid of the trouble of finding too much information, and hope to provide enterprises with information content that is more in line with their interests and consumption habits. The recommendation system of the shopping website provides customers with products, and automatically realizes the humanized product selection process to achieve the personalized needs of customers [3, 4].

With the birth of personalized recommendation system, personalized recommenda tion technology has gradually become a research hotspot in the industry and academia,

J. H. Abawajy et al. (Eds.): ICATCI 2022, LNDECT 169, pp. 344–351, 2023.
https://doi.org/10.1007/978-3-031-28893-7_41

and the personalized recommendation technology development has been continuously promoted. Collaborative filtering is widely used due to its advantages such as unlimited range of service resources, easy application in technology, and small restrictions on users. However, it also suffers from data sparsity, scalability, and cold starts. The problem of data sparseness is mainly solved by methods such as zero-padding and the use of new similarity [5, 6]. Some scholars use median, mode, etc. methods to input unevaluated points on sparse scoreboards, and recommend simultaneous filtering in fully dense tables [7, 8]. Some scholars are using new similarity measurement methods to solve the problem that the processing power of sparse data is lower than that of traditional similarity measurement methods [9]. Some researchers use grouping to reduce the nearest neighbor search space and improve the scalability of the algorithm [10]. Some researchers also use matrix factorization ideas, such as singular value decomposition (SVD), to decompose the high-scoring board into several lower sub-matrices and perform personalized recommendations in the lower matrices [11]. Some researchers use appropriate methods to compress user rating panels and reduce the size of datasets to address scalability issues [12].

On the basis of consulting a large number of relevant references, this paper combines the collaborative filtering algorithm in the personalized recommendation algorithm to improve its shortcomings. The algorithm combines the singular value decomposition dimensionality reduction and the BP neural network to predict the unscored items.

2 Business Personalized Automatic Recommendation Algorithm Based on AI Technology

2.1 Personalized Recommendation Technology

(1) Recommendation technology based on collaborative filtering.

Collaborative filtering technology is currently the largest personalized product recommendation technology. It is based on the principle of nearest neighbors. It counts the distances between users through information such as user descriptions, and then predicts product evaluations by taking the weighted average of the distances of the nearest neighbors among users. It satisfies the user's preference for a certain product, and finally the recommendation system pushes it to the user according to this preference.

(2) Recommendation technology based on association rules.

The technology of association rules uses association rule detection algorithms, which can be introduced to users through the current market situation. The general market correlation calculation using association rules can include two processes: establishing association rules and forming market recommendations. In the process of implementing recommendation through association rule technology, searching for association rules is the most time-consuming and bottleneck in the algorithm, but offline can be considered.

(3) Content-based recommendation technology.

Content-based recommendation is the inheritance and development of information filtering technology. Elements or objects are defined by properties of associated properties. The system can learn usage interest by using the characteristics of the evaluation object and provide opinions on the degree of correspondence between usage data and predicted data.

(4) Utility-based recommendation.

Utility-based recommendations are based on an evaluation of the usefulness of a product or object. First, we generate the utility function f, and then calculate how well the user needs correspond to the sentence set. Supplier reliability and product availability can be used as usefulness factors.

(5) Recommendation based on user statistics.

Recommendations based on user statistics do not require the use of historical data, categorizing users based on personal characteristics, or recommending users by category.

2.2 Collaborative Filtering Algorithm Optimization

2.2.1 Algorithm Optimization Description

The collaborative filtering recommendation algorithm in the traditional sense directly calculates the similarity of users or elements in the user element matrix when creating recommendation results, without training or preprocessing. Based on the reviews provided for these products, products of interest are recommended to users.

To address the sparsity issue, that combines the reduction of the decaying dimension of unique values with a BP neural network to predict unscored scores. The algorithm decomposes the user's behavioral scoreboard through single-value decomposition and assigns it to a low-dimensional table, making the scoreboard relatively dense without affecting data loss. This paper will introduce the following two methods in detail, and show the algorithm improvement strategy.

2.2.2 SVD Decomposition

SVD decomposition is a method of orthogonal matrix decomposition, which is a more efficient matrix decomposition method in MATLAB.

The user ratings of the product are represented by the table m × n. where m is the number of users, n is the number of products, and the items in the table are user i's preference for product j, which can be represented by numerical values. as shown in the matrix R.

$$R = \begin{bmatrix} R_{11} & \cdots & R_{1j} & \cdots & R_{1n} \\ \vdots & & \vdots & & \vdots \\ R_{i1} & \cdots & R_{ij} & \cdots & R_{in} \\ \vdots & & \vdots & & \vdots \\ R_{m1} & \cdots & R_{mj} & \cdots & R_{mn} \end{bmatrix} \tag{1}$$

If r is a classification of the table m × nR, when the table R is decomposed, the result is shown in Eq. (2).

$$R = U * S * V^T = U * \begin{bmatrix} \sum_r & 0 \\ 0 & 0 \end{bmatrix} * V^T \tag{2}$$

Among them, U is a m × m array of Emirates array, which can represent a typical user array. s is a positive half-defined diagonal array m × n.

2.2.3 Introducing Neural Networks

The recommendation system of e-commerce is constantly changing, user behavior is also changing, and product evaluations are also improving. This new evaluation method can be further added to BP's neural network to enrich experimental data and enhance the accuracy of future product predictions.

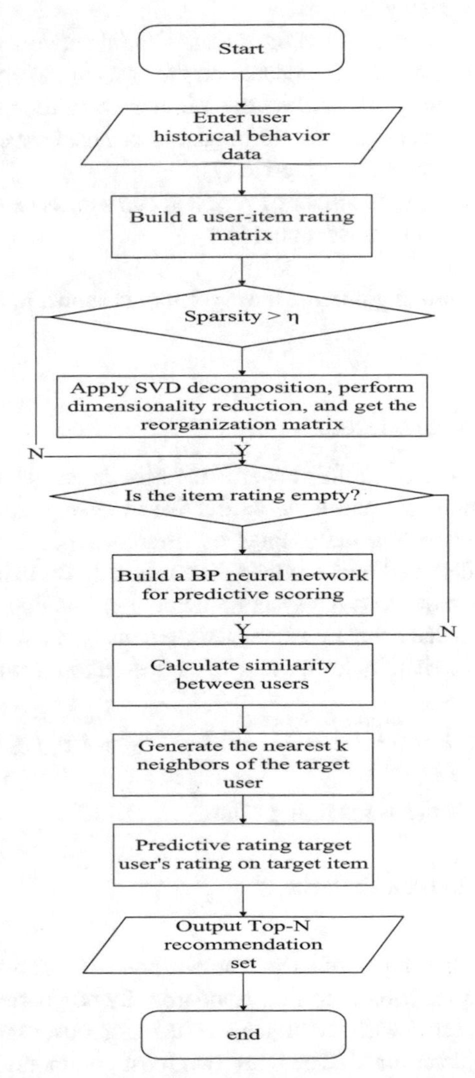

Fig. 1. The process of improving the algorithm

BP neural network training continues to evolve, and repetition is an important feature. The original neural network of BP is considered as net(i), and the updated network is considered as net(i + 1). Training results using thc latest n data as continuous samples are the best because the latest data has the best real-time performance. The training process is as follows.

(1) The algorithm starts;

(2) Define the original sample set, update the original sample training set to obtain a new sample set, and use the new sample set for training;

(3) Define and initialize the data of the three-layer network structure, including input parameters, learning accuracy, bias, etc.

(4) Set the weight and threshold of the updated initial network as the initial network value, so that it can be continuously updated on the basis of the original network;

(5) Repeat the training online, and set the initial training time as needed;

(6) Determine whether the network learning accuracy has converged to the minimum value, if it converges, go to (8), if not, go to (7);

(7) Determine whether the number of repeated steps exceeds the specified number of steps. If yes, go to (8), otherwise go to (5);

(8) End the algorithm.

In summary, the improving process the algorithm is shown in Fig. 1.

3 Experiment

3.1 Source of Experimental Data

The experimental data source for this paper is the MovieLens dataset, which is mainly used to analyze research data and is very effective in testing algorithmic recommendations. This dataset collects actual ratings of different types of movies from people of different ages, genders and backgrounds. The rationale for using this experimental dataset is to scrutinize movies on the system, including providing detailed information on users and people, so that scholars and experts can analyze and study various factors affecting the future. The sparsity level s can be expressed as formula (3)

$$s = 1 - \frac{num\ of\ R_{ij} \neq \varphi}{num\ of\ R_{ij}}, 1 \leq i \leq m, 1 \leq j \leq n \tag{3}$$

In the formula, R_{ij} is the user's rating value.

3.2 Experiment Evaluation Criteria

(1) Accuracy.

There are roughly two types of measurement precision standards: data precision standards and decision function precision standards. By comparing and analyzing the object's expected user score with the object's actual target user score, the target's data accuracy level can be determined. There are two most commonly used ways to judge the correctness of recommender system algorithms, namely MAE and RMSE. Among them, MAE is also the mean absolute error. If user u's average expected rating set for n items is assumed to be $\{p_1, p_2, ..., p_n\}$, then if user u's average actual rating set for n items is $\{q_1, q_2, ..., q_n\}$, the user's predicted mean absolute MAE error is:

$$MAE = \frac{\sum_{i=1}^{n} |p_i - q_i|}{n} \tag{4}$$

In recent years, in the International Netflix Grand Prix, the Root Mean Square Error (RMSE) method is used as the main measure of collaborative filtering accuracy. The RMSE algorithm is as follows:

$$RMAE = \sqrt{\frac{\sum_{i=1}^{n} (p_i - q_i)^2}{n}} \tag{5}$$

Both MEA and RMSE can be measured by the deviation between the predicted score and reality. The smaller the magnitude of MAE and RMSE, the smaller the error.

The accuracy standard of decision support mainly lies in prediction and prediction, which is especially suitable for zero-minus-one system. That is, if the target user is assumed to like or dislike the item, then the recommended result is true or false. In information retrieval technology, the recommendation system takes into account the precision and recall effect. Assuming that R_u is the recommended result set, and T_u is the item that user u likes in the test set, then:

$$\text{Re}call = \frac{\sum_u |R_u \cap T_u|}{\sum_u |T_u|} \tag{6}$$

$$\Pr ecision = \frac{\sum_u |R_u \cap T_u|}{\sum_u |R_u|} \tag{7}$$

(2) Coverage.

Coverage is used to introduce items to users. A good introduction system must introduce all items of interest to users, and introduce as many items as possible without producing high-quality items. The scope of a recommender system is mainly based on the percentage between the number of elements recommended and all elements in the system. If this percentage is too low, it may affect user satisfaction. If it is assumed that the user set is U, the item set is I, or the recommended result set is R_u, then the formula for calculating coverage is

$$Coverage = \frac{|\bigcup_{u \in U} R_u|}{|I|} \tag{8}$$

4 Discussion

Under different data sparsity, 30 neighbor users are taken, and the recommendation performance of the improved algorithm is tested below.

Table 1. MAE values of three algorithms under different sparsity

Sparsity	Collaborative filtering algorithm without SVD and BP neural network	Matrix Factorization SVD Collaborative Filtering Algorithm	An Algorithm Based on Singular Value Decomposition and BP Neural Network Optimization
0.7481	0.3637	0.4230	0.4897
0.8120	0.4293	0.4720	0.5451
0.8392	0.4711	0.5343	0.5788
0.8795	0.5933	0.5611	0.6309
0.9195	0.5455	0.6385	0.6804
0.9530	0.5940	0.7000	0.7624
0.9700	0.6752	0.7788	0.8583

The experimental results are shown in Fig. 2. The dimensionality reduction of the matrix is accompanied by the reduction of the sparsity. The dilution formula can be used to obtain the dilution in the dimension table, and the obtained results can be represented by the data in Table 1. Obviously, the larger the error, the lower the accuracy. Figure 2 shows that the trends of the three algorithms are the same, indicating that the recommendation error of the algorithm increases as the data becomes more diluted, but the magnitude of the trend is not the same, the improved algorithm in this paper produces the best recommendation, and the error value is always lower than the other values, which also confirms that the optimization algorithm in this paper makes sense.

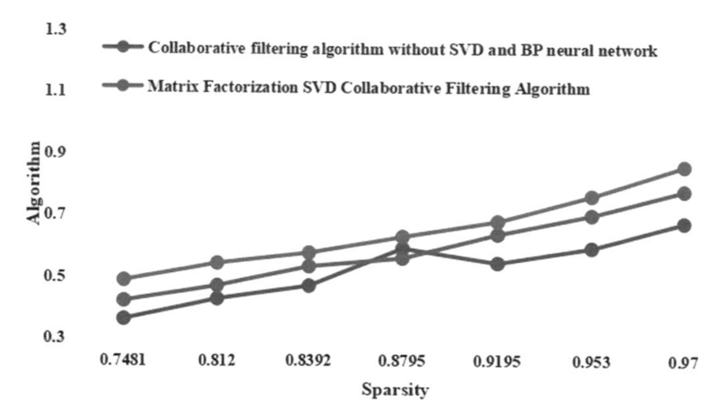

Fig. 2. MAE values of three algorithms under different sparsity

5 Conclusions

Personalized recommendation system is the product of the common development of e-commerce and Internet technology. The personalized recommendation algorithm is the core of the entire personalized recommendation system. The quality of the recommendation algorithm will directly affect the recommendation quality of the entire recommendation system. At present, because the collaborative filtering recommendation algorithm is relatively mature and used in the personalized recommendation system, it provides a recommendation technology with excellent recommendation results. Therefore, the main purpose of this paper is to study the collaborative filtering recommendation algorithm in detail, and make corresponding technical improvements to improve the quality of the collaborative filtering recommendation algorithm, thereby improving the development and implementation of the personalized recommendation system.

References

1. Alazab, M., Al-Nemrat,A., Shojafar, M.: et al. Foreword: Special Issue on Trends in AI and Data Analytics for an Ethical and Inclusive Digitalized Society. International Journal of Uncertainty, Fuzziness and Knowledge-Based Systems, 2021, 29(Suppl 2):v-vii
2. Thrall, J.H., Fessell, D., Pandharipande, P.V.: Rethinking the Approach to AI for Medical Image Analysis: The Case for Precision Diagnosis. J. Am. Coll. Radiol. **18**(1), 174–179 (2021)
3. Daisy, P.S., Anitha, T.S.: Can AI overtake human intelligence on the bumpy road towards glioma therapy? Med. Oncol. **38**(5), 1–11 (2021)
4. Rahimi, S. A., Légaré, F., Sharma, G., et al.: Application of AI in Community-Based Primary Health Care: Systematic Scoping Review and Critical Appraisal. J. Med. Internet Res., 23(9), e29839 (2021)
5. Barrera, A., Gee, C., Wood, A., et al.: Introducing AI in acute psychiatric inpatient care: qualitative study of its use to conduct nursing observations. Evid. Based Ment. Health **23**(1), 34–38 (2020)
6. Pandiyan, V., Shevchik, S., Wasmer, K., et al.: Modelling and monitoring of abrasive finishing processes using AI techniques: A review. Journal of Manufacturing Processes, 57(5) 114–135 (2020)
7. Ström, P., Kartasalo, K., Olsson, H., et al.: AI for diagnosis and grading of prostate cancer in biopsies: a population-based, diagnostic study. The Lancet Oncology, 21(2) 222–232 (2020)
8. Tewari, A.S., et al.: Sequencing of items in personalized recommendations using multiple recommendation techniques. Expert Systems with Application, 97, 70–82 (2018)
9. Pla Karidi, D., Stavrakas, Y., Vassiliou, Y.: Tweet and followee personalized recommendations based on knowledge graphs. J. Ambient. Intell. Humaniz. Comput. **9**(6), 2035–2049 (2017). https://doi.org/10.1007/s12652-017-0491-7
10. Pielak, R., Wei, P., Peyret, H., et al.: 14188 Wearable UV/HEV light sensor and smartphone application for personal monitoring and personalized recommendations. Journal of the American Academy of Dermatology, 83(6), AB133 (2020)
11. Divyaa, L.R, Pervin, N.: Towards generating scalable personalized recommendations: Integrating social trust, social bias, and geo-spatial clustering. Decis. Support Syst. **122**, 1–17 (2019)
12. Danaf, M., Becker, F., Song, X., et al.: Online discrete choice models: Applications in personalized recommendations. Decis. Support Syst. 119(APR.), 35–45 (2019)

Low-Carbon Garden Landscape Design Based on Image Processing Technology

Junhua Xiao[1(✉)], Fanling Chen[1], and Jumshaid Ullah Khan[2]

[1] Gongqing College of Nanchang University, Jiujiang, Jiangxi, China
xiaojunhua630630@163.com
[2] Commune d'Akanda, Akanda, Gabon

Abstract. The process of urbanization is accelerating, the scale of cities is expanding, and the requirements for the quality of garden landscape are getting higher and higher. Low-carbon garden landscape design is a green and environmentally friendly action. In order to strengthen the design effect of low-carbon garden landscape, it is necessary to use image processing technology for analysis. Therefore, this paper proposes a low-carbon garden landscape design based on image processing technology, in order to improve the feasibility and rationality of the design. This paper mainly uses the methods of investigation and comparison to analyze the related problems and strategies of low-carbon garden landscape design. The survey data shows that 78.35% of people agree with the design of low-carbon garden landscape, and combine it with regional characteristics, landscape characteristics, and low-carbon materials and low-carbon behavior, intending to promote the construction of low-carbon garden landscape.

Keywords: Image processing technology · Low-carbon garden · Landscape design · Applied research

1 Introduction

Due to the continuous expansion of life-like atmosphere, its pollution is becoming more and more serious. In order to improve the quality of living environment, low-carbon design of some garden landscapes is a way to protect the environment. In greening planning, not only factors such as green area, building density, and land use should be considered, but also the coordination and safety between environmental conditions and surrounding buildings.

There are many related studies on the application of low-carbon garden landscape design based on image processing technology. For example, R. Geetha Ramani et al. proposed magnetic resonance imaging (MRI) to perform computer analysis of these images to identify abnormal areas. In this work, high-grade glioma images were used to identify tumor regions in the brain using image processing and data mining techniques [1]. Raman Perumal and others proposed that the steel is a precious treasure of the world cultural heritage. Digital image processing is one of the powerful methods of lossless technology. They introduced the use of k-means ensemble clustering and intelligent

J. H. Abawajy et al. (Eds.): ICATCI 2022, LNDECT 169, pp. 352–359, 2023.
https://doi.org/10.1007/978-3-031-28893-7_42

edge detection for non-invasive detection of moss and cracks in monuments [2]. The purpose of the study by Viviana Yarel Rosales-Morales et al. was to propose and describe ImagIngDev. This is a new approach to developing automated cross-platform mobile applications using image processing techniques [3]. There are many researches on image processing technology, but there are not many researches on applying it to low-carbon garden landscape design. Therefore, this paper discusses about it.

This paper firstly studies low-carbon garden landscape and describes it in detail. Secondly, it analyzes the low-carbon design methods and principles of garden landscape. Then the image processing hardware platform system in landscape design is elaborated. Finally, through the form of questionnaire survey, the strategy of low-carbon garden landscape design is counted and analyzed, and relevant conclusions are drawn.

2 Application Research of Low-Carbon Garden Landscape Design Based on Image Processing Technology

2.1 Low-Carbon Garden Landscape

As a sustainable development model of urban garden landscape, low-carbon garden has a good promotion trend. Low-carbon gardens choose more energy-efficient materials and technical means in terms of content and construction, and minimize carbon dioxide emissions during construction and subsequent use by reducing the large-scale use of reinforced concrete. Based on the following garden aesthetics, the earthworks on the existing natural conditions of the planned land should be minimized, and recycling methods such as rainwater collection should be adopted to promote the sustainable development of gardens [4, 5].

Landscaping generally improves the climate in two ways: energy saving and emission reduction and carbon capture and storage. Low-carbon gardens pay more attention to the use of three-dimensional plants, which can maximize environmental sustainability.

2.2 Low Carbon Design of Garden Landscape

Landscape has spatial differentiation and biodiversity, that is to say, there are not only interrelationships among various elements, but also a tendency to comprehensively integrate and converge to develop into the same theory.

The theory of ecological location refers to the use of natural resources in accordance with local conditions, scientific and rational use of natural resources through the optimal ecological utilization and allocation of elements and elements in the landscape, according to the ecological laws and the interests pursued by human beings. Landscape ecological construction refers to the introduction of new elements through combination on the basis of maintaining the original landscape elements [6, 7].

Modern landscape planning and design theory, with the ecology of the landscape as the core, realizes the theory of satisfying human needs and the sustainable development of natural environment ecology, which can be summarized into three levels:

First of all, it is the planning and design for the visual perception of natural and artificial forms based on the visual perception of the landscape. Secondly, the level of ecological resources, including the investigation, analysis, planning and protection of natural ecological elements. Thirdly, it includes the humanistic aspect of customs and history, and these parts related to human spiritual life belong to spiritual landscape planning [8, 9].

"People-oriented" is the basic principle that landscape architects must follow in the work process. The principle of "people-oriented" requires that landscape architects must design works from the perspective of users, fully measure the feeling of the landscape, and also need to fully consider the use, nature, and function of the landscape project, as well as the age and grade of the audience.

Landscape construction must follow the ecological principle, which is the trend of the times. The main purpose of landscape construction is to build a sustainable and healthy living environment for people, which is also the greatest value of this industry.

Each region has its geographical features, customs, regional characteristics, and cultural environment with representative significance and characteristics. Therefore, the design of the landscape must comprehensively analyze the local environment, and accurately and scientifically measure various quantitative elements to comprehensively synthesize the above elements, and design the landscape according to the local topography and topography.

The concept of low-carbon landscape involves the meaning of saving, that is to say, in the process of constructing and constructing the landscape, the issue of saving must be taken into account [10, 11].

The first thing to consider in landscape design is to satisfy the audience's use functions. With the overall improvement of people's living standards, the requirements for landscapes are also constantly evolving. Not only the landscape needs to have leisure and entertainment functions, but also the landscape is required to improve the living environment, popular science education, Functions such as fitness exercise, gathering and communication [12].

The construction of landscape must pay attention to the principle of economy, but not blindly saving, but to maximize the value of each material, so that all kinds of energy and resources can be fully utilized.

In the process of building the landscape, the green soft landscape is increased, while the hard pavement landscape is reduced. The low-carbon landscape design concept advocates expanding the area and types of green space and building high-quality green space soft landscapes.

Landscape designers must follow the laws of nature in the process of project planning and design, in order to achieve the expected design effect, which can also fully express the concept of low-carbon landscape.

2.3 Image Processing Hardware Platform System in Landscape Design

With the continuous advancement of computer technology, artificial intelligence technology is an inevitable trend of development. At the same time, in the photoelectric detection industry, higher artificial intelligence automation requirements are put forward for the photoelectric detection equipment platform. The main subsystems of the photoelectric detection system include control system, calibration system, sensor system, power supply system, etc. Among them, the control system in the photoelectric detection equipment is basically equipped with a machine vision acquisition subsystem. In the acquisition system, the acquisition of image data signal is extremely important. The image data signal is input to the control system, and then the data information is transmitted to the upper computer. The application of artificial intelligence technology to photoelectric detection equipment will put forward higher standards for the transmission of its data information to realize data processing. For example, in the photoelectric ranging system, in order to avoid human error in the detection process, the intelligence and automation of the photoelectric detection system is necessary, and the high-speed data transmission is an inevitable trend in the development of the photoelectric detection industry. The digital technology of video monitoring technology needs to realize the rapid transmission of video images, which provides more opportunities and challenges for the photoelectric detection industry. In the field of video monitoring, due to some problems in video image processing technology, many difficulties are encountered in practical application. In this way, image stabilization processing is performed on the video image wobbles caused by swaying or shaking, and to obtain a stable video image, it is necessary to greatly improve the data transmission speed. This paper mainly designs the design of the processing system for fast image transmission based on FPGA, mainly to obtain the compressed video image through the configuration of the acquisition device, and quickly transmit it to the host computer through the Gigabit Ethernet interface to achieve the real-time display of the image.

In the low-carbon garden landscape design, image data can be processed from the aspects of its design, activity development and garden operation, so as to improve the design scheme of low-carbon garden landscape. Therefore, the research on image processing technology in this paper is also necessary.

There are two types of image transmission, one is analog and the other is digital. This paper adopts the digital image acquisition method, which can effectively compress and encode the digital image signal. The design of the system mainly includes: CMOS image sensor configuration, image acquisition, image storage, Ethernet communication, and global clock management.

From the overall design scheme, the main hardware subsystems of this system are: Ethernet communication module, image acquisition module, DDR3 storage module, JTAG module, power supply module. The hardware system structure is shown in Fig. 1:

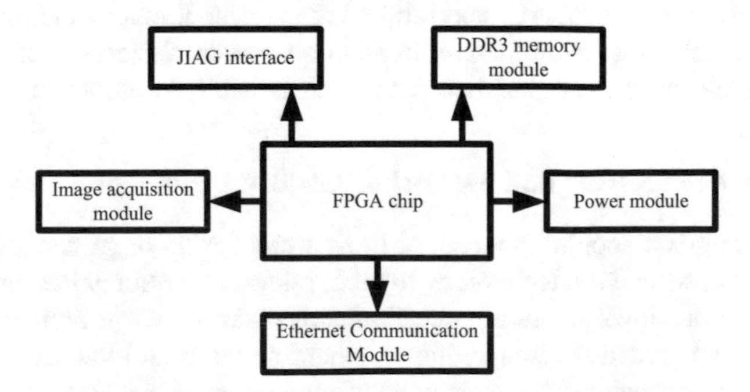

Fig. 1. Image Processing System Hardware System Structure

Ethernet communication module: Ethernet transmission uses Ethernet UDP communication protocol to achieve fast transmission of video image data.

Image acquisition module: Controlled by SCCB bus, it can input 8-bit data images of various resolutions such as whole frame, sub-sampling, and window.

DDR3 storage module: storage of video image data.

JTAG module: The JTAG interface is to download the compiled program (.bit) to the FPGA chip.

Power supply module: Provide the required voltage to the entire hardware system.

Image matching refers to the relative relationship between image content, representation, layout, texture and grayscale, the comparison of similarity, and the method of finding similar images.

Set $O_s(a, b)$ as the reference target, set $O_p(a, b)$ as the observation target, and set the matching method as a functional relationship between two sets of data, which is expressed as:

$$O_p(a, b) = h(O_s(g(a, b))) \tag{1}$$

g refers to the coordinate transformation of the image, and h refers to a certain mapping relationship between the two images. If g_a, g_b is a transformation function, then formula (1) can be expressed as:

$$O_p(a, b) = O_s(g_a(a, b), g_b(a, b)) \tag{2}$$

So, in simple terms, image matching is a method of using a matching function to identify the same feature points in two or more images.

3 Necessity Investigation of Low-Carbon Garden Landscape Design

3.1 Purpose of Low-Carbon Garden Landscape Design

There are still some problems in the level of landscape design in our country. In many urban green space construction, the ecological environment is not paid enough attention. Landscape design lacks scientific and rational management mode, and the application of low-carbon technology is immature. In addition, garden design sometimes does not consider the concept of ecological benefits and sustainable development.

3.2 Questionnaire Survey on Low Carbon Garden Landscape

In order to strengthen people's concept of low-carbon design of garden landscape, this paper intends to use network technology to fill in relevant questionnaires online. Among them, the questions involved in the questionnaire mainly revolve around the existing problems of local garden landscape, improvement methods and low-carbon landscape design strategies that cater to the awareness of low-carbon environmental protection.

3.3 Questionnaire Process

The object of this questionnaire is the necessity of low-carbon garden landscape design. The protagonists of the questionnaires are local citizens and netizens. A total of two questionnaires were designed, with two different types of questions designed for adults and minors. A total of 500 questionnaires were distributed, and netizens were invited to answer. The questionnaire lasted for one week. After the questionnaire was over, it was entered into the computer for data processing. A total of 400 valid questionnaires were recovered, with an effective recovery rate of 80%.

4 Analysis of Questionnaire Results

4.1 Low-Carbon Landscape Planning and Development Strategy

According to the survey of this paper, the opinions of netizens on the low-carbon landscape planning of gardens mainly focus on four aspects: people-oriented, interpenetration of regional characteristics and landscape characteristics, low-carbon landscape elements and low-carbon tourism behavior. Its support is shown in Table 1:

Table 1. Low-Carbon landscape planning and development strategy

	Agree	Disagree	General
People oriented	76	11	10
Regional characteristics and landscape characteristics	85	10	12
Low carbon landscape elements	83	11	8
Low carbon tourism behavior	69	12	13

As shown in Fig. 2, we can see that 76 netizens are very important to the people-oriented concept in low-carbon landscape planning and development. There are 85 netizens who believe that the interpenetration of regional characteristics and landscape characteristics is very important in low-carbon landscape planning and development. There are 83 netizens that low-carbon landscape elements are very important in low-carbon landscape planning and development. There are 69 netizens that low-carbon tourism behavior is very important in low-carbon landscape planning and development.

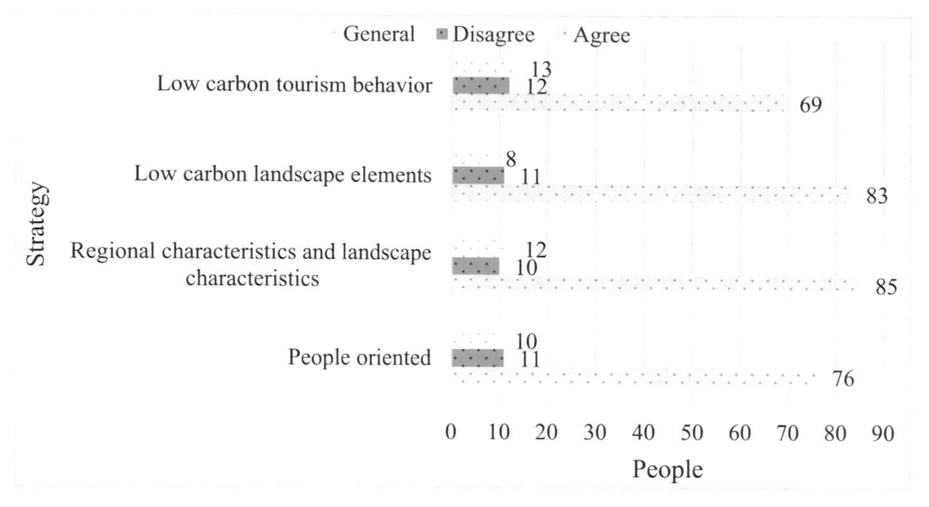

Fig. 2. Low-Carbon landscape planning and development strategy

5 Conclusion

With the development of society, the scale of urban construction has gradually expanded, and people have higher and higher requirements for environmental quality. Therefore, it is an inevitable trend to introduce image processing technology in garden landscape design, find problems in garden landscape, and propose low-carbon landscape design scheme. Landscape design based on image processing technology can effectively solve the problem of urban environmental pollution, which is of great significance to the construction of ecological civilization. Through in-depth research on low-carbon garden landscape design, it provides an effective reference for urban construction in my country, and is also conducive to the realization of sustainable development.

References

1. Geetha Ramani, R., Faustina, F., Shalika Siddique, K. Sivaselvi: automatic brain tumour detection using image processing and data mining techniques. Int.. Inf. Technol. Manag. **20**(1/2), 49–65 (2021)
2. Perumal, R.: Subbiah Bharathi Venkatachalam: non invasive detection of moss and crack in monuments using image processing techniques. Ambient Intell. Humaniz. Comput. **12**(5), 5277–5285 (2021)
3. Rosales-Morales, V.Y., Sánchez Morales, L.N., Alor-Hernández, G., García-Alcaraz, J.L., Sánchez-Cervantes, J.L., Rodríguez-Mazahua, L.: ImagIngDev: a new approach for developing automatic cross-platform mobile applications using image processing techniques. Comput. **63**(5), 732–757 (2020)
4. Singh, S., Sharma, A., Aggarwal, A.: Performance comparison of image processing techniques on various filters: a review. Int. Secur. Priv. Pervasive Comput. **13**(3), 34–42 (2021)
5. Al-Maitah, M.: Analyzing genetic diseases using multimedia processing techniques associative decision tree-based learning and Hopfield dynamic neural networks from medical images. Neural Comput. Appl. **32**(3), 791–803 (2020)

6. Abd El-Samie, F.E., et al.: Enhancement of infrared images using super resolution techniques based on big data processing. Multimed. Tools Appl. **79**(9–10), 5671–5692 (2019). https://doi.org/10.1007/s11042-019-7634-0
7. Bhalerao, R.H., Raval, M.: Automated tabla syllable transcription using image processing techniques. Multim. Tools Appl. **79**(39–40), 28885–28899 (2020)
8. Safaida, H.O., Aksasse, B., Ouanan, M., El Amraoui, M., Azrour, M.: Clay-based brick porosity estimation using image processing techniques. Intell. Syst. **29**(1), 1226–1234 (2020)
9. Saudagar, A.K.: Biomedical image compression techniques for clinical image processing. Int. Online Biomed. Eng. **16**(12), 133–154 (2020)
10. Prashar, N., Sood, M.: Shrutiain: novel cardiac arrhythmia processing using machine learning techniques. Int.. Image Graph. **20**(3), 2050023:1–2050023:17 (2020)
11. Paspalakis, S., Moirogiorgou, K., Papandroulakis, N., Giakos, G.C., Zervakis, M.E.: Automated fish cage net inspection using image processing techniques. IET Image Process. **14**(10), 2028–2034 (2020)
12. Kahlon, Y., Fujii, H.: Framework for metaphor-based spatial configuration design: a case study of Japanese rock gardens. Artif. Intell. Eng. Des. Anal. Manuf. **34**(2), 223–232 (2020)

Dynamic Economic Scheduling Optimization Based on Particle Swarm Optimization Algorithm

Guoqing Du[1](✉) and Fawaz Almulihi[2]

[1] Xi'an Traffic Engineering Institute, Xi'an, Shaanxi, China
dgq69_69@aliyun.com

[2] Taif University, Ta'if, Saudi Arabia

Abstract. Due to the excellent characteristics of resource conservation and environmental protection, wind energy has been developed on a large scale, and grid-connected operation has been gradually realized. However, there are severe interruptions in wind power generation, so it is difficult to make accurate predictions, which inevitably leads to many problems in the power system when generating electricity. The previous model is no longer applicable to the current situation, and a new model needs to be established to improve it. The economic dispatch model and method meet the needs of large-scale wind power grid integration under the current situation of rapid development of new energy in my country. Therefore, this paper takes the dynamic economic dispatch of power system as the research object, and uses particle swarm optimization algorithm to solve the problem, in order to ensure reliable power supply, the micro gas turbine power generation system has the largest output, and the power supply operation cost is lower when considering the complementary characteristics of wind and solar.

Keywords: Particle swarm optimization algorithm · Power system · Dynamic economic dispatch · Wind-solar hybrid

1 Introduction

The power supply characteristics in a microgrid are completely different, with not only grid cells that are easily exposed to the weather, but also energy storage elements that can be in charge and discharge modes. They can also exchange power externally, which undoubtedly increases the difficulty of designing the economic dispatch optimization model, making distributed optimization an important means to solve the dynamic economic dispatch problem of microgrids.

Many scholars at home and abroad have conducted in-depth discussions on the optimization of dynamic economic dispatch based on particle swarm optimization algorithm. For example, Jia Y H, Chen W N, Gu T and others integrated VESS into the microgrid model to reduce daily operating costs, and adopted a two-tier planning method to study the multi-objective optimal configuration of microgrids. The upper layer is the optimal configuration, whose goal is to minimize the daily fixed cost of investment, the rate of

J. H. Abawajy et al. (Eds.): ICATCI 2022, LNDECT 169, pp. 360–367, 2023.
https://doi.org/10.1007/978-3-031-28893-7_43

load loss, and the rate of excess energy, thereby determining the capacity of each supply. The lower layer is economic dispatch, whose goal is to minimize the cost of operation management and pollutant treatment cost, and it determines the output power of each distributed generator [1]. Wang H, Jiang Z, Wang Y et al. proposed an improved particle swarm optimization algorithm to solve the power dispatching optimization model, studied two dispatching scenarios in grid connection mode, and discussed the dispatching results of different optimization objectives. The results show that, optimizing operation for economic management of electric vehicle batteries [2]. Although there have been good research results on the dynamic economic dispatch optimization of the power system, in order to stabilize the power grid operation and minimize the economic cost, it is necessary to combine the two to perform multi-objective optimization to obtain the optimal result.

In this paper, the theoretical basis of particle swarm optimization algorithm is firstly introduced, and several kinds of power and wind power systems are listed. Then, mainly for the dynamic economic load distribution problem of wind power grid-connected power system, it is proposed to apply particle swarm optimization algorithm to the analysis of power grid economic dispatch. It is concluded that the use of wind-solar complementary characteristics for economic dispatch is a better dispatch scheme.

2 Particle Swarm Optimization Algorithm and Dynamic Economic Dispatch of Power Grid

2.1 Particle Swarm Optimization Algorithm

The particle swarm optimization algorithm simulates the predation and migratory behavior of birds or fish according to the objective function and constraints, and corrects its travel position according to the individual optimal solution and the overall optimal solution, and finally finds the optimal solution [3]. Particle swarm optimization has undergone more than 20 years of improvement and development, and many scholars have combined it with other algorithms and achieved good results in practical applications.

The particle swarm algorithm regards each bird in the flock as a particle, and quantifies the position of each bird in the current period to the position of the particle in the search space by mathematical methods. The search space of the flock corresponds to the movement interval of the particle, and the search space The position of the food in the middle corresponds to the optimal solution of the objective function to be optimized, and the distance between the bird and the food is used to judge the pros and cons of each particle solution [4]. Therefore, the closer the particle is to the optimal solution of the objective function, the better the quality of the solution in the particle, which is usually called fitness [5]. The currently found historical optimal position of each particle in the population is taken as the individual optimal position, and the currently found historical optimal position of all particles in the population is taken as the global optimal position. In the overall optimization process, through the continuous competition and cooperation of each particle, the group is constantly approaching the optimal solution [6].

When the PSO algorithm solves the optimization problem, it is equivalent to capturing the position information of the flying bird in the air and performing a functional mathematical model calculation, which is the so-called particle. The particle update speed and position formula are as follows:

$$A_{i,j}^{l+1} = A_{i,j}^{l} + e_1 d_1 (p_{i,j}^{l} - B_{i,j}^{l}) + e_2 d_2 (g_{j}^{l} - B_{i,j}^{l}) \tag{1}$$

$$B_{i,j}^{l+1} = B_{i,j}^{l} + A_{i,j}^{l} \tag{2}$$

Among them, e_1 and e_2 are normal numbers, e_1 is the individual learning factor, e_2 is the social learning factor, d_1 and d_2 are random numbers uniformly distributed in the period [0,1], l represents the number of iterations, and p and g are the maximum The optimal position solution, A represents the particle update speed, and B represents the update position.

2.2 Electric Power Generation System

(1) Wind power generation system.
Wind energy is a clean energy source, which is favored by more and more countries because of its environmental protection and huge energy output. According to the survey, the wind energy that can be exploited and utilized in the world is more than ten times that of the water energy [7]. It is especially suitable for developing wind energy in places with no fuel and inconvenient transportation, such as islands, mountainous plateaus, etc., and can meet the energy needs of the local population.
The wind power generation system consists of a wind turbine generator, a controller, a battery, and a power electronic device, and is then merged into a large power grid through a power transformer for energy transmission [8]. However, the power produced by wind turbines is not constant, so the regulator must check the quality of the output power and plug in batteries to make the output voltage as smooth as possible. The AC output current is then converted to DC, which is processed by the controller and battery, and then converted to AC by the electrical equipment. Finally, the transformer is converted into a large grid for use by other loads in the network [9]. In the microgrid transmitter model, the influence of wind speed, junction height and peripheral deviation of wind turbine output in the area where the wind farm is located is generally ignored.

(2) Photovoltaic power generation system.
The photovoltaic power generation system is composed of photovoltaic panels, control devices, batteries, and power electronic devices, which are then transformed into large power grids by transformers [10]. The working principle is that the semiconductor material of the photovoltaic panel is irradiated by sunlight. There is a voltage difference between the two ends, and the energy of the solar radiation is directly converted into electric energy according to the photoelectric effect. Its radiation intensity directly affects the output power of the photovoltaic panel, so the output power is unstable. The control system and batteries need to be upgraded to balance the output voltage and then turn it into an electronic AC device. Connect to large networks through transformers [11].

(3) Micro gas turbine power generation system.
A microturbine is a small heat generator with hundreds of kilowatts of power that uses radiant turbines and a recovery cycle to increase efficiency. Fuels are diesel, gasoline, natural gas, etc. Micro gas turbines have the advantages of high reliability, high efficiency and good controllability, and are widely used in distributed power generation, peak load supplementation and standby power stations. In addition, because it is suitable for power generation in remote areas, it also has great advantages in military applications such as sea, land and border defense. Therefore, many companies in various countries are also actively developing related equipment.
Compared with the thermal power units in the traditional centralized large power grid, the micro gas turbine units in the micro grid have a long service life, the power generation fuel is clean energy, the emission of polluting gases is less, the pollution to the environment is relatively small, and there is no pollution. Like the shortcomings of traditional thermal power units that are cumbersome to start and stop, micro-gas turbines have the characteristics of rapid start and stop. The power generation efficiency of micro-gas turbines in ordinary micro-grids can reach 30%. It can reach 75% or even higher [12].

(4) Fuel cell.
Since there is no energy loss in intermediate links, it is not limited by the Carnot cycle effect, so the efficiency is higher, and the actual efficiency can reach 40% to 60%. Furthermore, there is no mechanical transmission mechanism, so there is very little noise pollution and zero pollution. The fuel cell power generation system consists of four major parts: fuel conversion device, battery body, converter device, and waste heat recovery device. Its working process is as follows: natural gas and other fuels enter the fuel conversion device and undergo a chemical reaction to decompose hydrogen, and part of the heat generated by the chemical reaction will flow back to the waste heat recovery device; then the hydrogen and air generated by the fuel conversion device are electrochemically generated in the battery body. The reaction generates direct current, which is then converted into alternating current through the converter device for use by other users. The heat generated by the electrochemical reaction is also collected by the waste heat recovery device and converted into hot water, steam and other forms for heat supply [13].

(5) Battery.
The reaction speed of fuel cells and micro-gas turbines in the micro-grid is slow, and it is difficult to cope with the large fluctuations of wind power and photovoltaic power generation and load. It can be used to ensure the short-term power balance of the system and reduce the number of unit starts and stops.

(6) Diesel engine.
The proportion of new energy (such as wind energy and solar energy) units in the microgrid is relatively high, and its output accounts for a large proportion of the system output, and the output of new energy units has strong intermittent and volatility. When the output of the unit is small, in order to meet the load demand of the microgrid operation, a diesel engine (Diesel Engine, DE) is often added to the microgrid. DE generates electricity by burning fossil energy, which has good thermal efficiency and economy.

2.3 The Impact of New Energy Access to the Grid on Dynamic Economic Dispatch

(1) Randomness: From a microscopic point of view, this type of energy is sometimes non-existent, sometimes large and sometimes small, and it is difficult to control; from a macroscopic point of view, long-term data records show that it still has certain statistical laws.
(2) Difficulty in storage: Although this type of energy is widely distributed, inexhaustible and inexhaustible, it is difficult to store, resulting in the grid-connected system being unable to dispatch this part of the energy, resulting in abandonment of wind and light.
(3) Dependency: the weather in a specific area is basically the same, resulting in the change of wind energy, light energy, etc. in the area is also tending to be consistent, then the wind farms and photovoltaic farms in this area have specific spatiotemporal dependencies.

3 Experimental Research

3.1 Research Significance

With the continuous expansion of the energy system scale, the security requirements of the power network are also constantly increasing. The emergence of batteries and wind power systems has made the economic dispatch model of the power grid more accurate, such as charge and discharge state constraints, charge and discharge power size limits, limit charge and discharge amount at the beginning and end, and limit the number of charge and discharge times. The characteristics of the unit itself cause it to take a certain time from start-up to normal operation and from normal work to shut down. In addition, frequent startup and shutdown have a great impact on the performance of the unit. Therefore, it is necessary to consider the constraints on the start-up and shutdown of the unit, such as Start-stop state constraints, start-stop time's limit, etc. In addition, the electric energy generated by the power plant is transmitted to the load or other regional power grids through the transmission line. Excessive transmission capacity will cause flow blockage, which cannot meet the requirements of safe and high-quality power grid operation. In short, with the continuous development of the power grid, the dynamic economic dispatch model must meet the actual discharge demand, so that the guiding significance of the dispatch results to the actual operation is also more in-depth.

3.2 Research Methods

Comparison method: Compare the output of controllable units such as micro gas turbines, fuel cells, and diesel engines in 24 h a day and the impact of wind-solar complementation on dispatching costs.

4 Dynamic Economic Dispatch Results of Microgrid Based on Particle Swarm Optimization Algorithm

4.1 Comparison of Controllable Unit Processing

Wind-solar complementarity means that when the wind energy output decreases or even zero, the light energy output increases; when the wind energy output increases, the light energy output decreases or even zero. The complementary characteristics between wind energy and solar energy make the power output of the power generation system more stable and reduce the randomness of wind and solar output. With the continuous development of new energy sources, the proportion of wind energy and solar energy as their representatives connected to the power grid continues to increase. When conducting research on economic dispatch problems, in addition to considering their randomness, the relationship between wind energy and solar energy has to be considered. Complementary properties between them are taken into account. Due to the immaturity of the related technologies, there are many applications in which one of the wind farms or photovoltaic power plants is connected to the distribution network at present, and there are very few instances where the two are connected to the distribution network at the same time. Penetration is also lower. As a model of comprehensive utilization of wind energy and solar energy with high penetration, microgrid is suitable for the research on the influence of wind and solar complementary characteristics on the dynamic economic dispatch of power grid. As shown in Fig. 1, the output of the controllable unit in one day, it can be seen that the micro-turbine of the power system has the largest output when the energy is supplied, followed by the fuel cell, and finally the diesel engine, and between 20–22 h, The power generation output of the three is the highest, and these two hours can be said to be the time when the electricity consumption is the largest in a day, so it will reach its peak state.

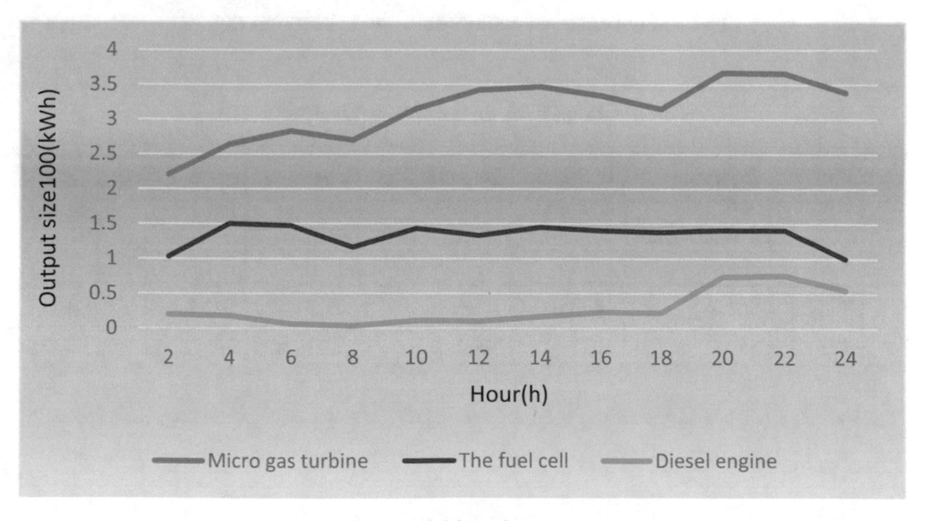

Fig. 1. Controllable unit output results

4.2 Considering the Influence of Wind and Solar Complementary Characteristics on Scheduling Results

As shown in Fig. 2 and Table 1, if the wind-solar complementary characteristics are considered in the dynamic economic dispatch of the power system, the power generation output of the three controllable generating units will be smaller, which will reduce the fuel use and the pollution gas. In addition, the sum of various costs incurred is also smaller, indicating that the impact of wind-solar hybridization on economic dispatch needs to be considered when constructing an economic dispatch model, which will reduce fuel costs, maintenance costs, and other power system operation and power supply costs. Save economic costs while reducing environmental pollution.

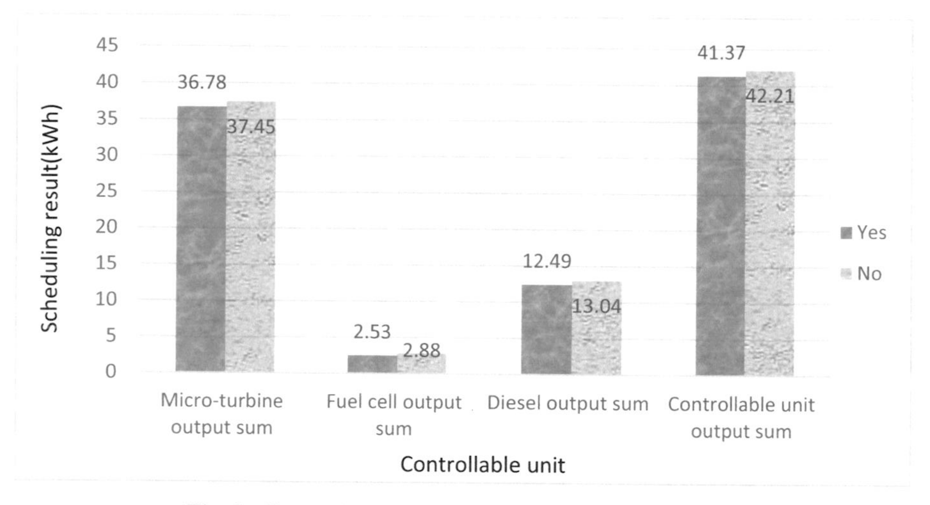

Fig. 2. Comparison of controllable unit scheduling results

Table 1. Comparison of various cost results (yuan)

	Fuel cost	Operation and maintenance costs	Pollution charges	Depreciation expense	Purchase and sale costs	rescheduling fee	Total cost
Yes	3957.6	368.2	1473.8	1596.1	-3031.5	545.9	4910.1
No	4034.5	384.0	1521.6	1667.4	-3210.8	783.2	5179.9

5 Conclusion

This paper uses the particle swarm optimization algorithm to calculate the power generation power of the power microgrid power generation system in each hour of the day, and then analyzes the impact of wind and solar complementation on economic dispatch, and obtains a microgrid system dispatch plan with lower economic and environmental costs, that is, to consider The wind-solar complementary characteristics show that the particle swarm optimization algorithm can solve the dynamic economic dispatching problem of the microgrid.

References

1. Nabi, S., Ahmed, M.: PSO-RDAL: particle swarm optimization-based resource- and deadline-aware dynamic load balancer for deadline constrained cloud tasks. J. Supercomput. **78**(4), 4624–4654 (2021). https://doi.org/10.1007/s11227-021-04062-2
2. Gabi, D., Ismail, A.S., Zainal, A., et al.: Hybrid cat swarm optimization and simulated annealing for dynamic task scheduling on cloud computing environment. J. Inf. Commun. Technol. **17**(3), 435–467 (2018)
3. Pahnehkolaei, S.M.A., Alfi, A., Machado, J.A.T.: Convergence boundaries of complex-order particle swarm optimization algorithm with weak stagnation: dynamical analysis. Nonlinear Dyn. **106**(1), 725–743 (2021)
4. Verma, P., Parouha, R.P.: Non-convex dynamic economic dispatch using an innovative hybrid algorithm. J. Electr. Eng. Technol. **17**(2), 863–902 (2021)
5. Kumari, R., Gupta, N., Kumar, N.: Cumulative histogram based dynamic particle swarm optimization algorithm for image segmentation. Indian J. Comput. Sci. Eng. **11**(5), 557–567 (2020)
6. Raheem, F.A., Hameed, U.I.: Heuristic D* algorithm based on particle swarm optimization for path planning of two-link robot arm in dynamic environment. Al-Khwarizmi Eng. J. **15**(2), 108–123 (2019)
7. Phommixay, S., Doumbia, M.L., Cui, Q.: Comparative analysis of continuous and hybrid binary-continuous particle swarm optimization for optimal economic operation of a microgrid. Process Integr. Optim. Sustainability **6**(1), 93–111 (2021)
8. Valarmathi, R., Sheela, T.: Ranging and tuning based particle swarm optimization with bat algorithm for task scheduling in cloud computing. Clust. Comput. **22**(5), 11975–11988 (2017). https://doi.org/10.1007/s10586-017-1534-8
9. Talaat, F.M., Ali, H.A., Saraya, M.S., et al.: Effective scheduling algorithm for load balancing in fog environment using CNN and MPSO. Knowl. Inf. Syst. **64**(3), 773–797 (2022)
10. Pattanaik, J.K., Basu, M., Dash, D.P.: Dynamic economic dispatch: a comparative study for differential evolution, particle swarm optimization, evolutionary programming, genetic algorithm, and simulated annealing. J. Electr. Syst. Inf. Technol. **6**(1), 1–18 (2019). https://doi.org/10.1186/s43067-019-0001-4
11. Bilal, R.D., Pant, M., et al.: Dynamic programming integrated particle swarm optimization algorithm for reservoir operation. Int. J. Syst. Assurance Eng. Manage. **11**(2), 515–529 (2020)
12. Gupta, V., Singh, B.: Study of range free centroid based localization algorithm and its improvement using particle swarm optimization for wireless sensor networks under log normal shadowing. Int. J. Inf. Technol. **12**(3), 975–981 (2018). https://doi.org/10.1007/s41870-018-0201-5

Student Behavior Analysis System in Smart Campus Based on Data Mining Algorithm

Wei Han[1(✉)] and Khadijah Mansour[2]

[1] Xi'an Fanyi University, Xi'an 710105, Shaanxi, China
582008886@qq.com
[2] University of Garden City, Khartoum, Sudan

Abstract. Colleges and universities must generate a lot of student behavior data during the informatization construction, such as card usage data, grades and grade point data, etc. These data can fully reflect the trajectory of students' activities. Therefore, strengthening the analysis and research of student behavior data will help the school to better manage it, so as to better build a smart campus. The purpose of this paper is to design a smart campus student behavior analysis system based on data mining algorithm. The relevant theories and technologies involved in the functional design and implementation of the student behavior analysis system based on data mining are studied. The software architecture, the design of the system operating system, the physical development architecture for storing system data, and the physical and logical design of the database are introduced. The detailed design process of system modulesis analyzed, including integrated vertical behavior module, student behavior early warning module, student big data analysis report module, student behavior trajectory module, early warning rule design, and interface design module. 82% of students borrowed books recommended by the association analysis book model within three months.

Keywords: Data mining · Smart campus · Student behavior · System design

1 Introduction

With the rapid progress of human informatization, from the first computer age to the Internet age, from the Internet age to the current artificial intelligence age, every change in the information technology age has brought about changes in expert opinions [1]. The teaching methods, management skills and information technology of smart campus are very diverse and advanced [2]. In the era of big data, in the process of writing intelligent systems, data is formatted into various forms with complex features. The purpose of comprehensively eliminating student behavior data with big data technology is to fundamentally change the school management system and seek more understanding and more humanized technology [3].

Conduct in-depth research on student behavior data, explore the student research, lifestyle and psychological state behind these behaviors, discover and solve potential problems, and promote student health [4]. Ikawati Y understands students' preferences

J. H. Abawajy et al. (Eds.): ICATCI 2022, LNDECT 169, pp. 368–375, 2023.
https://doi.org/10.1007/978-3-031-28893-7_44

in the learning process by understanding how each student learns. To identify appropriate student learning styles, students' behaviour was analysed based on the frequency of visits to Moodle E-learning and completed the Index Learning Style (ILS) questionnaire. A learning style prediction model using an ensemble tree approach, namely Bagging and Boosting-Gradient Boosted Tree, is proposed. Evaluate classification results using stratified cross-validation, and use accuracy to measure performance. The results show that the classification efficiency of the Ensemble Tree method is higher than the accuracy of the single tree classification model [5]. Pang C combines the traditional cluster analysis algorithm and random forest algorithm to improve the traditional algorithm, and combines the human skeleton model to identify students' classroom behavior in real time. In addition, combined with the needs of students' classroom behavior identification, a network topology model is constructed. The error rate of feature reconstruction using spatiotemporal features is lower than that of single features. The validity of the spatial angle features extracted based on the human skeleton model is verified by experiments. The algorithm performance test results show that the network structure of the proposed algorithm is better than the network structure of a single feature extraction algorithm [6]. Big data student behavior is a treasure trove worth digging into. Student behavior data analysis system has important practical significance for school teaching management, student management and sustainable development [7].

This paper starts with user needs and project needs, researches and designs the analysis of themes such as student portraits, behavioral trajectories, library management, and early warning modules such as social relations, life rules, and lateness. Corresponding data tables are designed to extract student behavior data from school system, curriculum, one-card system and access control system for data purification and storage. Error detection and early warning, student misbehavior, social interaction analysis and cryptanalysis algorithms, development and implementation of student behavior data analysis programs, and analysis and impact prediction through test case determination, providing key data support for school management and student development.

2 Research on the Design of Student Behavior Analysis System in Smart Campus Based on Data Mining Algorithm

2.1 Data Mining Technology

Apriori algorithm is the first type of legal mining algorithm, and it is also the most classic and effective algorithm in data mining, and it is widely used [8]. The basic idea is to use the method of layer-by-layer retrieval to explore the relationship between data systems in the database and formulate rules. This algorithm class can be used to capture maximum data within the minimum supported size allowed. Under the framework of group rules, the Apriori algorithm does not need to read the transaction data in the memory, but can be used directly on the hard disk data, with powerful data processing capabilities, which is one of the biggest advantages of the Apriori algorithm [9].

2.2 Student Behavior Data

There are many types of student behavior data, and domestic and foreign experts and scholars have different understandings of student behavior data from different perspectives. The student behavior data mentioned in this study refers to the data generated by college students in their study and life on campus, including campus One-card data, library data, student test score data, and student participation in postgraduate entrance examination and scholarship data [10]. One-card data mainly includes canteen consumption data, bus consumption data, and dormitory hot water consumption data; library data mainly includes library access control data and book borrowing data [11].

2.3 System Design Goals

The specific goals of the development of the student behavior analysis system are as follows:

Create the database required by the system. Clear, export, merge data from different data sources, select appropriate data schema, and create data tables [12]. The processed data is stored in the system database, which is convenient for the system to efficiently analyze the data, and retrieve and store the analysis results.

Perform efficient system user management, complete user registration and login functions. Users can register normally to ensure that the user account is unique, and there will be no problem of multiple registrations for the same account. Users log in to the system through their accounts to ensure system security and ensure that system data will not be arbitrarily used by others.

Implement functions for split correlation analysis and student behavior. Complete the cluster analysis of students from the perspectives of grades, consumption and other behaviors, and get the classification of students with different behaviors. Assist in the administration and education of students.

Visualize analysis results. Ensure that users can see the results of system analysis efficiently and accurately, so that users can observe and analyze analysis results in a more intuitive and clear way.

3 Investigation and Research on the Design of Smart Campus Student Behavior Analysis System Based on Data Mining Algorithm

3.1 System Physical Deployment

The collection layer is mainly business data collection: visual collection system based on big data analysis system platform, FTP collection; external data collection: Apache open source crawler tool. The storage layer mainly includes static data: the storage of static data is mainly based on the cluster database based on Mysql; the unstructured dynamic data is mainly stored in the distributed file system HDFS, and the Spark platform is used for behavioral data storage for unstructured dynamic data. Analysis; hotspot data is stored in the HBase column database, which is convenient for quick search and retrieval of hotspot data; the background management of the display layer mainly adopts the current mainstream SpringMvc + Mybatis framework structure, and the front-end data presentation mainly adopts the stable operation. The Jquery + BootStrap + Echarts component.

3.2 System Software Architecture

The overall structure of the student behavior analysis system is divided into seven layers: data support layer, data preprocessing layer, data storage layer, data analysis layer, data display layer, business application and school management. The design of the overall structure is guided by the school's business needs, the big data analysis and processing process is the method, and the school management level is improved as the purpose, and the application of the frontier technology of big data and the informatization of colleges and universities.

3.3 Database

The entity relationship diagram of the system is also called the E-R diagram, which mainly designs the system data information table and the relationship between the tables. The data table design of the student behavior analysis system is designed on the basis of various business systems. The major business systems are the data sources. This system is based on the extraction of data from all business systems. The database layer used for data statistical analysis is the business presentation layer database. Because this database involves many tables, the data tables include student information table, poor student information, student consumption early warning information table, student safety early warning information table, and early warning strategy. Information, early warning strategy - information on each early warning type, student score sheet, student online segment information sheet, etc.

3.4 Description of Behavioral Data Analysis

Correlation analysis is carried out on the data of student canteens, supermarkets, public transportation, borrowing books, etc. Finally, unearth valuable correlations hidden between disparate data. Apriori algorithm is a common dataset algorithm for mining association rules.

(1) Find all frequent datasets with frequencies greater than or equal to the default minimum support
(2) Find all frequent itemsets through frequent item groups
(3) Create strong association rules from frequent datasets that must satisfy minimum trust and minimum support
(4) The support degree is shown in formula 1:

$$(A \to B) = P(A \cup B) \tag{1}$$

The confidence level is shown in Eq. 2:

$$(A \to B) = P(A|B) \tag{2}$$

4 Analysis and Research on the Design of Smart Campus Student Behavior Analysis System Based on Data Mining Algorithm

4.1 System Functional Structure Design

The intelligent student behavior management system is initially divided into four functional modules, as shown in Fig. 1, which consists of four parts: student picture module, behavior early warning module, behavior training module, and big data analysis report. It analyzes students' early learning behaviors, lifestyles, and spiritual behaviors during school, and develops individual and group portraits; the behavioral warnings section deals primarily with student learning, science, and strength warnings. In the case of economic problems, the behavior tracking module mainly reflects the behavior of students and understands the performance trajectory and social status of students; through the monthly and semi-annual big data analysis website, students can print data analysis reports.

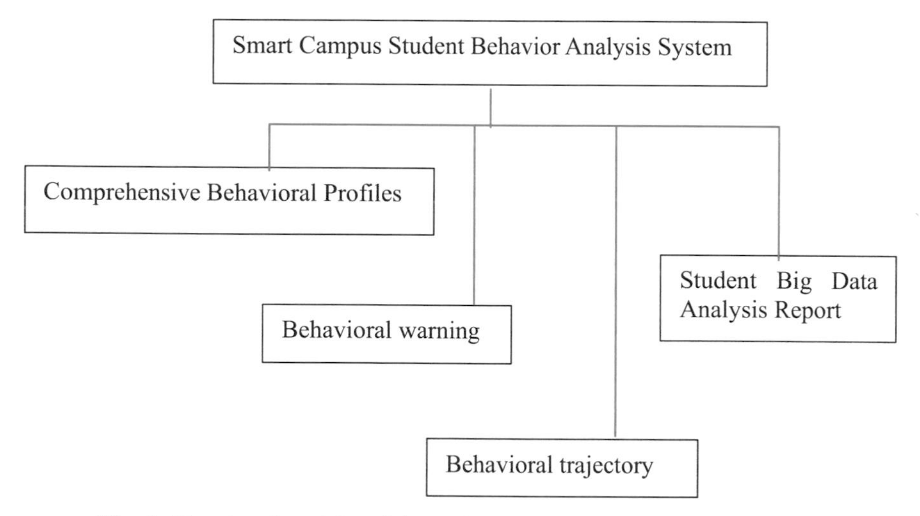

Fig. 1. Functional modules of the student behavior big data analysis system

4.2 Association-Based Book Analysis

Through statistical calculation, the most popular books in the school, the most popular books in grades, and the most popular books in majors can be obtained as shown in Table 1. According to these, the school can add books of this type as appropriate; of course, it can also According to students' majors and interests, different types of books are pushed to students, so that students can read useful knowledge during school. Figure 2 shows the relationship between the library borrowing data of undergraduate students and master students in the past year and the total number of books borrowed and months.

Apriori correlation analysis was performed on the obtained data, and the minimum support was set to 0.5. The sorted call numbers are encapsulated into an independent

Table 1. Monthly borrowing of books by students in the past year

Month	Undergraduate school	Master student	The total amount of books borrowed
1	3541	1872	5413
2	3348	1921	5269
3	4729	2343	7072
4	3940	2367	6307
5	3842	2092	5934
6	4032	1761	5793
7	3920	1922	5842
8	4112	1872	5984
9	4877	2322	7199
10	4321	2203	6524
11	3860	2081	5941
12	3533	1933	5466

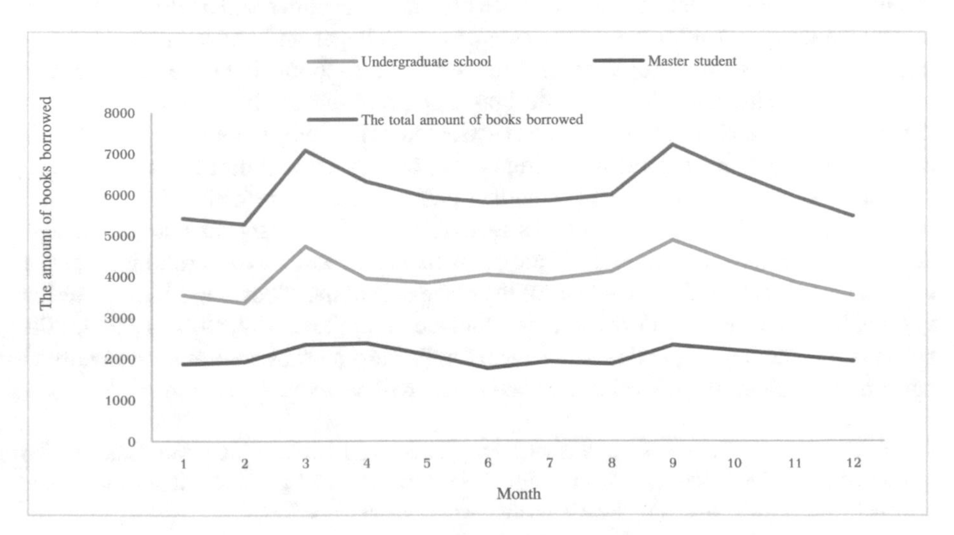

Fig. 2. Monthly borrowing of books by students in the past year

data set, which is used as the input set of the association algorithm. First, the algorithm will generate a list of itemsets for all single call numbers in the dataset, that is, each call number is an independent set that exists as a subset in the generated item list; secondly, the algorithm scans the itemset list And calculate the support of each subset, and compare it with the set minimum support of 0.5, and delete the item sets whose support is less than 0.5; then, delete the call numbers that do not meet the minimum support. A new itemset list, the new itemset list is composed of itemsets with two different call numbers;

finally, the algorithm scans the data records here, and calculates the support of each two-element itemset, and deletes those that do not meet the requirements one by one. Itemset with a minimum support of 0.5. The whole process of scanning, calculating, judging, deleting, and generating is repeated until all itemsets are deleted, and the algorithm ends, returning all frequent itemsets. By setting the minimum confidence level to 0.8, the related information of books borrowed by students can be obtained.

Through book association analysis, it recommends relevant books for students who have borrowed books, and provides high-quality reading services. The model realizes the self-recommendation function of students' borrowed books, recommends books for students through the history of students' borrowed books, and greatly improves the efficiency of students' borrowing books. Books are recommended to students through book associations, and 82% of the students have borrowed books recommended by the association analysis book model within three months.

5 Conclusions

Facing the era of big data, how can colleges and universities integrate valuable data generated by professional management and services such as construction industry, human training, and campus management, and use big data to organize and analyze these data to provide important information for enterprises. Colleges and Universities. Cultivating school-running knowledge for decision-supporting schools is the basis of campus information production. This paper designs a program that analyzes student behavior data through research, making maximum use of the basic data generated by the school's business information systems, providing corresponding data support for school teaching and student management, and providing data support and reference for leadership decision-making. The main work results are as follows: The student behavior data analysis system establishes an analysis and judgment model for various abnormal situations of student behavior through the analysis of the grade data of the educational administration system, the basic student information data and the library loan data, so as to analyze the students' grades and academic achievements. Behaviors such as abnormality, abnormal consumption, abnormal life rules, and late return will be warned.

Acknowledgement. This work was financially supported by Xi'an Fanyi University, Construction project of counselor's studio of Xi'an Fanyi University "IPE and guidance studio for college students -- ideological and theoretical education and value guidance".

References

1. Rizal, M., Bulan, P.L., Amilia, S.: Analysis of factors affecting student Bidik Misi savings behavior. Jurnal Manajemen Motivasi **14**(2), 65–72 (2018)
2. Choi, S.W.: Influence of college student's exercise regularity to health improvement behavior and connected recognition for life-time sports. Korean J. Sports Sci. **27**(2), 117–124 (2018)
3. Malinda, S., Nangoy, O., Susilo, G.: Analysis on Small Spaces with Special Behavior: Study Case Student Housing at Kost Provokatif in Gading Serpong, Tangerang. Jurnal Muara Ilmu Sosial Humaniora dan Seni **4**(1), 75–87 (2021)

4. Rawlins, M., Bacon, C., Tomporowski, P., et al.: A qualitative analysis of concussion-reporting behavior in collegiate student-athletes with a history of sport-related concussion. J. Athl. Train. **56**(1), 92–100 (2021)
5. Ikawati, Y., Rasyid, M., Winarno, I.: Student behavior analysis to predict learning styles based felder silverman model using ensemble tree method. EMITTER Int. J. Eng. Technol. **9**(1), 92–106 (2021)
6. Pang, C.: Simulation of student classroom behavior recognition based on cluster analysis and random forest algorithm. J. Intell. Fuzzy Syst. **40**(2), 2421–2431 (2021)
7. Albluwi, I., Salter, J.: Using static analysis tools for analyzing student behavior in an introductory programming course. Jordanian J. Comput. Inf. Technol. (JJCIT) **6**(3), 215–233 (2020)
8. Susanto, D., Qurani, N.R., Rasyid, M.: Develop a user behavior analysis tool in ETHOL learning management system. EMITTER Int. J. Eng. Technol. **9**(1), 31–44 (2021)
9. Lednicky, J.A., Shankar, S.N., Elbadry, M.A., et al.: Collection of SARS-CoV-2 virus from the air of a clinic within a university student health care center and analyses of the viral genomic sequence. Aerosol Air Quality Res. **20**(6), 1167–1171 (2020)
10. Lerche, T., Kiel, E.: Predicting student achievement in learning management systems by log data analysis. Comput. Hum. Behav. **89**(DEC), 367–372 (2018)
11. Mathews, A., Patten, E.V., Stokes, N.: Foodservice management educators' perspectives on nutrition and menu planning in student-operated restaurants. J. Nutr. Educ. Behav. **53**(3), 223–231 (2021)
12. Haura, A.T., Ardi, Z.: Student's self esteem and cyber-bullying behavior in senior high school. Jurnal Aplikasi IPTEK Indonesia **4**(2), 89–94 (2020)

A Research Status of 3D Printing of Different Metal Forms

Jianxiu Liu[1(✉)], Yi Li[1], Jianglei Fan[1], Shen Wu[1], Ying Li[1], and Jun Wang[2]

[1] School of Mechanical and Electrical Engineering, Zhengzhou University of Light Industry, Zhengzhou, Henan, China
jianxiuliu@126.com

[2] College of Environmental Arts and Engineering, Henan Polytechnic, Zhengzhou, Henan, China

Abstract. 3D printing of metal materials plays an important role in the ranks of 3D printing and is one of the core development directions of 3D printing in the future. According to the different forms of metal raw materials, it can be roughly divided into powder printing, solid-state printing and liquid printing. Printed with powder metal as raw materials include laser selective sintering technology (SLS), laser selective melting technology (SLM), 3D printing technology (3DP), etc. printed with solid metal as raw materials include arc fuse deposition molding(WAAM)and layered solid manufacturing(LOM), etc. according to the different forms of 3D printing raw materials, this paper explores the applicable printing processes of different raw materials, and discusses the processing methods, process characteristics and research status of each process.

Keywords: Metal 3D printing · Different forms · Research status

1 Introduction

3D printing technology (Three Dimension printing) is a kind of additive manufacturing technology. Based on the principle of discretization and stacking, according to the digital model of the object, it uses the method of material stacking layer by layer to form three-dimensional entities. It can use molten materials for continuous stacking, quickly create complex modeling, can achieve the traditional process is difficult to complete the structural modeling, with great design freedom, without taking into account the difficulties of any manufacturing method.it has important performance in machinery manufacturing, medical treatment, aerospace, military industry, automobile and other fields [1, 2]. There are many different technologies for metal 3D printing, and different technologies are used for different forms of metal materials. For example, the processing technologies of powder materials include selective laser sintering (SLS), 3D printing technology (3DP) and selective laser melting (SLM) etc. in recent years, metal 3D printing has gradually taken shape, and commercial 3D printers have begun to appear in the market. This paper will introduce several common 3D printing technologies of different forms of metal materials in the market, and discuss their forming methods, process characteristics and research status.

J. H. Abawajy et al. (Eds.): ICATCI 2022, LNDECT 169, pp. 376–384, 2023.
https://doi.org/10.1007/978-3-031-28893-7_45

2 Powder Metal Manufacturing Technology

2.1 Process Principle, Characteristics and Research Status of Laser Selective Sintering Technology

Laser selective sintering technology (SLS) using the three-dimensional data of the model, the metal particles are selectively melted and bonded through the CO2 laser beam in the machine. When parts processing begins, the powder processing bed and the powder supply cylinder shall be raised to the specified position, the roller rolls the raw material powder from the feeding cylinder to the powder processing bed, and the metal powder in the designated area is sintered by a laser beam. After the sintering of powder, the powder processing bed decreases by one slice thickness, and the powder spreading roller continues to work to spread powder. The above steps are processed layer by layer until the whole three-dimensional part is completely formed. Metal powders processed by SLS are divided into two categories: high melting point metals and low melting point metals. Generally, high melting point metal is used as the final structure, and low melting point metal is easy to melt as adhesive. The metal powder in the formed parts processed by SLS is not fully bonded, so further high-temperature sintering is needed to enhance the mechanical properties of the formed parts. The common SLS forming process is shown in Fig. 1.

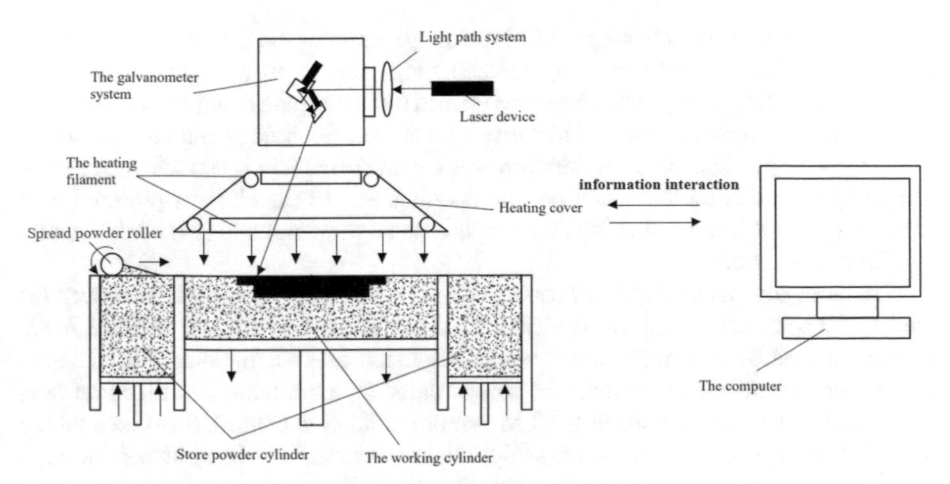

Fig. 1. SLS process principle

The advantages of SLS process are as follows: (1) the process is simple and can produce complex parts; (2) the powder material is not completely melted, and the sintering mechanism of semi-solid liquid phase can reduce the reaction heat inside the material; (3) there is no support structure, the cantilever structure is supported by unsintered metal powder particles; (4) The material utilization rate is high, and the unsintered metal powder can be recycled [3]. However, SLS process also has some shortcomings: (1) unsintered metal particles will lead to defects such as high porosity, low density and low tensile strength, which requires secondary sintering; (2) high use and maintenance costs. The equipment covers a large area and requires a special laboratory environment; (3) High surface roughness. The surface roughness will be caused by the incomplete melting of metal powder particles and the floating dregs during processing; (4) holes to remove excess powder need to be reserved when printing fully enclosed parts [4].

Xiao et al. Proposed a method to predict the accuracy of SLS molded parts. This method carries out orthogonal experiments on five process parameters: laser power, scanning speed, scanning temperature, preheating temperature and scanning spacing to obtain the sample parameters. The search strategy is determined by SOA algorithm to obtain the optimal solution of BP neural network. Then, the optimized BP neural network prediction results are established and compared by MATLAB. The experimental results show that the prediction model based on SOA-BP neural network has high prediction accuracy, and plays a guiding role in improving the accuracy of SLS molded parts and selecting process parameters. Zhu et al. Proposed a network model for predicting and controlling the temperature of sintering node. Based on the GA-BP neural network model, the simulation experiment of continuous sintering node temperature prediction and control is carried out. The results show that the sintering point temperature of the model can remain stable and improve the quality and accuracy of molded parts.

2.2 Principle, Characteristics and Research Status of Laser Selective Melting Technology

Laser selective melting technology (SLM) forming technology is developed on the basis of SLS technology. Limited by computer technology and expensive high-power lasers, the early Metal 3D printing technology was formed by coating and bonding metal powder. With the gradual development and maturity of computer technology and laser manufacturing technology, Fraunhofer in Germany was the first to explore laser technology SLM technology for fully molten metal powder forming. SLM technology is powered by a high-energy laser beam that completely melts a metal powder, making it stick together and form a solid body.

The main advantages of SLM forming are: (1) suitable for most metal powders; (2) the formed parts have good mechanical properties. The microstructure of SLM metal is characterized by fine microstructure and dispersion of strengthening phase [5]; (3) capable of manufacturing lightweight porous parts; (4) high material utilization rate, save material. Compared with SLS, SLM is more difficult to control. SLM technology requires printing support structures to avoid collapse when the laser beam sweeps through the thick metal powder. The support structure can effectively suppress the part warping deformation caused by shrinkage stress during the forming process. The large energy input of laser beam and the complete melting of particles are easy to cause problems such as metal particle spheroidization, residual stress and deformation, which may lead to deformation, delamination or cracking of formed parts, resulting in part failure. The optimization of laser pulse energy, powder layer thickness, printing speed and other parameters can improve the quality of the molded parts [6].

Nguyen et al. studied the pretreatment optimization of Ti-6Al-4V products processed by SLM process. Through python programming language and TensorFlow library, the laser power, laser scanning speed, layer thickness and filling distance of SLM are optimized, which effectively improves the density of formed parts [7]. Vandenbrucke et al.

Studied the possibility of producing medical and dental parts by SLM technology. The characteristics of small volume, complex structure and strong personalization can be quickly manufactured by SLM technology, It proves the development prospect of SLM technology in medical and health care [8].

2.3 Principle, Characteristics and Research Status of 3D Printing Technology

The process principle of 3D printing technology (3DP) is based on powder materials. The nozzle of the printer sprays adhesive along the track generated by the slicing software according to the instructions. After each section is completed, the printing platform reduces a slice size, and the powder roller spreads another layer of powder material. This process is repeated, and finally the whole formed part is processed. Then the blank is sintered or heat treated to remove the adhesive to bond the metal particles and powders together and improve its mechanical properties. The forming process is shown in Fig. 2.

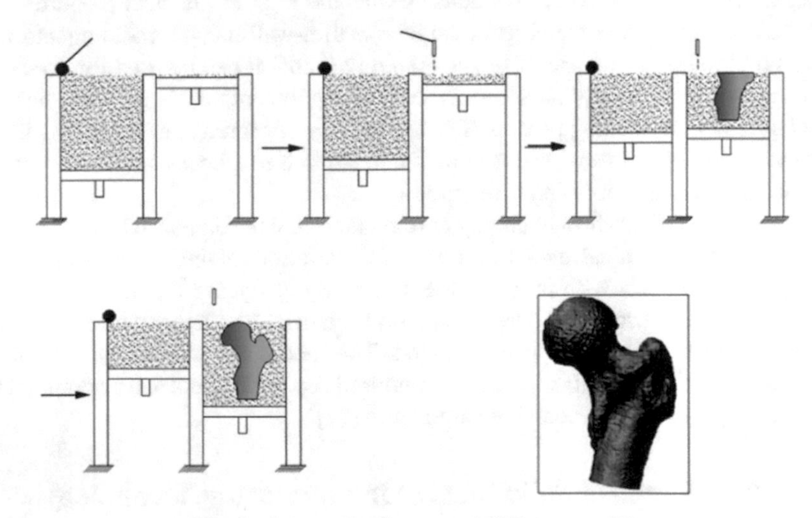

Fig. 2. Schematic diagram of 3DP process [9]

The main advantages of 3DP molding are: (1) wide selection of materials, applicable to black and non-ferrous metal casting. (2) high degree of design freedom, which can manufacture molded parts with complex shapes. (3) the equipment is cheap. Although all of them are powder printing, the 3DP printing process does not require expensive laser transmitters. However, the 3DP method also has some shortcomings. Compared with the traditional process, the molded parts manufactured by bonding need post-treatment to improve their strength. The forming accuracy also needs to be improved. The preparation accuracy of 3DP is related to the characteristics of powder materials, printing speed, nozzle and the parameters such as the distance between powder beds, material thickness and slice row spacing have a great relationship with the printing accuracy, and sintering and heat treatment will also affect the forming accuracy [10].

Yang et al. prepared porous 316L stainless steel using 3DP process, mixed viscous powder polyvinyl alcohol (PVA) and starch into 316L stainless steel powder, sprayed water-based solution to the mixed powder, and studied the effects of PVA content and sintering temperature on the size shrinkage, porosity, mechanical properties and other properties of printed parts.

2.4 Principle, Characteristics and Research Status of Laser Direct Melting Deposition Process

The basic principle of laser direct melting deposition (LDMD) is to deposit metal powder layer by layer. It uses the laser beam as the energy source to directly melt the metal powder to form a molten pool. The auxiliary device synchronously feeds the powder and gradually sends the alloy powder into the melted metal. In the process of processing, a laser beam is shot down the axis of the print head and focused on the convergence point of the powder, which melts and solidifies under the protection of an inert gas.

The advantage of laser direct melting deposition technology is that it can manufacture large-size parts. The parts have dense tissue and high mechanical properties. It is widely used in high-value metal parts, large-size thin-wall integral molding and metal injection mold. The advantage of this technology is that it can be used for processing and manufacturing a variety of materials, but the equipment cost is high and has large internal stress in the forming process. The technology of printing and annealing has not been developed. The dimensional accuracy and surface roughness of printed parts are poor, and subsequent processing is required.

Wang et al. used LDMD technology to remanufacture the damaged blades. Liu et al. established a mathematical model based on the variation of the motion trajectory of molten pool and nozzle with process parameters. By analyzing the interaction model between laser and powder, the laser attenuation intensity and temperature distribution of heated powder on the base were obtained. The model can effectively simulate the geometric dimensions of Ti-6Al-4V Alloy monorail deposition pool and cladding, which is helpful to optimize the processing parameters [11].

3 Research Status of Solid Metal Manufacturing Technology

3.1 Principle, Characteristics and Research Status of Arc Fuse Deposition Forming Process

Wire Arc Additive Manufac-ture (WAAM) is a manufacturing technology that takes the arc as the energy beam, uses the layer by layer cladding principle, uses the arc generated by welding machines such as melting electrode inert gas shielded welding and tungsten electrode inert gas shielded welding as the heat source, adds wire, and gradually forms metal solid components according to the three-dimensional model under the control of the program. This technology is mainly based on TIG, MIG and SAW (submerged arc welding) and other welding technologies have been developed. The formed parts are made up of welds with high density and uniform structure. Compared with the traditional powder metal printing, the wire feeding process is safer and cleaner, avoiding the dangerous powder environment for the experimenters as much as possible. In addition, the material utilization rate of wire feeding printing is higher, and the raw materials are easier to obtain.

The advantages of WAAM: (1) low price of finished product; (2) fast stacking speed; (3) The manufacturing size and shape are free; (4) all metal materials can be used.

Ding et al. Established the single weld contour model through a variety of curve fitting methods. Based on the single bead model, established the multi bead overlap model, and analyzed the critical center distance to realize the stable multi bead overlap process, so as to produce stable deposits and improve the surface quality and dimensional accuracy of molded parts [12].

3.2 Principle, Characteristics and Research Status of Melt Deposition Molding Process

Fused deposition modeling (FDM) has been widely used because of its simple operation and low cost. In metal printing, it is mostly used to manufacture eutectic extruded metal. In this manufacturing process, the hot-melt wire material is wound on the feeding roller, and the roller is driven by the servo motor to rotate. The wire material is sent down the extruder nozzle driven by the roller friction, and sent to the nozzle by the wire feeding mechanism. There is a set of guide sleeve between the feeding roller and the nozzle. The guide sleeve is composed of materials with low friction coefficient, which can accurately send the silk material to the nozzle. There is a set of resistance wire for heating above the nozzle. The resistance wire is heated to a certain temperature to melt the wire to the molten state, and then extruded by the nozzle. After the material cools, the cross-section profile of the workpiece is formed. The key of FDM forming process is to keep the temperature of semi flow forming material just above its melting point. The thickness of each layer is determined by the diameter of extrusion wire, usually 0.25–0.50 mm.

FDM has the following advantages: (1) low equipment cost and no expensive laser and galvanometer system; (2) good toughness of formed parts; (3) Low material cost and high material utilization; (4) the process operation is simple. The disadvantages are: (1) low heating temperature, which is only suitable for low melting point metals; (2) cantilever parts need support, so it is difficult to build complex components; (3) the strength in the direction perpendicular to the section is low; (4) it is not suitable for manufacturing parts with small features and thin-wall features.

Mimsa et al. Explored the possibility of using FDM technology to manufacture metal automobile brake pedal. BASF ultrafuse 316L stainless steel was used as raw material to print automobile brake pedal. After degreasing and sintering, their microstructure and mechanical properties were studied and tested. The results successfully proved the application reliability of FDM technology in manufacturing complex metal automobile brake pedal [13]. Qin et al. Proposed a near unsupported printing method. In this experiment, a three degree of freedom printing platform was designed to increase the degree of freedom of printing. During the printing process, the printing angle of parts and nozzles was changed to realize unsupported printing, which provided a solution to the problems such as poor surface quality, material waste and long printing time caused by support in the FDM printing process.

4 Research Status of Liquid Metal Manufacturing Technology

With the increasing popularity of 3D printing, people began to use liquid metal to print flexible circuits. Liquid metal, usually refers to the metal that remains liquid at room temperature, such as metal ink made of common low melting point metals or alloys such as mercury, gallium and indium. They not only have excellent electrical conductivity of metals, but also have good ductility. This technology comes from the use of liquid metal ink to print flexible circuit, change the adhesion of ink, overcome its high surface tension through dispensing machine, and prove that slightly oxidized alloy ink can be printed flexibly on coated paper [14]. In addition, the technology uses RTV silicone rubber as isolation ink and packaging material to ensure the stability of the circuit and provide a path for printing 3D electromechanical hybrid devices.

At present, the printing method of liquid metal is to put the metal ink into the printing needle, extrude it downward with the pressure, and print it layer by layer according to the specified route. The direct writing printing technology is simple to operate. It can quickly print metal circuits only by inputting the drawn model into the computer. It has great potential in the fields of electronic product manufacturing, circuit discipline education, industrial manufacturing and personal personalized design.

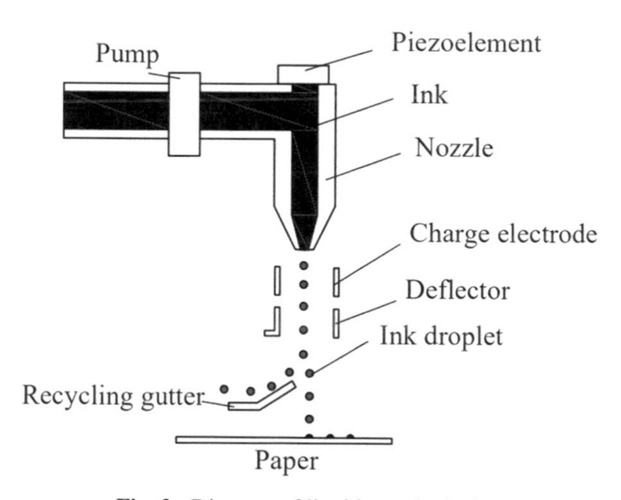

Fig. 3. Diagram of liquid metal printing

Zheng et al. Developed a tapping composite fluid conveying system, which revealed the conveying and adhesion mechanism of liquid metal ink, making it possible to print liquid metal. Using liquid metal 3D printing to manufacture smaller conductive components makes the development and application of circuits simple and efficient, and has great application potential in global intelligent electronic manufacturing. Its unique advantages are the most important in the growing field of personalized consumer electronics. It can freely manufacture all kinds of integrated circuits and functional electronic products to meet the needs of personalized customization. However, due to the problems of excessive surface tension, high evaporation temperature and easy oxidation of metal ink, the 3D printing of liquid metal ink is still in the early stage of development and has great development prospects in Fig. 3.

5 Conclusions

This paper introduces several metal printing methods based on different raw materials, and draws the following conclusions: (1) due to the high melting point of most metals, metal 3D printing is mainly solid metal and powder metal, and laser and electron beam are the main energy sources; (2) Liquid printing of low melting point metals is rare. In addition to linear direct writing technology introduced in this paper, mask deposition technology and liquid phase 3D printing technology are also under research and development. At present, metal printing has great development prospects. Metal 3D printing has high resource utilization, environmental friendliness and high design freedom, which is in line with the development trend of future science and technology. With the development of science and technology, metal printing will face a further breakthrough in the technical bottleneck, so that metal printing has a wider range of applications.

References

1. Goyanes, A., et al.: 3D printing of medicines: engineering novel oral devices with unique design and drug release characteristics. Mol. Pharmaceutics **12**(11), 4077–4084 (2015)
2. Azad, M.A., Olawuni, D., Kimbell, G., et al.: Polymers for extrusion-based 3D printing of pharmaceuticals: a holistic materials-process perspective. Pharmaceutics **12**(2), 1–34 (2020)
3. Khazaee, S., Kiani, A., Badrossamay, M., Foroozmehr, E.: Selective laser sintering of polystyrene: preserving mechanical properties without post-processing. J. Mater. Eng. Perform. **30**(4), 3068–3078 (2021)
4. Zhao, G., Liu, X., Zhang, Z., Zhang, Q.: Research progress of metal part manufactured by 3d printing. New Chem. Mater. **46**(08), 42–45+50 (2018)
5. Bourell, D., Kruth, J.P., Leu, M., Levy, G., Rosen, D., Beese, A.M., et al.: Materials for additive manufacturing. CIRP Ann. Manuf. Technol. **66**(05), 659–681 (2017)
6. Kruth, J.P., Froyen, L., Van Vaerenbergh, J., Mercelis, P., Rombouts, M., Lauwers, B.: Selective laser melting of iron-based powder. J. Mater. Process. Technol. **149**(11), 616–622 (2004)
7. Nguyen, D.S., Hong, S.P., Chang, M.L.: (2020) Optimization of selective laser melting process parameters for ti-6al-4v alloy manufacturing using deep learning. J. Manuf. Process. **55**, 230–235 (2005)
8. Vandenbroucke, B., Kruth, J.: Selective laser melting of biocompatible metals for rapid manufacturing of medical parts. Rapid Prototyping J. **13**(4), 196–203 (2013)
9. Chen, X., Yang, J., Huang, D., Chen, J., Tang, Y.: Development status of 3dp method in preparation of orthopedic implants. Hot Working Technol. **47**(04), 35–39 (2018)
10. Cader, H.K., Rance, G.A., Buanz, A., et al.: Water-based 3D inkjet printing of an oral pharmaceutical dosage form. Int. J. Pharm. **564**(04), 359–368 (2019)
11. Liu, J., Stevens, E., Yang, Q., Chmielus, M., To, A.C.: An analytical model of the melt pool and single track in coaxial laser direct metal deposition (ldmd) additive manufacturing. J. Micromechanics Mol. Phys. **02**(04), 1750013 (2018)
12. Ding, D., Pan, Z., Cuiuri, D., Li, H.: A multi-bead overlapping model for robotic wire and arc additive manufacturing (waam). Robot. Comput.-Integr. Manuf. **31**, 101–110 (2015)

13. Mimsa, C., Shm, B., Sp, B., Ej, A., Ak, B.: Additive manufacturing of an automotive brake pedal by metal fused deposition modelling. Mater. Today Proc. **45**(10), 4601–4605 (2021)
14. Lee, S., Kim, J.H., Wajahat, M., et al.: Three-dimensional printing of silver microarchitectures using newtonian nanoparticle inks. ACS Appl. Mater. Interfaces. **9**(22), 18918–18924 (2017)

Design of Remote Intelligence Monitoring System Based on LabVIEW for the Greenhouse

Xiaoyong Bo[1,2](✉)

[1] Electrical and Information Engineering College, Jilin Agricultural Science and Technology University Jilin, Jilin, China
110647710@qq.com

[2] Smart Agricultural Engineering Research Center of Jilin Province, Jilin, China

Abstract. In order to improve the efficiency of greenhouse cultivation and reduce management costs, an intelligent control system is designed to control the temperature, the soil moisture, the light intensity and other environmental factors in the greenhouse. The system allows the growers to monitor the crop growth in the greenhouse environments remotely on the computer and adopts proportional integral differential (PID) algorithm to realize the automatic and effective control for the temperature, the soil moisture, the light intensity in the greenhouse based on LabVIEW software and associated devices & peripheral circuits. When the crop species and the corresponding growing stages are typed in the front panel, the system will automatically generate the most suitable parameters such as soil moisture, temperature and light intensity so that the soil moisture, the temperature and the light intensity in the greenhouse will be controlled in the suitable range. And the automatic/manual buttons were set on the front panel by the system in order to make the system in the automatic/manual control. It was demonstrated in the laboratory condition. The results showed that the maximum error of the parameter is 8.7% and the effect is not ideal. But for the areas such as the greenhouse, cold chain logistics, aquaculture and so on with low control accuracy requirement, it still has certain reference significance.

Keywords: LabVIEW · Greenhouse · Remote monitoring · Automatic/manual switch · Intelligent control

1 Introduction

In agricultural production, greenhouse planting technology has been continuously promoted across the country, and the number of greenhouses has continued to increase. With the wide application of greenhouses, how to improve the production efficiency of growers and reduce management costs has attracted people's attention. Domestic scholars have proposed various design schemes of intelligent control systems for greenhouses [1–3]. The so-called intelligent control of greenhouses is to adjust various environmental conditions required for crop growth through advanced science and technology, such as temperature, soil humidity, light and other environmental parameters, so that crops are in the best growth environment to improve production efficiency. Because my country's

J. H. Abawajy et al. (Eds.): ICATCI 2022, LNDECT 169, pp. 385–392, 2023.
https://doi.org/10.1007/978-3-031-28893-7_46

greenhouse automatic control technology has developed relatively late, and most of our farmers' cultural level is not high, it is necessary to design an easy-to-operate and easy-to-understand control system. In response to this problem, this paper designs a remote intelligent monitoring system for greenhouses based on LabVIEW [4].

The intelligent monitoring system uses LabVIEW as the main control software, combined with sensor technology [5], measurement and control technology and computer technology, to achieve intelligent and scientific greenhouse environmental control and management [6]. Using the powerful graphics environment of the computer, using the visual graphics programming language and platform, the system has a friendly human-computer interaction interface, making the user operation simpler and more convenient. The system has good practical value and application prospect.

2 Overall System Design

The hardware of this design system mainly includes a computer, a camera, a temperature sensor, a soil humidity sensor, a light sensor, a roller blind lifting module, a solenoid valve on-off module, a ventilation module, and a data acquisition card. The computer uses the LabVIEW software platform to monitor and display the greenhouse environment in real time, and performs data processing, storage and intelligent control. The overall design scheme is shown in Fig. 1.

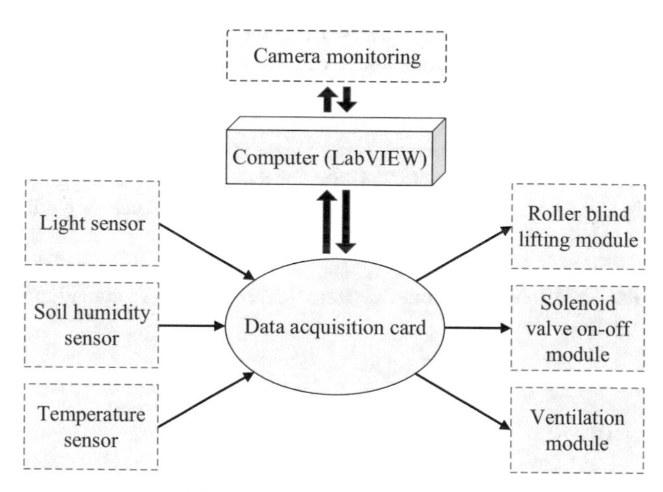

Fig. 1. Overall design scheme

The camera installed in the greenhouse is connected to the computer through the USB port, and the captured images are displayed on the front panel of LabVIEW, so that the user can directly see the situation in the greenhouse. The temperature sensor and photosensitive sensor are installed in the greenhouse, and the soil humidity sensor is inserted into the soil. The data collected by all sensors are input into the computer through the data acquisition card, their dynamic curves are displayed on the front panel, and the sensor data is displayed in real time. There is also an automatic/manual switch button on the front panel of the system. When the front panel is set to automatic, LabVIEW processes the read temperature value, humidity value, and light value, etc., and sends it to the corresponding actuator through the digital output port of the data acquisition card to complete the purpose of controlling these parameters. When the front panel is set to manual, the output value can be manually set on the front panel to realize manual control.

In addition, the front panel of LabVIEW has option boxes for crop type and growth stage. Enter the crop name and growth stage in the corresponding option box, and the system will automatically generate the humidity, temperature, and light parameter values that are most suitable for the growth of the crop at this time, making the system more operable.

3 System Hardware Design

The system hardware mainly includes computers, sensors, data acquisition cards, cameras, stepper motors (including drive modules), solenoid valves, etc.

3.1 Signal Conditioning and Control of Temperature Quantities

Temperature is one of the important factors related to the growth and development of crops, and suitable temperature [7] is conducive to the accumulation of photosynthetic products of crops. The system adopts thermal resistance Pt100 to detect temperature.

The temperature in the greenhouse is collected by the temperature sensor, and the A/D conversion is performed by the data acquisition card. The LabVIEW control collects and processes the converted digital quantity, and outputs the appropriate control quantity to the data acquisition card, so as to control the stepper motor connected to the vent. By adjusting the opening and closing of the vents, the temperature in the greenhouse is adjusted. The temperature control block diagram is shown in Fig. 2.

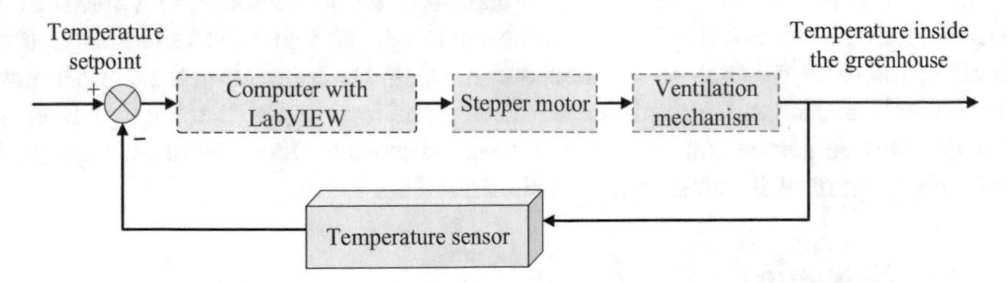

Fig. 2. Block diagram for temperature control

3.2 Acquisition and Control of Humidity Quantities

Soil moisture plays a vital role in the water absorption of crop roots and the transport of mineral nutrients, and also affects the reproduction of pathogens. Appropriate humidity can make crops grow better. The system uses a humidity sensor (YL-69) that can be directly inserted into the soil to measure the humidity. Its surface is treated with nickel plating, the sensing area is widened, the conductivity is improved, and rust is prevented from contacting the soil.

The output signal of the sensor module is connected with the USB port of the computer through the data acquisition card. The LabVIEW control analyzes and decides the monitoring data, and realizes the switch of the solenoid valve by outputting the high and low level control relay, so as to control the opening and closing of the irrigation equipment and adjust the soil moisture value in time. The humidity control block diagram is shown in Fig. 3.

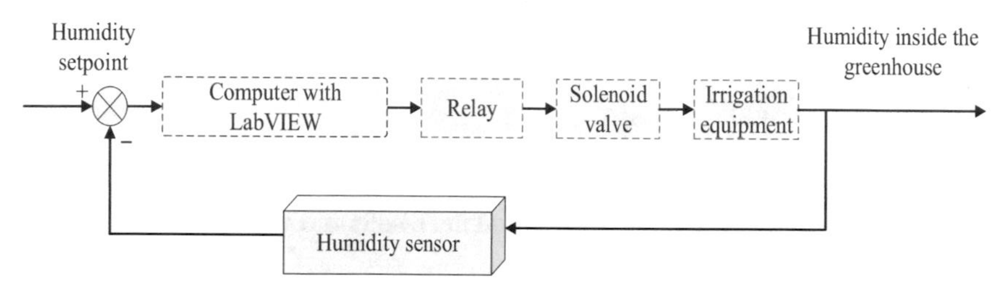

Fig. 3. Block diagram for humidity control

3.3 Signal Conditioning and Control of Light intensity

Light is not only a necessary condition for crop seeds to germinate, but also an indispensable condition for crops to perform photosynthesis. Therefore, in the greenhouse, the light intensity must be controlled within a certain range, otherwise too strong or too weak light will affect the normal growth of crops.

This system adopts HA2003 light sensor, and then uses photoelectric conversion module to convert light intensity value into voltage value.

The light sensor collects the light information in the greenhouse, and also through the data acquisition card, displays it on the computer, and sets the range through the LabVIEW interface for analysis and decision-making. By controlling the stepper motor on the roller blind, the roller blind can be rolled up and lowered, so that the light intensity in the greenhouse can be stabilized in the range suitable for the growth of crops [8, 9]. The block diagram of lighting control is shown in Fig. 4.

3.4 Video Monitoring

Knowing the growth of various crops in the greenhouse and the parameters of various indicators in time can enable farmers to make fast and accurate decisions and quickly adjust the parameter settings of various indicators of crops. At the same time, parameter changes can also be recorded.

The camera installed in the greenhouse is connected to the computer. Users can directly understand the growth of crops in the greenhouse by watching the real-time images collected by the camera on the front panel of LabVIEW. Users save a lot of time and effort.

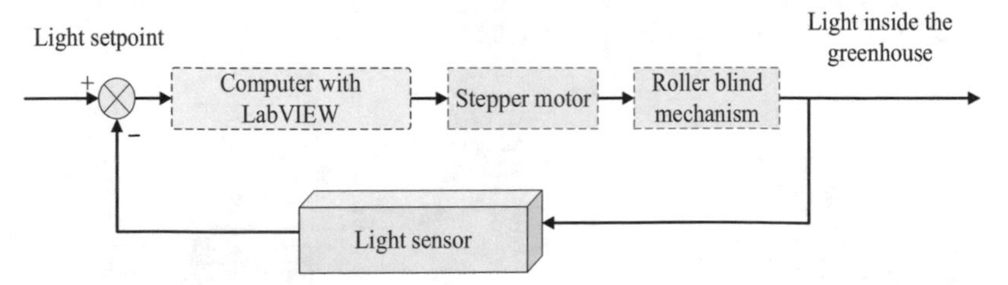

Fig. 4. Block diagram for light control

4 System Software Design

LabVIEW software is a graphical programming language [10] that uses icons instead of text lines to create applications. Similar to the C language and BASIC development environment, it uses the graphical programming language G to write programs. The main program flow and the subprogram flow are shown in Fig. 5 and Fig. 6, respectively.

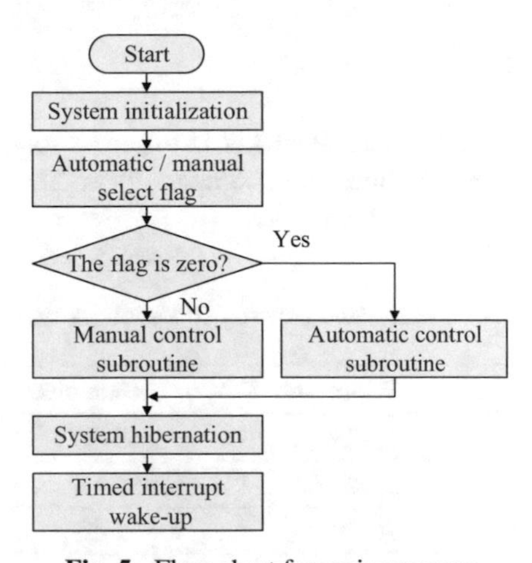

Fig. 5. Flow chart for main program

When the program starts running, the camera monitors the growth of the crops in the greenhouse. After selecting the type of crops and the growth stage, the system automatically generates the set values of temperature, humidity and light. For temperature and illumination, the temperature value and illumination value collected by the sensor are converted by the data acquisition card and compared with the set value to determine the forward and reverse rotation of the motor. The system adopts Proportional Integral Differential (PID) control, and outputs Pulse Width Modulation (PWM) with different duty ratios to control the speed of the motor, so as to realize the stable opening and closing of the ventilation mechanism and the stable lifting of the rolling shutter mechanism. For humidity, when the collected humidity is less than the set value, open the solenoid valve for irrigation. Otherwise, there is no action.

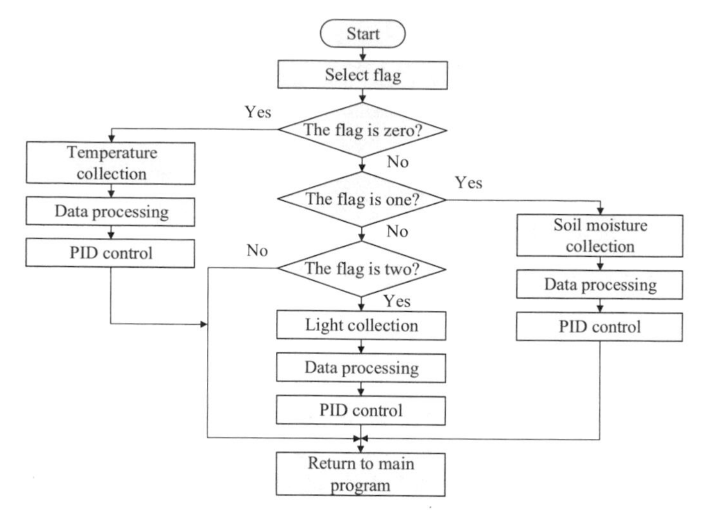

Fig. 6. Flow chart for subprogram

5 Result and Analysis

First, collect some growth indicators of plants at different stages. Then, set the various parameter values in the greenhouse and compare them. The most suitable growing conditions for typical crops are shown in Table 1.

Table 1. The most suitable growing condition for typical crops

Plant	Growth stage	Temperature/°C	Humidity/%	Light/kLux
Peanut	Stage one	12–15	40–60	5–8
	Stage two	22–24	50–60	5–8
	Stage three	26–30	45–55	5–8
Eggplant	Stage one	25–30	65–75	7–10
	Stage two	25–30	60–80	7–10
	Stage three	26–33	70–80	7–10
Tomato	Stage one	25–30	65–80	7–12
	Stage two	20–25	60–80	7–12
	Stage three	20–30	75–85	7–12

The acquisition of experimental data is carried out under the actual growth environment of crops, and the measurement range of the sensor is: temperature 0–100 °C. Humidity 20% to 90%. Light 0 ~ 20 kLux. These measurement ranges fully meet the needs of normal greenhouses. The second growth stage of each crop was tested and the resulting measurements were compared with control values.

Table 2. Test data

Parameter	Unit	Peanut	Eggplant	Tomato
Set temperature	°C	23.0	30.0	25.0
Indoor temperature	°C	33.0	34.0	35.0
Temperature after control	°C	25.0	28.0	27.0
Error rate	%	8.7	6.7	8.0
Set humidity	%	55.0	70.0	75.0
Soil humidity	%	45.0	60.0	63.0
Humidity after control	%	59.0	74.0	79.0
Error rate	%	7.3	5.7	5.3
Set light	kLux	7.0	8.5	10.0
Indoor light	kLux	10.5	15.0	16.0
Light after control	kLux	6.6	8.9	9.5
Error rate	%	5.7	4.7	5.0

The test data are shown in Table 2.

6 Conclusion

The intelligent greenhouse monitoring system designed in this paper takes LabVIEW as the development platform, and makes full use of the software resources of LabVIEW and the hardware resources of the computer system by replacing the "hard" with "soft", and realizes the monitoring and control of greenhouses suitable for various conditions. Control and management. The system has two control modes, automatic and manual, so that the system can run well during the startup process or under special conditions. At the same time, because of the friendly man-machine interface of LabVIEW, it is convenient for operators to use the greenhouse monitoring system, which reflects the advantages of LabVIEW in practical applications. In addition, the system creatively puts the growth index parameters of various crops at various stages into the system in the form of modules, and automatically selects them through the system interface, which improves the reliability of the system in practical aspects and has good promotion in practical applications.

Acknowledgements. This work was financially supported by Scientific Research Project of Jilin Province Education Department (No. JJKH20210416KJ), "Digital Agriculture" Emerging Cross Key Discipline of Jilin Province, Smart Agricultural Engineering Research Center of Jilin Province, 2021 Jilin Agricultural Science and Technology University Students' Innovative Training Program (No. 202111439007) and Jilin Province Educational Science "13th Five-Year Plan" Project (No. GH19251).

References

1. Geng, X., Zhang, Q.L., Wei, Q.G., et al.: A mobile greenhouse environment monitoring system based on the internet of things. IEEE Access **7**, 135832–135844 (2019)
2. Guan, S.P., Fang, Q.Y., Guan, T.Y.: Application of a novel PNN evaluation algorithm to a greenhouse monitoring system. IEEE Trans. Instrum. Meas. **70**, 2510712 (2021)
3. Ouammi, A., Choukai, O., Zejli, D., et al.: A decision support tool for the optimal monitoring of the microclimate environments of connected smart greenhouses. IEEE Access **8**, 212094–212105 (2020)
4. Espejo, S.C.C., Souza, S.S.D., Junior, O.H.A.: Development of a biochemical oxygen demand incubator prototype based on thermoelectric effect with monitoring system. IEEE Lat. Am. Trans. **18**(12), 2037–2046 (2020)
5. Subahi, A.F., Bouazza, K.E.: An intelligent IoT-based system design for controlling and monitoring greenhouse temperature. IEEE Access **8**, 125488–125500 (2020)
6. Hafshejani, E.H., TaheriNejad, N., Rabbani, R., et al.: Self-aware data processing for power saving in resource-constrained IoT cyber-physical systems. IEEE Sens. J. **22**(4), 3648–3659 (2022)
7. Vazquez-Carmona, E.V., Vasquez-Gomez, J.I., Herrera-Lozada, J.C.: Environmental monitoring using embedded systems on UAVs. IEEE Latin Am. Trans. **18**(2), 303–310 (2020)
8. Martin, J., Ansuategi, A., Maurtua, I., et al.: A generic ROS-based control architecture for pest inspection and treatment in greenhouses using a mobile manipulator. IEEE Access **9**, 94981–94995 (2021)
9. Fernando, S., Nethmi, R., Silva, A., et al.: Intelligent disease detection system for greenhouse with a robotic monitoring system. In: 2020 2nd International Conference on Advancements in Computing (ICAC), Malabe, Sri Lanka. IEEE (2020)
10. Kitpo, N., Kugai, Y., Inoue, M., et al.: Internet of things for greenhouse monitoring system using deep learning and bot notification services. In: 2019 IEEE International Conference on Consumer Electronics (ICCE), Las Vegas, NV, USA. IEEE (2019)

The Minimum Impact of the Space Manipulator Based on the Intelligent System When Grabbing the Load

Xiaojie Hu(✉)

Department of Mechanical Engineering, Dalian University of Science and Technology, Dalian, Liaoning, China
844975062@qq.com

Abstract. As an important execution component of on-orbit services, the space manipulator plays an important role in the process of exploring the universe. A lot of work has been done in the research and application of space manipulators at home and abroad. The purpose of this paper is to study the minimum impact when the space manipulator grasps the load based on the intelligent system. Firstly, using the equivalent mass method and the multi-body collision dynamic model, the relationship between the magnitude of the collision force and the configuration of the space manipulator during the collision is analyzed, and the optimization objective of the minimum collision force on the manipulator is proposed, and the minimum collision force is solved. Then, using the impact effect of the collision on the space manipulator, the combined optimization objective of the minimum impact moment on the joint and the minimum change of the joint angle is proposed, and the particle swarm optimization algorithm is used to obtain the configuration that satisfies the minimum impact effect of the collision. The experimental results show that the maximum error of the pose does not exceed 2.5%. Using this configuration as the initial configuration of the contact collision can effectively reduce the maximum impact moment received by the joint and effectively suppress the sudden change of the joint angle, which improves the stability of the system.

Keywords: Intelligent system · Robotic arm grasping · Minimal impact · Configuration optimization

1 Introduction

When the spacecraft is in orbit, aerospace robots are required to complete the replacement of on-orbit refueling and maintenance units, thereby increasing the on-orbit life of the spacecraft; while the space station requires aerospace robots to assist astronauts to complete the handling and assembly of large space structures and complete the space shuttle and space station functions such as connection and separation [1]. Therefore, the development of space robots is of great significance for space applications [2]. When a space robot is orbiting, it will inevitably come into contact with and collide with the target, the external environment and the working platform. Space robots and ground robots

J. H. Abawajy et al. (Eds.): ICATCI 2022, LNDECT 169, pp. 393–400, 2023.
https://doi.org/10.1007/978-3-031-28893-7_47

have different dynamic characteristics and limitations: such as operator and base, flexible boom vibration, Dynamics, limited fuel supply, limitations of stop control systems, and collisions during operation [3].

Space robots have encountered many challenging problems during on-orbit operations, and put forward more constraints and requirements for the dynamic modeling and control of on-orbit operations [4]. Liu W uses LfD, BC, and DDPG to improve sample utilization. Use multiple critics to integrate and evaluate input actions to address algorithmic instability. Finally, inspired by Thompson's sampling idea, the input actions are evaluated from different angles, which increases the algorithm's exploration of the environment and reduces the number of interactions with the environment. Simulation results show that under the same number of interactions, the success rate of grasping 1000 random objects by the robotic arm is more than doubled, reaching the state-of-the-art (SOTA) performance [5]. Sepulveda D introduced a dual-arm eggplant harvesting robot consisting of two robotic arms configured in an anthropomorphic manner to optimize the dual workspace. To automatically detect and locate eggplants, an algorithm based on a support vector machine (SVM) classifier was implemented, and a planning algorithm was designed to schedule efficient fruit harvesting, coordinating two arms throughout the harvesting process [6]. The successful development and partial successful application of space manipulators fully prove the necessity of using manipulators to replace astronauts to perform complex tasks, and also strengthen people's confidence in using manipulators to perform on-orbit missions [7].

The purpose of this paper is to discuss the path planning of the space manipulator to capture the target and the collision dynamics during capture. Through the study of path planning and collision dynamics, the arm shape and path planning conditions for minimum collision requirements are obtained, and it is pointed out that due to the huge difference in inertia parameters between the manipulator and the cabin, the optimization goal is not suitable for large-load self-assembly task for this particular work scenario. Then, an optimization scheme is proposed to reduce the impact effect of collision on the manipulator system. It has guiding significance for the capture of engineering practice, especially the moving target.

2 Research on the Minimum Impact When the Space Manipulator Based on the Intelligent System Grabs the Load

2.1 Space Robotic Arm

The on-orbit service system of the space manipulator is mainly composed of three parts: the base (serving spacecraft), the n-degree-of-freedom manipulator installed on the serving spacecraft, and the target spacecraft being served [8]. According to whether there is information communication between the service and the spacecraft being served, the service objectives of the on-orbit service mission can be divided into two categories: cooperative objectives and non-cooperative objectives. Cooperative targets can usually obtain information such as position and attitude directly through communication, and a convenient service spacecraft with a grab handle is installed on the target to work. Non-cooperative targets usually do not provide such information [9, 10].

2.2 Dynamic Simulation Analysis Software

ADAMS, multi-body dynamics simulation analysis software, is mainly used to study the kinematics and dynamic characteristics of structures. It is a computer-aided design software favored by researchers in various related industries around the world [11]. The ADAMS software itself contains a number of professional modules, and can complete functions such as preprocessing, solution and result analysis in an interactive graphical interface. The three core modules of the software are ADAMS/View, ADAMS/Solver and ADAMS/Post Process in turn, and most conventional mechanisms can be simulated and analyzed by using these three modules. In addition to this, the software includes some special specialized modules for technicians to complete specific analyses.

The graphic parts library of ADAMS is very complete, so its modeling function is relatively powerful, and ADAMS can be competent for the creation of general parts [12]. In ADAMS, not only the dynamic simulation of rigid body structures can be completed, but also the mechanical simulation of flexible body structures.

2.3 Collision Impulse

In the current collision theory, only the relationship between the relative velocity changes at the collision point is included. We express the velocity and angular velocity relationships separately. For the single-point collision problem, the collision conditions are first identified, and the point P closest to the target surface to the end of the manipulator at each moment in the motion process, the unit normal direction η of the tangent plane, and the distance between this point and the end of the manipulator are obtained. The relative velocity $\eta \cdot vhp$ in the direction of the line. In practice, in order to facilitate grasping, first of all, the speed and angular velocity of the space target are generally estimated, and the most suitable parts for grasping are selected on the target object, such as handles, propeller nozzles, etc., and then the robotic arm is extended. To a very close range to the target grasping part, and move with the part. In this way, when a collision occurs, the collision point area is generally limited to a small variation range of the part of the target, that is, the known collision point h, and the position rth relative to the center of mass. For single-point collision, the collision normal direction η is the normal direction of the point h tangent plane, and the outward direction is defined as the positive direction. Assuming that the magnitude of the impulse generated in the collision process is A, the impulse received by the end of the manipulator is $I = A\eta$.

3 Investigation and Research on the Minimum Impact When the Space Manipulator Based on the Intelligent System Grabs the Load

3.1 Experimental Subjects

The simulation object is a typical spatial seven-degree-of-freedom manipulator. The ZYX Euler angles of the base cabin and the cabin to be assembled at the moment of contact collision in the self-assembly task are set to be [0.24, −0.11, 0.21] (rad), [04, −0.5, 0.1] (rad), thc actual pose of the space manipulator while maintaining the current angle is [2 m, 4 m, 3 m, 0 rad, 0 rad, −120 rad].

4 Load Model

In the inverse dynamics, the base cabin of the end rod is fixedly connected to form a new end rod. At this time, as a new end link, its center of mass coordinate system is $\sum C$, and the mass and inertia parameters can be expressed as:

$$^{1}I_{cabin} = ^{1}R_{C}[I_{cabin} + m_{cabin}(p_{cabin}^{T} p_{cabin} E_{3} - p_{cabin} p_{cabin}^{T})]^{1} R_{C}^{T} \tag{1}$$

Among them, mcabin represents the mass of the cabin to be assembled, mc. Indicates the mass of the composite; Pcabin, and pc. Are the representation of the center of mass vector of the cabin to be assembled and the combined body in the inertial frame, respectively; Icabin and 'Icabin represent the inertia tensor of the cabin to be assembled in its own coordinate system and inertial frame, respectively; 'pc represents the The position vector of the coordinate system in the inertial system; E3 is the third-order unit matrix.

4.1 Configuration Optimization Based on Particle Swarm Optimization

In the optimization process of the particle algorithm, a group of populations are randomly set as the initial value, and the position and flight speed of the particles are updated according to the following formulas according to the optimal point:

$$v_{i}^{d}(s+1) = wv_{i}^{d}(s) + c_{1}\wp(p_{i}^{d} - x_{i}^{d}(s)) + c_{2}\wp(p_{g}^{d} - x_{i}^{d}(s)) \tag{2}$$

$$x_{i}^{d}(s+1) = x_{i}^{d}(s) + v_{i}^{d}(s) \tag{3}$$

Among them, s is the number of iterations of the particle swarm algorithm, w is the inertia factor, C1, C2 are the acceleration factors, its value affects the weight of the particle's flight acceleration towards Pbest and gbest, $\wp$ is a positive random number less than 1, pid Pgd is the historical optimal point that has been searched for the i-th particle, and Pgd is the historical optimal point that has been searched for the entire population.

The pose constraint equation is an equation. For optimization problems with equality constraints, the particle swarm optimization algorithm generally cannot directly calculate and solve, and usually convert the equality constraint equation into a fitness function:

$$f_{fitness}(\theta) = f(\theta) + k\|\theta\|_{2} \tag{4}$$

Among them, K is the coefficient of the penalty factor, and the larger the coefficient of the penalty factor, the better the result can satisfy the constraint equation.

5 Analysis and Research of the Minimum Impact When the Space Manipulator Based on the Intelligent System Grabs the Load

5.1 Configuration Optimization Strategy Aiming at Minimizing the Impact of Collision

Using the equivalent mass method, the space machine is equivalent to an imaginary ellipsoid object. At this time, the large-load self-assembling space manipulator system can be approximately regarded as a three-sphere collision system, and the space manipulator is analyzed independently. At this time, the manipulator arms are respectively Under the force from both directions of the base and the cabin to be assembled, the installation positions of the space manipulator and the base and the cabin to be assembled are fixed, so the direction of the force is also fixed. At this time, the magnitude of the collision peak force is related to the contact configuration of the manipulator. When only the space manipulator is the research object, this problem can be transformed into which configuration of the space manipulator receives the smallest resultant external force when the force direction is certain. The schematic diagram of the model is shown in Fig. 1.

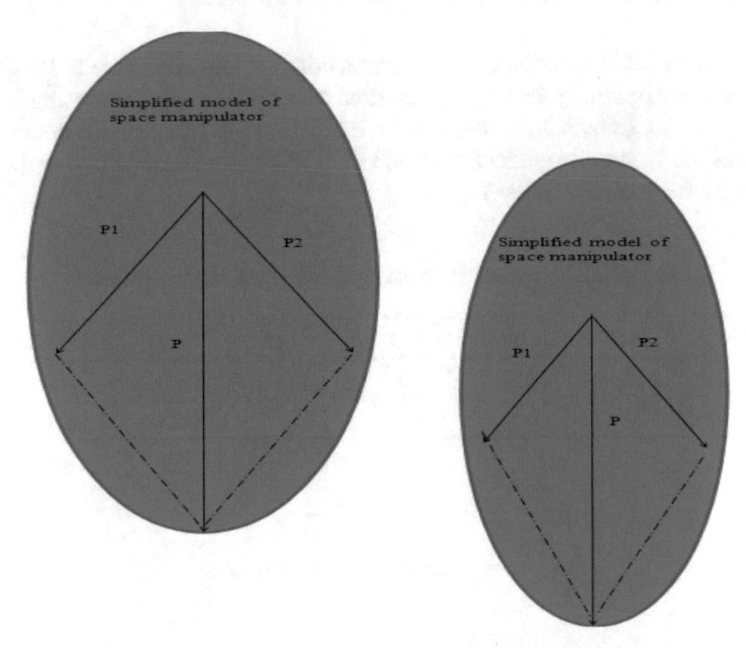

Fig. 1. Configuration optimization model

For the determined collision contact point and the determined collision direction, there are countless contact configurations of the space manipulator. If the configuration of the manipulator is changed, the magnitude of the collision force peak will change, which provides the possibility to minimize the collision force. The collision direction has been determined. The smaller the end equivalent mass of the manipulator in the configuration corresponding to, the smaller the collision force peak generated by the collision will be.

5.2 Configuration Optimization Results Aimed at Minimizing the Impact of Collision

The specific implementation steps of the particle swarm algorithm used in this paper are as follows:

Step 1: Initialize parameters, such as particle swarm scale, acceleration factor, inertia factor constraint index set, etc.;
Step 2: Calculate the global optimal value gbest;
Step 3: Calculate the fitness value of each particle;
Step 4: Compare the fitness value of each particle with its corresponding lbest, and the two are better as the new lbest, and update the optimal position.
Step 5: Compare the lbest of each particle with the global optimal value gbest of each particle, the better of the two is used as the new gbest, and the global optimal position is updated.
Step 6: Update the particle's position and velocity information;

Finally, after 1000 iterations, the minimum comfort function value is 103.01. After optimization, the optimal joint angle at the time of assembly of the space manipulator is [−0.6547, 0.0128, 1.6899, −2.0564, 1.7654, 0.8675, 0.0583] rad, and the corresponding end pose of the space manipulator is [0.988 m, 3.98 m, 0.970 m, 0.000 rad, 0.001 rad, −1.567 rad]. As shown in Table 1.

Table 1. Comparison table between actual pose and target pose

Target pose	2 m	2 m	4 m	0 rad	0 rad	−1.20 rad
Actual pose	1.95	1.99	3.98	0	0.002	−1.17
Error%	2.5	0.5	0.5	0	0.2	2.5

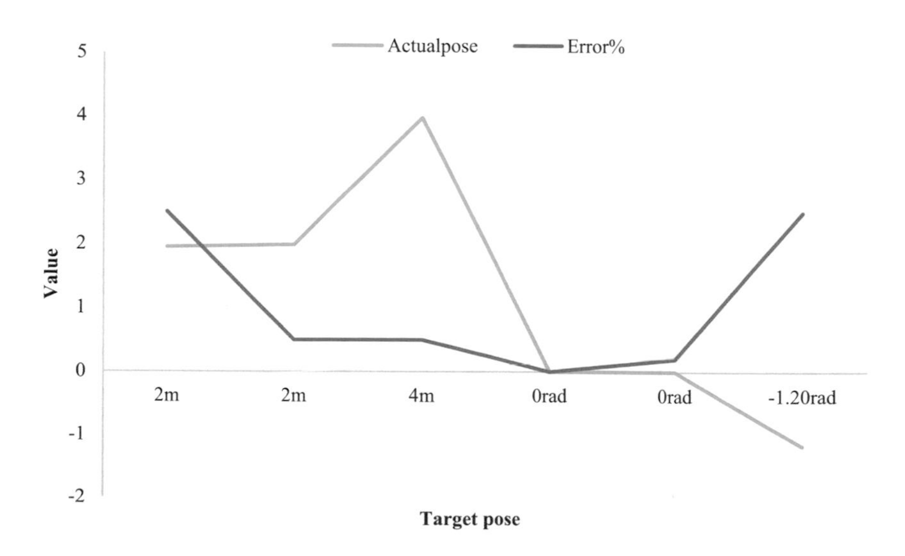

Fig. 2. Comparison of actual pose and target pose

It can be seen that the error of the pose is very small and the maximum error does not exceed 2.5%, as shown in Fig. 2. It is proved that the algorithm can meet the pose requirements. The main error is that the fitness function of the particle swarm optimization algorithm has a two-norm in the target function. When the fitness function is not zero, there is always an error in the constraint equation.

After optimization, the peak value of the impact moment of the collision on the space manipulator joint was reduced from 1925N to 887N, and the maximum value of the joint angle mutation caused by the collision was reduced from 0.258 deg/s before optimization to 0.187 deg/s. The optimized configuration can effectively reduce the disturbance of the collision impact to the manipulator.

6 Conclusions

With the rapid development of aerospace science and technology and the continuous growth of aerospace demand, more and more spacecraft are launched into space every year around the world. In-orbit spacecraft has gradually developed in a complex and large-scale direction, and various large-scale space structures will be produced in the future. In this paper, for the collision problem of capturing moving targets, the collision process when the target has a geometric shape is analyzed. From the point of view of impulse, the dynamic equation of collision is deduced by using the momentum expression of the space manipulator, and the linear velocity between the platform and the target is obtained. The dynamic Eq. 1 of the relationship with the angular velocity change, the collision dynamic Eq. 2 of the linear velocity and angular velocity change relationship between the end and the target, and the dynamic Eq. 2 is used to calculate the impulse generated in the collision process. Analysis of the impulse expression shows that the magnitude of the impulse generated in the collision process is related to the arm shape, the relative velocity of the collision point, and the position of the target capture point, and the local minimum collision arm shape is solved. The configuration optimization based on particle swarm optimization is designed. It can be seen from the simulation that the optimized configuration can effectively reduce the disturbance of the collision impact to the manipulator.

References

1. Gheorghe, A.C.: Robotic humanoid arm. Sci. Bull. Electr. Eng. Faculty **20**(1), 37–39 (2020)
2. Fattal, C., Leynaert, V., Laffont, I., Baillet, A., Enjalbert, M., Leroux, C.: SAM, an assistive robotic device dedicated to helping persons with quadriplegia: usability study. Int. J. Soc. Robot. **11**(1), 89–103 (2018). https://doi.org/10.1007/s12369-018-0482-7
3. Jiang, S., Guo, W., Liu, S., et al.: Grab and heat: highly responsive and shape adaptive soft robotic heaters for effective heating of objects of 3d curvilinear surfaces. ACS Appl. Mater. Interfaces. **11**(50), 47476–47484 (2019)
4. Chen, I.Z., Chang, J.T.: Applying a 6-axis mechanical arm combine with computer vision to the research of object recognition in plane inspection. J. Artif. Intell. Capsule Networks **2**(2), 77–99 (2020)
5. Liu W, Peng L, Cao J, et al. Ensemble Bootstrapped Deep Deterministic Policy Gradient for Vision-Based Robotic Grasping. IEEE Access **PP**(99), 1 (2021)

6. Sepulveda, D., Fernandez, R., Navas, E., et al.: Robotic aubergine harvesting using dual-arm manipulation. IEEE Access **PP**(99), 1 (2020)
7. Russell, P.R.: Robotic trash pickers. Eng. News-Record **281**(16), 60 (2018)
8. Banerjee, D., Yu, K., Aggarwal, G.: Robotic arm based 3D reconstruction test automation. IEEE Access **6**(1), 7206–7213 (2018)
9. Banger, M., Doonan, J., Rowe, P., et al.: Robotic arm-assisted versus conventional medial unicompartmental knee arthroplasty: five-year clinical outcomes of a randomized controlled trial. Bone Joint J. **103-B**(6), 1088–1095 (2021)
10. Ng, N., Gaston, P., Simpson, P.M., et al.: Robotic arm-assisted versus manual total hip arthroplasty: a systematic review and meta-analysis. Bone Joint J. **103-B**(6), 1009–1020 (2021)
11. Asokan, A., Baawa-Ameyaw, J., Kayani, B., et al.: Nursing considerations for patients undergoing robotic-arm assisted joint replacements. British J. Nursing (Mark Allen Publishing) **30**(10), 580–587 (2021)
12. Fahruzi, A., Agomo, B.S., Prabowo, Y.A.: Design Of 4DOF 3D robotic arm to separate the objects using a camera. Int. J. Artif. Intell. Robot. (IJAIR) **3**(1), 27–35 (2021)

Computational Prediction Approaches for Predicting Mutation Impact on Protein-Protein Interactions

Yi Ping(✉), Laura Hoekstra, and Anton Feenstra

Vrije University Amsterdam, De Boelelaan 1105, 1081 HV Amsterdam, The Netherlands
pingyi8010@163.com

Abstract. Protein-protein interactions (PPIs) involves in many significant mechanisms for human. The mutation impacts on PPIs can lead to the differences in conformational stability of proteins, as well as the PPIs kinetics and thermodynamics. Revealing these impacts is essential for understanding the underlying mechanism and designing new therapies. Therefore, there is a need to develop a reliable predictor for free energy changes of protein-protein binding affinity upon mutants. There have been many methods built on this purpose in the past two decades. So, we aim to conclude several aspects of these computational predictors for mutation impacts on PPIs, including several kinds of features, databases and measures used by predictors. A comparison is also conducted but not so accurate as the performances on measures are not on the same dataset. So, we got a conclusion that a benchmark database is needed for general validation and comparison of approaches related to mutant impact on PPIs. Besides, we categorized previous computational predictors into three groups, named energy-based approaches, structure-based methods and sequence-based methods. After comparison, sequence-based methods are a bit better than structure-based methods on the measure of PCC, especially the method SAAMBE-SEQ and MUPIPR. This comparison also illustrates a trend to predicting the impact of mutations on more PPIs with only sequence information.

Keywords: Impact of mutations · PPIs · Computational predictors · Sequence-based · Structure-based

1 Introduction

Protein-protein interactions (PPIs) are significant parts involved in numerous mechanisms such as cell metabolism, immune system, cellular processes and signalling transductions inside the human body [1]. Genetic mutations of all organisms lead to changes in proteins, which includes the mutants' influence on the conformational stability of proteins as well as the PPIs kinetics and thermodynamics. Revealing these impacts is essential for understanding the underlying mechanism, developing novel medicine and designing new therapies [2].

In recent years, a large variety of approaches to predict mutation effect on the protein-protein interactions have been developed and we review the current status of the field,

J. H. Abawajy et al. (Eds.): ICATCI 2022, LNDECT 169, pp. 401–409, 2023.
https://doi.org/10.1007/978-3-031-28893-7_48

including previous computational prediction approaches, datasets needed in training, testing, and validating the models, features usually used and assessment techniques. A deep understanding of the methodology applied in this area is needed for further progress, for example, finding a general benchmark for methods comparison, using which specific algorithm or which kind of features can lead to better prediction results later.

Until now, the experimental approaches remain pillars of measuring several characteristics of PPIs, for example, PPIs binding affinity can be observed by isothermal titration calorimetry [2]. Although they are accurate, they spend a lot of time and are expensive to conduct, which obstruct their steps to do such large-scaled studies. Besides, they need the physical existence of wild and mutant types for prediction. PPIs with mutants are not always accessible for experiments.

With the development of next-generating sequence techniques and advanced bioinformatics nowadays, computational approaches are developed and becoming reliable to employ useful information related to folding free energy and PPIs binding affinity more quickly.

There are several kinds of computational approaches, some are formed by machine learning algorithms like 'SAAMBE-SEQ' [1], 'MUPIPR' [2], while others are based on other theories such as one method 'CC/PBSA' [3]. The datasets used by these methods are also diversed including SKEMPI, SKEMPI 2.0, and many other databases collected by the researchers for their own researches [4, 5]. Therefore, the comparison between models is a bit hard to conduct.

So, we aim to conclude several aspects of computational predictors for mutation impacts on PPIs, databases used by predictors. Then, we focus on several computational predictors in every group of this field, revealing their basic underlying algorithms, and comparing their performances together with several databases about PPIs binding affinity. Some suggestions about efficient benchmark database development and feature selection will be provided and we wish to show that community standards are important to achieve development in the field for testing, training, and performance measurements.

2 Protein-Protein Interactions: A Major Concern

Protein-protein interactions (PPIs) represent the effects between proteins during the process of two or more proteins binding together with molecular docking to make complexes for conducting their biological functions. Usually, proteins do not function separately in cells, so PPIs are underlying mechanisms for molecular recognition [2].

Besides, proteins bind with others specifically can results in different PPIs patterns [3]. And this can be used to design proteins complexes for human on purposes such as producing specific medicines and developing therapies.

3 Databases Used by Predictors

Most of the databases employed for developing and validating of the predictors for PPI free energy changes upon mutations (Table 1) are extracted from the Structural database of Kinetics and Energetic of Mutant Protein Interactions (SKEMPI) [4] and SKEMPI v2.0 [5]. The SKEMPI and SKEMPI v2.0 datasets contain experimental data about mutant protein-protein complexes' free energy changes with their structures accessible in the Protein Data Bank (https://life.bsc.es/pid/skempi2/).

Table 1. Several databases which are frequently used by predictors.

No.	Database	Remarks	Models also use this database
1	SKEMPI	This database includes binding affinity data for 3047 mutations from 85 protein complexes	BindProfX [6], mCSM [7], BeAtMusic [8], Li et al. [3], MutaBind [9]
2	SKEMPI v2.0	This database includes binding affinity data for 7085 mutations from 389 protein complexes	SAAMBE-SEQ [1]
3	BindProfX database	Extracted from SKEMPI including 114 protein complexes and 1402 mutations	mCSM [7], MUPIPR [2], TopNetTree [10]
3	CC/PBSA database (or NM set)	Extracted by Alexnder et al. [3] contains 367 mutants of 9 protein complexes	Li et al. [3], iSEE [11]
4	BeAtMusic	This dataset contains 2007 mutations, subset of SKEMPI	MutaBind [9]
5	iSEE	The dataset was extracted from the SKEMPI 1.1 database, 1102 single point mutations in 57 protein complexes were selected	MUPIPR [2], ProAffiMuSeq [12]

4 Features of Predicting Mutation Impact on PPIs

Computational tools are built based on the following features (in Table 2).

5 Evaluation Measures Used by Predictors

For all the computational approaches in following section, their prediction performances are mainly evaluated by two measures. One is the Pearson Correlation Coefficient (PCC) for correspondence change between experimental data information and the computational predicted results. Another measure is the Root Mean Square Error (RMSE), which is calculated by the root square of total value of the differences between experimental and predicted values.

Table 2. Different kinds of features for predictors

Kind of features	Features inside the group
Structural features	Backbone angles, hydrophobic area, electrostatic interactions
Sequence features	Conserved sequences scores, amino acid position
Energy features	Van der Waals forces, solvation energy, extra-stabilizing free energy like water bridges, hydrogen-bond related free energy, electrostatic contribution of charged groups interactions et al.
Molecular features	Solvent accessible surface area of the interface, hydrophobic and hydrophilic area

6 Predictors for Mutation Impacts on PPIs

6.1 Previous Predictors

A diversity of computational techniques (Table 3) are developed for predicting. Based on the types of the features they used, we categorized these methods into three groups: energy-based approaches, structure-based models and sequence-based methods.

Table 3. Several methods for predicting mutation impacts on PPIs of three groups.

NO	Groups	Tool	Remarks	Publish time
1	Energy-based models	FoldX [13]	FoldX uses the physical energies to build an empirical linear method	2002
2		BindProfX [6]	BindProfX is based on random forest with the combination of FoldX score and interface conservation profile	2016
3		BeAtMuSiC [8]	BeAtMuSic uses the protein structure coarse-grained representation which dependents on the statistical energy potentials	2013
4	Structure-based models	mCSM [7]	mCSM takes the neighboring atomic-distance pattern as features	2014

(*continued*)

Table 3. (*continued*)

NO	Groups	Tool	Remarks	Publish time
5		iSEE [11]	iSEE is built with a random forest model based on combining structure-based features with energy-based features	2018
6		MutaBind [9]	MutaBind uses fast side-chain optimization algorithms, molecular mechanics force fields, and statistical potentials	2016
7		MutaBind2 [14]	MutaBind2 uses the combination of energy and structure information for the single point mutants and the multiple mutations	2020
8		mCSM-PPI2 [15]	The mCSM-PPI2 model improves the performance of mCSM method by combining other features like evolution scores, complex network with the original prediction properties	2019
9		CC/PBSA [3]	CC/PBSA uses the physical chemistry and the Poisson-Boltzmann equation to calculate free energy	2009
10		TopNetTree [10]	TopNetTree is based on topological structure properties	2020
11		SAAMBE-3D [16]	SAAMBE-3D uses some knowledge-based features to represent the environment of specific mutant	2020

(*continued*)

Table 3. (*continued*)

NO	Groups	Tool	Remarks	Publish time
12	Sequence-based models	ProAffiMuSeq [12]	It is a sequence-based approaches but still need some structural information	2020
13		SAAMBE-SEQ [1]	SAAMBE-SEQ predicts with 80 sequence-based features based on GBDT algorithm	2020
14		MUPIPR [2]	It is an end-to-end deep learning using deep contextualized representation learning	2020

Because almost all models mentioned here use PCC as an evaluation, we also conclude the detailed results of the PCC for all methods in Table 4. To compare the PCC results between these three model groups, Table 4 illustrates that sequence-bases models outperform the other two groups, with energy-based models performs worse than the others.

However, it is difficult to come up with a conclusion that one model is the perfect one having the best performance of PCC with the values present, because there is no general used benchmark database for models to compare with each other. So a validating or comparing database is needed for computational techniques which are utilized to predict free energy difference of mutants impact on PPIs.

Table 4. The PCC results for different kind of models on several databases. The context in brackets are the specific subsets extracted by the model.

No	Groups	Tools	PCC on SKEMPI	PCC on SKEMPI v2.0	PCC on other datasets
1	Energy-based models	FoldX [13]	NA	NA	0.64(FoldX); 0.72(CC/PBSA)
2		BindProfX [6]	0.625(BindProfX)	NA	NA
3		BeAtMuSic [8]	0.40(BeAtMuSic)	NA	NA
4	Structure-based models	mCSM [7]	0.801(mCSM), 0.58(BeAtMuSic)	NA	NA
5		iSEE [11]	NA	NA	0.73(CC/PBSA)
6		MutaBind [9]	0.68(BeAtMuSic)	NA	NA

(*continued*)

Table 4. (*continued*)

No	Groups	Tools	PCC on SKEMPI	PCC on SKEMPI v2.0	PCC on other datasets
7		MutaBind2 [14]	NA	0.82(MutaBind2)	NA
8		mCSM-PPI2 [15]	0.82(mCSM-PPI2)	NA	NA
9		CC/PBSA [3]	NA	NA	0.84(CC/PBSA)
10		TopNetTree [10]	0.85(TopNetTree)	NA	NA
11		SAAMBE-3D [16]	NA	0.78(SAAMBE-3D)	NA
12	Sequence-based models	ProAffiMuSeq [12]	NA	0.73(PROXiMATE)	NA
13		SAAMBE-SEQ [1]	NA	0.83(SAAMBE-SEQ)	NA
14		MUPIPR [2]	0.883(BindProfX)	NA	0.742(CC/PBSA)

6.2 Important Steps Forwards in This Field - Detailed Information of Several Predictors

The following part reveals the progression of algorithms predicting mutation impact on protein-protein interactions specifically.

6.2.1 FoldX

First of all, we focus on the method FOLD-X energy function (FOLDEF) [13]. FoldX is developed on an empirical linear method. It is developed for predicting mutants impacts on protein stability. And it is used to estimate the free energy changes ($\Delta\Delta G$).

This FoldX model attempts to have a progress for predicting the structure of the mutated complexes. It uses rotamers to carry out the conformational search, making the model fast to finish calculating a 300 residues long protein within 5 min [13]. However, they are unable to accurately predict backbone conformational changes caused by mutation, which lead to limitations of model FoldX.

6.2.2 Concoord/Poisson-Boltzmann surface area (CC/PBSA)

Then the second method is Concoord/Poisson-Boltzmann surface area (CC/PBSA) [3] which is a structure-based method. It predicts by two aspects, one is conformational stability, another is the effect of mutations on protein-protein binding affinity.

And it gets a great performance for mutated protein-protein binding affinity comparing with other methods of structure-based as shown in Table 4 in 2009. However, the limitation for this method is that it is time consuming as it needs more than 4 h for calculating every mutant.

6.2.3 SAAMBE-SEQ

The SAAMBE-SEQ [1] only uses sequence data based on the gradient boosting decision (GBD) machine learning algorithm. The underlying algorithm for this model is Gradient Boosting Decision Tree (GBDT). The advantages of SAAMBE-SEQ are not limited to the only using of sequence information but also keeping the accuracy for prediction similar or even better than structure-based methods and improving the computational time.

6.2.4 MUPIPR

Another sequence-based model is MUPIPR (Mutation Effects in Protein–protein Interaction Prediction Using Contextualized Representations) [2], which is based on end-to-end deep learning. MUPIPR has several advantages: It collects more accurate features on amino-acid-level; Every small signal of mutations are magnified even in a long protein sequence. Besides, this model only uses the sequence-based features.

7 Conclusion and Critical Discussion

PPIs involves in a significant number of mechanisms for human. Therefore, developing a reliable predictor for free energy changes of protein-protein binding affinity upon mutants is essential to inspire people to design proteins by introducing mutants for researches and therapies use. This review concludes several aspects of computational predictors for mutation impacts on PPIs. Firstly, these predictors contain several kinds of features mainly in structure features, sequence features, energy features and molecular features groups. Then databases used by predictors are different from each other but a majority of them are extracted from SKEMPI and SKEMPI v2.0. Furthermore, the predictors are shifting from methods with empirical and energy features to structure-based methods and sequence-based approaches.

Although, there are more than ten different approaches in this field. They can be validated using two general evaluation measures, PCC and RMSE. However, the variability of databases employed for different predictors' processes makes it hard to compare all predictors to get the best methods. Therefore, a benchmark database is needed for general validation and comparison of approaches related to mutant impact on PPIs. This benchmark dataset can be the testing or validation datasets, which are not used by training processes of all models' developing. Or new datasets extracted from the existing databases excluding the data used for training by previous methods.

In addition, obviously structure-based and sequence-based approaches outperform energy-based approaches. Sequence-based approaches are comparable or even a bit better than structure-based methods on the measure of PCC, especially the method SAAMBE-SEQ and MUPIPR, without considering the structural features.

Besides, even though the detailed structure information of proteins is adequate, the known human interactome with structural information is lower than 10%. So, trying to use sequence information only to conduct models may be a trend for predicting the impact of mutations on more PPIs, and the utilization for high-throughput information.

The field Machine learning as well as deep learning methods are participating in the basic algorithms for building the predictors.

For the further work, a general benchmark should be made for models' comparison, and a method with shorter computational time and better performance is needed for predicting mutant impacts on PPIs.

References

1. Li, G., et al.: SAAMBE-SEQ: a sequence-based method for predicting mutation effect on protein-protein binding affinity. Bioinformatics **37**(7), 992–999 (2021)
2. Zhou, G., Chen, M., Ju, C.J.T., Wang, Z., Jiang, J.-Y., Wang, W.: Mutation effect estimation on protein–protein interactions using deep contextualized representation learning. NAR Genomics Bioinforma. **2**(2), 1–12 (2020)
3. Benedix, A., Becker, C.M., de Groot, B.L., Caflisch, A., Böckmann, R.A.: Predicting free energy changes using structural ensembles. Nat. Methods **6**(1), 3–4 (2009)
4. Moal, I.H., Fernández-Recio, J.: SKEMPI: a structural kinetic and energetic database of mutant protein interactions and its use in empirical models. Bioinformatics **28**(20), 2600–2607 (2012)
5. Jankauskaite, J., Jiménez-García, B., Dapkunas, J., Fernández-Recio, J., Moal, I.H.: SKEMPI 2.0: an updated benchmark of changes in protein-protein binding energy, kinetics and thermodynamics upon mutation. Bioinformatics **35**(3), 462–469 (2019)
6. Xiong, P., Zhang, C., Zheng, W., Zhang, Y.: BindProfX: assessing mutation-induced binding affinity change by protein interface profiles with pseudo-counts. J. Mol. Biol. **429**(3), 426–434 (2017)
7. Pires, D.E.V., Ascher, D.B., Blundell, T.L.: MCSM: predicting the effects of mutations in proteins using graph-based signatures. Bioinformatics **30**(3), 335–342 (2014)
8. Dehouck, Y., Kwasigroch, J.M., Rooman, M., Gilis, D.: BeAtMuSiC: prediction of changes in protein-protein binding affinity on mutations. Nucleic Acids Res. **41**(Web Server issue), 333–339 (2013)
9. Li, M., Simonetti, F.L., Goncearenco, A., Panchenko, A.R.: MutaBind estimates and interprets the effects of sequence variants on protein-protein interactions. Nucleic Acids Res. **44**(W1), W494–W501 (2016)
10. Wang, M., Cang, Z., Wei, G.-W.: A topology-based network tree for the prediction of protein–protein binding affinity changes following mutation. Nat. Mach. Intell. **2**(2), 116–123 (2020)
11. Geng, C., Vangone, A., Folkers, G.E., Xue, L.C., Bonvin, A.M.J.J.: iSEE: Interface structure, evolution, and energy-based machine learning predictor of binding affinity changes upon mutations. Proteins Struct. Funct. Bioinforma. **87**(2), 110–119 (2019)
12. Jemimah, S., Sekijima, M., Gromiha, M.M.: ProAffiMuSeq: Sequence-based method to predict the binding free energy change of protein-protein complexes upon mutation using functional classification. Bioinformatics **36**(6), 1725–1730 (2020)
13. Guerois, R., Nielsen, J.E., Serrano, L.: Predicting changes in the stability of proteins and protein complexes: a study of more than 1000 mutations. J. Mol. Biol. **320**(2), 369–387 (2002)
14. Zhang, N., et al.: MutaBind2: predicting the impacts of single and multiple mutations on protein-protein interactions. iScience **23**(3), 100939 (2020)
15. Rodrigues, C.H.M., Myung, Y., Pires, D.E.V., Ascher, D.B.: MCSM-PPI2: predicting the effects of mutations on protein-protein interactions. Nucleic Acids Res. **47**(W1), W338–W344 (2019)
16. Pahari, S., et al.: SAAMBE-3D: predicting effect of mutations on protein-protein interactions. Int. J. Mol. Sci. **21**(7), 2563 (2020)

Passive Tracking Principle Based on Millimeter Wave Sensing

Wen Li[1](✉) and Haonan Xing[2]

[1] School of Information and Communications Engineering, Institute for Electronic and Information Science, Xi'an Jiaotong University, Xi'an 710000, Shaanxi, China
liwenaa788@163.com

[2] Baoji Oilfield Machinery Co. Ltd., Baoji 721000, Shaanxi, China

Abstract. With the rapid popularization of 5G technology from China in 2019, a large number of mobile devices have integrated 60 Ghz millimeter wave modules. Millimeter wave also has the advantage of high accuracy, 60 GHz millimeter wave will be able to perceive millimeter-level movement. The use of millimeter wave to realize the edgeless tracking can realize the trajectory tracking with high portability and high precision for small objects.

At present, radio-frequency-based active sensing and tracking technology has been extensively researched and applied, but the tracking technology based on active sensing needs to modify the target object or even equip special equipment as the target object to be tracked, which will damage the mobile device's performance. Portability, and will incur additional material costs, so active sensing technology is not an ideal solution. This article introduces the trajectory tracking of small target objects based on the principle of passive sensing of millimeter waves.

Keywords: Passive tracking · Millimeter wave · Principle

1 Research Status of Passive Sensing

In the field of passive sensing, video and image-based methods have been applied in many practical applications, such as leap motion, a somatosensory controller developed by leap company, and xboxkinect, a somatosensory device launched by Microsoft. These sensing devices will be disturbed by ambient light, and there are certain privacy problems, which will hinder their further promotion and application. Passive sensing based on microwave can avoid these two problems [1].

As a ubiquitous information medium without affecting people's normal life, RF microwave has been favored in the field of passive sensing. In the passive sensing location based on RF and microwave, the transmitting antenna is used to actively transmit the microwave signal, and then the receiving antenna receives the signal reflected by the relevant object. After a series of signal processing and analysis, the motion state of the target object can be perceived. Because the electromagnetic wave will be reflected by the object whose size is larger than its wavelength, we analyze the reflected signal of the target object by selecting the microwave with appropriate wavelength as the sensing

J. H. Abawajy et al. (Eds.): ICATCI 2022, LNDECT 169, pp. 410–417, 2023.
https://doi.org/10.1007/978-3-031-28893-7_49

medium, so as to extract the distance information and angle information, so as to realize the positioning of the target object. In the field of passive location and tracking, traditional radar is mainly used to track large moving objects, and the radio wave with sampling rate of GHz is used to ensure its granularity. In contrast, due to the limitation of hardware devices, the sampling rate of millimeter wave modules equipped on mobile devices is difficult to reach the GHz level [2].

In terms of application scenarios, what we need to achieve is high-precision positioning and tracking of small objects in close range, which cannot be solved by traditional radar methods. Pulse radar needs high-speed pulse generator, and its analog-to-digital converter needs to work at GB/s rate [3]. Such analog-to-digital converter has high power, high price and low bit resolution, which makes this method unattractive in practice. With the help of inverse synthetic aperture radar technology, Fadi et al. Realized through wall positioning through WiFi, and could track people's movement with IM granularity. Witrack uses FMCW radar to track the human body. These tracking methods use low-frequency WiFi signals to realize coarse-grained large-scale object tracking, which is mainly used for indoor positioning. Moreover, due to the long wavelength of WiFi signal, it limits the tracking method based on WiFi signal to track small objects. We are facing a small object tracking scene, so we choose a 60 GHz millimeter wave with a wavelength of 5mm, which can be reflected by small objects such as pens, which is a characteristic that traditional WiFi signals do not have. Mtrack, which realizes the tracking of small objects based on 60 GHz millimeter wave, and realizes accurate positioning by using the characteristics of phase change and beam maneuverability of 60 GHz millimeter wave. Its phase tracking module continuously tracks the target by measuring and operating the phase increment of each receiving antenna. In this paper, FMCW technology is used to locate the target object. All hardware components can be integrated into a small chip, which is more similar to the millimeter wave communication module equipped on mobile devices in the future, and can be directly applied to mobile devices. For the unique challenge of burst noise in passive tracking of small objects, this paper proposes a burst noise processing algorithm to solve this problem [4, 5].

2 Millimeter Wave Radar

2.1 Radar Parameters

The electromagnetic wave will be blocked or reflected by the object whose size is larger than its wavelength, and will be bypassed by the object whose size is smaller than its wavelength, which limits the application of the electromagnetic wave with longer wavelength in the positioning and tracking of small objects. The wavelength of the millimeter wave signal is in the millimeter level, so the millimeter wave can be reflected even if the target object is small. This is the unique advantage of millimeter wave in the field of passive tracking of small objects.

Millimeter wave also has the advantage of high accuracy. 60 GHz millimeter wave will be able to sense millimeter level movement. In addition, the size of hardware components (such as antenna, analog-to-digital converter, etc.) required to process millimeter wave signals can be designed very small and easy to be integrated into a chip as small as a few centimeters, which provides the possibility for its application in mobile devices, which is another advantage of millimeter wave [6].

At any time point, the baseband signal received by frequency modulated continuous wave (FMCW) radar is narrowband, so FMCW radar does not need high-speed analog-to-digital converter, which meets the needs of low power consumption of our mobile devices. Therefore, this paper uses FMCW millimeter wave radar as the sensing equipment of passive tracking system. FMCW radar transmits linear frequency modulation pulse signal through transmission (TX) antenna. Its signal amplitude remains unchanged, but the frequency changes periodically with time. The reflection of the object on the linear frequency modulation pulse is received by the receiving (Rx) antenna, and its reflected signal is also a linear frequency modulation pulse signal. For convenience of description, the linear frequency modulation pulse signal transmitted through the transmitting antenna is hereinafter referred to as the transmitting signal, and the frequency modulation pulse signal reflected by the object and received by the receiving antenna becomes the receiving signal [7–10].

The mixer combines the transmitted signal and the received signal to generate a new signal, extracts and amplifies the IF signal using an analog-to-digital converter, converts it into a digital signal, and outputs it to the track tracking system for subsequent processing.

The millimeter wave equipment adopts radar chip, and the setting of one frequency sweep time is jointly determined by the sampling rate set by the radar and the number of sampling points in a frame. The formula is as follows:

$$T = \frac{N_{\text{sampie}}}{F_s} \tag{1}$$

where: T is the time of one frequency sweep; N_{sampie} is the number of sampling points in a frame and F_s is the sampling rate.

2.2 Time Delay of Radar Transmitted Signal

The principle of FMCW radar detecting object position is based on the analysis of transmission signal delay. The time when the signal is sent from the transmitting antenna and returned to the receiving antenna after being reflected by the object. After analyzing this time delay, determine the distance between the reflected object and the radar. FMCW radar transmits a signal called FM pulse. Because the signal has different frequencies at different times, the receiving antenna can determine when the received signal is sent from the transmitting antenna according to the frequency [11, 12].

FMCW radar receives signals while transmitting signals. There is time delay due to the propagation of signals in space. The frequency of FM pulse signal at the transmitting end changes regularly with time. Therefore, at the same time, the frequency of radar transmitting signal and receiving signal is not equal, and there is a difference between them. As shown in Fig. 1.

The frequency difference f_τ can be obtained directly by subtracting the received signal from the transmitted signal at that time. The time delay of the received signal is

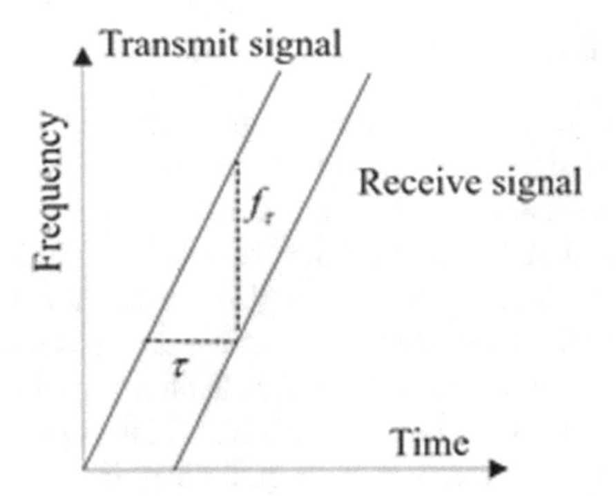

Fig. 1. FMCW transmit and receive signals

represented by τ, then

$$\frac{f_\tau}{\tau} = S \quad (2)$$

where S is the frequency modulation rate of the frequency modulation pulse signal, and this value is the fixed value, which is determined by the bandwidth B and the frequency modulation time T. So,

$$\tau = \frac{f_\tau}{S} = f_\tau * \frac{T}{B} \quad (3)$$

It can be seen from the above formula that the value of time delay τ can be determined by the frequency difference and is directly proportional to it.

The above description is the signal reflection in the ideal state, that is, there is only one reflector in space. In this case, the signal output by FMCW radar is a sine wave signal with unique frequency [13]. However, in the actual environment, there may be multiple reflective objects, which are mixed with background reflection and burst noise. At this time, the radar output signal is a mixed and superimposed signal of multiple sine wave signals. After fast Fourier transform, sine waves of multiple frequencies can be separated. Each frequency represents a signal reflection at a corresponding distance [14].

3 Position Calculation

This paper only describes the tracking scene in two-dimensional plane, which not only meets the needs of practical application, but also facilitates the description of trajectory. Tracking in two-dimensional plane can approximately project the target object into a point, and the corresponding spectrum is obtained by FFT processing of the received signal. From the spectrum, the distance and angle of arrival of the reflected object relative to the FMCW radar can be calculated, so as to determine the position of the object on the two-dimensional plane.

3.1 Distance Calculation

Since the signal is continuously transmitted and received, the frequency difference between the transmitted signal and the reflected signal is constant at any time during transmission. This shows that when there are multiple reflectors, the IF signal obtained after mixing is a mixed signal containing several single tone signals with constant frequency. Now assume that there is only one single tone signal in the IF signal. The single tone signal frequency is the fixed frequency difference between the transmitted signal and the reflected signal, expressed in f_τ. The reflected signal received by the FMCW radar is the delay signal of the transmitted signal, which is determined by the distance between the radar and the reflected object. Signal delay is:

$$\tau = \frac{2R}{c} \tag{4}$$

where C is the speed of light and R is the distance between the radar and the reflected object.

By introducing Eq. (3) into Eq. (4):

$$R = f_\tau * \frac{cT}{2B} \tag{5}$$

Since T and B are the working parameters set by the radar in advance, their values are fixed, and the light speed C is also a fixed value, the distance R is the frequency of the IF signal output by the FMCW radar, which is uniquely determined. When reflected in the

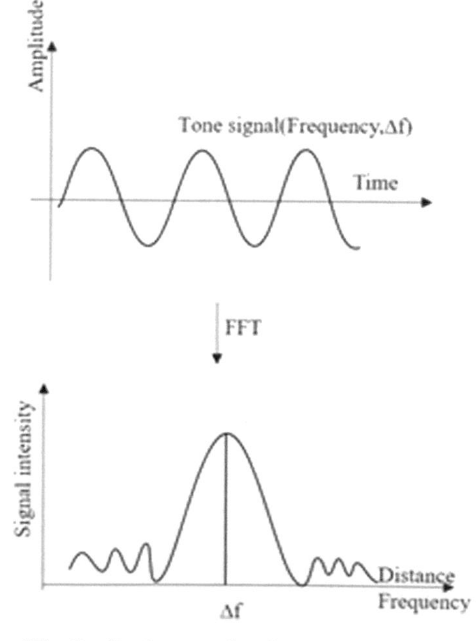

Fig. 2. Single tone signal processed by FFT

FFT spectrum of the signal, the frequency corresponds to the distance. The embodiment of single tone signal in the spectrum after FFT transformation is shown in Fig. 2.

In practice, there are often multiple single tone signals in the IF signal, and multiple peaks will appear in the spectrum after FFT processing. The signal strength of a single tone signal is different, and each peak point of signal strength indicates the existence of a reflective object. If the IF signal containing multiple single tone signals is processed by FFT, different single tone signals can be separated. The spectrum generated after FFT processing has different peaks. Each peak corresponds to an object at a specific distance, and its distance can be known from the single tone signal frequency corresponding to the peak.

3.2 Angle Calculation

In the horizontal plane, the angle between the object and the node normal is called the angle of arrival (AOA). The small change of distance will cause the phase change of FFT peak. Based on this feature, we can use the phase change to calculate the angle. This calculation process requires at least two receiving antennas, as shown in Fig. 3.

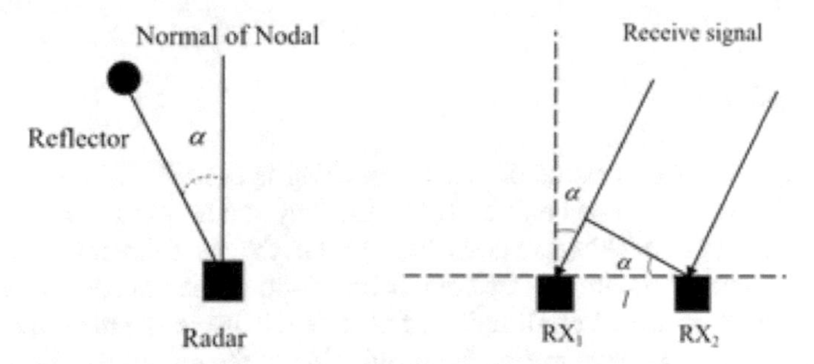

Fig. 3. AOA calculation method

As shown in Fig. 3, according to the basic geometric principle, the distance change Δd can be represented by the distance l between two receiving antennas and the angle of arrival α:

$$\Delta \mathrm{d} = l \sin\alpha \tag{6}$$

The phase change due to distance change can be deduced as:

$$\Delta\Phi = \frac{2\pi f_0 \Delta \mathrm{d}}{c} = \frac{2\pi f_0 * l \sin\alpha}{c} \tag{7}$$

Finally, the angle of arrival a can be calculated using the results obtained in (7):

$$\alpha = \sin^{-1}\left(\frac{\Delta\Phi c}{2\pi f_0 l}\right) \tag{8}$$

After the distance R calculated in (5) and α calculated in (8) are converted into Cartesian coordinates, the position can be determined on the two-dimensional plane to realize the positioning of the object.

The comparison between the millimeter wave passive location technology used in this paper and the existing technology is shown in Table 1. The tracking accuracy of this method has been further improved, and it is easier to transplant to mobile devices than other methods. In applications such as virtual touch panel on mobile devices, the detection range of 0.01 M has been able to meet the accuracy requirements of many devices.

Table 1. Comparison with existing technologies

	RFID raw	Tagoram	Paper method
Tracking type	Active	Active	Passive
Signal characteristics	Phase	Phase	Phase
Number of antennas	8	32	N3
Tracking range	2–5 m	1–10 m	0.1–0.5 m
Average error	49 mm	14 mm	<5 mm

4 Conclusion

In this paper, the research status of passive tracking is described. The characteristics of millimeter wave radar based on FMCW technology and how to use millimeter wave radar to realize passive tracking are described. That is, the spectrum of the two receiving antennas is obtained through the fast Fourier transform of the obtained signal, and the distance and angle of arrival of all reflected objects relative to the radar are calculated through the spectrum, so as to realize the positioning of the reflected objects.

At present, facing the trajectory tracking scene of two-dimensional plane, this method can be realized. How to realize the trajectory tracking in three-dimensional space has high requirements for the analysis of the physical structure of the target object, so it is necessary to find a stable positioning center point. This issue still has many challenges.

References

1. Lien, J., Gillian, N., Karagozler, M.E., et al.: Soli: ubiquitous gesture sensing with millimeter wave radar. ACM Trans. Graph. (TOG) **35**(4), 1–19 (2016)
2. Kim, S., Nguyen, C.: On the development of a multifunction millimeter-wave sensor for displacement sensing and low-velocity measurement. IEEE Trans. Microw. Theory Tech. **52**(11), 2503–2512 (2004)
3. Alloulah, M., Huang, H.: Future millimeter-wave indoor systems: a blueprint for joint communication and sensing. Computer **52**(7), 16–24 (2019)
4. Brown, E.R.: Fundamentals of terrestrial millimeter-wave and THz remote sensing. Int. J. High Speed Electron. Syst. **13**(04), 995–1097 (2003)
5. Meng, Z., Fu, S., Yan, J., et al.: Gait recognition for co-existing multiple people using millimeter wave sensing. In: Proceedings of the AAAI Conference on Artificial Intelligence, vol. 34(01), pp. 849–856 (2020)

6. Wu, C., Zhang, F., Wang, B., et al.: mSense: towards mobile material sensing with a single millimeter-wave radio. In: Proceedings of the ACM on Interactive, Mobile, Wearable and Ubiquitous Technologies, vol. 4(3), pp. 1–20 (2020)
7. Wan, Z., Gao, Z., Shim, B., et al.: Compressive sensing based channel estimation for millimeter-wave full-dimensional MIMO with lens-array. IEEE Trans. Veh. Technol. **69**(2), 2337–2342 (2019)
8. Barneto, C.B., Riihonen, T., Turunen, M., et al.: Radio-based sensing and indoor mapping with millimeter-wave 5G NR signals. In: 2020 International Conference on Localization and GNSS (ICL-GNSS), pp. 1–5. IEEE (2020)
9. Yang, X., Liu, J., Chen, Y., et al.: MU-ID: multi-user identification through gaits using millimeter wave radios. In: IEEE INFOCOM 2020-IEEE Conference on Computer Communications, pp. 2589–2598. IEEE (2020)
10. Kimionis, J., Georgiadis, A., Tentzeris, M.M.: Millimeter-wave backscatter: a quantum leap for gigabit communication, RF sensing, and wearables. In: 2017 IEEE MTT-S International Microwave Symposium (IMS), pp. 812–815. IEEE (2017)
11. Melgarejo, P., Zhang, X., Ramanathan, P., et al.: Leveraging directional antenna capabilities for fine-grained gesture recognition. In: ACM International Joint Conference, pp. 541–551. ACM (2014)
12. Wang, H., Sen, S., Elgohary, A., et al.: No need to war-drive: unsupervised indoor localization. In: International Conference on Mobile Systems, pp. 499–500. ACM (2012)
13. Sen, S., Radunovic, B., et al.: You are facing the Mona Lisa: Spot Localization using PHY Layer Information. ACM MobiSys, pp. 183–196 (2012)
14. Liu, H., Gan, Y., Yang, J., et al.: Push the limit of WiFi based localization for smartphones. ACM MobiCom, pp. 305–316 (2012)

Application of Data Mining in the Design of English TRAnslator's Speech Recognition System

Fenglan Gui[1(✉)] and Zhengmao Yang[2]

[1] Party School of Liaoning Provincial Committee of the Communist Party of China, Shenyang, Liaoning, China
guifenglan6666@126.com

[2] Henan University of Economics and Law, Zhengzhou 450046, Henan, China

Abstract. With the further improvement of basic equipment, more and more data can be discovered and used. With the continuous improvement of computer computing power and the perfection of machine learning and other theories, speech recognition technology gradually tends to be practical. Research on the application of data mining in speech recognition and constructing an English translator speech recognition system has practical significance in our country. Therefore, this article mainly studies the application of data mining in the design of English translator speech recognition system. This paper analyzes the speech data of data mining technology, speech recognition technology and machine translation technology, and proposes a speech error correction algorithm based on the Jaro-winkler algorithm, which improves the accuracy of speech recognition. This paper designs a platform development plan for an English translator speech recognition system, and tests the translation performance of the system. The results show that the overall recognition rate of the system is 89.45%, and the translation accuracy rate is 91.64%. This shows that the accuracy of the system is high, and the input sentence can be translated more accurately, which meets the requirements of the English speech translation system in specific situations.

Keywords: Data mining · Speech recognition · Machine translation · System testing

1 Introduction

Voice is the most common way for people to transmit and exchange information [1]. With the current development of science and technology, more and more products have added voice recognition function [2, 3]. In today's economic globalization, English has become an important tool for people to communicate with each other, and people's demand for the speech recognition system of English translators is becoming more and more exuberant [4, 5]. In the era of big data, the number of users of English translation systems and hardware products has reached hundreds of thousands, millions, tens of millions or even more, and the traditional methods of artificially analyzing voice data

J. H. Abawajy et al. (Eds.): ICATCI 2022, LNDECT 169, pp. 418–425, 2023.
https://doi.org/10.1007/978-3-031-28893-7_50

are no longer applicable. It is particularly important to mine and analyze English speech data with the help of data mining technology [6, 7].

Regarding the research of data mining and speech recognition, many scholars have conducted multi-angle explorations. For example, Long Y studied the method of acoustic data enhancement for speech recognition in Chinese-English translation [8]; Wei L studied cloud computing and Speech recognition technology in English teaching in English [9]; Dorothy ND has studied the English speech recognition system for students in the English classroom [10]. Therefore, the research on the application of data mining in the design of English translator speech recognition system is innovative.

This article mainly analyzes the speech data of data mining technology, speech recognition technology and machine translation technology, and firstly proposes a speech error correction algorithm based on the Jaro-winkler algorithm. Then design the platform development plan of the English translator speech recognition system, and briefly introduce the platform function modules. Finally, the translation performance of the system is tested, which verifies that the system has good speech recognition and translation accuracy.

2 Application of Data Mining in the Design of English Translator Speech Recognition System

2.1 Speech Error Correction Algorithm Based on Data Mining Technology

The recognition rate of English speech is not 100% at present, so the speech log stored in the database has some recognition errors. English voice log error correction uses natural language processing technology to correct the wrong text. On the basis of the original system, the recognition accuracy is improved, which has a positive effect on the improvement of the system. In the process of using the voice recognition function, the system will the user's voice input is converted into text and stored in the database, and the voice recognition data is analyzed according to the text analysis algorithm and artificially set rules [11, 12]. This part is based on the Jaro-winkler algorithm to correct the content of speech recognition errors.

The Jaro-winkler algorithm is an algorithm for calculating the similarity between two strings. By calculating the similarity between the strings, it finds the string with the greatest similarity to the target string, and treats it as a corrected string to achieve voice correction. The calculation method of similarity between strings is shown in formula (1):

$$d\begin{cases} 0, m = 0 \\ \frac{1}{3}\left(\frac{m}{|s1|} + \frac{m}{|s2|} + \frac{m-t}{m}\right), m \succ 0 \end{cases} \tag{1}$$

Among them, d represents the similarity between string s1 and s2, and the value range is [0,1]; |s1| and |s1| represent the length of string s1 and s2 respectively, m represents the number of matching strings, t Indicates the number of transpositions between strings.

The calculation method of t is related to the matching window. The calculation method of the defined matching window is shown in formula (2):

$$MW = \left\lfloor \frac{max(|s1|, |s2|)}{2} - 1 \right\rfloor \tag{2}$$

When the characters in the two strings are the same but the positions are different, if the distance between the two characters is not greater than MW, the two characters are matched characters, t is half of the number of matched characters. In order to give a higher score to the same string at the beginning, P is defined to be no greater than 0.25, and the string similarity calculation method is shown in formula (3):

$$\mathrm{D} = \mathrm{d} + lp(i - d) \tag{3}$$

In the formula, P is 0.1 by default, and l is the same length as the prefix. In the actual experiment process, the recognition results of the speech recognition system have the following characteristics: the processing content is the content that the system does not correctly recognize, so many words have no practical meaning; the speech recognition results are prone to fewer words; the user's speech content and actual trends The relevance is great, and there are some popular words among them.

Therefore, the process of the speech error correction algorithm is as follows: first, create a dictionary used as a template for speech error correction; then clean the data to find out the wrong sentences, and remove the verbs, adverbs, pronouns, adjectives, etc. from the word segmentation.

2.2 Voice Translation Model of English Translator

The system uses a multi-task approach to improve the performance of the speech translation model. In multi-task learning, the speech recognition task serves as an auxiliary trainer and the end-to-end speech translation model shares certain parameters to train these two tasks together. The input is the source language speech feature sequence x = (x1, x2,..., xt), and the output encoder coded speech signal intermediate vector sequence h = (h1, h2,..., hu).

For the decoder of the speech translation task, first use the attention mechanism to generate a context vector for the output of the shared encoder, and then use a multi-layer perceptron with a non-linear activation function softmax to output the predicted probability of the target language to translate the word Distribution, the final decoding generates the translated target language word sequence $y^1 = (y_1^1, y_2^1, \ldots, y_n^1)$, and the formalization of the process is shown in formulas (4), (5), (6):

$$c^1{}_i = Attention_{st}(s^1{}_i, H) \tag{4}$$

$$s^1{}_i = LSTM_{st}(s^1{}_{i-1}, c^1{}_{i-1}, y^1{}_{i-1}) \tag{5}$$

$$y^1{}_i = generate_{st}(s^1{}_i, c^1{}_i) \tag{6}$$

In the process of multi-task training, the error is propagated back to the input through the two decoders, so the loss function of the model is the weighted sum of the losses of the speech recognition task and the speech translation task.

$$\mathrm{L} = \lambda L_{asr} + (1 - \lambda)L_{st} \tag{7}$$

In formula (7), L_{st} is the negative log-likelihood loss of the speech translation task, L_{asr} is the negative log-likelihood loss of the speech recognition task, and λ is the hyperparameter, which represents the importance of the speech recognition subtask in the model. In the corpus D = (x, y1, y2), their calculation methods are as shown in formulas (8) and (9):

$$L_{st} = \sum_{x,y^1 \in D} \log P(y^1|x; \theta) \tag{8}$$

$$L_{asr} = \sum_{x,y^2 \in D} \log P(y^2|x; \theta) \tag{9}$$

Among them, X is the speech signal, y^1 is the transcribed text in the source language, y^2 is the translated text in the target language, and θ is the parameter of the model. When the model is inferring, the decoder of the speech recognition subtask will no longer be used, and only the decoder of speech translation is used to decode and generate the translation of the target language.

2.3 Detailed Design of English Translator Speech Recognition System

(1) Voice data collection
The first task to be completed before the system design is to collect the phonetic samples of English isolated words. The voice data sample collection of this system is recorded by English professionals. When audio recording, choose a space with less environmental noise as much as possible to reduce the interference factors of voice data.

(2) Voice data preprocessing
This module is the operation of preprocessing the collected voice sample data. First, use the wavread function in the MATLAB system to sample and quantize the voice information. The voice signal is converted into a digital signal, and then highlighted pre, framing, window, and finally endpoint detection.

(3) Feature extraction module
The system will select the frequency head coefficient as a characteristic parameter of speech to extract characteristics. The first step is to preprocess the voice data to obtain valid data; the second step is to perform feature extraction on the valid data and pass the Mel filter The group completes the discrete cosine transform of the voice data, and obtains the cepstrum coefficients and the cepstrum difference coefficients; finally, head coefficients and head difference coefficients are obtained from the 24-dimensional transformation of a characteristic MFCC parameter.

(4) Codebook generation module
In order to ensure the correct degree of recognition of the system, it is necessary to improve the training quality of the best code book. First, the collected initial English speech data is divided into several speech segments and combined into a multidimensional speech data set. Then perform cluster analysis on each data set, calculate the central feature of each voice data segment, and repeat the iteration until the result meets the error range of the recognition accuracy rate. Finally, the obtained central features are combined to generate the best codebook.

3 English Translator Speech Recognition System Test

3.1 Experimental Environment

The experimental operating system uses Windows 10, uses Python as the programming language, and uses the TIMIT voice data set for experiments.

3.2 Experimental Program

(1) Experiment 1
Speech recognition and Chinese-English machine translation: Let 5 testers read Chinese digits 0–9 and good morning 20 times through the microphone to check the recognition and English translation results.

(2) Experiment two
Choose CCMT2019-BSTC Chinese to English speech simultaneous interpretation corpus as the speech translation corpus, select 50 source language sentences, each source language sentence corresponds to 4 reference translations, and use Google online translation, Baidu online translation and the offline translation system of this article For translation source language, BLEU is used as the evaluation index. The higher the score, the better the translation effect, and the full score is 1 point.

4 Speech Recognition and Machine Translation Performance

4.1 Chinese-English Translation Results

The results of speech recognition and Chinese-English machine translation are shown in Table 1: The overall recognition rate of the system is 89.45%, and the translation accuracy rate is 91.64%.

From thc results of Fig. 1, we can see that the accuracy of the system is high, the recognition of English isolated words is good, and the input sentence can be translated more accurately, which meets the requirements of the English speech translation system in specific situations.

Table 1. Chinese-English translation results

Input voice	Average recognition rate	Translation accuracy
0	85	88
1	92	92
2	89	87
3	93	93
4	90	90
5	87	92
6	83	93
7	89	94
8	90	94
9	94	93
Good morning	92	92
average	89.45	91.64

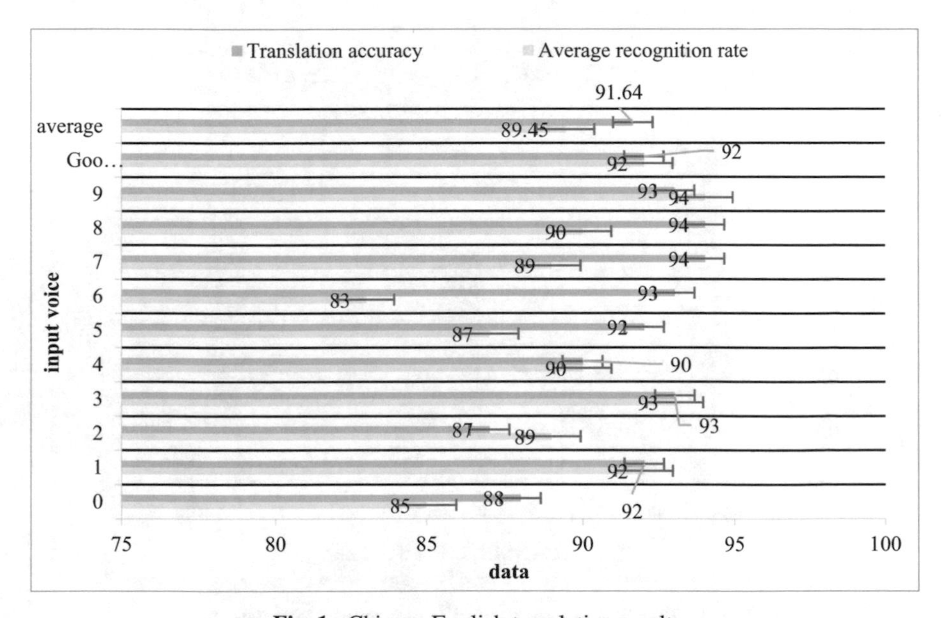

Fig. 1. Chinese-English translation results

4.2 Translation Results of Different Translation Systems

The translation test results of the three speech translation systems are shown in Table 2: The BLEU values of Google Translate and Baidu Translate are 0.5772 and 0.4417, and the error ratcs are 8.1% and 14.6%.

Table 2. Translation results of different translation systems

Translation system	BLEU	Error rate
Google translate	0.5772	8.1%
Baidu translator	0.4417	14.6%
This article	0.5143	8.4%

It can be seen from the translation results in Fig. 2 that the translation of individual sentences by different translation platforms in the online environment is slightly different, but the overall translation effect is almost the same. The BLEU score in the table objectively reflects the translation effect of each translation system. The BLEU value of the English translation system in this article is 0.5143, and the error rate is 8.4%, which has a certain gap with Google translation. Mainly because of the limitations of computer hardware and other aspects, it is difficult to train large-scale corpora. However, the translation effect has a lot to do with the scale of the corpus. Therefore, in the future research, we will upgrade the hardware equipment, increase the scale of the corpus, and achieve better translation results through continuous training and optimization.

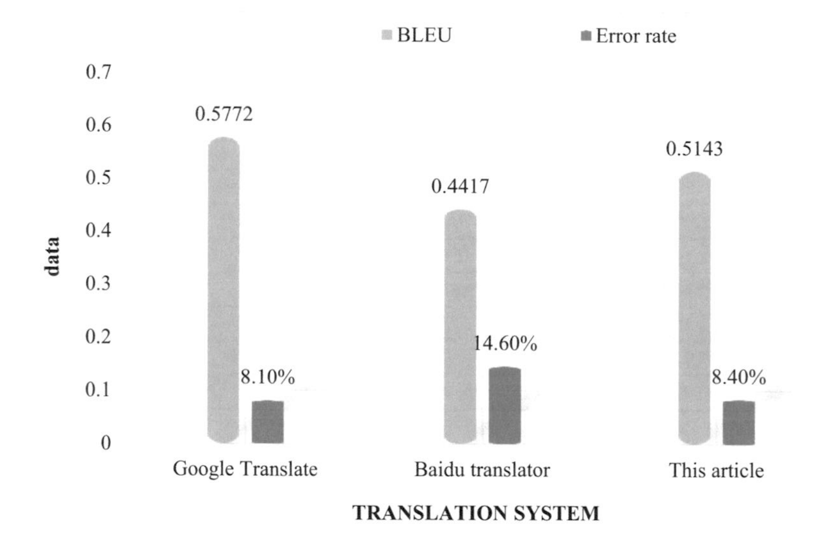

Fig. 2. Translation results of different translation systems

5 Conclusions

The era of big data is accompanied by massive amounts of user data. If this data is not used, it will be a stagnant water. How to find the interconnection between data in these data, data mining technology came into being. At the same time, speech translation technology, as a research hotspot in the field of artificial intelligence, can to a large extent solve the problem of communication barriers between people of different races. Therefore, it is very necessary to study an English translator speech recognition system based on data mining technology. This article mainly starts from three aspects of speech recognition technology, data mining technology and machine translation technology, completes the research of the final speech translation system, and expands the application of data mining technology.

References

1. Karimi, M.B., Isazadeh, A., Rahmani, A.M.: QoS-aware service composition in cloud computing using data mining techniques and genetic algorithm. J. Supercomputing **73**(4), 1–29 (2017)
2. Giczela-Pastwa, J.: Developing phraseological competence in L2 legal translator trainees: a proposal of a data mining technique applied in translation from an LLD into ELF. Interpreter Translator Trainer **1**, 1–18 (2021)
3. Sethi, K.K., Ramesh, D.: HFIM: a Spark-based hybrid frequent itemset mining algorithm for big data processing. J. Supercomput. **73**(8), 1–17 (2017)
4. Wood, D.A.: Net Ecosystem Carbon Exchange Prediction And Insightful Data Mining With An Optimized Data-Matching Algorithm. Ecol. Ind. **124**(82), 107426 (2021)
5. Krivov, M.V., Ulyanova, A.V.: Application of the built-in translator of scenarios in the automated training system of training complexes. Modern Technol. Sci. Technol. Progress **1**(1), 150–151 (2019)
6. Bharara, S., Sabitha, S., Bansal, A.: Application of learning analytics using clustering data mining for students' disposition analysis. Educ. Inf. Technol. **23**(2), 957–984 (2018)
7. Chiba, K., Nakata, M.: From extraction to generation of design information - paradigm shift in data mining via evolutionary learning classifier system. Procedia Comput. Sci. **108**, 1662–1671 (2017)
8. Kundu, S., Garg, M.L.: Analysis, design and implementation of multi agent system in web data mining. Int. J. Future Gener. Commun. Networking **10**(12), 59–70 (2017)
9. Bhardwaj, V., Kukreja, V.: Effect of pitch enhancement in punjabi children's speech recognition system under disparate acoustic conditions. Appl. Acoust. **177**(3), 107918 (2021)
10. Dorothy, N.D., Adrienne, R., Neuman, A.C.: Speech recognition in nonnative versus native english-speaking college students in a virtual classroom. J. Am. Acad. Audiol. **28**(5), 404–414 (2017)
11. Motlaq, M., Sepora, T.: The attrition of common and distinctive english words in translator trainers - published in the cypriot journal of educational sciences (CJES). Cypriot J. Educ. Sci. **15**(3), 575–586 (2020)
12. Kwon, H., Kim, Y., Yoon, H., et al.: Selective audio adversarial example in evasion attack on speech recognition system. IEEE Trans. Inf. Forensics Secur. **PP**(99), 1 (2019)

The Construction of Dynamic Model of College English Blended Learning Based on Informational Data Analysis

Jianhua Yu(✉)

Shandong Transport Vocational College, Weifang, Shandong, China
cassie1217@126.com

Abstract. With the development of the Internet, the education sector has realized resource sharing and information interaction, which has promoted the pace of modern education into the information age. The development of the Internet has opened up an online learning model in the education field, and has also created a foundation for the subsequent dynamic model of blended learning. In simple terms, blended learning is to combine traditional education with Internet education, and promote student-oriented teaching ideas. The dynamic mode of college English blended learning constructed in this article integrates traditional offline teaching and Internet online teaching, and analyzes the changes in students' English scores after using this mode of teaching. From the research results, it can be seen that the failure rate of English scores and the passing rate the increase in the student's average grade reflects a significant improvement.

Keywords: Information-based data analysis · College English · Blended Learning Dynamic Model · Online learning

1 Introduction

The traditional English teaching model has no breakthrough in improving students' academic performance, and the research hotspots seeking breakthroughs in teaching under the background of the information age make blended learning enter the educational vision. Under the guidance of the technological age, various interactive learning software provides a platform for the innovation of new educational concepts. The blended learning mode is the learning mode most suitable for modern teaching discovered in the process of educators' continuous reform of teaching thinking.

So far, many scholars have conducted research on the construction of the dynamic model of college English blended learning in informationized data analysis, and some results have been achieved. For example, a scholar believes that traditional classroom teaching does not focus on cultivating students' creative thinking, students passively accept knowledge, and lack of interaction between teachers and students prevents academic discussions on the same platform. Modern education based on online learning breaks communication barriers and eliminates teachers. Student communication barriers have established a strong learning environment for students [1]. A university uses

J. H. Abawajy et al. (Eds.): ICATCI 2022, LNDECT 169, pp. 426–433, 2023.
https://doi.org/10.1007/978-3-031-28893-7_51

computer network technology to carry out online English teaching, providing students with easy-to-use and rich learning resources. Students can choose their favorite English teachers to teach, and fully mobilize students' learning initiative. Teachers also develop personalized teaching plans for students, to achieve educational equality [2]. Although there have been significant results in the study of English blended learning models, in the face of the current situation of college English teaching that teachers are the main body and students are the learning objects, blended teaching can change this teaching situation and make English teaching more active and overall sexual play out.

This article explains the advantages of the dynamic mode of college English blended learning, introduces several common blended learning models, and proposes a blended learning method suitable for English teaching based on the principles of the model construction. The application of this model in a college English major makes students' English scores have been significantly improved, greatly reducing the probability of failing grades.

2 Construction of a Dynamic Model of English Blended Learning

2.1 The Main Advantages of the Blended Learning Model

(1) The complementary advantages of classroom teaching and online learning

Blended learning combines the advantages of classroom teaching and online learning. In classroom teaching, students acquire knowledge under the influence of teachers and teachers, but multiple students teach in one classroom. This kind of collective teaching focuses on the student community. The uniqueness of students is easily overlooked, and each student responds very much. It is difficult to focus on the individual needs of students, which is not conducive to the development of students' individualization, and the teaching time is limited. It is necessary to complete the teaching content within the predetermined time and give students less time to think about problems. Online learning can provide students with personalized guidance, and teachers can do a good job of teaching planning. In general, blended learning maximizes strengths and avoids weaknesses, and achieves the best learning goals to the maximum extent [3].

(2) Combination of collective teaching, independent learning and collaborative learning

Collective learning originates from traditional school teaching, and its teaching method is mainly by teachers passing knowledge to students. Teachers are the teaching center. Students blindly accept knowledge instillation and are in a state of passive learning. Blended learning is a universal learning method. It not only teaches knowledge, but also teaches students what to do and how to do it, so that students can take the initiative to learn and finally master the learning skills. Collaborative learning in blended learning tends to use teaching methods such as research, discovery, and anchoring, so that students change from passive acceptance to active and effective teaching [4]. Group learning is suitable for students with no learning experience, which can ensure faster and more efficient knowledge transfer, and effectively achieve learning effects and performance learning. Autonomous learning is a self-centered learning activity that can stimulate students' enthusiasm for learning and allow students to complete their learning tasks independently. In addition, students can choose a convenient time and place to study, with full autonomy [5].

(3) Diversity of teaching resources

Blended learning can provide students with rich and diverse learning opportunities. In school teaching, students can acquire knowledge through different types of resource channels, while being exposed to English practice. Slides and micro videos are also widely used in school teaching. The online learning environment can provide students with more resources, and students can directly obtain teaching materials through related research or through short videos and courses [6]. In addition, according to the characteristics and needs of the school, teaching resources that meet the needs of the school can be developed. A variety of teaching skills can enable students to improve their academic performance with keen experience, cognition and positive and enthusiastic motivation [7].

(4) Diversity of interaction

Blended learning provides a variety of interactive teaching courses for teachers and students. Students can communicate with the teacher face to face, or communicate with the teacher through the Internet, such as Internet platforms or group social programs. A variety of interactive programs can help students solve most of the problems and puzzles in the learning process, and can promote a better understanding of students and their learning effects [8].

2.2 Construction Principles of Blended Learning Model

(1) Goal orientation

The hybrid learning system is used to optimize learning performance. Pay attention to the goal-oriented role in blended learning. Therefore, blended teaching does not follow the tendency of provocation and abuse, and the recommended teaching should be served from beginning to end, and the most suitable learning method should be selected according to the blending. Goal orientation should run through the entire learning process. In each teaching link, the students' ability to receive education and the goals that should be achieved should be clarified. The realization of goals should be the most important evaluation [9]. In addition, teachers should bring these goals to students in a timely and accurate manner through explanations, demonstrations, network reminders, and communication, so that students know what they are doing.

(2) Integrity

Although from a macroeconomic point of view, the mixed curriculum model includes two completely different dogmatic models: the school teaching model and the online learning model. The two are not mutually exclusive or unrelated, nor are they merely a 1 + 1 integration. However, with the support of learning theory, the substantive acceptance of constructivist learning theory enables students to organically integrate and review the important learning factors of the entire learning process. This integrity should include the unity of learning, systematic learning, diversified educational forms and the integrity of knowledge structure [10].

(3) Flexibility

Constructivist learning theory is one of the theoretical foundations of blended learning, emphasizing the essential learning process of students in the construction of knowledge. In terms of learning skills, learning is no longer the only source of knowledge acceptance for students. Abundant resource networks, textbooks carefully prepared by teachers or resources discovered and shared by students on the Internet make them highly efficient, flexible, diverse and full of learning vitality. At the same time, with the support of online platforms or social media, teachers can always understand the current level of students, and conduct dynamic assessments and evaluations of course learning effects after the completion of the teaching. In summary, the use of the Internet can help teachers guide, inspire and control the learning process of students, allowing students to learn and construct knowledge independently and improve their learning skills [11].

2.3 Several Dynamic Modes of English Blended Learning

(1) Hybrid learning dynamic mode based on MOOC

In this mode, front-line teachers formulate learning goals, learning skills and teaching content in the first stage of teaching, and then plan school learning activities, allowing students to watch MOOC teaching videos in advance before class. Teachers use online learning when teaching in class. The advantage is for students to answer their doubts and make a summary evaluation of the teaching process after learning.

(2) Hybrid learning dynamic mode based on SPOC

This model still requires pre-teaching analysis, but unlike MOOC teaching, SPOC teaching analyzes learning content, learning environment and learning resources. For the entire teaching process, it is necessary to design knowledge transfer before, during and after class, and then form it. Summative evaluation of teaching.

(3) Dynamic model of blended learning based on flipped classroom

Flipped classroom is to allow interactive communication between teachers and students in the classroom. This model allows students to study and preview courses online before class, and through offline teaching during class. Teachers instruct students to digest the knowledge of this lesson and consolidate it after class.

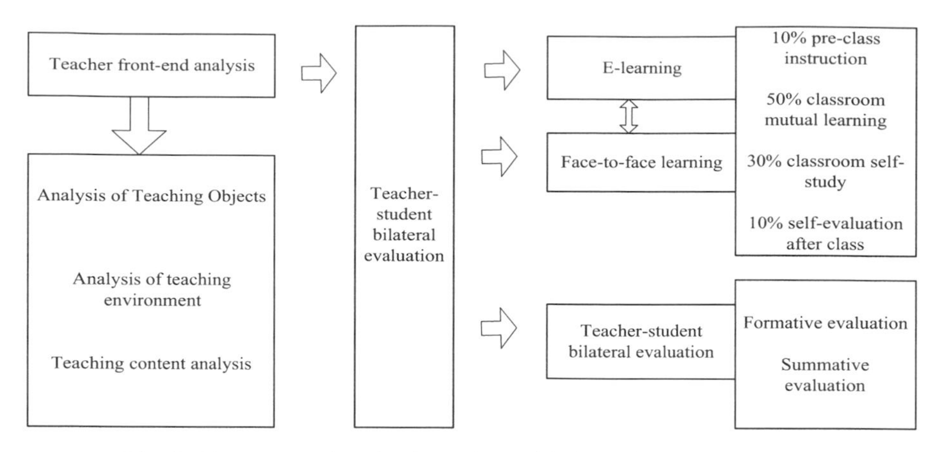

Fig. 1. The construction of a dynamic model of English blended learning

As shown in Fig. 1, the dynamic mode of English blended learning constructed in this article, which combines the characteristics of the above three modes, integrates online learning, face-to-face learning, and teacher-student bilateral evaluation, and incorporates 10% of pre-class instruction, 50% of classroom mutual learning, 30% of classroom self-study and 10% of after-class self-evaluation. The self-evaluation teachers and students both conduct formative and summative evaluations of the teaching process [12].

2.4 Information Data Analysis Algorithm

The data resource search success rate is related to the expected value of the total number of information, data information coverage, etc. Assuming that the total number of data remains unchanged during a data information search process, the data search success rate and the expected value of the total number of messages can be expressed as:

$$\text{Success}_i = 1 - (1 - r_i)^{K \bullet TTL} \tag{1}$$

$$\text{E}(O_i) = \frac{K}{r_i}\left(1 - (1 - r_i)^{TTL}\right) \tag{2}$$

Among them, r_i is the data information coverage rate, $\text{E}(O_i)$ is the expected value of the total number of messages, TTL is the query information value, and K is the information parameter.

3 Research on English Blended Teaching

3.1 Research Purpose

The blended learning model takes learners as the main role, and teachers play a leading role. The implementation of this model stimulates learners' enthusiasm for learning, and greatly improves academic performance, moral education, etc., so that most students and teachers can Accept this model.

3.2 Research Methods

This article takes teachers and students of an English major in a university as the research object. First, the students and teachers will be issued a questionnaire satisfaction survey aimed at improving students' learning autonomy, learning interaction, diversity of teaching resources, and flexibility of teaching in the dynamic teaching model of blended learning. After the statistical survey results, the informationized data is used to analyze the satisfaction results, and then mixed English teaching is implemented for English majors. The monthly test is conducted every month, and the results of mixed English teaching are tested by comparing the monthly test scores.

4 Analysis of Research Results

4.1 Analysis of Satisfaction Survey Results

Table 1. Satisfaction with the dynamic model of blended learning (%)

	Learning autonomy	Learning interactivity	Resource diversity	Teaching flexibility
Teacher's attitude	87	93	96	88
Student's attitude	92	91	95	85

Table 1 shows the results of the survey of satisfaction with the dynamic mode of English blended learning among teachers and students of this school. Among them, teachers and students are 87% and 92% satisfied with the blended English teaching model to improve students' learning autonomy, 93% and 91% are satisfied with the improvement of learning interactivity, and are satisfied with the diversity of teaching resources. They were 96% and 95%, respectively, and the satisfaction with teaching flexibility was 88% and 85%, respectively. From the above data, it can be seen that teachers and students have a higher degree of satisfaction in all aspects of the English blended learning model, indicating that this model is helpful for college English learning.

4.2 Analysis of English Learning Performance

Figure 2 shows the students' monthly test scores after applying the mixed teaching model for a period of English teaching. From the data in the figure, it can be seen that the failing rate of the monthly test is gradually decreasing from the first to the fifth. The passing rate, good rate and excellent rate are on the rise. From the drop in the failing rate, it can be seen that the longer the learning cycle of the blended teaching, the greater the drop in the failing rate, indicating that the longer the blended teaching is used, the faster the student's academic performance will rise. The scores of most students are

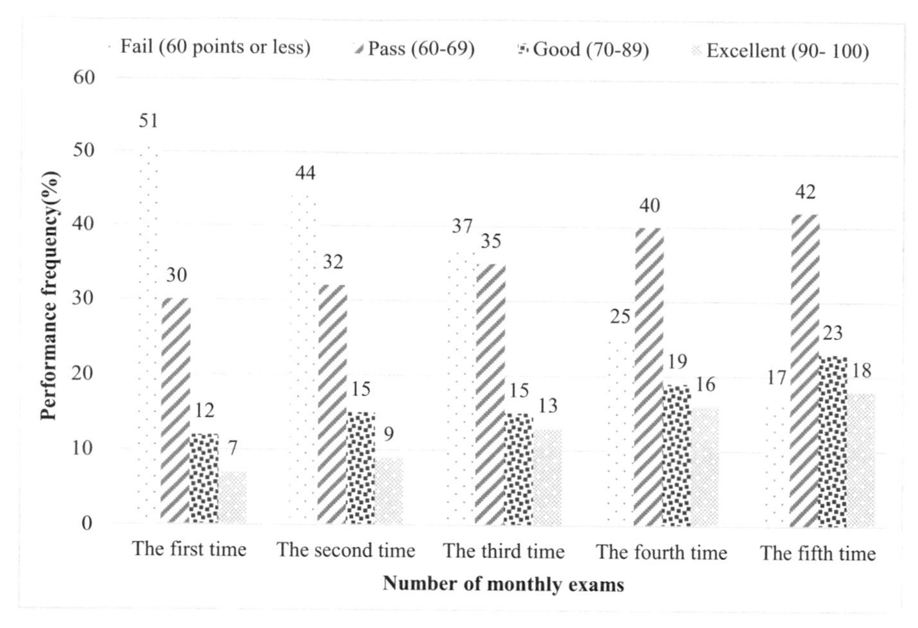

Fig. 2. English performance distribution under the dynamic mode of blended learning (%)

concentrated in the range of 60–89, which means that the pass rate and the good rate are the most students, the English scores are gradually improved, and the teaching is developing towards a good side.

5 Conclusion

When students enter the university, they may ignore English learning and think that English is no longer the focus of learning. The main reason is that students lack interest in English learning, and colleges and universities have not spent energy on English education thinking about how to make students have the idea of active learning. However, the dynamic model of blended learning follows the student-centered concept of teaching and adapts to the characteristics of modern education, allowing students to relax and immerse themselves in the process of English learning, improving their academic performance while having learning fun.

References

1. Bamoallem, B., Altarteer, S.: Remote emergency learning during Covid-19 and its impact on university students perception of blended learning in KSA. Educ. Inf. Technol. **27**(1), 157–179 (2022)
2. Ibrahim, F., Padilla-Valdez, N., Rosli, U.K.: Hub-and-spokes practices of blended learning: trajectories of emergency remote teaching in Brunei Darussalam. Educ. Inf. Technol. **27**(1), 525–549 (2022)

3. Stanislaus, I.: Forming digital shepherds of the church: evaluating participation and satisfaction of blended learning course on communication theology. Interact. Technol. Smart Educ. **19**(1), 58–74 (2022)
4. Ameloot, E., Rotsaert, T., Schellens, T.: The supporting role of learning analytics for a blended learning environment: exploring students' perceptions and the impact on relatedness. J. Comput. Assist. Learn. **38**(1), 90–102 (2022)
5. Li, D.: A study on the construction of college English project-based learning platform in the context of information. Boletin Tecnico/Tech. Bull. **55**(11), 540–547 (2017)
6. Martínez, S., Guíñez, F., Zamora, R., Bustos, S., Rodríguez, B.: On the instructional model of a blended learning program for developing mathematical knowledge for teaching. ZDM Math. Educ. **52**(5), 877–891 (2020). https://doi.org/10.1007/s11858-020-01152-y
7. Kumar, A., et al.: Blended learning tools and practices: a comprehensive analysis. IEEE Access **9**, 85151–85197 (2021)
8. Hamad, M.M.: Pros & Cons of using blackboard collaborate for blended learning on students learning outcomes. High. Educ. Stud. **7**(2), 7 (2017)
9. Hadjiandreou, M., Conejeros, R., Vassiliadis, V.S.: Towards a long-term model construction for the dynamic simulation of HIV infection. Math. Biosci. Eng. **4**(3), 489–504 (2017)
10. Kim, E., Cho, M.: Equivalent model construction for a non-linear dynamic system based on an element-wise stiffness evaluation procedure and reduced analysis of the equivalent system. Comput. Mech. **60**(1), 1–16 (2017)
11. Viktorova, M., Hentschke, R., Fleck, F., et al.: Mesoscopic model for the simulation of dynamic mechanical properties of filled elastomers: model construction and parameterization. ACS Appl. Polymer Mater. **2**(12), 5521–5532 (2020)
12. Castellaro, S., Russo, S.: Dynamic characterization of an all-FRP pultruded construction. Compos. Struct. **218**(JUN.), 1–14 (2019)

Construction of Chinese Knowledge Graph Based on Multiple Data Mining Algorithms

Xiaohong Li(✉)

Beihua University Teacher's College, Jilin 132013, Jilin, China
jilinxue2021@126.com

Abstract. The importance of language research is self-evident. In modern society, people have higher requirements for language learning. In order to enable information and data to serve humanity and national construction faster and better. How to dig out the knowledge graph from the text is one of the topics worthy of consideration and in-depth discussion. This article will study this through a variety of data mining algorithms. This article mainly uses experimental methods and examples to conduct sampling experiments on the established language knowledge graph. Through data analysis, data mining is carried out on the text, and the accuracy rate and recall rate are obtained. The experimental results show that the accuracy of map sampling is above 90%, and the design can meet the basic requirements.

Keywords: Multiple data mining · Mining algorithm · Language knowledge · Map construction

1 Introduction

In the Internet age, the development of computer technology has become an indispensable and indispensable part of human life. With the continuous acceleration of the process of social informatization, various application software such as network digitization and massive information processing tools have also been popularized. As one of the most important carriers of Chinese cultural transmission, Chinese has a strong attraction.

There are many theories about knowledge graphs and data mining algorithms. For example, Some scholars uses CiteSpace, a new information visualization technology, to perform keyword co-occurrence analysis, keyword mutation detection, and cluster analysis to reflect scientific knowledge in education and English vocabulary research [1]. Some scholars uses techniques such as entity recognition, relationship extraction, and visual analysis to create a medical knowledge graph, which can realize systematic and efficient knowledge retrieval, organization and management [2]. Some scholars uses data mining technology to extract and calculate the relationship between scientific research entities, create a distributed index based on the entity knowledge graph, and realize multi-dimensional knowledge retrieval and presentation and related navigation [3]. Therefore, this article also uses a variety of data mining methods to construct the Chinese knowledge graph, hoping to design a system that meets the requirements.

J. H. Abawajy et al. (Eds.): ICATCI 2022, LNDECT 169, pp. 434–442, 2023.
https://doi.org/10.1007/978-3-031-28893-7_52

This article first studies the application of data mining methods in knowledge graphs. Secondly, it analyzes and researches the construction of Chinese knowledge graph. Then the related technology of knowledge graph is studied. After that, the key technology of knowledge reasoning is analyzed. Finally, a sampling experiment was carried out and the data results were obtained.

2 Chinese Knowledge Graph Construction Based on Multiple Data Mining Algorithms

2.1 Application of Data Mining Methods in Knowledge Graph

Entity matching, as one of the core technologies in the data integration and cleaning process, has different performances in different application fields. The input data for the entity matching process are two different knowledge base data and labeled training data. The entity matching calculation is carried out under the control of some manual intervention, and the entity matching is judged according to the result of the entity matching calculation. Measuring the similarity of features generally requires multiple similarity functions, which is called feature comparison based on similarity functions. There are two similarity functions commonly used to calculate the similarity of long text strings, one is the similarity function based on token, and the other is the similarity function based on distance for editing [4, 5].

The language knowledge graph program conceived in this paper combines the advantages of various tools, displays the development status of the subject in a multi-dimensional and accurate way, and realizes the verification of several methods for specific problems. Methods can be divided into vertical cross integration and horizontal integration. The first type of permutation and combination aims at the visualized association types and methods of specific knowledge units, and integrates the relationships between knowledge units and their processing methods, such as method analysis and grouping of co-words, co-occurrences, and co-citation relationships (reflecting the macro structure). The strategic coordinate method (reflecting meaning) and the social network visualization method (reflecting micro-connections) are cross-integrated. The latter focuses on the visual relationship of the knowledge units involved in the discipline, presenting the development of the discipline structure, research focus, and professional knowledge system respectively, with multiple fusions and a single source. A variety of visualization methods to establish relationships (based on keyword grouping, multi-dimensional scales, emergent words, time series grouping, etc.), reveal the maturity and core of the topic, analyze and reflect the microstructure of the external and topic [6, 7].

The basic principle of data mining is to summarize and sort out a large amount of original text information, and extract the implicit connections among them. According to the characteristics of things, they can be divided into two parts: "class" and "group"; then the classification standards, methods and rules are determined according to specific issues. Finally, according to the needs, further refine the object or target attributes and form a new data type. Choosing the corresponding mining strategy from the mining algorithm to achieve the purpose of prediction is a complex system process, which involves the mutual restriction of many factors [8, 9].

2.2 Construction of Chinese Knowledge Graph

The knowledge graph is a dynamic, continuous, and comprehensive data set, which can be automatically generated according to the user's input and output information, has a certain regularity, and can be understood and remembered by humans. So we define it as a time series or a space series. Chinese Knowledge Graph is a new type of data mining method based on text, multimedia and network. It uses computer technology to extract and extract a large amount of topic information to discover the useful content hidden in it. The Chinese Knowledge Graph is a dynamic and open data set, with a topic or time as a unit, covering various topics in the entire teaching process. When constructing Chinese, we should pay attention to the following issues: The same object type has polymorphism and co-structure [10, 11].

It has the following advantages: high recovery efficiency, implicit relational knowledge can be obtained, and multiple exploration modes are provided. It can have advantages in text display and semantic expression skills. First, conduct a demand survey in the corpus to determine its type and quantity. Secondly, according to the specific requirements of technical information and users, a text classification model is created, and appropriate methods are proposed to realize the output function of model prediction, semantic retrieval and recommendation results [1].

When creating a language resource library, determine the subject based on the text content, word count, period and other information. Then, according to the main characteristics of related concepts and the article types involved in the text, a corresponding class structure model is created, and each sub-library is sorted. Before establishing a complete language resource database, it is necessary to make statistics on the subject, type and author of the text, and use it as the basic data.

The traditional web organizes web pages through hyperlinks between web pages, while ignoring the rich semantic relationship between web content, while the knowledge graph classifies web content in a structured manner and discovers the diversity of web content. At present, the construction of knowledge graph mainly uses information extraction, fusion and other technologies to combine these data into structured data to generate a semantic network. Different from traditional search engines, the search engine query based on the knowledge graph not only returns the content of the keyword query, but also returns the search results of other relevant information returned after understanding the user's intention at the semantic level. This improves the quality of the search.

The knowledge graph modifies the existing method of finding information. Thanks to the knowledge graph, the network can be transformed from a network link to a conceptual link. Users can search by subject, not just string. The search engine based on the knowledge graph can understand the user's intention, use the structured knowledge graph of the classification standard to realize accurate positioning and in-depth knowledge acquisition, and realize the real semantic retrieval. Building a knowledge map is a huge project, involving technologies such as knowledge extraction, knowledge representation, knowledge fusion, knowledge review, and knowledge verification. Each technology is a research hotspot.

Data source: The knowledge graph is organized by extracting valuable information from the Internet.

Knowledge acquisition: There are two main methods: one is knowledge extracted from the outside world, and the other is intermediate knowledge generated by reasoning.

Knowledge representation: The main method is to express knowledge as triples based on the representation of the complex network graph, and transform the entities and relationships in the knowledge graph into low-dimensional vector representations.

Knowledge fusion: Knowledge fusion needs to unify data into a unified framework and standard through entity matching, knowledge processing and knowledge update, and then store it in the knowledge graph, including entity binding, entity disambiguation and other technologies.

Knowledge reasoning: Knowledge reasoning is to extract hidden knowledge on the basis of the existing knowledge graph, and use the known knowledge stock to derive potential but unknown knowledge to supplement and perfect the knowledge graph.

Knowledge verification: The reason for this inconsistency may be the poor accuracy of the original new knowledge or the unclear expression of the new knowledge.

The basic framework of knowledge graph construction is shown in Fig. 1:

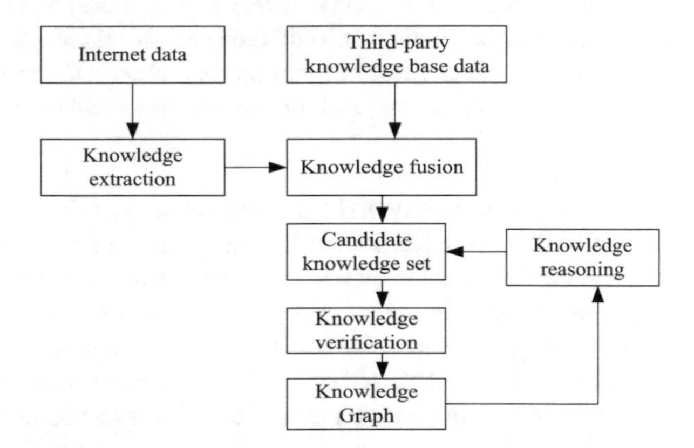

Fig. 1. The basic framework of knowledge graph construction

The structure of the knowledge graph consists of two parts: logical structure and system structure. The logical structure represents the form of data storage in the knowledge graph, which can be divided into a model layer and a data layer. The content of the data layer is mainly composed of knowledge and several facts. Triples are a common way to present knowledge and facts in a knowledge graph.

The related technologies of knowledge graph include the following aspects. The main task of entity extraction is to automatically identify named entities from the original text corpus. The earliest identification of named entities focused on certain fixed and unique information, such as names of persons, organization names, percentages, time and place names. Relation extraction is a natural language processing technology that automatically detects the factual relationship between two entity concepts. Attribute extraction refers to extracting multi-directional description information of the same thing from the same or different data sources.

2.3 The Key Technology of Knowledge Reasoning

(1) Pathfinding strategy

Breadth-first search was originally used to find the shortest path through the maze. The algorithm for traversing the graph width first is similar to traversing the tree width first. The difference is that the graph can contain cycles so that the same node can be visited multiple times. In order to ensure that the adjacent corners of the corners visited first are also preferentially accessed, a queue (first in, first out) is required to record the order of the visited corners, so that the adjacent corners of these corners can be visited in the order of leaving the queue The first breadth-first search can find the shortest path from the start point to the end point.

(2) Random walk search strategy

There are two methods for migration research. One of them is the point verification method, in which the coordinates and probabilities of all walking steps and actions are determined. It can only be used when the total number of steps is small, and it is difficult to use a large number of calculations when the total number of steps is large. The Monte Carlo method can effectively solve the arithmetic problems in the walking process, and the Monte Carlo method samples the walking process in several steps. In principle, the accuracy of the approximate solution can be achieved by increasing the number of samples, and this method has a wide range of roaming functions.

(3) Reasoning algorithm

Graph-based thinking is mainly divided into three stages: One is path recognition, which is to find the connected path of the entity pair under the target relationship. The second is the feature selector, which converts path expressions into feature expressions. Feature engineering is an important step in transforming graph data into machine learning, and it is also the research focus of algorithm optimization. The third is to calculate the weight. The classification model is trained by machine learning, and the weight of the path feature is calculated to obtain the inference model.

1) PRA algorithm

The PRA algorithm involves finding the edges of the node labels in the graph to calculate the feature matrix and predict the relationship of the graph. This method has strong reasoning logic.

Given a set of paths, calculate the path distribution of each connecting node t to f as the characteristic path value of node f, and use the machine learning classification model to train the weight value of the path characteristic. The node can be defined by a linear model and can be sorted:

$$\ell_1 g_{t,p_1}(f) + \ell_2 g_{t,p_2}(f) + \cdots + \ell_m g_{t,p_m}(f) \tag{1}$$

which ℓ_m represents the weight of the path.

2) SFE algorithm

For a given set of node pairs in the graph, the SFE algorithm first performs a local search to get the graph structure around each node. On this basis, a set of feature extraction is performed on the local graph structure, and the path feature vector is obtained from it, thereby obtaining each pair of nodes. Since the SFE algorithm does not calculate the random walk probability distribution value of the path in the feature matrix, which reduces the amount of calculation, the breadth first search can be used to obtain more path features.

(4) Classification model

Classification is the basic and core problem in machine learning and data mining. Logistic regression is mainly used to evaluate the possibility, can be used for classification and regression, and is mainly used to solve two classification problems. Support vector machines can solve the linear inseparability problem of sample data, and are mainly used for binary classification. Through the kernel function, the samples of the two-dimensional planc are mapped to the multi-dimensional space, and the hyperplane that can classify the samples in the multi-dimensional space is found, that is, the decision limit. SVM maximizes the distance between the sample data point closest to the decision limit and the classification decision limit. If the data point is farther from the limit, the classification result is more reliable.

3 User Demand Pattern Mining Method Based on Fusion Graph

3.1 Demand Mode

In the context of user needs, demand patterns refer to common elements in user needs and are the embodiment of community user needs. The demand model is derived from the historical record of user demand, and a large number of demand data sets need to be used for extraction and analysis. In this article, the user demand data set is converted into a demand map set, and the user demand is displayed as a graph. Therefore, the demand model in this article is similar to a common subgraph model. The method of model mining is to extract a large number of demand maps by merging maps. The common pattern, and finally the subgraph pattern of the merged graph, is composed of entities with higher co-occurrence.

3.2 Pattern Mining Method Based on Frequent Graphs

This paper uses probabilistic graph models to study user demand models based on common graphs. The probability graph model introduces the conditional probability based on the graph to reflect the relationship between elements. According to whether the edges of the graph are directed, it can be divided into directed and undirected. Therefore, since the relationship is directional in our demand graph, refer to the directed graph model. The joint probability of the directed probability graph is:

$$G(a_1, a_2, a_3, \ldots, a_m) = \prod_{i=1}^{m} g(a_i \mid \tau(a_i)) \tag{2}$$

where $\tau(a_i)$ is the parent node of a_i.

3.3 Field Term Extraction Experiment

This article conducts experiments to extract domain terms from multiple domains on the website. Randomly extract 500 user demand description texts from each field, split them according to their respective field categories, and use the method of information entropy to filter out the words that meet the conditions. Randomly select 10 texts that participate in the extraction of domain words in each field from the data set, label the domain words in these texts according to the extracted domain vocabulary, and compare them with the manually labeled domain words. The accuracy and recognition value of the extraction algorithm are calculated based on the correct number of algorithm labels, the number of label errors, and the number of unlabeled entries. The experiment was carried out in the Java language under the Windows environment.

4 Field Entity Relationship Prediction Experiment Results

4.1 Experimental Results of Domain Term Extraction

Table 1 shows the results of the extraction experiments in various fields. The comprehensive results of the experiment are precision = 92.75% and recall = 80.25%. This article searches from several language terms such as unit, text, topic, and goal.

Table 1. Domain terms draw the experimental results

	Precision	Recall
Unit	93	74
Text	92	85
Theme	95	87
Target	91	75

As shown in the experimental results in Fig. 2, the accuracy of the extraction results for each entry is relatively balanced, but the memory is quite different. The accuracy rate of each word is significantly higher than the recall rate, and the recall rate is linked to the threshold. The higher the threshold, the fewer keywords are retained, and the higher the accuracy rate, but at the same time some edge keywords are eliminated, resulting in a lower recall rate.

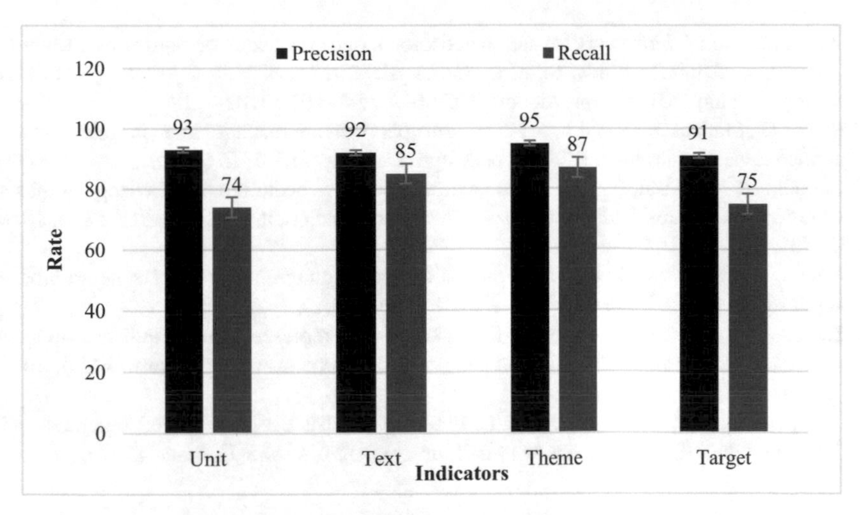

Fig. 2. Domain terms draw the experimental results

5 Conclusion

In modern society, we have entered the information age, and the knowledge graph refers to the processing of data through computers so that it can meet the needs of users. Therefore, how to use a variety of mining techniques to build a Chinese digital text database has become an important issue in the current research field. This article analyzes various mining algorithms and their advantages and disadvantages, and selects appropriate methods based on actual conditions. And established a corpus map and knowledge expression system based on different models. Through a series of technical analysis and finally a systematic experiment, this article concludes that the accuracy of the Chinese term extraction is higher. This is in line with the basic needs of the language knowledge map.

References

1. Sigalov, K., Knig, M.: Recognition of process patterns for BIM-based construction schedules. Adv. Eng. Inform. **33**(Aug.), 456–472 (2017)
2. Martinez-Rodriguez, J.L., Lopez-Arevalo, I., Rios-Alvarado, A.B.: OpenIE-based approach for knowledge graph construction from text. Expert Syst. Appl. **113**(DEC.), 339–355 (2018)
3. Hartmanns, A., Junges, S., Katoen, J.-P., Quatmann, T.: Multi-cost bounded reachability in MDP. In: Beyer, D., Huisman, M. (eds.) TACAS 2018. LNCS, vol. 10806, pp. 320–339. Springer, Cham (2018). https://doi.org/10.1007/978-3-319-89963-3_19
4. Budde, C.E., Dehnert, C., Hahn, E.M., Hartmanns, A., Junges, S., Turrini, A.: JANI: quantitative model and tool interaction. In: Legay, A., Margaria, T. (eds.) TACAS 2017. LNCS, vol. 10206, pp. 151–168. Springer, Heidelberg (2017). https://doi.org/10.1007/978-3-662-54580-5_9
5. Lara-Navarra, P., Falciani, H., Sánchez-Pérez, E.A., et al.: Information management in healthcare and environment: towards an automatic system for fake news detection. Int. J. Environ. Res. Publ. Health **17**(3), 1066 (2020)

6. Weng, H., et al.: A framework for automated knowledge graph construction towards traditional Chinese medicine. In: Siuly, S., et al. (eds.) HIS 2017. LNCS, vol. 10594, pp. 170–181. Springer, Cham (2017). https://doi.org/10.1007/978-3-319-69182-4_18
7. Russo, C., Madani, K., Rinaldi, A.M.: An unsupervised approach for knowledge construction applied to personal robots. IEEE Trans. Cogn. Dev. Syst. **13**, 6–15 (2020)
8. Choudhury, N., Faisal, F., Khushi, M.: Mining temporal evolution of knowledge graphs and genealogical features for literature-based discovery prediction. J. Informetrics **14**(3), 101057 (2020)
9. Koutra, D., Faloutsos, C.: Individual and collective graph mining: principles, algorithms, and applications. Synth. Lect. Data Min. Knowl. Discov. **9**(2), 1–206 (2017)
10. Zangeneh, P., Mccabe, B.: Ontology-based knowledge representation for industrial megaprojects analytics using linked data and the semantic web. Adv. Eng. Inform. **46**(6), 101164 (2020)
11. Boas, H.C., Lyngfelt, B., Torrent, T.T.: Framing constructicography. Lexicographica – Int. Annu. Lexicogr./Internationales Jahrbuch für Lexikographie **35**(1), 41–85 (2019)

Searchable Encryption Scheme Based on Multiple Access Control Authority Attributes

Fanglin An[1,2], Long Su[1,2], Yin Zhang[1,2], and Jun Ye[1,2](✉)

[1] School of Computer Science and Cyberspace Security, Hainan University, Haikou, China
yejun@hainanu.edu.cn

[2] Key Laboratory of Internet Information Retrieval of Hainan Province, Haikou, China

Abstract. Searchable encryption technology searches data without disrupting the user's encrypted ciphertext. Existing searchable encryption technology schemes has too little research on the control of access rights of users. In order to focus on protecting data security, the existing security schemes have to sacrifice data sharing, which loses most of the value of using cloud server (CS) for storage. This paper constructs an encryption scheme that can control access rights. It uses the key to quantifying the data to ensure the security of the data in the cloud. It uses public access control structure and the shared key encryption keyword to realize the query and verification of the data. Unauthorized users can do not get and decrypt the ciphertext data on the CS. Finally, the security proof shows that this scheme can ensure the security of user keywords and trapdoors.

Keywords: Access control · Encryption · Trap-door

1 Introduction

Cloud computing can reduce users' local resources, increase data sharing. Cloud server providers use large-scale serial storage technology and high concurrency and large bandwidth network technology to provide service for users. Users can easily upload the data to provide to the cloud server. In this process, users lose ownership control over the data stored in the cloud. The cloud environment is complex, and the data stored by users is likely to be leaked, which is contrary to the original intention of users to share data within a limited range, at present, the most effective solution is to encrypt the data before uploading it. When the cloud ciphertext data is stored in the cloud service, when the user (here the user may also be another person uploaded by the data user) wants to access the data, let the cloud server search the desired data according to the ciphertext keyword provided. However, in order to guarantee the privacy of the keyword in the user data, the plaintext data query scheme can be utilized according to the general situation, that is, search with the keyword corresponding to the plaintext of the data. In this case, the cloud service also needs to decrypt the file to match. Even the most vigorous cloud server can't query and calculate a large amount of data, and this scheme also loses the original intention of users to encrypt data to protect data security. Considering that CS as a data storage Party, cannot believe that by users, there is a danger of being tracked at any time

J. H. Abawajy et al. (Eds.): ICATCI 2022, LNDECT 169, pp. 443–452, 2023.
https://doi.org/10.1007/978-3-031-28893-7_53

when the relevant privacy information of data is stored in CS. For this subject, searchable technology based on attribute encryption is developing continuously. The attribute encryption scheme can realize the superior authority control of each query user, so the functional requirements of user data security and sharing.

In the existing search encryption scheme for controlling authority, the user encrypts data and send to the server in the open network channel. When the user needs to retrieve the required data, he uses keywords and his own key to generate a trap (with authority characteristics and target data characteristics), then send the trap door data to the CS for verification. CS processes the trap door so that it can be equipped with the generation rule form of keyword ciphertext data in the repository, and then traverse the corresponding data in the repository. Once the trap door is paired successfully, it indicates that the trap door has access to a certified data. The subsequent operation of CS is to sort out the ciphertext data, package it and send it to the requesting user, and make records. The user decrypt the plaintext of the data by use key according to the ciphertext data returned by the CS. The encryption scheme does not have high requirements for users' operation, and a large amount of computing costs to cloud, and server is the party with strong computing power and university data management ability, so it is very suitable for the existing cloud server customer architecture. However, the existing encryption schemes still have some problems: the computational complexity is prohibitive, and the cloud server's control over the calculation results can't be trusted. So far, many encryption schemes with keyword search [1–4] proposed by researchers can not completely control the query results of ciphertext data. Although the privacy of keywords and data itself can be, the generation efficiency of attribute keys in actual situations is too low. These solutions cannot be applied to the urgent need of large-scale permission control in the current Internet environment. In addition to the above problems, the application of encryption technology with access rights in reality also faces: the private key held by the user may only access a certain part of the data, and when the amount of data is huge, the user needs to access the private key more. When the number of private keys is large, the private keys are also easily leaked.

2 Related Work

Andola et al. [1] conducted a research summary: searchable encryption techniques should guarantee the balance between the number of encrypted keywords, search accuracy and search efficiency under the premise of guaranteeing the system to accomplish data searchability and data security functions, and the existing attribute encryption can be deployed with the support of trusted third parties to perform searchable encryption schemes on data, and the existing research directions mainly focus on achieving fuzzy search, fine-grained search and revocable permissions. Gupta et al. [2] deployed the scheme on a blockchain in order to find the ideal trusted third party, and for frequent partial search token generation the task is performed by consensus nodes in the blockchain to try to eliminate the need for central authority, but the resources required to deploy the federated blockchain are even more enormous, resulting in a waste of computational resources. Oya et al. [3] proposed an attack for the security of search patterns, i.e., for private information such as search keywords and user search habits, to collect information by leaking the capacity information calculated for access patterns and the frequency information obtained from search pattern leaks, which poses a greater challenge for the research of searchable cryptography. Senouci et al. [4], in order to describe the attacker's guessing attack on keywords, analyzed the attack probability and proposed a certificate-free searchable encryption scheme with relevant optimization in terms of ciphertext and trapdoor computation, making the scheme secure and efficient, but using a large number of hash operations, which cannot support the subsequent decryption of the ciphertext and cannot resist the attacker and cloud server This operation cannot support the subsequent decryption of the ciphertext and cannot resist the attacker's collusion with the cloud server.

Zhang K et al. [5] proposed that the computational power requirements corresponding to IoT devices in industrial IoT environments should not be too strong and proposed a lightweight multi-keyword search scheme and extended it to multi-authority scenarios to suit the scenario requirements, but the scheme uses a bilinear mapping that imposes a high load on the cloud server. Zhang M et al. [6] proposed a fuzzy search scheme constructed from asymmetric scalar product-protected encryption and Hadamard product operations and used it for machine learning, and experimental results showed that the scheme has high accuracy, but the scheme did not prove the strong security of the matrix computation used by the scheme by performing a rigorous formal proof. Liu et al. [7] proposed a searchable private encryption scheme for protecting data and search patterns in distributed systems and designed a new subset decision mechanism to resist keyword guessing attacks, but the scheme does not have a formal description of the process. Varri et al. [8] analyzed symmetric searchable encryption schemes, public-key searchable encryption schemes, and attribute-based searchable encryption schemes and suggested that future research needs to focus on the key escrow problem. Xu et al. [9] used attribute encryption and number theory principles to construct a multi-client dynamic searchable encrypted medical database, which focused on fine-grained access, but the scheme utilized a large number of power operations, which increased the computational overhead. Lu et al. [10] proposed a pairing-free and privacy-preserving CBEKS scheme with significant advantages in computational performance in order to avoid the use of computationally intensive bilinear pairings, which are incompatible with performance-limited industrial IoT smart devices, which also provided inspiration for our construction scheme.

Shang et al. [11] propose a SSE scheme called OSSE (fuzzy SSE) for hidden patterns, which can be independently fuzzy access patterns for each executed query, but achieves optimal performance only when the keywords are evenly distributed in the dataset. Fan et al. [12] MSIAP, a searchable encryption scheme for multi-keyword subset retrieval, achieves efficient multi-keyword subset retrieval and dynamic update with insignificant information disclosure in smart grid, which is more suitable for scenarios with minimal privacy protection requirements in smart grid. Liang et al. [13] proposed scheme DMSE, which pre-classifies outsourced documents in DMSE and then performs separate schemes and pre-processes them according to their privacy to achieve categorical search, which is useful for improving search efficiency, but the cloud server operation of pre-processing to support keyword updates requires a large computational overhead. Ali et al. [14] proposed a distributed database blockchain based on homomorphism cryptography for secure database search and keyword-based access to provide users with access control policies for privacy and security of patient health data in PHR systems, but this deployment scheme is not strictly secure. Olakanmi et al. [15] developed a lightweight searchable public key encryption scheme for generating and distributing symmetric encryption keys for secure key distribution in order to combat internal keyword guessing attacks, and the certificate free encryption method of this scheme solves the key escrow problem associated with key management. Wang et al. [16] proposed an efficient lattice-based searchable public-key encryption algorithm for traditional hardness assumptions that are vulnerable to quantum attacks, which utilizes the learning with error (LWE) assumption and subset predicate encryption to achieve post-quantum security and highly flexible access control policies for multi-user applications, but the search phase of the scheme requires a high computational complexity. Awais et al. [17] used a trapdoor obfuscation technique to deceive third-party adversaries as a means to secure access patterns, but the scheme not only requires more communication overhead and computational overhead, but also the scheme itself is not described in enough detail to be easily deployed and implemented in practical scenarios. Wang K et al. [18] proposed a C-ITS multi-party encryption sample alignment forward privacy protection scheme AFFIRM using collaborative learning. By introducing searchable encryption method, the sample alignment with collaborative learning in multi-party encrypted data space is achieved, but the scheme does not consider the honesty of cloud servers, and the disadvantage of the scheme when cloud servers are treated as trusted verifiers. Wang H et al. [19] used 0,1 encoding theory to reduce the number of numeric keywords and thus improve the range search mechanism, but the scheme can be implemented to scenarios where the number of keywords increases. Su et al. [20] addressed the problem that entrusted cloud servers may return false search results or results related to some special commercial purposes by implementing a ranked multi-keyword search that returns only the top k documents that best meet the user's needs, but this scheme uses a large number of bilinear operations for searching, which elongates the computation time. Yang et al. [21] proposed an improved certificate free public-key searchable encryption scheme to secure the privacy of multiple trapdoors by outputting the trapdoor output results into the hash function domain, which not only requires more computation but also exposes the privacy information on the cloud server side.

2018, Cui et al. [22]'s scheme satisfies a huge attribute space with dedicated access policy expression, but the scheme cannot achieve complete security. Jiang et al. [23] proposed a scheme to solve the problem of key delegation abuse by constructing a traceable mechanism that satisfies the new security requirements, which prevents malicious attackers from obtaining keys illegally and has strong access policy expressiveness, but the scheme key setting is too cumbersome, making the overall computational overhead too large. The scheme of Zhang Y et al. [24] supports large attribute Spaces. In 2021, Meng et al. [25] introduced the hidden access policy CP-ABE scheme with keyword search function, which constructs a dual access policy based on the hard problem assumption, i.e., one public and the other hidden, making the access policy endowed with stronger expressiveness and compactness, which is suitable for devices with limited resources, but the scheme requires 13 data for trapdoor generation, which increases the computational overhead and communication overhead and does not meet the practical verification requirements.

3 Contributions

Our contribution is to construct a more secure, efficient and reasonable ciphertext retrieval scheme:

Access control technology is combined with searchable encryption technology to classify users, generate access structure by using construction rules, Integrate keyword generation key with attribute key, and search encrypted data on CS through specific keywords.

Ciphertext retrieval technology that supports access control: when large amount of privacy data (such as personal health records, etc.) must be stored on the server in the form of ciphertext through encryption technology, and access control is introduced Technology enhances data security. However, if files are encrypted, their search and use will be seriously affected. Searchable encryption can search encrypted data through specific keywords, but in the current research. It will bring a lot of redundancy to the search results, which increase the user's resource consumption in the decryption process. We will deeply study the retrieval technology of ciphertext data. Design an efficient and practical retrieval plan, and remove the redundancy in the search process, which is one of the main contents of this project.

4 Preliminaries

4.1 Discrete Logarithm Problem

For finite cyclic group $G = \{g^k, k = 0, 1, 2, ...\}$ and its generator g and order $n = |G|$.

a) Given the integer a, it is easy to calculate $g^a = b$;
b) Given element b, it is very difficult to calculate that integer a satisfies $g^a = b$.

4.2 Access Structure

Let u represent users, then the set of users is $U = \{u_1, u_2, ..., u_n\}$.

Let att represent user attributes, and p represents the total number of attributes, then the authorized attribute set is $S = \{att(1), att(2), ..., att(p)\}$.

Access structure: Assuming that the user's attribute set is $P = \{P_1, P_2, ..., P_n\}$, set $A \subseteq 2^{\{P_1,P_2,...,P_n\}}$, choose any B and C, if $B \in A, B \subseteq C$, then $C \in A$, then A is said to be monotonous. $A \subseteq \{P_1, P_2, ..., P_n\} \wedge A \neq \emptyset$, namely $A \subseteq 2^{\{P_1,P_2,...,P_n\}} \big| \{\emptyset\}$.

5 System and Security Model

The access control model is shown in Fig. 1.

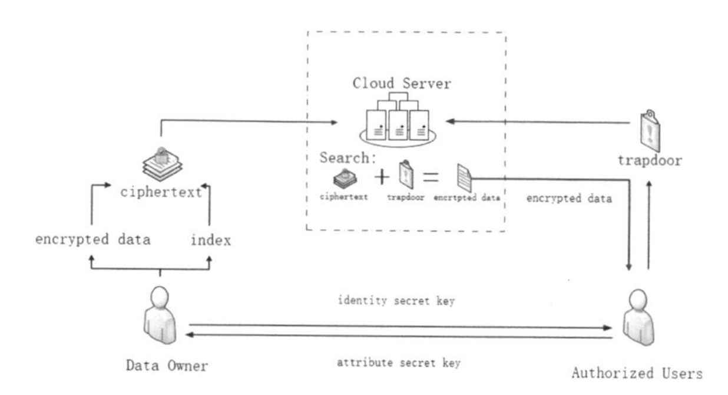

Fig. 1. System model

Data Owner

(A) Define access rights and encrypt data.
(B) The data owner gives access to the attributes of the specified data set and sends it to the access user as a key.

Cloud Server (CS)

When the access user submits a query request, it verifies it according to the access user's trap door.

Access User

Generate trapdoor access data.

6 Scheme Description

6.1 System Initialization

For the smooth operation of the scheme, the system selects a prime p and generates two cyclic multiplication groups G and G_T with p, finite field Z_p, in which the generator of G is g, and selects a bilinear mapping function: $e : G \times G \rightarrow G_T$. Generate random function: 8 and expose parameters:

$$P = <p, g, G, G_T, Z_p, e, F>$$

6.2 Key Generation

In order to obtain their unique identity, each access user selects a random number $y_i \in Z_p$ as their private key $MK_i = y_i$, and calculates $Y_i = e(g, g)^{y_i}$ as their public key $PK_i = Y_i$.

Data owners generate private keys:

a) Select two random numbers $r_1, r_2 \in Z_p$;
b) Use random functions to generate access control structure $F(\prod Y_i) = h$;
c) Generate your own private key $SK_u = g^{r_1 + r_2 \sum_{i=1}^{N} h}$.

6.3 Encryption

Data owners categorize their plaintext data D_i and its corresponding keyword C_i at different levels, and divides them into several levels according to different attributes $att(j)$. The data $D_{i,j}$ is encrypted, and the keyword $C_{i,j}$ of the data is calculated for the corresponding ciphertext data. The index:

$$index_{i,j} = SK_u \prod_{j=1}^{n} e(Y_j, C_{i,j})$$

And will generate an authorization key:

$$ASK_j = \frac{\prod_{j=1}^{n} e(Y_j, C_{i,j})}{e(Y_j, C_{i,j})}$$

6.4 Generated Trap

When an access user wants to request data with the clear text of the query keyword C_i from the cloud server, the following steps are required to calculate the trap door sent to the cloud server for validation, calculate trap $Tr : <T_1, T_2>$ as follows:

$$T_1 = \text{SK}_u \cdot ASK_j, T_2 = e(Y_j, C_{i,j})$$

6.5 Cloud Server Validation

When an access user requests the ciphertext data needed for queries to the cloud server using Trap Gate $Tr : <T_1, T_2>$, the cloud server computes:

$$index'_{i,j} = T_1 \cdot T_2$$

If the cloud server traverses to find the index that meets the criteria: $index'_{i,j} = index_{i,j}$, if the search succeeds, the corresponding ciphertext is sent to the data visitor, and the data clear text obtained by the decryption is executed on the ciphertext with its own private key.

7 Security Analysis

Theorem 1: A malicious attacker can't disclose user data and keyword privacy.

Proof: As long as the attacker is not authorized, even if the trap value is calculated according to the program, the trap cannot pass data validation on the cloud server. Even if an attacker obtains cryptographic data by illegal means (such as colluding with or attacking the cloud server), the decryption program cannot be executed, that is, clear information of the data cannot be obtained.

Theorem 2: A malicious attacker cannot fake a trap that can be verified by a cloud server.

Proof: Assuming that a malicious attacker accidentally acquires the keyword plaintext of a user data and wants to fake a trap to obtain the ciphertext. If the attacker does not have a private key with access, there is no possibility to execute a trap generation program to fake the correct trap value.

8 Conclusion

In order to address the problem that cloud cipher data cannot be shared with complex rights control, this paper presents a public key encryption scheme with access control rights. This scheme does not use a secure channel shared key and eliminates the possibility of key leaking for each user. At the same time, the data owner has a straightforward operation to encrypt the data. As long as the calculation is done according to the specified program, the data he owns can be successfully encrypted into cipher data that can be accessed by a certain range of accessing users. Access user's trap generation program is easy to compute and allows access to cryptographic data within all permission without frequently changing keys. The scheme in this paper can realize the access control and security guarantee of shared data.

Acknowledgements. This work is partially supported by the National Natural Science Foundation of China (62162020), Hainan Province Science and Technology Special Fund (ZDYF2021GXJS216), the Science Project of Hainan University (KYQD(ZR)20021).

References

1. Andola, N., Gahlot, R., Yadav, V.K., et al.: Searchable encryption on the cloud: a survey. J. Supercomput. **78**(7), 9952–9984 (2022)
2. Gupta, B.B., Li, K.C., Leung, V.C.M., et al.: Blockchain-assisted secure fine-grained searchable encryption for a cloud-based healthcare cyber-physical system. IEEE/CAA J. Automatica Sinica **8**(12), 1877–1890 (2021)
3. Oya, S., Kerschbaum, F.: Hiding the access pattern is not enough: exploiting search pattern leakage in searchable encryption. In: 30th USENIX Security Symposium (USENIX Security 2021), pp. 127–142 (2021)
4. Senouci, M.R., Benkhaddra, I., Senouci, A., et al.: An efficient and secure certificateless searchable encryption scheme against keyword guessing attacks. J. Syst. Architect. **119**, 102271 (2021)

5. Zhang, K., Long, J., Wang, X., et al.: Lightweight searchable encryption protocol for industrial internet of things. IEEE Trans. Industr. Inf. **17**(6), 4248–4259 (2020)
6. Zhang, M., Chen, Y., Huang, J.: SE-PPFM: a searchable encryption scheme supporting privacy-preserving fuzzy multikeyword in cloud systems. IEEE Syst. J. **15**(2), 2980–2988 (2020)
7. Liu, X., Yang, G., Susilo, W., et al.: Privacy-preserving multi-keyword searchable encryption for distributed systems. IEEE Trans. Parallel Distrib. Syst. **32**(3), 561–574 (2020)
8. Varri, U., Pasupuleti, S., Kadambari, K.V.: A scoping review of searchable encryption schemes in cloud computing: taxonomy, methods, and recent developments. J. Supercomput. **76**(4), 3013–3042 (2020)
9. Xu, L., Xu, C., Liu, J.K., et al.: Building a dynamic searchable encrypted medical database for multi-client. Inf. Sci. **527**, 394–405 (2020)
10. Lu, Y., Li, J., Wang, F.: Pairing-free certificate-based searchable encryption supporting privacy-preserving keyword search function for IIoTs. IEEE Trans. Industr. Inf. **17**(4), 2696–2706 (2020)
11. Shang, Z., Oya, S., Peter, A., et al.: Obfuscated access and search patterns in searchable encryption. arXiv preprint arXiv:2102.09651 (2021)
12. Fan, K., Chen, Q., Su, R., et al.: MSIAP: a dynamic searchable encryption for privacy-protection on smart grid with cloud-edge-end. IEEE Trans. Cloud Comput. (2021)
13. Liang, Y., Li, Y., Zhang, K., et al.: DMSE: dynamic multi keyword search encryption based on inverted index. J. Syst. Architect. **119**, 102255 (2021)
14. Ali, A., Almaiah, M.A., Hajjej, F., et al.: An industrial IoT-based blockchain-enabled secure searchable encryption approach for healthcare systems using neural network. Sensors **22**(2), 572 (2022)
15. Olakanmi, O.O., Odeyemi, K.O.: A certificateless keyword searchable encryption scheme in multi-user setting for fog-enhanced Industrial Internet of Things. Trans. Emerg. Telecommun. Technol. **33**(4), e4257 (2022)
16. Wang, P., Chen, B., Xiang, T., et al.: Lattice-based public key searchable encryption with fine-grained access control for edge computing. Futur. Gener. Comput. Syst. **127**, 373–383 (2022)
17. Awais, M., Tahir, S., Khan, F., et al.: A novel searchable encryption scheme to reduce the access pattern leakage. Futur. Gener. Comput. Syst. **133**, 338–350 (2022)
18. Wang, K., Chen, C.M., Shojafar, M., et al.: AFFIRM: provably forward privacy for searchable encryption in cooperative intelligent transportation system. IEEE Trans. Intell. Transport. Syst. (2022)
19. Wang, H., Li, Y., Susilo, W., et al.: A fast and flexible attribute-based searchable encryption scheme supporting multi-search mechanism in cloud computing. Comput. Stand. Interfaces **82**, 103635 (2022)
20. Su, J., Zhang, L., Mu, Y.: BA-RMKABSE: blockchain-aided ranked multi-keyword attribute-based searchable encryption with hiding policy for smart health system. Futur. Gener. Comput. Syst. **132**, 299–309 (2022)
21. Yang, G., Guo, J., Han, L., et al.: An improved secure certificateless public-key searchable encryption scheme with multi-trapdoor privacy. Peer-to-Peer Network. Appl. **15**(1), 503–515 (2022)
22. Cui, H., Deng, R.H., Lai, J., et al.: An efficient and expressive ciphertext-policy attribute-based encryption scheme with partially hidden access structures, revisited. Comput. Netw. **133**, 157–165 (2018)
23. Jiang, Y., Susilo, W., Mu, Y., et al.: Ciphertext-policy attribute-based encryption against key-delegation abuse in fog computing. Futur. Gener. Comput. Syst. **78**. 720–729 (2018)

24. Zhang, Y., Zheng, D., Deng, R.H.: Security and privacy in smart health: efficient policy-hiding attribute-based access control. IEEE Internet Things J. **5**(3), 2130–2145 (2018)
25. Meng, F., Cheng, L., Wang, M.: Ciphertext-policy attribute-based encryption with hidden sensitive policy from keyword search techniques in smart city. EURASIP J. Wirel. Commun. Network. **2021**(1), 1–22 (2021)

Design and Implementation of PSO Based on Ternary Optical Computer

Lijun Fang[1], Qiangqiang He[1], Man Ling[1], Jie Zhang[1], Kai Song[2], Zhehe Wang[3], and Xianchao Wang[1](✉)

[1] Fuyang Normal University, Fuyang 236037, Anhui, China
wxcdx@126.com
[2] East China Jiaotong University, Nanchang 330013, Jiangxi, China
[3] Hainan Tropical Ocean University, Sanya 572022, Hainan, China

Abstract. As particle swarm optimization is widely used in communication, electronic system design and other fields, it cannot meet the high real-time requirements. To this end, this paper intends to use the giant bitness and parallelism of the ternary optical computer, and adjust the effect of fixed parameters on particles to make it dynamic adaptive, so as to speed up the convergence speed of the particle and the ability to find the global optimal solution. The simulation results show that, compared with other literature, the convergence speed of the particle swarm optimization algorithm based on ternary optical computers is significantly improved.

Keywords: Particle swarm optimization algorithm · Ternary optical computer · Parallelism · Time complexity

1 Introduction

In Particle Swarm Optimization (PSO), each particle represents a possible solution to the problem and has two properties of position and velocity [1].

However, due to the traditional PSO algorithm, it needs to consume a lot of computing time, which affects its real-time performance. In the literature [2], the fitness value range is determined, and then according to the current calculated fitness value of the particle, it is decided whether to discard the particle. Although this improvement can reduce the amount of computation, it also affects the optimization performance of the objective function.

Therefore, it is necessary to find an algorithm that can reduce the time complexity of PSO while maintaining its performance without loss. This paper utilizes the megabitness and parallelism of Ternary optical computer (TOC) to reduce the time complexity of the algorithm [3, 4].

In this paper, an optimized particle swarm algorithm is designed to verify the effectiveness of the algorithm by using the dynamically adaptive inertia weight and an improved learning factor, which greatly improves the convergence speed of the particles and reduces its time complexity. The contents of each section of this paper are

J. H. Abawajy et al. (Eds.): ICATCI 2022, LNDECT 169, pp. 453–461, 2023.
https://doi.org/10.1007/978-3-031-28893-7_54

distributed as follows: Sect. 2 mainly introduces the development and characteristics of ternary optical computers. Section 3 introduces the PSO algorithm flow and optimization of learning factors. Section 4 introduces the design of the TOC-based particle swarm algorithm and analyzes the time complexity of its implementation. Sections 5 and 6 are mainly based on TOC simulation to realize the parallel operation of particles in three different test functions, which effectively improves the convergence speed of particles.

2 Introduction to Ternary Optical Computer

Although the performance of electronic computers is constantly improving, they still cannot solve practical, complex problems [5]. The main problems lie in the parallel ability, the line time constant problem, and so on. For this reason, it is proposed to replace the electrical signal with an optical signal and realize a certain algorithm with optics.

A ternary optical computer has up to a million processor bits. An optical processor can be divided into several small parts, and each part can work independently of other programs and can be grouped arbitrarily. These characteristics enable massive data parallel computing [3, 6], and its computing power is significantly greater than that of current electronic processors, improving the ability to use computing to study and solve complex problems.In 2017, the first parallel addition operation of MSD numbers was successfully completed at Shanghai University, which was also the time when the ternary optical computer processor was born [4].

The ternary optical computer is based on the background of the current electronic computer. The SD11 gets in touch with the user through a PC, and the included optical devices are invisible to the user. In the resource management software of the ternary optical computer, it can help the user complete the data conversion, and there is no big difference between the user and the PC. But when solving more complex tasks, the ternary optical computer can highlight its advantages. Because of the giant bitness of TOC, it can run hundreds of small data sets at a time, which greatly reduces the amount of calculation compared with electronic computers.

Currently, ternary optical computers have low power consumption for high-performance classical computing applications using ternary optical computer processors for verification. Such as cellular automata, fast Fourier transforms have obtained relatively good results. It can be seen that the current TOC has a parallel implementation of particle swarm optimization.

3 Particle Swarm Optimization Algorithm

This algorithm was mainly proposed by Kennedy and Eberhart in 1995 [7]. In the D-dimensional target space, there are T particles, and it iterates M times, where X_{id} represents the position of the ith particle in the D-dimensional space vector.

$$X_i = (X_{i1}, X_{i2}, \ldots, X_{iD}) \quad i = 1, 2, \ldots, T \tag{1}$$

where V_{id} represents the moving speed of the ith particle in the D-dimensional space vector.

$$V_i = (V_{i1}, V_{i2}, \ldots, V_{iD}) \quad i = 1, 2, \ldots, T \tag{2}$$

where pbest_{id} indicates that the ith particle finds the local optimal position in the D-dimensional space vector.

$$\text{pbest}_i = \left(\text{pbest}_{i1}, \text{pbest}_{i2}, \ldots, \text{pbest}_{iD}\right) \quad i = 1, 2, \ldots, T \tag{3}$$

where gbest_{id} means that the entire particle swarm particle finds the global optimal position in the D-dimensional space vector.

$$\text{gbest}_i = (\text{gbest}_{i1}, \text{gbest}_{i2}, \ldots, \text{gbest}_{iD}) \quad i = 1, 2, \ldots, T \tag{4}$$

According to the above description, the velocity and position of the updated particle i in the d-th dimension can be expressed by a mathematical formula.

$$V_{id}^k = W * V_{id}^{k-1} + c_1 r_1 (pbest_{id}^{k-1} - {X_{id}}^{k-1}) + c_2 r_2 (gbest_{id}^{k-1} - {X_{id}}^{k-1}) \tag{5}$$

$$X_{id}^k = X_{id}^{k-1} + V_{id}^k \tag{6}$$

W is the inertia weight of the particle, and W_Start and W_End are the inertia weight coefficients of the starting point and the ending point.

$$W = W\_Start - W\_End * i/M \tag{7}$$

PSO is mainly divided into the following steps, as shown in Fig. 1:

Input: V max, X max, W: 0.4–0.9, C1, C2: 0.5–1.5

Output: output the optimal value.

Step 1: Randomly initialize the speed and position of the particles.

Step 2: Calculate the fitness value of the particle according to the fitness function and record the position of the local optimal solution of the individual particle and the global optimal solution of the particle group.

Step 3: Compare the position of the local optimal solution of the individual particle and the global optimal solution of the particle group obtained according to Step 2 with the historical record value. If it is better than the historical record value, update the local optimal value of the particle and recalculate the particle's velocity and position.

Step 4: Determine whether the given maximum number of iterations is reached or the global optimal value satisfies the minimum limit, and if not, jump to Step 2.

4 Design of PSO Algorithm Based on TOC

4.1 Design of PSO Based on a TOC

In particle swarm optimization, a large number of particles transmit information to each other through collective cooperation [8], and each particle is independent of each

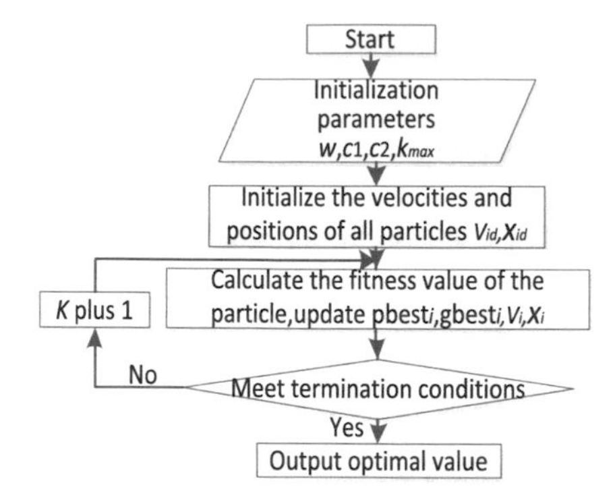

Fig. 1. Flow chart of serial particle swarm.

other in the process of iterative update. The parallelism of particle swarm optimization algorithms is analyzed. (1) To achieve independent computing space for each particle on TOC, allocate the same optical processor as the number of particles. (2) According to the parallelism of TOC, when calculating the fitness value of a particle swarm, the time calculated by several serial particles can be parallelized into the running time of a single particle. (3) Obtain the speed of each particle and the distance between the current position and the local optimal position. (4) Recalculate the particle's global optimal position. As shown in Fig. 2.

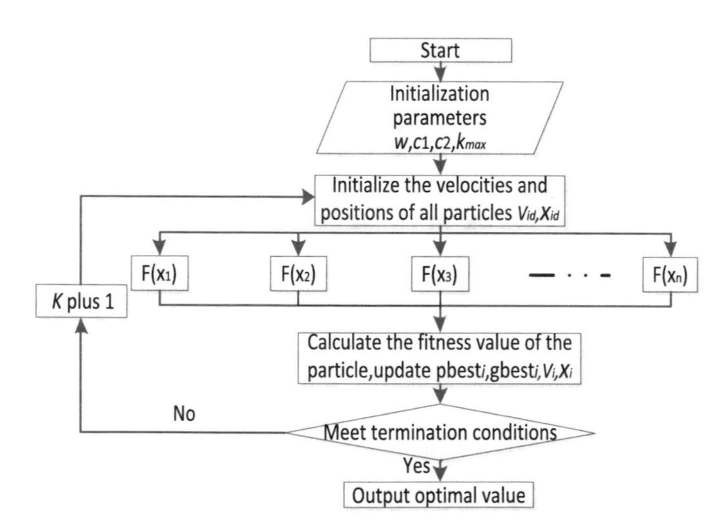

Fig. 2. Parallel particle swarm flow chart

4.2 Time Complexity Analysis

When the particles perform serial operations, assuming that the number of particles is T and the number of iterations is M, it can be seen that the time complexity of one particle iteration is O(T), and the time complexity of M iterations is O($T * M$). Based on the giant bitness and parallelism of the ternary optical computer, an optical processor is assigned to each particle in turn, so that the particles can be independently calculated. The time complexity of running one particle iteratively in parallel can be reduced to one particle iteratively running once. The time complexity is O (1), and the time complexity of iteratively running M times is O (M). The parallel operation procedure for particle fitness is as follows:

Allocate a processor bit for each particle and initialize the particle's velocity and position

```
for i → 1 to particle dimension
do
    Calculate the fitness value f(x_i) of each particle in parallel
    Store the fitness value f(x_i) corresponding to the i dimension
end
```

The fitness value $f(x_i)$ (i = 1, 2, 3…T) of each particle obtained according to the above calculation is stored in the array Fit, and the local optimal position and local optimal position of the particle are updated optimal fitness value. And store the global optimal value of each iteration in the array Gfit. The above pseudo-code of the particle computing parallel program is as follows:

```
if Fit(i) < PFit (i)
   PFit(i) = Fit(i)
   for all dimensions D of the particle
      do saves the calculated X position of each dimension to the corresponding P_pb ,
   end for
end if
if GFit (i) < PFit(i)
   GFit (i)= PFit (i)
   do each iteration the optimal value GFit(i) is stored in the array Gfit
end if
```

When the particles are operated in series, when the particles search for the global optimal value, the particles are compared in order of size, and the comparison is T minus 1 times in total. The time complexity is O (T). In the ternary optical computer, each particle calculates the fitness value in parallel and updates its local optimal value, and then compares them by the local optimal value, iterating in turn until the global optimal value is found. It can be seen that in the process of finding the optimal value, the time complexity of the particle is reduced from O (T) to O ($\log_2 T$).

According to the above analysis, the particle swarm optimization algorithm is mainly manifested in serial and parallel optimization. When designing the algorithm, each particle is assigned a total of n processors so that they can search for the optimal fitness in parallel and improve the search efficiency of the particle. The particle algorithm has a parallel computing time T_l in the process of sequential execution. The serial computation time is T_2, and the communication time between particles is T_3. According to the given fitness function, the corresponding fitness value is calculated in parallel, and then the position and velocity of the particle are updated. The parallel time in the algorithm is $T_b = T_l/N + T_2 + T_3$, and the serial time is $T_c = T_l + T_2$. Then speed up $a = T_c/T_b$.

5 Implementation of Particle Swarm Algorithm Based on TOC

5.1 PSO Algorithm Related Parameter Settings

In this paper, three test functions are used for simulation experiments. The experimental parameters are $M = 200$, $D = 20$, and $T = 200$. According to formulas (8) and (9) and $C1 = C2 = 1.5$, the search ability of the particle swarm is tested 10 times in an experiment cycle to get the average optimal value. The results are shown in Table 1 below:

1) Sphere function:

$$f(x) = \sum_{i=1}^{D} X_i^2 \qquad x_i \in (-10, 10)$$

2) Rastrigin function:

$$f(x) = \sum_{i=1}^{D} X_i^2 - 10\cos(2\pi X_i) + 10 \qquad x_i \in (-10, 10)$$

3) Rosenbrock function:

$$f(x) = \sum_{i=1}^{D} 100(X_{i+1} - X_i^2)^2 + (X_i - 1)^2 \quad x_i \in (-10, 10)$$

The figures of these functions are shown in Fig. 3.

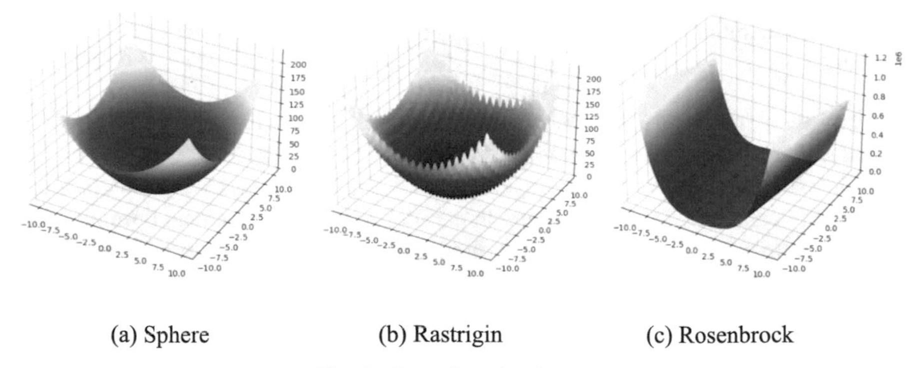

(a) Sphere (b) Rastrigin (c) Rosenbrock

Fig. 3. 3 test function images.

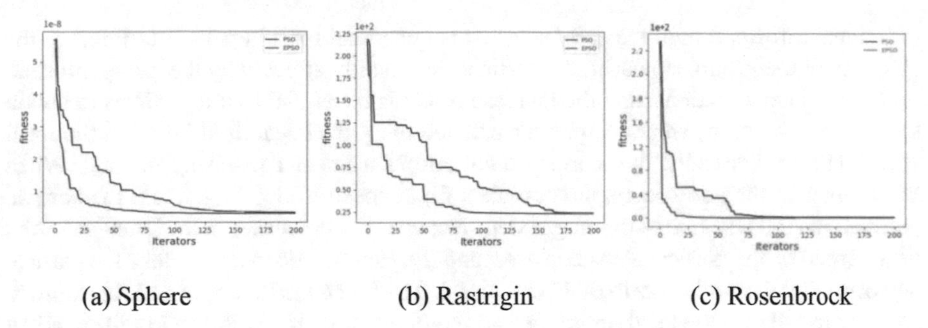

(a) Sphere (b) Rastrigin (c) Rosenbrock

Fig. 4. Convergence curves of two algorithms.

5.2 Simulation Experiment of EPSO Algorithm

The experimental results are shown in Fig. 4.

According to Fig. 4, it can be seen that the standard PSO algorithm has a slow convergence speed in finding the global optimum. Adjusting the dynamic adaptation of multiple parameters significantly improves the performance of the PSO algorithm and speeds up the convergence speed of particles to find the global optimum. The method shows the feasibility of the solution.

Table 1. Test results of two algorithms.

Evaluation criteria	Algorithm	Sphere	Rastrigin	Rosenbrock
The optimal value	PSO	4.339244e−9	5.153629	5.153629
	EPSO	68.463414e−9	2.319468	2.319468
Average optimum	PSO	0.251472e−9	8.728636	8.728636
	EPSO	3.716143e−9	6.380836	6.380836

Table 2. PSO test control group

Evaluation criteria	Algorithm	Sphere	Rosenbrock
The optimal value	PSO	6.107499e−9	12.767528
Average optimum	PSO	0.435726e−9	15.247575

Comprehensive analysis of the experimental results in Table 1 shows that the sphere function, as shown in Fig. 3, is a simple single-peak function. Compared with the literature [9], as shown in Table 2, it can be found that the optimization performance of the improved PSO algorithm has a great effect. Improvement. The reason is that the improved C1 and C2 are the step sizes of the particles flying towards the local optimum and the global optimum, respectively. During the optimization process of the particles, with the progress of the number of iterations, the particles are mainly affected by their individual information in the early stage. It is beneficial for a larger C1 to improve the diversity of the group. However, the particles are mainly affected by the group information in the later iterations, and the increase of C2 is beneficial to the particles to obtain the optimal solution, which can better balance the global search ability and the local search ability. Secondly, the inertia weight adopts a linear decreasing method. When W is larger in the early stage, the particle's flight speed is also larger, and the particle performs a global search with a larger step size; when W decreases in the latter stage, the flight speed of the particle also decreases, and the step size decreases. Make the particle perform a local search so as to find the optimal. The functions Rastrigin and Rosenbrock are easy to fall into the local optimum, and the function image is shown in Fig. 3. When dynamic adaptive learning factors C1 and C2 and inertia weight W are used, the local optimal solution can be jumped out during the search process, and the global optimal solution can be found.

6 Summary

Existing PSO algorithms are executed in a serial manner, which is less efficient. This experiment makes full use of the bit-assignment, megabits, and parallelism of TOC and effectively reduces the time complexity of particle swarm optimization. The improved PSO algorithm enhances the global optimization ability of the particles, and the convergence speed of the particles is greatly improved compared with the standard PSO algorithm. The experimental results show that the particle swarm optimization with high parallelism simulated by ternary optical computer simulation is reasonable and feasible. This paper has made some progress by improving the parameters of PSO, and there is still a lot of research space in other aspects. Using the improved PSO algorithm can solve more complex system optimization problems, and its application can be further expanded.

Acknowledgements. This research was supported in part by the Project of National Natural Science Foundation of China under Grant 61672006 and 6186202, and Hainan Provincial Natural Science Foundation of China under Grant 622MS084. And we would like to thank Jin from Shanghai University for their careful guidance on my thesis writing and the reviewers for their beneficial comments and suggestions, which improves the paper.

References

1. Cai, Y., Li, G.Y., Wang, H.: Design and implementation of parallel particle swarm optimization algorithm based on CUDA. Journal **30**(08), 2415–2418 (2013)
2. Soudan, B., Saad, M.: An evolutionary dynamic population size PSO implementation. In: 2008 3rd International Conference on Information and Communication Technologies, pp. 1–5 (2008)
3. Jin, Y., Ouyang, S., Song, K., Shen, Y.F., Peng, J.J., Liu, X.M.: Data bit management theory and technology of ternary optical processor. China Sci. (Inf. Sci.) **43**(03), 361 373(2013)
4. Zhang, S.L., Peng, J.J., Shen, Y.F., Wang, X.C.: Programming model and implementation mechanism for ternary optical computer. Opt. Commun. **428**, 26–34 (2018)

5. Song, K., Zhang, B.Y., Li, W., Yan, L.P., Wang, X.C.: Research on parallel principal component analysis based on ternary optical computer. Optik. **241**, 167–176 (2021)
6. Wang, X.C., Yao, Y.F., Jin, Y.: Parallel carry-free addition based on ternary optical computer. Comput. Sci. **37**(02), 290–293 (2010)
7. Lu, J.L., Miao, Y.Y., Zhang, C.X., R, H.: Optimal scheduling of power systems with wind farms based on improved multi-objective particle swarm optimization. Power Syst. Protect. Control **41**(17), 25–31 (2013)
8. Mohanty, S.D., Fahnestock, E.: Adaptive spline fitting with particle swarm optimization. Comput. Stat. **36**(1), 155–191 (2020). https://doi.org/10.1007/s00180-020-01022-x
9. Niu, Z.X., Hu, M.: A particle swarm algorithm for dynamically adjusting inertial weights. Industr. Control Comput. **33**(03), 28–30 (2020)

Design and Implementation of Simulated Annealing Optimization Algorithm Based on TOC

Qiangqiang He[1], Lijun Fang[1], Man Ling[1], Jie Zhang[1], Kai Song[2], Xianchao Wang[1(✉)], and Sulan Zhang[3(✉)]

[1] Fuyang Normal University, Fuyang 236037, Anhui, China
wxcdx@126.com
[2] East China Jiaotong University, Nanchang 330013, Jiangxi, China
[3] Jiaxing University, Jiaxing 314000, Zhejiang, China
zhangsl000111@zjxu.edu.cn

Abstract. When the traditional simulated annealing algorithm solves the traveling salesman problem (TSP), with an increase in the size of the problem, the time required increases and the quality of the solution decreases. To this end, this paper takes advantage of the parallelism and bit-wise assignability of ternary optical computers and proposes a parallelization strategy that divides an optical processor into many small optical processors and then distributes the initial solution to the problem to each small optical processor. On the optical processor, the optimal solution is independently searched for. Due to the parallel search of many processors, the convergence speed of the algorithm and the global optimization ability are effectively improved. Experimental tests show that the parallel strategy can achieve a better convergence effect and effectively improve the quality of understanding.

Keywords: Simulated annealing algorithm · TSP problem · TOC · Parallelism

1 Introduction

In the TSP problem, as the number of cities increases, the amount of time spent solving the problem increases exponentially. For solving the TSP problem with the traditional simulated annealing algorithm (SAA), in order to obtain accurate results, it will not only consume a lot of search time, but also, because the SA algorithm will accept the deteriorating solution with a certain probability, it is easy to miss the global the optimal solution is [1]. Using the parallelism, megabit and bit-assignable characteristics of the ternary optical computer (TOC), the optical processor on the TOC is first distributed through the equalization strategy [2], and the average will be It is divided into multiple small optical processors and then parallelizes thc traditional SA algorithm when solving the TSP problem, distributing the initial solution of the problem to each small optical processor that has been divided, and then using the proposed method in this paper. Two search strategies are used to obtain the global optimal solution to speed up the search and improve the quality of the global optimal solution.

J. H. Abawajy et al. (Eds.): ICATCI 2022, LNDECT 169, pp. 462–470, 2023.
https://doi.org/10.1007/978-3-031-28893-7_55

2 Introduction to TOC

In 2003, Professor Jin proposed the idea of TOC for the first time [3]. After nearly 20 years of research, he successively proposed a series of achievements, such as the devaluation design theory [4] and the reconfigurability theory [5]. In 2016, the TOC team successively launched ShangDa 16 (SD16) [6], marking the maturity of theory and technology. So far, many algorithms such as vector matrix multiplication, cellular automata, and fully parallel matrix [7] have been implemented on TOC.

The three-valued information representation of TOC is two mutually perpendicular polarized lights and a no-light state when there is a light state. Compared with electronic computers, ternary optical computers have millions of processor bits and can arbitrarily divide the processor into many small parts according to their own needs, and each small part can serve an application program independently. Not only that, each processor bit of the ternary optical computer can reconfigure the computing function of its hardware at any time, so that it not only has strong parallelism, but also reduces the time to solve the problem and the power consumption of the device [8].

3 SA Algorithm

The SA algorithm is a simulated solid annealing process that belongs to a greedy algorithm. The difference is that it introduces random factors into the process of searching for the optimal solution. Based on the Metropolis criterion, it accepts a solution that is worse than the current solution with a certain probability. It makes the algorithm have the possibility to jump out of the local optimum, so as to find the global optimum solution as much as possible. The simulated annealing algorithm has a wide range of applications today, such as the 0–1 knapsack problem, the scheduling problem, and the TSP problem discussed in this paper, etc. [9] (Fig. 1).

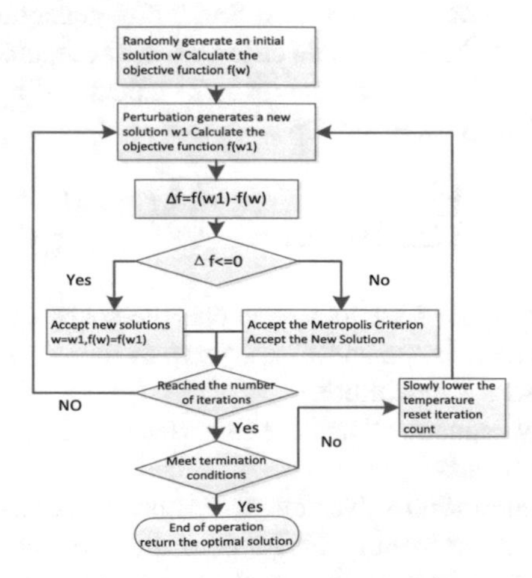

Fig. 1. Flowchart of simulated annealing algorithm

4 Solving TSP Problem Based on SA Algorithm

4.1 TSP Problem Description

The TSP problem can be described as follows: there are n cities, the coordinates of each city and the distance between the cities are known, a traveling salesman starts from a certain city and passes through each city, and each city can only pass through once. Finally, return to the starting city and ask to choose a path with the shortest total distance [10]. That is, set $S = \{s_1, s_2, \ldots, s_n\}$ (the elements in S represent the number of the corresponding city), and the total distance can be expressed as:

$$K = \sum_{i=1}^{n} d(s_i, s_{i+1}) + d(s_n, s_1) \qquad i = 1, 2, \ldots, n \tag{1}$$

Find an optimal path so that K takes the minimum value, where $d(s_i, s_{i+1})$ represents the distance from city s_i to city s_{i+1}.

4.2 Algorithms for Solving TSP Problems

The algorithm of SA can be described as:

Step 1 Initialization parameters, given the initial control temperature T_0 and termination temperature T_1, generate the initial city sequence $S_0 = \{s_1, s_2, \ldots, s_n\}$, the temperature decay coefficient is α, , the Markov chain length is L (that is, the number of iterations), $i = 0$, the current state $S_i = S_0$, calculate the total distance $K(S_i)$ of the current city sequence.

Step 2 $i = i + 1$, if $i \leq L$.

Step 3 Randomly generates a new sequence S_a from the current city sequence S_i. The new sequence is generated by the roulette method, and three strategies of crossover, shift and inversion are randomly selected to generate the new sequence.

Step 4 if $K(S_i) > K(S_a)$, then $S_i = S_a$, $K(S_i) = K(S_a)$, that is, accept the new state as the current state, and jump to step 3; else Randomly generate a decimal θ between (0,1), based on the Metropolis acceptance criterion, the acceptance probability p under the current city sequence S_i and temperature is T_i calculated by formula (2), if $\theta < p$, then $S_i = S_a$, $K(S_i) = K(S_a)$ Jump to Step 3.

$$p = \begin{cases} 1 & K(S_i) \geq K(S_a) \\ \frac{\exp([K(S_i) - K(S_a)])}{T_i} & K(S_i) < K(S_a) \end{cases} \tag{2}$$

Step 5 if $T_i \leq T_1$, terminates the loop, executes the next step Step6; else $T_{i+1} = \alpha * T_i$, that is, decay to the next temperature T_{i+1}, jump to Step 4, and continue to find the optimal solution at the next temperature.

Step 6 The final city sequence S and the total distance S are obtained, the result is output, and the algorithm ends.

The traditional SA algorithm solves the TSP problem by accepting the deteriorating solution with a certain probability. However, as the number of cities increases, the solution time gradually increases, and the quality of the solution is also closely related to the initial temperature T and the attenuation coefficient. Poor parameter settings will cause the loss of the global optimal solution to a certain extent.

5 Parallel Design of SA Algorithm Based on TOC

5.1 Parallel Strategy

Based on the SA algorithm [10, 11], an improved strategy in the TOC environment is proposed. First, on the optical processor of TOC, an optical processor is equally divided into $N + 1$ small optical processors, and one of the optical processors is used to collect and allocate the optimal solution space, which is temporarily abbreviated as CDTOC. The initial city sequence is generated on the CDTOC, and then the initial sequence is distributed to the remaining N optical processors, and then each of the N optical processors independently iterates at this temperature to find their own local optima. After the iteration, CDTOC collects the final solution space sequence obtained by each optical processor, distributes the optimal one to each optical processor at the next temperature, and repeats the above operations until the cooling is over, and the optimal solution is obtained the solution space sequence of.

5.2 Parallel Algorithm Steps

According to the above algorithm description, the following algorithm steps can be obtained:

Step 1: Divides the optical processor of TOC into $N + 1$ small optical processors equally, in which CDTOC is used to collect and distribute the obtained optimal solution space.

Step 2: Generates initialization parameters on the CDTOC, given the initial control temperature T_0 and termination temperature T_1, generates an initial city sequence $S_0 = \{s_1, s_2, \ldots, s_n\}$, the temperature decay coefficient is α, and the Markov chain length is L (that is, the number of iterations), $\underline{i} = 0$, the current state $S_i = S_0$, and the total distance $K(S_i)$ of the current city sequence is calculated.

Step 3: Copy the current sequence to N optical processors by CDTOC as the initial solution space on each optical processor at the current temperature.

Step 4: $i = i + 1$, if $I \leq L$. Each optical processor searches for the optimal solution space at the current temperature independently at the same temperature and at the same time.

Step 5: if $T_i \leq T_1$, If N optical processors obtain the city sequence at the current temperature and copy it to the CDTOC, the CDTOC finds an optimal solution space sequence and distributes it to each optical processor at the next temperature, and jumps to Step 3.

Step 6: Obtains the final city sequence S and the total distance K (S), outputs the result, and the algorithm ends (Fig. 2).

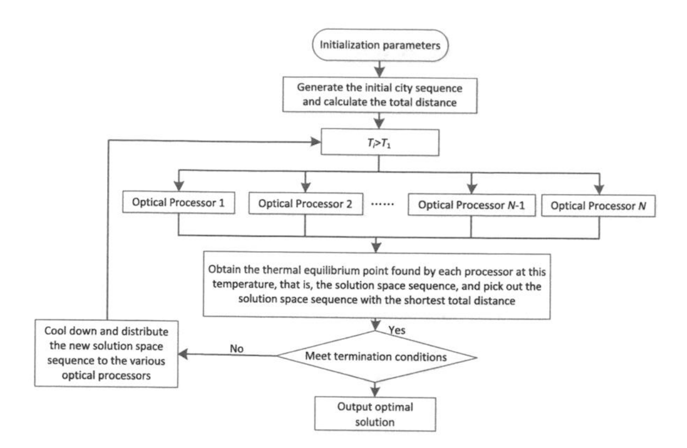

Fig. 2. Concurrent design flowchart for the TSP Problem

5.3 Improved Parallel Strategy

Compared with the common SA algorithm to solve the TSP problem, the parallel strategy proposed above has improved convergence and global optimization, but only because the optimal solution is found at each temperature and distributed to the next temperature. The optical processor and the SA algorithm itself have strong randomness, so the worst-case and best-case fluctuations of the obtained optimal solution are large.

On this basis, an improved parallel strategy is proposed. The algorithm steps are as follows:

Step 1: Divides the optical processors of the TOC into $N + 1$ small optical processors, one of which is only used to collect the best solution space sequence, which is temporarily abbreviated as CTOC.

Step 2: Generates initialization parameters on CTOC, given the initial control temperature T_0 and termination temperature T_1, generates the initial city sequence $S_0 = \{s_1,s_2,\ldots s_n\}$, the temperature decay coefficient is α, and the Markov chain length is L (that is, the number of iterations), $i = 0$, the current state $S_i = S_0$, calculate the total distance K (S_i) of the current city sequence, CTOC records the initial city sequence S_0, and the total distance K (S_i), and copy the initial parameters to N optical processors.

Step 3: Each optical processor independently searches for the solution space sequence. No information exchange is required during the search process, and CTOC is responsible for comparing the current value after each cycle of each processor. If it is, then copy the solution space sequence and total distance onto this processor.

Step 4: if $T_i \leq T_1$, to terminate the loop, CTOC records an optimal solution space sequence obtained in the search process of the N optical processor, outputs the result, and the algorithm ends.

The improved parallel strategy has the same initial parameters on N optical processors. In the subsequent iteration process, the processors independently search for the global optimum, and one optical processor records the best solution in the entire search process. Since the SA algorithm has a certain probability of accepting a deteriorating solution during the search process, to a large extent, the solution obtained after the algorithm is finished is not optimal, but with CTOC, the solution will be improved to a certain extent. Moreover, since the N optical processors independently search for the global optimal solution, the global optimization capability is also enhanced.

6 Experiment

6.1 Experimental Environment

This experiment mainly uses the TOC simulation software of Prof. Shen [11] to simulate the running environment of the TOC prototype system SD 16 (as shown in Fig. 3), which was originally 192 digits. This platform can be used to develop TOC-based applications. This platform includes four arithmetic operations on MSD numbers. Experimental environment: a simulation program written in the C language on code::blocks 20.03.

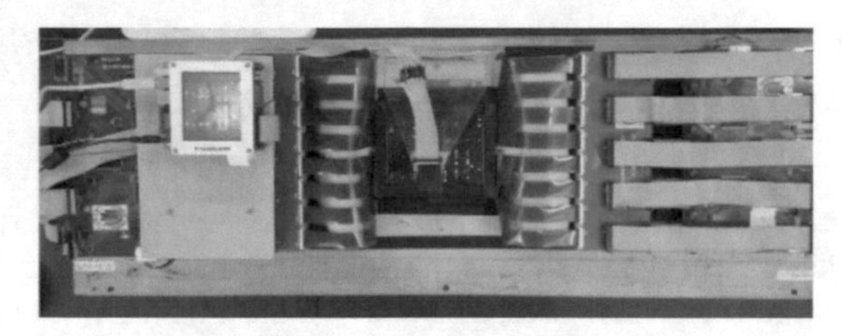

Fig. 3. SD 16 physical map

6.2 Experiment and Analysis

Att48 [12] in the TSP test library TSPLIB as an example, and the best known result is 33522. This experiment solves the three algorithms of the ordinary SA algorithm, the parallel algorithm and the improved parallel algorithm respectively. Experiment with the TSP problem. The parameters of the algorithm are set as: initial temperature $T = 5000$, termination temperature using outer loop $L_1 = 500$, temperature decay coefficient $\alpha = 0.95$, memory loop $L_2 = 200$, allocated optical processor bits $M = 10$. The experimental results are shown in Fig. 3. The number of iterations in the figure is the product of the outer loop and the inner loop. Figure 3 (a) is the route and convergence diagram of the common SA algorithm, and Fig. 3 (b) is the route and convergence of the parallel SA algorithm. Convergence graph, Fig. 3 (c) is the route and convergence graph of the improved parallel SA algorithm (Fig. 4).

As can be seen from Fig. 3 (a) and (b), when the SA algorithm solves the TSP problem, its convergence speed is much slower than that of the TOC-based parallel SA algorithm, and the best solution obtained by the SA algorithm is the same as the worst solution range is [34144, 35009], while the best and worst solutions obtained by the TOC-based parallel SA algorithm are in the range [33723, 34590]. Obviously, in the process of solving the global optimum, the parallel SA algorithm the quality of the obtained solution is significantly better than that of the SA algorithm. This is because of

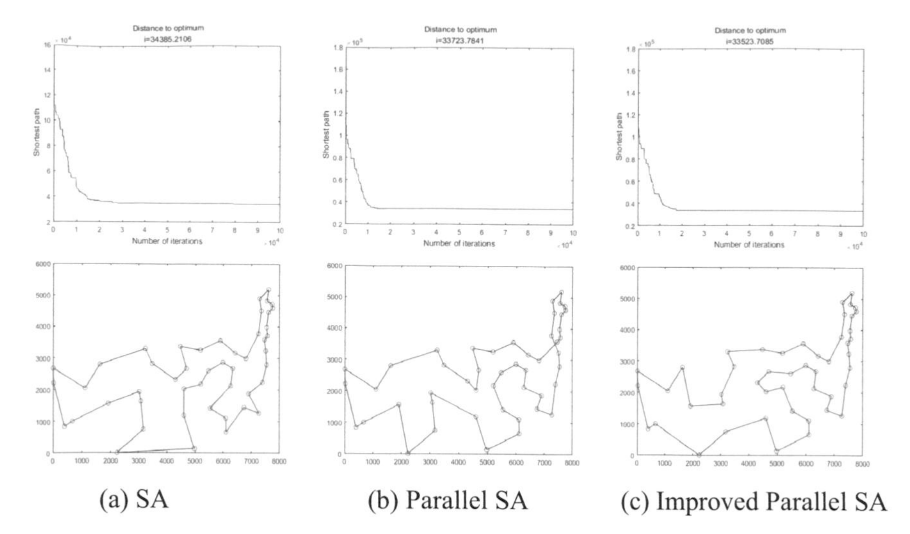

(a) SA (b) Parallel SA (c) Improved Parallel SA

Fig. 4. Roadmap and convergence graph of three different algorithms for solving tsp problems

the parallel SA algorithm in the search process. Due to the parallel search of N optical processors, the theoretical search time is much lower than the ordinary SA algorithm, and because the search range increases, the quality of the final solution becomes obvious improve.

In Fig. 3 (c), the range of the best solution and the worst solution of the improved parallel algorithm is [33523, 34076], and the best solution route is:

$43 \rightarrow 30 \rightarrow 36 \rightarrow 46 \rightarrow 33 \rightarrow 20 \rightarrow 47 \rightarrow 21 \rightarrow 32 \rightarrow 39 \rightarrow 48 \rightarrow 5 \rightarrow 42 \rightarrow 24 \rightarrow 10 \rightarrow 45 \rightarrow 35 \rightarrow 4 \rightarrow 26 \rightarrow 2 \rightarrow 29 \rightarrow 34 \rightarrow 41 \rightarrow 16 \rightarrow 22 \rightarrow 3 \rightarrow 23 \rightarrow 14 \rightarrow 25 \rightarrow 13 \rightarrow 11 \rightarrow 12 \rightarrow 15 \rightarrow 40 \rightarrow 9 \rightarrow 1 \rightarrow 8 \rightarrow 38 \rightarrow 31 \rightarrow 44 \rightarrow 18 \rightarrow 7 \rightarrow 28 \rightarrow 6 \rightarrow 37 \rightarrow 19 \rightarrow 27 \rightarrow 17.$

As can be seen from Fig. 3 (b) and (c), although the convergence speed of the improved parallel algorithm is relatively longer, because it is equipped with an optical processor, it is used to collect the optimal solution in the entire iterative process. Each optical processor is independently optimized, and the quality of the obtained solution is obviously greatly improved. Compared with the current best solution, the optimal results obtained by the improved parallel algorithm are almost the same. In order to further illustrate the effectiveness of the given strategy, Oliver30 in the TSP test library TSPLIB is selected as an example in the experiment, and the obtained results are shown in Table 1, which shows that the TOC-based parallel SA algorithm and the improved parallel SA algorithm proposed in this paper are both effective.

Table 1. The solution result of Oliver30

Algorithm	Worst solution	Best solution	Average value
SA	476	424	450
Parallel SA	440	422	429
Improved Parallel SA	431	420	423

7 Summarize

The simulated annealing algorithm has strong randomness when solving the TSP problem, so it is likely to miss the global optimal solution. The TOC-based parallel SA algorithm not only enhances the global optimization ability, but also improves the convergence speed of the algorithm and the algorithm. The overall running time has been greatly improved. Based on the parallel processing characteristics of the TOC optical processor, the problem of carry delay in the search process of the SA algorithm on the electronic computer processor is further solved. In a word, the SA parallel algorithm based on TOC shows strong superiority in simulation experiments for different numbers of cities.

Acknowledgements. This research was supported in part by the Project of National Natural Science Foundation of China under Grant 61672006 and 61862023. And we would like to specially thank to Mr. Jin from Shanghai University for his careful guidance and the reviewers for their beneficial comments and suggestions, which improves the paper.

References

1. Zhang, Y., Han, X., Dong, Y., Xie, J., Xie, G., Xu, X.: A novel state transition simulated annealing algorithm for the multiple traveling salesmen problem. J. Supercomput. **77**(10), 11827–11852 (2021). https://doi.org/10.1007/s11227-021-03744-1
2. Wang, X.C., Zhang, J., Gao, S., Zhang, M., Wang, X.C.: Performance analysis and evaluation of ternary optical computer based on a queueing system with synchronous multi-vacations. IEEE Access **8**, 67214–67227 (2020)
3. Jin, Y., He, C.H., Lv, Y.T.: Ternary optical computer principle. Sci. China (Ser. F: Inf. Sci.) **46**, 145–150 (2003)
4. Jin, Y., Jin, J.Y., Zuo, K.Z.: Decrease-radix design principle for carrying/borrowing free multi-valued and application in ternary optical computer. Sci. China (Ser. F: Inf. Sci.) **51**, 1415–1426 (2008)
5. Jin, Y., et al.: The principle, basic structure and implementation of a reconfigurable ternary optical processor. Sci. China Inf. Sci. **42**(06), 778–788 (2012)
6. Jin, Y., Gu, Y.Y., Zuo, K.Z.: The theory, technology and implementation of TOC decoder. Sci. China Inf. Sci. **43**(02), 275–286 (2013)
7. Song, K., Zhang, Y., Yan, L.P., Jin, Q.Q., Chen, G.: Research on fully parallel matrix algorithm of ternary optical computer for the shortest path problem. Appl. Opt. **59**(16), 4953–4963 (2020)

8. Li, S., Jiang, J.B., Wang, Z.H., Zhang, H.H.: Basic theory and key technology of programming platform of ternary optical computer. Optik. **178**, 327–336 (2018)
9. Omer, C.: Parallelizing simulated annealing algorithm for tsp on massively parallel architectures. Havacılık ve Uzay Teknolojileri Dergisi. **11**(1), 75–85 (2018)
10. Sharma S., Jain V.: A novel approach for solving TSP problem using genetic algorithm problem. IOP Conf. Ser. Mater. Sci. Eng. **1116**(1), 012194 (2021)
11. Shen, Y.F., Zhang, S.L., Wang, Z.H., Li, W.M.: Design and implementation of parallel radix-4 MSD iterative division of ternary optical computer. Opt. Commun. **501**, 127360–127368 (2021). Prepublish
12. Xu, X.P., Zhu, Q.Q.: Improved simulated annealing algorithm for solving TSP. Appl. Comput. Syst. **24**(12), 152–156 (2015)

Recommendation Algorithm Based on Product Category Path

Zhenyuan Fu, Xianchuan Wang(✉), Baofeng Qi, Xiuming Chen, and Xianchao Wang(✉)

Fuyang Normal University, Fuyang 236037, Anhui, China
xchwang@fynu.edu.cn, wxcdx@126.com

Abstract. In recent years, due to the rise of deep learning, most researchers are trying to combine the traditional recommendation system with deep learning, however the traditional recommendation model is still suitable for most scenes or be widely used as an auxiliary model of deep learning recommendation model because of its strong interpretability, fast training and deployment. When the recommendation algorithm calculates the recommended goods for users, this paper establishes the category path of goods, then find the goods according to the category path of goods and recommend them. At the same time, according to the characteristics of the data obtained in this paper, the similarity of goods is calculated by the combination of TF-IDF and Jaccard similarity coefficient, so as to infer the list of goods recommended to users.

Keywords: Recommendation algorithm · TF-IDF algorithm · Knowledge graph

1 Introduction

As an information retrieval tool in the Internet era, recommendation system has made great progress since the 1990s. With the rapid development of the Internet, people's shopping behavior through the e-commerce platform is also increasing. The commodity information and user behavior data show an explosive growth trend. As a method that can effectively alleviate this problem, the recommendation system plays an important role in the network service by analyzing the user's historical behavior data to determine the user's preferences and recommend personalized content [9].

Before the deep learning model blooms in 2015, the traditional recommendation algorithms mainly include content-based recommendations, coordinated filtering recommendations and hybrid recommendations [9]. In recent years, due to the rise of deep learning, most researchers are trying to combine the traditional recommendation system with deep learning, however the traditional recommendation models such as collaborative filtering, logistic regression and factor decomposition machine are still applicable to most scenes or be widely used as an auxiliary model of deep learning recommendation model because of their irreplaceable advantages of strong interpretability and rapid training and deployment.

J. H. Abawajy et al. (Eds.): ICATCI 2022, LNDECT 169, pp. 471–479, 2023.
https://doi.org/10.1007/978-3-031-28893-7_56

The collaborative filtering recommendation algorithm based on users' purchase intention introduced in Paper [2] Can effectively recommend relevant products for users according to users' purchase behavior, In some e-commerce platforms that people often use, such as Taobao and JingDong, when users click to buy a commodity, the e-commerce platform can also make personalized recommendations to users according to their behavior of purchasing commodities and some attributes of commodities. Although most of these recommended goods are in line with the user's purchase intention, but the recommendation algorithm does not filter out the attributes of the goods that the user has purchased when calculating the goods that need to be recommended to the user. For example, users buy a computer on the e-commerce platform, although the e-commerce platform will recommend computer related accessories such as mouse and keyboard to users after users purchase computers, but most of the goods in the recommendation list are still computers.

In Paper [1], Paper [6], Paper [7] and Paper [9], the knowledge graph is combined with the recommendation system to obtain better recommendation results. However, most of the algorithms introduced in these papers focus on the goods or users themselves and ignore the relationship between different goods. In order to better provide users with personalized recommendation services, this paper establishes the category path of goods. When the recommendation algorithm calculates the recommended goods for users, first find the goods according to the category path of goods, and then recommend them. This can not only filter out the goods that the user has purchased, but also recommend the goods that the user is likely to buy according to the category path of the goods.

2 Recommendation Model Based on Product Path

2.1 Data Acquisition

The data set used in this paper is the commodity data set on Alibaba cloud Tianchi., and the specific data characteristics are described in the Table 1.

Table 1. Data feature description

Feature	Feature explanation	Size
item_id	Commodity ID	499981
user_id	User ID	13611038
cat_id	Commodity category ID	281
property	Commodity property	85488
day	User purchase time	13611038

2.2 Save Product Category Path by Neo4j

Neo4j is a distributed and open source native graph database based on Java. It not only has good graph storage and graph processing engine capabilities, supports clustering, backup, ACID and failover functions, but also contains visual management tools. It has two versions: community version and enterprise version. Compared with other graph databases, Neo4j has some obvious advantages, such as efficient query, wish graph calculation, open source Network oriented database and adaptive development.

Based on the above information, this paper uses Neo4j graph database to store the path of goods. First, classify the commodities according to the primary labels of commodities, such as digital, men's clothing, mother and baby products, and establish the commodity path according to the secondary labels of commodities.

Take mother and baby products as an example. Mother and baby products include six categories: infant and child products, children's clothes, children's shoes, snacks, toys and maternity clothes. These six categories are used as nodes to establish contact in pairs. The specific effect is shown in Fig. 1.

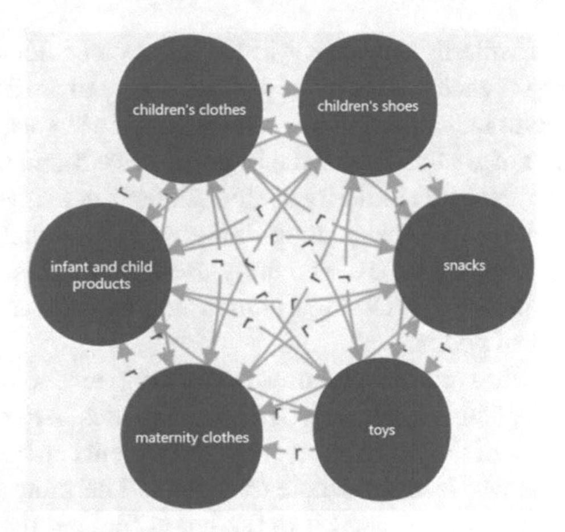

Fig. 1. Route map of mother and baby products

As shown in Fig. 1, since there are only six secondary labels in the primary label of mother and baby products in the dataset, six nodes are established. These six secondary labels are related to each other, so the relationship between the six nodes is established, and any node can find all other nodes according to the path.

2.3 Jaccard Similarity Coefficient

After a user purchases a commodity, the recommendation system filters out the candidate recommended commodities according to the commodity path obtained in Sect. 2.2. Next, it needs to calculate the similarity between the candidate commodities and the commodities purchased by the user to determine which commodities to recommend for the user.

There are many methods to calculate commodity similarity, the most commonly used are European Distance, Manhattan Distance and Cosine Similarity. In this paper, the similarity of goods is calculated according to the attribute characteristics of goods. As shown in Sect. 2.1, although the attribute characteristics of goods have been digitized, it is obvious that these numbers can not be applied to numerical calculation. Therefore, this paper uses Jaccard similarity coefficient to calculate the similarity of goods.

We can regard the characteristics of goods as some sets, and Jaccard similarity coefficient is a method to calculate the similarity of these sets. Assuming that there are two sets A and B, the ratio of the intersection of A and B and the union of A and B is called the Jaccard similarity coefficient of the two sets, which is expressed by J(A, B). The corresponding formula is as follows.

$$J(A, B) = \frac{|A \cap B|}{|A \cup B|}$$

2.4 TF-IDF Algorithm

Although the Jaccard similarity coefficient in Sect. 2.3 can calculate the similarity of goods. But the meaning of each feature of the goods is likely to be different. For example, the features of a computer include brand and screen size. If users want to buy accessories related to this computer, it is obvious that the importance of brand is greater than screen size for users, However, Jaccard similarity coefficient does not consider giving different weights to different features when calculating similarity. Although the TF-IDF algorithm introduced in Paper [5] and Paper [10] is mainly used for text classification, but in the case discussed in this paper, TF-IDF algorithm is undoubtedly a good choice to deal with the characteristics of goods.

TF-IDF algorithm is a commonly used weighting technology for information retrieval and text mining. This algorithm is based on statistics. After counting the words, the discrimination degree of the words to the target document can be evaluated according to the proportion of the words in the whole document. The more words appear in the target file, the more important the word is to the target file, but the more words appear in the whole document, the less important the word is to the target file.

The main idea of TF-IDF algorithm is to try to find a word or phrase with high frequency in the target article but low frequency in other articles. These words or phrases have good classification ability for the target article. Corresponding to the commodity data in this paper, it is the differentiation of each characteristic of the commodity from the commodity.

TF (Term Frequency) represents the frequency of occurrence of feature t in commodity Di. TF is calculated as follows:

$$TF = \frac{count(t)}{|Di|}$$

In the formula, count(t) represents the number of occurrences of feature t, and |Di| represents the number of all features in commodity Di.

IDF (Inverse Document Frequency) represents the distinguishing ability of the feature t in the whole commodity for the target commodity. IDF is calculated as follows:

$$IDF = lg\frac{N}{1 + \sum_{i=1}^{N} I(t, Di)}$$

In the formula, N is the total number of all goods. I(t,Di) is a judgment value indicating whether feature t is included in product Di. If it is included, the value is 1, and if not, the value is 0. In order to prevent feature t from appearing in all goods, add 1 for smoothing.

The TF-IDF value of final word feature t in commodity Di is:

$$IF - \mathrm{IDF}_{\mathrm{t,Di}} = \mathrm{TF}_{\mathrm{t,Di}} * \mathrm{IDF}_t$$

From the calculation process of TF-IDF value, it can be seen that when the frequency of feature t in commodity Di is higher and the freshness in the whole commodity set D is lower (i.e. low universality), the higher the corresponding TF-IDF value, which means t has a greater degree of differentiation of commodity Di.

2.5 Time Decay Function

The jackard similarity coefficient in Sect. 2.3 and the TF-IDF algorithm in Sect. 2.4 can calculate the similarity of goods according to the user's purchase behavior. However, in practice, users' interests are constantly changing, and for e-commerce platforms, users' interests are timely. For example, in the mother and baby products path shown in Sect. 2.2, if a user purchases children's clothing products under the mother and baby products category, the recommendation system based on the commodity path will recommend the other five types of products under the mother and baby products category to the user according to the user's purchase behavior. It is assumed that the user has purchased all kinds of products recommended by the recommendation system, it means that the product category purchased by this user has covered all kinds of products under the category of mother and baby products. Therefore, it can be inferred that this user will not consider purchasing all kinds of products under the category of mother and baby products in the short term. However, according to the above recommendation algorithm, the recommendation system will still recommend mother and baby products for users, so it needs to add a time decay function to make the recommendation results change constantly, so as to meet the real-time needs of users.

The formula of time decay function used in this paper is as follows:

$$f(|t_0 - t_{ui}|) = \frac{1}{1 + \alpha|t_0 - t_{ui}|}$$

In the formula, α is the time decay factor, t_0 represents the current time, and t_{ui} represents the time when the user u acts on the item i.

According to this formula, a time threshold can be set for the user's purchase behavior. When the user's purchase time is greater than this threshold, other categories of products will be added to the candidate recommendation list, otherwise the products to be recommended will be selected according to the product path.

3 Experiment and Analysis

In addition to the accuracy evaluation indicators such as accuracy and recall, the recommendation system also has non accuracy evaluation indicators such as personalization, richness and diversity. In the experiment, this paper does not use the commonly used accuracy evaluation indicators such as accuracy and recall, but uses the non accuracy evaluation indicators such as personalization, richness and diversity to evaluate the experimental results.

In the test experiment, this paper randomly selects the purchase records of 10000 users for the experiment. Figure 2 and Fig. 3 are the experimental results of predicting the recommendation results based on the recommendation algorithm of calculating the commodity similarity by simply using the Jaccard similarity coefficient and the recommendation algorithm of calculating the commodity similarity by TF-IDF without using the commodity path to screen the commodities, and recommending 200 commodities most similar to the commodities purchased by the user for the user.

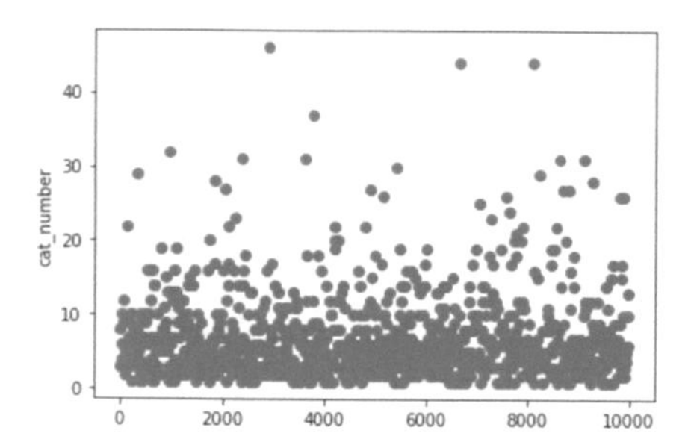

Fig. 2. Only Jaccard similarity coefficient

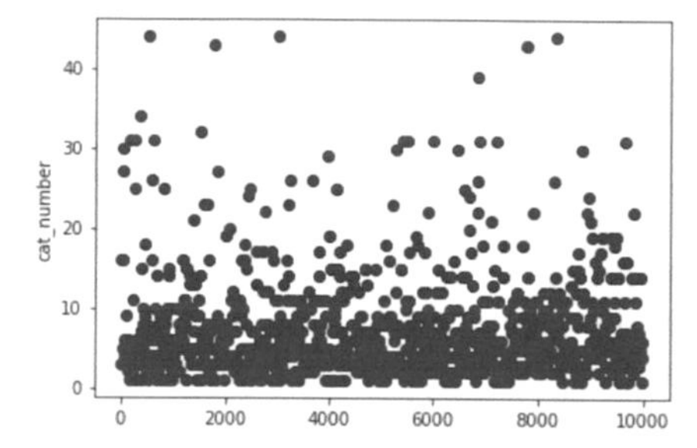

Fig. 3. Without commodity path

As shown in Fig. 2 and Fig. 3, the scattered points in the graph are mainly concentrated between 0 and 10, indicating that the number of recommended commodity types is relatively small. This result is predictable. Because the two algorithms do not screen the goods, they will calculate the similarity among all the goods, and the goods most similar to the goods purchased by the user must be the same type of goods. Therefore, most of the 200 goods recommended by the algorithm belong to the same category as the goods purchased by the user.

According to the recommendation algorithm based on commodity path introduced in this paper, the recommendation results are shown in Fig. 4.

It can be seen that for the commodity path based recommendation algorithm, as shown in Fig. 4, the scatter distribution is more dispersed than Fig. 2 and Fig. 3. For the commodity path based recommendation algorithm, the average number of 200 commodity types recommended is 11.211, The average number of 200 kinds of goods recommended by the recommendation algorithm that only uses the Jaccard similarity coefficient to calculate the commodity similarity and the TF-IDF recommendation algorithm that does not use the commodity path to screen the commodities are 6.968 and 7.328 respectively, which shows that the recommendation algorithm based on the commodity path does effectively solve the problem of single commodity recommended by the traditional recommendation system.

However, it can be observed that some points in Fig. 2 and Fig. 3 have exceeded 40, but there are no more than 35 points in Fig. 4, This means that for some commodities, the number of commodity types recommended by the commodity path based recommendation algorithm is not as large as the number of commodity types recommended

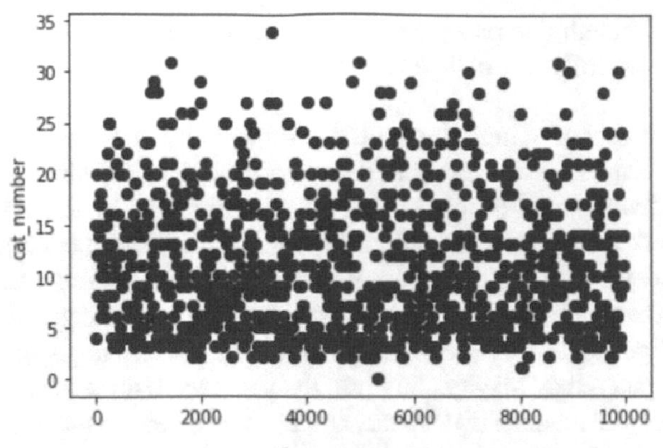

Fig. 4. Using commodity path

by the recommendation algorithm of calculating commodity similarity by simply using Jaccard similarity coefficient and the TF-IDF recommendation algorithm of calculating commodity similarity without using commodity path to screen commodities, This paper speculates that it is because some commodities do not have strongly related commodities, resulting in little difference in the similarity between other commodities and the commodity, so it is impossible to establish an effective path for the commodity. In fact, there is a result that the number of recommended categories is 0 in the recommendation results of the recommendation algorithm based on commodity path, which also verifies the speculation of this paper.

4 Summarize

With the rapid development of the Internet, people's shopping behavior through e-commerce platform is also increasing. Commodity information and user behavior data show an explosive growth trend. The emergence of personalized recommendation system alleviates this problem to a great extent. Although in recent years, due to the rise of deep learning, most recommendation systems are trying to combine the traditional recommendation system model with deep learning, the role of the traditional recommendation algorithm in the recommendation system can not be ignored.

In the application scenario of e-commerce, combined with the real e-commerce data, in order to better provide users with personalized recommendation services, combined with knowledge map, this paper studies the recommendation algorithm based on commodity category path. This paper establishes the category path of goods. When the recommendation algorithm calculates the recommended goods for users, first find the goods according to the category path of goods, and then recommend them. At the same time, according to the characteristics of the data obtained in this paper, the similarity of goods is calculated by the combination of TF-IDF and Jaccuard similarity coefficient, so as to infer the list of goods recommended to users. Of course, in order to maintain the real-time of user interest, this paper introduces a time decay function to control the recommendation system.

Of course, although this paper has made some research results, but because of the short time, there are still some deficiencies in this paper, which need to be further studied. Such as:

The establishment of commodity path has subjectivity. In the research of this paper, the commodity path is set according to the classification of commodities, but although some commodities belong to different categories, they have internal relations, and relying on people to mark the commodity path, it is easy to ignore these internal relations.

The time decay of commodity recommendation needs to be improved. Although this paper controls the timeliness of user interest through the time decay function, in fact, the time decay threshold should be different for different kinds of goods. For example, if users buy a computer, they may not buy it in the next two to three years, but if users buy snacks, it will only take one to two weeks, and users will need to buy it again. Therefore, different time decay functions are needed for different categories of goods.

Acknowledgements. This research was supported in part by the Planning Youth Project of Philosophy and Social Sciences of Anhui under Grant AHSKQ2021D47. And we would like to specially thank to the reviewers for their beneficial comments and suggestions, which improves the paper.

References

1. Wen, F., Cao, X., Huang, X., Yan, X.Y.: Research on recommendation algorithm based on knowledge map. J. Shenyang Univ. Technol. (06), 13–17 (2021)
2. Wang, G.X., Liu, H.P.: Overview of personalized recommendation system. Comput. Eng. Appl. 70–80 (2012)
3. Huang, C.H., Yin, J., Hou, F.: A text similarity measurement method combining word item semantic information and TF-IDF method. J. Comput. Sci. 98–106 (2011)
4. Zhang, L., Jiang, Y., Sun, L.: An improved TF-IDF text clustering method. J. Jilin Univ. (Sci. Ed.) 203–208 (2021)
5. Chen, Y.Y., Feng, W.L., Huang, M.X., Feng, S.L.: Behavior path collaborative filtering recommendation algorithm based on knowledge map. Comput. Sci. 182–189 (2021)
6. Zhu, M.Q., Wen, M.: Design of filtering recommendation algorithm based on knowledge map. Electron. Technol. Softw. Eng. 152–153 (2021)
7. Qi, J., Liu, Y., Liu, Y.X., Hu, M.Z., Yue, H.F.: Research on collaborative filtering recommendation method based on tag. J. Beijing Union Univ. 53–58 (2021)
8. Ning, Z.F., Sun, J.Y., Wang, X.J.: Recommendation algorithm based on knowledge map and label perception. Comput. Sci. (11), 192–198 (2021)
9. Yan, Y.Y.: Comparative study of word bag model and TF-IDF in text classification. Comput. Knowl. Technol. 144–146 (2021)
10. Hazrati, N., Ricci, F.: Recommender systems effect on the evolution of users' choices distribution. Inf. Process. Manag. **59**(1) (2022)
11. Seo, Y.-D., Kim, Y.-G., Lee, E., Kim, H.: Group recommender system based on genre preference focusing on reducing the clustering cost. Expert Syst. Appl. **183** (2021)
12. Asani, E., Vahdat-Nejad, H., Sadri, J.: Restaurant recommender system based on sentiment analysis. Mach. Learn. Appl. **6** (2021)

13. Gómez, E., Shui, Z.C., Boratto, L., Salamó, M., Ramos, G.: Enabling cross-continent provider fairness in educational recommender systems. Futur. Gener. Comput. Syst. **127**, 435–447 (2022)
14. Twyman, M., Newman, D.A., DeChurch, L., Contractor, N.: Teammate invitation networks: the roles of recommender systems and prior collaboration in team assembly. Soc. Netw. **68**, 84–96 (2022)

Cyber Intelligence for Health and Education Informatics

Digital Marketing System Based on Improved Teaching Optimization Algorithm

Yanmei Zeng(✉)

Gongqing Institute of Science and Technology, Gongqingcheng, Jiangxi, China
zym851927973@126.com

Abstract. The advent of the digital age has changed people's traditional marketing models, and has also caused companies to face huge challenges. The construction of a digital marketing system enables companies to integrate, utilize and develop digital media. They can also understand their information from the perspective of customers for the reference of decision makers and make corresponding countermeasures, so as to provide users with more satisfactory services. This is very necessary to promote sales growth and improve market competitiveness. At the same time, compared with other optimization algorithms, the improved teaching optimization algorithm has better convergence accuracy, faster convergence speed, and better results in solving optimization problems with complex parameters. Based on the improved optimization learning algorithm, the content and system of digital marketing are studied. Through experimental analysis and data analysis, we can better understand the performance of digital marketing system. The experimental results show that the system has shorter response time and higher performance. At the same time, as the number of concurrent users increases, the system response time also increases.

Keywords: Teaching optimization algorithm · Digital marketing · Big data · Network economy

1 Introduction

With the acceleration of informatization, the digital marketing model has been widely used in the production and management of enterprises. This marketing model combines traditional sales methods with network technology, and effectively integrates them on this basis, so as to improve market competitiveness and maximize economic benefits. At present, many companies are facing these problems, one is how to make information transmission faster, and the other is to use various methods in the marketing process to reduce costs and improve customer satisfaction. At the same time, compared with other optimization algorithms, the improved teaching optimization algorithm has better convergence accuracy, faster convergence speed, and better results in solving optimization problems with complex parameters. Based on the improved teaching optimization algorithm, this article has carried out research on digital marketing related content and system design.

J. H. Abawajy et al. (Eds.): ICATCI 2022, LNDECT 169, pp. 483–489, 2023.
https://doi.org/10.1007/978-3-031-28893-7_57

Currently, many scientists have studied the algorithm of teaching optimization. Zhai Junchang pointed out that the teaching optimization algorithm is a new type of heuristic optimization algorithm, aiming at the problem that the teaching optimization algorithm is easy to fall into the local optimum, and proposed an improved teaching optimization algorithm [1]. Hou Jingwei believes that the construction of a multi-objective nonlinear high-dimensional evaluation model is conducive to the strategic improvement of the teaching and learning optimization algorithm, and finally the algorithm can solve the problem [2]. Jiang Jiayan et al. proposed that the improved teaching and learning algorithm based on the elite strategy can obtain better optimization results and provide a new method and idea for solving the reactive power optimization problem of the distribution network [3]. This article starts from a new perspective, based on the improved teaching optimization algorithm, to carry out research on digital marketing and its related content.

First, the teaching optimization algorithm and related research are explained. Then, a detailed introduction to digital marketing. Finally, an experimental study was carried out around the digital marketing system, and the corresponding experimental results and analysis conclusions were drawn.

2 Related Theoretical Overview and Research

2.1 Teaching Optimization Algorithm and Related Research

The teaching optimization algorithm was proposed by an Indian scholar. The algorithm achieved the goal of finding the best by simulating the teacher's teaching to students in the classroom and the students' self-study process after class.

Compared with other optimization algorithms, the teaching optimization algorithm has the advantages of good convergence accuracy and fast convergence speed, especially in solving the problem of optimizing complex parameters. The algorithm has been applied in many fields, and has attracted the attention of researchers at home and abroad, especially in function optimization, multi-objective optimization and other issues that have achieved good results.

The teaching and learning optimization algorithm is mainly based on the teaching process of teachers and the process of learning and communication between students. The teacher's teaching process is to bridge the gap between the teacher's personal knowledge level and the average level of the class, thereby improving the overall level of the class. In this algorithm, the best person in the class acts as the teacher for the teaching process. Once the teacher starts the classroom phase, students will communicate and learn. The learning process of students also follows the principle of being close to the best individual. Individuals improve their personal level by learning from neighbors who are better than them [4, 5].

In the process of class initialization, each learner in the class is randomly generated in the search space. The generation of initial learners is shown in formula (1).

$$A_u = A_u^{min} + \text{rand}(A_u^{max} - A_u^{min}) \tag{1}$$

Among them, Au is the u-th in the class, A_u^{min} is the lower limit of the control variable, A_u^{max} is the upper limit of the control variable, and rand is a random number in the range of 0–1.

In the classroom stage, the students with the highest scores are selected as teachers to improve the learning level of the students in the class and move the average level of the class toward the optimal solution. The gap between the average level of the classmates (population average) and the teacher is shown in formula (2).

$$\text{Differ} = \text{rand}(0, 1) * (A_{\text{teacher}} - R_e * A_{\text{mean}}) \tag{2}$$

Among them, Differ is the gap between the average level (population average) and the teacher, A_{teacher} the individual with the best fitness value of the current population, and Re is the teaching factor, A_{mean} is the average individual of the classmates.

Since it was proposed, the teaching optimization algorithm has received extensive attention and research from many researchers with its best optimization performance, and has been applied to the optimization of many technical problems. But it also has some shortcomings, and it is easy to fall into a local minimum when solving some complex problems. Based on this, researchers have done a lot of research to improve it. The following describes a typical improved teaching optimization algorithm-the teaching optimization algorithm based on elite strategy [6, 7].

In order to retain the best individuals in the population, many population-based intelligent optimization algorithms follow an elite strategy to improve their optimization performance. In order to further improve the optimization performance of the teaching optimization algorithm, Rao et al. proposed an elite strategy teaching optimization algorithm. If an individual is better than the worst individual in the elite set in each iteration, replace the worst individual in the elite set, so that the best elements are always kept in the population and ensure that all individuals in the population are constantly searching for the best, The algorithm also converges quickly. The teaching optimization algorithm that adopts the elite strategy effectively improves the optimization performance of the algorithm. The algorithm also uses random mutation for the same individuals in the population to create new individuals instead of repeating individuals. This reduces the invalid iterations in the algorithm optimization process, and also increases the possibility of the algorithm jumping out of the local optimum [8, 9].

2.2 Digital Marketing and Related Research

As a product of the rapid development of digital technology and the Internet, digital marketing has updated and improved the marketing content and structure, entered a new marketing era, and become the trend of marketing development.

Digital marketing has effectively mobilized enterprise resources through modern electronic means and communication technology. Special emphasis is placed on the effective coordination and standardization of logistics, capital and information flows to meet the needs of customers and ensure their profitability. The core of digital marketing is marketing, and digitalization is just a way, that is, the development of modern marketing theory under the new economic conditions.

The reason why digital marketing can prevail and present an uncontrollable trend is that digital marketing conforms to the current economic development trend and can help companies establish their own competitive advantages in the fierce market competition. The marketing methods involved in digital marketing are shown in Fig. 1.

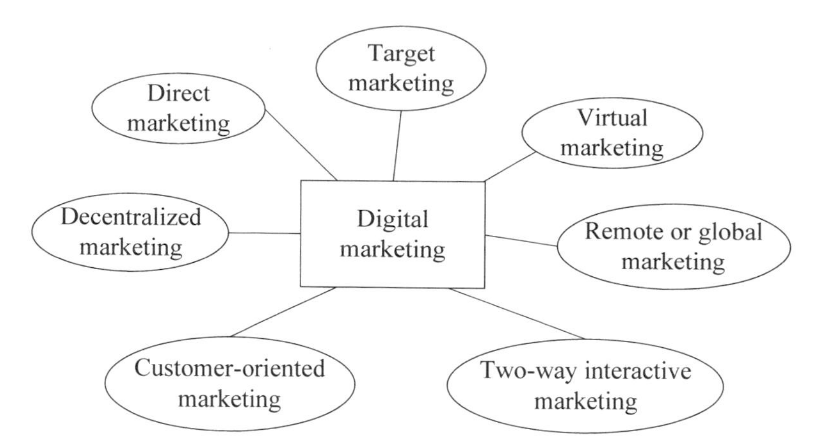

Fig. 1. Marketing methods involved in digital marketing

The basic goals and marketing concepts of digital marketing are the same as those of traditional marketing, but they have their own unique advantages and characteristics in the process of implementation and operation.

First, digital marketing is highly interactive. In the network environment, companies can collect consumer information and feedback in real time through online chat and e-mail, which can help companies achieve their marketing goals.

Second, digital marketing helps reduce marketing costs, including material costs and communication costs. It also reduces research costs. Online research saves investigators a variety of expenses and improves the efficiency of after-sales service.

Third, digital marketing can help companies increase sales and market share. It cooperates with the international market through the Internet, breaks market barriers, and promotes fair marketing of SMEs.

In short, in the network marketing environment, digital marketing can add more value to enterprises and customers, and maximize customer benefits.

The main difference between digital marketing and e-commerce is that e-commerce focuses on electronic transactions, while digital marketing focuses on the use of computer networks to achieve commercial marketing goals, e-commerce is only a means to achieve it [8, 10].

This connection exists between digital marketing and other marketing models. Internet marketing, digital marketing, and e-commerce are all results of informatization that enterprises must accept to some extent. In addition, these methods focus on different effects. For example, online marketing focuses on building your own website on the Internet, increasing click-through rates by connecting to search engines and constantly updating the user interface and website content, thereby generating potential demand and ultimately facilitating transactions, while digital marketing focuses on extracting and analyzing various Information and data to draw conclusions that guide the direction of marketing. Focus on business digitization, mainly to solve the problem of success rate of e-commerce transactions. Digital marketing combines the advantages of previous marketing methods and has the uniqueness of its own system marketing. It can allocate resources systematically to maximize marketing effects.

In addition, digital marketing is a strategic choice for corporate marketing informatization, which has long-term significance for enhancing corporate competitiveness. Digital marketing is a kind of integration of information from different marketing resources of a company. It is neither a simple combination of several information systems nor a transformation of a certain marketing method, but a comprehensive overall planning of the company's marketing resources from a systematic perspective. The planning and further development of computerized corporate marketing with a strategic vision is in line with the overall marketing concept. The future competition of enterprises is not only the competition of products and services, but also the competition of ideas and concepts. The purpose of digital marketing is not only to establish a digital marketing management system, but also to establish relevant rules and regulations within the company to continuously adapt and motivate employees to adapt to new marketing methods and continuously improve the concept of digital marketing, and to continuously strengthen to form the company's core competition force [11].

3 Experiment and Research

3.1 Experimental Background

As an integral part of the business information system, the digital marketing system must work with other systems to achieve business goals. This experiment attempts to build a digital marketing system based on an improved course optimization algorithm. The digital marketing system is a comprehensive information management system for digital users. It provides various services to customers through the Internet platform and realizes daily business operation and management. The system mainly includes the following functional modules: marketing strategy management, statistical analysis and data forecasting, customer relationship management, etc. The dynamic system structure diagram is used to describe the behavior of the system in detail. After users log in to the digital marketing system, they can view and manage basic personal information, as well as various activities such as purchasing goods.

3.2 Experimental Process

In this experiment, 100, 300, 500, and 800 real users were simulated to enter the system, and some functions of the system were tested. The test items include user login, placing an order, and receiving goods. After-sales service, etc.

4 Analysis and Discussion

Table 1. Test results

Concurrent number	User login	Place an order	Receipt	After-sales service
100	0.9	1.1	1.7	1.1
300	1.3	1.8	2.4	1.9
500	2.1	2.5	3.1	2.3
800	2.9	3.1	3.8	2.9

Some functions of the system were tested. The test items included user login, placing orders, receiving goods, and after-sales service. The test results are shown in Table 1.

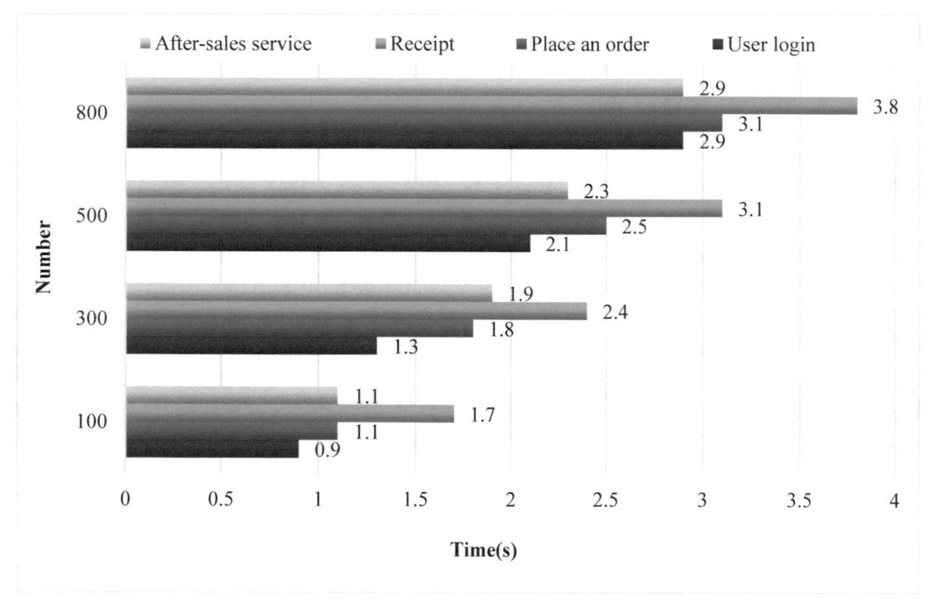

Fig. 2. Test results of the system performance

It can be seen from Fig. 2 that when the number of concurrent users is 100, the response time of the system for user login, placing an order, receiving goods, and after-sales service are 0.9, 1.1, 1.7, and 1.1 s, respectively. It can be seen that the response time of the system is shorter and the performance is better. At the same time, as the number increases, the response time is correspondingly extended.

5 Conclusion

Digital marketing is a series of activities that companies use network technology as a means to analyze customer needs and use various electronic tools to carry out product promotions, market research and other activities. The digital marketing system is a complete system, which includes many aspects such as enterprises, consumers and distributors. By collecting and analyzing digital information in these aspects, we can provide effective services for the entire market. Therefore, this article combines the improvement of teaching optimization algorithm to carry out research on digital marketing and related content.

References

1. Shukla, A.K., Singh, P., Vardhan, M.: An adaptive inertia weight teaching-learning-based optimization algorithm and its applications. Appl. Math. Model. **77**(Jan.), 309–326 (2020)
2. Prasad, A., Chokshi, S., Khan, S.: Predictive programmatic re-targeting to improve website conversion rates. J. Phys. Conf. Ser. **1714**(1), 012027 (15p.) (2021)
3. Abbassi, A., Mehrez, R.B., Abbassi, R., et al.: Improved off-grid wind/photovoltaic/hybrid energy storage system based on new framework of moth-flame optimization algorithm. Int. J. Energy Res. **46**(5), 6711–6729 (2022)
4. Onana, K.: Effect of digital marketing on consumer based brand equity of micro enterprises in Nigeria. Int. J. Manag. Soc. Sci. **2**(2), 1–37 (2021)
5. Dairo, A., Szcs, K.: Analytical approach to digital channel performance optimization of mobile money transactions in emerging markets. Innov. Mark. **16**(3), 37–47 (2020)
6. Suganthi, S.T., Devaraj, D.: An improved teaching learning–based optimization algorithm for congestion management with the integration of solar photovoltaic system. Meas. Control-Lond.-Inst. Meas. Control **53**(7–8), 002029402091493 (2020)
7. Thawkar, S.: A hybrid model using teaching–learning-based optimization and salp swarm algorithm for feature selection and classification in digital mammography. J. Ambient. Intell. Humaniz. Comput. **12**(13), 1–16 (2021)
8. Rabhi, S., Semchedine, F., Mbarek, N.: An improved method for distributed localization in WSNs based on fruit fly optimization algorithm. Autom. Control. Comput. Sci. **55**(3), 287–297 (2021). https://doi.org/10.3103/S0146411621030081
9. Aeloor, D.: Fruit-fly optimization algorithm for disability-specific teaching based on interval trapezoidal type-2 fuzzy numbers. Int. J. Fuzzy Syst. Appl. **9**(1), 35–63 (2020)
10. Slim, S.O., Elfattah, M., Atia, A., et al.: IoT system based on parameter optimization of deep learning using genetic algorithm. Int. J. Intell. Eng. Syst. **14**(2), 2021 (2021)
11. Kumar, A., Verma, S., Das, R.: Eigenfunctions and genetic algorithm based improved strategies for performance analysis and geometric optimization of a two-zone solar pond. Sol. Energy **211**, 949–961 (2020)

Research on Pension Service Medical Optimization Quality System Based on Machine Learning Algorithms

Yu Zhang(✉)

School of Management, Wanjiang University of Technology, Maanshan 243031, Anhui, China
zhangyu20210106@163.com

Abstract. The development of mobile Internet and big data technology has made the medical practice of elderly care services more practical and feasible. The application of big data mining technology in medical communications and elderly care services affects the production and distribution of medical resources. This paper clarifies the concepts of machine learning algorithms and medical optimization quality systems through combing related concepts, and provides a solid foundation for the subsequent research on the establishment of medical optimization quality systems for elderly care services based on machine learning algorithms. The article then elaborated on the current lack of diversity in elderly care services and the incomplete feedback system of the medical system. It points out that under the trend of modern information technology, combining machine learning algorithms to research on the optimization of quality system platform for elderly care services and medical care is the way to solve the problems.. The thesis analyzes the medical optimization quality system from the perspective of machine learning algorithms, with a view to making contributions to the further development of the elderly service medical field.

Keywords: Machine learning · Elderly care services and medical care · Medical optimization quality · Digital medical care

1 Introduction

The medical optimization client for elderly care services based on machine algorithms integrates elderly care services into the medical service system, and is recognized by a broad audience with the advantages of intelligent reading. The outstanding change of this new old-age service medical system is that the construction of the old-age service medical system is jointly undertaken by "manpower" and "machine algorithm." Through observation and research on elderly care, it can be found that artificial intelligence has improved the accuracy of data sources in the optimized quality system of elderly care services, the audience's personalized information needs have been met, the diversification of information has been promoted, and the elderly care services based on machine algorithms The medical optimization quality system has an undeniable positive effect in the research of social information transformation.

J. H. Abawajy et al. (Eds.): ICATCI 2022, LNDECT 169, pp. 490–497, 2023.
https://doi.org/10.1007/978-3-031-28893-7_58

2 Understanding of Related Concepts

2.1 Machine Learning Algorithm

Machine learning research aims to use computers to simulate human learning activities. This is a computer research method used to identify existing knowledge and acquire new knowledge based on data to continuously improve its performance. With the continuous improvement of learning new knowledge, frequently used machine learning algorithms such as artificial neural networks and support vector machines have evolved [1]. Due to the related characteristics and benefits, many scholars have conducted research and discussions based on different disciplines. At present, the analysis and research methods based on machine learning are supervised learning, semi-controlled learning and unsupervised learning. Supervised learning means that the training data set has input and output data, and each data is characterized and labeled [2]. By learning the input-output relationship of the training data set, the appropriate functional relationship matches the input data. Obtain the result data such as classification and regression. Unsupervised learning means that only data is input into the training set and each data is unique. Semi-controlled learning needs to maximize the gap between different types of data, such as grouping, which means that some data in the training set only contains attributes, and the rest of the data contains both attributes and labels. The two jointly establish an appropriate quantitative relationship.

2.2 Medical Quality System

In 1966, the three-dimensional implication theory of the concept of medical quality was first proposed: structure-process-result. To some extent, medical quality is regarded as a characteristic of medical institutions. Therefore, medical quality control models are widely used in various countries. The Joint Health Certification Committee believes that this kind of medical quality system research is in line with the current professional knowledge system. Health systems, services, and procurement can provide opportunities for individuals and groups to benefit [3]. Traditional medical quality is limited to clinical impact assessment, which is a narrowly defined concept of medical optimization quality. Clinical impact assessment mainly refers to the accuracy, comprehensiveness, and rapidity of diagnosis, and whether it is related to whether effective and thorough hospitalization brings patients There is nothing to do with unnecessary pain or injury. With the continuous advancement of medicine and society, as well as changes in medical models, the significance of medical quality continues to expand. The current concept of medical quality is not limited to the timeliness, safety and efficiency of traditional medical services. Can the cost of patient consultation not only satisfy the efficiency of healthcare professionals? Is the service process convenient? At the same time as patient satisfaction, the focus of traditional medical quality control models is gradually shifting from final quality assessment to process control management and associated quality [4]. Currently, medical quality control is based on three main levels of structural control. Link quality control and final quality control. Basic quality control refers to the quality control of components. Also called element quality control, it is the basic condition of quality because it contains five quality elements: medical staff, technology, equipment, materials and information. The main management methods of basic quality control include the formulation, actions, rewards and sanctions of control rules and regulations. Link quality management refers to all aspects of work from patient inspection to hospitalization, diagnosis, treatment, and performance appraisal. The quality of the discharge link defines the medical quality and is the focus of medical quality control. End point quality control refers to the sorting, calculation, and step-by-step scientific operation of the digital data collected from hospital medical end points. The first is to use numbers as facts to provide a more reliable basis for improving the quality of medical quality control. The second is to use the final quality statistical data to provide a reliable basis for quality control planning, decision-making, content, measurement and evaluation. In order to improve the health of patients, the commonly used quality indicators on the evaluation screen include improvement rate, mortality rate, cure rate, hospital diagnosis accident rate, nurses, average hospitalization rate, and bed turnover rate. Unified quality control such as medical expenses means that each patient's steps in the hospital are recorded and tracked [5].

3 Current Status of the Medical Quality System for Elderly Care Services

3.1 There is Room for Improvement in the Quality of Elderly Care Services

Most nursing homes mainly provide simple life care services, with little or no medical services. For example, about 40% of nursing homes in Beijing do not have an intelligent nursing home system or cooperate with public institutions. Most large hospitals focus on the early treatment of acute illnesses for the elderly. Patients with convalescence or chronic diseases cannot enjoy meticulous care and services. In addition, many elderly people in the hospital believe that their condition is still unstable, ignoring the doctor's advice, and insisting on hospital observation to ensure follow-up treatment and avoid the risk of recovery after discharge. This situation has also exacerbated the shortage of medical resources in large general hospitals. The existing large-scale hospitals cannot meet the real medical needs of all the elderly, and cannot give full play to the hospital's proper treatment functions. This causes a waste of medical resources. Therefore, the old-age care model of the combination of medical and elderly care has received extensive attention from the state, emphasizing the connection between treatment and rehabilitation, and better caring for the elderly. However, no matter what kind of old-age care model, there is a lack of attention to the quality of medical services for the elderly, and the elderly have a low degree of satisfaction with the medical services for the elderly. Therefore, the construction of a medical quality system for elderly care services is an issue that must be faced in response to an aging society.

3.2 Digital Healthcare is the Future Trend of Optimization of Elderly Care Services

As shown in Fig. 1, the construction of medical informatization refers to the construction of a medical information system in accordance with the level of medical informatization.

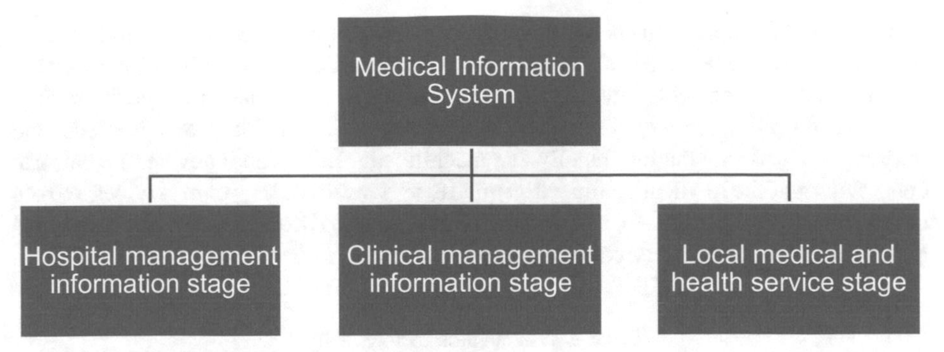

Fig. 1. Classification of medical information system construction stages

It has gone through three stages: Hospital Management Information (HMIS), Clinical Management Information (HCIS) and Community Health and Medical Services (GMIS) [6].

At present, reports on digital healthcare claim that European and American countries have entered the forefront of the development of the digital healthcare industry, and the most promising area for the development of digital healthcare today is the service platform for the medical optimization quality system. The medical optimization quality system aims to use mobile communication technology and electronic equipment to provide people with necessary medical information management and medical transaction processing. Its biggest feature is to provide national medical data analysis anytime and anywhere, so that the medical system can allocate resources, so that patients can obtain more effective medical services and obtain quality feedback, and timely repair problems that occur to improve patient satisfaction.

4 Machine Learning Algorithms Optimize the Medical Quality System of Elderly Care Services

4.1 The Key Advantages of Machine Learning Algorithms

The current world trend is the era of big data, in the context of rapid expansion of data volume, enhancement of cloud computing capabilities, and advancement in modeling technology. Artificial intelligence technology (especially machine learning) is promoting the development of global productivity in new and dynamic ways, shaping the face of cities, and enabling government agencies to improve their governance capabilities. The main body of social governance is to implement effective government management under the basic leadership of the party committee. The combination of state responsibility, citizen participation and legal protection can maintain harmony and social stability. In order to use the powerful analysis and prediction capabilities of machine learning technology to improve the overall governance level [7], more and more scholars today have moved machine learning technology to various fields of social governance such as economic and social development, especially the quality of elderly care services and medical optimization. In the system, the diversity of elderly care services can be improved first, so that the elderly can choose home care, nursing home care, and hospital care, all of which can enjoy the necessary care services for the elderly; secondly, it improves the elderly's satisfaction with care services. The introduction of machine algorithms into the medical optimization quality system can further predict and improve problems and improve medical satisfaction. Finally, the machine algorithm can innovate the platform construction of the medical optimization quality system for elderly care services, reduce human resource expenditures, and invest human resources instead. Go where it deserves to improve elderly care services (see Table 1).

Table 1. System architecture design

Data application	Medical health system	Active care system	Community service system	Security rescue system
Data support	Elderly information	Health file	Business data	Video data
Infrastructure	Smart elderly terminal equipment	The internet	Server	Sensor

4.2 Platform Innovation for the Optimization of Quality System for Elderly Care Services and Medical Care

(1) Personalized medical optimization quality system

Machine learning algorithms are mainly used to interact with the interface of the medical improvement quality system. The core logic includes three dimensions: the audience's personal attention label, context dimension and content dimension. The personal attention label of the target audience refers to the attention obtained from the user's clicks, browsing or social behavior. The contextual dimension is related to the environment in which the audience is located, and the function includes the dimensions of time and space. The text content related to the elderly medical services and results on the user's mobile phone, as well as the category, keywords, keyword optimization of the medical quality system, length, and text format (text, image, video, etc.) [8]. Medical improvement is also an aspect of content analysis. Using a special algorithm program to combine the above dimensions, users may click on the content multiple times. When the specified value is reached, the medical improvement quality assurance system can analyze the relevant data and perform corresponding processing, and push the corresponding questionnaire to the user's client interface. This is the core of the project. The system is responsible for establishing a medical information service platform from las, which provides services such as central database, data storage, and network security. At the same time, the Paas medical information service platform was built to cover services such as network backup and voice calls from third-party branches [9]. Introducing barcodes for unique centralized verification of patient identity and visual inspection numbers. Medical information services, Cloud-FLS, Cloud-LIS, Cloud-PACS, Cloud-HER and other hospitals provide remote consultation and business systems that can be built on the SaaS platform. The automated dispensing and platform logistics management system is responsible for building a logistics network, transferring samples from public hospitals to high-end hospitals, and supplementing consumables and over-the-counter drugs in public hospitals. The operation and maintenance of the operation and maintenance medical platform responsible for this investigation project can be mediated by a third-party company[10].

(2) Analysis of data sources for elderly care services

The elderly community health data file is the health record of the elderly's life process, including the sum of the data obtained from various health records. The medical optimization quality system can create a time axis to display multiple levels of health events. The horizontal axis represents the time series, and the vertical axis represents the specified time. Identify different types of medical and health events for community health agencies. Community health management can be based on residents' electronic health records, such as community chronic disease management [11]. Doctors with electronic medical records in general hospitals can refer to past medications to help diagnose patients' records. Doctors can thus provide more accurate diagnosis results and prescribe better treatment plans. Based on these huge sources of clinical information, if it can be used in scientific research, clinical, drug management and other departments, it can also be used as scientific research and educational resources such as medical research and clinical pathways. For new application management departments such as drug interaction research and monitoring, the system can create a monitoring and tracking system, create an optional decision-making system for management, and finally provide users with various aspects such as work monitoring, and conduct medical quality scientific research and analysis. [12]. The first is the establishment of elderly data collection and the establishment of a regional information system to realize the sharing of electronic medical records in network hospitals across the country; second, the health assessment of the elderly is carried out. The chronic elderly patients in the community collect blood samples from healthy cases, and relevant sites can be set up for collection. Including monitoring reports such as electronic medical records and blood glucose levels, and uploading them to the data platform. The background assessment physician can also issue corresponding alarms and suggestions through system data analysis. Third, for elderly patients with chronic diseases, if the medical service system shows that a hospital is recommended for diagnosis, you can use the HIS system to make an appointment. The medical process is much simpler, and the elderly can save treatment time and human resources. In terms of prescriptions for the elderly, the elderly are the main group of patients with chronic diseases [13]. Chronic diseases are very inconvenient. The old-age medical system can be used to advance the system to apply for drugs in advance according to their own drugs, so as to avoid inconvenience to the normal medical treatment of the elderly; finally, public health data collection. Through the establishment of a medical institution resource database, it is possible to monitor the statistical data and health information of the local elderly. Provide timely and accurate health and disease management information for the elderly. Of course, the old-age medical service system can also promote the health education of the elderly, by pushing a whole set of knowledge construction system to carry out chronic disease education for the elderly [14], and other suggestions to improve diet, exercise and quality of life, so as to achieve the purpose of interaction between patients and medical staff. The elder care service medical optimization platform construction process uses the most advanced Internet of Things technology, and the medical optimization quality system provides follow-up services for the analysis and evaluation of the medical device platform of medical institutions. Through the gradual acquisition of information, the integration of personal management and public management is realized (see Fig. 2) [15].

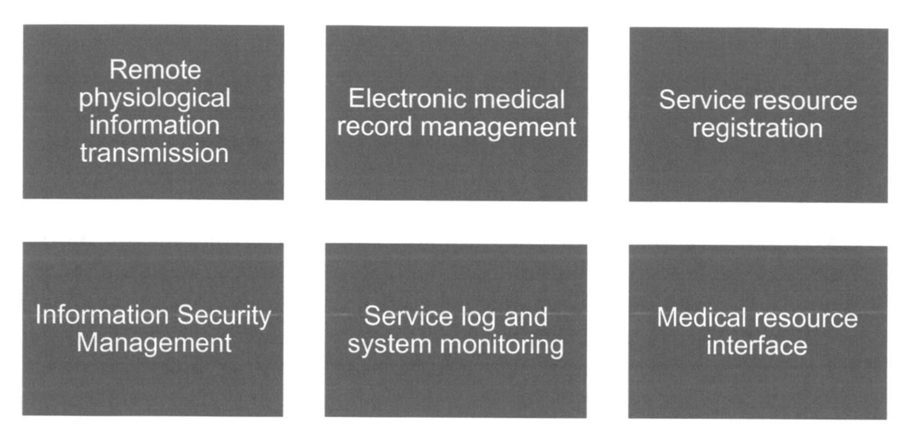

Fig. 2. The basic elements of an optimized quality system for elderly care services

5 Conclusion

Machine algorithm is a modern technology used to improve the quality system of medical optimization of elderly care services. The research of medical optimization quality system is different from other medical systems. It mainly emphasizes the development of big data, artificial intelligence and other aspects to improve the quality of medical optimization system of elderly care services. The importance of the status quo. The optimized quality system of elderly care services has become an important channel for users to feedback real information in the elderly care system. Based on the understanding of machine learning algorithms, this article analyzes how to build an optimized quality system for elderly care services from the perspective of machine learning algorithms. At the technical level, it relies on artificial intelligence and big data to screen data sources, and uses platforms such as Paas to establish interactive pages. Logic, and ultimately realize the extensive use of the optimized quality system for the elderly care services.

Acknowledgments. This work was financially supported by Anhui Province University excellent talents support program "Study on the quality evaluation and optimization path of medical care combined with elderly care service supply" (gxyq2021056) and Anhui Province University Humanities and social sciences research project "Research on collaborative construction and shared development of medical and elderly care service system in the Yangtze River Delta" (SK2021ZD0033) fund.

References

1. Ding, Q., et al.: An overview of machine learning-based energy-efficient routing algorithms in wireless sensor networks. Electronics **10**(13), 1539 (2021)
2. Rghioui, A., et al.: A smart architecture for diabetic patient monitoring using machine learning algorithms. Healthcare **8**(3), 3–5 (2020). Multidisciplinary Digital Publishing Institute
3. Zhu, M., et al.: Using machine learning algorithms to guide rehabilitation planning for home care clients. BMC Med. Inform. Decis. Mak. **7**(1), 1–13 (2007)
4. Panch, T., Szolovits, P., Atun, R.: Artificial intelligence, machine learning and health systems. J. Glob. Health **8**(2), 2 (2018)
5. Thakur, N., Han, C.Y.: A study of fall detection in assisted living: identifying and improving the optimal machine learning method. J. Sens. Actuator Netw. **10**(3), 39 (2021)
6. Belhor, M., et al.: A new MIP model and machine learning approach for home health care: optimization of cancer treatment process by chemotherapy. In: 2020 5th International Conference on Logistics Operations Management (GOL). IEEE (2020)
7. Kumar, S.M., Majumder, D.: Healthcare solution based on machine learning applications in IOT and edge computing. Int. J. Pure Appl. Math. **119**(16), 1473–1484 (2018)
8. Gianfrancesco, M.A., et al.: Potential biases in machine learning algorithms using electronic health record data. JAMA Intern. Med. **178**(11), 1544–1547 (2018)
9. Jin, Y., Sendhoff, B.: Pareto-based multiobjective machine learning: an overview and case studies. IEEE Trans. Syst. Man Cybern. Part C (Appl. Rev.) **38**(3), 397–415 (2018)
10. Yacchirema, D., de Puga, J.S., Palau, C., Esteve, M.: Fall detection system for elderly people using IoT and ensemble machine learning algorithm. Pers. Ubiquit. Comput. **23**(5–6), 801–817 (2019). https://doi.org/10.1007/s00779-018-01196-8
11. Fatima, M., Pasha, M.: Survey of machine learning algorithms for disease diagnostic. J. Intell. Learn. Syst. Appl. **9**(01), 1 (2017)
12. Snoek, J., Larochelle, H., Adams, R.P.: Practical Bayesian optimization of machine learning algorithms. In: Advances in Neural Information Processing Systems, vol. 25 (2015)
13. Silver, D.L., Yang, Q., Li, L.: Lifelong machine learning systems: beyond learning algorithms. In: 2013 AAAI Spring Symposium Series (2013)
14. Bradley, A.P.: The use of the area under the ROC curve in the evaluation of machine learning algorithms. Pattern Recogn. **30**(7), 1145–1159 (2017)
15. Ibrahim, I., Abdulazeez, A.: The role of machine learning algorithms for diagnosing diseases. J. Appl. Sci. Technol. Trends **2**(01), 10–19 (2021)

Two-Dimensional Code Information Security Design Based on Hash Function and Encryption Algorithm

Xingwei Chen(✉)

Jinhua Polytechnic, Jinhua 321007, Zhejiang, China
uzak42cc@163.com

Abstract. With the development of information networks, encryption technology has also been improved to a large extent. In the research of information security, encryption is a very important subject. It is not only related to the integrity, confidentiality and availability of information and data. It has practical significance for the information security design of the two-dimensional code. Therefore, this article conducts related research on hash function, encryption algorithm and QR code information security, with the purpose of improving network security. This article mainly uses the experimental method and the comparative method to study the application of the hash function encryption algorithm in the information security design of the two-dimensional code. Experimental data shows that its safety performance can reach more than 90%, which basically meets actual needs.

Keywords: Hash function · Encryption algorithm · QR code information · Security design

1 Introduction

Information security is a very important subject. In today's society, with the rapid development and popularization of applications such as Internet technology, mobile communication networks, and multimedia storage, encryption technology has also developed rapidly. Among them, the main aspects are cryptography and data access control. The most commonly used in the field of modern communications is to keep secrets based on hash algorithms.

There are many research theoretical results on the design of two-dimensional code information security based on hash functions and encryption algorithms. For example, Some scholars said that image encryption adopts hash function and splicing mode, which can resist a large number of attacks [1]. Some scholars said that the current chaotic encryption scheme mainly uses iterative periodic sequence and one-way broadcast mechanism to achieve pixel confusion, and designed an encryption algorithm to solve the problem of poor anti-decryption performance [2]. JSome scholars feels that the QR code has played an important role as an information carrier and identity authentication tool, but the security problems it brings are also endless [3]. Therefore, this article

J. H. Abawajy et al. (Eds.): ICATCI 2022, LNDECT 169, pp. 498–505, 2023.
https://doi.org/10.1007/978-3-031-28893-7_59

has a more in-depth understanding of QR code information security design, and a deeper understanding of hash functions and encryption algorithms.

This article first studies the image encryption technology and discusses its basic theory. Secondly, the design of the hybrid encryption security protocol is analyzed in detail. Then the related theories of the two-dimensional code are explained, and the function of the two-dimensional code is proposed. After that, the realization of data encryption on the QR code is analyzed. Finally, the performance of information security is tested through experiments and relevant conclusions are drawn.

2 Two-Dimensional Code Information Security Design Based on Hash Function and Encryption Algorithm

2.1 Image Encryption Technology

The purpose of the image encryption algorithm is to change the color, brightness and other information of the original image through certain calculations and operations, so that the image is composed of noise-like information randomly.

Other theoretical knowledge and tools such as hash functions are often used in the design of image encryption algorithms. Because it is irreversible, it can resist known plaintext attacks and selected plaintext/ciphertext attacks. Therefore, this article chooses SHA1 and MD5 hash algorithms [4, 5].

The hash function can be simply understood as a message digest. Changing one or more digits in the message will change the digest value of the message. Usually the hash function algorithm is public. Hash functions can be used for digital signatures and data integrity testing, password protection, message verification codes, etc. The design of one-way hash function is based on the idea of compression function, and the hash function constructed using this method is called iterative hash function. Hash functions can be further subdivided into directly constructed hash functions and hash functions based on block ciphers. Compared with the directly constructed hash function, the hash function based on block cipher, because in the iterative process, the key will change every time it is encrypted, which affects the speed [6, 7].

For hash function attack methods, there are mainly search collision method and exhaustive method. MD5 and SHA1 currently have no effective collision detection methods, but the exhaustive methods are applicable to some computational objects in a region. The hash function is used to calculate these calculation objects and compare the calculation results with their equal strings. The operation object must be a part of the original object, thereby attacking system encryption. Although the hash algorithm is irreversible, some encryption systems have very simple keys. Therefore, this article proposes to use a multiple hybrid hash algorithm, that is, use different hash functions plus an additional control parameter key to encrypt the key multiple times. If the key is complex, the exhaustive method becomes extremely difficult [8, 9]. This article uses the following mixed formula for hashing:

$$r = sha1(hex1dec(MD5(z)) \oplus hex2dec(MD5(key))) \tag{1}$$

Among them, $\oplus$ stands for exclusive-or operation, hex2dec($\cdot$) represents the conversion of hexadecimal to decimal.

Assume that for a plaintext image Q of size $X \times Y$, X is the number of rows and Y is the number of columns. Its average pixel value:

$$m_1 = floor(\sum_{p=1}^{16} Q_m/16) \tag{2}$$

$$m_2 = floor(\sum_{p=1}^{8} Q_m/8) \tag{3}$$

where floor(·) means round down.

2.2 Design of Hybrid Encryption Security Protocol

The current RFID system mainly adopts the following security protocols: Hash Lock protocol, Random Hash Lock protocol, Hash String protocol, Hash Based ID Conversion protocol, LCAP protocol, Hash Authentication distributed query, RFID authentication protocol and library RFID protocol.

It can be seen from the log recording process that the hash lock protocol only executes a hash function on the tag, and the value K is maintained in the background database. In this way, the hash lock protocol can also be extended to multi-user access, provide access controls, and can also be used as an object identifier in the locked state. However, during the entire protocol execution process, the ID value is not dynamically updated; in particular, the ID value is also transmitted in clear text in the communication channel to facilitate its monitoring. An attacker can pretend to be a reader to obtain the meta ID and K value of the tag, thereby unlocking the tag and obtaining tag information [10, 11].

The random hash lock protocol is an improved form of the hash lock protocol. The implementation of this protocol indicates that the back-end database sends all its stored credentials to the reader without identity checking, and, like the hash lock protocol, it is always completed after the reader is authenticated. Encrypted transmission, once acquired by an attacker, can be disguised as a legal label, making it unable to resist tampering and forwarding attacks [12, 13].

Commonly used hash functions include MD series, SHA, etc.. This protocol guarantees high security while occupying less resources. The MD5 function is used to encrypt communication data. The hash algorithm has the following advantages:

Check the file. Currently, the most widely used check algorithms are the CRC check algorithm and the parity check algorithm. Although these two algorithms can detect and correct errors in data information transmission to a certain extent, they cannot prevent data from being maliciously tampered with.

Electronic signature. The hash algorithm is a very important part of the modern cryptosystem, especially in the digital signature protocol, the hash function plays a key role. In the analysis and statistics, the digital signature of the hash value can be recognized as the digital signature of the file itself.

Authentication agreement. The authentication protocol is also called challenge authentication mode. Assuming that the data transmission channel can be monitored but not manipulated, then the hash function can be described as a very simple and safe method.

The two-dimensional code information security design takes into account both the encryption algorithm and the key, and proposes a cryptographic coding scheme. First, use the hash function to generate the ciphertext and send it. Then use the hash function to express and store the necessary information in the ciphertext as plaintext symbols. Secondly, by entering the content to be transmitted in the buffer port of the receiving end, the ciphertext and the decryption result are obtained, and it is judged whether it is a private key thief.

2.3 QR Code

The possibility of encountering similar graphics elsewhere in the symbol is extremely low, so any QR code symbol can be quickly identified in the field of view. Recognizing the three position recognition patterns that make up the viewfinder pattern can clearly determine the position and direction of the symbol in the field of view.

There is a 1-module wide separator at each position between the recognition pattern and the coding area.

The horizontal and vertical positioning graphics each occupy one row and one column wide module. The module is composed of alternating light and dark modules, and the dark modules are at the beginning and the end. The horizontal positioning pattern is located between the two upper position sensing patterns on the 6th line of the symbol. The vertical positioning map is located between the two position detection maps on the left side of the sixth column of the symbol. Their function is to determine the density and version of the symbol, and provide a reference position to determine the modulus coordinates.

The coding area includes symbols representing data codewords, error correction codewords, version information, and format information.

According to the data type, select the appropriate encoding mode and the level of error detection and correction required.

According to the version used, Finder graphics, separators, positioning graphics, correction graphics and codeword modules are formed in the matrix.

The mask pattern is in turn used for the coded area of the symbol. Evaluate the result and choose the result that optimizes the dark and light modulus ratio and minimizes the unwanted graphics.

Finally, the format and version information are generated.

2.4 Implementation of Data Encryption on QR Code

In recent years, the use of QR codes has become more and more widespread. Combining the two-dimensional code encryption technology, overcome the two-dimensional code problem. The information it contains is a code transmitted through the Internet and other physical spaces, and the error that is easy to decrypt and copy ensures the security of the QR code. Therefore, in order to strengthen the security of the two-dimensional code, it is necessary to improve the security of the selected encryption and ensure the effectiveness of its promotion and application. Encrypt and re-encrypt the source information, the specific process is shown in Fig. 1:

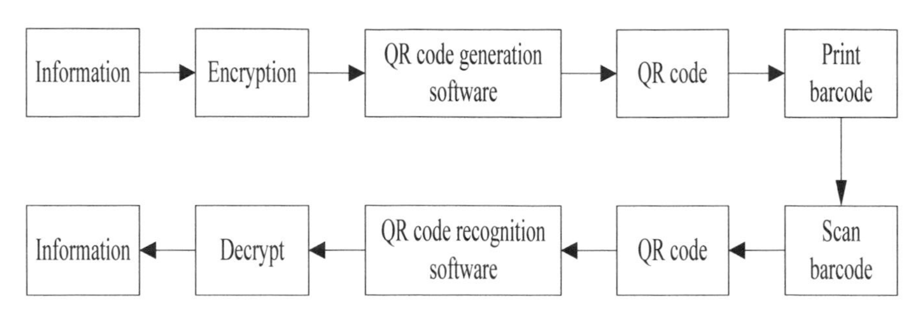

Fig. 1. The specific process of first encryption and then encoding

3 Implementation of QR Code Information Security System

3.1 Architecture Design

The data layer is the lowest layer of the architecture, responsible for network communication, as well as retrieving data and sending data to upper layers. The data layer can be divided into network layer, local data layer and supply layer. These three layers respectively perform three tasks: network communication, data retrieval, and data transfer to the upper layer. The main function of the network layer is to implement certain network API calls and manage communication strategies under different network states. The main job of the local data layer is caching. The design of the caching strategy includes what data should be cached and what file format should be cached. In order to reduce server-side pressure and improve user experience, you can save some important data or frequently requested pages as local data without waiting for a certain amount of time for each request. The delivery layer is composed of some open interfaces and protocols to process data requests sent from the top-level structure to the data layer. The client uses the HTTP protocol to exchange data with the server, and sends and receives data in Json format. After receiving the Json data from the server, the client parses the data layer in a Java instance, and then parses the data in Json format.

The main function of the business layer is to process business logic. The business layer is above the data layer and below the display layer. Interaction with the data layer is the interface for the business layer to call the data layer, and interaction with the display layer is the interface for the display layer to call the data layer and management layer. A single-activity multi-segment model was adopted, and certain types of tools were written to supplement the business design.

The main function of the display layer is to carry out human-computer interaction, comprehensively considering various factors such as surface layout, image selection and display, and font embellishment.

3.2 Development Environment Construction

The experimental environment of this article: 64-bit Windows8 operating system, Intel Core i7 3.70 GHz processor, 12 Gb memory, JDK 2.8.0 version, Android Studio 4.0 version, AndroidSDK r24.4.1 version. Android application development requires the installation of JDK, Android SDK and Android Studio.

3.3 The Overall Framework of the Information System

The watermark embedding subsystem is mainly based on electronic documents, adding the confidential information of the file and the user's identity information to the corresponding paper documents.

The copy control subsystem mainly controls whether users are allowed to copy confidential documents through the correspondence between user authorization and document secret level, and adds the identity information of the copy user and copy-related information to the watermark, records log information and tells the copy process at the same time.

The audit and configuration platform configures equipment usage policies, manages and analyzes the equipment usage logs of the copy users imported from the replication control subsystem, and can also extract information from the watermark for efficient document tracking.

4 Analysis of System Test Results

System testing is to find errors in the program and make improvements. By analyzing the existing encryption algorithm and its supporting information security design, it can achieve the purpose of protecting users' private data, improving encryption and decoding efficiency, and reducing costs.

4.1 Analysis of Performance Results of Multiple Test Information

According to the experimental test survey in this article, four samples were selected for analysis. In these four samples, their security, integrity and credibility data are similar, as shown in Table 1:

Table 1. Analysis of the performance results of multiple test information

	Safety	Integrity	Creditability
1	90%	92%	90%
2	91%	94%	91%
3	94%	95%	95%
4	91%	92%	92%

As shown in Fig. 2, we can see that the highest safety and integrity is the No. 3 test. In addition, the indicators of the remaining groups can meet the basic requirements. The average value of safety data is 91.5%, the average value of completeness is 93.25%, and the average value of credibility is 92%.

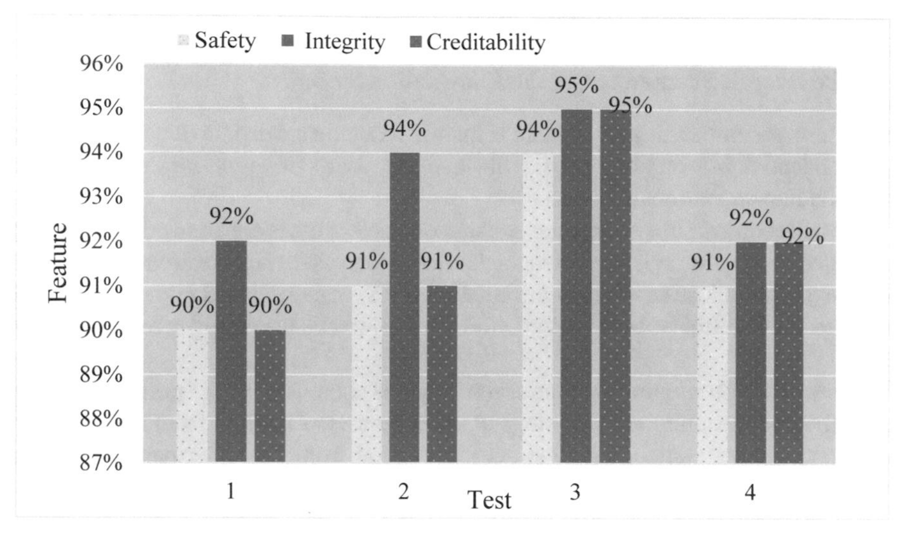

Fig. 2. Analysis of the performance results of multiple test information

5 Conclusion

A two-dimensional code is a plane structure composed of one-dimensional or two-dimensional character codes, which can be organized according to certain requirements. The encryption algorithm in this article is based on the security design of the QR code information itself. Therefore, when implementing password cracking, we only need to use different types of plaintext to replace the original code. In this system, in order to verify the security of the model, an encryption algorithm—hash function is used. This paper proposes a scheme based on hash function and matrix to protect the encoding and hiding and decryption, and verify the effectiveness and security of the method by verifying the experimental results.

References

1. Panwar, K., Purwar, R.K., Srivastava, G.: A fast encryption scheme suitable for video surveillance applications using SHA-256 hash function and 1D sine-sine chaotic map. Int. J. Image Graph. **21**(02), 50–54 (2021)
2. Abbasi, A.A., Mazinani, M., Hosseini, R.: Evolutionary-based image encryption using biomolecules operators and non-coupled map lattice. Optik – Int. J. Light Electron Opt. **219**(12), 164949 (2020)

3. Sbaytri, Y., Lazaar, S.: A design of a new hash function based on cellular automata. J. Theor. Appl. Inf. Technol. **99**(10), 2280–2289 (2021)
4. Abed, S., Waleed, L., Aldamkhi, G., et al.: Enhancement in data security and integrity using minhash technique. Indones. J. Electr. Eng. Comput. Sci. **21**(3), 1739–1750 (2021)
5. Fitriyanto, R., Yudhana, A., Sunardi, S.: Implementation SHA512 hash function and Boyer-Moore string matching algorithm for Jpeg/exif message digest compilation. Jurnal Online Informatika **4**(1), 16 (2019)
6. Pieprzyk, J., Suriadi, S.: A QR code watermarking approach based on the DWT-DCT technique. In: Pieprzyk, J., Suriadi, S. (eds.) ACISP 2017. LNCS, vol. 10343, pp. 314–331. Springer, Cham (2017). https://doi.org/10.1007/978-3-319-59870-3_18
7. Sharma, V., Joshi, A.M.: VLSI implementation of reliable and secure face recognition system. Wirel. Pers. Commun. **122**(4), 3485–3497 (2021). https://doi.org/10.1007/s11277-021-09096-6
8. Soltani, M., Bardsiri, A.K.: A new secure hybrid algorithm for QR-code images encryption and steganography. APTIKOM J. Comput. Sci. Inf. Technol. **2**(2), 86–96 (2017)
9. Mathivanan, P., Ganesh, A.B., Venkatesan, R.: QR code–based ECG signal encryption/decryption algorithm. Cryptologia **43**(3), 233–253 (2019)
10. Maetouq, A., Daud, S.M.: HMNT: hash function based on new Mersenne number transform. IEEE Access **8**, 80395–80407 (2020)
11. Avrylova, A. H., Korol, O., Milevskyi, S.: Mathematical model of authentication of a transmitted message based on a McEliece scheme on shorted and extended modified elliptic codes using UMAC modified algorithm. Cybersecur. Educ. Sci. Tech. **1**(5), 40–51 (2019)
12. El-Latif, A.A., Abd-El-Atty, B., Venegas-Andraca, S.E., et al.: Efficient quantum-based security protocols for information sharing and data protection in 5G networks. Futur. Gener. Comput. Syst. **100**(Nov.), 893–906 (2019)
13. Rosales-Roldan, L., Chao, J., Nakano-Miyatake, M., Perez-Meana, H.: Color image ownership protection based on spectral domain watermarking using QR codes and QIM. Multimedia Tools Appl. **77**(13), 16031–16052 (2017). https://doi.org/10.1007/s11042-017-5178-8

Method and Implementation of Vehicle Body Attitude Detection Based on Beidou Satellite

Lin Wang(✉)

Wuhan Institute of Shipbuilding Technology, Wuhan 430050, Hubei, China
tintinmini@163.com

Abstract. Attitude measurement parameters are important information to characterize the safety of aircraft carrier operations, such as measuring and controlling the attitude of spacecraft, measuring the deflection angle of aircraft, and obtaining orientation information of land vehicles. The detection of attitude measurement parameters is the basis for the development of active traffic condition safety monitoring. The Beidou satellite navigation system can not only receive carrier position, velocity and time information, but also attitude parameters that can be applied to a wide range of applications. This paper studies the attitude detection method of the vehicle planning problem on the Beidou satellite, and understands the relevant knowledge theory of the vehicle body attitude detection on the basis of the literature. The designed system is tested, and the test results show that the error between the two output headings of the Beidou satellite vehicle body attitude system in this paper is less than 0.2mil.

Keywords: Beidou satellite · Vehicle attitude · Attitude detection · Navigation system

1 Inductions

Attitude measurement is one of the key technologies of air navigation, aerospace, navigation and land navigation systems, and is the basis for active air traffic safety surveillance [1, 2]. By obtaining the stopover information, not only the attitude of the fuselage itself can be understood, but also the trajectory with the future can be predicted, which is widely used in the fields of aircraft and ship navigation, automatic driving of land vehicles, and aerial photogrammetry [3, 4]. Attitude measurement can not only quickly and accurately obtain stop and azimuth results, but also can artillery placement, stop measurement and radar control, as well as fast general azimuth measurement of maneuvering rocket launches. If there is an error in attitude measurement, it will affect the accurate attack and lead to the appearance of fighters. Errors even affect entire armies [5, 6]. With the development of IT equipment and the expansion of application scenarios, attitude measurement has become a research hotspot in the fields of navigation and intelligent control in various countries [7, 8].

In attitude measurement research, there is a view that different current attitude measurement methods have different environments and application fields. The monitoring

J. H. Abawajy et al. (Eds.): ICATCI 2022, LNDECT 169, pp. 506–513, 2023.
https://doi.org/10.1007/978-3-031-28893-7_60

of low dynamic and static platforms is mainly optical and infrared devices. While such sensors are very accurate, they are susceptible to weather, soil or other objective factors, making it difficult to provide real-time disruption or direction information. GPS attitude measurement systems mainly rely on complex measurement systems such as inertial navigation systems and platform compasses to provide conveyor position information. Both are self-contained navigation systems with excellent concealment and robust high-frequency dynamic measurements, but drift dynamics can degrade measurement accuracy over time [9]. Some researchers have used the Beidou B1 and B2 dual-frequency vector carrier centerline observations to design and study the software of the Beidou dual-frequency carrier measurement system, and combined the LAMBDA and TCAR algorithms to solve the integer ambiguity problem [10]. To sum up, with the development of aviation technology, the attitude measurement of Beidou satellites has attracted more and more scholars' attention.

This paper studies the vehicle body attitude detection method of Beidou satellite, analyzes the significance of vehicle body attitude detection and the method of Beidou satellite vehicle body attitude detection on the basis of literature data, and then analyzes the vehicle body attitude of Beidou satellite. The detection system is designed, and the designed system is tested, and relevant conclusions are drawn from the test results.

2 Research on Vehicle Body Attitude Detection

2.1 Significance of Vehicle Body Attitude Detection

The field of intelligent driving perception mainly studies the acquisition of information and the understanding of the world environment by intelligent driving systems. There are many kinds of information that can be used for intelligent driving, such as image information recorded by cameras similar to human eyes, GPS location information, map information, road condition information, vehicle condition information, etc. However, vehicle parking information is abstract information that cannot be directly collected from the environment, and must be further calculated using original information, such as image information [11].

Body posture information is also one of the important determinants when people drive. The human brain uses visual information to determine the position of the vehicle ahead. If its orientation is not parallel to the lane, expect it to inevitably turn and move sideways. The larger the steering angle, the more the vehicle turns. At the same time, observing whether the vehicle ahead is changing direction is also an important basis for judging whether the vehicle can safely overtake.

At the same time, the vehicle location information can be used as important dynamic semantic information on the semantic map. With the participation of a sufficient number of online map crowdfunding vehicles, the real-time semantic information of vehicle stops in the city can be obtained relatively completely. This kind of semantic information provides a basis for the macroeconomic flow forecast of vehicles in the traffic control department, and at the same time helps to divert traffic flow and improve the operation efficiency of urban traffic.

The vehicle attitude detection algorithm uses the image information collected by the camera to detect the vehicle recorded by the camera to determine the location and parking information. It is one of the solutions to automatically identify the target vehicle, which does not require additional sensors and is inexpensive to deploy. The challenge is accuracy on the one hand and computational speed on the other. Accuracy determines the reliability of information,and1sufficiently accurate attitude information can provide a reliable basis for driving decisions and achieve safe driving, but major misinformation may also lead to wrong decisions and accidents. In terms of computing speed, this is the factor that determines how quickly the intelligent drive system reacts. If the algorithm runs too slowly, the intelligent driving system of course cannot detect changes in the environment in time, nor can it react in time. This is extremely fatal in high-speed application scenarios such as driving. On the other hand, if the time complexity of the algorithm is low, the algorithm can be developed on an inefficient and cheap hardware platform without sacrificing real-time performance, saving material cost and consumption. Such low-cost platforms tend to have low prices and strong adaptability to harsh environments (for example, industrial computers are hardware platforms that reduce performance and improve environmental adaptability). This makes the entire smart drive system cheaper and improves ease of use [12].

The perception field of intelligent driving mainly studies the information acquisition and understanding of the surrounding world by the intelligent driving system. There are various kinds of information that can be used by intelligent driving, such as: image information captured by cameras similar to those observed by the human eye, GPS positioning information, map information, road condition information, vehicle state information, and so on. However, the vehicle attitude information, as an abstract information that cannot be directly collected from the environment, needs to be further calculated by using the original information such as image information.

2.2 Beidou Satellite Vehicle Body Attitude Detection Method

Compared with the traditional station system, the station system based on the Beidou system has the advantages of low cost, high precision, convenient installation, and high stability of research and development products. However, with the development of modern military aircraft and weapons and equipment, higher requirements have been placed on the accuracy and reliability of navigation technology. Based on the analysis and investigation of the status quo of the Beidou satellite signal system, the main problems currently faced by the Beidou station technology are:

(1) The single-satellite navigation station system has low short-term stability, low data rate, and is easily limited by satellite signals, while the inertial navigation data rate is high, the interference is strong, and the inertial navigation and inertia are large. In view of the above shortcomings, the traditional Kalman filter is easy to diverge due to the accumulation of rounding errors during data fusion, and other filtering algorithms can suppress the problem of filtering deviation while ensuring accuracy.
(2) The measurement of direction and phase is the basis of Beidou navigation position measurement. Vector phase measurements are output from the vector monitoring loop. Crystal oscillators are an important standard source of frequency stability for satellite navigation systems, and their performance can suffer. When the Beidou navigation system is used in a high dynamic environment, it is easily affected by

complex environments such as acceleration shock, and the crystal oscillator is very sensitive to acceleration, resulting in the continuous shift of the output frequency of the crystal oscillator. The acceleration effect of the crystal oscillator has an impact on the performance of the Beidou measurement system. The Doppler shift caused by carrier-satellite motion within the carrier tracking loop can be effectively supported by estimating inertial information. Although this technology has matured, the impact and compensation of the crystal oscillator acceleration effect on the carrier tracking loop has not been studied in detail at home and abroad.

(3) There are many types of orbits of Beidou satellites, such as geostationary orbit, IGSO, MEO, etc. The orbits are generally higher than those of GPS satellite systems, which are not suitable for eliminating the ambiguity of the entire cycle. In addition, the traditional process of solving fuzzy frequency loops is complex and time-consuming, but inertial information can be input to solve loop ambiguities, compress the search space of loop ambiguities, and improve search efficiency and accuracy.

3 Design of Vehicle Body Attitude Detection Based on Beidou Satellite

3.1 Main Scheme Design of Beidou Navigation Attitude Measurement System

The core part of the Beidou data acquisition platform used in this paper consists of a receiver connected to two Beidou antennas. Two Beidou antennas are connected at both ends of the carrier, and the connection between the two Beidou antennas is the baseline. Figure 1 shows the main scheme of the Beidou navigation attitude measurement system.

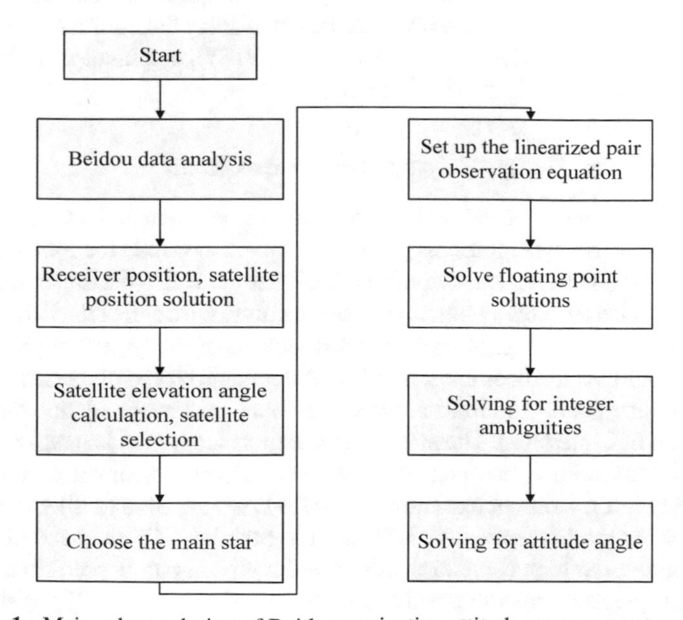

Fig. 1. Main scheme design of Beidou navigation attitude measurement system

Assuming that the Beidou-1 antenna acts as the main antenna, and the Beidou-1 antenna and the Beidou-2 antenna receive the Beidou satellite data at the same time, it is possible to obtain false orange and carrier phase observations. Due to the current overall operation of the Beidou system, the navigation mode is relatively stable, and the number of satellites observed by the receiver is usually the number of satellites required to measure the satellite position in an unobstructed position. Therefore, according to the characteristics of the Beidou data acquisition platform, a specific Beidou attitude measurement algorithm scheme is formulated. The design idea of this scheme is: first, according to the Beidou online data received from the main and auxiliary antennas, calculate the observation position of the satellite; the position of the receiver; and then the double difference in the pseudo-distance. Defined by the vector phase observation model and the double-difference model, the double-difference observation equation is linearized to obtain the linear double-difference observation equation, and then the LAMBDA algorithm using the basic constraints of the distributed table point solution and the floating point solution is used to find the integer ambiguity degrees and obtain a fixed solution for integer ambiguities.

3.2 Data Collection

The data subsystem includes the main controller, GPS receiver unit. The FPGA is selected as the main controller, the GPS module is used as the position and speed information retrieval module, and the GPS module is responsible for obtaining the position and speed data. The FPGA collects the position and velocity data output by the GPS and sends the collected data to the stop computing subsystem through the bus interface.

The attitude calculation subsystem uses the DSP as the main controller. The DSP mainly receives the attitude information through the interface calculation, and sends the received attitude information to the host interface to display the information in real time. The host data display interface is designed by LabVIEW, and communicates with the DSP host controller through the UART interface.

3.3 Overall Scheme Design of System Hardware Circuit

The system is mainly composed of data acquisition subsystem and parking calculation subsystem. The data acquisition subsystem is mainly responsible for collecting the output data of the GPS receiver, and outputs the DSP instructions to the Ethernet and RS232 interfaces through the decoding function. More accurate selection of the IMU as the inertial measurement unit for high-precision measurements of the system. A programmable field gateway (FPGA) has been chosen as the master controller for the subsystem, allowing parallel collection of data from multiple GPS channels by simulating the timing of the UART and SPI interfaces. The attitude calculation subsystem is mainly responsible for calculating the attitude measurement system algorithm and communicating with the host to display information in real time. The ADI DSP (ADSP-21489) was selected as the processor for legal computation. DSP has the function of high-speed digital signal processing, and it has high computing power and high computing precision, which can meet the high-precision measurement requirements of the system. The system design is divided into two parts: the data acquisition subsystem and the attitude calculation subsystem, so that the DSP processor focuses on real-time data and attitude data calculation, and the FPGA is used for data acquisition,and take full advantage of system performance.

3.4 Resolving Integer Ambiguity

Carrier-based position and position measurements are the primary way that precision satellite navigation systems perform position and position measurements. When using carrier-phase observations to measure position and orientation with high accuracy, the most difficult problem is solving integer ambiguities.

Since thc ambiguity of the integer itself must be an integer, the square of the distance between the integer vector and the floating-point solution is used as the objective function to find the ambiguity of the integer that minimizes the objective function.

$$\min_{a} \|a - \hat{a}\|^2_{Q_{\hat{a}}^{-1}} \tag{1}$$

For the least squares problem of Eq. (1), if the coefficient table $Q_{\hat{a}}^{-1}$ of $\hat{a}$ is a diagonal table, the minimum solution $\hat{a}$ of a is directly equal to the rounded value of i. But in reality, $Q_{\hat{a}}^{-1}$ is usually not a diagonal array, so rounding $\hat{a}$ to the nearest integer won't work. Therefore, it is necessary to find the best solution when applying it. With χ^2 as a suitable search limit, the search space for integer solutions of integers can be expressed as:

$$\|a - \hat{a}\|^2_{Q_{\hat{a}}^{-1}} \leq \chi^2 \tag{2}$$

Equation (2) is actually a multi-dimensional ellipsoid. Double difference ambiguity table $Q_{\hat{a}}^{-1}$ has different weights for each measurement. Due to the constant correlation between the ambiguities of the double-differenced integers in the double-difference observation model, the weights of different pairs of measurement values vary greatly, and the multidimensional elliptical search space becomes narrower and narrower. The optimal solution a may not be close to or far from the floating-point solution a. Faced with these problems, the LAMBDA algorithm provides two basic contents. First, reduce the fuzzy correlation of two-digit integers, so that the multi-dimensional ellipse search space tends to be a sphere, and the volume is unchanged, and the search algorithm is used to search the space in the reduced space, and the optimal solution $\hat{a}$ is obtained.

4 System Test

The 5-m baseline of the GPS stop-measure system is mounted on the top of the vehicle and placed as centrally on the roof as possible. There is an error between the theoretical central axis of the vehicle and the actual installation position. In the actual use process, the placement error needs to be adjusted. For densities above 0.1, GPS stop measurement system and standard instrument data are recorded each time the parking lot is passed. The experimental results are shown in Table 1.

Table 1. System test results

	GPS attitude measurement system	Standard instrument
1	1645.98136	1645.9387
2	438.66612	438.536689
3	3455.75545	3455.672978
4	1644.53467	1644.496434
5	1616.62934	1616.555078

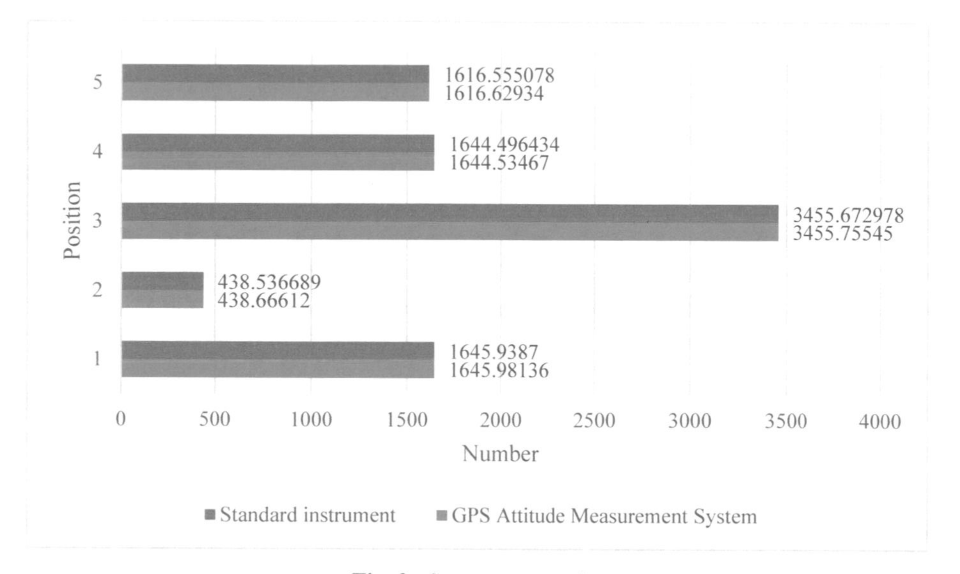

Fig. 2. System test results

As can be seen from Fig. 2, the error between the two outputs of the vehicle stop measurement system is less than 0.20mil, and the two output headings are the dynamic sports car and the instrument, respectively. Many application goals proposed prior to system design have been successfully tested, including accurate measurement accuracy and excellent dynamic performance, which can be used to significantly guide vehicle carriers.

5 Conclusions

This paper studies the vehicle body attitude detection method of Beidou satellite. After analyzing the relevant knowledge theory, the Beidou satellite vehicle body attitude monitoring system is designed, and the designed system is tested. The test results show that this The error between the two output headings of the vehicle attitude measurement system is less than 0.2mil, and the two output headings are the dynamic sports car and the instrument respectively.

References

1. Svenningsson, J., Hultén, M., Hallström, J.: Understanding attitude measurement: exploring meaning and use of the PATT short questionnaire. Int. J. Technol. Design Educ. **28**(1), 67–83 (2016). https://doi.org/10.1007/s10798-016-9392-x
2. Liu, J., Ma, Q., Zhu, H., et al.: Non-scanning measurement of position and attitude using two linear cameras. Optics Lasers in Eng. **112**(JAN.), 46–52 (2019)
3. Batista, P.: Robustness to measurement noise of a globally convergent attitude observer with topological relaxations. Nonlinear Dyn. **98**(1), 589–600 (2019). https://doi.org/10.1007/s11071-019-05214-z
4. Zhou, M., Wang, D.: Generational differences in attitudes towards car, car ownership and car use in Beijing. Transportation Research Part D Transport and Environment, **72**(JUL.), 261–278 (2019)
5. Ao, Y., Yang, D., Chen, C., et al.: Exploring the effects of the rural built environment on household car ownership after controlling for preference and attitude: Evidence from Sichuan, China. Journal of Transport Geography, **74**(JAN.), 24–36 (2019)
6. Malekzadeh, M., Sadeghian, H.: Attitude control of spacecraft simulator without angular velocity measurement. Control Eng. Pract. **84**(MAR.), 72–81 (2019)
7. Kautish, P., Sharma, R.: Value orientation, green attitude and green behavioral intentions: an empirical investigation among young consumers. Young Cons. Insight Ideas Responsible Market. **20**(4), 338–358 (2019)
8. Jacotă, V.-G., Negruş, E.-M., Toma, M.F.: Evaluation and measurement the recovered energy from automobile suspension in the operation conditions. Int. J. Automot. Technol. **19**(6), 1049–1054 (2018). https://doi.org/10.1007/s12239-018-0102-4
9. International S. Rolling resistance measurement procedure for passenger car, light truck, and highway truck and bus tires. Tire Technology International, (Mar.), 106–106 (2018)
10. Blokland, W., Koppel, S., Lodewijks, G., et al.: Method for performance measurement of car companies from a stability-value leverage perspective The balancing act between investment in R&D, supply chain configuration and value creation. Int. J. Lean Six Sigma **10**(1), 411–434 (2019)
11. Severgnini, E., Vieira, V.A., Galdamez, E.: The indirect effects of performance measurement system and organizational ambidexterity on performance. Bus. Process Manage. J. **24**(5), 1176–1199 (2018)
12. He, Z., Bu, X., Cao, Y., et al.: Infrared attitude measurement method for spinning projectile under snow background. Infrared Phys. Technol. **111**(1–4), 103528 (2020)

Computer Image Recognition Technology Based on Deep Learning Algorithm

Yuan Jiang[1], Yongjun Qi[2(✉)], and Junhua Wang[3]

[1] South China Business College, Guangdong University of Foreign Studies, Guangzhou 510545, Guangdong, China

[2] Faculty of Megadate and Computing, Guangdong Baiyun University, Guangzhou 510450, Guangdong, China

qyj120040878@126.com

[3] Institute of Engineering, Guangzhou College of Technology and Business, Guangzhou 510850, Guangdong, China

Abstract. In the field of computer vision, deep learning is a very important research topic. It can not only help us analyze images, but also understand images into recognizable language. This paper uses deep learning algorithms for computer image recognition. On the basis of traditional recognition, a brand-new machine vision system is proposed. This paper analyzes the prediction accuracy of several different algorithms in computer image recognition through experimental methods. And through analysis and comparison, the algorithm has been explored, and the recognition method has been further understood. Through experimental data, we found that among the deep learning algorithms, RBM-SVM has the highest prediction accuracy, which is above 93%. Moreover, the recognition of static objects is easier and more accurate than dynamic objects.

Keywords: Deep learning algorithm · Computer image · Recognition technology · Technology research

1 Introduction

Computer vision is a new subject developed in recent years. Its main research is the algorithm of computer image recognition. It is widely used in artificial intelligence and pattern recognition. At present, a relatively mature theoretical system has also been formed for deep learning. Deep learning algorithms can quickly and efficiently process the relationship between the various parts of a complex data set.

There are many theoretical results in the research of computer image recognition technology for deep learning algorithms. For example, Zhou Lin verified the accuracy of SAR image recognition through the improvement of the convolutional neural network and the analysis of the network propagation process [1]. Sun Pingan has proposed an improved iterative deep learning algorithm fused with convolutional neural network for the problem of the loss of discriminative information in the image recognition algorithm [2]. Lv Jiaosheng considers that traditional image recognition is susceptible to noise

J. H. Abawajy et al. (Eds.): ICATCI 2022, LNDECT 169, pp. 514–521, 2023.
https://doi.org/10.1007/978-3-031-28893-7_61

interference during processing, so it proposes a robust sparse representation method on multiple scales [3]. Therefore, this article intends to conduct in-depth research on computer image recognition technology, analyze deep learning algorithms, and link them with recognition technology to produce conclusions.

This article first studies the related theories of image recognition, including several methods and specific steps of image recognition. Then the deep learning is elaborated and the deep Boltzmann machine is proposed. After that, the structural analysis of the discriminative image feature fusion and computer image recognition system is carried out. Finally, design and experiment on the image recognition model.

2 Computer Image Recognition Technology Based on Deep Learning Algorithms

2.1 Image Recognition

As a new image processing technology, computer vision has been researched by professionals in the fields of computer science, information science and artificial intelligence in recent years. With the continuous expansion of network communication and digital TV system applications. People use it more and more widely. This makes the traditional identification methods unable to meet the requirements, but also brings great challenges to our daily work. On the other hand, human beings cannot accurately understand the essential characteristics of things due to the limitations of human understanding of nature, which leads to misjudgments [4, 5].

Feature extraction is to extract useful information from the image, such as shape edge contour features and texture features, to distinguish different types of images, minimize the similarity of the same category, and maximize the similarity of different categories. The extracted features directly affect the efficiency of image recognition. Classification and recognition are based on the output of extracted features, and the algorithm forms a certain classification standard after certain training. Commonly used image recognition methods include Bayesian classification, model comparison, kernel methods, artificial neural network methods, etc. [6, 7].

Bayesian classification: The disadvantage of this method is that it cannot extract and describe the features of the image well at this stage, or cannot identify the extracted features.

Pattern matching method: Determine whether the sample matches the pattern. In order to build a suitable model, it is necessary to have some prior knowledge of the shape of the detection target. The disadvantage is that when the model and unknown samples are constructed, the consistency of the corresponding unit in the model and the sample is determined.

Kernel method: Integrate the original data into a large-dimensional feature space through some nonlinear mapping, and use a linear learner to analyze and process the model. The typical kernel method is a support vector machine for classification tasks.

Artificial neural network method: construct a plane neural network and adjust the parameters to improve the recognition rate of the image [8, 9].

2.2 Deep Learning

Computer image recognition technology based on deep learning has good effects in solving complex worlds, can effectively deal with complex real-world problems, and provide people with simple, practical, efficient, easy-to-understand, and widely used functions. Deep learning is a type of machine learning. Its essence is to explore the learning of unsupervised features and the analysis and classification of patterns through the hierarchical structure of multi-party information processing, and use multi-layer neural networks to achieve machine learning learning functions. The different high-level information of deep learning mainly includes two important pieces of information: the model is composed of non-linear information processing and supervised or unsupervised learning feature representation. As the number of network layers increases, the extracted information becomes more abstract. Deep learning uses abstract layered ideas and greedy layered training algorithms to extract useful learning functions. Deep learning algorithms usually take the form of unsupervised learning and are usually applied to unlabeled data. Compared with labeled data, the data is richer and easier to obtain [10, 11].

Most traditional machine learning and signal processing techniques use flat architectures. These structures usually contain one or two non-linear feature transformation layers, which can be viewed as structures with or without hidden layers.

As a new method, deep learning has many advantages over traditional artificial neural networks. It reduces the requirements and complexity of the learning set. Because deep learning can perform complete, correct and unsupervised output control and other functions that can be run in different environments. This makes it more suitable for complex systems [12, 13].

The computer image recognition framework based on deep learning is mainly composed of three parts, which are deep learning training set, feature extraction and neural network. First, pre-processing is performed to classify and sort the different types of character samples in the original data one by one. Secondly, according to the classification criteria, the text and pictures to be recognized are classified into the corresponding categories, and a suitable model is established. The required parameters are calculated by the convolution method and the final result is obtained. The final result is output to the text box for inputting the size information of each template in the box for convenience the content of the next step.

The computer image recognition framework based on deep learning mainly includes: a data acquisition system. After acquiring feature information in the original database, these features are described and extracted, and the obtained information is converted into machine language through preprocessing. Design of input and output interface for visual analysis and recognition. In order to be able to easily realize the automatic image classification function and the fast and accurate reading, storage and other operational requirements, it is necessary to establish a complete and efficient computer network structure system to support the interconnection and interaction process between the various layers of data in the deep learning system.

Deep Boltzmann machine is a more useful method in image recognition. Assuming that the nodes are all binary values (0 or 1), the overall probability distribution P(v, h) satisfies the Boltzmann distribution. Therefore, when w is input, the hidden layer g can be obtained by the formula P(h|w), and then the parameters can be adjusted, and the visible layer w_1 can be obtained by P(w|g). When w and w_1 is the same, g can be considered as w Characteristic performance. The joint configuration energy of the RBM model is:

$$F(w, g, \mu) = -\sum_{k} x_k w_k - \sum_{i} y_i w_i - \sum_{ik} Q_{ik} w_i g_k \tag{1}$$

Among them, $\mu = \{Q, x, y\}$ the joint probability distribution of v and h can be determined by Boltzmann distribution:

$$T_\mu(\mathrm{w}, g) = \frac{1}{c(\mu)} \exp(-F(w, g, \mu)) \tag{2}$$

Among them, the normalization factor is $c(\mu) = \sum_{\mathrm{w},g} \exp(-F(w, g, \mu))$.

2.3 Discriminative Image Feature Fusion and Recognition System Composition

In digital image processing, different descriptions of target image characteristics can be considered as different modalities. Different categories may contain different information, so the focus of describing the target image is different. For some problems, even powerful feature descriptors may be inapplicable or invalid. Multi-modal feature description methods have many different manifestations and uses. The deep neural network model can easily extract the overall attributes of the image, and has a good understanding of the semantic information and spatial structure information of the image, but this technology is generally not good at describing very detailed images with characteristics. The multi-modal learning model minimizes the difference in information between different modalities to produce a representation of shared features between multi-modal information. Even when information loss occurs in some modalities, it can also work well. However, in this model, the information is directly transformed and manipulated in different modes, and features are not selectively eliminated, so the effect of the model may be weakened by introducing bad features. The common convolutional neural network and long short-term memory network model combine the performance of automatic feature extraction and the function of time domain information storage to improve the recognition accuracy of the image recognition system.

The hardware composition of the computer image recognition system mainly has the following parts. The human-computer interaction interface must be simple to use, easy to use, friendly and easy to operate during the entire identification process. And it can control and monitor the input and output in real time so as to process the error information in time and make adjustments to meet people's needs. At the same time, it should have good anti-interference ability and high reliability and other performance indicators to ensure that the system can work normally. In visual platform part, we use digital template matching technology to extract features by processing the image. The advantage of this method is that it can quickly download pictures from massive databases to devices such as presets and camera model databases to obtain the required identification information. At the same time, functional modules such as higher accuracy, stable recognition rate and accuracy detection system requirements can also be realized. Network platform, this part mainly includes computer network transmission, as well as communication interface and server.

3 Image Recognition Model Experiment

3.1 Experimental Thinking

Due to the insufficient generalization ability of traditional limited-sample machine learning methods, RBM-SVM is based on the VC dimension and structural risk minimization, and the generalization ability is better. The basic idea of RBM-SVM is to solve the optimal classification range by maximizing the classification interval. A vector machine is similar to a three-layer neural network. The number of nodes in its hidden layer is determined by the support vector, and its purpose is to solve the convex quadratic programming optimization problem.

3.2 Experimental Data Selection

This part of the experiment is based on the MATLAB R2009a operating system platform, using the SVM classifier. The MINIIST data set contains 100 training samples, 20 test samples, 500 training samples, or 100 test samples. Take two types of training sample numbers or test sample numbers.

3.3 Experimental Method

Although support vector machine has many advantages in theory and application, it is a flat structure and can be regarded as a hidden layer. The constrained Boltzmann machine is used to construct a multi-layer model for classification, the deep learning method is combined with the support vector machine, and the RBM-SVM method is used for notation.

4 Analysis of Experimental Results

4.1 The Influence of the Number of Samples, the Number of Nodes, and the Number of Layers

Take two training sample numbers or test sample numbers, choose different node numbers and hidden layer offset numbers, and compare the effects of the number of training samples, node numbers, and offset numbers. The results are shown in Table 1:

Table 1. The Prediction Accuracy of Different Methods in the Hidden Layer

	100	200	300	400
DBN	91	92	93	93.7
RBM-SVM	92	93	94	94.6
SVM	81	83	84	85

As shown in Fig. 1, we can see that in the three algorithms, the accuracy of SVM is above 80%. The second is the DBN algorithm, which has an accuracy rate of over 91%. RBM-SVM has the highest prediction accuracy.

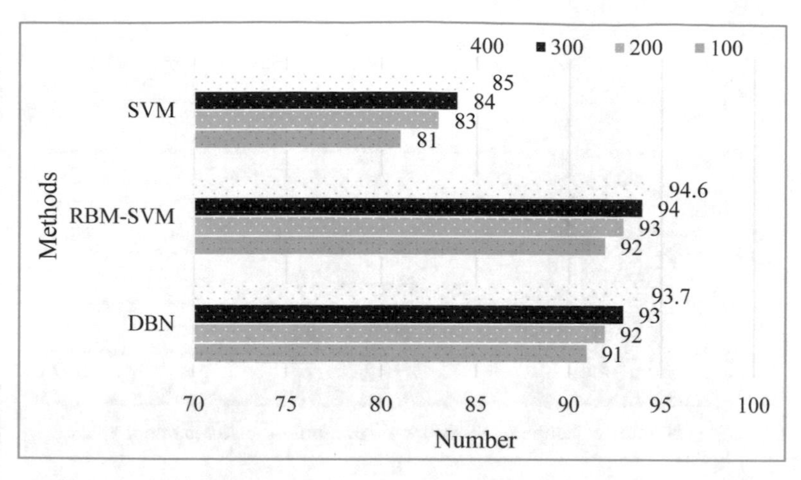

Fig. 1. The Prediction Accuracy of Different Methods in the Hidden Layer

4.2 Recognition Rate of Various Samples

This paper selects 100 five kinds of objects for image recognition, namely airplanes, cars, birds, boats and dogs. The specific recognition rates obtained are shown in Table 2:

Table 2. Identification rate of all types of samples

Type	Airplane	Car	Bird	Ferry	Dog
Number of samples	18	20	23	25	14
Number of recognition	12	13	12	15	5
Recognition rate	66.7%	65%	52.2%	60%	35.7%

As shown in Fig. 2, we can see that the recognition rate of cars can reach 65%, the recognition rate of airplanes can reach 67%, and the recognition rate of dogs is the lowest, only 36%. Due to the large number of ground animals, identification is prone to errors.

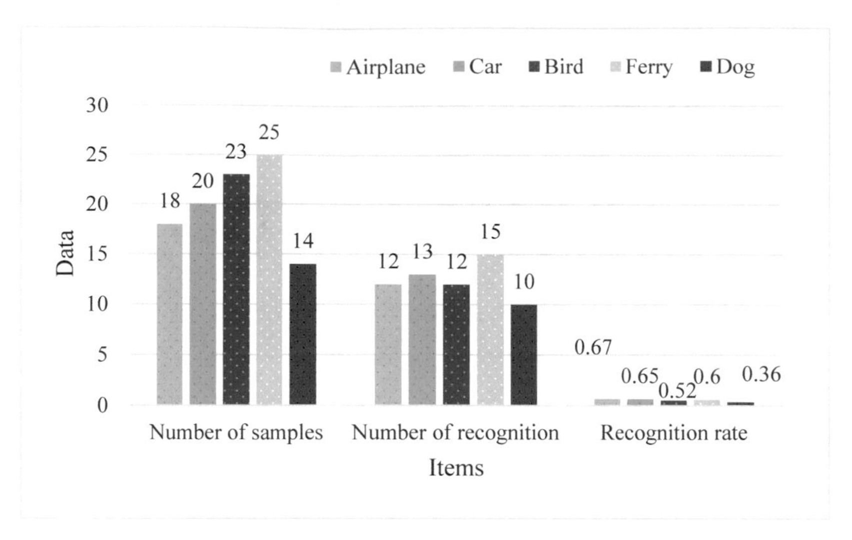

Fig. 2. Identification rate of all types of samples

5 Conclusion

With the continuous development of computer technology, deep learning has also become a new research direction. It uses artificial neural networks for information processing and analysis by inputting image data into the machine. This method is simple and fast. Compared with traditional manual recognition, deep learning has obvious advantages. The experiments in this article tell us that the RBM-SVM algorithm in deep learning has the highest accuracy in image recognition systems. Of course, the accuracy of the other two methods is also above 80%, which is faster and more accurate than traditional manual identification methods. Therefore, it is a good choice to study computer image recognition methods and start with deep learning technology.

Acknowledgements. This work was supported by Feature innovation project of colleges and universities in Guangdong province (No.2020KTSCX163) and Feature innovation project of colleges and universities in Guangdong province (No.2018KTSCX256) and Guangdong baiyun university key project (No.2019BYKYZ02) and Special project in key fields of colleges and universities in Guangdong Province (No.2020ZDZX3009).

References

1. Lin, Z.: Research on deep learning algorithm in SAR image target recognition. Electron. Prod. 408(22), 29+67–68 (2020)
2. Pingan, S., Jun, Q., Qiuyue, T.: Research on image recognition method using convolutional neural network to improve iterative deep learning algorithm. J. Comput. Appl. Res. **036**(007), 2223–2227 (2019)
3. Jiaosheng, L.: Research on image recognition algorithm based on sparse representation and deep learning. J. Xinxiang Univ. (Natural Sci. Ed.) **035**(009), 31–34 (2018)

4. Ilyas, B.R., Beladgham, M., Merit, K., et al.: Illumination-robust face recognition based on deep convolutional neural networks architectures. Indonesian J. Electr. Eng. Comput. Sci. 18(2), 1015–1027 (2019)
5. Sharma, C., Singh, R.: A performance analysis of face and speech recognition in the video and audio stream using machine learning classification techniques. Int. J. Comput. Appl. **183**(13), 975–8887 (2021)
6. Kim, H., Lee, W., Kim, M., et al.: Deep-learning-based recognition of symbols and texts at an industrially applicable level from images of high-density piping and instrumentation diagrams. Expert Syst. Appl. **183**(3), 115337 (2021)
7. Batchuluun, G., Jin, K.K., Nguyen, D.T., et al.: Deep learning-based thermal image reconstruction and object detection. IEEE Access (99), 1 (2020)
8. Rahman, M.T., Jabiullah, M.I., Sultana, N., et al.: Computer vision-based plant leaf disease recognition using deep learning. Int. J. Innov. Technol. Exploring Eng. **9**(5), 622–626 (2020)
9. Cohen, I., David, E.O., Netanyahu, N.S.: Supervised and unsupervised end-to-end deep learning for gene ontology classification of neural in situ hybridization images. Entropy **21**(3), 221 (2019)
10. Mourad, C., Akhtar, Z., et al.: 3D palmprint recognition using unsupervised convolutional deep learning network and SVM classifier. IET Image Processing 13(5), 736–745 (2019)
11. Jittawiriyanukoon, C.: Proposed algorithm for image classification using regression-based pre-processing and recognition models. Int. J. Electr. Comput. Eng. **9**(2), 1021 (2019)
12. Hieu, N.V., Hien, N.: automatic plant image identification of Vietnamese species using deep learning models. Int. J. Emerg. Trends Technol. Comput. Sci. **68**(4), 25–31 (2020)
13. Madduri, A.: Human gait recognition using discrete wavelet and discrete cosine and transformation based features. Int. J. Comput. Trends Technol. **69**(6), 22–27 (2021)

Path Planning and Task Scheduling of AGV System Based on Digital Twin Technology

Xingyu Luo and Hua Zhou(✉)

Southwest Forestry University, Kunming, Yunnan, China
zhua301@yeah.net

Abstract. In recent years, with the development of digital technology, higher requirements are put forward for the performance of AGV system in various complex environments. This paper mainly takes the realization of path planning as the objective function, establishes the mathematical model and solves the algorithm, and completes the task generation process design and allocation strategy research. Firstly, this paper introduces the concept and characteristics of digital twin technology, then studies the related technologies of path planning, then introduces the role of task scheduling, and designs the path planning framework of AGV system. Then the performance of digital twin technology is verified by simulation. Finally, the test results show that the performance of digital twin technology in AGV system path planning and task scheduling is good. It can find the solution points in the solution by parallel computing, and optimize the path when the constraints are met, so as to obtain the global feasible scheme and realize the global task planning.

Keywords: Digital twin technology · AGV system · Path planning · Task scheduling

1 Introduction

In today's era, various resources are interdependent, interconnected and interdependent. With the continuous improvement of socio-economic and scientific and technological level and the increasing demand of various departments for AGV system, its performance requirements are also increasing [1, 2]. It is a series of strategies, means and method systems adopted by simulating nature and human cognitive world and transforming it into visual spatial form to describe things or process relations and state transformation, so as to provide the optimal path to complete complex tasks and achieve the desired purpose for global optimization [3, 4]. Therefore, how to plan efficient, stable and cost-effective is one of the urgent problems to be solved in the research of modern path problems.

Many scholars have done relevant research on AGV system. AGV appeared in the United States in the 1950s, and it was not put into use until the late 1970s. The development status abroad is that military technology is used in the military field, but its function is limited to simply completing national defense tasks. With the rapid rise of science and technology industry and civil aviation and people's increasing demand for intelligent life, countries all over the world recognize that AGV system brings great benefits to

J. H. Abawajy et al. (Eds.): ICATCI 2022, LNDECT 169, pp. 522–529, 2023.
https://doi.org/10.1007/978-3-031-28893-7_62

themselves, and vigorously study high performance, high reliability and operability. At the same time, many countries, such as Japan, have started independent research and development and put into application [5, 6]. At present, China has begun to develop the cooperative working mechanism between functional modules in a complex system with computer as the core unit (including hardware and software), and on this basis, a complete, scientific and reasonable network structure, resource allocation and management system have been established. Some scholars have established a multi-objective optimization programming problem model based on genetic algorithm (ant colony algorithm, simulated annealing). Other scholars have carried out a series of simulation experiments on parameter path optimization in complex systems [7, 8]. The above research has laid the foundation for this paper.

With the continuous development of science and technology in recent years, various advanced manufacturing technologies have been gradually applied to production and life, making industrial robots an indispensable part of the development of the times. The most common and important one is AGV system. In recent years, scholars at home and abroad have studied and developed it and achieved certain results and application results, and began to turn to other fields to carry out experimental exploration. This paper establishes an object-oriented task allocation and path planning problem model based on resource constraints to realize the path optimal configuration and task scheduling of AGV, which provides reference ideas for subsequent chapters.

2 Discussion on Path Planning and Task Scheduling of AGV System Based on Digital Twin Technology

2.1 Digital Twin Technology

2.1.1 Principle

The application of digital twin technology was originally designed to solve the problem of maintaining the health and safety of space vehicles. The U.S. Department of defense, the U.S. Air Force Research Laboratory and NASA are all using the concept of digital twins. By integrating some closely related models and data analysis, such as heat transfer model, dynamic model, finite element model type analysis, stress analysis and fatigue mode analysis, it is used to simulate and judge whether the body itself needs maintenance and meets the requirements of task conditions. The digital twinning method is a special algorithm based on the hashhofstein equation. It is of great significance in the reduction of x value and Z value. The model linearizes a pair of multi chromosomes as the basic unit to generate the initial trajectory of the population, solves it based on the classical algebraic equation, and gives the feasible solution of the optimization problem, Or the optimal solution of the objective function relationship that may appear under different types and different evolutionary algebra numbers in optimal path planning. Digital twin algorithm is based on the principle of natural evolution and achieves its goal by simulating the biological evolution process in nature. In terms of evolutionary problem solving [9, 10]. It decomposes a complex system into multi-stage sequences or continuous functions such as sub, population and local, generates population parameters according to randomly selected fitness values, and designs corresponding strategies for each generation of individuals to form a cluster structure, in which each chromatin produces the densest combination of individuals with different probabilities.

2.1.2 Features

Digital twinning refers to a multi-disciplinary and multi-scale simulation process that makes full use of physical models, sensors, operation history and other data. In order to promote the understanding of digital twin, digital twin uses information technology to digitally define and model the composition, attributes, functions and performance of physical entities. Digital twin refers to the information model completely corresponding to the physical entity in the virtual machine room, which can simulate, analyze and optimize the physical entity based on digital twin. Digital twin is technology, process and method, and digital twin is object, model and data. It has the following characteristics: (1) high robustness. After digitization, the execution efficiency of the system is greatly improved, and it has good fault-tolerant performance, which makes the whole algorithm more acceptable. At the same time, because of its high flexibility, visualization and easy implementation, it has been widely used in practical production. (2) Strong parallel processing ability. For a complex system with multiple processors for control and management, each module is required to complete the task requirements without affecting the overall operation efficiency [11, 12].

2.2 AGV System Path Planning and Task Scheduling

2.2.1 Definition of Path Planning

AGV is an autonomous guided robot. In order to realize the autonomy of AGV, three problems must be solved: 1) where am I? 2) Where am I going? 3) How do I get there? The goal of AGV path planning is to solve the third problem on the basis of these two problems before solving them. The principle of AGV trajectory planning is to provide map and destination location, that is, to solve "where am I?" And "where am I going?" Once the destination of AGV is determined, the trajectory planning of AGV is carried out. AGV trajectory planning aims to generate collision free optimal or quasi optimal trajectory from the current AGV position to the target AGV position. AGV path planning can be divided into single AGV path planning and multi AGV path planning according to the number of AGVs. Planning a unique AGV path aims to make the AGV move along the generated path and reach the target position as soon as possible. The goal of optimization is usually to maximize the sum of AGV steering angles from the starting position to the target position according to the generated trajectory. The goal of multi AGV travel planning such as the shortest and shortest distance is to maximize the overall operation efficiency of AGV system, guess and consider the possible conflict and blocking problems between multiple AGVs. The optimization goal is usually completed by several AGVs in batches, which has the shortest time, the shortest total duration, the shortest total distance, etc.

2.2.2 Local Path Planning Algorithm

The idea of artificial potential field method comes from the field theory in physics. It is a virtual force method. The principle is that the movement of the AGV in the space environment is regarded as the movement of the AGV in the virtual artificial force field. A gravitational field is set around the target point and a repulsive field is set around the obstacle. The gravitational field will guide the AGV to move towards it, and the repulsive field will hinder the AGV to move towards it. The combined force of the AGV is equal to the sum of the gravitational force and the repulsive force, and the AGV is driven by the combined force field, and drive to the target position. The moving direction of AGV depends on the direction of the resultant force field. The potential energy function UATT of gravitational field is:

$$U_{AIT} = (X) = \frac{1}{2}\Re\rho^2(X_R, X_G) \tag{1}$$

where $\Re$ is the gravitational scale factor, P (XR, XG) represents the distance between the current position of AGV and the target node. In the gravitational field, the gravitational function fat is the derivative of the potential energy function of the gravitational field to the distance, such as:

$$FAIT(X) = -\nabla U\text{att(X)} = \Re\rho(X_R, X_G) \tag{2}$$

This means that when the distance between the AGV's current position and the obstacle exceeds a certain distance, the repulsion field of the obstacle has no repulsion effect on the AGV.

2.3 The Role of Task Scheduling

The role of task scheduling is to reasonably allocate the objects involved in the system, so that it can successfully complete the established objectives and execute according to the predetermined requirements within the specified time. When implementing a specific goal, we need to determine its trajectory first, and then decide whether to start work according to the actual situation. If a specific operation is completed, the next action can be ended and returned to the initial position to stop the operation, or there is no response in the next program, a new round of task cycle will be restarted, and the above process will be repeated until all operations are terminated. The whole system is in a stable state, waiting for all operators to execute the specific target again. Task scheduling is helpful to improve the overall performance of the system. Due to hardware and software resources, when performing different degrees of operations, you can reduce the running speed to complete higher-level target tasks. It is also conducive to reducing the equipment failure rate, timely and effective maintenance of the machine, so as to prevent unrecoverable losses caused by failures, realize the principle of minimizing multi-level maintenance costs, and the technology with the highest functional priority required in scheduling optimization problems.

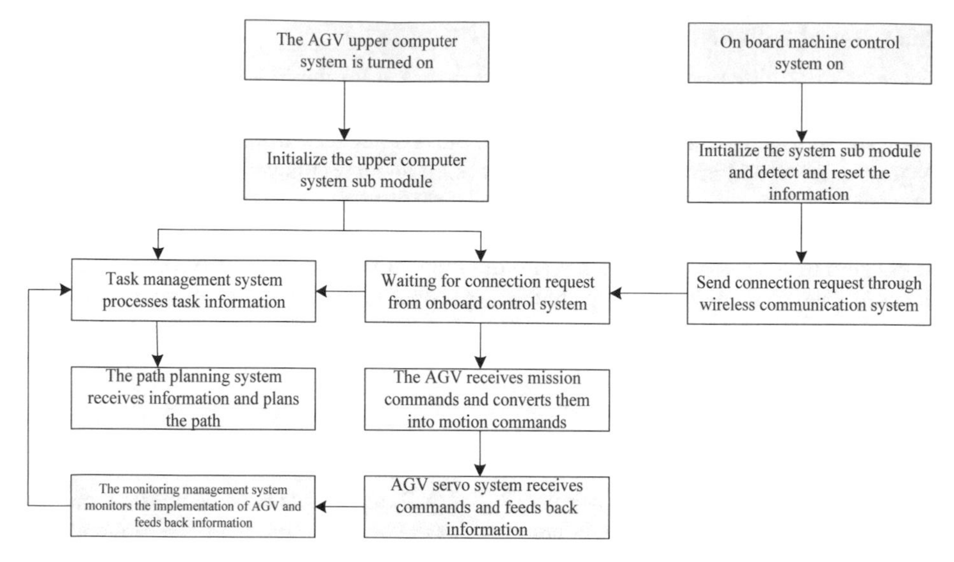

Fig. 1. AGV system construction

3 Experiment of AGV System Based on Digital Twin Technology

3.1 AGV System Based on Digital Twin Technology

This paper designs a path planning algorithm for parallel motion based on multi robot cooperation platform. In this method, multiple tasks are assigned to a single execution unit, and then all priorities are combined and crossed by genetic operation function. When generating a new chromosome, first define an initial start time and calculate its length as the penalty value. Secondly, the position information corresponding to the vehicle in the parent population is called to determine the optimal backward speed solution with the minimum distance between the obstacles in the parking space, so as to make the global tend to the optimization state. At the same time, the parameters such as the shortest path width and the final shortest path are determined according to the speed difference between the two adjacent vehicle routes. As can be seen from Fig. 1, the AGV system software startup process is as follows: first initialize the navigation and positioning module, autonomous obstacle avoidance module, motion control module and communication module, and then reset the sensor detection frame to ensure the normal operation of each module. Then, the on-board computer control system sends a binding request to the superior computer management system through wireless communication, and establishes a TCP / IP communication link with the AGV management system. If the corresponding connection is successful, the current AGV accepts the task command sent by the superior computer and executes the corresponding task. If the corresponding connection fails, the AGV is currently in standby mode and has been waiting for the connection and operation command of the main control computer.

3.2 Experimental Test of AGV System Based on Digital Twin Technology

In this paper, the simulation is carried out in AGV, and the path planning and task scheduling results are analyzed. There are several sets of data for the objectives to be completed in this experiment. It includes: route length, starting time, speed, acceleration and other parameter information, as well as the function value of each part, etc. The trajectory planning algorithm adopts the iterative method based on black box constraints, that is, when a certain initial state is satisfied, it starts to solve and stops optimization. In AGV system, the program is initialized first, and then the path planning method is used to generate an initial state travel. Then the constraint conditions and required time at each position are obtained through the running track. Then, after returning to the starting point, start the iterative route and output the blackboard coordinate data to each task execution module, so as to continuously update the parameters and calculation amount required for solving new problems in the later stage. Finally, the global optimal solution is obtained according to the objective function, and all possible schemes and final results (minimum or maximum optimization) under the final expected path are given.

4 Experimental Analysis of AGV System Based on Digital Twin Technology

4.1 Performance Test and Analysis of Digital Twin Technology in AGV System Path Planning and Task Scheduling

Table 1 shows the performance test of digital twin technology in AGV system path planning and task scheduling.

Table 1. Performance testing of the digital twinning technology

Test times	Route length	Starting time (s)	Speed (m/s)	AGV number
1	5	2	110	4
2	6	1	114	4
3	5	3	128	6
4	11	2	132	5
5	8	1	110	7

In AGV system, the test system needs to track different operation conditions, and record the corresponding objective function value under each execution path. Task scheduling is a continuous, dynamic and static process. As can be seen from Fig. 2, the performance of digital twin technology in AGV system path planning and task scheduling is good. It can find out the solution points in the solution by parallel computing, and optimize the path when the constraints are met, so as to obtain the global feasible scheme and realize the global task planning.

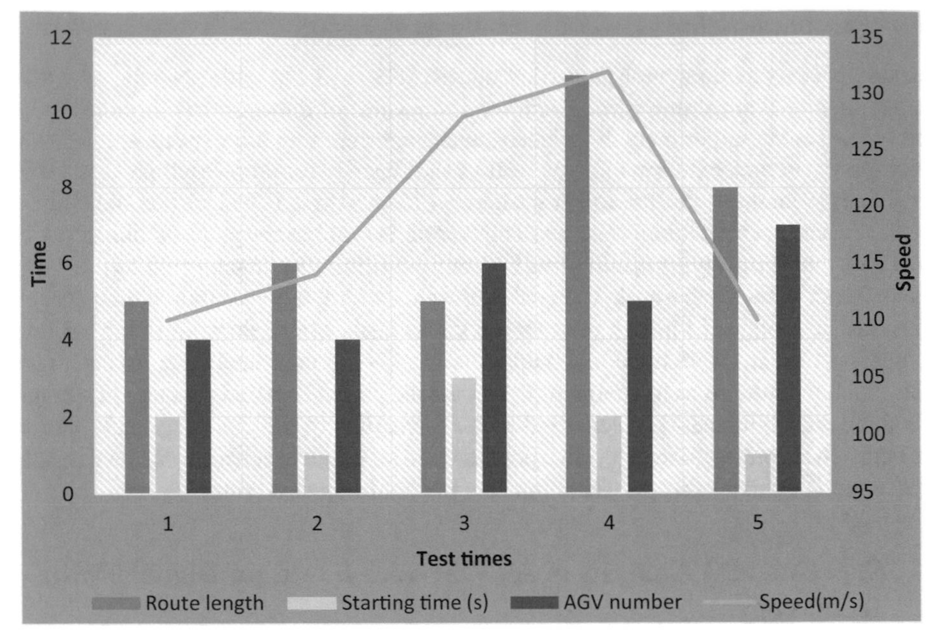

Fig. 2. Performance testing of the digital twinning technology

5 Conclusion

AGV system is widely used in modern industry. It has the advantages of high efficiency, strong real-time, high flexibility, universal and open. This paper analyzes its constraints and path planning. Firstly, the mathematical model of target assignment problem based on digital twin technology is introduced. Then the object-oriented task execution optimization algorithm is established, and the corresponding system simulation experiments are designed to verify the feasibility and practical value of the planning method in the industrial field. Finally, the shortest path is achieved by minimizing the time required for AGV scheduling.

References

1. Kwok, P.K., Yan, M., Qu, T., Lau, H.Y.K.: User acceptance of virtual reality technology for practicing digital twin-based crisis management. Int. J. Comput. Integr. Manuf. **34**(7–8), 874–887 (2021)
2. Mohammadi-Moghadam, H., Foroozan, H., Gheisarnejad, M.: Mohammad Hassan Khooban: A survey on new trends of digital twin technology for power systems. J. Intell. Fuzzy Syst. **41**(2), 3873–3893 (2021)
3. Preetha Evangeline, P.: Anandhakumar: Chapter two-digital twin technology for smart manufacturing. Adv. Comput. **117**, 35–49 (2020)
4. Anderl, R., Haag, S., Schützer, K., De Senzi Zancul, E.: Digital twin technology - an approach for industrie 4.0 vertical and horizontal lifecycle integration. IT Inf. Technol. **60**(3), 125–132 (2018)

5. Riazi, S., Bengtsson, K., Lennartson, B.: Energy optimization of large-scale AGV systems. IEEE Trans Autom. Sci. Eng. **18**(2), 638–649 (2021)
6. Matsumoto, H., Shibako, Y., Neba, Y.: Contactless power transfer system for AGVs. IEEE Trans. Ind. Electron. **65**(1), 251–260 (2018)
7. Almadhoun, R., Taha, T., Seneviratne, L.D., Zweiri, Y.H.: Multi-Robot hybrid coverage path planning for 3D reconstruction of large structures. IEEE Access **10**, 2037–2050 (2022)
8. Gee, M., Vladimirsky, A.: Optimal path-planning with random breakdowns. IEEE Control. Syst. Lett. **6**, 1658–1663 (2022)
9. Puente-Castro, A., Rivero, D., Pazos, A., Fernandez-Blanco, E.: A review of artificial intelligence applied to path planning in UAV swarms. Neural Comput. Appl. **34**(1), 153–170 (2021). https://doi.org/10.1007/s00521-021-06569-4
10. Lathrop, P., Boardman, B.L., Martínez, S.: Distributionally safe path planning: wasserstein safe RRT. IEEE Robotics Autom. Lett. **7**(1), 430–437 (2022)
11. Akbar, R., Prager, S., Silva, A.R., Moghaddam, M., Entekhabi, D.: Wireless sensor network informed UAV path planning for soil moisture mapping. IEEE Trans. Geosci. Remote. Sens. **60**, 1–13 (2022)
12. Al-Kaseem, B.R., Taha, Z.K., Abdulmajeed, S.W., Al-Raweshidy, H.S.: Optimized energy-efficient path planning strategy in WSN with multiple mobile sinks. IEEE Access **9**, 82833–82847 (2021)

Research on Computer Information Security Defense Methods Based on PCA and BP Neural Network Underneath the Backdrop of Large Set of Information

Rong Chen(✉)

Shanghai Customs College, Shanghai 201204, China
longpuren299571@163.com

Abstract. Based on the characteristics of high dimension and huge data of abnormal data in the background of big data, traditional detection methods aim at the phenomenon that the detection efficiency of abnormal data is not high and the identification progression of unknown atypical data is negligible. In this paper, data safety detection technique comprising of PCA and the proposal for the BP neural system is presented, which is suitable for big data environment. By training and learning the known abnormal data, this method can effectively identify the set abnormal data, and can also identify the abnormal data without training and learning. This PCA based BP neural network abnormal data recognition method is applied to information security defense in the backdrop of large set of information, which has the advantage of proactive defense.

Keywords: Big data · PCA · BP neural network · Information security · Defense methods

1 Preface

The so-called big data, as the name implies, is the use of computer technology to search, process, and analyze large-scale data, and filter out data that is valuable for economic development. Thanks to constant growth of the social economy and the further betterment of social productivity, the Internet has shown a rapid development momentum. While bringing huge economic value, it also demands greater conditionalities for cyber safety control capabilities. Continuous improvement of network information security control technology can effectively reduce the theft and leakage of network information and data, and improve the confidentiality and stability of network data. Therefore, based on the backdrop of large set of information, the analysis of the network data security control mechanism and evaluation system holds significance for on the ground utility for promoting the long-term progression of the Internet industry. As the process of globalization continues to deepen, the Internet industry is also showing a trend of globalization. People's demand for network information is increasing, and the times of large set of information has begun.

J. H. Abawajy et al. (Eds.): ICATCI 2022, LNDECT 169, pp. 530–537, 2023.
https://doi.org/10.1007/978-3-031-28893-7_63

The arrival of the era of big set of information is of great significance to promoting the development of my country's social economy. At this stage, our country mainly applies data and information processing software with high-efficiency and intelligent features to realize the processing of large-scale data and information, and to screen out valuable data and information. Big data contains all the peculiarties of, large quantity as well as complex structure. It can process and analyze data information such as video, audio, picture and text, and has the advantages of economy and efficiency. Simultaneously, big information technology can efficiently improve Internet technology, thereby enriching information and data processing methods [1]. Using large information can actualize the collection, refining with analysis of information, and can store valuable information, so as to play the economic value of data information. At the same time, in the face of the complex characteristics of big data, new requirements are also put forward for the control methods and methods of network information security in the context of big data. The research in this direction mainly includes the following directions, network information and data security management personnel, building a secure environment, technical support, etc. Fig. 1 [2].

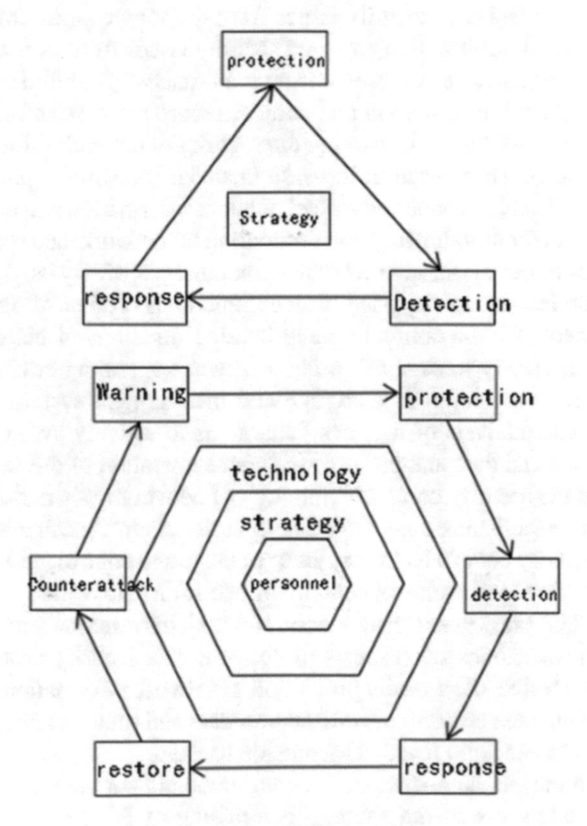

Fig. 1. Network security model

Network information and data security management personnel are the foundation and core of the system information safety oversight method. System data and information safety problems are caused by network personnel's improper operations. Maintaining system information safety is also the key to ensuring the legitimate rights and interests of network users. Therefore, the network system security management personnel should perform well for the division of personnel levels, and put the network mechanism security supervising staff to begin with, which is related to the system information safety oversight strategy. Put network users, network data providers and network hackers at the personnel volume of the system data safety control mechanism. At the same time, in the process of controlling network data security, the personnel layer can also be specifically divided into network equipment controllers and network information data security managers. As managers of network information and data security, they should establish a correct awareness of network information and data security and fully actualize the importance of network cyber security. Improve self-management and control capabilities, formulate unified network behavior specifications, and continuously improve the management and control of network data security. As a network user, we must continue to raise awareness of network information and data security, reduce the occurrence of network information and data security issues from the source, and achieve the purpose of preventing network information and data security issues from occurring. Building a secure network information environment is an important way to build a security control mechanism for network information and data. A secure network information environment plays an important role in improving the security of network information and data [3]. Therefore, the government must increase financial investment, purchase advanced network equipment, and introduce advanced network information and data management technologies and methods to form a good foundation for network data security. Comprising of large set of information, Internet cloud computing is widely used in various fields of society. While advancing the pace of economic construction, it also puts forward further requirements for the computing and bearing functions of network equipment. Therefore, it is necessary to establish a good network operating environment, build a complete network information data storage and management system, and realize the mining, sorting and analysis of network data, so as to actively avoid the safety risks about network data and data and promote the stable operation of the Internet. Continuously improving the security control technology of network information and data is an important part of establishing a network data security control mechanism. Making full use of various security control technologies to manage and control network information data can give full play to the value of network information data, which is beneficial to the development and progress of social economy. Network information data security control technology mainly includes two aspects of network data security protection and real-time monitoring. Realize the security protection of network information data, which can effectively guarantee the security of information data and prevent network information data from being attacked and invaded by outside hackers.

According to the characteristics of the data in the big data environment, this study proposes cyber safety disclosing strategy comprising of PCA combined with neural system of BP. After training and learning the known abnormal data, this method can not only efficiently identify the known abnormal data, but also recognize the atypical data which has not been reinforced and acquired. This PCA-based BP neural network abnormal data recognition method can provide big data security defense to solve the defects of traditional abnormal data recognition defense [4].

2 PCA Data Statistical Method

Principal Component Analysis (PCA) is a statistical method. Transform a group of potentially correlated variables into a group of linearly uncorrelated variables through orthogonal transformation. This group of variables after conversion is called principal component. In actual subjects, in order to analyze the problem comprehensively, many variables (or factors) related to this are often put forward, because each variable reflects some information of the subject to varying degrees. Principal component analysis was first introduced by K. Pearson (Karl Pearson) for non-random variables, and then H. Hotelling extended this method to the case of random vectors. The size of the information is usually measured by the sum of squared deviations or variance [5].

The network data processed by the PCA method in statistics not only greatly reduces the amount of data and the dimensionality of the data, similarly it retains the peculiarites of the actual set of information. Certain quantity of data and the dimensionality is reduced, so the detection performance of abnormal data is improved. Pursuant to troubleshooting the glitches in cyber safety identification display, a significant roadblack regading cyber safety is the search for atypical set of information emanating from recently developed viral onslaughts. Traditional system security equipment compares and checks and kills abnormal data by updating the virus signature database, which always brings the problem of security lag, and cannot protect information security in an active, real-time, intelligent, and self-educating way [6]. This neural system has precisely this characteristic of discovering and identifying abnormal data in real time via constant autonomous discovery. The acquired neural system has significantly higher identification speed for the recongnized atypical set of information. Similarly, it has positive identitificaiton effect for undisclosed atypical set of information. Therefore, neural network has an excellent research utility for the application of cyber safety [7].

Under normal circumstances, when we deal with some high-dimensional datagenerally we use feature selection and feature extraction these two methods to reduce the data dimensionality. Although these two methods can reduce the dimensionality of the data, there are essential differences in the dimension reduction method. PCA is a dimension reduction technique for numerical examination which can delineate various qualities of considerable dimentional set of information into scanty representative values of source data. These few feature values can mostly reflect the feature attributes of the source data, and they are not related to each other. In addition to the normal data, the abnormal network data also contains a lot of redundant information. It is precisely because of the redundant information that these attributes are irrelevant that they not only interfere with the accuracy of the detection work. Moreover, analyzing these large amounts of irrelevant redundant information will also greatly increase the amount of data training and detection operations, and reduce the efficiency of the system in analyzing abnormal data in Fig. 2 [8].

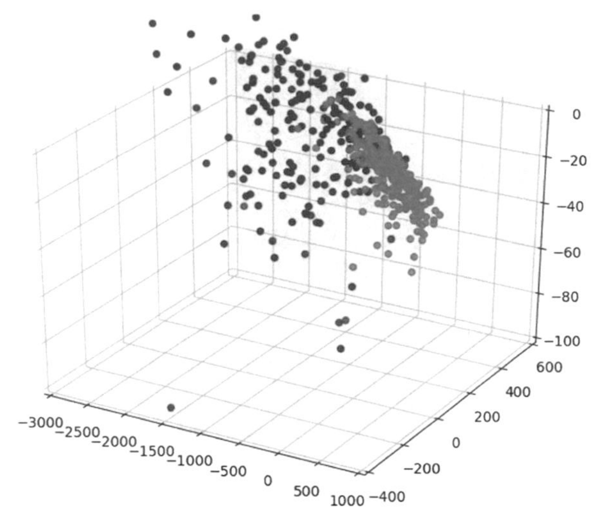

Fig. 2. PCA statistical method application of principal component analysis

Before detecting abnormal data, it is necessary to "clean" the data, that is, to preprocess the data, to remove irrelevant attributes from the information containing redundant attributes, to extract the data containing the original information characteristics, and to reduce the dimensionality of the original data. It is necessary to detect abnormal data in information security. The PCA method in statistical methods can map multiple features in the data to a few main features [9]. These main features can not only retain the features of the source data, but also reduce the dimensionality of the data. Compared with other dimensionality reduction methods, PCA is more than simply deleting some data to achieve a purpose of reducing dimensionality. It provides a feature with a relatively high contribution rate compared to the original feature information, so that the features of the original data are worth keeping, and guarantees the accuracy of abnormal data detection. This analysis method is very suitable for solving the problem of large traffic, high dimensionality, and strong real-time network data to make it more efficient and accurate to detect abnormal data [10].

3 BP Neural Network

Artificial neural network is a parallel interconnection network that is composed of the most basic unit group neurons and is adaptive and can learn and adjust independently. Through training and learning, artificial neural networks can simulate biological nervous systems. The system gives feedback to specific objects, and its basic unit neuron is a simplification and simulation of biological neurons. Artificial neural networks are composed of these artificial neurons that simplify and simulate biological neurons [11].

The implementation of fake neural network inside field of system security has solved the defect that the new abnormal data cannot be blocked due to the incomplete feature database of security equipment [12]. With its self-learning and self-adaptive capabilities, as well as efficient computing performance, it can quickly and accurately make judgments in the face of intrusion risks and abnormal data. The neuron model of the BP neural network is modeled on the structure of the biological neuron as demonstrated in Fig. 3.

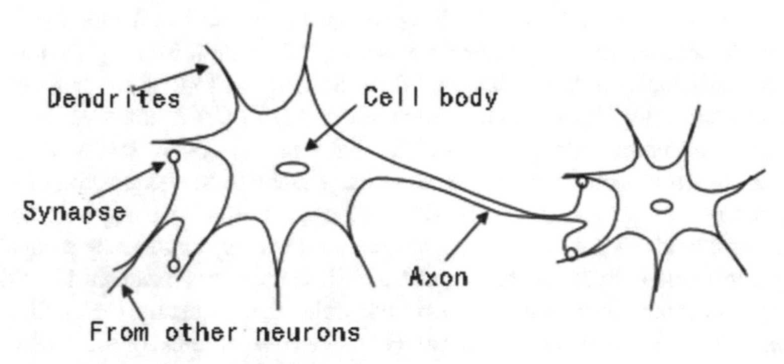

Fig. 3. The formation of neurons inside biology

From the point of view of its composition structure, the commonality of various neurons is composed of three main parts: axons, dendrites and cell bodies. The dendrites of biological neurons are the bridge of information communication between neurons and neurons, and are the information input terminals of neurons. The main function of dendrites is to receive the information transmitted by the last connected neuron axon and deliver the information to the cell body. In biological neurons, axons have the function of transmitting information processed by the cell body to the dendrites of the next neuron. There are many branched nerve endings at the ends of the axons. The functional contacts between these nerve endings and the dendrites of other neurons are used to transmit signals and play a role in signal transmission in neurons.

The learning rules of artificial neural networks are actually a way of network training. Its purpose is to modify the weight of the neural network and adjust the threshold of the neural network to better complete some specific tasks. At present, neural network supervised learning and autonomous learning are two different learning methods. Neural network has the ability of fault-tolerant processing, and has a certain general abstraction completion function for incomplete input information. During the procedure of network data transfer, information is often lost and incomplete or deformed. BP neural network is very suitable for handling such situations. Neural network has strong autonomous learning ability. Through continuous training and learning of input normal samples, the neural network can not only identify the known abnormal data in the training samples, but also has a good recognition effect on the unknown abnormal data and the known abnormal data deformation forms. Neural network has extremely high execution efficiency. The calculation and data storage of BP neural network is a unified whole. The working mode of each neuron is essentially the weight transfer process, and the weight transfer process completes the information storage process at the same time.

4 PCA and BP Combined Information Data Detection Model in Reference to Large Set of Information

Using PCA cuts down this dimensionality of considerable dimensional set of information. It also guarantees the data which has peculiarities of the actual data. First, extract the main elements of the actual abnormal set of information into big data and lowers the dimensionality of these abnormal data. Subsequent to that, the main elements of data acquired after dimensionality reduction through PCA is employed as inserted covering of BP neural system for education. False neural system is a multi-covering analogous formation consisted of fake neurons. Thanks to the acquisition of training data, obtainment of memory can be installed in the sytem itself by using weights as well as other threshholds. A proficient neural system has the capacity to easily recognize the training information which has been gathered. Similary, it has excellent identification effects over urtrained information with similar traits. Consequently, the application of aforementioned characteristic in cyber safety sphere conspensates for the defects which the conventional safety equipment identificiation technique cannot recognize. It also affects the undisclosed atypical data, and vastly refines the identification performance and precision of the said network in Fig. 4.

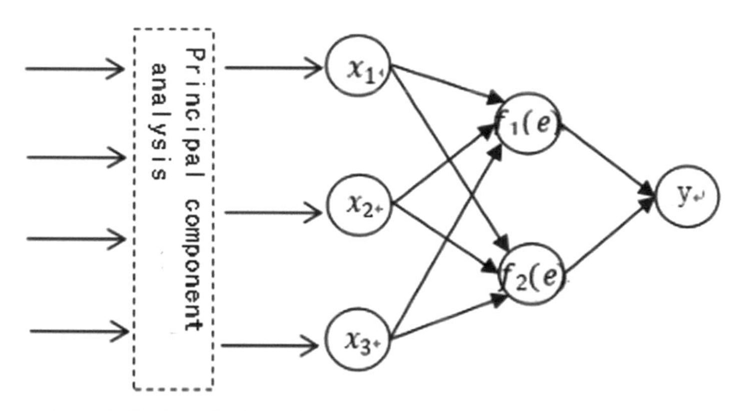

Fig. 4. PCA and BP combined information data detection analysis

PCA and BP combined with the information data detection and analysis model through principal component analysis to reduce the total amount of data while retaining all the information characteristics of the original data. Using a small amount of principal components as the input of the BP neural network, with the help of the advantages of the BP neural network in processing nonlinear problems, the generalization ability and robustness of the system can be improved. Therefore, this detection method that combines the statistical method and the BP neural network has greatly improved the accuracy and detection performance compared with the traditional single detection method.

5 Conclusion

According to the experimental simulation results in MATLAB, the performance of computer information security defense methods comprsing of PCA as well as neural system of BP in the network training can be greatly improved in the context of big data. The false alarm rate dropped to about 10%. Incorporationg PCA as well as neural system of BP information security models to improve the identification performance as well as the progression of the identification of information safety regarding abnormal set of information. It provides a new research reference for computer information security defense methods in the background of big data.

References

1. Xiaoyan, R., Danwa, S.: Research on cyber-attack defense system based on big data and threat intelligence. Info. Secur. Res. **5**(5), 383–387 (2019)
2. Lei, Z., Yong, C., Jing, L., Jiang Yong, W., Jianping.: Application of machine learning in cyberspace security research. Chinese J. Comput. **41**(9), 1943–1975 (2018)
3. Siqin, W., Biao, W.: Computer network security evaluation simulation model based on neural network. Mod. Electron. Technol. **40**(3), 89–91 (2017)
4. Raguseo, E.: Big data technologies: an empirical investigation on their adoption, benefits and risks for companies. Int. J. Inf. Manage. **8**(1), 187–195 (2018)
5. Custers, B., Bachlechner, D.: Advancing the EU data economy: conditions for realizing the full potential of data reuse. Inf. Polity. **22**(4), 291–309 (2017)
6. Sheng, J., Amankwah-Amoah, J., Wang, X.: Implications of COVID-19 pandemic on lung cancer management: a multidisciplinary perspective. Critical Reviews in Oncology/Hematology 156 (2020)
7. Yang, C.-C.: The integrated model of core competence and core capability. Total Quality Management Business Excellence (1–2) (2015)
8. Tien, J.M.: Big data: unleashing information. J. Syst. Sci. Syst. Eng. **22**(2), 127–151 (2013)
9. Habib, M.A., Mella-Barral, P.: Skills, core capabilities, and the choice between merging, allying, and trading assets. J. Math. Econ. (2012)
10. Blomster, M., Koivumäki, T.: Exploring the resources, competencies, and capabilities needed for successful machine learning projects in digital marketing. ISEB **20**(1), 123–169 (2021). https://doi.org/10.1007/s10257-021-00547-y
11. Yang, H., Zeng, R., Wang, F., Guangquan, X., Zhang, J.: An unsupervised learning-based network threat situation assessment model for internet of things. Secur. Commun. Netw. **26**(4), 345–355 (2020)
12. Yeom, S., Shin, D., Shin, D.: Scenario-based cyber attack·defense education system on virtual machines integrated by web technologies for protection of multimedia contents in a network. Multimedia Tools Appl. **80**(26–27), 34085–34101 (2020). https://doi.org/10.1007/s11042-019-08583-0

Modeling Method of 3D Environment Design Based on Genetic Algorithm

Weidong Zhao(✉) and Na Zou

Jilin Engineering Normal University, Changchun, Jilin, China
falunwen0418@163.com

Abstract. At present, mobile robots have been applied in various fields, and technologies such as autonomous research and environmental modeling have also received attention. Although this technology has been researched according to certain rules, it still faces certain problems when it comes to solving more complex 3D environment applications. Therefore, this paper studies the 3D environment design modeling method of genetic algorithm, understands the relevant knowledge of 3D environment design modeling on the basis of literature data, and then constructs the 3D environment design modeling method based on genetic algorithm. The algorithm is tested, and the test results show that the algorithm in this paper can eliminate noise very well. In the case of 5% noise, the final fuzzy system has 10 division points (5X5 rules).

Keywords: Genetic algorithm · Environmental design · Modeling method · Fuzzy system

1 Introduction

Intelligent robots must be practical, and intelligence is also a key factor in determining the prospects of its implementation [1, 2]. The main indicators for evaluating the intelligence level of mobile robots also include conditions such as adaptability, autonomy and interactivity. In particular, adaptability shows that intelligent robots can cope with complex and changeable working environments, not only can identify and detect surrounding objects, but also recognize the surrounding environment, and make appropriate choices and implement appropriate functions [3, 4]. Adaptability refers to the ability of autonomous robots to choose their own work procedures and methods in accordance with work tasks and environmental requirements, and interactivity is the cornerstone of human wisdom. There are mainly the following interactions between the robot and the environment: robots, humans and robots, which mainly include the collection, processing and understanding of information [5, 6]. In the field of mobile machines and human intelligence research, due to the rapid progress of sensor technology, electronic computers, artificial intelligence and other related fields, active environmental adaptation has been the main symbol of human intelligence [7, 8].

Regarding the study of 3D environment modeling, other researchers have developed and completed a new type of comprehensive topology model and navigation system

J. H. Abawajy et al. (Eds.): ICATCI 2022, LNDECT 169, pp. 538–545, 2023.
https://doi.org/10.1007/978-3-031-28893-7_64

based on optical camera technology. The system generates travel information with a local imaging area, and uses a hidden Markov model to identify the relationship between the received router signal in the current view and the topographic product peak, and adds a learning method to determine the environment using graphical attributes and their relationships. This is the most reasonable way to monitor environmental navigation [9]. Some designers use complex data fusion systems with independent sensors to point and deploy mobile robot systems seen in the command environment. The system can analyze the ground level and recognize the ability of mobile robots, and support the robot model in the environment and the design of the walking distance of the mobile robot through the navigation system [10]. Some researchers believe that geometric map environmental modeling has the advantage of generating environmental interference information. A variety of advanced graph search algorithms can be used to find the best solution in static and offline programming environments, making it a better platform for correcting new algorithms. The biggest advantage of field mode modeling lies in its actual performance. This method works well and is very popular among robot designers. Mathematical modeling has advantages based on dynamic goals, safety, reliability, optimized operation and overall system control. Modeling based on biological intelligence can manage high complexity and various structural obstacles, avoiding unrealistic dynamic obstacles. However, the main disadvantage is that the programming cycle is very long, so it is not commonly used in mobile robot systems, and the process needs to be approved through the review process. Therefore, for practical problems, different methods can be used to solve the real-time path planning problem [11]. In summary, although there are many researches on 3D environment modeling, its specific construction methods need to be further studied.

This paper studies the 3D environment design modeling method of genetic algorithm, analyzes the process of 3D environment modeling and some existing problems on the basis of literature data, and then analyzes the 3D modeling method based on genetic algorithm, and the designed algorithm performs detection [12].

2 Research on 3D Environment Design Modeling

2.1 3D Environment Design Modeling Process

(1) High-speed acquisition of DEM data

The ground data here mainly refers to digital elevation model (DEM) data. DEM is an important key to realizing 3D high-speed design. At present, the main methods of obtaining ground data are: 1) directly use total station, GPS and other equipment for measurement; 2) directly use DEM in 2DGIS; 3) digital photogrammetry to process aerial images; 4) use aerial laser scanning system and tracking scan directly during the process; 5) Use synthetic aperture radar to receive the elevation of the numerical model.

(2) Collection of large amounts of image data

Another important part of 3D spatial information is image data. It is not only used to realize soil texture data in 3D scenes, but also can be used as an image source to quickly capture urban buildings, roads and rivers. At present, the main technologies for urban image data collection are: 1) The use of aerial survey and remote sensing technology; 2) Digital photogrammetry technology, including short-range photogrammetry technology. Among them, the image resolution obtained by aerial survey and remote sensing technology is lower than that of digital photogrammetry technology, which is the most widely used technology in 3D modeling today.

(3) Acquisition of architectural model data

Architectural model data is also another key element of three-dimensional urban spatial information, and most of the three-dimensional scenes of China's high-speed three-dimensional urban railways are also composed of a large number of three-dimensional architectural models. Building model data is mainly divided into elevation data, geometric elements and building texture data. This work has achieved the fastest way to obtain building model information through free high-resolution orthophotos obtained from Google. This is also a way to achieve free data sources, reduce workloads, and be highly intelligent. It is possible to use high-resolution pixels for automated modeling.

(4) Texture processing of architectural model

At present, in 3D modeling, the editing of architectural textures is mainly divided into three formats: shooting with standard digital cameras, shooting with standard digital cameras, and shooting architectural textures with aerial and high-resolution satellite images. The three methods have the characteristics of high shooting and high cost, and are more suitable for high-precision digital city applications. The modeling in this article does not require high accuracy of the original texture of the building. What is needed is the ability of the building to execute quickly. The corresponding texture map allows you to quickly create urban architectural landscapes. For the purpose of this function, this article draws on the automatic mapping method based on texture library.

2.2 Difficulties in 3D Environment Design Modeling

The mobile robot accurately models its environment, which can provide a reliable guarantee for the layout and route planning of the mobile robot itself. Compared with 2D maps, 3D maps usually have important research significance for autonomous decision-making of mobile robots because they can provide rich environmental information. At present, vision-based (especially monocular vision) mobile robot environment reconstruction schemes have more or less problems in the robustness and application of algorithm model estimation, and further research is needed.

When it comes to the accuracy of the mobile robot's visual positioning, the image feature matching method is usually used to create data correlation, and then the pose estimation model is created and analyzed to perform the positioning of the mobile robot itself. Generally, robust estimation methods such as M estimation and RANSAC are needed to improve the positioning accuracy, because it limits the accuracy of image detection and mapping. However, due to the lack of the above robustness evaluation method for accurate evaluation of samples, its performance efficiency is relatively low.

For the single-sided visual positioning ruler of mobile robots, it is difficult to determine the positioning result through the metric system due to the lack of environmental depth information in the single-sided visual positioning process. In order to effectively solve the above problems, areas with obvious collective attributes such as the ground and roof are usually selected as the prior information, and the geometric relationship between them and the robot is obtained by the method of visual calibration. As mentioned above, the currently proposed algorithm is relatively inferior in terms of robustness in acquiring features of a specific region.

When it comes to the 3D modeling of the monocular vision map of a mobile robot, the process of creating the monocular vision map is more complicated due to the limited mobility of the mobile robot. At the same time, there are a lot of difference information in the 3D environment map created by the single view. How to improve the placement accuracy and construction of mobile robots based on unilateral vision, and how to effectively use the above information to effectively construct a high-density reconstruction of the mobile robot environment is an urgent problem in this field.

3 3D Environment Design Modeling Based on Genetic Algorithm

3.1 Data Collection

In this article, wheel odometer, inertial sensor and optical odometer will be used to form a multi-sensor system to collect environmental data. The measurement characteristics of each sensor are different. For example, optical odometer has high relative accuracy but low reliability, inertial unit has good real-time performance but large displacement error, and wheel odometer has high reliability, but its accuracy is affected by sliding motion. In order to improve the fusion effect of multiple sensors under different working conditions and adaptively adjust the fusion parameters, this paper uses a fuzzy inference system to calculate and complicate the reliability of each sensor in real time, optimize the fusion effect, and improve the positioning system under flexible conditions, to fault tolerance rate (Fig. 1).

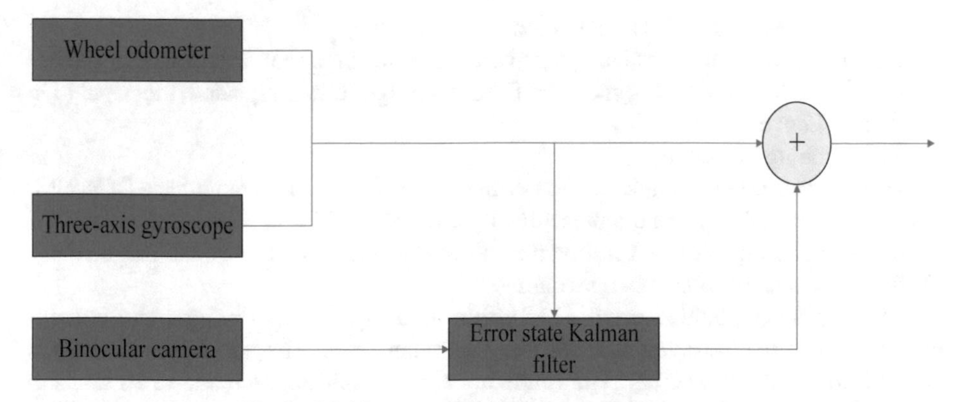

Fig. 1. Multi-sensor positioning system framework

Based on the built-in image acquisition function, the image data is combined with various drawings and other map data of the project area to meet the needs of 3D imaging, and combined with image processing and street view image multi-source measurement data. Not only reflects the real geometric characteristics of the scene, but also takes into account the design elements. After automatic image matching and editing, the drawing and the real scene model are completely integrated to accurately represent the spatial objects and the characteristics of the objects contained in the project area. The creation of 3D models is mainly based on images taken multiple times and the camera lens used in the multi-sensor system, and other image files and multi-source drawings, including model and texture files. There are many types of files and a lot of data. Therefore, when reconstructing a 3D scene, it is necessary to adjust the accuracy of the object representation and reasonably control the data scale.

3.2 Modeling

In this article, the boundary method will be used to build a 3D model. With this model, all you have to do is to get the spatial coordinates of all feature points, define two element lines, and define the boundary of the surface from the lines. A closed surface can form an "individual", and then a real 3D model can be created.

3.3 Application of Genetic Algorithm

This paper draws on the fuzzy system modeling algorithm based on genetic algorithm, which has a dynamic structure. That is, different numbers of member functions are allocated according to the value of the input variable. A new performance index test method is proposed to determine the total number of member functions and the compensation relationship between system complexity and approximation accuracy. This avoids the shortcomings of using cumbersome and time-consuming heuristics to determine GA parameters.

The specific genetic algorithm operation is as follows:

(1) Identification of the initial population

According to the number of language variables in the fuzzy control rule table, 100 initial populations are randomly generated. Even if the algorithm is repeated, the population number will not change.

(2) Fuzzy coding of rules

Use decimal notation to encode fuzzy rules. A set of ambiguous languages {NB, NM, NS, ZE, PS, PM, PB} can be represented by a set of {1, 2, 3, 4, 5, 6, 7}. It expands it into a one-dimensional form, forming the chromosomes required by genetic algorithms.

(3) Improved adaptive genetic algorithm

The improved adaptive genetic algorithm used in this paper adjusts the individual fit between the average value and the maximum value. The adaptation strategy is derived from Eqs. (1) and (2). The following adjustments can be made to P1 and P2: These changes have also improved the quality group. Excellent groups were effectively retained during the search process. This increases the possibility of subordinate individuals mutating and avoids limiting the algorithm to local optimal solutions.

$$p_1 = \begin{cases} \frac{p_{11}\left(f-f'\right)+p_{12}\left(f'-f_{min}\right)}{f-f_{min}} \\ \frac{p_{12}\left(f-f'\right)+p_{13}\left(f'-f\right)}{f_{min}-f} \end{cases} \tag{1}$$

$$p_1 = \begin{cases} \frac{p_{21}\left(f-f'\right)+p_{22}\left(f'-f_{min}\right)}{f-f_{min}} \\ \frac{p_{22}\left(f-f'\right)+p_{13}\left(f'-f\right)}{f_{min}-f} \end{cases} \quad (2)$$

Among them: 2, respectively represent the maximum and minimum values of individual fitness of the group, f represents the average fitness value, and 2 represents the larger fitness value of the two individuals that need to be crossed.

(4) Select operation

This article uses the roulette algorithm. The specific procedure is that the probability of a person being selected is proportional to his ability. Assuming that the population is n and the ability of person i is denoted by f, the probability p of that person being hired is expressed as follows.

$$P = f/\sum\nolimits_{i=1}^{n} f_i \quad (3)$$

4 Algorithm Test

In order to test the robustness of the new algorithm, the output data is interpolated with additional noise ±5% and 9%. Since the triangle split structure and point conclusion of the fuzzy system are the same as the new algorithm, the parameters required for each rule are also the same as the new algorithm. The fuzzy system uses four rule structures (3X3, 4X4, 5X5, 6X6), and each MSE is shown in Table 1.

Table 1. Algorithm test results

	5% noise	9% noise
3X3	0.014	0.017
4X4	0.0042	0.0078
5X5	0.0021	0.0077
6X6	0.0019	0.0082

It can be concluded from Fig. 2 that the new fuzzy system can successfully remove the additional noise. Under 5% noise, the final fuzzy system has 10 division points, 35 chromosomes and 20 generations of evolution. Under 9% noise, the final fuzzy system has 11 division points, 25 chromosomes and 10 generations of evolution.

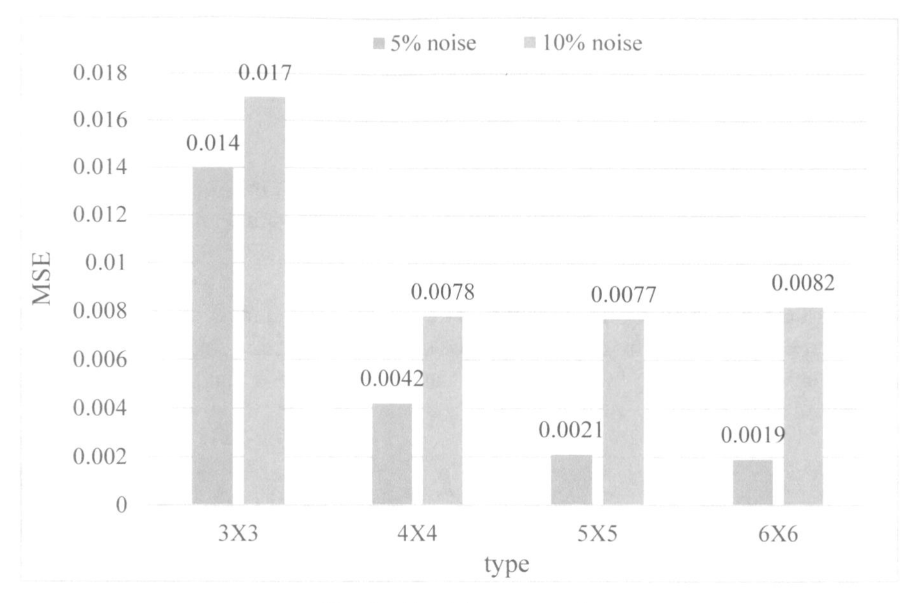

Fig. 2. Algorithm test results

5 Conclusion

This paper studies the 3D environment design modeling method based on genetic algorithm. After understanding the relevant theories, the 3D environment design modeling method based on genetic algorithm is constructed, and then the algorithm is tested, and the detection result shows that the new fuzzy system performance Successfully eliminate additive noise.

Acknowledgements. This article is part of the research results of the 2021 Jilin Engineering Normal University doctoral research start-up funding project "Research on the Construction of University Campus Image from the Perspective of Space Aesthetics" (Project No.: BSGC202133).

References

1. James, A.G., Maciej, S., Abbas, S., et al.: Simulation of single-species bacterial-biofilm growth using the glazier-graner-hogeweg model and the compucell 3D modeling environment. Math. Biosci. Eng. MBE **5**(2), 355–388 (2017)
2. Perez-Perez, Y., Golparvar-Fard, M., El-Rayes, K.: Segmentation of point clouds via joint semantic and geometric features for 3D modeling of the built environment. Autom. Constr. **125**(7), 103584 (2021)
3. Bejaoui, B., Solidoro, C., Harzallah, A., et al.: 3D modeling of phytoplankton seasonal variation and nutrient budget in a southern mediterranean lagoon. Mar. Pollut. Bull. **114**(2), 962–976 (2017)
4. Kim, M., Hwang, D.G., Jang, J.: 3D pancreatic tissue modelingin vitro: advances and prospects. BioChip J. **14**(1), 84–99 (2020)

5. Safhalter, A., Vukman, K.B., Glodez, S.: The effect of 3D-modeling training on students' spatial reasoning relative to gender and grade. J. Educ. Comput. Res. **54**(3), 395–406 (2016)
6. Neiß, M., Sholts, S.B., Wärmländer, S.K.T.S.: New applications of 3D modeling in artefact analysis: three case studies of viking age brooches. Archaeol. Anthropol. Sci. **8**(4), 651–662 (2014). https://doi.org/10.1007/s12520-014-0200-9
7. Lee, D.G., et al.: Key point extraction from LiDAR data for 3D modeling. J. Korean Soc. Surv. Geodesy, Photogrammetry and Cartography 34(5), 479–493 (2016)
8. Sanduleac, I., Casian, A.: Nanostructured TTT (TCNQ) 2 organic crystals as promising thermoelectric n-type materials: 3D modeling. J. Electron. Mater. **45**(3), 1–5 (2016)
9. Popovski, F., Spasov, N., Mijakovska, S., et al.: Comparison of rendering processes on 3D model. Int. J. Comput. Sci. Inf. Technol. **12**(5), 19–28 (2020)
10. Perego, R., Guandalini, R., Fumagalli, L., et al.: Sustainability evaluation of a medium scale GSHP system in a layered alluvial setting using 3D modeling suite. Geothermics, 59(JAN.PT.A), 14–26 (2016)
11. Singh, A., Dehiya, R., Gupta, P.K., et al.: A MATLAB based 3D modeling and inversion code for MT data. Comput. Geosci. 104(JUL.), 1–11 (2017)
12. Hocine, A., Poncet, S., Fellouah, H.: CFD modeling and optimization by metamodels of a squirrel cage fan using OpenFoam and Dakota: ventilation applications. Build. Environ. **205**(1), 108145 (2021)

Pattern Recognition Based on Variant Reinforcement Meta-learning

Zeyang Zheng[1(✉)], Haiyong Chen[1], Shijie Wang[1], and Ramya Radhakrishnan[2]

[1] School of Artificial Intelligence, Hebei University of Technology, Tianjin, China
abc22343244@outlook.com
[2] Jimma University, Jimma, Ethiopia

Abstract. Pattern recognition is the study of automatic processing and interpretation of patterns using computers through mathematical techniques, where the environment and the objects are collectively referred to as "patterns". It studies the problem of classifying samples by their attributes using computational methods. The main research directions of pattern recognition are computer vision, image processing, and natural language processing. In this paper, we discuss the application of a variant of reinforcement meta-learning, which combines experience replay and meta-learning in reinforcement learning, to the problem of continuous learning in pattern recognition. It also compares its accuracy with current solutions such as online learning for the continuous learning problem in pattern recognition. It mainly consists of reinforcement learning, reptilian meta-learning and experience replay in SGDM optimizer. It is finally concluded that for the problem of continuous learning in pattern recognition, its ability to preserve past learned knowledge with rapid learning of current knowledge is higher than the current online learning. Based on this, the effect of different parameters on the performance of this variant of reinforcement meta-learning is explored.

Keywords: Pattern recognition · Experience replay · Meta-learning · Reinforced meta-learning

1 Introduction

For biological intelligence, the ability to recognition of patterns is the most basic cognitive ability in the human brain and the cornerstone of other advanced creature policy-making. It is evident from the history of human evolution that this ability provides us with security in the face of bad weather and fierce beasts. On the other hand, pattern recognition is also an important goal in the study of machine learning and artificial intelligence, and the solution of many advanced intelligence problems relies heavily on the success of automatic and accurate pattern recognition [1].

In recent years, with the continuous progress in computer performance and recognition algorithms, the accuracy of many patterns' recognition tasks has rapidly improved continuously, reaching or even surpassing human performance in some places. However, when solving some practical problems about continuous learning of pattern recognition,

J. H. Abawajy et al. (Eds.): ICATCI 2022, LNDECT 169, pp. 546–554, 2023.
https://doi.org/10.1007/978-3-031-28893-7_65

highly accurate pattern recognition systems often become unsteady and undependable, and end up with less than satisfactory accuracy. One reason for such a phenomenon is catastrophic forgetting [2]. That is, after learning new knowledge, one almost completely forgets what has been learned before. Because at each stage of learning, the current mainstream algorithms use different data or tasks, the problem that learning a new task or data can cause a significant degradation in the performance of the old task. Meta-learning is a new algorithm currently proposed, also known as learning to learn. Its essence is to increase the generalization ability of the learner across multiple tasks to address catastrophic forgetting. Experience replay was proposed long ago as the basic data generation mechanism for non-policy deep reinforcement learning and is still in constant use today [3]. Therefore, in this paper, we propose pattern recognition based on variant reinforcement meta-learning.

2 Ability to Preserve Past Knowledge and Learn Present Knowledge in Continuous Learning

The better a model is at preserving past knowledge and learning from current knowledge, the less likely it is to interfere with future gradients and the more likely it is to transfer future gradients [4]. With the parameters w and loss L, be able to define two practical parts of transferred and hindrance between two separate samples (x_i, y_i) and (x_j, y_j) when trained with SGD. The 2 figures below show the future gradients of transfer and interference phenomena, respectively (Figs. 1 and 2).

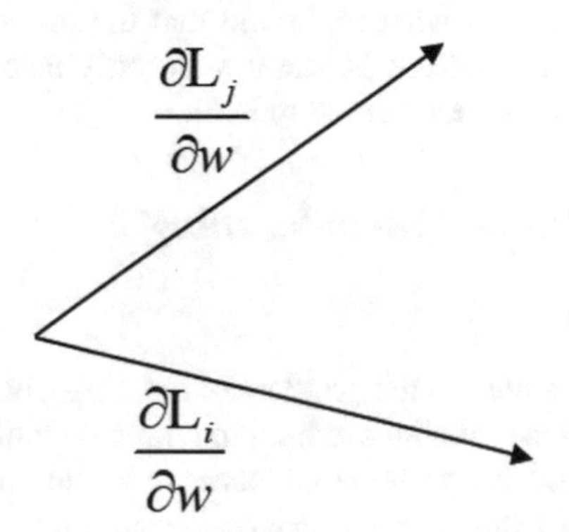

Fig. 1. The emergence of transfer

Transfer appears when:

$$\frac{\partial L(x_i, y_i)}{\partial w} \cdot \frac{\partial L(x_j, y_j)}{\partial w} > 0 \tag{1}$$

This means that the performance of learning example i is improved if example j is not a duplicate, and vice versa.

Interference appears when:

$$\frac{\partial L(x_i, y_i)}{\partial w} \cdot \frac{\partial L(x_j, y_j)}{\partial w} < 0 \tag{2}$$

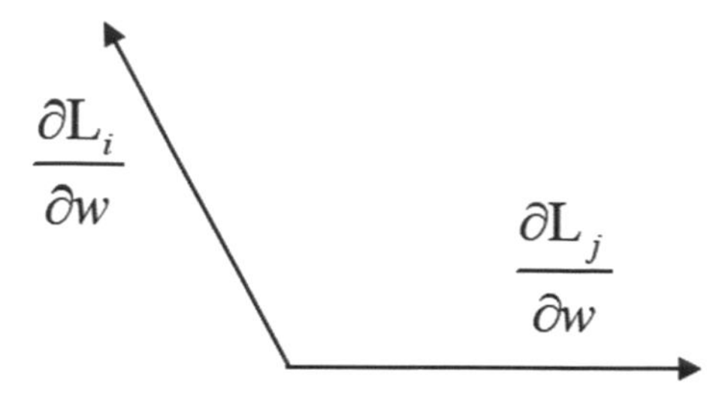

Fig. 2. The emergence of Interference

Contrary to the case above, here, studying sample i will give rise to forgetting sample j and vice versa. When learning between i and j with overlapping parameters, there is weight sharing between them. Therefore, the transfer potential is maximum when weight sharing is maximum, and the interference potential is minimum when weight sharing is minimum.

Former remedies to the difficulty of dealing with stability and plasticity in continuous learning usually take place in an uncomplicated temporal environment, where studying is separated into two stages: The whole bypast experiences are classified as aged memory and the data presently being learned belong to new studying. In that case, the objective is simply to minimise the disturbance backwards projected in time, which is usually achieved by mitigating the extent to which weights are shared explicitly or implicitly. However, this solution encounters an important problem that cannot be ignored, namely that the system is still studying what to do and that the future may produce a system that is largely unaware of this. That is, the system only mechanically learns what is happening at the time, and has not learned to learn.

3 Experience Replay and Meta-learning

3.1 Experience Replay

Reinforcement learning centres on the problem of learning a behavioural policy, a mapping from states or situations to actions, which maximizes cumulative long-term reward [5]. Reinforcement learning is a system that learns in a continuous trial-and-error manner. The system communicates with the external environment continuously, and after each communication with the external environment, the external environment will give the system feedback, which will evaluate how good or bad the system just behaved. If the feedback thinks that the system just behaved in accordance with the requirements, then the system will move forward in the direction of the previous behavior. If the feedback that the system just behavior is not in line with the requirements, then the system will move in the direction of the opposite of the previous behavior. So on and so forth, until the training is complete.

Experience replay is an old technique proposed in 1992. It still plays an important role today and is an important part of reinforcement learning. It allows agents to remember and reuse past experiences. Experience replay has been shown to greatly stabilise the training process and increase the efficiency of samples by breaking temporal correlations

[6]. The core of Experience Replay is to keep the memory of the examples you have already seen.

$$w = \arg\min_{w} E_{(x,y)\sim H}[L(x, y)] \tag{3}$$

L is the loss function and it is chosen to tackle the problem. H approximates a dataset with present dimension H_{size} and maximum dimension $H_{\max}$. a reservoir sampling algorithm is used to update the buffer. The reservoir sampling algorithm allows all data in the dataset to be selected with equal probability. This ensures that at each time step, this chance of any O samples being seen in the buffer is the same as H_{size}/O. Also, we have to care more about the current example to ensure that it intersects with the examples in the replay buffer.

3.2 Meta-learning

The previous training system worked well for a certain situation after training with a large amount of data. But as soon as the environment changed, the effect of using the system again was not acceptable, and then the system produced a catastrophic forgetting. Meta-learning was created to solve the problem of catastrophic forgetting after training. Let the system learn to learn, just like people. Meta learning examines how the learning system understands learning itself in relation to the domain or task being studied. Meta learning is about dynamically choosing the appropriate bias based on the current situation [7].

Model agnostic meta-learning (MAML) is a very classical meta-learning algorithm. The meta-learning algorithm consists of a method with two optimization loops, an outer loop to find a meta-initialization and an inner loop from which a new task can be efficiently learned [8]. The basic principle of MAML is to find the best set of initialization parameters, so that it can achieve better results on any new case with only a few updates to the gradient. After that, the model can be fine-tuned to achieve better results even on small samples by analysing specific situations.

In addition, there is a meta-learning algorithm called Reptile, which can be considered as an improvement of the MAML algorithm. It can also be called as first-order MAML. The MAML algorithm needs to partition each learned task into a test set and a training set. However, the reptile algorithm does not need to partition each learned task data into training and test sets. Reptile can sample from the same distribution to which the training task belongs and quickly learn tasks that have not been seen before [9]. It is easy to implement and does not require an optimization process for differentiation. It is implemented by sequentially optimizing the data across s batches Using the stochastic gradient descent optimization method and studying rate β. After training these batches, we obtain the starting parameters for training w_0 and update them to $w_0 + \chi * (w_k - w_0) \rightarrow w_0$, where χ is the studying rate of the meta-learning amend. Reptile generally optimizes the aim over a series of z batches:

$$w = \arg\min_{w} E_{B_1,\ldots,B_z\sim D}[2\sum_{i=1}^{i-1}[L(B_i) - \sum_{j=1}^{i-1}\beta\frac{\partial L(B_i)}{\partial w}\cdot\frac{\partial L(B_j)}{\partial w}]] \tag{4}$$

where $B_1, ..., B_z$ are batches within D. The gradients generated on these batches approximate stationary distributed samples.

4 Variant Reinforcement Meta-learning

The organic combination of the modified crawler algorithm, experience replay module and SGDM produces a variant of reinforcement meta-learning. This variant combines the advantages of the crawler algorithm, experience replay, and SGDM. It effectively retains the knowledge learned in the past and learns from the current knowledge. The model has a high probability of transferring future gradients and a low probability of interfering with future gradients. It is advantageous for solving the problem of continuous learning in pattern recognition (Table 1).

The algorithm keeps an empirical replay memory H with reservoir sampling and draws $k - 2$ random samples from the buffer at each time step to form a batch, which is then trained together with the current example. Where the current example is placed at a random position in the batch. After processing the batch, Reptile is performed for meta-updating. Finally, the reservoir sampling method is applied to update H again, and so on and so forth until the end of the cycle. The optimizer used in this algorithm is SGD with Momentum (SGDM) instead of SGD, which has the disadvantage that its update direction depends entirely on the gradient computed from the current batch and is therefore very unstable. It simulates the inertia of an object in motion, i.e., the update preserves the direction of the previous update to a certain extent, while fine-tuning the final update direction using the gradient of the current batch. In this way, it can increase the stability to a certain extent, thus learning faster, and there is a certain ability to get rid of the local optimum. After continuous testing, it was found that SGD with a driving volume works better than SGD when the momentum value of SGDM is 0.5 and the buffer size is small. Therefore, the following experiments were conducted with a momentum value of 0.5 for SGDM. The following is the detailed flow of the algorithm.

Now to test if the algorithm is better than the previous model in the continuous learning domain. Online learning was chosen to compare with this algorithm to determine if the algorithm identifies the MNIST rotated dataset with better accuracy. The MNIST dataset is a very classical handwritten digit dataset. The dataset was split into two parts: a 60,000-row training dataset and a 10,000-row test dataset. Such a cut is important in that there must be a separate test dataset not for training but for evaluating the performance of this model at the time of the study, thus making it easier to generalize the designed model to other datasets i.e., generalization. MNIST rotation is a variant of MNIST proposed by Lopez-Paz & Ranzato. As the name suggests, each of these tasks contains numbers rotated at a fixed angle from 0 to 180 degrees. Therefore, it is more difficult to identify the rotation of this variant of MNIST than to identify the original MNIST. In the experiments, the final retention accuracy of all tasks was used as a metric for comparing methods and is referred to as retention accuracy (RA). The results of the experiments are presented in the following Table 2.

From the experimental results, it can be seen that variant reinforcement meta-learning is superior and more accurate than online learning in solving problems with continuous learning like the MNIST Rotations dataset.

Table 1. Variant reinforcement meta-learning

Algo 1: Variant reinforcement meta-learning

Algo ($D, w, \alpha, \gamma, z, k$)
  $H \leftarrow \{\}$
  for $t = 1, ..., T$ do
    for (x, y) in D_t do
      $B, index \leftarrow sample(k-2, H)$
      $w_0 \leftarrow w$
      for $i = 1, ..., k-2$ do
        $x_c, y_c \leftarrow B_i[j]$
        if $j = index$
          $w_{k-1} \leftarrow SGDM(x, y, \theta_{k-2}, z\alpha)$
        else
          $w_{i-1} \leftarrow SGDM(x_c, y_c, w_{i-2}, \alpha)$
      end for
      $w \leftarrow w_0 + \gamma(w_{k-1} - w_0)$
      $H \leftarrow H \cup \{(x, y)\}$
    end for
  end for
  return w, H
end Algo

Table 2. Comparison of variant-enhanced meta-learning and online learning

Model	Online	Variant of reinforcement meta-learning
RA (%)	52.76	76.67

5 The Effect of Different Parameters on the Performance of this Variant of Reinforcement Meta-learning

5.1 Buffer Size

Buffer is the area used to store previous examples, that is, it stores the knowledge previously learned by the system. Its size has an important impact on the system's recognition accuracy. Generally speaking, the larger the size of the buffer, the more previous knowledge it holds and the higher the recognition accuracy of the system. So of course, we would like the buffer size to be close to infinity, but this is unrealistic because we cannot find a space that can store an infinite amount of previously learned knowledge. So, we want the size of the buffer to be as small as possible with little impact on the recognition accuracy. The following table shows the system's recognition accuracy for MNIST Rotations with buffer from 200–700, holding all other parameters constant (Table 3).

Table 3. Variant reinforcement meta-learning recognition accuracy under different buffer sizes

Buffer Size	RA (%)
200	76.67
300	79.73
400	81.66
500	82.19
600	82.43
700	83.55

As can be seen from the table, the recognition accuracy of the system keeps increasing as the size of the buffer keeps going up. However, the resources consumed also increase. To weigh the resource consumption and recognition accuracy, the larger the buffer size does not mean the better the final result. Therefore, it is better to choose a smaller buffer size because it can reduce the resource consumption and get good results.

5.2 Lr

Lr is called the learning rate. It is a hyperparameter that controls how much the optimizer adjusts the network weights in terms of the loss gradient. The lower the value, the slower the loss function will be along the downward slope [10]. Therefore, when the learning rate is very low, training takes a lot of time. Then, if the learning rate is high, then the training may not converge or even diverge. The change in weight can be so large that the optimizer exceeds the minimum value, thus making the loss even worse [11]. Therefore, choosing a suitable learning rate is very important to improve the recognition accuracy of the system. The following table shows the recognition accuracy of the system for

Table 4. Variant reinforcement meta-learning recognition accuracy under different learning rates

Lr	RA (%)
0.01	74.75
0.02	76.67
0.03	77.06
0.04	75.94
0.05	11.56

MNIST Rotations when the learning rate is from 0.01 to 0.05 with other parameters held constant (Table 4).

As can be seen from the table, the recognition accuracy of the system keeps improving as lr keeps increasing from 0.01 to 0.03. However, when lr goes from 0.04 to 0.05, the recognition accuracy of the system is decreasing compared with the previous one. Especially, when lr is 0.05, the recognition accuracy flies down to about 11%. Therefore, the learning rate is chosen reasonably so that it is not too big or too small to help improve the recognition accuracy.

6 Conclusions

In the field of pattern recognition, each task to be recognized is not completely unrelated to the other tasks. We would like to have a system model that can learn one task and then learn the subsequent tasks without forgetting the previous learning experience. But the previous system models were not as accurate in recognizing a task after learning it as we had hoped. When using them to recognize another task, the recognition accuracy drops dramatically compared to the previously learned task, and catastrophic forgetting occurs. In this paper, we propose a hypothetical system model: variant reinforcement meta-learning. The model mainly consists of experience replay in reinforcement learning, reptile meta-learning and SGDM optimizer. It does not just learn the task by rote like previous system models but can learn to learn. The experimental results also show that variant reinforcement meta-learning is superior in solving continuous learning problems such as MNIST rotated datasets. In summary, it has been our goal to make the system learn, but until now the system has not been able to learn as well as it should. However, this should not stop us from trying and researching. Looking ahead, as science and technology continue to advance, i.e., mathematical theory continues to advance and machine computing power increases dramatically, the system will certainly fully grasp what it has learned before and use it for the task of learning other knowledge.

References

1. Zhang, X.Y., Liu, C.L., Suen, C.Y.: Towards robust pattern recognition: a review. Proc. IEEE **108**(6), 894–922 (2020)
2. Kirkpatrick, J., et al.: Overcoming catastrophic forgetting in neural networks. Proc. Natl. Acad. Sci. **114**(13), 3521–3526 (2017)
3. Fedus, W., et al.: Revisiting fundamentals of experience replay. In: International Conference on Machine Learning (pp. 3061–3071). PMLR (2020, November)
4. Riemer, M., et al.: Learning to learn without forgetting by maximizing transfer and minimizing interference. arXiv preprint arXiv:1810.11910 (2018)
5. Botvinick, M., Ritter, S., Wang, J.X., Kurth-Nelson, Z., Blundell, C., Hassabis, D.: Reinforcement learning, fast and slow. Trends Cogn. Sci. **23**(5), 408–422 (2019)
6. Zha, D., Lai, K.H., Zhou, K., Hu, X.: Experience replay optimization. arXiv preprint arXiv: 1906.08387 (2019)
7. Vilalta, R., Drissi, Y.: A perspective view and survey of meta-learning. Artif. Intell. Rev. **18**(2), 77–95 (2002)
8. Raghu, A., Raghu, M., Bengio, S., Vinyals, O.: Rapid learning or feature reuse towards understanding the effectiveness of maml. arXiv preprint arXiv:1909.09157 (2019)
9. Nichol, A., Schulman, J.: Reptile: a scalable metalearning algorithm. arXiv preprint arXiv: 1803.02999, 2(3), 4 (2018)
10. Zulkifli, H.: Understanding learning rates and how it improves performance in deep learning. Towards Data Sci. **21**(23) (2018)
11. Surmenok, P.: Estimating an optimal learning rate for a deep neural network. Towards Data Sci. 5 (2017)

Online Crisis Learning Alert and Intervention Based on OU Analyse

Shuting Liu[1(✉)] and Muhammad Hamam[2]

[1] College of Engineering Technology, Xi'an Fanyi University, Xi'an, Shaanxi, China
xafy01@126.com

[2] Uni de Moncton, Moncton, Canada

Abstract. In the Internet plus education environment, big data and learning analysis technology have been applied to educational reform, providing a convenient and effective tool for solving the online education quality. Online learning Alert can improve students' learning effect, accurately identify learning crisis students, send warning information and intervene in advance, provide help and guidance for students, and promote students' academic success. Taking OU Analyse as the research object, we use learning analysis technology to identify students with academic crisis, and implement effective intervention measures to improve students' retention rate and promote students' academic progress. By discussing the successful practice of the British Open University, it has great enlightening significance for the development of online education institutions plagued by high dropout rate and education quality in China, and provides experience and practices that can be used for reference.

Keywords: Learning crisis · Learning alert · Online learning · OU analyse

1 Introduction

China's online education has played an important role in promoting the deep integration of modern information technology and education and teaching, serving the popularization of higher education, building an education system serving the whole people's lifelong learning, and building a learning society [1]. Although the development of online education has made remarkable achievements, it still faces many problems, such as high dropout rate in online learning, low participation of students in learning, lack of monitoring and management in the online learning process, untimely feedback and help from teachers, etc., which makes it difficult for students to complete the course learning smoothly. According to relevant survey data, although the public has different opinions on the quality of online education, most of them are skeptical and worried on the whole. With the development of Internet plus education, big data and learning analysis technology are applied more and more in the field of education. Learning Alert is the in-depth application of big data and learning analysis technology in the field of education, providing a convenient and effective tool for solving the problems faced by online education [1].

J. H. Abawajy et al. (Eds.): ICATCI 2022, LNDECT 169, pp. 555–563, 2023.
https://doi.org/10.1007/978-3-031-28893-7_66

Students' learning behavior data is recorded and stored on the online learning platform. Learning Alert through mining and analyzing students' online learning data, it can accurately identify students with learning crisis, send warning information in advance and intervene, which can not only provide decision-making reference for teaching management, but also provide help and guidance for students, but also improve the quality of online teaching through online education, It is an inevitable choice to promote students' academic success.

However, according to the literature research, there are few studies on learning Alert in the field of online education in China, the theoretical and practical research is still in its infancy, and there are no mature application cases. The research on learning Alert abroad started earlier and developed more deeply. Some online education institutions have developed different learning Alert systems according to their actual needs, and achieved remarkable results in practice. Among them, the UK Open University, as a model in the field of international online education, has been at the forefront of developing learning analysis methods [2]. In recent years, it has carried out in-depth exploration in learning Alert, launched OU analyze online learning Alert project, used learning analysis technology to early identify students with academic crisis, and implemented effective intervention measures to improve students' retention rate and promote students' academic progress. The successful practice of the British Open University has great enlightening significance for the development of online education institutions plagued by high dropout rate and education quality problems, and provides experience and practices that can be used for reference.

2 Introduction to OU Analyse

Since its establishment, the UK Open University has been committed to providing adult learners with opportunities to participate in higher education. It adheres to the open admission policy and online teaching form, and has helped more than 2 million students obtain degrees. However, in recent years, due to changes in the internal and external situation, the UK Open University, like most online education institutions, is still facing the realistic dilemma of low retention rate and high dropout rate. In order to meet the challenge, the UK Open University has been thinking and exploring how to use advanced information technology to improve the retention rate of students [3].

At present, the application of learning analysis technology makes it possible for online education to realize large-scale personalized learning. Through the measurement, collection, analysis and report of learners and learning situation data, it can help teachers deeply understand learners' learning status and provide higher quality and personalized learning support. Therefore, the UK Open University has invested a lot of resources to develop and evaluate learning analysis methods and launched the OU analyze learning Alert project, which is sponsored by the knowledge Media Institute of the UK Open University The Institute of Educational Technology and Student Data and Analytics Team jointly carry out cooperative research and practice to help students successfully complete their studies and improve retention rate. Therefore, the developed online learning Alert system uses machine learning technology to establish an early warning model to quantitatively predict students' development to determine the possibility of learning failure.

According to the visual results output by the system, teachers can accurately identify students with learning crisis and provide them with timely and effective intervention to promote students to successfully complete their studies, Make learning progress.

3 Construction of OU Analyze Online Learning Alert System

3.1 Data Acquisition

Data collection and analysis is the basis and premise of learning Alert. The data collected by OU analyze online learning Alert system mainly comes from the student information data and learning behavior data stored on the learning management system and learning platform of the UK Open University, mainly including the following four categories: the first category is demographic information data, such as gender, age, major, region Family background, etc.; The second category is to prepare data for learning, such as the highest level of education, the number of courses taken, and the grades obtained; The third category is online learning behavior data, such as learning time, page access, forum discussion, resource retrieval, etc.; The fourth category is learning result data, such as test scores, task completion, etc. In order to improve the early warning effect, the project team integrates multiple types of data to achieve common early warning. Whether learners' learning background, past learning achievements or current learning behavior are possible factors affecting students' learning results. Therefore, only by collecting and analyzing data from multiple dimensions can learning Alert become more effective and accurate.

3.2 Construction of Learning Alert Model

OU Analyze online learning Alert system uses machine learning method to build prediction model. The goal is to identify whether there is a learning crisis by predicting students who may not complete the next Tutor Market Assessment in advance. Previous studies have shown that when the data types are rich and the amount of data is large, the prediction model formed by machine learning method is more accurate and reliable. In machine learning methods, classification is the key algorithm to judge the success of learning [4]. At present, the most used classification algorithms include K-nearest neighbor, Bayesian, decision tree, support vector machine, linear discriminant analysis and logistic regression. In order to achieve better prediction effect, the online learning Alert system comprehensively uses three classification algorithms: K-nearest neighbor, naive Bayes and classification regression tree. Combined with the collected student information and learning behavior data, four early warning models are developed: Naive Bayes early warning model; K-nearest neighbor algorithm early warning model using student demographic information data; An early warning model of K-nearest neighbor algorithm based on online learning behavior data; Classification regression tree early warning model. These models consider the different attributes of input data and complement each other. Their conclusions are integrated for final prediction, and the four early warning models have the same weight. If a student meets the prediction results of two or more models, it can be determined that the student will not complete the next TMA assessment and has a potential learning crisis. The specific process is shown in Fig. 1.

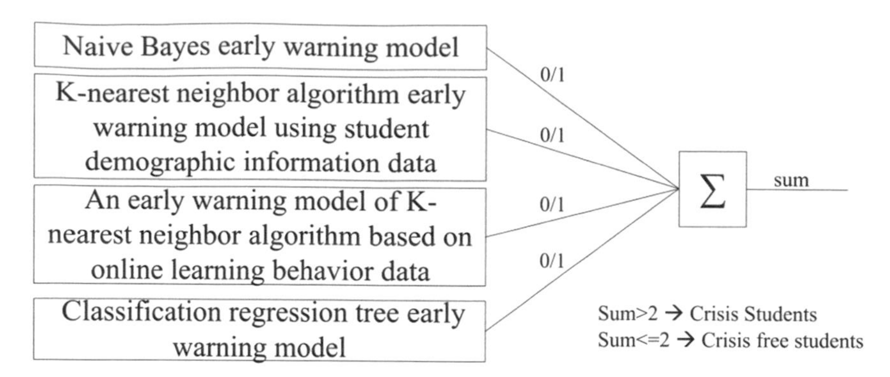

Fig. 1. Early warning model determination process

In order to evaluate the prediction effect of the model, it is necessary to comprehensively use metrics such as Precision, Accuracy, Recall and F-Measure. Among them, the accuracy rate represents the proportion of students correctly predicted among all students predicted by the model as "not completing the next TMA assessment". Recall rate refers to the proportion of students correctly predicted by the model among all students who actually "will not complete the next TMA assessment". The accuracy rate represents the proportion of all correctly predicted students in the total number of students. The comprehensive evaluation index is the harmonic average of accuracy and recall. The early warning model has been applied and evaluated in an undergraduate course of the Open University of the UK. It predicts whether students will complete the assessment within 30 learning weeks of the course. The specific parameter values are shown in Table 1. It can be seen from the data in the table that the accuracy and accuracy of the model are maintained at a high level, and continue to improve with the increase of prediction time and prediction times. The accuracy increases from 0.78 in the first week to 0.88 in the thirtieth week. The recall rate increased from 0.39 in the first week to 0.52, and the comprehensive evaluation index increased from 0.46 in the first week to 0.64 in the thirtieth week, and has been at a stable level. Practice shows that the model has high accuracy and accuracy, stable recall rate and comprehensive evaluation index, and has good prediction effect.

3.3 Visual Result Presentation of Learning Alert

OU Analyse project has developed The Early Alert Indicator Dashboard (EAI) to enable teachers, managers and students of the UK Open University to obtain the latest model prediction results in a visual way. For tutors, using the dashboard can quickly observe students' learning performance, obtain early warning information in advance and take intervention measures to help students successfully complete the course learning. For teaching managers, using the dashboard can find out the crisis students of various majors and courses as soon as possible, make scientific and reasonable teaching decisions and provide more accurate teaching services. For students, using the dashboard can more clearly understand their learning performance, obtain timely feedback, and actively adjust their learning behavior.

Table 1. Model Accuracy Rate, Recall Rate and Comprehensive Evaluation Index

Evaluation Index	Accuracy Rate	Recall Rate	Comprehensive Evaluation Index
Week 1	0.78	0.39	0.46
Week 5	0.81	0.51	0.62
Week 10	0.79	0.47	0.57
Week 15	0.77	0.48	0.59
Week 20	0.79	0.57	0.61
Week 25	0.84	0.54	0.67
Week 30	0.88	0.52	0.64

In the warning interface of EAI dashboard on the overall learning situation of the class, the top shows the average score of class students in the three TMA assessments completed in the form of histogram, and the activity of students on the platform in the form of broken line chart. At the same time, it also shows the historical performance data of the course, which is convenient for teachers to compare and analyze. At the bottom of the dashboard, the prediction results of whether students will complete the fourth TMA assessment are presented in the form of a list [4]. The results are divided into two states: submitted assessment (low risk) and not submitted assessment (high risk), which are represented by green and red circles respectively. The more red indicators, the higher the learning crisis of students, and the lower the possibility of successfully passing the course. In the warning interface of students' individual learning situation, the dashboard displays the prediction results and prediction basis of whether the student will complete the next TMA assessment, as well as the neighboring students whose situation is most similar to that of the student, and uses three colors to represent the risk status of neighboring students: green indicates low risk (normal), yellow indicates medium risk, and red indicates high risk. Teachers can obtain detailed prediction information by clicking the adjacent student icon, so as to realize group intervention.

4 OU Analyze Online Learning Crisis Intervention Strategy

In the process of learning Alert, learning intervention, as a part directly related to the teaching process, is an important link to help students solve the learning crisis and achieve academic success. After identifying students' learning status through the dashboard, the tutors and student support team of the UK Open University will take further intervention measures to provide various support and help at the key nodes where students may have academic crisis. There are two main ways to implement intervention. One is manual intervention, that is, teachers and student support teams provide personalized learning guidance and suggestions to students manually according to the early warning information of the dashboard. In order to further improve the effect of manual intervention, OU Analyse project provides a complete set of processes to guide teachers. The process includes five steps: monitoring data, investigating problems, taking actions, selecting methods and evaluating results [5].

First, teachers log in to the dashboard to understand students' learning performance and identify whether students have learning crisis. After finding the problem, teachers can determine the root cause of the problem by obtaining more and more detailed data. Then, teachers take targeted intervention actions and choose appropriate intervention methods to help students solve the learning crisis. The project also helps teachers implement intervention by providing intervention strategy guidelines, as shown in Table 2. The guidelines mainly include providing teaching support and adjusting teaching design. Specific intervention strategies are given for students' cognitive dimension, interactive dimension, teaching dimension and emotional dimension. Finally, teachers should track and investigate the intervention effect and carry out reasonable evaluation.

Table 2. Guidelines for Intervention Strategies

	Provide teaching support	Adjust teaching design
Cognitive dimension	(1) timely provide homework correction or performance feedback (2) strengthen pre examination counseling	(1) redesign learning materials (2) redesign course assessment
Interaction dimension	(1) add video conference link (2) conduct individual conversation (3) recommend learning partners	(1) organize forum discussion and online communication (2) carry out group cooperative learning (3) introduce problem solving / Project-based cooperative learning activities
Teaching dimension	(1) add video conference link (2) contact students with learning crisis by telephone / email / video (3) provide guidance for students' difficult problems	(1) hold an online video conference every two weeks (2) prompt and analyze the key learning contents (3) strengthen the learning guidance in the first two weeks (4) provide teaching scaffold
Emotional dimension	(1) carry out individual conversation and psychological counseling (2) send email or message to encourage students to make progress	(1) use the questionnaire to understand students' learning emotion (2) establish a learning group

Another intervention method is system intervention, that is, the intervention engine or intervention system automatically generates and pushes personalized learning suggestions and learning resources to students, including automatic message pop-up, e-mail reminder, learning risk report, personalized Resource Recommendation, etc. The system can send notification information to students in the form of automatic message pop-up window or e-mail to remind students to log in to the learning platform on time, complete learning assessment in time, submit course homework, and guide students to complete the course learning at this stage [6]. By sending the learning risk report, students can better understand their own learning situation and promote self-monitoring and self-reflection. It can also push appropriate resources to students through personalized resource recommendation to reduce the time for students to find learning resources and meet personalized learning needs. In the system intervention, the diagnosis and intervention of students' learning crisis are carried out circularly with the continuous development of learning activities. With the continuous development of intervention engine in the direction of intelligence, the continuous increase of students' learning data and the continuous enrichment of intervention strategies, the effect of system intervention will be more and more scientific, timely and accurate.

5 Enlightenment to Online Education in China

5.1 Building an Organizational Environment for Learning Analysis

At present, China's online education institutions generally lack the organizational environment for learning analysis, so they must make a comprehensive deployment from the perspective of top-level design [7]. Firstly, relevant management departments should be established within the organization as soon as possible to formulate relevant planning and Implementation Rules for the application of learning analysis, guide the application of learning analysis and provide guarantee from the organizational level. Secondly, establish a professional technical department and team to be responsible for the project development of learning analysis. On this basis, gradually promote the application of learning analysis.

5.2 Promote the Development and Application of Learning Alert System

At present, the technical development and practical application of learning Alert system in the field of online education in China is still in its infancy. Online education institutions should build a learning Alert system with complete functions in combination with their own reality. In terms of data collection, online education institutions have accumulated a large amount of student information and learning data, but most of the data are stored in different formats and granularity levels, which makes it difficult to collect [8]. The effectiveness of learning Alert data collection can be guaranteed by unifying the data format and data structure or establishing a data center. In the aspect of model construction, multiple prediction models are constructed by using a variety of algorithms for comprehensive evaluation, and the optimal model is selected. In the aspect of dashboard development, we should reasonably plan and design the early warning indicators and early warning interface of the dashboard to help teachers and managers better understand, analyze and apply data. Finally, the system is tested and improved through practical application.

5.3 Promote the Improvement of Teachers' Data Literacy and Ability

In the application of learning analysis technology and methods, China's online education institutions should pay attention to strengthening the training of teachers and teaching managers to help them improve data literacy and data use ability. Teachers should know and understand the application of learning analysis technology, accurately understand data and student behavior, and provide accurate teaching services. Teaching managers should be able to gain insight into the students' learning growth trajectory and the law of educational development reflected behind the data, so as to improve the foresight and scientificity of decision-making. Online education institutions can carry out coherent training programs at different levels from primary to senior according to the actual situation of personnel, so as to promote their data literacy and ability development, so as to meet the practical demands of teachers and teaching managers for learning analysis and application.

5.4 Improve the Efficiency and Effect of Learning Alert

In the field of online education in China, the research on learning Alert is still in its infancy, so it is urgent to strengthen the research on learning Alert. In terms of theory, we should fully learn from the theories of different disciplines such as learning science, educational psychology and social psychology, deeply explore the cognitive basis and operation mechanism of learning Alert, and enhance the interpretability of early warning results. In terms of technology, strengthen the research on basic technologies such as educational data mining and machine learning, and constantly improve the accuracy and accuracy of early warning models. In terms of application, actively carry out the application research of learning Alert system, accumulate experience on the basis of practical research, summarize the application mode, evaluate the application effect, and continuously improve the efficiency and effect of learning Alert [9].

5.5 Pay Attention to the Protection of Students' Privacy Data

China's online education institutions should pay full attention to the privacy and ethical issues of data, strengthen the normative requirements of data use, formulate relevant documents to standardize the application standards and ethical norms of educational data, seek the balance between personal privacy and learning analysis, and reasonably integrate the teaching and ethical issues in the application of learning analysis.

6 Conclusion

With the advent of the Internet plus education era, data support and data driven role are becoming more and more obvious. Online education institutions must adapt themselves to the requirements of technological change and innovate the organizational mode, service mode and teaching mode of online education, in order to provide more high-quality, flexible and personalized teaching services for online learners. As the pioneer and leader of online education in the world, the UK Open University has always attached importance to using new technologies and tools to provide innovative ways for teaching. By implementing online learning crisis early warning based on big data learning analysis, it has effectively resolved students' learning crisis, improved students' retention rate and promoted academic progress [10]. By studying and learning from the successful experience of the British Open University, it will help to promote China's online education institutions to strengthen the research and practice of learning analysis technologies such as learning Alert. With the more and more extensive and in-depth application of big data and learning analysis technology in the field of online education, online education will eventually realize large-scale personalized learning and provide learners with higher quality learning services.

Acknowledgements. This work was supported by the Educational Reform Project of Xi'an Fanyi University (Project Number: J21B16). Open Online Course Construction Project of Xi'an Fanyi University (Project Number: ZK2001).

References

1. Herodotou, C., Rienties, B., Boroowa, A., Zdrahal, Z., Hlosta, M.: A large-scale implementation of predictive learning analytics in higher education: the teachers' role and perspective. Educ. Tech. Res. Dev. **14**(5), 3515–3521 (2020)
2. Gutiérrez, F., et al.: LADA: a learning analytics dashboard for academic advising. Comput. Hum. Behav. **10**(4), 105826–105837 (2020)
3. Walker, S., Olney, T., Wood, C., Clarke, A., Dunworth, M.: How do tutors use data to support their students? Open Learn. J. Open, Distance e-Learning **5**(1), 351–342 (2019)
4. Herodotou, C., Rienties, B., Boroowa, A., Zdrahal, Z., Hlosta, M.: A large-scale implementation of predictive learning analytics in higher education: the teachers role and perspective. Educ. Technol. Res. Dev. **67**(5), 1273–1306 (2019)
5. Hlosta, M., Herrmannova, D., Zdrahal, Z., Wolff, A.: OU analyse: analysing at-risk students at the open university. Learn. Anal. Rev. **15**(5), 1–16 (2021)
6. Walker S., Olney, T., Wood, C., Clarke A., Dunworth, M.: How do tutors use data to support their students? Open Learn. J. Open, Distance e-Learning **34**(1) , 118–133 (2019)
7. Bin, X., Dan, Y.: Motivation classification and grade prediction for MOOCs learners. Comput. Intell. Neurosci. **34**(1), 214–233 (2021)
8. Moreno-Marcos, P.M., et al.: Generalizing predictive models of admission test success based on online interactions. Sustainability **11**(18), 4940–4950 (2020)
9. Duru, I., et al.: A case study on English as a second language speakers for sustainable MOOC study. Sustainability **11**(10), 2808–2818 (2019)
10. Chen-Hsiang, Y., Jungpin, W., Liu, A.-C.: Predicting learning outcomes with MOOC clickstreams. Educ. Sci. **9**(2), 104–114 (2020)
11. Nkhoma, M., et al.: Examining the mediating role of learning engagement, learning process and learning experience on the learning outcomes through localized real case studies. Educ. Train. **56**(4), 287–302 (2020)

Prediction Model of Mental Health (MH) Based on Apriori Algorithm

Feifei Sun(✉)

Mental Health Education and Counseling Center, Zao Zhuang University, Zaozhuang 277160, Shandong, China
uzz_paper@163.com

Abstract. Today's database provides us with a large amount of rich information, but inevitably, a large amount of information will also have a lot of negative and negative effects on us. People hope to analyze the generated data more deeply, explore and refine the hidden valuable information, so as to make better use of these data and obtain higher value. In this context, Apriori algorithm, a data mining technology, was born, which is considered to be a typical algorithm. At the same time, mental health prediction is a very complex and systematic process. The interaction between environment and environmental factors, genes and genes, environment and genes will have a significant impact on mental health. This paper investigates the mental health status of college students by questionnaire. The results show that gender, grade, parents' marital status and whether they are the only child have a certain impact on College Students' mental health; The influence of these factors on the results needs a method that can reflect the corresponding relationship between the factors affecting mental health and the results to establish a prediction model. In this paper, the model prediction of mental health through Apriori algorithm can just reflect the corresponding relationship between this influencing factor and the result, so as to establish the prediction model of mental health.

Keywords: Apriori algorithm · Mental health · Health status prediction · Prediction model

1 Introduction

With the rapid development of society, the improvement of people's living standards and material satisfaction, the MH status shows a downward trend. The MH status has become an invisible "killer". China's youth groups play a role in promoting development in various fields, A student with all-round development of morality, intelligence, physique, art and labor will produce more social benefits to make up for the social costs previously consumed. From the perspective of marginal benefits, this can be in line with the sustainability of higher education. However, nowadays, various social events such as suicide and self-mutilation caused by psychological diseases or problems and bad emotions continue to occur, these are just the tip of the iceberg. This paper uses Apriori

J. H. Abawajy et al. (Eds.): ICATCI 2022, LNDECT 169, pp. 564–571, 2023.
https://doi.org/10.1007/978-3-031-28893-7_67

algorithm to test and analyze the MH status, and establishes a prediction model of MH status.

Many scholars at home and abroad have studied the MH prediction model based on Apriori algorithm. Harvey s B studied the first comprehensive and systematic meta-review of evidence linking the development of work, common mental health problems, especially depression, anxiety and / or work-related stress, and considered how to identify risk factors that may be interrelated [1]. Lu P H explored the potential association rules in the treatment of diabetic gastroparesis (DGP) by using Apriori algorithm and another partition based algorithm, frequent pattern growth algorithm. Apriori algorithm is an analysis method based on data mining, which is widely used in business, medicine and other fields to mine frequent patterns in data sets [2].

In this paper, thc Apriori algorithm is used to study and analyze the questionnaire data. The advantage is to reduce the frequency of database scanning and optimize the a priori rules of the algorithm. However, the application effect of Apriori algorithm is not ideal, and its application limitations are high. For a large number of data sets with complex structure, the calculation efficiency is low, which cannot well meet the data analysis of this survey, so the data will inevitably have some deviations and need to be improved [3, 4].

2 Concept and Research Methods of MH

2.1 Concept and Significance of MH

Who proposes that health is not only the absence of disease and pain in the body, but also the absence of psychological deficiencies and defects, normal psychology and full integration into society. In the state of MH, a person can clearly understand himself, others and the world, can correctly deal with all kinds of emergencies in work and life, and is willing and able to realize his own value [5, 6]. Different people have their own unique views on MH. MH does not mean that people are perfect in all aspects, but can show a balanced state under the conditions of their own characteristics and objective reality. Based on the current western mainstream view, MH is not a temporary psychological activity, which is closely related to their own needs and the society in which they live, In this state, it can have a positive impact on people's all-round development, stimulate personal potential, make their behavior accepted by the public, and make themselves feel satisfied. The views of domestic scholars are that MH includes having normal intelligence, being able to learn knowledge and skills conducive to themselves, being independent of others in daily work and life, having self-independence, knowing and accepting themselves Respond appropriately to external pressure, adjust in time when inner fluctuation occurs, and maintain a good relationship with others [7].

2.2 Research Methods

2.2.1 Questionnaire Survey Method

According to the relevant characteristics of schools and teenagers with bad behavior, investigate the school students, including the school students' family structure, atmosphere, upbringing style, performance in the original school, relationship with teachers and classmates, investigation of bad behavior before entering school, favorite entertainment methods, knowledge of law, performance in the original school, friend selection criteria, etc. [8, 9].

2.2.2 Measurement Method

(1) Basic information survey

Demographic variables of the subjects, including gender, grade, age, family situation, etc.

(2) Mental Symptom Self-Diagnosis Scale

The scale contains seven independent factors, namely anxiety, hysteria, schizophrenia, bipolar disorder, depression, neuroticism and fiction. The score of each item is 20–0. The higher the score, the more obvious the corresponding mental symptoms. If the score of fiction is greater than 10, it means that the answer to the questionnaire is not practical, and the score of each symptom is questionable [10, 11].

2.2.3 Test Score and Analysis

16–33 points: MH, no adverse signs. We should practice spiritual self-management and be good at adapting to all kinds of tense situations.

34–49: MH, but check whether the score of a certain symptom type is too high. When a symptom score is higher than 3 points, it is necessary to check the construction and health of a certain aspect and treat it symptomatically.

50–63 points: the MH status is general, which cannot be regarded as health. We should completely adjust our health status, find out the types and causes of symptoms with high scores and treat them in time.

64–78: if there are slight signs of mental illness, it is best to ask a specialist for diagnosis and careful analysis.

79–99 points: if you have suffered from some degree of mental illness, you must accept the diagnosis and reassuring treatment of a specialist.

(3) Questionnaire test

With reference to the self-made questionnaire, investigate the school students' family atmosphere, family education, school education, social experience, their own situation and so on. By issuing questionnaires to school students to understand their family atmosphere, family education, school education, social experience and themselves, we can more accurately analyze the causes of school students' MH problems from multiple angles.

Interview method

Design an interview outline and interview 12 teachers, managers and staff of management units of special schools, focusing on the current situation of students' MH education, the problems existing in school MH education and the development direction of school MH system construction in the next step.

2.3 Data Sorting and Analysis

A total of 105 questionnaires were distributed and 99 valid questionnaires were recovered. Sort out the survey results. In order to ensure the effectiveness, delete the questionnaires with incomplete information, incomplete answers or obvious rules. The Apriori algorithm is used to input the survey data, and the Apriori algorithm is used to calculate and analyze the data [12].

3 The Concept and Calculation of Apriori Algorithm

Apriori algorithm has three main characteristics. First, extract the data related to the calculation target, and reduce the three-dimensional data containing factors, results and the number of transactions into two-dimensional data containing only factors and the number of transactions. The second is to carry out calculation based on the reduced dimension data, which greatly reduces the amount of calculation. Third, in the calculation process, only the data column is pruned, and the number of transactions remains unchanged. Therefore, the calculation result of the available support number is regarded as the support degree.

Let Vr(r = 1, 2,...) Is a transaction, VK is a transaction set, its order is AXB, and the formed association rules are recorded as X → Q (k%, H%). Then the core calculation formula of Apriori algorithm is as follows:

$$Sup(X \Rightarrow Q) = K\% = \sum_{r=1}^{a} \{(X \Rightarrow Q) \subseteq V_r\} \tag{1}$$

$$Con(X \Rightarrow Q) = H\% = \frac{\sum_{r=1}^{a} \{(X \Rightarrow Q) \subseteq V_r\}}{\sum_{r=1}^{a} \{X \Rightarrow V_r\}} * 100\% \tag{2}$$

4 Sorting and Analysis of MH Survey Data Based on Apriori Algorithm

According to the students' test, the students' MH was analyzed. A total of 105 students participated in the MH test, and the data after removing 6 inaccurate questionnaires were analyzed. In terms of the total number of students tested, the school currently has only 99 students, and the total number of samples cannot be increased, so the sample size may be too small; From the perspective of gender, the number of girls is far less than that of boys, and there may be some deviation in the representativeness of the sample. The population data of the sample is shown in Table 1 and Fig. 1.

According to the above survey and statistical structure, gender, grade, parents' marital status and whether they are only children have a certain impact on College Students' MH. It can be seen from the table that the MH status of boys is better than that of girls; The marital status of parents is also an important factor affecting the MH of

Table 1. Basic statistics of samples

Statistic	Category	Proportion of students (n = 99)
Age	13	15.3%
	15	61.5%
	17	5.1%
Gender	Male	86.2%
	Female	13.5%
Single parent	Yes	33.9%
	No	64%
Father's educational level	Primary school and below	32.6%
	Junior high school	54.7%
	High school (technical secondary school or technical school)	3.4%
	Junior college	7.8%
	Bachelor degree or above	5%
Mother's educational level	Primary school and below	35.6%
	Junior high school	37.7%
	High school (technical secondary school or technical school)	13.7%
	Junior college	4.3%
	Bachelor degree or above	3.6%

college students; Parents are the first teachers of children and the most exposed objects of children. Therefore, the impact of family on children's MH cannot be ignored.

According to the test situation, the sample data are analyzed by using Apriori algorithm. The structure is shown in Table 2 and Fig. 2.

It can be seen from the above table that the seven indexes of anxiety, schizophrenia, manic depression, depression, neuroticism, fiction and psychological symptoms in different gender samples will not be significant ($P > 0.05$), which means that the indexes of anxiety, schizophrenia, manic depression, depression, neuroticism, fiction and psychological symptoms in different gender samples are consistent and have no difference. In addition, the gender samples showed significant difference in one item of Hysteria ($P < 0.05$), which means that there are differences in hysteria between different gender samples. The specific analysis shows that gender has a significant level of 0.05 for hysteria ($F = 4.098$, $P = 0.054$), and the specific comparative differences show that the average value of girls will be significantly higher than that of boys.

It can be concluded that there is no significant difference in anxiety, schizophrenia, manic depression, depression, neuroticism, fiction and psychological symptom index

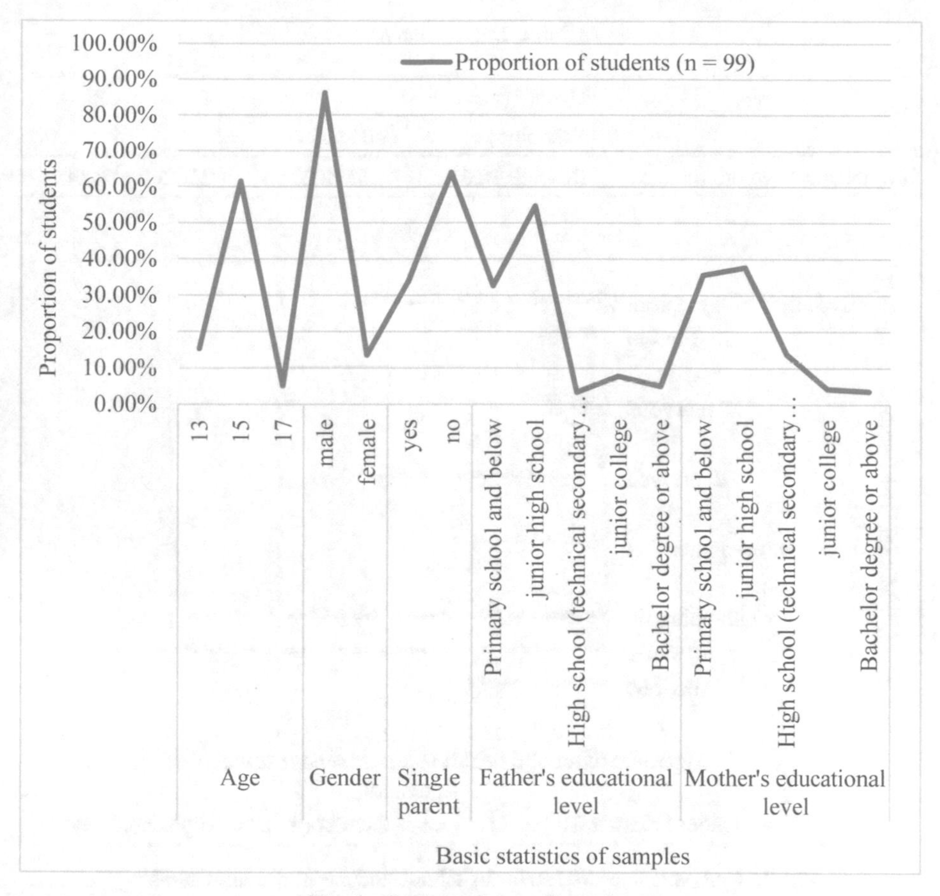

Fig. 1. Basic statistics of samples

Table 2. Comparison of MH status of school students of different genders

	Gender (mean ± SD)		F	P
	Schoolboy	Girl student		
Anxious	9.60 ± 2.60	8.79 ± 4.68	1.467	0.234
Hysteria	9.90 ± 3.54	6.I7 ± 4.81	4.098	0.054
Schizophrenia	9.40 ± 2.64	7.54 ± 3.80	2.776	0.075
Manic-depressive	9.20 ± 2.34	9.65 ± 4.07	0.009	0.922
Depressed	9.90 ± 3.54	7.65 ± 4.343	2.165	0.165
Nervous	8.70 ± 2.76	7.56 ± 3.45	0.334	0.554
Fiction	6.76 ± 1.86	6.32 ± 2.69	0.386	0.543

(*continued*)

Table 2. (*continued*)

	Gender (mean ± SD)		F	P
	Schoolboy	Girl student		
Psychological symptom index	57.75 ± 10.61	47.67 ± 20.53	2.276	0.145

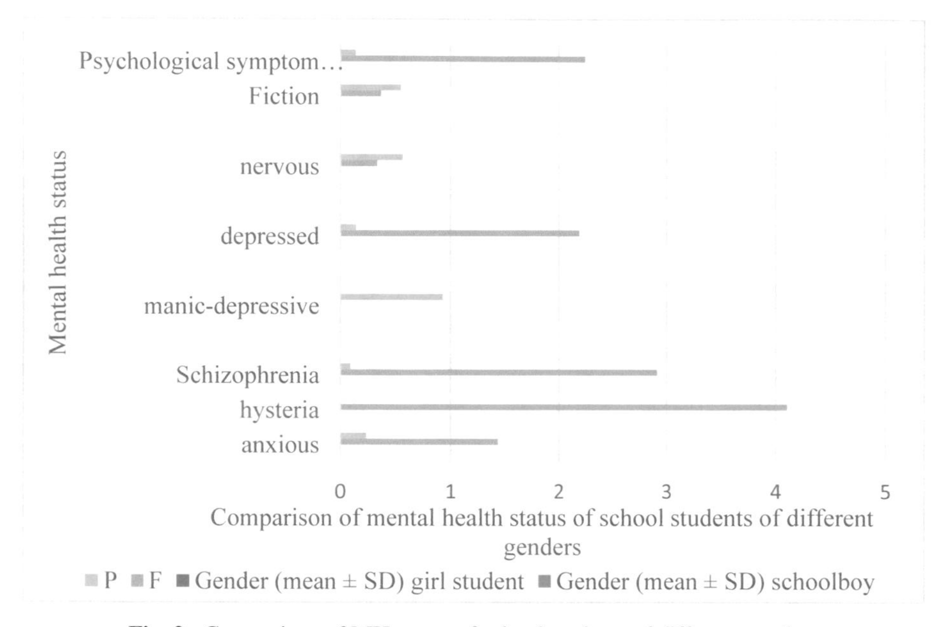

Fig. 2. Comparison of MH status of school students of different genders

among different gender samples, and there is significant difference in hysteria among gender samples.

5 Conclusion

With the wide application of modern enterprise computers and the rapid optimization development of enterprise database classification management technology system in China, the database technology has also undergone earth shaking changes. The previous classification database mainly contains and is used to manage various simple types of data, but now it has developed to such as text, image, audio many types of complex data such as video can be managed and processed by the database. In modern society, with the rapid development of productive forces, people have greatly improved their ability to create and collect data and rapidly expanded the scale of data through continuous changes in production information technology. The rapid growth of data and databases forces people to Adopt new technical means and tools to deal with massive data, automatically and independently help people manage, extract and analyze useful information, explore valuable knowledge and provide decision-making services for people. Thus, data mining was born under such a macro background. This study collects data through the questionnaire, uses the mining technology Apriori algorithm to make full use of the prediction of MH status for fitting, and analyzes the influencing factors of MH status. The innovation of this paper is to use Apriori algorithm to explore and study. This paper uses Apriori algorithm to explore the influencing factors of College Students' MH, and draws the final conclusion through continuous exploration, which provides a certain value for the further research of this subject in the future. The deficiency of this paper is that the experimental data collected is less and the sample data distribution is slightly uneven, which may have a certain impact on the research results. In the future study, we will continue to study relevant topics and strive for more in-depth research.

References

1. Harvey, S.B., Modini, M., Joyce, S., et al.: Can work make you mentally ill? A systematic meta-review of work-related risk factors for common mental health problems. Occup. Environ. Med. **74**(4), 301–310 (2017)
2. Lu, P.H., Keng, J.L., Tsai, F.M., et al.: An apriori algorithm-based association rule analysis to identify acupoint combinations for treating diabetic gastroparesis. Evid. Based Complement. Altern. Med. **2021**(17), 1–9 (2021)
3. Irfiani, E.: Determination of book loan association pattern using apriori algorithm in public libraries. Jurnal Techno Nusa Mandiri **17**(2), 137–142 (2020)
4. Sornalakshmi, M., Balamurali, S., Venkatesulu, M., et al.: An efficient apriori algorithm for frequent pattern mining using mapreduce in healthcare data. Bull. Electr. Eng. Informat. **10**(1), 390–403 (2021)
5. Nandini, G., Rao, N.: Utility frequent patterns mining on large scale data based on apriori mapreduce algorithm. Int. J. Res. **3**(8), 19381–19387 (2019)
6. Jeyakarthic, M., Singaram, S.: An efficient metaheuristic based rule optimization of apriori rare itemset mining for adverse disease diagnosis model PJAEE. PalArch's J. Archaeol. Egypt/Egyptol. **17**(7), 4763–4780 (2020)
7. Velmurugan, T., He, M.B.: Mining implicit and explicit rules for customer data using natural language processing and apriori algorithm. Int. J. Adv. Sci. Technol. **29**(9), 3155–3167 (2020)
8. Niksa-Rynkiewicz, T., Landowski, M., Szalewski, P.: Application of apriori algorithm in the lamination process in yacht production. Polish Maritime Res. **27**(3), 59–70 (2020)
9. Montoya, G.A., Donoso, Y.: A prediction algorithm based on markov chains for finding the minimum cost path in a mobile WSNs. Int. J. Comput., Commun. Control (IJCCC) **14**(1), 39–55 (2019)
10. Ariestya, W.W., Supriyatin, W., Astuti, I.: Marketing strategy for the determination of staple consumer products using fp-growth and apriori algorithm. Jurnal Ilmiah Ekonomi Bisnis **24**(3), 225–235 (2019)
11. Shambharkar, S.A., Bhajipale, R., Nagpure, N., et al.: Study on market basket analysis using apriori and classification rule based association algorithm. Int. J. Comput. Sci. Eng. **7**(3), 723–728 (2019)
12. Khan, S., Khan, M., Iqbal, N., et al.: A two-level computation model based on deep learning algorithm for identification of piRNA and their functions via Chou's 5-Steps rule. Int. J. Pept. Res. Ther. **26**(2), 795–809 (2019)

Psychological Counseling System Based on Association Rules Mining Algorithm

Feifei Sun(✉)

Mental Health Education and Counseling Center, Zao Zhuang University, Zaozhuang 277160, Shandong, China
uzz_paper@163.com

Abstract. With the rapid development of society, people's life and work pressure is increasing. How to relieve psychological pressure, express inner true emotions, and avoid the formation of mental illness is an urgent problem to be solved at present become the focus of society. The traditional way of psychological counseling usually adopts face-to-face communication. Some counselors do not want to share their thoughts with experts directly, which makes counseling impossible. When there are too many counselors, the counseling experts have no time to deal with them, resulting in low counseling efficiency. Therefore, the psychological counseling system designed in this paper can be consulted online, avoiding the embarrassment of communication, protecting the privacy of the counselors, and improving the efficiency of counseling services. The system combines the association rule algorithm, which can calculate the correlation degree of the information data of the consultants in the system, and generate consultation data reports for the consultants, so that experts can provide targeted suggestions for the consultants according to the report results.

Keywords: Association rule algorithm · Psychological counseling system · Mental illness · Counseling efficiency

1 Introduction

Due to the scarcity of psychological counselors, the lack of counseling platforms, and people's long-term resistance to the psychological counseling process, many people's psychological problems cannot be solved, which seriously affects their physical and mental health. Therefore, in order to solve the above problems, the psychological counseling system based on the association rule mining algorithm developed by the structured system analysis and design method has important practical significance.

At present, many scholars have conducted research on the design of psychological counseling system based on association rule mining algorithm, and have achieved fruitful research results. For example, a psychological counseling teacher in a university discusses students' mental health problems, and analyzes the correlation degree of students' psychological problems through association rule mining algorithm to formulate countermeasures for students to relax their mood. They proposed to ensure the

J. H. Abawajy et al. (Eds.): ICATCI 2022, LNDECT 169, pp. 572–579, 2023.
https://doi.org/10.1007/978-3-031-28893-7_68

all-round development of students, not only to supervise academic progress, but also to care about students' psychological problems, so as to be conducive to the formation of a sound personality [1, 2]. The psychological counseling system designed by a scholar has the function of posting counseling, users can publish counseling information, and the operation on the system is simple, and they can publish what they want to counsel in a few steps. Through the viewing function, the counselor can view the psychological information of the counselor or view the counseling events by category, so as to improve the counseling work efficiency of the counselor [3]. Although the design of the psychological counseling system based on the association rule mining algorithm has achieved good results, there are still many people with psychological problems in life. It is hoped that the use of the psychological counseling system can help people solve their psychological troubles.

This paper introduces the definition of association rule mining algorithm, designs a multi-functional module system after understanding the requirements of the psychological counseling system, introduces the mining algorithm to mine the data of the counselors stored in the system and analyzes the relationship between the data and information. After that, the system performance was tested and the user login module verification was completed to ensure the stable operation of the system.

2 Psychological Consultation System Design Requirements

2.1 Association Rule Mining Algorithm

Association rule mining can find the subtle relationship between some data from a bunch of seemingly disorganized data. Therefore, association rule mining can help us better understand the relationship between objective things [4, 5]. Assuming that there are N pieces of data in the transaction database, the calculation formulas of support and confidence are:

$$Sup(X) = \frac{Sum(X)}{N} \tag{1}$$

$$Conf(X \Rightarrow Y) = \frac{Sup(X \cup Y)}{Sup(X)} \tag{2}$$

where Sup represents support, Conf represents confidence, and X and Y are transactions.

2.2 System Requirements Analysis

(1) User needs

Login registration: It is required to have basic login and registration functions. After the user completes the registration, he can enter the system through the login operation, and can log out and exit the system after the operation is completed. The system can respond to the user's operation in time, give user-friendly prompts, and store the relevant information in the database [6, 7].

User information management: After logging in, users can view consultation results, change passwords and other functions. Psychologists have the authority to set users, delete users and other functions [8].

Psychological counseling management: Users can view the content of their own counseling. After logging in, the psychological consultant can view the consultation content of all users and give a reply [9].

(2) System performance requirements

Stability: It is required that the system should have a stable working state, be able to process and respond to user operations in a timely manner, and require a good network environment when the system is running [10].

Security: Good security is not only one of the needs of the system, but also one of the needs of users. Good security performance can effectively protect user privacy and data. The system can only log in after the user has entered the correct password, and each page has a password to determine whether the user is already logged in. The system adds a page that only administrators can access to determine if the ID is an administrator. If you are not an administrator, you cannot log in. Therefore, the design system needs to have good security performance [11, 12].

Reliability: The system needs to have high reliability in order to ensure the transmission speed of the network. When the system fails, the system will detect the failure and make a repair plan by itself to ensure the reliability of the system software and hardware. System reliability can be improved by selecting hardware equipment with high reliability, allowing appropriate redundancy of hardware systems, strengthening fault detection and other system security software and management measures [13].

(3) System feasibility requirements.

The operation of the system needs to avoid the embarrassment of face-to-face, protect the personal privacy of the consultants, and solve the consulting problems at the first time, saving the time of the consultants. The system is a psychological counseling system, which consumes very few resources, and the computer can run without a high configuration. Therefore, the system is feasible in operation [14].

3 System Design

3.1 System Function Module Design

Figure 1 shows the five modules of the psychological counseling system, and each module is introduced below.

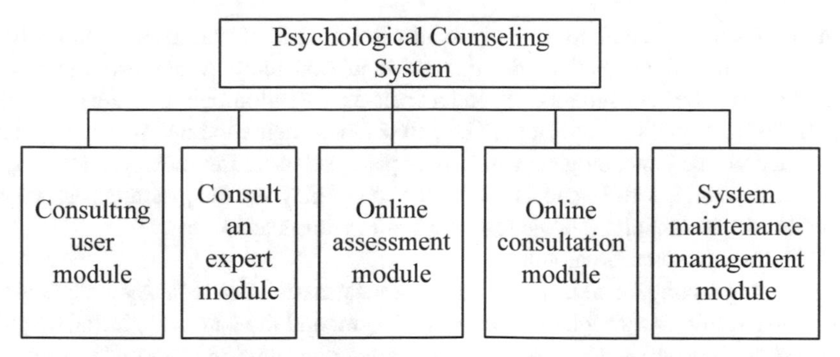

Fig. 1. Module structure of psychological counseling system

(1) Consulting user module

The user module is relatively large, mainly including user registration, questioning, query viewing, custom user information collection, and expert column collection. User registration is mainly to collect general user information. Users can also ask questions without registering, but must provide a valid email address to receive an automatic response from the system. The email address is used as the user's username, and the system automatically generates a password and sends it to the user's mailbox. User-defined user information basically allows users to select some of the services provided by the system and receive the latest information released by the system on a regular basis. The "Favorite Consultants" column essentially aggregates a list of frequently consulted experts so that a consultant can contact the same expert again.

(2) Consulting Expert Module

Consultants first register electronically and log in after verification. Consultants must provide their basic personal information, including name, gender, education, technical qualifications, consulting fields, contact numbers, personal profiles, etc. Visitors can click on the list of consultants to view their personal details, and visitors can select consultants based on their personal characteristics. Consultants can see the consultation questions they need to answer, and client questions that experts cannot answer can refer the matter to other consultants for their help. When the consultant answers questions, the system sorts the questions, saves them in a database, and emails the answers to the visitor. If possible, the answer to the question can be sent to the visitor in the form of a mobile phone text message. But this will increase the cost of the entire consultation, so it is not used for the time being.

(3) Online assessment module

The online assessment module mainly completes the psychological test of the counselors. Online psychological assessments are widely used as self-examinations for psychological counseling. The online psychological assessment system adopts association rule mining algorithm and software assessment method, and formulates assessment standards based on psychological knowledge. Through the test results, the psychological state of the counselor is displayed, and the counseling conclusion is drawn and guidance is given. The counseling and evaluation data are included in the system database to provide valuable suggestions for the smooth development of psychological counseling in the future. Using this method not only avoids the trouble of classifying and analyzing the information of the consultant during traditional consultation, but also improves the work efficiency. The consultant can conduct self-examination online at any time, which protects the privacy of the consultant. The online assessment module needs to log in first, log in to the online psychological assessment page, select or randomly select test questions, and conduct a self-diagnosis test. After completing the test, submit your answers and review the test results, file the test results and complete the test.

(4) Online consultation module

Consulting management is the exchange of information between psychological consultants and inquirers through the network platform, and the automatic initial diagnosis of the system through question and answer. Online consultation is a true "face-to-face" form of questioning, also known as real-time reference consultation. In this system, short-term consultation services can be completed through text chat, and "face-to-face" consultation can also be completed through the video function.

(5) System maintenance management module.

This module mainly includes changes to some general information, such as changes to information output, site announcements, and contact information. User management needs to manage registered user information and use rights, manage experts to apply for registration verification, delete invalid data, and manage problem databases, which mainly include regular inspection and sorting of databases, removal of invalid data, ensuring database integrity, and regular automatic backup of the database, etc.

3.2 Database Table Design

Table 1. Data table field settings

Field Type	length	Is it empty?
Username	6	No
User password	10	No
Consult expert information	50	No
Consultant Information	50	No
Consultation time	20	No
Consultation content	100	No

As shown in Table 1, the user name, user password, consulting expert information, consultant information, consulting time, consulting content and other data information that need to be retained in the system database, set the length of these information fields, respectively 6, 10, 50, 50, 20, 100, and cannot be empty in the final reservation.

4 System Implementation

4.1 System Test

As shown in Fig. 2, the algorithm is used to test the stability, security and reliability of the psychological counseling system. According to the test results, the test values of stability, security and reliability are 94%, 97% and 95% respectively, the set theoretical values are 88%, 92%, and 90%, respectively, and the performance test values are all greater than the theoretical values, indicating that when the inquirer uses the system to conduct real-time dialogue with consulting experts, the system can respond in time to ensure the consultation and communication process. It is smooth, meets the needs of consultants and consultants, and also verifies the success of the test.

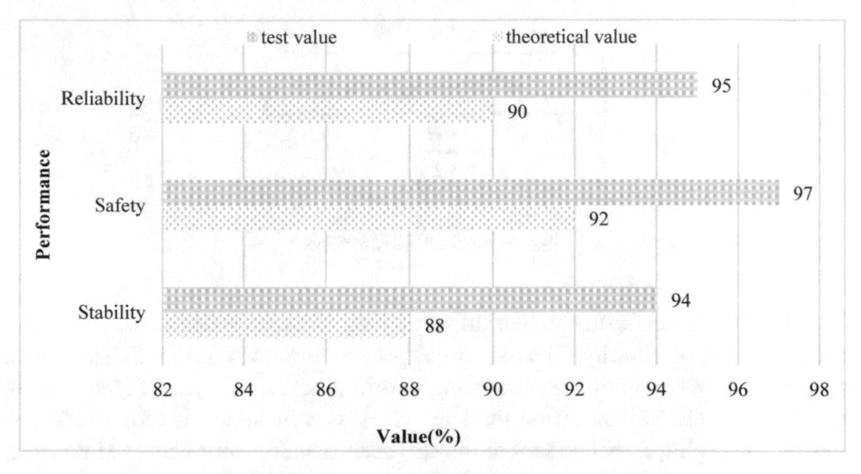

Fig. 2. System performance test results

4.2 Consulting User Login Module Implementation

The login registration module is mainly related to user login and registration. If the user is not registered, provide the user with a registration page. When registering, click "Register New User" to enter the system registration interface to register. Follow the instructions to enter the registration details correctly. Log in again after registration is complete. To ensure that users are logged in from the login page, a piece of code is added to every page of the system to determine if they are logged in. If they are not logged in, they will be automatically directed to the login page. The system login window is the first interface to connect to the system, so when creating an application, common variables are defined to determine the permissions of the connecting user. After the user has successfully logged in, assign the username to a global variable. User permissions determine the actions a user can take after logging in. The realization flow chart of the login module is shown in Fig. 3.

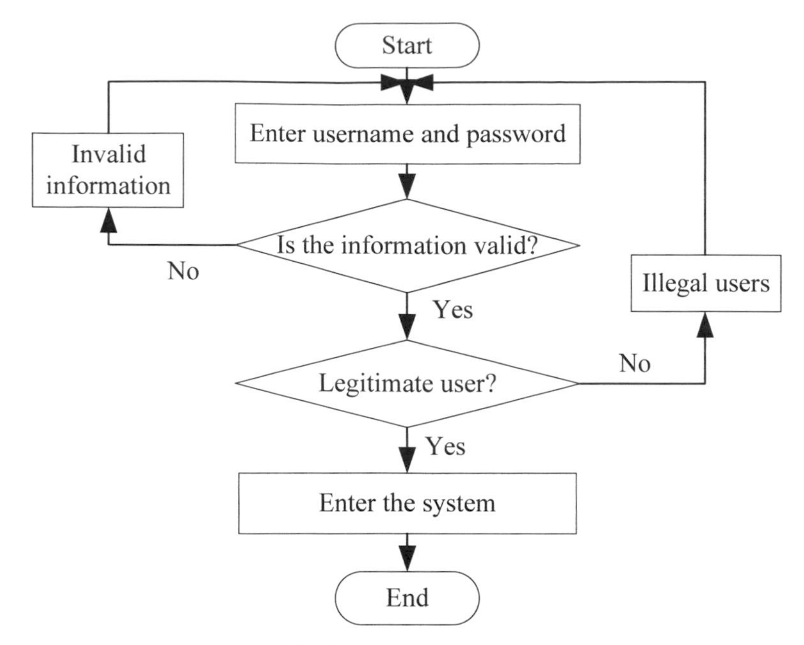

Fig. 3. User login process

It is necessary to perform effective authentication on users, and it is also an important guarantee for system security. The user login process is shown in Fig. 3. The user first selects different system entrances according to different identities, enters the user login page, and fills in the authentication information. If it is invalid, it is required to return and refill. If it is valid, it will be sent to the server for legality verification. If it is a legal user, it is allowed to enter the system. Otherwise, a prompt message will be given and a refill is required.

5 Conclusion

In the traditional counseling model, the work of entering the information and data of the counselors is cumbersome and cannot meet the needs of users. Therefore, this paper designs a psychological counseling system based on the association rule algorithm. Through this algorithm, the counseling data is counted and the reasons for the psychological problems of each counselor are analyzed. The relationship is convenient for the consultant to quickly obtain the information of the consultant. Moreover, the system allows people to communicate with experts online, and the relaxed online environment makes the counselor free from the pressure of counseling and promotes the smooth progress of the psychological counseling process. This paper also verifies that the system can complete the consulting work after testing the performance of the system, indicating that the system design is successful.

References

1. Nie, A.: Design of english interactive teaching system based on association rules algorithm. Secur. Commun. Netw. **2021**(s1), 1–10 (2021)
2. Nguyen, L.T.T., Vo, B., Nguyen, L.T.T., Fournier-Viger, P., Selamat, A.: ETARM: an efficient top-k association rule mining algorithm. Appl. Intell. **48**(5), 1148–1160 (2017). https://doi.org/10.1007/s10489-017-1047-4
3. Wu, Z., Chen, Y.: Digital art feature association mining based on the machine learning algorithm. Complexity 2021(1), 1–11 (2021)
4. Wei, B.: On scientificity and artistry of psychological counseling. Theory Pract. Psychol. Couns. **3**(3), 148–152 (2021)
5. Kutsenko, N.: Phenomenology experience in the context of the existential approach to psychological counseling. Bull. Baikal State Univ. **29**(1), 39–47 (2019)
6. Lee, E.J.: Disability and psychological counseling from a multicultural perspective. Korean J. Couns. Psychother. **32**(1), 197–224 (2020)
7. Kai, Z., Ning, Y., Wu, F.: Case analysis for psy-chess: a tool for psychological counseling and organizational counseling. Theory Pract. Psychol. Couns. **2**(12), 855–865 (2020)
8. Bi, Y.: Application of psychological counseling of drawing in psychological adjustment of college students during COVID-19. Theory Pract. Psychol. Couns. **2**(8), 451–457 (2020)
9. Wang, Y.: A psychological counseling case of a college student with narcissistic personality. Theory Pract. Psychol. Couns. **2**(6), 355–365 (2020)
10. Lee, S.M.: The need for statutes in the field of psychological counseling in South Korea. Korean J. Couns. Psychother. **32**(1), 547–557 (2020)
11. Cao, M., Wan, Z.: Psychological counseling and character analysis algorithm based on image emotion. IEEE Access (99), 1 (2020)
12. Karyagina, T.D.: Scientific Heritage of F.Ye. Vasilyuk: from psychological counseling to counseling psychology. Cult. Hist. Psychol. 15(1), 4–14 (2019)
13. Getachew, A.: Assessment of psychological counseling service for higher education institution students. Int. J. Educ. Literacy Stud. **7**(4), 53–61 (2019)
14. Aydn, D., Aytekin, C.: Controlling mathematics anxiety by the views of guidance and psychological counseling candidates. Eur. J. Educ. Res. **8**(2), 421–431 (2019)

Design of Intelligent Recognition Corpus System Based on Ternary N-gram Algorithm

Shiyang Zhang[1(✉)] and Amar Jain[2]

[1] Institute of Foreign Languages, Jiangxi Science and Technology Normal University, Nanchang, Jiangxi, China
zsy202017@163.com

[2] Madhyanchal Professional University, Bhopal, India

Abstract. With the rapid development of computers and the Internet, mankind has entered a new information age. Language processing plays an irreplaceable role in the information age. The emergence of language processing research mainly solves the problem that computers cannot understand human language. However, there are still deficiencies in the field of intelligent recognition language today, and it is very necessary to further optimize the intelligent recognition corpus. Therefore, based on the ternary N-gram algorithm, this paper studies and designs an intelligent recognition corpus system. First, this article explains the composition of the corpus and the concept of the intelligent recognition corpus; then, the application of the ternary N-gram algorithm is studied, and on this basis, the framework and modules of the intelligent recognition corpus system are designed. Finally, the system is tested, and the test results show that the accuracy of the ternary N-gram algorithm is higher than that of the traditional algorithm, and the response speed of the optimized system is faster than that of the traditional algorithm.

Keywords: Ternary N-gram · Intelligent Recognition · Corpus · System Design

1 Introduction

From the late 1940s to the early 1950s, the field of research in the field of language understanding began to develop. Computers have just appeared at this time, because computers can process symbols, making it possible to understand and process languages on computers. This early research laid the foundation for the later formation of formal language theory [1, 2].

Many experts and scientists have begun to discuss the research of intelligent recognition corpus. GPU technology currently enables scientists to simulate very large artificial neural networks. Under the combined effect of deep learning and GPU technology, the accuracy of speech recognition technology and image recognition technology can be increased by 91%. Due to the relatively slow development of national language processing research, there are few systematic researches on language understanding in our country. Due to the different language foundations at home and abroad, if Chinese is the main body of national research, the language of literature research is English, but at the national

J. H. Abawajy et al. (Eds.): ICATCI 2022, LNDECT 169, pp. 580–588, 2023.
https://doi.org/10.1007/978-3-031-28893-7_69

level, it should be based on Chinese. Therefore, there are many challenges in intelligently recognizing Chinese. In recent years, with the continuous efforts of researchers, the research of Chinese speech recognition technology has made considerable progress. Among them, the direction of progress is as follows: one is the application field of the technology; the other is the original foundation of the technology. Because many experts have done a lot of research work on the theory of Chinese character combination and characteristics of Chinese characters, and achieved localization, they can more closely integrate the characteristics of the country and better serve the country [3, 4].

Although there are many previous results, there are still some limitations such as unclear recognition of professional vocabulary, incomplete vocabulary grammar, and lack of semantics. Therefore, based on the ternary N-gram algorithm, the research on the intelligent recognition corpus system is to complement the current deficiencies in the field of intelligent recognition [5, 6].

2 Overview of the Intelligent Recognition Corpus System Based on the Ternary N-gram Algorithm

2.1 The Composition of the Corpus

Although the corpus existed before the advent of computers, a computerized corpus can be considered a real corpus. Therefore, what people call a corpus refers to a computerized corpus. The 1960s and 1970s marked the beginning of the development of the human body. The corpus is relatively small and unmarked, mainly used for description, analysis and research of English grammar. In 1964, Brown University in the United States created the first corpus, the Brown Corpus, with a capacity of 1 million words and a large collection of language materials of various styles of American English at that time. The United Kingdom created the LOB (Lancaster-Oslo-Bergen) corpus in 1978, which can collect 1 million words of British English documents. At the same time, the UK also established Spoken Language Corpus LLC (London-Lund Corpus) with a capacity of 500,000 words. A corpus is a set of languages stored in a computer, used to reflect the state of the language and modify the attributes of the language. The above two definitions reflect people's understanding of corpus in different periods of corpus development. With the development of a corpus, its most comprehensive definition should be: a corpus is a collection of language materials created, stored and used by a computer, which can be retrieved, analyzed, and processed to give students a harmonious, correct or beneficial language feature Language response and teaching. Simply put, it is a computerized language database created with specific goals and clear design standards. The content of the corpus is closely related to its purpose, and the content of the corpus is guaranteed by the collection of the corpus. If the collected corpus is inaccurate, incomplete, unsystematic, and unrepresentative, it will lead to differences in the results of corpus use, and there is no universal rule. Therefore, the collection of the corpus is the key to the construction of the corpus. With the development of corpora, there are more and more types of corpora [7, 8].

2.2 Basic Concepts of Intelligent Recognition Corpus

The intelligent recognition corpus is an intelligent system for computers to process human language, which includes mainstream languages in today's society, such as English, Chinese, Spanish, Arabic, Korean, Japanese, etc. As an important tool for communication and learning, the intelligent recognition corpus system can process and check the type of language input into the system, create a corresponding language model framework, transfer language data into the language framework, and finally apply the output improved language recognition results To various practical occasions. In short, the use of computer technology to study and process language information and transform it into words that people can understand is an intelligent recognition corpus system. The study of language comprehension strategy aims to deal with the knowledge base in a specific field in response to the problems raised by users. The system applies appropriate strategies to achieve specific understanding and analysis, the statistical function of understanding user needs and the control function of processing specific situations. At the same time, Language processing is transferred to the Internet of Things, so it becomes smarter. Whether it is human-computer interaction or intelligent control, the process and level of the computer processing system make the language have a higher standard [9, 10].

2.3 Processing of Corpus

The research on corpus processing is divided into two parts: written corpus and spoken corpus, in which computer processing is easier to deal with written language comprehension. Computer analysis and understanding of language is generally a hierarchical process. Linguists divide this process into three types: pragmatic analysis, phonetic analysis, and semantic analysis [11, 12].

(1) Word segmentation processing: divide the article into phrases.
(2) Lexical analysis: segmentation of vocabulary types through word segmentation, indicating the part of speech of the vocabulary to determine the type of words, including nouns, verbs, adjectives, adverbs, prepositions, etc.
(3) Grammatical analysis: Analyze the grammatical components of sentences.
(4) Semantic analysis: It refers to allowing computers to understand language.

2.4 The Basic Principle of the Ternary N-gram Algorithm

The language model is generally constructed as the probability distribution p(s) of the string S, where p(s) reflects the probability that the string s appears as a sentence. For example, if in an article to be translated, there is about one sentence among 100 sentences in the article "Chinese standing", we can assume that more (Chinese standing) is 0.01. Because we can consider some sentences with a probability of 0, such as "rabbit eat snake", because almost no such sentence appears in the article.

At this point, it should be noted that, unlike linguistics, the language model has nothing to do with whether a sentence is grammatical or not. Even if a sentence is completely grammatical and logical, we believe that its occurrence rate is always close to zero.

For a sentence composed of 1 primitive (the primitive can be a word, word or sentence, etc., for the sake of brevity, it is agreed that only "word" will be used in the future) S = A1, A2,... Ai, the formula for calculating the probability can be expressed for:

$$P(S) = P(A1)P(A1/A2)P(A1/A2A3)\ldots P(A1/A2\ldots Ai) \quad (1)$$

In formula (1), the probability of generating the i-th (1i1) word is determined by the i-1th generating words A1, A2...Ai. Usually we call the first word i-1 A1, A2... Ai the "story" of word i. Of course, with this method, as the length of the history increases, the number of different histories increases exponentially. Example: L = 8000 (assuming L is the size of the vocabulary set), i = 6, then the number of free parameters is 80006. In this way, it is almost impossible to correctly estimate these parameter data from a large amount of training. In fact, most of the history just cannot appear in the training data. For the above reasons, we can say that the history (A1, A2...Ai) is mapped to an equivalence class E (A1, A2...Ai) according to the judgment rule. The data of the equivalence class is much smaller than the different historical data. The specific calculation is shown in formula 2:

$$P(A1/A1\ldots Ai-1) = P(A1A2A3\ldots Ai) \quad (2)$$

Then the number of free parameters is drastically reduced. Here we divide history into equivalence classes by mapping two histories to the same equivalence class. If and only if the nearest n-1 words of the two histories are the same, the language model that satisfies the above conditions is called n-gram or n-gram.

3 The Establishment of an Intelligent Recognition Corpus System Based on the Ternary N-gram Algorithm

3.1 System Structure

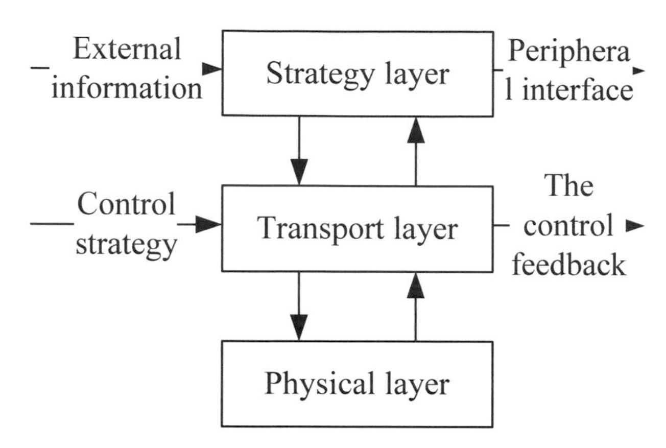

Fig. 1. Structure of a system

As can be seen from Fig. 1, the general system structure of the intelligent recognition corpus system is divided into the strategy layer, which collects and processes corpus data; the transmission layer, which transmits the corpus data to the speech recognition chip for speech recognition; the physical layer, which is an external device of the system and can be connected to other devices to assist in the recognition of the corpus.

3.2 Detailed Functional Module Design

As can be seen from Fig. 2, building a complete intelligent recognition corpus system should include the following 7 functions:

(1) User management function. The user management function includes two parts: adding users and deleting users. On the one hand, the system must be able to collect and store the voice characteristics of new users; on the other hand, it can also delete users;
(2) Voice login function. Only registered legal users can use this system;
(3) Microphone function. The microphone can transmit sound into the device for better recognition of the corpus.
(4) Voice recognition. Voice recognition can process the sound input into the microphone;
(5) Identify the chip. The recognition chip is the core of the entire intelligent corpus system, and the recognition chip can achieve a system recognition rate of more than 95%;

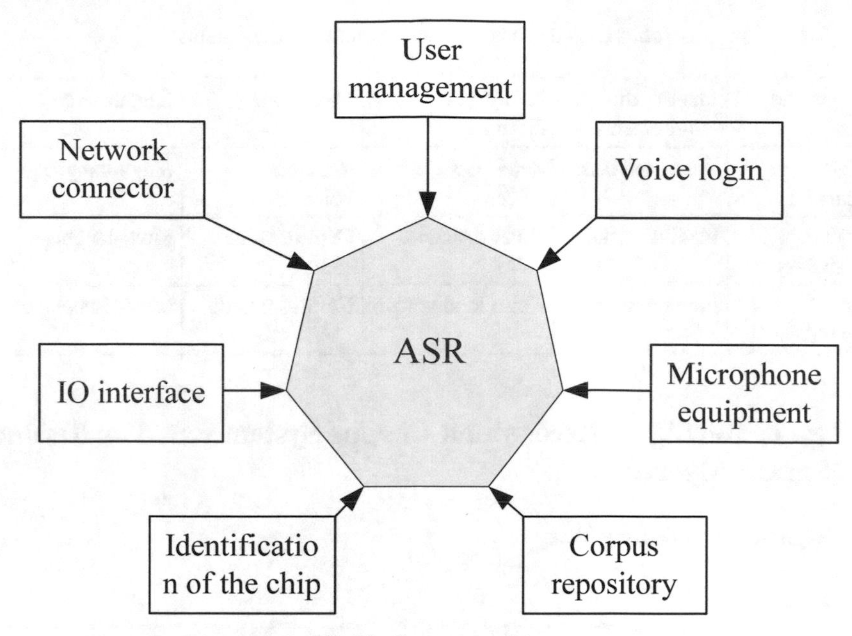

Fig. 2. System module

(6) IO interface. The IO interface can provide users with a variety of input and output devices for voice collection and voice recognition;
(7) Network interface. The network interface can connect to the network, conduct online activities, and search for more corpus resources;

Through the above-mentioned functions, an intelligent recognition corpus system can be constructed. As a practical system, the recognition rate and processing speed are required. Whether the user's voice can be recognized correctly affects the reliability of the system, and the system should control the time for voice registration and voice recognition processing within the user's acceptable range.

3.3 Intelligent Recognition of the Database of the Corpus System

It can be seen from Table 1 that in the speech recognition system, the recognition system based on a single word and small vocabulary of a specific character is still the simplest. Due to the changes in today's social tasks, human language recognition is still the most complicated, because human languages have a huge vocabulary. Noise, speech, and mixed languages are much more difficult than the tasks to be solved in the past.

Table 1. A database divided by actual requirements

Corpus and quantity	Identify the requirements	Identify the distance	Voice tone	Language type
Small vocabulary	Restricted tasks	Zero-distance voice	Read out the voice	Single-language
Large vocabulary	Flexible tasks	Quiet-distance voice	Careful voice sounds	Multi-language
Huge vocabulary	Free mission	Noise loud voice	Natural sounds	Mixed languages

4 Test of Intelligent Recognition Corpus System Based on Ternary N-gram Algorithm

4.1 Algorithm Accuracy Test

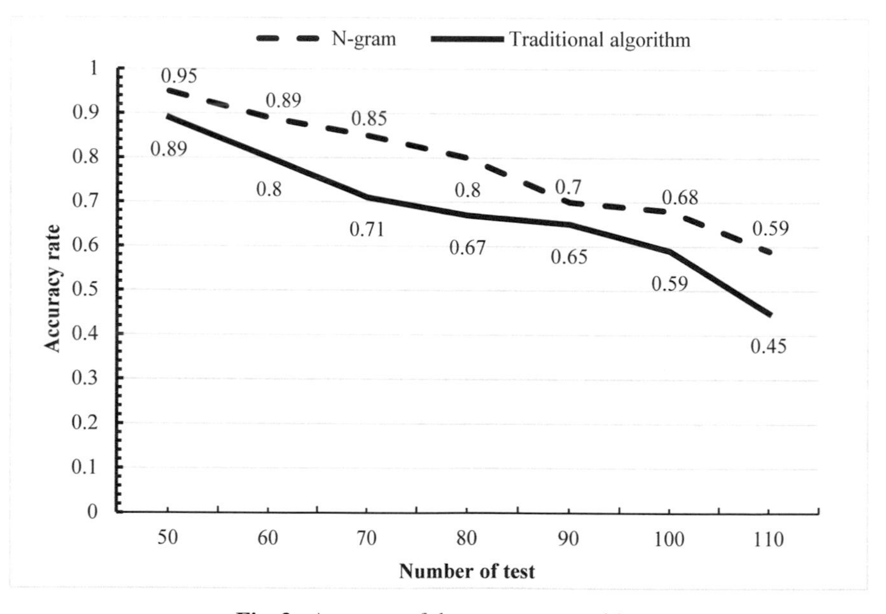

Fig. 3. Accuracy of the corpus recognition

Figure 3 compares the traditional algorithm with the algorithm used in this article. The test results are divided into two directions. One is to compare the accuracy of the word segmentation algorithm; the other is to compare the response time of the small vocabulary set and the original sentence. Let's use data analysis to observe the test results. It can be seen that the accuracy of the ternary N-gram algorithm is higher than that of the traditional algorithm. After experimental testing, the accuracy is indeed higher.

4.2 System Response Time Test

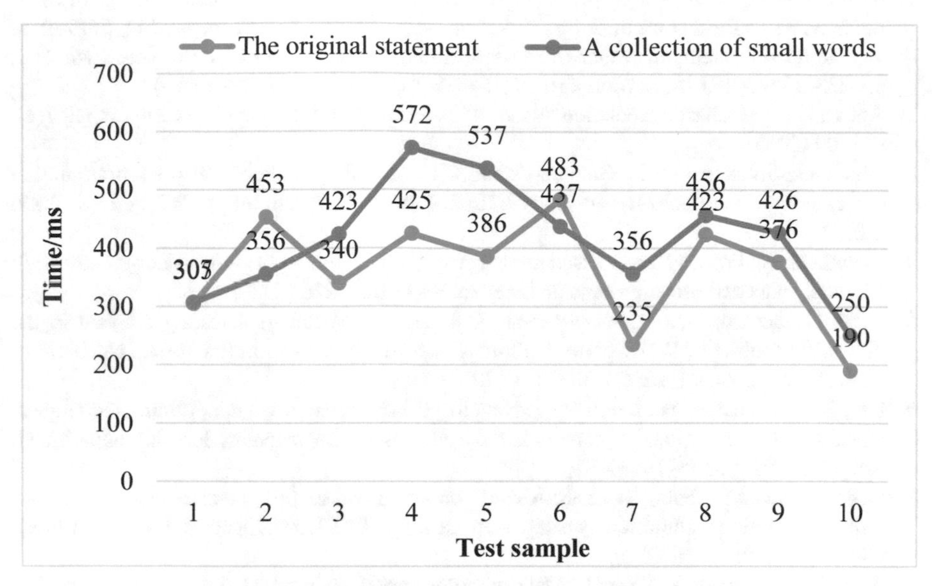

Fig. 4. Response time comparison

In Fig. 4, we can see that these sentences can express the same meaning. After language processing, the combination of small vocabulary collection and sentences without language processing are tested in the system respectively, and the response time of the original sentence is higher than that of the small vocabulary collection. This shows that small vocabulary sets are more efficient than non-small ones.

5 Conclusion

In today's world, there are more and more language exchanges, and we cannot rely on ourselves to use words alone to express what they need to convey. Therefore, the transformation of text into information that is readable, highly flexible, highly accurate, and can meet the needs of different groups is an urgent problem to be solved. These problems have prompted researchers to research and develop automatic intelligent recognition systems. In this paper, the ternary N-gram algorithm is used to solve the problem of large storage and search in traditional manual translation of words, and it is not easy to damage the original sentence structure of the sentence and affect the word frequency.

Acknowledgements. This work was supported by research Project of Humanities and Social Sciences in Colleges and Universities of Jiangxi Province in 2020.

References

1. Melin, P., Castillo, O., Kacprzyk, J.: [Studies in Computational Intelligence] Design of Intelligent Systems Based on Fuzzy Logic, Neural Networks and Nature-Inspired Optimization. || An Improved Intelligent Water Drop Algorithm to Solve Optimization Problems. vol. 601, pp. 233-239 (2015). https://doi.org/10.1007/978-3-319-17747-2 (Chapter 19)
2. Abbas, Q.: A stochastic prediction interface for urdu. Int. J. Intell. Syst. Technol. Appl. **7**(1), 94–100 (2015)
3. Olgun, G., Sokmensuer, C., Gunduz-Demir, C.: Local object patterns for the representation and classification of colon tissue images. IEEE J. Biomed. Health Inform. **18**(4), 1390–1396 (2017)
4. Mettouchi, A.: From a corpus-based to a corpus-driven definition of clefts in Kabyle (Berber): Morphosyntax and prosody. Faits de Langues **52**(1), 207–226 (2021)
5. Paoli, C., Segard, S., Burel-Vandenbos, F.: Is H3 K27M mutation testing relevant in the diagnostic routine of WHO grade 4 gliomas exclusively located in the corpus callosum in adults? J. Neurooncol. **155**(3), 383–384 (2021)
6. Lee, H., Warschauer, M., Lee, J.H.: Advancing CALL research via data-mining techniques: unearthing hidden groups of learners in a corpus-based L2 vocabulary learning experiment. ReCall, **31**(PT.2), 135–149 (2019)
7. Nagayama, I., Miyahara, A., et al.: A study on intelligent security camera system based on sequential motion recognition by using deep learning. IEEJ Trans. Electron. Inf. Syst. **139**(9), 986–992 (2019)
8. Alhassan, A., Sabtan, Y., Omar, L.: Using parallel corpora in the translation classroom: moving towards a corpus-driven pedagogy for omani translation major students. Arab World Engl. J. **12**(1), 40–58 (2021)
9. Saichaolu.: Development of intelligent automatic recognition system of breadboard function circuit based on image processing. Open J. Circ. Syst. **08**(3), 35–40 (2019)
10. Akinwonm, A.E.: Development of a prosodic read speech syllabic corpus of the yoruba language. Commun. Appl. Electron. **7**(36), 13–32 (2021)
11. Almuqren, L., Cristea, A.: AraCust: a saudi telecom tweets corpus for sentiment analysis. PeerJ Comput. Sci. **7**(6), e510 (2021)
12. Trautmann, C., Versen-Hynck, F.V.: Corpus luteum, vaskulre gesundheit und preklampsierisiko nach ART. Journal für Gynäkologische Endokrinologie/ Österreich **31**(3), 94–101 (2021)

Meta-analysis of the Prediction Model of Obstructive Sleep Apnea Based on Image Fusion Algorithm

Ying Fu(✉)

People's Hospital of Guangxi Zhuang Autonomous Region, Fangchenggang, Guangxi, China
1516866968@qq.com

Abstract. Sleep apnea syndrome (SAS) is a sleep disorder that seriously affects sleep quality and human health. Common symptoms of sleep are snoring, short-term asthma, respiratory arrest, body contractions, and even shock. This paper studies the obstructive sleep apnea prediction model of the image fusion algorithm, understands the relevant theoretical knowledge of obstructive sleep apnea on the basis of the literature, and then establishes the obstructive sleep apnea prediction model of the image fusion algorithm, and then conducted data analysis of actual cases. According to the results of model analysis, it was concluded that the descending area of rCBF in obstructive patients was mainly located in the posterior lobe of the left cerebellar cells, the left temporal lobe, the right middle frontal gyrus, and the bilateral middle and parahippocampal gyrus; RCBF the increased level of the midbrain is mainly in the bilateral superior frontal gyrus.

Keywords: Image Fusion · Obstructive Sleep Apnea · Fusion Algorithm · Sleep Disorders

1 Inductions

Sleep is a key physiological need of the human body in the process of life, an important link in the body's memory repair and integration functions, and an important part of physical and mental health [1, 2]. The best time for a healthy adult to fall asleep is seven to nine hours, and newborns need at least twelve hours of sleep a day to adapt to their own growth and development needs [3, 4]. However, due to the improvement of life rhythm and the pressure of social competition, the number of people with irregular life and insufficient sleep time has increased rapidly. Those with poor skin quality are the main ones [5, 6]. Research in my country has shown that sleep disorders not only affect obesity, slow thinking, poor memory, myopia and other problems, but also affect the human immune system and the development and intellectual development of adolescents, and more than 38.2% of our residents have sleep disorders. Therefore, the academic community has paid more and more attention to sleep quality and sleep-related diseases [7, 8].

In order to study sleep apnea, some researchers use 3D upper airway reconstruction as an important tool for the diagnosis of obstructive sleep apnea, and most hospitals have

J. H. Abawajy et al. (Eds.): ICATCI 2022, LNDECT 169, pp. 589–596, 2023.
https://doi.org/10.1007/978-3-031-28893-7_70

used spiral computed tomography (MSCT) or cone beam CT (CBCT) for obstructive sleep apnea. Patients with sleep apnea syndrome (OSHAS) are imaged as the basis for treatment, however, if physical changes affect the patient's upper airway morphology and whether there are differences between upper airway measurements and long-distance cross-sectional measurements, there is still a lack of practical answers [9, 10]. Although some researchers have identified obesity as an important risk factor for sleep apnea syndrome (SAHS), craniofacial morphology, along with the maxilla, is also considered an important risk factor for sleep apnea syndrome (OSAS). Therefore, it is unclear if there are differences between patients with obsessive- compulsive apnea syndrome (OSAHS) with normal craniofacial morphology and normal individuals with normal craniofacial morphology in 2D and 3D airway monitoring [11, 12]. To sum up, due to the importance attached to sleep quality by human society, more and more researches have been conducted on sleep apnea syndrome in academia.

This paper studies the obstructive sleep apnea prediction model of image fusion algorithm, analyzes the characteristics of obstructive sleep apnea and the application of image fusion on the basis of literature data, and then analyzes the image fusion algorithm for obstructive sleep apnea prediction. Models are constructed, and examples are used for analysis, and the results are used to draw relevant conclusions.

2 Obstructive Sleep Apnea Research

2.1 Features of Obstructive Sleep Apnea

Obstructive sleep apnea syndrome (OSAHS) is a common condition that can cause serious damage to the body. Studies have shown that the average incidence of OSAHS in different regions ranges from 2% to 9%, with a maximum of 0.15% and a minimum of 1.4%. Recent reports indicate that the prevalence of OSAHS is on the rise. The clinical characteristics of OSAHS are: airflow obstruction in the nasopharynx, mid-pharynx, and larynx leads to frequent apnea, hypoxia and complete destruction of sleep structure during nighttime sleep, which makes cardiovascular and brain diseases easy to combine and increase, and memory leads to other diseases appear.

Although the primary cause of OSAHS is the upper respiratory tract, it affects many organs and systems, is an independent risk factor for many diseases, and has a profound impact on patients' quality of life and life expectancy. It is a systemic disease. Such as causing or aggravating early hypertension, it may also cause or aggravate coronary heart disease through autonomic nerve dysfunction, endothelial cell damage, chronic inflammation or oxidative stress, and increase blood viscosity through oxidative stress, factors and lack of sleep, etc., which lead to impaired glucose tolerance. Studies have shown that up to 45% to 48% of people with SAHS have high blood pressure, compared to 20% in the general population. The relationship between the two is so strong that OSAHS may be one of the independent risk factors for hypertension. Another epidemiological study showed that up to 20% to 30% of OSAHS suffer from coronary heart disease, and that elevated AHI is strongly associated with coronary heart disease mortality and prognosis. Other studies have shown that patients with coronary heart disease with OSAHS have a 62% improved 5-year survival rate compared to controls. At present, the mechanism of OSAHS causing diabetes and insulin resistance is still unclear, but related studies have shown that OSAHS is actually related to the occurrence and development of diabetes. OSAHS can cause many diseases and accidents, and impose a heavy burden on socioeconomic and medical resources, thus attracting the attention of clinicians and snorers, and active monitoring of snorers, treatment is essential.

At present, the main treatment methods for OSAHS include etiological treatment, effective weight management, weight loss, smoking cessation, alcohol treatment, non-invasive positive airway pressure treatment, oral devices, etc., and surgical treatment according to the basic conditions of patients.

2.2 Application of Image Fusion

Image difference is a difference graph obtained by subtracting two pixels in adjacent areas at different times, according to the gray value of the corresponding pixel, to describe the change between the two time-phase pixels. Ideally, if a region changes, the pixel value of the received difference map should be greater or less than zero, and if the region does not change, the pixel value of the received difference map should also be zero. If the image is remote or the image is noisy and the image is blurred, the effect of image difference on the remote image is better than the actual image effect. Ideally, the user only needs to set the corresponding time limit on the generated difference map, and then get a measure of the change in the result. This method is very concise, intuitive and simple. The disadvantage is that it can give the signals of changing and constant graphs, but cannot give the specific signals of object changes, and is very sensitive to noise. At the same time, the finite difference method that selects the threshold of the incremental graph in the graph It is also very critical, but the threshold selection is a rather troublesome problem, and the current threshold method cannot determine the result.

The image ratio method has good anti-calibration and radiation error ability, and its basic principle is the same as that of the aberration method. The basic method of image aspect ratio is to compare two pictures of different periods in adjacent areas, so as to obtain different graphics pixel by pixel. But once the area is changed, the pixel value of different graphics should be much higher than one or even more. If it is equal to one, the pixel value is close to one under the condition of constant area, but due to the existence of noise, the difference image obtained will be quite different from the actual situation. In addition, the aspect ratio method usually requires the preprocessing of the radiation compensation information. Although it is simple and convenient, the efficiency still needs to be further improved. However, since the image aspect ratio method is more resistant to long-term image magnification noise than the image aspect ratio method, the image aspect ratio method is used to build a difference map to monitor long-term changes. The logarithm method is to apply the ratio operator to the long-term image, and then obtain the logarithm (usually based on the natural logarithm), this operator has gained more common use in recent years. Although this operator has strong anti-noise properties, it uses the log-ratio operator to process the received difference map to convert the multiplier noise into additional noise. Since the obtained pixel values do not follow a normal distribution, the analysis is very difficult.

Sorting comparison method: Sorting and comparison method is to sort the two images respectively, and then perform a one-to-one comparison and analysis on the sorting graphs of the acquired pixel values of the two images, and finally obtain the information obtained by changing the characteristic scene. Images with different phases are sorted separately and marked using a sorting and comparison method, so the user can determine the number of categories in the final result by looking at the number of categories in the two polyphase images. Although classification and comparison methods provide information about the location of modified regions and the nature of features within regions. However, this method requires very accurate classification, otherwise, it will directly affect the accuracy of the final change detection results.

Principal Component Analysis: Orthogonal (linear) transformations can be performed within the image through principal component analysis. Principal component analysis is about redundant compression and redundant execution of signals in multivariate random variables, focusing on simultaneously receiving multiple uncorrelated transform components. The PCA method is also widely used to measure multiple pixel changes, because if the area changes greatly, then the correlation between pixel data is very low, otherwise the correlation between image data is very high. However, the PCA method can isolate linearly correlated data point nodes. This usually causes errors in data nodes related to constitutive nonlinearity.

3 Obstructive Sleep Apnea Prediction Model Based on Image Fusion Algorithm

3.1 Image Acquisition

All subjects were examined by a Siemens Magnetom Trio Tim3.0TMR system (an 8-channel phased array coil), and upon completion of the examination, the individual's head was reinforced with a special foam pad and kept still. The subjects were then asked to calmly close their eyes and remain awake. The relevant scan sequences and parameters were: first, routine head and high-resolution T1-3D sequence scans were performed to eliminate organic lesions, followed by routine scan reference: T1WI axial position.

3.2 Image Fusion

Figure 1 shows the main framework of the fusion-based image detection algorithm proposed in this paper. First, a more accurate surface volume is constructed based on superpixel segmentation and fusion, and then an initial projection map is created based on the color histogram and peripheral contrast, respectively. Finally, using the constructed exact convex hull, the two original projection maps updated under the Bayesian framework are merged to obtain the final map.

(1) Image segmentation

The general idea of the SLIC algorithm is to first convert the image from the RGB color space to the CIELAB color space. The position coordinates (x, y) of each pixel in the image and the color value (La, b) corresponding to the pixel in the CIELAB color space form a five-dimensional vector [L, a, b, x, y]. Also, the euclidean distance of the 5D vector is used to represent the similarity between two pixels. The greater the distance between the vectors corresponding to two pixels, the greater the distance. The less similar two pixels are, the less likely they are to belong to the same superpixel.

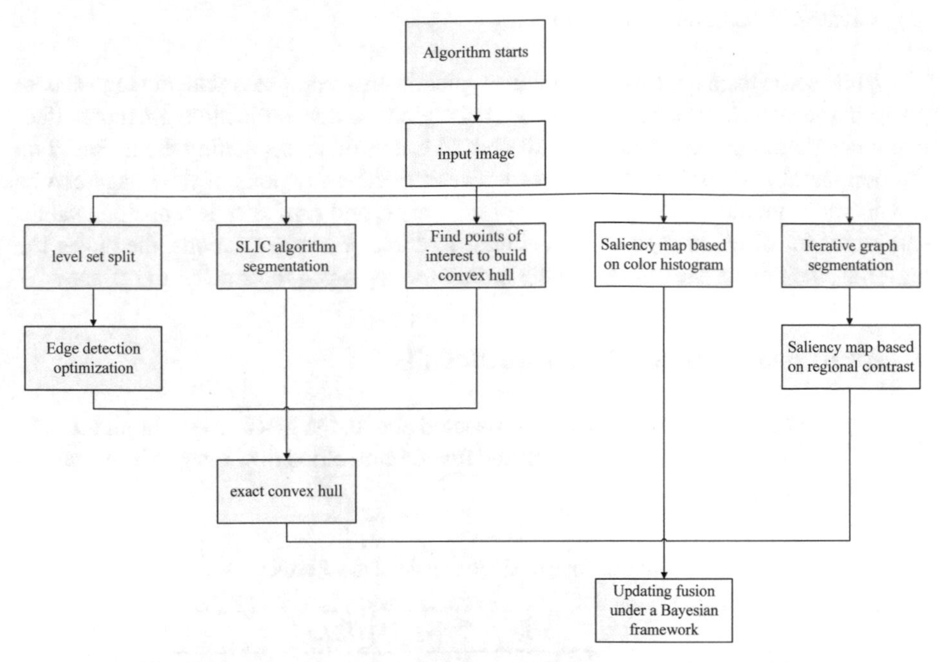

Fig. 1. The main framework of fusion-based image detection algorithm

The SLIC algorithm computes the color feature difference and the spatial position difference between the pixel and the cluster center, and uses different weights to overlap the two differences.

$$d_{lab} = \sqrt{(l_k - l_i)^2 + (a_k - a_i)^2 + (b_k - b_i)^2} \tag{1}$$

$$\mathrm{d_{xy}} = \sqrt{(\mathrm{x_k} - \mathrm{x_i})^2 + (\mathrm{y_k} - \mathrm{y_i})^2} \tag{2}$$

Among them, k represents the cluster center of the k-th superpixel; i represents the ith pixel around the cluster center; d_{lab} represents the color difference of the pixel; $\mathrm{d_{xy}}$ represents the spatial position difference of the pixel.

(2) Build a convex hull to detect visible parts in the image

Perspective technology is an important way to extract feature information when editing images. Interest points help people pay attention to the area where the most interesting objects are located in the image and reduce unnecessary computation, so in this article, Harris operator will be used to detect points in the image. Meanwhile, the interest points obtained by the Harris operator contain more background pixels, which have a significant impact on the detection results of key regions, so this task is a simple linear SLIC iterative grouping to reduce redundancy.

(3) Create a local comparison map

While some local contrast projections typically only compute relatively high salient position values near the target contour, global region contrast projections are more effective when targeting areas wider than the actual background, separating them. Based on the peripheral contrast, the differences between different regions of the image can be considered from the overall perspective of the image, and similar salient position values can be set for different regions with similar features. For other regions, the closer the region is, the greater the effect, and the farther the region is, the smaller the effect.

4 Analysis of Image Algorithm Results

Through the image fusion algorithm constructed above, the MRI images of all the subjects collected were compared and analyzed for cerebral blood flow images. The analysis results are shown in Table 1:

Table 1. Image Algorithm Analysis Results

	GSs	OSAs
L.CPL	27	18
L.PG	20	8
R.PG	32	21
L,SFG	12	28
R.SFG	17	40
L.TL	38	21
R.MFG	45	39

As can be seen from Fig. 2, compared with the normal group, the decrease in rCBF in the obstructive sleep apnea group was mainly seen in the left posterior cerebellar lobe, left temporal lobe, right middle frontal gyrus and bilateral atrioventricular nodes.

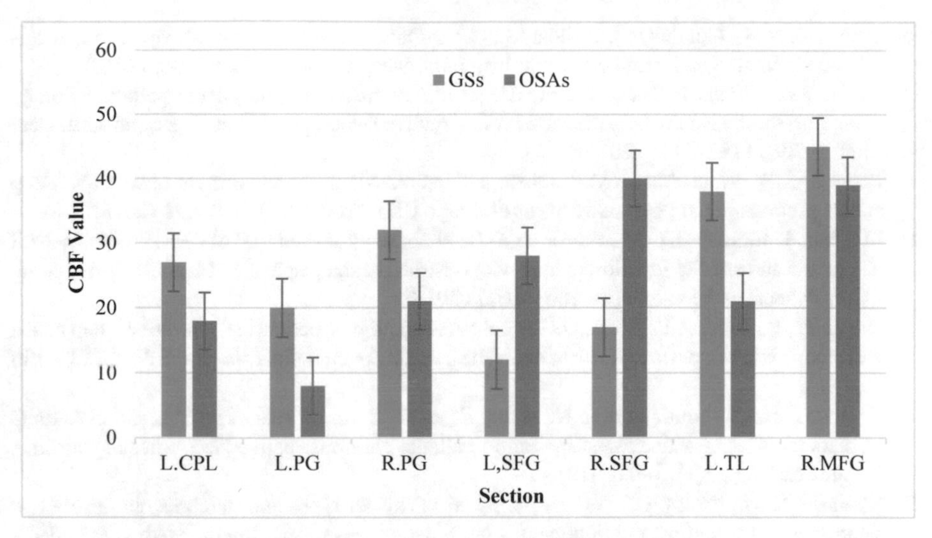

Fig. 2. Image Algorithm Analysis Results

5 Conclusions

In this paper, the obstructive sleep apnea prediction model of image fusion algorithm is studied. After analyzing the relevant theoretical knowledge, the obstructive sleep apnea prediction model of image fusion algorithm is constructed, and an example is analyzed. In the OSAHS group, the decreased rCBF was mainly located in the left posterior cerebellar lobe, the left temporal lobe, the right middle frontal gyrus, and the bilateral parahippocampal gyrus; the brain regions with increased rCBF were mainly located in the bilateral superior frontal gyrus.

Acknowledgements. Project Name: self funded scientific research project of Guangxi Health Commission "Research on multidisciplinary joint sleep apnea prediction model for snoring population", Project No.: 20212178.

References

1. Jin, S., Jiang, S., Hu, A.: Association between obstructive sleep apnea and non-alcoholic fatty liver disease: a systematic review and meta-analysis. Sleep Breathing **22**(1), 1–11 (2018)
2. Kang, D.Y., Deyoung, P.N., Malhotra, A., et al.: A state space and density estimation framework for sleep staging in obstructive sleep apnea. IEEE Trans. Biomed. Eng. **65**(99), 1201–1212 (2018)
3. Haddock, N., Wells, M.E.: The association between treated and untreated obstructive sleep apnea and depression. Neurodiagnostic J. **58**(1), 30–39 (2018)
4. Marrone, O., Bonsignore, M.R.: Obstructive sleep apnea and chronic kidney disease: open questions on a potential public health problem. J. Thorac. Dis. **10**(1), 45–48 (2018)
5. Hedner, J., Zou, D.: Drug therapy in obstructive sleep apnea. Sleep Med. Clin. **13**(2), 203–217 (2018)

6. Tasbakan, M.S., Gunduz, C., Pirildar, S., et al.: Quality of life in obstructive sleep apnea is related to female gender and comorbid insomnia. Sleep Breathing **22**(9), 1–8 (2018)
7. Sands, S.A., Edwards, B.A., Terrill, P.I., et al.: Phenotyping pharyngeal pathophysiology using polysomnography in patients with obstructive sleep apnea. Am. J. Respir. Crit. Care Med. **197**(9), 1187–1197 (2018)
8. Prakash, K.B., Mansukhani, M.P., Olson, E.J., et al.: Medical cannabis for obstructive sleep apnea: premature and potentially harmful. Mayo Clin. Proc. **93**(6), 689–692 (2018)
9. Melehan, K.L., Hoyos, C.M., Hamilton, G.S., et al.: Randomised trial of CPAP and vardenafil on erectile and arterial function in men with obstructive sleep apnea and erectile dysfunction. J Clin Endocrinol Metab **25**(4), 1601–1611 (2018)
10. Goyal, A., Pakhare, A.P., Bhatt, G.C., et al.: Association of pediatric obstructive sleep apnea with poor academic performance: a school-based study from India. Lung India **35**(2), 132–136 (2018)
11. Voulgaris, A., Archontogeorgis, K., Nena, E., et al.: Serum levels of NGAL and cystatin C as markers of early kidney dysfunction in patients with obstructive sleep apnea syndrome. Sleep Breathing **23**(1), 1–9 (2018)
12. Kunisaki, K.M., Nancy, G., Wajahat, K., et al.: Provider types and outcomes in obstructive sleep apnea case finding and treatment: a systematic review. Ann. Intern. Med. **168**(3), 195–202 (2018)

Labeling Algorithms for Sensitive Areas of Visual Images in Multimedia Environment

Huihuang Wu(✉)

Xiamen Institute of Software Technology, Xiamen 361024, Fujian, China
wuhuihuang666@163.com

Abstract. With the development of computers, multimedia information has become an indispensable part of people's lives. In the field of visual image recognition, how to process and detect images is very important and difficult. Therefore, studying image features and their annotation algorithms is the basis of image recognition and processing. This article mainly uses experimental method and comparison method to detect the standard rate, over-standard rate and miss-standard rate of the selected images. Experimental results show that for the calibration of birds and human bodies, the missing and over-standard rates are significantly higher than those of other static objects, and the standard rates are only 26% and 35%. The EM algorithm can be used to identify and calibrate relatively static objects, but the calibration of dynamic objects needs further improvement.

Keywords: Multimedia Environment · Visual Image · Sensitive Area · Labeling Algorithm

1 Introduction

Visual recognition technology is a kind of advanced information technology that is widely used in the fields of computer and communication. It uses image collection, motion feature extraction and the process of human eyes to recognize external information. Identifying sensitive areas means that a certain point or certain part of an image is marked as specific content, has a certain color attribute, or has characteristics such as meaning for certain information. The identification object plays a very important role in the visual system, and is a way of information exchange and transmission. However, people still do not have a clear, unified and effective method for how to effectively identify image quality problems. Therefore, research on labeling algorithms for sensitive areas can improve the reliability and accuracy of recognition.

There are not a few researches on the labeling algorithms for sensitive areas of visual images in multimedia environments. For example, some people proposed a visual image annotation algorithm for the problem of low accuracy of the sensitive area division algorithm [1, 2]. Some people have also proposed that the labeling of sensitive areas of multi-dimensional visual images is a very important step in image analysis [3, 4]. In addition, some scholars said that massive image data inevitably puts forward higher requirements on image processing technology. How to quickly and efficiently find the

J. H. Abawajy et al. (Eds.): ICATCI 2022, LNDECT 169, pp. 597–604, 2023.
https://doi.org/10.1007/978-3-031-28893-7_71

pictures you want from the huge picture library has become an urgent and arduous task [4, 5]. Therefore, this article is based on the multimedia background, the study of the visual image sensitive area labeling algorithm mainly starts with the semantic-based algorithm to detect the recognition rate of the image.

This article first studies some basic theories and methods of image retrieval and image calibration. Secondly, it analyzes the image automatic semantic annotation algorithm based on the sensitive area. Afterwards, the image quality evaluation method is described. Finally, the data set is collected, and the experimental analysis is carried out, and conclusions are drawn.

2 An Algorithm for Labeling Sensitive Areas of Visual Images in a Multimedia Environment

2.1 Image Retrieval and Image Calibration

Multi-label image calibration is also one of the important applications of content-based image restoration technology. At the same time, better calibration results can very effectively improve the efficiency and accuracy of image retrieval to a certain extent [6].

Cognitive gap. Perception gap mainly refers to the gap between real objects in the real world and the information we use to describe objects in the scene. The main problem caused by this defect is that regardless of the type of information description method, the recognition of image content is always restricted by the information description method [7].

Semantic gap. The semantic gap mainly refers to the possible inconsistency between the information that the viewer of the image can extract from the visual data of the image and the information that the image author can perceive or express for the same visual data in a given situation. The main problem caused by this defect is that the viewer of the image may not be able to understand or perceive the information description of the image by the image author through the visual data of the image alone [8, 9].

Most content-based image restoration techniques use feature extraction as a preprocessing step in the algorithm. Once the description of the image features is obtained, the visual features are used as input for some subsequent image analysis tasks (such as similarity estimation, concept recognition, or image calibration) [10].

Image segmentation. Image segmentation is an important step to obtain image feature description based on sensitive areas. Reliable image segmentation is particularly important for extracting shape features from images. If there is no reliable image segmentation, the estimation of most shape features is meaningless [11].

Visual characteristics refer to specific visual attributes that can describe global or local pixel sets in an image. The most commonly used visual features include color, texture, shape, and key points in the image. In the extraction of global features, some features of the image are described as a whole through feature calculation.

2.2 Image Automatic Semantic Annotation Algorithm Based on Sensitive Areas

The algorithm process in this article has two stages: learning and testing. The training phase consists of three main modules: segmenting the sensitive area of the image, extracting the underlying visual features of the image, and establishing the concept class described by the Gaussian mixture model. In the test phase, first perform image segmentation and extraction of the lowest-level features, and then combine the lowest-level features with the probability distribution model of each semantic category obtained by l'learning to calculate the posterior semantic probability. Finally, based on the probability and posterior decision algorithm, the annotation results are given. The process of image annotation method based on image sensitive area is shown in Fig. 1:

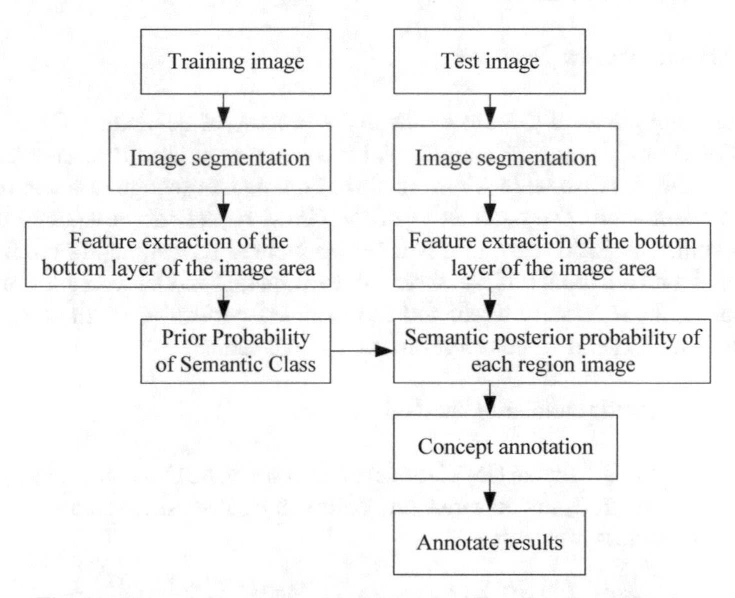

Fig. 1. Process of Image Annotation Method Based on Image Region

(1) Image segmentation

The semantics of the image is closely related to the sensitive area of the target in the image. For semantic-based image retrieval, one of the key steps of semantic automatic image annotation is to segment the image and identify multiple image sensitive areas with semantic content meaning. In order to effectively express the content of the image, the first preprocessing to be carried out is the segmentation of the image. Each sub-image obtained after segmentation can be regarded as a "visual word", which is used for model generation and category evaluation of the subsequent image labeling system. These sensitive areas do not overlap, and each sensitive area corresponds to the consistency of a specific sensitive area. Image segmentation is the basis of target expression and has an important influence on the measurement of underlying features. The pros and cons of the image segmentation results directly affect the extraction of the underlying features of the hotspot, which in turn affects the effect of automatic image restoration.

(2) Low-level feature extraction

The color characteristics are closely related to the objects or scenes contained in the image. In addition, the color attributes have little dependence on the size and viewing angle of the image. Usually, it reflects the uniformity of the image and has nothing to do with color and brightness. The sensitive areas of different objects in the image have different texture attributes. Therefore, texture functions are widely used in content-based image retrieval and annotation.

(3) GMM concept class generation

The learning phase of the algorithm is the process of generating GMM concept classes. GMM or Gaussian Mixture Model is a commonly used mathematical model in statistics, and it is also widely used in clustering, pattern recognition and multivariate density estimation. The parameters of the GMM model can be learned using the EM algorithm. The EEM algorithm is a classic method for estimating the maximum likelihood of a mixed model. If the observed data contains hidden variables or the data is incomplete, the maximum likelihood estimation is performed on the data, and the probability that the iteration converges to the optimal value.

(4) Concept semantic annotation method

After training to obtain the GMM model of each semantic class, in the testing phase, according to Bayes' rule, the posterior probability of each sensitive area corresponding to each semantic class is calculated:

$$G_{v|a}\left(v_i\middle|y^k\right) = G_{v|a}(v_l|a) = \frac{G_{ap}(a|v_l)\cdot G_v(v_l)}{G_a(a)} \tag{1}$$

where a is the underlying feature vector extracted from the sensitive area y^k, and $G_v(v_l)$ is the probability that the l-th semantic concept appears in the entire training set. $G_a(a)$ is the prior probability of the underlying feature vector, assuming a uniform distribution, that is, a constant.

In the traditional labeling method based on posterior probability, the conceptual semantic posterior probability $G_{v|a}(v|M)$ of each image is composed of the semantic posterior probability of the sensitive area of the image after segmentation, which represents the possibility that the current image contains the semantics of each concept.

$$G_{v|a}(v_l|M) = \sum\nolimits_{y^k=M} G_{v|a}\left(v_l|y^k\right)\cdot g\left(y^k|M\right) \tag{2}$$

The latter probability $g\left(y^k|M\right)$ represents the proportion of the sensitive area y^k to the entire image area. Arrange the posterior probabilities corresponding to all the semantics in descending order, and take thes semantics corresponding to the first s posterior probabilities as the final tagging words of the image.

(5) Annotation method combining semantic relevance between sensitive areas

In the aforementioned traditional annotation method, the posterior semantic probability of an image is weighted by adding the posterior semantic probability of each sensitive area to the area of the sensitive area. Therefore, a larger sensitive area has a greater impact on the final result of the marking. But in fact, some smaller sensitive areas can also correspond to important semantics. The principle of the algorithm is as follows: Due to the imprecision of segmentation, there are generally multiple sensitive areas in an image corresponding to the same semantic concept. In addition, the semantics of each sensitive area corresponds to another semantics, which also has a certain relevance to natural images. The basic idea of the algorithm is: on the one hand, for each sensitive area, the candidate semantics of the sensitive area is obtained from the posterior probability. Secondly, the similarity between each candidate semantics and the semantics of other sensitive areas is calculated. Finally, the similarity is calculated Sort by degree to determine the final semantic annotation of the entire image.

2.3 Image Quality Evaluation Method

Image quality is the subjective perception of images by humans through the visual system, and it also reflects the ability of images to provide information to humans. In the case of a standard reference image, the quality of the image is related to the degree of difference between the target image and the reference image in human eyes.

Generally speaking, the picture quality is mainly reflected in the picture fidelity and picture clarity. The smaller the difference between the two, the better the image quality. The comprehensibility of the image is related to the comprehensibility of the information in the image, that is, the more the image can reflect the different information contained in the image, the higher the quality of the image. The overall evaluation methods of image quality can be divided into subjective evaluation methods and objective evaluation methods. The subjective evaluation result of image quality can be regarded as a realistic attribute of the image quality test environment and test duration. The physical and psychological conditions of each observer at different times are different. This makes the accuracy of the subjective results of quality evaluation vulnerable to the external environment.

Generally speaking, objective image quality evaluation methods can be divided into full reference image quality evaluation, semi-reference image quality evaluation and standard image quality evaluation. In fact, this is only a classification method of the objective image quality evaluation method. The so-called classification method based on the original image is classified according to the information dependence of the original image when calculating the quality of the distorted image. The other two classification methods are based on human vision based on specific applications. The second classification method is based on the application of classification methods. Each distorted image has its own main distortion type. For example, compressed block-based images are distorted due to block effects in the compression process. Then, the method of judging image quality for blocking effect will give good results. The third classification method is based on the characteristics of human perception: it is divided into bionic method and technical method.

3 Experimental Analysis

3.1 Data Set

In the experiment, we used the more commonly used MSRC data set, and extracted 60 images of 6 categories for analysis. The resolution of its image has reached a high level. For some data sets, if the image resolution changes greatly, the resolution may be normalized before processing.

3.2 Evaluation Criteria

In order to present the experimental results more clearly and better evaluate the accuracy of various label calibrations, in the experiment we outline the approximate sensitive areas of the image for each image. Here we mainly quote the evaluation criteria of the V0C Challenge. Here we mainly refer to its evaluation criteria for classifying and identifying objects. At the same time, we revised and broadened the evaluation criteria accordingly. We decompose the final result of calibration into the following three situations in detail: missing standard, accurate calibration and over-calibration.

3.3 Experimental Preprocessing

In the specific experimental process, we carried out appropriate pre-processing for the following different situations. First, there is a label calibration error in the MSRC record, which means that the corresponding image does not contain the object at all. There is only a small part of the objects in the corresponding image. The concept of labels in related images is too large or too abstract to be accurately calibrated. In these cases, we do not make reference.

4 Experimental Results and Analysis

4.1 Calibration Accuracy, Missing Standard Rate and Over-Standard Rate of Some Labels

Table 1. Calibration Accuracy, Missing Standard Rate and Over-Standard Rate of Some Labels

	Slag rate	Standardized rate	Overbid rate
Car	0.12	0.81	0.01
Bike	0.04	0.85	0.09
Bird	0.35	0.26	0.31
Body	0.11	0.35	0.52
Book	0.05	0.87	0.02

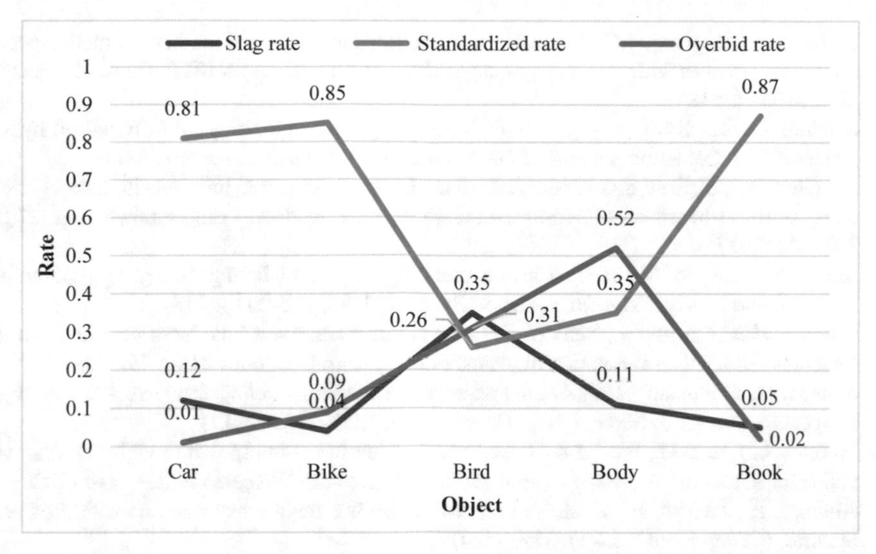

Fig. 2. Calibration Accuracy, Missing Standard Rate and Over-Standard Rate of Some Labels

After the corresponding processing, the calibration accuracy rate, the missing standard rate and the over-standard rate of the label obtained from the experiment are shown in Table 1. From the table, we can see that the standard rate has the largest data, and there are fewer cases of missing standard.

As shown in Fig. 2, various labels such as car, bike, and book have distinct characteristics. After EM iteration, good calibration results are obtained in this experiment. For dynamic objects such as bird, the calibration effect is not very good. For its calibration situation, the standard rate appears to be slightly lower than that of others.

5 Conclusion

Feature extraction based on sensitive areas is a way to classify and recognize images. It is mainly divided according to its threshold, and pictures with similarity or similar location attributes (such as brightness, saturation, etc.) are used as key markers. This method can well solve the problem of vague edge information caused by the existence of a single sign in the same area. Therefore, research on the labeling algorithm for sensitive areas of visual images in a multimedia environment can further improve the image calibration technology. The experiments in this article tell us that the EM algorithm has a certain role in the calibration of sensitive images, but it needs to be improved.

References

1. Junior, W., Oliveira, E., Santos, A., et al.: A Context-sensitive offloading system using machine-learning classification algorithms for mobile cloud environment. Futur. Gener. Comput. Syst. **90**, 503–520 (2019)

2. Prabhakaran, B., Jiang, Y.G., Kalva, H., et al.: Editorial ieee transactions on multimedia special section on video analytics: challenges, algorithms, and applications. IEEE Trans. Multimedia **20**(5), 1037 (2018)
3. Satellite, Jia, L., Xiao, S., et al.: Multi-view image generation algorithm based on hybrid generative confrontation network. Acta Autom. Sin. **47**(11), 2623–2636 (2021)
4. Subramaniam, S., Haw, S.C., Soon, L.K., et al.: QTwig: a structural join algorithm for efficient query retrieval based on region-based labeling. Int. J. Software Eng. Knowl. Eng. **27**(2), 321–342 (2017)
5. Marin, A., Pelegrin, M., et al.: Towards unambiguous map labeling - integer programming approach and heuristic algorithm. Expert Syst. Appl. **98**, 221–241 (2018)
6. Cagirici, H.B., Galvez, S., Sen, T.Z., et al.: LncMachine: a machine learning algorithm for long noncoding RNA annotation in plants. Funct. Integr. Genomics **21**(2), 195–204 (2021)
7. Arun, K.S., Govindan, V.K.: A Context-aware semantic modeling framework for efficient image retrieval. Int. J. Mach. Learn. Cybern. **8**(4), 1259–1285 (2017)
8. Weon, I.S., Lee, S.G., Ryu, J.K.: Object recognition based interpolation with 3D lidar and vision for autonomous driving of an intelligent vehicle. IEEE Access PP(99), 1–1 (2020)
9. Andreas, R., Maxim, R., et al.: A practical algorithm for the external annotation of area features. Cartographic J. **54**(1), 61–76 (2017)
10. Nobuaki, T., Koichi, I., Koji, T., et al.: Automated ablation annotation algorithm reduces re-conduction of isolated pulmonary vein and improves outcome after catheter ablation for atrial fibrillation. Circ. J. **81**(11), 1596–1602 (2017)
11. Shishido, H.Y., Estrella, J.C., Toledo, C., et al.: Optimizing security and cost of workflow execution using task annotation and genetic-based algorithm. Computing **103**(6), 1281–1303 (2021)

Algorithm Design of Dynamic Course Recommendation Model Based on Machine Learning

Chunrong Yao(✉)

Wuhan Qingchuan University, Wuhan, Hubei, China
ycrong1209@sina.com

Abstract. With the development of Internet technology, dynamic course recommendation systems based on network learning models and cloud computing have been gradually applied and developed in various disciplines. In order to improve the recommendation accuracy of dynamic courses, this paper uses the recommendation algorithm of machine learning to analyze and design. This article mainly uses experimental method and comparison method to compare and analyze the computing power of different algorithms. The experimental results show that under the two evaluation indicators of RMSE and MAE, the minimum value of NeuCNN is 1.2836 and 0.9765. These data are sufficient to show the advantages of the intelligent algorithm.

Keywords: Machine Learning · Dynamic Curriculum · Recommendation Model · Algorithm Design

1 Introduction

With the rapid development of Internet technology, online learning has become a common and effective teaching method. In the field of distance education, more and more researchers pay attention to the data mining problem in the course resource library based on the Web environment. With the popularization and application of the Internet, people are increasingly using online course resources in the learning process, so it is necessary to improve the system's data analysis.

There are many research theoretical results on the algorithm design of the dynamic course recommendation model of machine learning. For example, it has been suggested that autonomous learning systems have become an important subject of modern distance learning. By analyzing the characteristics of autonomous learning, a personalized learning model is proposed, and then the organization of curriculum resources and recommendation algorithms are sought [1, 2]. Some researchers integrate and create university curriculum resources from the perspective of information resource construction, and provide services for university curriculum customization and recommendation [3, 4]. In addition, in the field of course recommendation, some researchers have proposed an improved recommendation algorithm based on collaborative filtering. Based on the

J. H. Abawajy et al. (Eds.): ICATCI 2022, LNDECT 169, pp. 605–612, 2023.
https://doi.org/10.1007/978-3-031-28893-7_72

original course similarity calculated based on the user's historical price selection behavior record, the study of the content of the course text is introduced [5, 6]. Therefore, this article intends to design and algorithm research on the recommendation model of dynamic courses from the perspective of machine learning.

This article first studies some basic theories and categories of data mining and machine learning. Secondly, it studies the analysis and modeling of the dynamic course recommendation system. Then analyze and explain the recommendation system. Finally, through experiments, several different algorithms are compared and analyzed, and conclusions are drawn.

2 Algorithm Design of Dynamic Course Recommendation Model Based on Machine Learning

2.1 Data Mining and Machine Learning

The main task of data mining is to discover the potential connection between hidden knowledge and reveal the hidden connection or inner law. The learning process involves a series of operations, from obtaining a large number of training sets to using real data sets to classify, compare, and filter prediction objects. Finally, train a new set of rules, and then run the specific algorithm to achieve the expected results. At the same time, it is possible to evaluate whether the model is feasible by analyzing the original data in the test set and designing new methods on this basis [7, 8].

Data Mining (DataMachine) is a science of research methods, tools and models. He is building a new system for many computers to find the information he needs in the future. With this in mind, data mining technology can be defined as: using existing machine learning theories and algorithms and other related knowledge to model, analyze, and predict the implicit relationships that exist in the database. Neural network is a learning tool that simulates the information processing system of the human brain. It mainly models the human brain and nervous system from low-dimensional and high-dimensional perceptual data structures. After preliminary training and decision-making, it approaches its minimization in the expected direction and realizes it in reality. These can be very smart behavioral characteristics (that is, the ability to understand things or behaviors, etc.). Neural networks can be used as models for parallel computing, pattern recognition, knowledge classification and optimization to process large amounts of complex information. Data mining technology is about extracting potentially valuable unstructured information from massive information, and transforming it into a model that can be used for calculation through mechanisms such as knowledge discovery and pattern recognition. Online learning usually requires high efficiency. And that efficiency mainly depends on the accuracy and correlation value of the algorithm on the original input data set (training sample) to determine whether the recommended result output matches the expected correct probability. In addition, it is based on rule base and parameter mining. The structure of the argument is also an important factor [9, 10].

2.2 Analysis and Modeling of Dynamic Course Recommendation System

Dynamic course recommendation is based on the user behavior model, and its algorithm will go through many experiments and parameter estimations from input data to output results. Therefore, a suitable machine learning tool is needed to support it. The traditional method will produce a lot of redundant information and consume long time in the calculation process. On the other hand, it also limits the modeling of the system [11, 12].

The design of the dynamic course recommendation system is based on the number of grades and comments, and the scores recorded by each student when they are not graded are divided into two sub-libraries between each level. Then the SVM classifier is used to identify objects of the same category with similar label values in the corresponding sets of all labels on the high-dimensional data set.

The dynamic course recommendation system is an algorithm based on artificial neural network. It filters and models different supply and demand information by combining existing knowledge bases (such as My) and database management systems.

This system is mainly composed of server side and course library. Among them, the client implements a hash-based method related to the learning category for similarity calculation of similar query items. Extract a part of the data from each score as the training set for weighted average and get the corresponding score; and for the questions that already exist but cannot be familiar to the students or refuse to provide help for their content scores, return to the group set (that is, the question is From the same mixed text) to save to the database for later review. In the course library, organize the collection of items that need to be recommended into a message file. After uploading, select a certain storage space as the model parameter. A BP neural network for processing the database is established in the application service software, which maps the user input vector to the output, and simulates the human brain and other information processing processes through algorithms. Create machine learning tools and control methods under the system support framework, and implement related functions based on model construction and training management unit modules.

Firstly, by writing a program, the static data is converted into video files and stored in the database; secondly, when the real-time requirements are high and the content is more complex, methods such as tDDT classification model and K-nearest neighbor feature are used to extract the relevant attribute information required by the experiment. Used as a training set for prediction score calculation. The dynamic course recommendation system based on machine learning algorithms mainly uses the C# language development framework and the SSH format. It is necessary to input its own information and basic knowledge points for the user name, password and other dimensions. And generate a training set based on the feature item. Then it is mapped to the output terminal through the neural network method, and the appropriate weight range is selected according to different situations to achieve high-dimensional space cross-validation, polynomial calculation and other operations.

The dynamic course recommendation system based on machine learning is mainly designed to solve the problem of statistical analysis of students' learning in traditional classrooms based on reading scores on the current online examination platform. By comparing with the existing traditional paper-based test papers, it can be more intuitive See the data flow between the modules, the efficiency of collaborative work, and the correction opinions given by different teachers and parents when answering questions face-to-face.

2.3 Recommendation System

(1) Content-based recommendation

The content-based recommendation algorithm models users and products separately, extracts their respective attributes and performs correlation calculations. The method of constructing product features generally requires the participation of experts in the field, and feature extraction must be completed before the product is included in the recommendation system. User characteristics can be created in the system based on the user's browsing, clicking, and purchasing behaviors. The recommendation system uses an association algorithm to match users and products, and generates and recommends a list of product recommendations that are most likely to be purchased by multiple users.

Collaborative recommendation generates recommendations for users based on their previous behavior information without extracting user or product features. The collaborative filtering process first analyzes the interaction between the two displayed in the collected user product preferences, and uses these functions to generate personalized recommendations for users. The hybrid filtering process combines content-based filtering and collaborative filtering to overcome their respective shortcomings and combine their respective advantages at the same time.

Recommendation systems generally refer to the use of collected information, such as specific intelligent system problems. The merge filter layer usually merges some callback sets, adds appropriate feature vectors, and filters some user/product pairs, thereby increasing the recall success rate and reducing negative samples. A typical recommendation system consists of three modules. At the same time, according to the relationship between the attributes and elements of the project, pick up the project template, and match the two. Through a specific recommendation algorithm, filter out which users are most interested or most interested in a specific product or course, and provide personalized recommendations. The specific model is shown in Fig. 1:

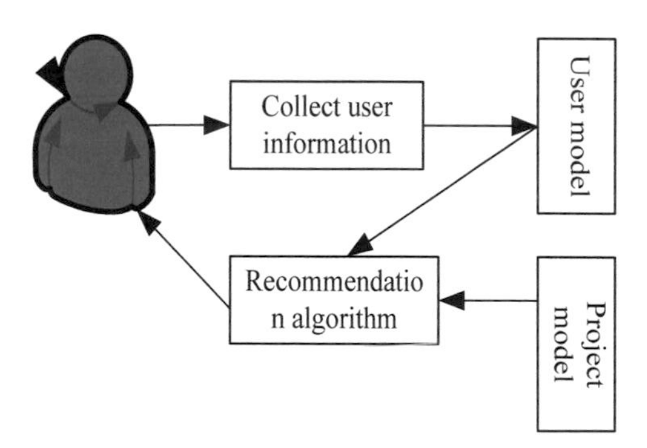

Fig. 1. Typical Recommender System Model

(2) Result prediction system

If the system or website can provide users with a course scoring service, the recommendation system can predict the user's scoring of new articles based on the user's scoring history of articles, and use the recommendation model to predict whether users will be interested in scoring new articles when the article is generated. The rating indicators of the rating prediction system are relatively simple and clear. There are generally two indicators: Mean Square Error (RMSE) and Mean Absolute Error (MAE):

$$\text{RMSE} = \sqrt{\frac{\sum_y \sum_k \left(s_{y,k} - \hat{s}_{y,k}\right)^2}{m}} \tag{1}$$

$$\text{MAE} = \frac{\sum_y \sum_k \left(s_{y,k} - \hat{s}_{y,k}\right)}{m} \tag{2}$$

Among them, MAE measures the average error of the predicted value deviating from the true value, which can intuitively reflect the accuracy of the predicted value. Since RMSE contains a quadratic term, it is more sensitive to terms whose predicted value is different from the true value, and may reflect the variance of the error value.

(3) TopN recommendation system

The TopN recommendation system uses certain pattern training algorithms and ranking algorithms to recommend the most interesting things to users and make it easier for users to find. The main identification indicators of TopN recommendation calculation system include recall rate, accuracy rate, coverage rate, novelty, etc. Recall rate and accuracy rate measure the accuracy rate recommended by the recommendation system, and correspond to the coverage rate of RMSE and MAE indicators in the scoring system.

3 Experimental Test

3.1 The Purpose of the Experiment

In order to verify the effectiveness of the proposed method, we selected the Mukenet as the data source, and collected user information and course-related data available on the website. Including 1000 MOOC learners, 100 courses.

3.2 Experimental Setup

In order to be consistent with the standard record format, users with more than 15 records were selected, and 200 users and all 100 courses were randomly selected as records. The statistical data on the progress of the user's courses are compiled in the selected data set. Process basic user information and extract features. As part of the model input, a total of 10 types of user attributes were extracted according to the user's gender and occupation, including Java development engineers, PHP development engineers, and test engineers.

The MOOC platform selects 5 categories from the course information according to the course classification, including JavaScript, Python, Android, etc. Other experimental standard settings, the learning rate is 0.001, the number of iterations is 30, the batch size is 198, and the activation function of the output layer is Relu.

3.3 Benchmark Experiment Comparison

Experimental evaluation standard: Combine the characteristics of online resource learning to evaluate the performance of the model proposed in this article. The experiment in this chapter adopts the omission evaluation method, that is, for each user, the last learning course is used as a test set. Make predictions, and use courses other than the previous course as the learning set, that is, predict the user's learning situation in the course month through n-1 records of the previous course. In this article, two valuation indicators are selected, RMSE and MAE. This article mainly compares three commonly used reference models PMF, SVD and SVD++.

4 Experimental Results and Analysis

4.1 Experimental Results of the Data Set on Different Models

For the two evaluation indicators RMSE and MAE, the NeuCNN model combined with auxiliary information proposed in this paper has higher performance. The details are shown in Table 1:

Table 1. Experimental Results of the Dataset on Different Models

	RMSE	MAE
PMF	1.3198	1.1056
SVD	1.3105	0.9872
SVD++	1.3209	1.1098
NeuCNN	1.2836	0.9765

As shown in Fig. 2, this is a plot of the experimental results of different models. It can be seen that the NeuCNN model outperforms PMF, SVD and SVD++ under the two evaluation indicators of RMSE and MAE. The performance of the learning prediction model based on the convolutional neural network is better than other models of the reference system.

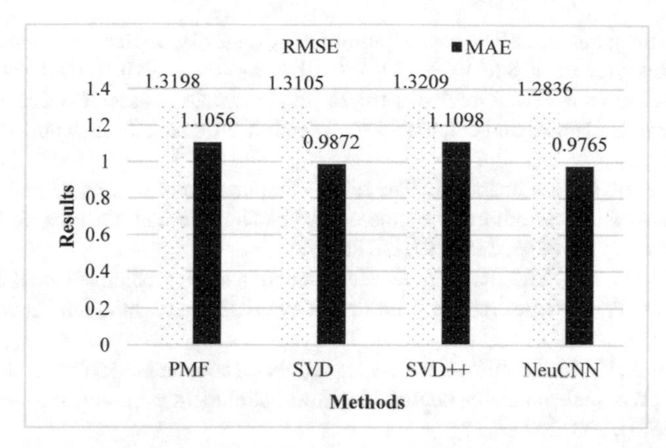

Fig. 2. Experimental Results of the Dataset on Different Models

5 Conclusion

According to the learning trends in recent years, we have found that the combination of machine learning and artificial neural networks will replace expert systems for a long period of time in the future. Recommendation model has been widely used in various occasions as a very practical and convenient tool to obtain information effectively. The dynamic course recommendation system (tag will be) is a more mature network learning algorithm in the future. The experiment designed in this paper tells us the importance of the NeuCNN model in the course recommendation system. The design of a dynamic course recommendation system based on machine learning is reasonable.

References

1. Mhawish, M.Y., Gupta, M.: Software metrics and tree-based machine learning algorithms for distinguishing and detecting similar structure design patterns. SN Appl. Sci. **2**(1), 11 (2020)
2. Abdo, A., Pupin, M.: Turbo prediction: a new approach for bioactivity prediction. J. Comput. Aided Mol. Des. **36**(1), 77–85 (2022)
3. Elkholy, M.M.: Steady state and dynamic performance of self-excited induction generator using facts controller and teaching learning-based optimization algorithm. COMPEL Int. J. Comput. Mathe. Electr. **37**(2), 00 (2017)
4. Radanliev, P., Roure, D.D., Page, K., et al.: Design of a dynamic and self-adapting system, supported with artificial intelligence, machine learning and real-time intelligence for predictive cyber risk analytics in extreme environments – cyber risk in the colonisation of mars. Saf. Extreme Environ. **2**(3), 219–230 (2020)
5. Alzubi, O.A., Alzubi, J.A., Alweshah, M., et al.: An optimal pruning algorithm of classifier ensembles: dynamic programming approach. Neural Comput. Appl. **32**(5), 1–17 (2020)
6. Buatois, S., Ueckert, S., Frey, N., Retout, S., Mentré, F.: Comparison of model averaging and model selection in dose finding trials analyzed by nonlinear mixed effect models. AAPS J. **20**(3), 1–9 (2018). https://doi.org/10.1208/s12248-018-0205-x
7. Tanaka, A., To, J., O'Brien, B., et al.: Selection of reliable reference genes for the normalisation of gene expression levels following time course LPS stimulation of murine bone marrow derived macrophages. BMC Immunol. **18**(1), 43 (2017)

8. Raffin, N., Seegmuller, T.: The cost of pollution on longevity, welfare and economic stability. Environ. Resource Econ. **68**(3), 683–704 (2016). https://doi.org/10.1007/s10640-016-0041-3
9. Sarwar, S., Qayyum, Z.U., García-Castro, R., et al.: Ontology based E-learning framework: a personalized, adaptive and context aware model. Multimedia Tools Appl. **78**(24), 34745–34771 (2019)
10. Guest, D., Kent, C., Adelman, J.S.: The relative importance of perceptual and memory sampling processes in determining the time course of absolute identification. J. Exp. Psychol. Learn. Mem. Cogn. **44**(4), 615–630 (2018)
11. Park, Y., Cheon, J.H., Yi, L.P., et al.: Development of a novel predictive model for the clinical course of crohn's disease: results from the CONNECT study. Inflamm. Bowel Dis. **23**(7), 1071 (2017)
12. Kyllingsbk, S., Bo, M., Bundesen, C.: Testing a poisson counter model for visual identification of briefly presented, mutually confusable single stimuli in pure accuracy tasks. J. Math. Psychol. **38**(3), 628–642 (2017)

English Differential Classification Teaching Based on Naive Bayes

Qi Zhang(✉)

English Department, Heilongjiang International University, Harbin, Heilongjiang, China
tomorrow202112@163.com

Abstract. In the new round of English curriculum reform, how to promote the individualized development of students has been pushed to the forefront of the reform of English teaching, and the essence of promoting the individualized development of students is the differentiated development of students. The rapid development of information technology and the popularization of big data have promoted the application of classification algorithms, especially the naive Bayes algorithm, in the teaching of English differentiated classification. Based on this, this article mainly studies the English differentiated classification teaching based on Naive Bayes. This article first investigates the current situation of the traditional English curriculum teaching mode, and analyzes the main problems that currently exist. This article proposes a student classification method based on Naive Bayes, and puts forward strategies and suggestions for the teaching of English differentiated classification, in order to provide a reference for the innovation of English teaching mode. The survey found that in teachers' understanding of differentiated teaching, 23.26% of teachers believed that "classification according to academic performance" accounted for 23.26%; 25.28% believed that "treat students differently in teaching"; 37.21% Believes that "respect the individual differences of students and carry out differentiated teaching activities". This shows that there are still some English teachers who do not have a deep understanding of differentiated classification teaching.

Keywords: Differential Teaching · Naive Bayes Algorithm · Teaching Strategy · Student Classification

1 Introduction

In the context of the continuous advancement of the new round of curriculum reform, the traditional teaching mode has undergone an essential change, from the original teacher-oriented to student-oriented [1, 2]. As independent individuals, students have differences in the needs and abilities of English learning. In order to allow students in each category to gain something, differentiated classification teaching has received attention as a new teaching strategy [3, 4]. Naive Bayes classification method, as a kind of data mining technology, has a very good effect in data classification because of its simple structure and fast data processing [5]. Therefore, the Naive Bayesian method is widely used in the teaching of English differentiated classification.

J. H. Abawajy et al. (Eds.): ICATCI 2022, LNDECT 169, pp. 613–620, 2023.
https://doi.org/10.1007/978-3-031-28893-7_73

Regarding the research of English teaching, many scholars have conducted multi-angle investigations. For example, Arulmozhi P developed a new adaptive linear regression classification method based on the learning of grade students using RFID technology [6]; Wei L research The application of cloud computing and speech recognition technology in English teaching [7]; Rao Z took East Asian primary school students as the survey object, and studied the challenges and future prospects of English as a foreign language teaching [8]. It can be seen that the research on English teaching has always attracted much attention. This article applies Naive Bayes to the classification of students, and then proposes a strategy for differentiated classification of English teaching, which has innovative and practical significance.

This article first analyzes the status quo of English differentiated classification teaching, taking students and teachers of H University Business English class as the survey object, reveals the problems of traditional English teaching mode and explores the teacher's understanding of differentiated classification teachers. Then, this article uses the Naive Bayes algorithm to classify students, and proposes strategies and suggestions for teaching English differentiated classification.

2 Investigation and Design of the Status Quo of English Differentiated Classification Teaching

2.1 Survey Design and Research Questions

This research aims to explore the status quo of differentiated classification teaching in college English audio-visual courses, understand the problems of traditional English teaching models, and propose corresponding teaching methods and strategies. This research mainly analyzes the data on the unreasonableness of the fixed teaching system and the teachers' understanding of differentiated teaching.

2.2 Research Objects

The survey subjects selected for this survey are sophomores in the business English class of H University, a total of 204 people. Because English teachers have just one year of understanding and understanding of sophomore students, it is universal to conduct a survey of the application status of differentiated classification teaching on this basis.

2.3 Research Tools

In this survey, questionnaire surveys and interviews will be adopted to explore the current status of the application of differentiated classification teaching in college English audio-visual courses. This questionnaire consists of two parts. The first part is the basic personal information of the teacher/student, including gender, age, and grade (professor class). The second part is the basic situation of college English teaching, including the problems of the current English curriculum model, the status quo of differentiated classification teaching, and teachers' understanding and suggestions on differentiated classification teaching.

2.4 Questionnaire Collection

The questionnaire survey issued 204 student questionnaires, excluding incomplete questionnaires, a total of 203 valid questionnaires were returned; the survey was also oriented to teachers, and a total of 43 valid questionnaires from teachers were issued and recovered. In order to ensure the authenticity of the answers to the questionnaire as much as possible, the questionnaire adopts an anonymous method.

2.5 Interview Method

After the investigation, this study randomly selected 16 students (12 students and 4 teachers) from students in different classes for interviews, so as to have a more comprehensive and objective understanding of the importance of differentiated classification teaching and teaching strategies. In order to ensure the effectiveness of face-to-face interviews as much as possible, face-to-face interviews are conducted in a one-to-one manner.

3 Survey Results of the Current Situation of English Differentiated Classification Teaching

3.1 Irrationality of the Fixed Teaching System

Table 1. Irrationality of the fixed teaching system

Options	Frequency	Proportion(%)
A. All students develop according to the uniform standard of the class	141	69.46
B. Student style and teacher style may not be appropriate	73	35.96
C. Lack of opportunities to learn and communicate with other teachers	98	48.28
D. Affect students' interest in learning	15	7.34
E. Other	20	9.85

From the perspective of students, the results of the questionnaire on the "unreasonable aspects of the fixed teaching system" are shown in Table 1: 141 people think that "all students develop uniformly in the class", accounting for 69.46%; 98 people think that "the lack of opportunities to learn and communicate with other teachers", accounting for 48.28%; 73 people think that "student style and teacher style are not necessarily appropriate", accounting for 35.96%; 15 people think "affecting students' interest in learning", accounting for 7.34%.

It can be found from Fig. 1 that the main irrationality of the fixed teaching system is the inability of students to carry out personalized development and the lack of opportunities to communicate with other teachers. At present, an English audiovisual and listening

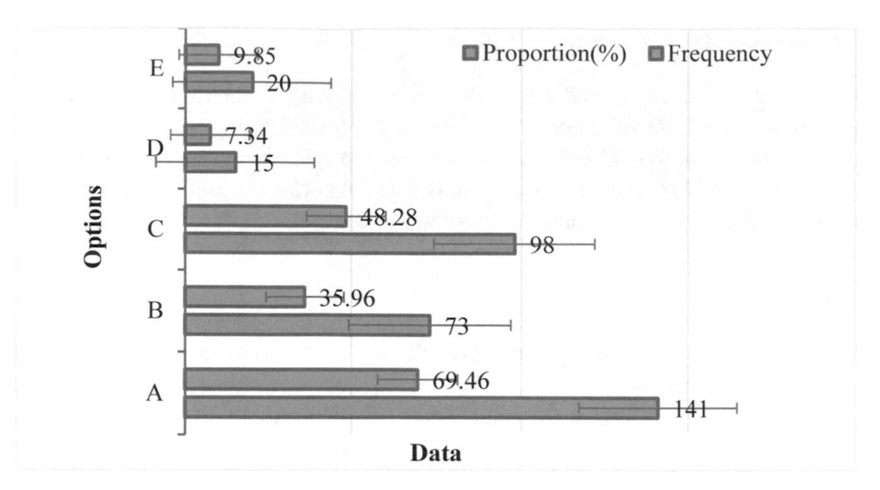

Fig. 1. Irrationality of the fixed teaching system

teacher needs to be responsible for the English curriculum of the entire class, so it is easy to regard all students in a class as a whole, ignoring the differences of students. But in fact, even in the same class, students' knowledge level, learning interest, personality performance, etc. are different. If a unified evaluation content and standard is adopted for all students, it will affect the improvement of students' academic performance in the light of the individual development of students has an adverse effect.

3.2 Teachers' Understanding of Differentiated Classification Teaching

Table 2. Survey results of teachers' understanding of differential teaching

Options	Frequency	Proportion(%)
A. Classes are divided according to the level of academic performance	10	23.26
B. Arrange different learning content for different students	6	13.95
C. Treat students differently in teaching	11	25.58
D. Respect the individual differences of students and carry out different teaching activities	16	37.21

From the perspective of teachers, a total of 43 English teaching teachers' questionnaires were collected. Among them, teachers' understanding of differential teaching is shown in Table 2: 10 people think that "classification is based on the level of academic performance", accounting for The ratio is 23.26%; 11 people think that "treat students differently in teaching", accounting for 25.58%; 6 people think that "arrange different learning content for students with different learning content," accounting for 13.95%;

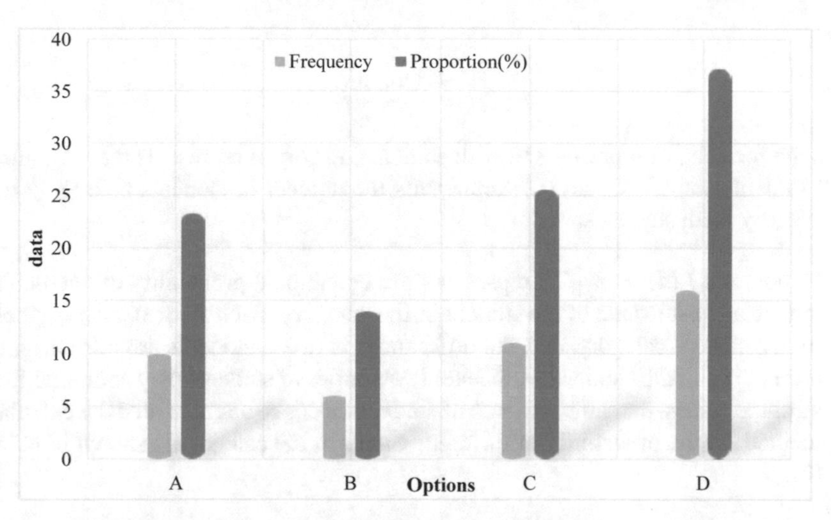

Fig. 2. Survey results of teachers' understanding of differential teaching

16 people think "Respect the individual differences of students and carry out different teaching activities", accounting for 37.21%.

It can be seen from Fig. 2 that teachers have different understandings of differentiated classification teaching. There are still many teachers who think that classifying or arranging different learning content based on academic performance is differentiated classification teaching. It can be seen that English teachers' understanding of differentiated classification teaching is not deep enough, and they lack higher-level differentiated classification teaching concepts.

4 English Differentiated Classification Teaching Based on Naive Bayes

4.1 Student Classification Method Based on Naive Bayes Algorithm

The core of the student classification method based on the Naive Bayes algorithm is to classify students according to their attribute information. Students in the same category are not simply judged by their grades, but combined with other attributes, and then integrated Judgment achieves the purpose of classification [9, 10]. Therefore, when using the Naive Bayes algorithm to classify students, first calculate the prior probability (category probability) of the collected student data. The calculation method is as follows:

(1) Hypothesis $P(C_i)$ represents the frequency of occurrence of student category C_i in the training sample set, that is, the category probability. For the student classification method of the naive Bayes algorithm, the probability of each category represents the proportion of the number of occurrences in the training sample set to the number of the entire sample. Among them, the calculation method of the frequency of the i-th students in all sample sets is shown in formula (1).

$$P(C_i) = \frac{\text{Count }(C_i)}{n} \tag{1}$$

In the formula, C_i represents the i-th student category. n represents the total number of all student samples. Count (C_i) represents the number of students belonging to the i-th category in all student samples.

(2) Hypothesis $P(A_j = a_j|C_i)$ represents the conditional probability of each characteristic attribute value of the student in the category. Each student can only belong to one student category, and the differences between students are attributes, each attribute has a different value, different categories of students may contain different attribute values, the same attribute of students needs to be calculated the calculation method for the probability of different values in the category is shown in formula (2).

$$P(A_j = a_j|C_i) = \frac{\text{Count }(A_j = a_j)}{\text{Count }(C_i)} \tag{2}$$

In the formula, A_j represents the j-th attribute. $A_j = a_j$ indicates that the value of the j-th attribute is a_j. Count $(A_j = a_j)$ represents the number of students whose attribute name A_j value is a_j in the i-th student category.

(3) Assumption $P(C_i|X_k)$ represents the conditional probability that student X_k to be classified belongs to the i-th category. The calculation formula is based on the Bayes theorem formula, as shown in formula (3).

$$P(C_i|X_k) = \frac{P(X_k|C_i)P(C_i)}{P(X_k)} \tag{3}$$

In the formula, $P(C_i)$ represents the occurrence probability of the i-th student category in the training sample set, $P(X_k|C_i)$ represents the conditional probability of the student to be classified X_k in the i-th student category, and $P(X_k)$ represents the probability of the student to be classified.

(4) The category that the student to be classified belongs to takes the maximum probability of each category, which is $P(C_{max}|X_k)$, which means that the student X_k to be classified belongs to the largest category probability in the student category. First, the probability that the student X_k to be classified belongs to each category is calculated. Select the category C_{max} with the largest value from it, which is the category to which the student to be classified belongs, as shown in formula (4).

$$P(C_{max}|X_k) = max\{P(C_1|X_k), P(C_2|X_k)\ldots, P(C_i|X_k)\} \tag{4}$$

4.2 English Differentiated Classification Teaching Strategy Based on Naive Bayes

(1) Implement difference classification strategies in English teaching and pay attention to individual differences among students

When preparing for English teaching, teachers need to pay attention to three different performances of students, which are differences in learning styles, differences in multiple intelligences, and differences in personality [11, 12]. Then, teachers should use different differentiated teaching strategies according to the characteristics of different students. Because students are a group that integrates thinking and emotions, their learning will be affected by all aspects, and students' differential performance will directly affect students' internalization of teaching content. The quality of teaching effect mainly depends on the degree of students' internalization of teaching content. Therefore, when preparing lessons, teachers need to fully understand the concept of differentiated classification teaching and prepare differentiated teaching strategies based on students' different performances, so as to effectively improve the quality of English teaching and improve teaching results.

(2) Strengthen the connection between teachers and teaching seminars

Strengthening the connection between teachers and teachers will help teachers improve their teaching literacy. Communication with teachers can enrich the teaching style of teachers. The new teacher team needs to strengthen the grasp of the course knowledge and class time in the English classroom, and the old teacher team needs to strengthen the use of multimedia and diversified teaching aids to enrich the English classroom. In particular, the subject of English has its own characteristics, which requires teachers to innovate in teaching resources and teaching models. The English course research team can promote communication and reflection among teachers by organizing teachers to prepare lessons collectively, to listen to each other's lessons, to hold regular English teaching seminars, and to organize English teaching seminars. The improvement of differentiated classification teaching ability requires teachers to continuously reflect on the basis of in-depth understanding of the differentiated classification teaching concept, verify it in the teaching time, and continue to cycle, so as to realize the independent development and teaching innovation of teachers' differentiated classification teaching ability.

(3) Change the mode of teaching classes to improve the overall level of teachers

The implementation of the concept of differential teaching has very high requirements for the comprehensive ability and professional quality of teachers. Therefore, schools should increase the theoretical training and practical skills training for the differentiated classification teaching of English teachers. On the one hand, they should train teachers to have a differentiated teaching concept, guide them to implement the differentiated classification teaching concept in their teaching, and implement differentiation for teachers. The effect of classified teaching is tracked and evaluated. In addition to the guidance of school leaders on differentiated teaching, it is also possible to rationalize the teaching unit and improve the quality of teaching and the teaching environment by appropriately reducing the size of the class. At the same time, it is necessary to pay attention to the balance between teachers and students, reduce teachers' pressure, and increase the quality of teaching monitoring.

5 Conclusion

Each student is an independent individual, and English differentiated classification teaching advocates people-oriented. It needs to be based on the differences and diversity of students and provide students with different teaching strategies. Through research, this article has completed the following tasks: using questionnaire surveys and interviews to investigate, it is found that teachers have insufficient understanding of differentiated English teaching, and traditional teaching models cannot promote students' individualized development; using Naive Bayes algorithm to achieve students' development Individualized classification promotes differentiated classification teaching of English; differentiated classification teaching of English requires teachers to implement different strategies in English, pay attention to individual differences of students, strengthen the connection and teaching research between teachers, improve teacher quality, change the teaching model, and improve the overall teacher level.

References

1. Fangfang, X., Chun, Z.: Critical Thinking-Oriented College Oral English Teaching Reform. American Scholars Press Inc, International Forum of Teaching and Studies (2020)
2. Zhang, F.: Quality-improving strategies of college English teaching based on microlesson and flipped classroom. Engl. Lang. Teach. **10**(5), 243 (2017)
3. Wolf, S.D., Smit, N., Lowie, W.: Influences of early English language teaching on oral fluency. ELT J. **71**(3), 341–353 (2017)
4. Brian, P.: World Englishes in English language teaching global Englishes for language teaching. ELT J. **3**, 3 (2020)
5. An, J., Macaro, E., Childs, A.: Classroom interaction in EMI high schools: do teachers who are native speakers of English make a difference? System **98**(2), 102482 (2021)
6. Arulmozhi, P., Hemavathi, N., Rayappan, J.B.B., Raj, P.: ALRC: a novel adaptive linear regression based classification for grade based student learning using radio frequency identification. Wireless Pers. Commun. **112**(4), 2091–2107 (2020). https://doi.org/10.1007/s11277-020-07141-4
7. Wei, L.: Study on the application of cloud computing and speech recognition technology in English teaching. Clust. Comput. **22**(4), 9241–9249 (2018). https://doi.org/10.1007/s10586-018-2115-1
8. Rao, Z., Yu, P.: Teaching English as a foreign language to primary school students in East Asia Challenges and future prospects. English Today **35**(3), 16–21 (2019)
9. Ho, W., Tai, K.: Doing expertise multilingually and multimodally in online English teaching videos. System **94**(3), 102340 (2020)
10. Shirakawa, M., Nakayama, K., Hara, T., et al.: Wikipedia-based semantic similarity measurements for noisy short texts using extended naive bayes. IEEE Trans. Emerg. Top. Comput. **3**(2), 205–219 (2017)
11. Jeremy, H.: The routledge handbook of English language teaching. ELT J. **3**, 3 (2019)
12. Jennifer, J.: International perspectives on english as a lingua franca: pedagogical insights new frontiers in teaching and learning english. ELT J. **1**, 99–104 (2017)

Applications and Case Study Session

Motivation Analysis and Prediction of Pension Institutions Based on CART Regression Prediction Tree Model

Jie Dong(✉), Xilin Zhang, and Said Muse Abdullahi

Shenyang Jianzhu University, Shenyang, Liaoning, China
dong_jie_28@163.com

Abstract. Under the background of China's aging population, the development of pension institutions has become an inevitable trend. In the 1970s, a large number of couples actively responded to the "family planning" under the influence of national conditions, so in the past two decades, the old-age care problem will become more and more intense. In this paper, the decision tree is used to analyze the trend and rules of institutional endowment choice for the aged over 35 years old with different personal characteristics. Questionnaire survey was conducted among the elderly by releasing questionnaires on the Internet, and a regression prediction model of elderly pension mode selection was constructed by using decision tree CART algorithm. The results show that the factors influencing the choice of institutional endowment mode mainly depend on the health status, age, income and number of children of the elderly.

Keywords: Institutional Endowment · Regression Prediction · CART Algorithm

Everyone will be old, how to spend our old age is the problem that each of us will face. China's elderly population reached 264 million in 2021 and is expected to keep growing in the next decade. With the development of urbanization, many children who work outside or settle down in other places have no way to take care of the elderly. Even if they live together in the same city with their children, conflicts and frictions will also affect family affection due to the big differences in ideology and living habits, and the traditional home-based care model is undergoing severe challenges. Facing the reality, in order to make their old age happy and their children happy, now more and more old people live in nursing homes. At the same time, many people do not understand the type of pension institutions, what kind of pension institutions are suitable for them, and what are the reasons for the old people to choose pension institutions [1]. This paper classifies and predicts the motivation of institutional endowment by using typical artificial intelligence algorithm, and determines the general law behind it, providing theoretical basis for the selection and development of institutional endowment.

J. H. Abawajy et al. (Eds.): ICATCI 2022, LNDECT 169, pp. 623–629, 2023.
https://doi.org/10.1007/978-3-031-28893-7_74

1 The Research Methods

1.1 CART Algorithm

CART is short for classification and regression tree, and the end result is a binary tree that can be used for classification as well as regression problems. The output of the classification tree is the category of the sample, and the output of the regression tree is a real number. The nodes are established from the root from top to bottom, and the best attribute should be selected at each node to split, so that the training set in the child nodes can achieve the highest purity [2].

In order to more clearly sort out the rules behind the elderly's choice of institutional pension, such as the influence of the elderly's age, health status, income status, marital status and other personal attributes on the elderly's choice of pension, this paper adopts the decision tree analysis method. Decision tree algorithm is the most commonly used inductive learning algorithm when it comes to classification problems. It can deduce a more intuitive decision tree from a group of irregular data samples to show classification rules from top to bottom. Depending on the problem, the basic purpose of classification research can be to produce an accurate classifier or to reveal the predictive structure of the problem. Classification And Regression Decision Tree (CART) is a typical binary decision Tree that can be used for both Classification And Regression. When the predicted data is discrete data, Classification and Regression Tree is used to deal with complex problems with unclear relationships between variables without assumptions on samples. The regression decision tree is used when the predicted data is continuous data. In this paper, the behavior of the elderly choosing the pension institution is analyzed by decision tree. The output value of the decision tree is the reason of the different categories mentioned above. Because of the discreteness of the data, the classified decision tree is adopted.

When CART is a classification tree, the attribute selection metric adopted is the attribute with minimum GINI and its attribute value as the basis for the optimal splitting attribute and the optimal splitting attribute value of node splitting. GINI can be selected as the purity index. The smaller GINI value is, the higher the purity of subsamples after dichotomy. The better the effect of selecting this attribute as the splitting attribute is, until each sample set is pure after division, the modeling of classification decision tree will stop [3].

1.2 Data Acquisition

The questionnaire was designed based on Anderson model's three factors. Non-scale questions were divided into single choice and multiple choice questions. The questions involved included descriptive statistics, influencing factors, and the willingness of pension institutions. In order to make the questionnaire more reliable, questionnaire investigators try to achieve a 1:1 ratio of men and women. A total of 1165 valid questionnaires were collected in this survey.

1.3 Data Statistical Method

In this paper, the likelihood ratio Chi-square test method is used to measure the importance of variables before constructing the prediction model [4]. The mathematical definition of likelihood ratio chi-square is as formula 1:

$$T = 2\sum_{i=1}^{r}\sum_{j=1}^{c} f_{ij}^{0} \ln \frac{f_{ij}^{0}}{f_{ij}^{e}} \tag{1}$$

where r represents the number of rows, C represents the number of columns, f_{ij}^0 represents the measured frequency of cell (i, j), and f_{ij}^e represents the expected frequency of cell (i, j). This paper uses cle-mentine to analyze the importance of input variables. Missing values of more than 60% of the classification variables are considered as unimportant variables, and their coefficient of variation is less than 0.1. The importance of this paper for influencing the motivation of the elderly to choose institutional endowment is shown in the following Table 1:

Table 1. Degree of importance of model variables

Variable	The importance
Degree of self-care	0.48
Age	0.21
Disposable income	0.16
The number of children	0.15

2 Model Building and Data Analysis

2.1 The CART Model

In this study, the CART decision tree model was selected to construct the root node of influencing factors affecting the elderly's choice of pension institutions, and the optimal data segmentation features were selected based on binary segmentation method and Gini index. Then, the classification was completed by dividing the data layer by layer through indicators until all the features were clearly described or the data set had only one dimension. Gini index, as the criterion for selecting classification attributes, refers to that in the segmentation node, if A value of A feature is selected to divide dataset D into two parts, D1 and D2 [5], then the Gi-Ni index of dataset D under A feature is expressed as formula 2:

$$Gini(D, A) = \frac{D_1}{D} Gini(D_1) + \frac{D_2}{D} Gini(D_2) \tag{2}$$

The questionnaire data will first select the best splitting attribute according to Gini index, and then perform recursive splitting for corresponding nodes until all nodes contain the same level of accident representation information. After the decision tree is generated, its specific form will be from the root node to each leaf node, and can formally correspond to the decision rules expressed as "if-then" [6].

2.2 Classification Index Evaluation

This paper uses the characteristic curve (ROC) curve of pension institution as a comprehensive evaluation index. The area of the ROC curve and the connecting segment of the two endpoints is AUC. The larger THE AUC value is, the higher the prediction accuracy of the classification model is. Each point on the ROC curve represents the feeling degree of a threshold, which is distributed from 0 to 1. The threshold at [0,0] corresponds to 1, and at [1,1] corresponds to 0. Values greater than the threshold point are classified as positive cases, while values less than the threshold point are classified as negative cases [7]. On a ROC curve, the closer it is to the curve point [0,1], the better its threshold division is.

3 Classification Prediction and Result Analysis

3.1 Attribute Assignment

The data should be preprocessed before the construction of the decision tree, and the individual category attributes of the elderly (gender, age, etc.) and the category attributes of family relatives (family pattern, number of elderly children, etc.) in the questionnaire of the elderly should be assigned values.

3.2 Attribute Classification

Ai (I = 1,2...... 10) Encode attribute categories as decision tree node names. At the same time, institutional endowment reasons are divided into three types: leisure endowment, transfer endowment and professional endowment, and they are numbered as the output of the decision tree. The classification numbers are as following Table 2:

Table 2. Select the institutional endowment reasons classification number table.

Serial number	Choice tendency	Reason
1	Leisure retirement	"More peers in the institution", "Good service quality and environment" and "Different living habits"
2	Transfer pension	"Sick, need professional care", "family is too busy to take care of", "afraid of troublesome children", "poor physical condition of family members"
3	Professional pension	"Sick, need professional care", "family busy with work, no time to take care of"

Leisure pension, need more companions, eliminate the loneliness of unaccompanied, prefer a diversified life, enrich the balance of other factors in their spiritual world, pay attention to the health of their diet, numbered 1; Transfer pension: In addition to reducing the burden of children and environmental factors, the elderly also consider the needs of professional nursing services and companionship after considering their own conditions and family members' conditions, no. 2; Professional pension, mainly to reduce the burden of children, dedicated care services, rich entertainment and companionship, the elderly need professional staff for all-weather care, number 3 [8].

3.3 Classification Prediction

The total sample data to be used is divided into two parts, one part as the training number set, the other part as the test number set. Test sets are used to verify the accuracy of decision genera constructed by training sets. The training set samples were imported into Matlab, and the CART classification decision tree operation program was run to construct the decision tree [9]. The selection results of partial model variables are analyzed according to the classification and prediction results of decision tree, as shown in Table 3 below:

Table 3. Model variable selection results

Serial number	The degree of care	Age	Children	Income	The decision results
1	Can	≤75	Yes	>4000	Leisure retirement
2	Basic can	>75	Yes	>4000	Leisure retirement
3	Basic can	>75	No	≤4000	Professional pension
4	Basic can	≤75	Yes	≤4000	Transfer pension

The test set was used to calculate the prediction accuracy of the model, and the prediction accuracy reached 0.731266, indicating that the prediction model has a high accuracy, indicating that the elderly's personal attributes and family attributes have a relatively clear impact path on the reasons for the elderly to choose institutional endowment.

3.4 Prediction Result Judgment

According to the Fig. 1 below if we can analyze the ROC curve for some point coordinates (0, 1), represents classification quite right, and named the perfect classifier, at the same time, the ROC curve is divided into two areas, the space in the line of point represents a good classification results, and below the line point represents the poor classification results [10]. Retrieved from The SPSSAU Project (2021).

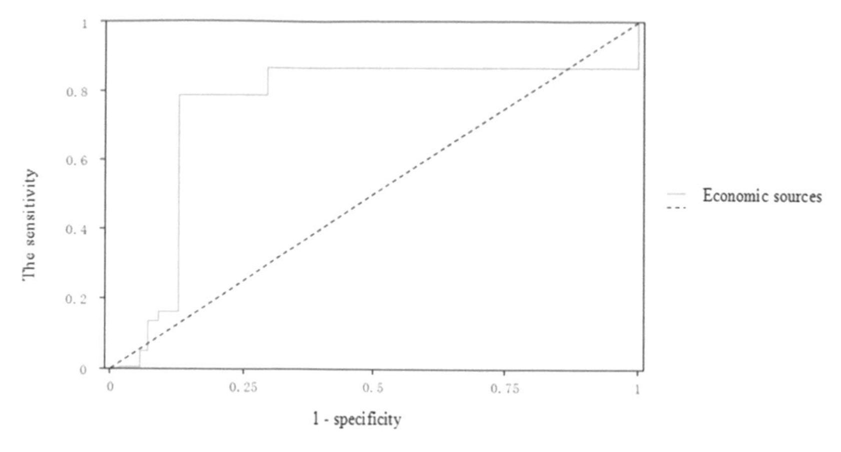

Fig. 1. ROC curve

The generated ROC curve shows that the economic source is the corresponding AUC value of 0.820, which has a high diagnostic value for institutional endowment selection, and the corresponding optimal boundary value is 0.659(sensitivity is 0.786 and specificity is 0.874 at this time).

4 Conclusion

At present, our country's endowment institution development is still in its beginning stage, most of them lack of awareness of and pension mechanism, and related industrial chain pension policy in China has not yet fully formed, on different groups for different types of pension institution or the endowment way is not a clear path, based on the CART back to the decision tree model was effective prediction, By the decision tree above we can see the elderly health status, age, income, children is the key to affect older people choose institutions endowment node number or key factors, and through the investigation material inductive reasoning out the old institution endowment way of decision tree in different state general rules to choose different way of pension. In addition, the likelihood ratio Chi-square test is used in this paper to measure the importance of variables. Compared with the traditional test, the difference is that the calculation formula is different, and it has a larger range and higher accuracy, and has greater advantages in dealing with multiple influencing factors, thus greatly improving the reliability of the analysis results in this paper.

References

1. Elgart, A., Klein, A.: Eigensystem multiscale analysis for the anderson model via the wegner estimate. Ann. Henri Poincaré **21**, 2301–2326 (2020)
2. Chae, Y.M., Ho, S.H., Cho, K.W., et al.: Data mining approach to policy analysis in a health insurance domain. Int. J. Med. Inform. **62**(2), 103–111 (2001)

3. Trendowicz, A., Jeffery, R.: Classification and regression trees. Softw. Project Effort Estim. 295–304 (2014)
4. Malik, M., Nehra, A.K., Saini, B.K.: A study on factors affecting job satisfaction of working women with Karl Pearson's Chi-Square test. Res. J. Human. Soc. Sci. **12**(2) (2021)
5. Maesya, A., Hendiyanti, T.: Forecasting student graduation with classification and regression tree (CART) algorithm. In: IOP Conference Series: Materials Science and Engineering, vol. 621 (2019)
6. Abdi, F., Khalili-Damghani, K., Abolmakarem, S.: Solving customer insurance coverage sales plan problem using a multi-stage data mining approach. Kybernetes **47**(1), 2–19 (2018)
7. Norton, M., Uryasev, S.: Maximization of AUC and buffered AUC in binary classification. Math. Program. **174**(1–2), 575–612 (2019)
8. Ermarth, A., Bryce, M., Woodward, S., et al.: Identification of pediatric patients with celiac disease based on serology and a classification and regression tree analysis. Clin. Gastroenterol. Hepatol. **15**(3), 396–402 (2017)
9. Carrington Andre, M., et al.: Deep ROC analysis and AUC as balanced average accuracy, for improved classifier selection, audit and explanation. IEEE Trans. Pattern Anal. Mach. Intell. **45**, 329–341 (2022)
10. Rao, G.: What is an ROC curve? J. Fam. Pract. **52**(9), 695 (2003)

The Integration of Artificial Intelligence Technology and Digital Media Art

Jie Liu[1](✉) and Neha Jain[2]

[1] Engineering and Technology College, Hubei University of Technology, Wuhan, Hubei, China
3356327828@qq.com
[2] Chitkara University, Punjab, India

Abstract. Given the progress of artificial intelligence technology, computer has been more popular. More, the digital media art and artificial intelligence technology have fused well to help modern cultures blossom with greater charm. So, the core of blending between AI techniques and digital media art will be discussed with the basis of the era of AI. More, current state of development of digital media art and technology will be analysed and finally innovative directions and future trends will be proposed.

Keywords: Artificial Intelligence · Digital Media Arts · Creative Development

1 Brief Description

Unlike conventional artistic creation, digital art and media is an interdisciplinary subject of information technology and art creation by means of computer communication technology. With the speed and stability of the Internet & computer communication technology, the application of digital media art has brought strong stimulation to people's vision and senses, and has gradually been regarded as a new trend of modern art expression. Under the background or context of this age for artificial intelligence, or the AI for short, the perfect integration of technological innovation & digital multimedia art is now one of the central industries, and which is bound to help China's development better and faster in the social and economic aspect [1].

2 Overview of the Artificial Intelligence

Artificial Intelligence (AI for short) Technologies has been employed in wide ranges of people's life, which also has produced great community value. To some extent, the technology of artificial intelligence (AI) has comprehensively altered of changing lifestyles of people and daily production mode of enterprises, and provided great convenience for the production of enterprises and helpful for people's life, learning and working. At present, artificial intelligence is widely used in physiology, linguistics and basic computer science. Artificial intelligence's great computing power is designed to is used to provide high-quality services for relevant industries. Artificial intelligence focuses on

J. H. Abawajy et al. (Eds.): ICATCI 2022, LNDECT 169, pp. 630–635, 2023.
https://doi.org/10.1007/978-3-031-28893-7_75

improving the intelligence degree of mechanical equipment, and then replace human beings to complete some complex work.

Artificial intelligence is far from computer science, but also involves natural science, social science and other fields, with neurophysiology, philosophy, art, mathematics and so on. The demand for artificial intelligence will be higher and wider in the future. However, artificial intelligence also has its uncertainties [2]. If intelligent robots exceed human intelligence and have autonomous consciousness in the future, it is very likely to cause security risks against human.

3 The Development Status of Technology and Art of Digital Media

3.1 Lack of Advanced Ideas When Creating

Digital media art and technology have been developing gradually, but compared with some more developed foreign countries, digital media art in China is at an immature stage in terms of innovation and ideas [3]. In an effort to produce digital media artworks in line with market demand, we should correctly and deeply understand the significance of digital media art, and also accept and understand some digital and artistic creation methods. In the times of AI, creative ideas are widely employed in a traditional way. In this case, the created masterpiece cannot be sustainable. In addition, due to the lack of innovative ideas in digital media art creation, it may make the works circulating in the market even if there are artistry, it will also cause the aggravation of the phenomenon of similarity.

3.2 Lack of Rich Understanding of Cultural Heritage

Only by properly spreading traditional culture, taking Chinese culture as the foundation of creation and avoiding blind worship of Western culture can we realize the long-term sustainable development of art and technology of digital media [4]. In an age of artificial intelligences (AI), there exists urgent need to develop digital media art and technology, and the rapid development also increases the diversification of art and technology. China is an ancient civilisation, which contains profound cultural heritage, plays a positive and effective role. Hence, the studies of digital media art and culture deposits should be combined right away. In addition, one of the realities we face is that it is difficult to integrate with the international community, so we need to take our culture as the foundation. Digital media art, full of cultural heritage, must be more profound, and can promote its own development.

3.3 Too Much Reliance on Technology Such as Computers

Digital media technology reflects the crystallization of human civilization wisdom. On the other hand, it also reflects the artistic thoughts of artists, which is the embodiment of the successful combination of art and science. However, although digital media art has made some progresses, it is still obvious that the development of technology is too dependent on the support of computer technology, which makes it difficult for art groups or individuals to truly improve their art and technology. Especially for the young artists in the social art community, some of them feel that it is very convenient to use computer technology, but also can greatly improve efficiency in time and practice [5]. As time goes by, the psychology of relying on computer technology comes into being. When a digital media artist relies on technology, it is easy to lose the new inspiration for innovation and creation, and the development of art is stalled.

4 The Integration of AI and the Art of Digital Media

At present, our global community has already reached the age filled with digital development., integration of AI technologies and digital media arts is deepening day by day, with initial results, thereby subverting the perception of what Art and Design are like upside down. New design means and artistic expression forms have come into being, while some insiders define this new digital media art as "digital intelligence art". According to a report on the Emerging Technology Maturity Curve 2017, published by a US consulting firm, AI accounted for more than 50% of 33 new technologies. In the following years, the use of artefactual intelligence technology has been widely promoted and adopted. The current AI technology is compatible with the intelligence system of human intelligence, things such as environmental perception, the storage of memory, thinking and reasoning, and the ability to learn. However, in the design world, artificial intelligence-assisted design has already been impressive. For example, "Luban", an artificial intelligence product which was introduced by Alibaba on November 11, designed its own 4×10^8 advertising banners by itself, a testimony to the powerful force of the collision between art and technology. Yet, in the replacement of basic art and design by artificial intelligence, people have also noticed the inadequacy of intelligent machines in aesthetic judgment and understanding, and the beauty of art could not be achieved by relying solely on computer technology. This time, artistic designers become "trainers" [6]. They are required to cultivate the aesthetic feeling of computers, so that computers can acquire the same aesthetic feeling and wisdom based on human beings during design and learning, so that they can formulate their own thoughts and opinions about the aesthetics of the design, and feed back into the relevant work.

Nowadays, with the emergence of "digital intelligence art", there are two different voices in the industry. Optimists believe that digital media art in the future has infinite possibilities, while pessimists see it as the ultimate end for art designers. As the global digital accelerates, digital media design in China should take the lead in embracing new knowledge and technology in our educational settings, and look objectively and rationally at the opportunities and challenges that presented by artificial intelligence and its high and new technologies. At the same time, we should always adhere to these guiding principles of integration of traditions and the moderns, close to ideal and realities, emphasis on practice and theory, reform and survival. By doing so, it is helpful to build a digital media design teaching and learning system in accordance with the contemporary contexts.

5 Core of the Technology of AI and Art of Digital Media

The fusion of artificial intelligence technologies and digital medium art is grounded on human's understanding and emotional experiences. Assuming that a single person has no idea of the creation of aesthetic thinking and consciousness, in variety of cultural backgrounds, it is not possible to resolve various design issues related to human life and work mode if only relying on calculation methodologies for intelligent machines [7]. Nowadays, the new advances in science and technologies are constantly enriching people's ways of life, improving their qualities of living and promoting people's aesthetic needs. The reason for the development of artificial intelligence technology is to provide better service to people, while art and design are integrated to happen to be based on people's emotional needs, among which art gives more emphasis to the expression of people's subjective concepts. Artificial intelligence in the future is more about thinking in terms of human needs and the ability to make changes in a different environment. From its current usage scenarios of artificial intelligence, artificial intelligence is seen to be working on some boundary issues, human will inevitably lose to the machine; However, there is also uncertainty in logical inference on the strength of artificial intelligence big data when it is needed to make decisions and judgments with sensibility. For the moment, artificial intelligence can only offer high and new technology, and the realization of brilliant works of art must be achieved by new technological methods. People emphasize that art give expression to subjective concept of human beings. Just like the saying of Mr. Qi Baishi, "if it is similar, it is common, if it is not similar, it is not", there is no limit to art. Therefore, artificial intelligence can never substitute human beings to achieve subjective concept. The core to the fusion or synthesis of artificial intelligence (AI) technology and the beaux-arts of digital media lies in the utilization of artificial intelligence technology to obtain new methods of artistic creation, for the sake of better enrich the forms of art design, the main points of the synthesis of the two depends upon people's values and cognitive competence of artificial intelligence [8].

6 The Innovative Development of Art and Technology of Digital Media in the Times of AI

To sum up, as we all know the notion of artificial intelligence, the way and core significance of the synthesis of artificial intelligence and digital media art. Under the background of artificial intelligence (AI), intelligent machinery can impulse or promote the sustaining progress & advance of society by right of their mighty computing capabilities. In the next phase, the in-depth synthesis and reformation of artificial intelligence and digital media art will directly show the following development characteristics (Fig. 1).

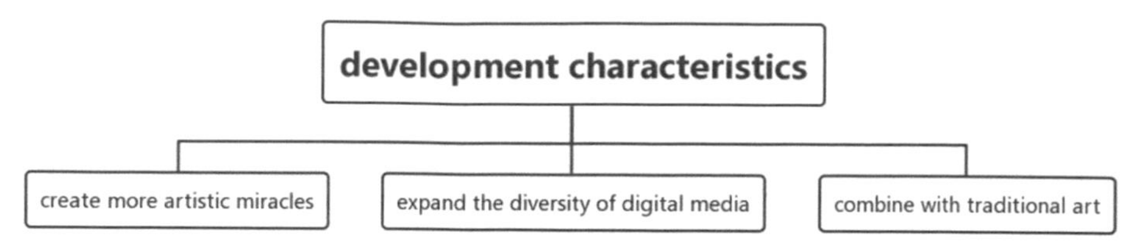

Fig. 1. Development Characteristics

6.1 In Favor of Creating More Artistic Miracles

Digital media art is characterized by openness, integration and interactivity, but also has the characteristics of artistic assessment of the situation, and changes with the change of times or means. It is ground on these particularities that make the recent digital media art more diversified. For the past few years, the fast progress of calculating machine & network technique in China has met the requirement of artistic invention, making art and artificial intelligence technology organic integration. Only by scientific and reasonable use of this integration, digital media art can make the audience get a more ideal visual experience and enhance the cognition and impression of artistic creation [9]. With all kinds of excellent and updated technologies as the creative support of digital media art, art and technology will blast out more artistic wonders.

6.2 Actively Expand the Diversity of Digital Media

The characteristics of digital media art are more diversified, but in fact the advance of digital media art and artificial intelligence (AI) technology presents obvious foible. During the period of the continuous advance of the art of digital media, China's digital media art will show more new models of art with richer and expressive forms, increase the multitudinous and diversified development of digital media, bring the audience a different consumption experience, and get a new experience and artistic perception in visual sense. Quoad hoc, the common fusion and synthesis of digital media art & artificial intelligence technology will definitely make the advance space of art and technology broader. From analysing the former progress of digital media art, the synthesis of digital media art and artificial intelligence technology in China can better present the technology and art through digital media only by constantly enhancing their own traditional culture and neoteric civilization [10]. Many people think that digital technology and computer technology is the activator for the advance of art and technology of digital media, but in nature, from another angle, it is grounded on the background of the era of artificial intelligence that the society has entered a new cycle of art and technology progress.

6.3 A More Perfect Combination with Traditional Art

With regard to the recent situation, China's digital media art is still in the primary stage of progress for a long time, facing numerous inevitable difficulties and questions that are difficult to solve. Under the background of AI, there will be many artistic and design geniuses for sustaining research and exploration, the appearance of these geniuses will promote the farther progress & improvement of digital media art and AI technology. In this course, it is peculiarly worth noting that we should not abandon our excellent cultural connotation. Only by excavating more beneficial elements of the art of digital media from traditional culturally essence can it often develop in a more significant method.

7 Conclusion

The powerful computing power of artificial intelligence can help them comprehensively explore people's needs by using historical big data, assist people handle actual problems through design, and get design thinking by using data. At the same time, there is an abundance of information in the world today, and this information encourages people to adopt updated products, technologies and digital lives in negative mode. The further study of esthetics and wisdom in digital media design order art designers to think about the changes of The Times from their own perspective and find the differences in the life of The Times.

References

1. Yang, L.: Research on application of artificial intelligence based on big data background in computer network technology. J. Jiujiang Vocat. Tech. Coll. **2**, 32–38 (2018)
2. Sousa, M.J.: Artificial intelligence: technologies, applications, and policy perspectives. In: Jeyanthi, P.M., Choudhury, T., Hack-Polay, D., Singh, T.P., Abujar, S. (eds.) Decision Intelligence Analytics and the Implementation of Strategic Business Management. EAI/Springer Innovations in Communication and Computing, pp. 68–73. Springer, Cham (2022). https://doi.org/10.1007/978-3-030-82763-2_6 (4):
3. Tsaih, R.-H.: Artificial intelligence in smart tourism: a conceptual framework. In: Proceedings of the International Conference on Electronic Business (ICEB), no. 5, pp. 100–104 (2018)
4. Papadopoulou, F.: Movement notation and digital media art in the contemporary dance practice: aspects of the making of a multimedia dance performance. In: ACM International Conference Proceeding Series, no. 7, 22–28 (2016)
5. Guzman-Serrano, R.: Where there are flies, media art you'll find: digital (Im) materiality, artistic medium, and media art decay. In: ACM International Conference Proceeding Series, no. 4, pp. 67–69 (2019)
6. Muthukumarana, S.: Jammify: interactive multi-sensory system for digital art jamming. In: Ardito, C., et al. Human-Computer Interaction – INTERACT 2021. INTERACT 2021. Lecture Notes in Computer Science, vol. 12936, no. 8, pp. 77–81. Springer, Cham (2021). https://doi.org/10.1007/978-3-030-85607-6_2
7. Ma, L.I.: Application of virtual reality technology and digital twin in digital media communication. J. Intell. Fuzzy Syst. (3), 66–69 (2021)
8. Sugiarto, E.: Virtual gallery as a media to simulate painting appreciation in art learning. J. Phys: Conf. Ser. (8), 200–205 (2019)
9. Chai-Arayalert, S.: A digital micro-game approach to improve the learning of hand-weaving art and history. Int. J. Emerg. Technol. Learn. (4), 44–48 (2021)
10. Leung, S.K., Choi, K.W., Yuen, M.: Video art as digital play for young children. Br. J. Educ. Technol. **51**(11), 88–93 (2020)

Research on Intelligent Application of Internet+ BIM Technology in Smart Site

Ting Han[1(✉)] and Sinem Alturjman[2]

[1] Fuzhou University of International Studies and Trade, Fuzhou 350202, Fujian, China
a272993345@126.com

[2] Artificial Intelligence Engineering Department, AI and Robotics Institute, Near East University, Mersin 10, Turkey

Abstract. Based on the model of "Internet+ BIM technology", Smart site can improve the construction Productivity and scientific management. This thesis mainly studied the effective method of building smart site in the context of "Internet+ BIM technology". It introduced the current situation of smart site, and the relationship between Internet, BIM technology and smart construction site. In view of the existing problems, the intelligent management mode based on Internet+ BIM technology was put forward. Applying BIM technology the concrete application of intelligent site management system was discussed.It provided a new idea for the realization of construction project management.

Keywords: Internet · BIM Technology · Smart Site

1 Research on Construction Status of Smart Site

"Smart site" refers to the use of BIM 3D design platform to carry out fine design of engineering projects and engineering construction process simulation [1]. Centering on construction process management, construction project information ecological circle of interconnection, cooperation and scientific management is constructed. Analyze and process engineering data based on Internet of Things. Provide construction forecast and treatment plan [2], realize visualization and intelligent management of construction process.

Housing urban and rural development in the several opinions about promote the development of construction and reform of the requirements, the competent department of construction and engineering companies to build construction market integration of quality and safety supervision work platform, the market and the field of subject dynamic recording project, released in a timely manner in the process of engineering construction supervision and law enforcement information, an effective social supervision mechanism [3]. In the outline of construction Industry Informatization Development from 2016 to 2020, it is specially pointed out that the digital achievement delivery system should be established and improved in terms of engineering construction informatization supervision [4].

J. H. Abawajy et al. (Eds.): ICATCI 2022, LNDECT 169, pp. 636–641, 2023.
https://doi.org/10.1007/978-3-031-28893-7_76

Smart site is a new management thinking, using data mining and modern technology to implement project management. According to the survey and research, the public's understanding of smart construction sites is shown in Fig. 1.

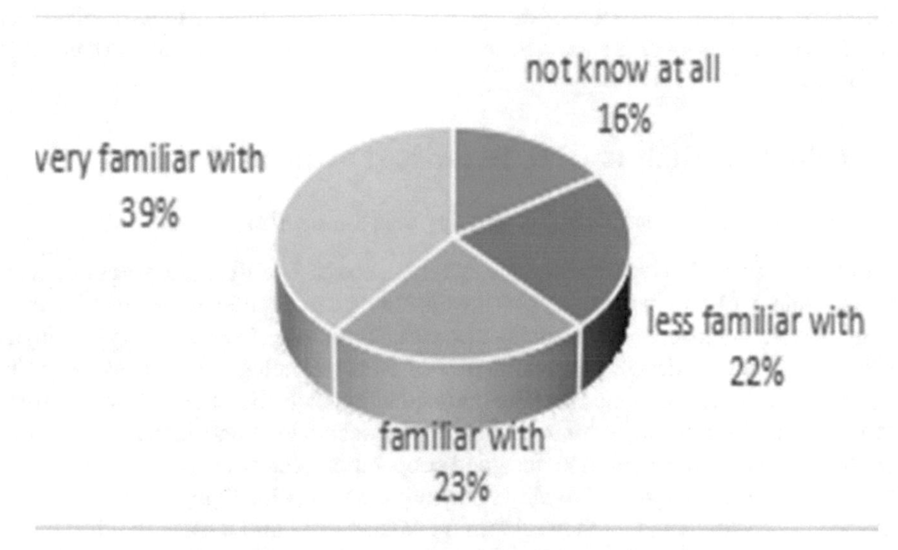

Fig. 1. Investigation and statistical results of the understanding degree of smart site

Smart site can create a special platform through the Internet to carry out information management work, and use BIM technology to carry out visualization and three-dimensional display, to promote its information development. Information technology can achieve the goal of information resource sharing, real-time dynamic monitoring of all aspects of construction, timely discovery of defects in management, so as to enhance management ability while reducing costs, shorten the time required for work. So as to achieve the goal of improving economic benefits while protecting the environment and promoting ecological construction.

2 Analysis of the Problems of "Internet+ Smart Site" in Grassroots Projects

Although the value of smart site promotion in construction enterprises is huge. However, at present, smart site construction is mainly applied in the leading promotion at the grassroots project level. There is no systematic construction from top to bottom, and there are still many problems in the lack of corresponding management system and construction standards as support [5]. The automation degree of construction site is low, and the integration of intelligence and project management is difficult. The comprehensive quality of employees is low, and the ability to accept intelligent operation is not strong [6].

Firstly, grassroots projects have insufficient informatization capability, limited understanding of informatization, and lack of overall demand analysis and planning. Secondly, grassroots projects lack the ability to negotiate and bargain with manufacturers [7]. All kinds of equipment in smart site applications are expensive and easy to be kidnapped. Thirdly, the smart site applications of external vendors mostly adopt the deployment mode of public cloud, which is prone to data leakage, and data security and business information are difficult to be effectively guaranteed [8]. Fourthly, the grassroots project level lacks the corresponding smart site management system and construction standards as support.

3 Application Study of BIM Technology in Smart Site

3.1 Relationship Between BIM Technology and Smart Site

As an important basic condition for the development of intelligent construction site, BIM technology plays a major role ensuring the efficient utilization of information technology. Before engineering construction, construction enterprises can carry out high-precision engineering design through the use of BIM technology, deeply excavate the value of collected data, and then provide corresponding data reference for the construction model [9]. The effective combination of BIM technology and Internet of Things technology can effectively promote the good cooperation between various equipment at the construction site, so as to give play to the advantages of intelligent management in engineering construction, and then relevant personnel can build a specific information management ecosystem on this basis. Combined with the data information obtained by BIM technology, relevant technical personnel can make scientific planning for the follow-up construction direction based on this, and then effectively guarantee the participation of personnel through the rational application of terminals. In addition, technicians can also manage the construction site visually by using the distributed network topology. This approach not only enhances the transparency of project management, but also enhances economic performance of firms in the establishment of smart site.

3.2 Establish Smart Site Management System Based on BIM

The BIM graphics engine is taken as the basic platform of visual and digital construction project management [10]. Through the establishment of standard data collection methods, sampling frequency and data output format, information application interface, the real-time collection of projects, personnel, equipment, material, quality, safety, environmental protection seven modules data seamless docking BIM model, visualization, digital management, based on BIM integration cloud platform, the wisdom of the site construction framework is shown in Fig. 2.

The intelligent site management system is composed of terminal layer, platform layer and application layer. Real-time monitoring, intelligent sensing and data collection of the construction site can be realized through the Internet of things data acquisition equipment. Through the cloud platform, the massive data collected by the system can be efficiently calculated, stored and provided with services. The new intelligent platform system can add real-name authentication of field workers, fingerprint or face recognition, intelligent access control, HIGH-DEFINITION monitoring, behavior identification analysis, intelligent transmission and information sharing system [11].The application layer

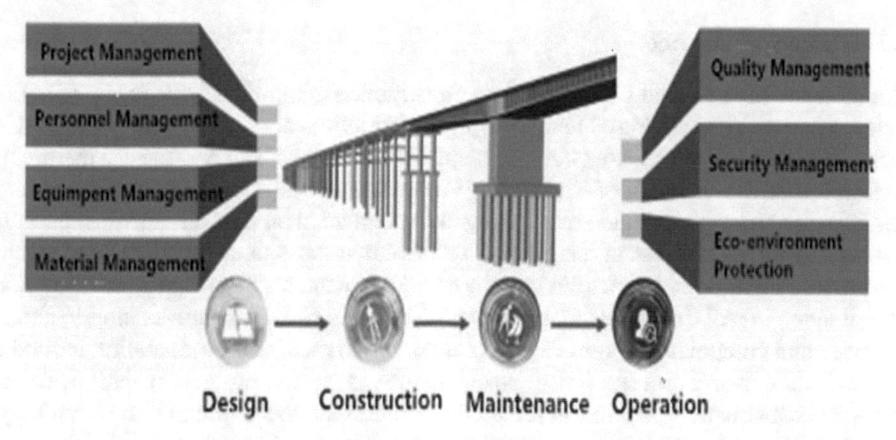

Fig. 2. Construction framework of smart site

provides the project management system for the field manager according to the project management requirements. BIM technology plays a key supporting role in smart site construction.Applying BIM technology, the architecture of intelligent field management system is shown in Fig. 3.

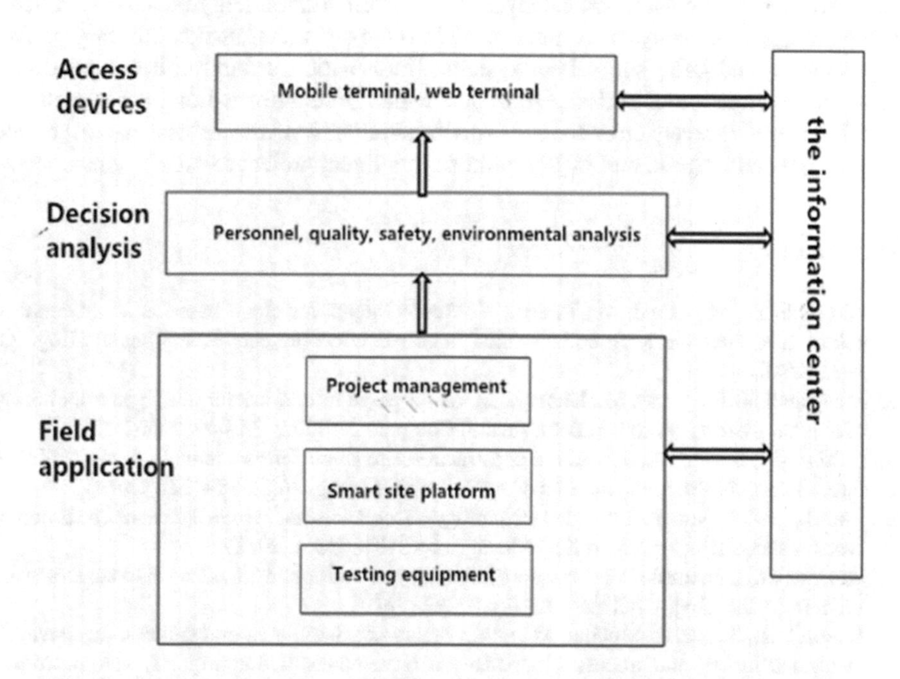

Fig. 3. Architecture diagram of intelligent field management system based on BIM

3.3 Video Surveillance

Combined with the actual situation, many construction enterprises have begun to apply video monitoring technology in engineering construction, and the introduction of BIM technology in engineering construction helps staff to better view and master the monitoring situation of each area of engineering construction. Under the background of smart construction site, remote video monitoring is implemented on the construction site. It is not only to set up cameras in the internal and external areas of construction and establish monitoring rooms, but also to assist various departments to grasp the construction site situation in real time by using information technology means such as mobile phone software and computer programs. For example, in the construction operation around a highway, the construction enterprise can obtain the video monitoring information of each point and construction position in the model through the use of smart site technology, so as to better understand and master the construction progress and quality information of the project.

4 Conclusion

Construction unit only by means of science and technology, relying on large data, to comprehensively improve the technical level of the information. Intelligent, cloud services, Web of things, data analysis are introduced into the construction site, building perfect project quality and safety supervision system of information, through collecting and analyzing the information of building site. Meanwhile, share information interconnection, raise the level of supervision on the construction site. The level of site information service can be higher and higher, and finally build an intelligent project to satisfy consumers.

References

1. Alreshidi, E., Mourshed, M., Rezgui, Y.: Requirements for cloud-based BIM governance solutions to facilitate team collaboration in construction projects. Requirements Eng. **23**, 1–31 (2018)
2. Mneymneh, B.E., Abbas, M., Khoury, H.: Vision-based framework for intelligent monitoring of hardhat wearing on construction sites. J. Comput. Civil Eng. **33**(040180662) (2019)
3. Edirisinghe, R.: Digital skin of the construction site smart sensor technologies towards the future smart construction site. Eng. Constr. Archit. Manag. **26**(2), 184–223 (2019)
4. Liu, D., Lu, W., Niu, Y.: Extended technology-acceptance model to make smart construction systems successful. J. Constr. Eng. Manag. **144**(040180356) (2018)
5. Stefanic, M., Stankovski, V.: A review of technologies and applications for smart construction. Proc. Inst. Civil Eng.-Civil Eng. **172**(2), 83–87 (2019)
6. Kuenzel, R., Teizer, J., Mueller, M., et al.: Smart site: intelligent and autonomous environments, machinery, and processes to realize smart road construction projects. Autom. Constr. **71**(SI1), 21–33 (2016)
7. Griffin, A., Hughes, R., Freeman, C., et al.: Using advanced manufacturing technology for smarter construction. In: Proceedings of the Institution of Civil Engineers-civil Engineering, vol. 172, no. 6SI, pp. 15–21 (2019)

8. Yu, Z., Peng, H., Zeng, X., et al.: Smarter construction site management using the latest information technology. In: Proceedings of the Institution of Civil Engineers -Civil Engineering, vol. 172, pp. 1–34 (2018)
9. Kim, J., Kim, J., Fischer, M., Orr, R.: BIM-based decision-support method for master planning of sustainable large-scale developments. Autom. Constr. **58**, 95–108 (2015)
10. Hu, X., Zhou, Y., Vanhullebusch, S., Mestdagh, R., Cui, Z., Li, J.:1CA1.smart building demolition and waste management frame with image-to-BIM. J. Build. Eng. **49**, 104058 (2022)
11. Li, C.Z., Xue, F., Li, X., Hong, J., Shen, G.Q.: An Internet of Things-enabled BIM platform for on-site assembly services in prefabricated construction. Autom. Constr. **89**, 146–161 (2018)

Intelligent Application of "Internet+" BIM Technology in Prefabricated Building Construction

Yanli Huang[1](✉) and Neha Jain[2]

[1] Fuzhou University of International Studies and Trade, Fuzhou 350202, Fujian, China
HYLmomo@163.com
[2] Hitkara University, Punjab, India

Abstract. With the promotion of national policies and the advocacy of the concept of green energy saving buildings, the development of prefabricated buildings ushered in the spring. How to vigorously develop prefabricated buildings and realize the leapfrog development of building industrialization and information technology has become the key problem that the construction industry needs to think about. In this paper, based on the "Internet+" BIM technology, relevant research is carried out in combination with the development needs of prefabricated buildings. Through research and analysis, this paper puts forward the favorable measures of "Internet+" BIM technology in the construction of prefabricated buildings, which is conducive to promoting the way of urban construction in China, reducing building consumption and promoting innovative industrialization of construction.

Keywords: Prefabricated Building · "Internet+" · BIM Technology · Intelligence Application

1 Introduction

On August 28, 2020, nine ministries and commissions of the state jointly issued several opinions on accelerating the industrialization development of new building. The opinions put forward that prefabricated concrete buildings should be vigorously promoted, green construction methods should be advocated, and intelligent construction and construction industrialization should be promoted [1]. The country's "14th Five-year" construction industry development plan proposed that by 2025, China's prefabricated buildings accounted for more than 30% of the proportion of new buildings, while building industrialization, digitalization, intelligent level will be greatly improved.

2 Overview of Prefabricated Building Development

The rapid development of prefabricated buildings has solved a series of problems such as high cost and high energy consumption of traditional buildings, improved construction

J. H. Abawajy et al. (Eds.): ICATCI 2022, LNDECT 169, pp. 642–647, 2023.
https://doi.org/10.1007/978-3-031-28893-7_77

efficiency, reduced resource loss in the construction process, and significantly reduced environmental pollution in the relevant process [2].

With the great support of national and local government policy dividend, prefabricated building develops rapidly, the industry policy standard system is increasingly improved, the research and development of prefabricated parts and production technology level is gradually improved, the permeability of prefabricated building is gradually increased, and the prefabricated building area started continues to increase [3]. Prefabricated buildings developed rapidly in China in 2018, with a floor area of about 290 million square meters. The total output of the prefabricated construction industry is about 105 billion yuan. The prefabricated building market size was 667 billion yuan, and the output value of related supporting industries was 333.5 billion yuan. In recent years, the total output value of China's assembly construction industry has shown a trend of rapid growth, as shown in Fig. 1.

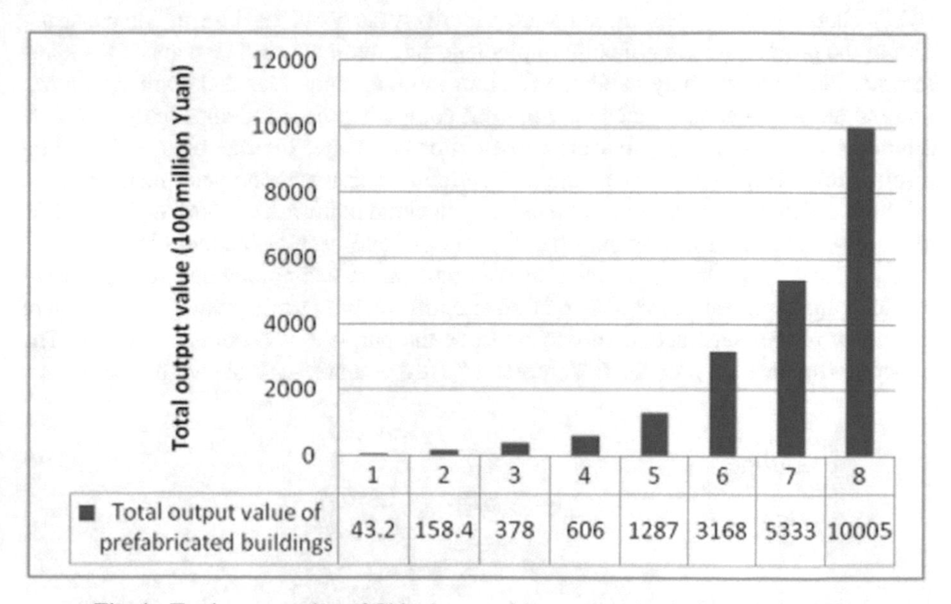

Fig. 1. Total output value of China's assembly construction industry 2011–2016

However, on the whole, prefabricated buildings in China are still large but not strong. The development of informatization and industrialization about prefabricated building is a historic opportunity to change the current situation of "big but not strong" in China's construction industry and achieve high-quality development. Prefabricated building is an important carrier of building information development [4]. At present, although there is some research and development and practice of building information modeling (BIM), the overall progress is slow, and it still stays at the level of design or simulation and display, and lacks the overall application of the whole industry chain of design, production, logistics and construction. There are significant differences in informatization level between urban and rural areas. In some areas, no management platform for building informatization has been established, and the overall informatization and intelligence level of the project implementation process is low [5].

3 "Internet+" BIM Technology Promotes the Information-Based Transformation of Construction Industry

BIM technology has strong integration ability [6]. The global unified construction software programming language makes it possible for the software of different functions and systems to work together. BIM technology realizes the collection of physical and mechanical parameters, time, cost, visualization, construction simulation and other factors, which is an unprecedented mode of production in the construction industry.

"Internet+" technology has the characteristics of rapid transmission, rapid analysis, mass storage and real-time sharing of information[7]. The application of "Internet+" BIM technology in engineering can more effectively carry out the information construction of the project and scientifically implement the intelligent management of the site." Internet Plus" has not only made the Internet mobile, ubiquitous and applied to a traditional industry, but also added omnipresent computing, data and knowledge to create omnipresent innovation [8]. For the construction industry, "Internet plus" is based on information technologies such as Internet communication and computer technology to establish an Internet cloud big data management cloud platform, or combined with BIM technology to form a new management mode of "end user + Internet cloud + BIM data platform" [9]. Or rely on Internet communication technology and remote monitoring equipment, realize the construction site information, visual management, ensure the safety of the construction site, to improve the purpose of economic benefits. The integrated application process of "Internet +" BIM technology is shown in Fig. 2.

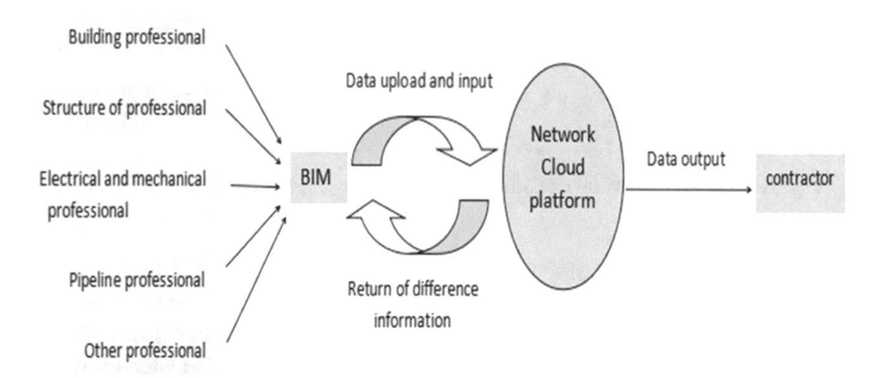

Fig. 2. Collaborative working diagram of "Internet+" BIM technology

4 Intelligent Application of "Internet+" BIM Technology in Assembly Building Construction

Intelligent engineering construction with advanced technologies such as "BIM" and "Internet+" meets the requirements of green development of the construction industry [10]. "Internet+" technique is based on computer information technology, the Internet with super cloud computing and cloud storage capacity, fast and efficient information transmission and the characteristics of not restricted by geographical space, combining it with BIM technology and application in the production of the construction industry, can significantly improve economic efficiency and the enterprise the competitive ability.

4.1 Accurate Matching of Components to Ensure the Accuracy of Construction

In prefabricated buildings, a large number of prefabricated components are used, such as wall panels, floors, stairs, balconies and so on. Assembly components need to be broken down in a factory. The visualization and simulation advantages of BIM technology can be fully utilized in the process of splitting production [11]. Through BIM technology, the size of the components to be used in this construction is confirmed. Based on the existing 3D model, the production, installation, transportation and construction scheme of prefabricated components are constantly optimized, and the information of each component is reasonably matched to ensure the accuracy of the construction project.

4.2 Build a Network Cloud Platform to Promote Multi-party Collaborative Participation

In architectural engineering of prefabricated, because components do not conform to the requirements of the building and construction problem requires parties to participate in solving the problems such as period extended phenomenon is not uncommon, integration of the production process can strengthen the communication and the exchange, makes the integration of design, production, construction, can adopt the integrated delivery mode, promote the building process progresses steadily [12]. Construction enterprises can data collected as part of the construction project related information via the Internet with BIM 3D conceptual model, uploaded to the exclusive cloud platform for declaration, examination and approval departments through the cloud platform for online assessment and audit of construction projects, and thereby reducing the examination and approval department for offline survey and the number of evaluation meeting, reach the purpose of save time and cost of construction project. Designers can upload design change information to BIM cloud in time through the Internet.

Construction units and construction units can realize online sharing and exchange of construction project information through the Internet, so that Party A can communicate and change the scheme at any time, making the design scheme more perfect. In the construction stage, the construction unit introduces BIM technology and establishes BIM 4D model [13]. By simulating the 4D construction process, the possible risks in the construction process are found out, construction difficulties are analyzed, and solutions are proposed. Then the BIM model and solutions are uploaded to the Internct interactive cloud platform. Construction personnel can use terminals to access the Internet cloud and guide construction online or download corresponding construction videos. At the same time, the problems encountered in the implementation of the project and its parameters can be quickly uploaded to the Internet cloud [14]. Designers or experts can obtain timely data and propose solutions, thus speeding up the actual construction progress and ensuring construction safety and quality. The participants of the assembly building can realize the intelligent management of the building by combining the three-dimensional model with the characteristics of real-time monitoring, fast transmission and remote operation of the Internet.

4.3 Promote Information Construction of Prefabricated Buildings

Through Revit software, we can create 3d information model of engineering projects, so as to make full use of the simulation characteristics of BIM technology, such as energy saving simulation, earthquake escape and fire evacuation analysis, sunshine analysis and heat conduction simulation, and finally to scientifically evaluate whether the construction project is reasonable. It can generate 3D animation to guide the actual construction, find the collision problems in advance, reduce the engineering rework and change in the construction stage, further optimize the engineering design and construction scheme, save costs and improve the quality of the project [15].

In the construction stage, it can also realize the visualization of the construction site, timely find the problems existing in the construction work, avoid causing more serious consequences. The connection between the family members of a single model or the family members of a multi-professional combination model can be checked for collision and conflict in space and time, so as to find conflicts between the members in the construction process in advance, modify the model and reduce the rework rate. Use BIM technology to create a library of various living facilities, construction materials and mechanical equipment, establish a 3D model of construction site layout, and carry out animation simulation of construction materials management and mechanical operation of the site in advance based on the construction status of each stage. Combined with Internet access control system, remote monitoring system and mobile communication, real-time construction monitoring and management are carried out on the access of construction personnel and the operation of construction site machinery, so as to improve the control and scheduling ability of the owner and construction technicians.

5 Conclusion

In a word, prefabricated building plays an important role in promoting China's future industrialization. The development of assembly building industry needs to further develop intelligent construction technology and strengthen the research and practice of Internet + BIM technology. In the implementation of assembly building projects, the powerful combination of Internet and BIM technology is conducive to the visual and collaborative operation of massive data of large-scale projects, as well as the establishment of smart construction sites and the application of collaborative platforms. This will have a significant impact on the construction industry, bring about the transformation of architecture, and lead the construction industry into the era of big data.

References

1. Li, S.C.: Green building technology of prefabricated building. In: E3S Web of Conferences, vol. 198, pp. 127–142 (2020)
2. Alcinia, Z.S., Diogo, G.S., Edgar, P.B.: BIM tools used in maintenance of buildings and on conflict detection. Sustain. Constr. **8**, 163–183 (2016)
3. Costa, G., Madrazo, L.: Connecting building component catalogues with bim models using semantic technologies: an application for precast concrete components. Autom. Constr. **57**, 239–248 (2015)
4. Jifei, W.: Application of assembly building intelligent construction technology in construction engineering management. Intell. Build. Smart Cities **11**, 89–94 (2021)
5. Ting, W.: Research on detailed design of prefabricated building based on BIM and Big Data. J. Phys: Conf. Ser. **2037**(1), 143–150 (2021)
6. Bataglin, F.S., Viana, D.D., Formoso, C.T.: 4D BIM applied to logistics management: implementation in the assembly of engineer-to-order prefabricated concrete systems. Ambient Constr **018**(01), 208–223 (2018)
7. Zhao, S., Wang, J., Ye, M., Huang, Q., Si, X.: An evaluation of supply chain performance of china's prefabricated building from the perspective of sustainability. Sustainability **14**(3), 101–123 (2022)
8. Yang, B., Fang, T., Luo, X., Liu, B., Dong, M.: A BIM-based approach to automated prefabricated building construction site layout planning. KSCE J. Civil Eng. **12**(8), 1–18 (2021)
9. Gunawardena, T., Mendis, P.: Prefabricated building systems—design and construction. Encyclopedia **2**(1), 70–95 (2022)
10. Zhong, J., Zhang, P., Wang, L.: An energy consumption calculation model of prefabricated building envelope system based on BIM technology. Int. J. Global Energy Issues **44**(2–3), 121–138 (2022)
11. Tong, S., Nan, Y.: Application of BIM based information management technology in prefabricated buildings. J. Sci. Res. Rep. **75**(2), 17–29 (2020)
12. Yong, L.: Research on prevention of common quality problems in prefabricated costruction based on computer. J. Phys.: Conf. Ser. **1992**(2), 82–97 (2021)
13. Tavares, V., Lacerda, N., Freire, F.: Embodied energy and greenhouse gasemissions analysis of a prefabricated modular house: the "Moby" case study. J. Clean. Prod. **212**, 1044–1053 (2019)
14. Wuni, I.Y., Shen, G.Q.: Critical success factors for management of the early stages of prefabricated prefinished volumetric construction project life cycle. Eng. Constr. Archit. Manag. **98**, 2315–2333 (2020)
15. Xiao-Juan, L., Ji-Yu, L., Cai-Yun, M., et al.: Using BIM to research carbon footprint during the materialization phase of prefabricated concrete buildings: a china study. J. Clean. Prod. **279**, 123454 (2021)

Financial Cyber Security Risk and Prevention in Digital Age

Hongyuan Duan[1](✉), Minghong Sun[1], and Ivy Jackson[2]

[1] Department of Information and Business Management, Dalian Neusoft University of Information, Dalian, Liaoning, China
duanhongyuan@neusoft.edu.cn
[2] Xiqin Technology Co., LTD, Dalian, Liaoning, China

Abstract. At present, global financial services are highly dependent on the Internet, and while service efficiency has been greatly improved, network risks have spread rapidly. With a wider spread and greater impact, network security and data security risks are huge. Due to unbalanced financial infrastructure construction of science and technology, new technology and business integration inadequate, standardization of financial supervision mechanism is not sound, the financial industry network security and supervision is facing new challenge, how to build and emerging technologies that meet the needs of the financial regulatory system, network security enhance network security management and risk prevention capacity. This paper aims to describe the types and sources of financial network risks, as well as financial network risk prevention measures.

Keywords: Digital age · Financial network security · Financial risk prevention

1 Introduction

With the development of technologies, such as big data, blockchain and AI, fintech has become the future trend of financial industry. Since 2017, annualized industry has been listed as the third-largest high-level sustainability threat of attack group, which is listed next to the government and energy [1].

As well as the scale of its adoption, fintech ecosystem bring business innovation in the financial sector and also increasing risks in financial network. Due to the security and maintenance of information services, financial institutions usually build their own data centers. Therefore, with the rapid development of banking and securities business, information security risks increase synchronously, service access efficiency and data security become the most concerned topic for network and application administrators. (see Fig. 1).

J. H. Abawajy et al. (Eds.): ICATCI 2022, LNDECT 169, pp. 648–653, 2023.
https://doi.org/10.1007/978-3-031-28893-7_78

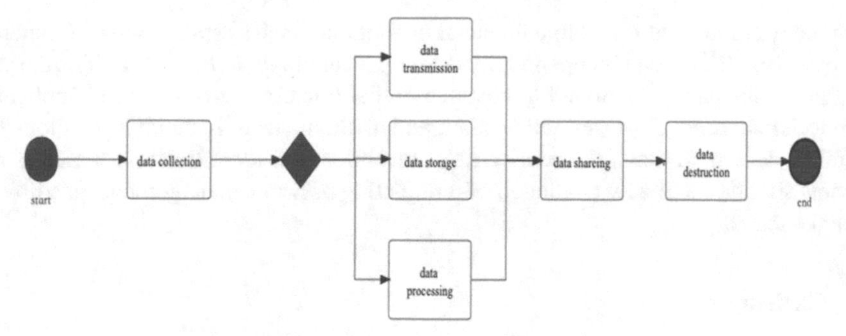

Fig. 1. Financial data life cycle management

2 Financial Industry Network Security Situation and Regulatory Challenges

2.1 Technical Level

On the one hand, financial business is becoming intelligent and digital after integrating with emerging technologies, fintech has became the most important direction of financial innovation in recent years. With the development of fintech in China, fintech will not only promote financial coverage and service efficiency, but also bring certain financial risk. Since different financial businesses and financial institutions are more correlative, technologies are integrated between different departments. On the other hand, the integration of the new technology brings network security about subversive influence and huge impact, because of more and more exposed loopholes, security problem is increasingly serious, the implementation of the safety protection difficulty improved significantly, regulatory need to redefine boundary. Because of the new technology, network security improve the regulatory framework, new system and the technical specification [2].

For example, CC attack use proxy server to simulate multiple users to constantly put forward access requests to the network system. DDOS attacks control multiple computers to send access requests to network system. Traffic attacks will cause CPU of the system server to reach the peak value, which will break down the network system and make the network system unavailable [3]. This kind of attack is most harmful, which may result in the bank customer information leaking massively, database tampering, bank system implanting back door. For example, on July 29, 2019, Capital One reported a massive theft of 106 million pieces of customer information, mainly from 2005 to 2019.

2.2 Management Level

At present, emerging technologies in the direction of fintech are mainly held by large technology companies and research institutions, Financial institutions usually engage in technology application innovation with third-party technology r&d institutions and then set up fintech subsidiaries. Because of new technology research and development is still on preliminary stage, the ability to integrate new technologies with business is not yet available [4]. Since the provision of fintech technology services mainly relies on technology companies, only large financial institutions, such as state-owned commercial banks such as ICBC, can independently develop technology. In the first half of the year, the four major banks continued to advance digital transformation, and the application of fintech increased. However, many small and medium-sized financial institutions face technical defects in network security risk prevention and identification, which is also the new situation and new challenge of completing network management faced by the financial industry.

2.3 Challenges

PBOC emphasized the supply chain security of financial institutions in the "Opinions on Regulating the Application and Development of Open Source Technology in the Financial Industry" on October 2021. The security of supply chain infrastructure provides a new way for traditional network attack. The suppliers of financial institutions are complicated, and at present the security review for suppliers' products is not mature. Once the supplier is invaded and utilized successfully, it will directly destroy the traditional network security defense trust system (Fig. 2).

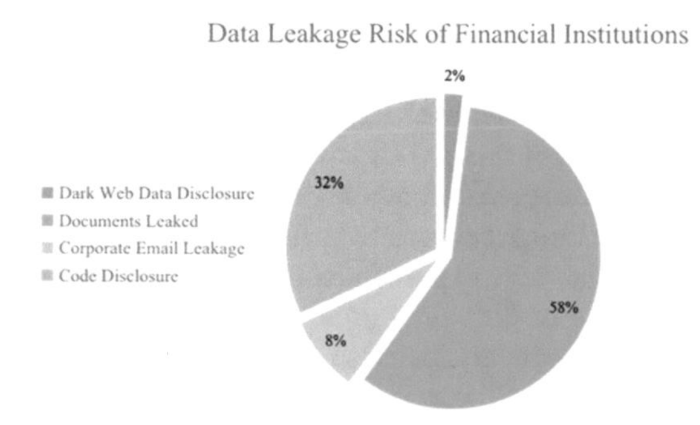

Fig. 2. Data leakage risk of financial institutions

According to IBM's 2021 Data Breach Cost Report, the average cost of a corporate data breach is $420,000, and the average cost $150.8 per lost or stolen record. For instance, on September 2021, the MISO crypto trading platform of SushiSwap community was attacked by a software supply chain, in which attackers hijacked platform transactions and stole about $3 millions of Ethereum coins [5].

Sensitive data leakage is also a high-risk security incidents of concern to the financial institutions and their regulators. For financial institutions, due to their special industry background, the characteristics of high data sensitivity and high black market value will promote the occurrence of malicious data leakage events, such as theft and other network attacks and intentional behavior of internal personnel. In addition, when internal personnel use network cloud disk, library, code warehouse and other network platforms for data transmission and storage, t they often leak sensitive internal data to the public network due to personal negligence or data insensitivity. In recent years, Internet exposure of sensitive data leakage incidents of financial institutions emerge one after another.

3 Financial Industry Network Security Countermeasures

As new technologies application in the field of financial deepening, the algorithm of defect, technical control, tampering with information, new technologies, such as leak risk and lead to financial problems get more and more attention, since 2021, the new technology of national regulators continuously strengthen the financial sector, risk prevention and control measures mainly against the risk from both technical and business aspects (Table 1).

Table 1. Regulatory policies on new technical risks

Country	Time	Main content
The United States	2021.3	The federal reserve and other five departments have announced consultation on AI applications in the financial sector, including fraud prevention and personalized services
South Korea	2021.7	FSC publishes the manual intelligence financial service guide to establish a risk monitoring and management system to prevent information leakage risk
UK	2021.4	Cma new digital marketing door, increase the supervision of large digital enterprises, especially large technology monopolies
Russian	2021.1	Russian cryptocurrency exchange Livecoin has been hacked

3.1 Build a Perfect Network Security Regulatory Framework

Countries continue to strengthen the top level of domestic financial data security, promote financial data safety standards, and some countries establish third-party institutions to monitor the risk assessment and vulnerability of financial institutions' data security capabilities [6]. For example, South Korea publishes the "my data" legislation, strengthening data protection and issuing licenses to several financial technology companies.

With the development of digital technology, digital currency and other encrypted assets have become the focus of financial network secure supervision [7]. Global issuance of non-sovereign digital currency regulation by private or enterprise is also more stringent.

The digital transformation of financial infrastructure has led to the continuous follow-up of financial network security supervision measures. With the deepening of the financial technology scenario, the digital transformation of the financial infrastructure areas such as the settlement of the clearing, the settlement of securities and the network system, and the continuous acceleration of the digital transformation of the financial sector, the international financial supervision is more concerned with the development of the financial infrastructure of the financial industry [8].

3.2 Focus on the Application of Technology in Financial Network Security Management

The network financial security technology includes firewall technology, intrusion detection technology ids, network encryption security protocol SSL and SET, which are progressing with the development of the network.

SSL provides client and server authentication, data integrity, and information confidentiality security for client/server applications based on TC/server. In addition to data encryption SSL also supports digital signature. SSL introduces many security mechanisms, such as asymmetric key exchange, data encryption, message authentication code, identity authentication. SSL also provides security notification function for exception cases, which includes error warning and connection closing.

Using the method of SSL and decryption, the log is processed by the use of the local traffic manager, which is used by the local traffic manager, which is used by the local traffic manager, and can be able to speed up the SSL acceleration of SSL, and uninstalls the pressure of the server to handle the SSL and decryption (Fig. 3). On the site, the load balancing product can simultaneously load the load balancing the web front server, application server, database server, cache server, and so on.

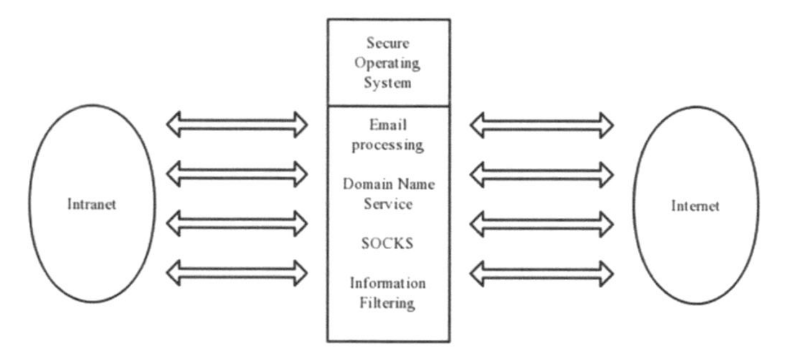

Fig. 3. Application of technology in network security management

The security protection of all kinds of servers is the key to the customer's most concern, and it is recommended that the web firewall be protected by the web firewall, and use the database security audit platform to audit and alarm all kinds of database operations, data table calls and modified actions, and ensure that the situation of a large number of users can audit the database [9]. The dynamic password authentication system ensures the password protection of the user account and password of the online trading system, and forms the "double factor" and the strong body is certified.

In the policy documents issued by the supervision and administration of the financial industry, the system log retention period is different from the risk level, at least for less than a year, the unified log audit platform can meet the requirements of various regulations, formulate the relevant strategies and processes, manage the activities of the productive system, and support effective audit, safe evidence analysis and prevention of fraud.

4 Conclusions

Financial network security needs well system support, and it is urgent to speed up the gap in the supervision of financial networks. Establish a sound risk the prevention and control system, promote prevention of systemic financial risks, with financial information infrastructure as the underlying construction, financial business-oriented, and jointly support financial network security.

References

1. McCallig, J., Robbb, A., Rohdec, F.: Establishing the representational faithfulness of financial accounting information using multiparty security, network analysis and a blockchain. Int. J. Account. Inf. Syst. **33**, 47–58 (2019)
2. Chinazzi, M., Fagiolo, G., Reyes, J.A., Schiavo, S.: Post-mortem examination of the international financial network. J. Econ. Dyn. Control **37**(8), 1692–1713 (2013)
3. Brignoli, M.A.: A distributed security tomography framework to assess the exposure of ICT infrastructures to network threats. J. Inf. Secur. Appl. **6**(59), 102833 (2021)
4. Hasan, S., Ali, M., Kurnia, S., Thurasamy, R.: Evaluating the cyber security readiness of organizations and its influence on performance. J. Inf. Secur. Appl. **58**, 102726 (2021)
5. Ma, C.: Smart city and cyber-security; technologies used, leading challenges and future recommendationse. Energy Rep. **7**, 7999–8012 (2021)
6. Kang, Q.: Do macroprudential policies affect the bank financing of firms in China? Evidence from a quantile regression approach. J. Int. Money and Finance **7**(115), 102391 (2021)
7. Varga, S., Brynielsson, J., Franke, U.: Cyber-threat perception and risk management in the Swedish financial sector. Comput. Secur. **6**(105), 102239 (2021)
8. Bat, J.V.: Bank-based versus market-based financing: implications for systemic risk. J. Bank. Finance **5**(114), 105776 (2020)
9. Olukoya, O.: Distilling blockchain requirements for digital investigation platforms. J. Inf. Secur. Appl. **11**(62), 102969 (2021)

Online Examination System of Japanese Course Based on WeChat Mini Program

Libao Niu[1], Zhenhui Wang[1](✉), and Deepak Kumar Jain[2]

[1] Xi'an Fanyi University, Xi'an 710105, Shaanxi, China
nlb9502@163.com
[2] Chongqing University of Posts and Telecommunications, Chongqing, China

Abstract. With the development of Internet technology and the popularity of mobile devices, the online and offline hybrid teaching mode in colleges and universities has gradually become popular. Online examination is not limited by time and place, which can effectively reduce the cost of examination and reduce the working intensity of teachers. By virtue of the advantages of the development and use of WeChat mini program, an online examination system for Japanese courses is designed and developed. The front end of the small program uses WeChat developer development tools, the back end uses Spring MVC framework, MySQL database to store business data. The system allows teachers to upload test papers and statistical scores, and students can complete online tests, exams and view scores. At the same time, the system supports automatic test paper group to avoid cheating. The system is simple to use and reliable in function. It has certain practicability for students to master Japanese knowledge points and carry out fragmented learning, and adapts to the development trend of "Internet + " education.

Keywords: Online examination · Japanese course · Mini program

1 Introduction

The sudden outbreak of COVID-19 has brought a lot of inconvenience to people's work and life, as well as a great obstacle to social and economic development. But on the other hand, it also promotes the rapid development of internet industry applications. Especially in the field of college education, the construction of online education resources is in full swing [1]. Online and offline hybrid teaching mode has become the main teaching mode [2, 3].The evaluation method of students' scores is not only dependent on the final exam, but more daily tests and attendance are increasingly valued by schools.

With the help of the online examination system, teachers can quickly understand students' learning, eliminating the cost of money and time such as printing papers, marking books and summarizing scores. Students can also test and consolidate the knowledge they have learned without time and space restrictions and realize fragmented learning [4, 5].Although there are many online foreign language test software in the market, 90% of them are developed for English majors. There are few software products for teaching and testing of small languages like Japanese.

J. H. Abawajy et al. (Eds.): ICATCI 2022, LNDECT 169, pp. 654–660, 2023.
https://doi.org/10.1007/978-3-031-28893-7_79

As a lightweight application, WeChat mini program has the advantages of fast development, low cost and free installation. Its "run out of time" concept can upgrade the traditional industry information service level and provide users with ubiquitous software services [6]. Due to the file size and performance requirements, mini programs can focus on users, subdivide services, better mining user functional needs, with more in line with the form of user preferences to carry out information services, enhance user experience, improve service level. Therefore, it is necessary to develop a practical, convenient online examination system for Japanese course based on WeChat mini program to meet the teaching and learning needs of Japanese major teachers and students in colleges and universities with the advantage of mini program.

2 Analysis of Online Examination System of Japanese Course

2.1 System Design Idea

1) MVC Design pattern

Mini program background uses Spring MVC framework technology. Spring MVC is a lightweight Web framework based on Java to realize the request-driven type of MVC design pattern. It uses the idea of MVC architecture pattern to decouple the responsibilities of the web layer. The purpose of the framework is to help simplify development and maintenance.

2) Micro service concept

Spring micro service technology is used to separate each module of the mini program background system into components to reduce system complexity. The services of each component communicate with each other through interfaces, so that a reusable system with high cohesion and low coupling can be constructed quickly, and service changes can be flexibly responded to the system. Mini programs can request restful API to access background micro service components.

3) Ajax + JSON asynchronous data communication

JSON file is the configuration file of the mini program and the data exchange standard between the mini program and the background Web service. It directly determines the response time of the mini program. Using Ajax by exchanging small amounts of data in the background with the server, Ajax enables the mini programs view to update data asynchronously, making it a good user experience [7].

2.2 System Architecture

The architecture of the online examination system (as shown in Fig. 1) is divided into three layers. The first layer is the front end of the system, namely WeChat small program, responsible for human-computer interaction, is the interface of students' examination operation; The second layer is the small program system background, that is, the online examination system Web application, provides Web service interface for the small program data, but also is the online examination system teachers and managers business processing platform, it provides data to the small program in JSON format; The third layer is the system data, that is, the examination system database, storage of teachers, students, administrators, question banks, papers, scores and other business data.

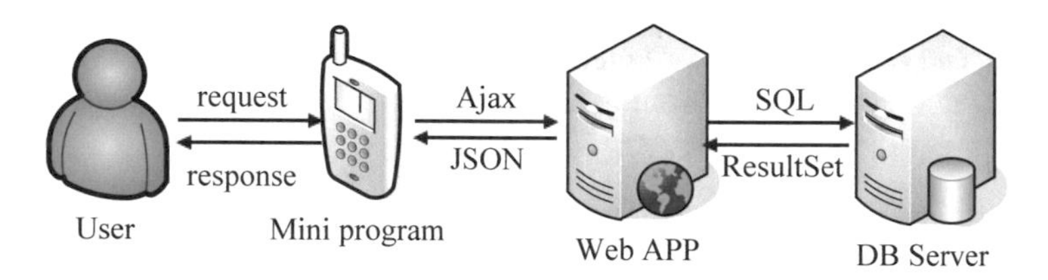

Fig. 1. Architecture of online examination system

2.3 System Function Analysis

The main users of the online examination system (as shown in Fig. 2) include students, teachers and administrators. The administrator is responsible for the management of all kinds of information (students, teachers, courses, papers, question banks, etc.). Teachers can set up their own courses, input the questions into the curriculum question bank, or upload the Excel paper. After a student has taken an exam, the teacher can check the statistics of the scores. An administrator uses a back-end Web system to manipulate data. Most of the functions of teachers can be completed by using mini programs, but

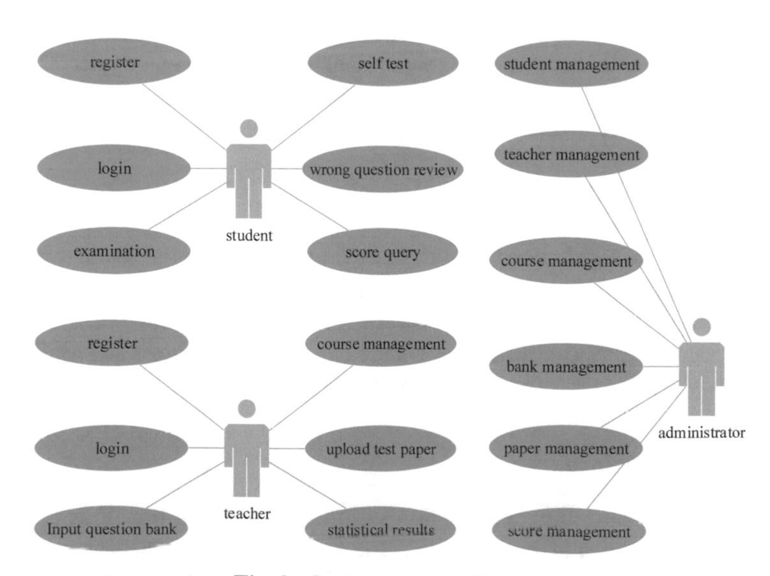

Fig. 2. System use case diagram

due to the amount of data and processing time requirements of input question bank and upload examination paper can be completed by using the back-end Web system. Student functions are all completed in the small program, mainly including login, registration, online test, online test, view results and view error records and other functions.

3 Data Storage Design

The data of the WeChat mini program of the online exam system is stored in the MySQL database [8–10]. Consider the three paradigms and performance requirements (with appropriate redundancy) at design time. The data stored in the small program is mainly teacher information, student information, administrator information, question bank information, test paper information, student answer information and test notice. This paper takes the paper information and answer information table as an example to introduce the structure of the table. Table 1 and Table 2 describe the table structure.

Table 1. Test paper information table

Field	Type	Null	Constraint
Paperid	varchar(20)	NO	Primary key
Nid	int(2)	NO	
Type	varchar(10)	NO	
Title	varchar(100)	NO	
Answer	varchar(20)	NO	
AnswerA	varchar(20)	NO	
AnswerB	varchar(20)	NO	
AnswerC	varchar(20)	NO	
AnswerD	varchar(20)	NO	

Table 2. Student answer sheet

Field	Type	Null	Constraint
Paperid	varchar(20)	NO	Primary key
Stuid	varchar(15)	NO	Primary key
Nid	int(2)	NO	Primary key
Answer	varchar(20)	NO	
Score	int	NO	

4 System Implementation and Key Technologies

4.1 System Implementation

MINA framework and micro service design ideas are used in system implementation to develop services with native APP experience, so as to bring better user experience. The mini program mainly consist of two modules that run independently: the View module (View layer) and the Service module (business logic layer). View module is responsible for the display of interactive interface and communicates with background through Weixin JSBridge. It is written by WXML and WXSS and displayed through components. The Service module is responsible for the logic operation of the background, and also communicates with the background through Weixin JSBridge object. JavaScript engine is used to provide the unique functions of the running environment and small programs. Therefore, the business logic of this project is completed by writing.js script files bound to each page. After the business logic layer processes the data, it sends it to the data layer and receives event feedback at the view layer.

4.2 Key Technologies

1) Importing Excel Files

In order to facilitate teachers to build the curriculum question bank, the system provides the test question uploading function, teachers can use the Excel test question template offline, and then upload to the system, so as to effectively improve the performance of the system, but also improve the efficiency of teachers to set questions. This system uses Commons -fileupload-1.2.2.jar, a free file upload component provided by Apache software foundation.

The principle of importing Excel test questions database is to import Excel test questions files with POI plug-in under Spring MVC framework. POI is a free, open source, cross-platform Java API written by the Apache Software Foundation in Java, which provides Java programs with the ability to operate files in Office format. POI plug-in operates Excel files through HSSFAPI, and the detailed steps of reading test data from Excel files and storing test data in the database are as follow steps:

(1) Generate the input stream of Excel test files

```
FileInputStream inputStream = new FileInputStream("/test/japancourse.xls");
```

(2) Get the Excel workbook object

```
HSSFWorkbook Testworkbook = new HSSFWorkbook(inputStream);
```

(3) Operate Excel worksheet obj ects

```
HSSFSheet sheetAt = Testworkbook.getSheetAt(0);
```

(4) Read test data in the Excel table and write it into the table

```
for (Row row:sheetAt) {
  if (row.getRowNum() == 0) {
   continue;
  }else{ //import
  }
}
```

(5) Combined with JDBC database operation, the above questions are stored in the question bank line by line.

2) Location service

Small program can achieve the function of examination positioning, combined with random unit volume algorithm and timed examination, can avoid cheating.

The location service function in the mini program can use Tencent location service, or third-party companies such as Baidu API interface. However, the third-party map API can only obtain data through the interface, and the map cannot be directly introduced into the small program, so using Tencent location service can develop more powerful functions, better compatibility, and more accurate location service application scenarios.

The steps of using Tencent location service for location service in the small program are as follows:

(1) Register a Tencent location service account
(2) Apply for developer key
(3) Download the WeChat small program JavaScript SDK and put it into the small program project
(4) Log in to WeChat mini program background and add the legitimate domain name of Request https://apis.map.qq.com
(5) Configure the permission field in app.json to obtain user location information
(6) Call wx.getLocation to obtain the geographical location of the current user (wechat returns longitude and latitude, speed and other parameters).
(7) Use the API function of JavaScript SDK reverse address resolution.

5 Conclusions

In this paper, an online examination WeChat mini program is designed and developed according to the daily teaching needs of Japanese courses. The program meets the needs of Japanese major students to learn Japanese knowledge and test the learning effect by using fragmented time, and also helps teachers to grasp students' grasp of knowledge points timely and accurately. It has realized automatic test paper group, online self-test, online examination, score query and so on. It provides a useful reference for small programs in the field of mobile learning. In the next step, based on the function of this software, we will use speech recognition technology to detect the accuracy of learners' pronunciation, and add a Japanese reading scoring module in the software to provide more powerful support for Japanese major students to improve their Japanese learning performance.

Acknowledgements. This work was supported by The 2021 Education and Teaching Reform Research Project of Xi'an Fanyi University (Project No.: J21B44).

References

1. Peimani, N., Kamalipour, H.: Online education and the Covid-19 outbreak: a case study of online teaching during lockdown. Education Sciences **11**(2), 72 (2021)
2. Yu, Y., Hao, T., Zhang, H.: Research and practice of hybrid teaching mode of "film and television production technology and art" course based on OBE concept. Creat. Educ. **12**(9), 2066–2073 (2021)
3. Singh, T., Afreen, S., Chakraborty, P., et al.: Automata simulator: a mobile app to teach theory of computation. Comput. Appl. Eng. Educ. **27**(5), 1064–1072 (2019)
4. Al-aqbi, A.T.Q., Al-Taie, R.R.K., Ibrahim, S.K.: Design and implementation of online examination system based on MSVS and SQL for university students in Iraq. Webology **18**(1), 416–430 (2021)
5. Mhaiskey, M.S.P., Thombare, M.R., Patle, M.N.M., et al.: Android APP for online examination and results systems. Int. J. Sci. Res. Eng. Trends **7**(3), 1914–1917 (2021)
6. Xiong, Y.: Research and design of campus bus reservation system based on WeChat mini-program. Int. Core J. Eng. **7**(6), 297–301 (2021)
7. Sharma, U., Singh, B.K.: Reintroduction of ajax using javascript libraries. J. Anal. Comput. **14**(11), 1–4 (2020)
8. Nugraha, A.H.: Making the casher application for the tonguewing meatball restaurant using Mysql and Netbeans. Int. J. Sci. Technol. Manag. **2**(5), 1793–1799 (2021)
9. Matallah, H., Belalem, G., Bouamrane, K.: Comparative study between the MySQL relational database and the MongoDB NoSQL database. Int. J. Softw. Sci. Comput. Intell. (IJSSCI) **13**(3), 38–63 (2021)
10. Klimek, B., Skublewska-Paszkowska, M.: Comparison of the performance of relational databases PostgreSQL and MySQL for desktop application. J. Comput. Sci. Inst. **18**, 61–66 (2021)

A Digital Supply Chain Financing Decision Algorithm Based on Blockchain Technology on Jiangxi New Energy Vehicle Industry as an Example

Hui Wang[1,2], Tiantian Xia[1], Yashi Che[2(✉)], Yinan Gao[1,2], and Josep Maria[3]

[1] Jiangxi Regional Development Research Institute, Jiangxi 330000, China
[2] School of Finance and Economics, Jiangxi University of Technology, Jiangxi 330000, China
kamui2022@163.com
[3] University of Pompeu Fabra, Tecnocampus Mataro Maresme, C Ernest Lluch 31, 08300 Mataro, Spain

Abstract. In this study, the conventional digital supply chain finance model and blockchain are separately portrayed by a game approach, and the game model is optimised and solved by using the inverse induction method to obtain the best wholesale price, retail store price and quantitative decisions for the supply chain of the two financing models in equilibrium; subsequently, sensitivity analysis is conducted on key parameters such as the initial amount of capital of retailers, the meaning of time of money rate of enterprises, the conventional digital supply chain A comparative analysis of the orthodox digital purvey chain finance type and the district digital a connected sequence purvey chain finance type was then conducted. The study showed that: as the initial capital volume of retailers decreases, the expected revenue of producers increases and the expected revenue of distributors and retailers decreases; compared to the orthodox digital purvey chain monetary model, blockchain. The numeral purvey chain monetary model is able to create value for the supply chain and district while the meaning of money in time rate is higher for companies. Digital purvey chain monetary provides better meaning for the purvey chain and achieves the result of victory at the same time. For producers, distributors and retailers. This deliberate attests the maneuverability advantages of the numeral purvey chain monetary model driven by district skill, and offer well-grounded complaints support and administrative insights for the recruit of district numeral purvey chain monetary.

Keywords: District technology · Numeral puevey chain monetary · Capital constraints

With the advancement of blockchain technology, the multi-million digital purvey chain monetary system relay on district skill can ensure value transfer and credit multi-layer penetration, so that the Nth level businesses in the purvey chain, which are far away from the core enterprises, can also dependent upon the corporate champions. to obtain credit backing, enabling digital purvey chain monetary to effectively cover the enterprises at the distal end of the core enterprises, realizing the transformation of traditional digital purvey

J. H. Abawajy et al. (Eds.): ICATCI 2022, LNDECT 169, pp. 661–667, 2023.
https://doi.org/10.1007/978-3-031-28893-7_80

chain monetary and greater serving the real economy [1]. This study aims to analyse the influnence of district credit transmission technology on digital purvey chain monetary from the perspective of quantitative research by establishing a three-level supply chain decision-making model of producer-distributor-retailer [2], so as to provide effective strategic policy directory to reasonably carry out the operation and administer of digital purvey chain monetary in the context of the application of blockchain credit transmission function [3].

1 Basic Assumptions

A complete supply chain usually consists of a core enterprise and its multi-level upstream and downstream enterprises [4]. This study focuses on the financing of multi-level enterprises downstream of the core enterprise and therefore constructs a supply chain consisting of a single core enterprise and multi-level distributors [5]. Without loss of generality this study considers a three-level pull supply chain distribution system ground consisting of a single producer A1 (core firm), a single distributor A2 (firm) and a single retailer A3. The retailer at the downstream end of the chain estimates the market demand and orders from the upstream distributor, who in turn orders from the upstream producer, who then produces the goods at cost C and sells them to the distributor at wholesale price JV (producer decision variable), who in turn distributes them to the retailer at wholesale price JE (distributor decision variable), who in turn sells them to the consumer market at retail price, with both the distributor's and retailer's The order quantity is Q (the retailer's decision variable) [6]. The market demand X is an uncertain quantity that obeys a probability distribution between [0, + oo] and a random probability distribution with density functions F(X) and f (X), respectively, and the probability distribution obeys the IGFR property.

This study also defines $F\Delta(X) = 1 - F(X)$, $H(X) = f(X)/F\Delta(X)$, and from the IGFR property of the probability distribution it is clear that both H(X) and $H\Delta(X)$ are increasing functions. When the retailer's order quantity is higher than the market demand, some of the goods are not sold and the retailer disposes of the surplus goods at a residual value O. The price of each commodity satisfies $P > JE > JV > C > O$. each participant can make a profit by selling the new product, but it is not profitable to dispose of the surplus product. Throughout the supply chain operation, the selling link occupies the longest time. At this point, the money-holding enterprise can invest the value of the funds held during that time, assuming that the time value of money rate is I. This study does not consider the time value of money in other links.

As a core company in the supply chain, the producer is well capitalized [7]. Distributors and retailers are both capital-constrained enterprises with initial capital amounts of uE and uR respectively, and since the capital required for their orders is JVQ and JEQ respectively, there is $uR < JEQ$, $uR + uE < JVQ$. In order not to take up white capital, producers are usually reluctant to provide trade credit financing for upstream and downstream enterprises, and prefer to provide credit guarantees for enterprises Consider two ways to solve the financing constraints of enterprises: the traditional digital supply chain finance model and the blockchain digital supply chain finance model. In this study, the superscripts α and β are used to denote the traditional digital supply chain finance and blockchain digital supply chain finance models respectively. The focus of this study is on the β financing model, while in the following, the α model will be used as the benchmark model for comparative analysis with the β model, so as to reflect the characteristics and advantages of the β model. This study also analyzes the value created by the financing tools (blockchain digital supply chain finance and traditional digital supply chain finance) for the supply chain by comparing the two scenarios of undercapitalised (β and α scenarios) and well-capitalised enterprises.

2 Model Comparison

The main differences and advantages of blockchain digital purvey chain finance compared with the orthodox digital purvey chain finance type as follows: (1) The blockchain digital supply chain finance model is more concise than the traditional digital supply chain finance model in terms of financing process. Traditional digital financial supply chain requires the combination of "digital financial supply chain and trade credit" to open up the entire supply chain capital flow, while after the introduction of the credit transfer function of blockchain technology, only a single financing mode of "blockchain digital financial supply chain" is required. (2) The blockchain digital supply chain finance model takes up less of the purvey chain's own capital and creates more time value of capital for the purvey chain. In the traditional digital financial purvey chain model, the distributor needs to provide trade credit financing for the retailer; (3) the risk bearers of the two financing models are different. In traditional digital financial supply chain, the producer and the distributor need to bear the bankruptcy and default risks of the distributor and the retailer respectively; whereas in blockchain digital supply chain finance, only the producer needs to bear the bankruptcy and default risks of the retailer.

α-case. The following is a modelling analysis of each funding scenario and a discussion of the changes to the financing model after the introduction of blockchain technology [8]. Firstly, the expression for the profit function of each member in the α case is analysed [9]. As mentioned above, the retailer pays its own funds uR from now on, acquires sales proceeds I3 at the eleventh hour and repays the loan I_2. $P = -u_E + [I_3\text{-}I_2]^+$. The distributor pays own funds as at the beginning of the period, receives distribution proceeds $\min\{I_1, I_4\}$ at the end of the period and repays the principal and interest on the loan I_6. $P = \text{-}u_E + [\min\{I_1, I_4\} - I_6]^+$. The producer pays production costs CQ^α at the beginning of the period, receives wholesale proceeds $J^\alpha{}_V Q^\alpha I$, and receives the meaning of money $J_\alpha V Q_\alpha I$ at the end of the period, repaying$[I_6 - I_1]^+$ for the distributor. Combining the above analysis, the expected returns of the producer, distributor and retailer in the α case can be expressed as

$$F^\alpha = \left(J^\alpha(1+I) - C\right)Q^\alpha - (P - O),\ F(X)dX \tag{1}$$

β-case. The expressions for the profit function of each member in the β case are analysed below [10]. Firstly, the changes in supply chain costs after the introduction of blockchain are explained: on the one hand, the introduction of blockchain technology will increase supply chain costs, such as blockchain construction, usage and operation costs; on the other hand, the introduction of blockchain technology will reduce supply chain costs, such as supply chain transaction costs, supply chain management costs, banking service costs, supply chain information identification costs, and supply chain business processing costs [11]. Considering that the introduction of blockchain technology will lead to both cost increase and cost reduction, and that the focus of this study is on the gossipy of the flow of district technology on the work flow of purvey chain raise funds, the changes in purvey chain costs due to the introduction of blockchain technology will not be considered in this study when modelling [12]. Combining the above analysis, the expected benefits of the producer, distributor and retailer in the β case can be expressed as

$$F^{\beta} = \left(J^{\beta}(1+I) - C\right)Q^{\beta} - (P - O),\ F(X)dX \tag{2}$$

3 System Simulation Comparison

A numerical analysis is carried out using a new energy vehicle as an example [13]. The product is produced by the manufacturer at a production cost of C = ¥60,000 and is sold to the distributor at a wholesale price of V. The distributor sells it to the retailer at a distribution price spelling E. The retailer further sells it to the consumer at P = ¥160,000 and the stalled product is disposed of at a price of O = ¥40,000. The distributor has an initial amount of capital uE = 0 and borrows from the bank at an interest rate of 0.05 with the manufacturer's guarantee. The market demand for this product obeys a normal distribution: X – Y (10000, 4000). Fixing the above parameters, a sensitivity analysis is performed on the amount the retailer started with of capital uY and the value of time about money rate/. The number of simulations is set to 30 and the results are averaged over 30 times. The results are as follows.

(1) Impact of retailers' own capital the one-stop purvey chain operational decision making and expected returns

In both the β and α cases, as the initial capital used by the sellers decreases, the distributor's distribution cost adds, the seller's order quantity augments, the initial capital used by goods to reduce, the producers wholesale price the producers prospective income adds, the distributor's prospective income decreases, the seller's prospective income decreases, and the overall prospective income of the one-stop upply chain increases [13]. The main reason for this is that the seller's original amount becomes less, the higher the amount of borrowing required and the higher the probability that sales proceeds will not fully cover the credit borrowing [14], which means that the distributor providing the trade credit facility will take more risk in doing so. This means that the distributors who provide them with trade credit finance will be exposed to a greater risk [15]. By increasing their distribution prices, the distributors are able to contain the risk of over-ordering by retailers and compensate them for possible default losses. However, the implementation of trade credit is tantamount to the distributor sharing part of the market risk for the retailer, and The seller's original amount becomes less, the higher the distributor's risk sharing, which provides an incentive for the retailer to order more. The above analysis yields the following management insights: the deepening of financing is beneficial to the core business, the provider of financing, but detrimental to the distributors and retailers, the demand side of financing. Look at it from the seller's perspective, which is capital constrained but holds some of its own funds, it should not rely too heavily on financing when it is not subsidised by the producer and should use as much of its own funds as possible in the transaction. From the producer's perspective, they should encourage retailers to use as little of their own capital as possible in the transaction through a reasonable subsidy mechanism. The subsidy mechanism should satisfy the incentive compatibility constraint, i.e. ensure that neither the producer nor the retailer benefits less from participating in the producer subsidy mechanism than before. It is worth noting that core firms should effectively screen the true value of the firm's initial capital to avoid losses due to information asymmetry and misrepresentation of information. The above analysis verifies the correctness of the hypothesis, and the following further compares the alpha and beta models.

(2) Comparison of the beta and alpha models and the advantages of the beta model

When the time value of money held by a company is $Z = 0$, the introduction of the blockchain credit transfer function increases the revenue of manufacturers and retailers, decreases the revenue of distributors, and increases the overall revenue of the one-stop supply chain. It follows that, compared with the traditional "digital provision chain finance 10 trade credit" financing method (α model), blockchain digital provision chain monetary (β model) can create value for the purvey chain, but for the sake of achieve the Pareto improvement of the interests of all participating members, the core enterprise should collect a certain participation fee from retailers to subsidize distributors. The producer, as the game leader, will then set a higher wholesale price to compensate for its risk losses, which is clearly to the detriment of the distributor. As the retailer no longer bears the risk of default, the blockchain digital provision chain finance model results in a significant benefit to the retailer as the distribution price set by the distributor falls. Ultimately, the overall supply chain revenue also increases as the producer and retailer revenue increases.

4 Conclusions

Blockchain technology triggered supply chain financial reform, making it truly serve Jiangxi new energy automobile enterprises and reshape the development of financial institutions Through the Stackelberg game analysis method and the three-level programming equilibrium solution algorithm, the orthodox purvey chain financial decision-making type and the district purvey chain financial decision-making type relay on credit transmission technology are established and compared respectively, and the susceptibility of hinge parameters such as the primary principal of retailers and the temporal value rate of business principal are analyzed.

Acknowledgements. We appreciate the financial support of Jiangxi Social Science Foundation Project (21GL51, 22ZXQH16); Science and technology research project of Jiangxi Provincial Department of Education (GJJ202013, GJJ212017); Jiangxi University Humanities and social sciences research project (GL21226); Collaborative education project of industry university cooperation of the Ministry of Education (202102013004, 202102348008); University excellence research project of Chinese Academy of Social Sciences (2021-KYLX02-02); Special project of Jiangxi Institute of regional development, Jiangxi University of science and technology(QYFZ2101, QYFZ2104).

References

1. Tiantian, X., Zhenduo, Z., Huan, X., Jing, X., Wentong, J.: The curvilinear relationship between job control and voice: role of emotional resistance to change and supervisor developmental feedback. SAGE Open **11**(2), 21582440211027960 (2021). https://doi.org/10.1177/21582440211027960
2. Raya, J.M., Vargas, C.: How to become a cashless economy and what are the determinants of eliminating cash. J. Appl. Econ. **25**(2), 543–562 (2022). https://doi.org/10.1080/15140326.2022.2052000
3. Ng, S.-H., Zhuang, Z.: Exploring herding behavior in an innovative-oriented stock market: evidence from ChiNext. J. Appl. Econ. **25**(1), 523–542 (2022). https://doi.org/10.1080/15140326.2022.2050992
4. Mulaga, A.N., Kamndaya, M.S., Masangwi, S.J.: Spatial disparities in impoverishing effects of out-of-pocket health payments in Malawi. Glob. Health Action **15**(1), 2047465 (2022). https://doi.org/10.1080/16549716.2022.2047465
5. Lee, S.T., Yang, E.B.: Factors affecting social accountability of medical schools in the Korean context: exploratory factor and multiple regression analyses. Med. Educ. Online **27**(4), 1–6 (2022). https://doi.org/10.1080/10872981.2022.2054049
6. Qiu, T., Ma, X., Luo, B., Choy, S.T.B., He, Q.: Land defragmentation in China: does rental transaction inside acquaintance networks matter? J. Appl. Econ. **25**(9), 259–278 (2022). https://doi.org/10.1080/15140326.2022.2043720
7. Jiang, M., Qi, J., Zhang, Z.: Under the same roof? The green belt and road initiative and firms' heterogeneous responses. J. Appl. Econ. **25**(7), 315–337 (2022). https://doi.org/10.1080/15140326.2022.2036566
8. Nester, M.S., Hawkins, S.L., Brand, B.L.: Barriers to accessing and continuing mental health treatment among individuals with dissociative symptoms. Eur. J. Psychotraumatology **13**(1), 5–12 (2022). https://doi.org/10.1080/20008198.2022.2031594
9. Zembe-Mkabile, W., Sanders, D., Ramokolo, V., Doherty, T.: I know what I should be feeding my child: foodways of primary caregivers of child support grant recipients in South Africa. Glob. Health Action **15**(1), 2014045 (2022). https://doi.org/10.1080/16549716.2021.2014045
10. Wildberger, J., Wenzel, K., Fishman, M.: Assessing clinical impacts and attitudes related to COVID-19 among residential substance use disorder patients. Substance Abuse **43**(4), 756–762 (2022). https://doi.org/10.1080/08897077.2021.2010249
11. Wang, Z., Wang, X., Yan, Xu., Cheng, Q.: Are green IPOs priced differently? Evidence from China. Res. Int. Bus. Financ. **61**, 101628 (2022). https://doi.org/10.1016/j.ribaf.2022.101628
12. Claudio, C.D.M.: A curiosa história do Real Collegio dos Nobres. Ensaio: Avaliação e Políticas Públicas em Educação **30**(8), 550–565 (2022). https://doi.org/10.1590/s0104-40362022003003764
13. Kim, H.M., Turesson, H., Laskowski, M., Bahreini, A.F.: Permissionless and permissioned, technology-focused and business needs-driven: understanding the hybrid opportunity in blockchain through a case study of insolar. IEEE Trans. Eng. Manage. **69**(10), 776–791 (2022). https://doi.org/10.1109/TEM.2020.3003565

14. Núñez, M.L., CáceresRuizDíaz, M., Correa, E.S.: Características de los proyectos financiados por el Consejo Nacionalde Cienciay Tecnologíade Paraguay Convocatorias 2013–2015. Poblacióny Desarrollo **28**(12), 55–67 (2022). https://doi.org/10.18004/pdfce/2076-054x/2022.028.54.055
15. Gulosino, C., Maxwell, P.: A comprehensive framework for evaluating Shelby County school district's voluntary preschool program: the challenges of equity, choice, efficiency, and social cohesion. Urban Educ. **57**(5), 779–813 (2022). https://doi.org/10.1177/0042085918801885

Data Quality Improvement Method for Power Energy Consumption Analysis in Customer-Side Management

Shiwang Yang[1], Qi Ding[2], Qingjuan Wang[2], Liuxin Yang[3], Wangying Jin[4], Peng Lu[5](✉), and Adil Israr[6]

[1] State Grid Zhejiang Electric Power Co., Ltd., Lishui Power Supply Company, Lishui, Zhejiang, China

[2] Marketing Service Center of State Grid, Zhejiang Electric Power Co., Ltd., Hangzhou, Zhejiang, China

[3] State Grid Zhejiang Hangzhou Yuhang District Power Supply Co., Ltd, Hangzhou, Zhejiang, China

[4] State Grid Zhejiang Electric Power Co., Ltd., Daishan Power Supply Company, Hangzhou, Zhejiang, China

[5] Institute of Computing Innovation, Zhejiang University, Hangzhou, Zhejiang, China
anthonyyang2021@163.com

[6] Balochistan University of Information Technology, Engineering and Management Sciences, Quetta 87300, Pakistan

Abstract. Data quality is of paramount importance for the data analytics of power consumption and behavior analysis. For the detection of missing data, firstly, according to the characteristics of missing data, the missing data is divided into typical deletion and atypical deletion, the location of typical missing segments is located through the absolute difference sequence of data, then the noise characteristics in the missing segments are learned, and then all suspected segments in the test data are generalized according to the noise characteristics, Finally, the linear correlation is used for secondary detection to generate the missing mask. The proposed solution is evaluated and its effectiveness is confirmed through simulation experiments.

Keywords: Missing data detection · Data recovery · Data analytics · Power marketing analysis

1 Introduction

Big data technology is a novel generation of information technology and paradigm. It is expected to extract value from different forms of super large-scale data with low cost and fast collection, processing and analysis technology. The continuous emergence and development of big data technology make it easier, cheaper and faster to deal with massive data, become a good assistant to use data, and even change the business model

J. H. Abawajy et al. (Eds.): ICATCI 2022, LNDECT 169, pp. 668–674, 2023.
https://doi.org/10.1007/978-3-031-28893-7_81

of many industries, e.g., the power industry. The development of big data-related technologies requires the timely collection and processing of massive data in power systems [1]. The statistics showed that the data collected in State Grid in China can exceed 60TB per day [2]. Big data in the power industry is considered with the following main characteristics:

(1) Large volume: a very large amount of collection, storage and calculation. The starting measurement unit of big data is at least at the P level (1000 T), E level (1 million T) or Z level (1 billion T).
(2) Variety: It includes structured, semi-structured and unstructured data, which are embodied in network logs, audio, video, pictures, geographic location information. Multiple types of data put forward higher requirements for data processing ability.
(3) Low-value density: the data value density is relatively low in practice or even not valuable in some cases. For the wide application of the power systems, the information is massive, but the value density is low.
(4) Fast speed: the data needs to be processed in a nearly real-time fashion, and hence it is required the fast processing speed, e.g., search engines generally demand that the few minutes to be queried by users.
(5) Data is online: the data needs to be available and calculated at any time. This is the biggest feature that big data is different from traditional data.

It should be noted here that there are often data missing occurs during the collection of the data in the application domain. In practice, the missing data for further processing and application can be simply categorized into two classes: continuous missing and discrete missing [3]. In practice, for the efficient utilization and application of the data in power system applications, the detection of missing data is of paramount importance and has been exploited in the literature (e.g., [4, 5]). In the traditional method, the detection based on residual and sudden change is proposed. Specifically, in the field of power systems, the existing study proposed a method combined with state pre-estimation to improve the performance [6]. However, due to the particularity of missing data, we can try to design a simpler but more targeted detection method only for missing data. As for the recovery of missing data in the power system, it can be regarded as a statistical interpolation problem. The mathematical methods consist of the mean filling method, polynomial interpolation, k-nearest neighbor method (e.g., [7, 8]). The aforementioned solutions are simple and easy to implement, but they are very sensitive to the neighboring data in small intervals before and after the missing data, which requires that the dataset itself varies gently. In addition, those methods cannot perceive the data features with a large time scale, and hence the performance on the long-time continuous missing data is not satisfactory.

With the recent advances of Artificial Intelligence (AI) technologies, the AI-based approaches for data recovery of power demand measurement in power systems have been exploited The authors in [9] proposed an unsupervised learning-based solution through using Wasserstein Generative Adversarial Network (WGAN) for the recovery of the missing data and information, including the measurements of active power, reactive power, voltage amplitude and phase measurements. In [10], a solution based on the Adaptive Neural Fuzzy Inference System (ANFIS) model was studied for the data recovery of wind power generation profiles.

In this work, the proposed solution developed a novel model for missing data detection and recovery of electric power load data. This is developed with the analysis and characterization of missing data in load data, a detection solution based on the absolute difference sequence and linear correlation is implemented to detect the missing mask and grade the mask as three levels according to the severity of missing. Secondly, since there is an evident periodicity in the load data, we can truncate the one-dimensional data into the two-dimensional matrix by proper cut width, and use the two-dimensional structure as data enhancement. Rather than using the feature vector as suggested in [11], this work considers these matrices as generalized images and transfer the original problem into image inpainting, then propose a cascaded convolutional neural network to hierarchically repair the graded missing data in different levels, which provides a new inspiration in this field.

The remainder of the work is organized as follows: Sect. 2 introduces the proposed missing data detection solution; Sect. 3 discussed the data measurement recovery method. The proposed solution is evaluated and the numerical results are presented in Sect. 4. Finally, Sect. 5 concludes this work with remarks.

2 Missing Measurement Data Detection

In this section, the mechanism in the generation of missing data is firstly overviewed, and a generic model for the missing mask is introduced. With the generative missing mask, we improved the present abnormal data detection methods to adapt to the missing data detection problem. Finally, we apply an improved normalization technique on the input data based on the detection results.

2.1 Missing Mask Generation

To describe the real situation when data missing, this paper first models the process of data missing and designs the generation model for the binary missing mask. Some relative variables are defined in Table 1, and (1) to (3).

$$m_i = \begin{cases} 1, x_i \ is\ missing\ data \\ 0, x_i \ is\ normal\ data \end{cases} \tag{1}$$

$$X^{`} = X - X \odot M + \aleph \odot M \tag{2}$$

$$\mathrm{NMD} = n\gamma \tag{3}$$

Here, it is considered that the data with missing as the normal data minus the element-wise product of a binary missing mask and normal data, then add the product of the mask and Gaussian noise in (2–2). The data in each position of the binary missing mask is a binary code, in which 1 represents the data in this position is missing, while 0 means normal.

Table 1. Variables to describe the missing process

Variables	Definition
$X(x_1, x_2, \ldots, x_{n-1}, x_n,)$	Normal data
$M(m_1, m_2, \ldots, m_{n-1}, m_n,)$	Missing binary mask
$X^{`}\left(x^{`}_1, x^{`}_2, \ldots, x^{`}_{n-1}, x^{`}_n,\right)$	Data under the missing mask
$\aleph(\mu, \sigma)$	Gaussian noise on missing data
$\gamma \in (0, 1)$	Missing rate
NMD	Num of missing data
$\odot$	Element-wise production operator

Under the given missing rate, if NMD missing points are randomly selected directly from n bits of M, the probability of discrete missing will be far greater than that of continuous missing; if we first slice the NMD missing points into several continuous segments, and then randomly place these segments in M, the collision might happen, which decrease the real missing rate, γ, as shown in Fig. 1.

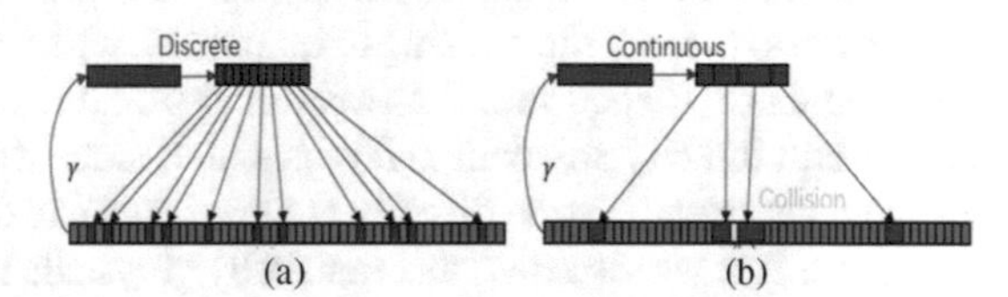

Fig. 1. The missing mask generation for missing data: (a) discrete missing case and (b) continuous missing case

2.2 Noise Feature Learning and Extended Detection

Compared with the problem of missing data recovery, the problem of missing data detection does not seem to get much attention. In the previous research on missing data recovery, scholars usually label the missing mask artificially, and then design algorithms for recovery. However, in a practical situation, how to automatically detect the missing mask is a kind of complementary part of the repair issue. Therefore, we design a missing detection algorithm based on the characteristics of missing data. In this work, the SNR between this normal data and the noise signal is defined in (3).

$$SNR = 10 \log \frac{P_{data}}{P_{noise}} \tag{4}$$

According to the previous analysis, another important feature is that the linear correlation coefficient r (2–9) between the missing segments and time is lower because the data in missing segments is replaced by unknown noise, while the correlation between the normal data and time is higher in the small window, as shown in Fig. 2. As the result, we can locate the typical missing segments by checking the sudden jump in the absolute difference sequence and the linear correlation.

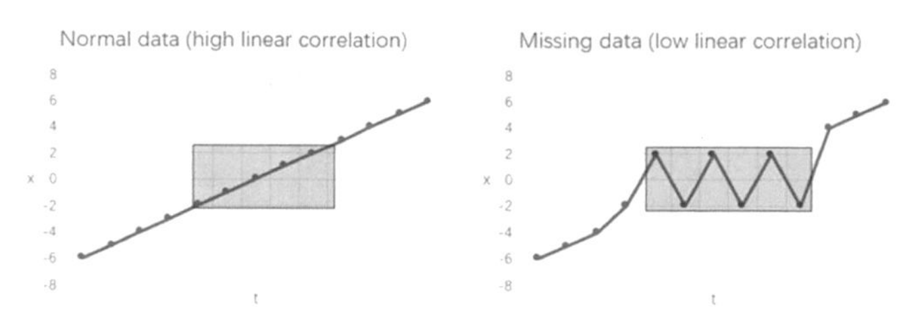

Fig. 2. The linear correlation for normal data and missing data (noise)

3 Recovery of Missing Data Measurement

In this work, a data enhancement algorithm is developed to reconstruct the data from one dimension to two dimensions that can be considered as "generalized" images. After this, the proposed solution divides the detected missing mask into three grades and adopts the image processing techniques to recover the missing data. In [11], when one-dimensional data is truncated by period and reshaped into the two-dimensional matrix, the original directly adjacent data are still directly adjacent and distributed alone the row, while the original indirectly adjacent data also become directly adjacent data and distributed along the column, as shown in Fig. 3. If we visualize this matrix by grayscale image, the direct adjacency in the rows and columns will constitute some regular texture on this image.

It can be seen that when the cut width is not an integral multiple of 96 (that is one day), the image will have horizontal and oblique stripes due to the displacement in the columns, and the image semantics will be confused seriously. Only when the cut width is an integral multiple of 96, the image will only have horizontal stripes, and the image texture is relatively regular and clear; Because the attenuation of the semantic information in this image is directly proportional to the cut width, we will choose 96, also one day, as the cut width, to ensure the image as clear as possible.

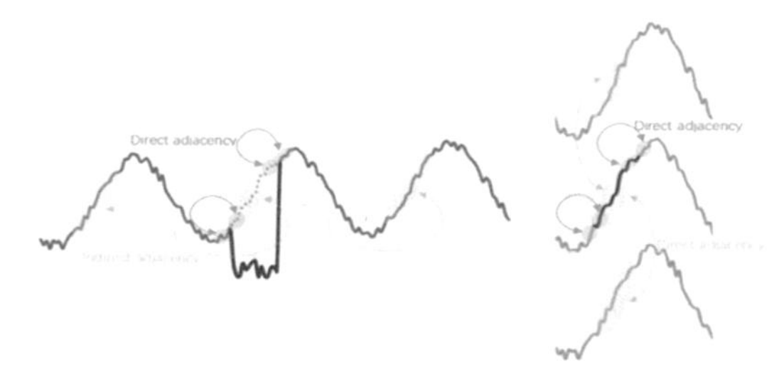

Fig. 3. Both the direct and indirect adjacency become direct adjacency

4 Simulation Experiments and Numerical Results

In this study, the power load data during 2014–2020 is adopted in the simulation experiments. For the missing recovery model, we will evaluate the algorithm under different padding depths p including zero. The step of the slice window is 4 and the total number of slices will be 4800, among which 4000 for training and 800 for the test, as presented in Table 2.

Table 2. Parameters for missing masks in training and test datasets

Parameters	Range
γ	{0.05, 0.1, 0.2}
(α, β)	{(0.1, 0.15), (0.1/2, 0.15/2), (0.1/4, 0.15/4)}

Here, the Cascaded Pure Convolutional Auto Encoder (CPCAE) based models on different padding depths from 0 to 11 are evaluated, corresponding to the padding ratio from 0% to 23%. During the training process, the batch size is set as 20 and the initial learning rate as 0.01 exponentially decreases along the epochs. We will stop the training until the loss function will not decrease. The numerical results are shown in Table 3 in detail and the Mean Absolute Error (MAE) is used as a complementary index.

Table 3. Recovery RMSE (MAE) for data normalized in (0.01, 1)

Padding depth	CPCAE $\left(\times 10^{-2}\right)$	Improved CPCAE $\left(\times 10^{-2}\right)$
$p = 0$	4.15 (2.72)	3.15 (2.11)
$p = 7$	3.94 (2.61)	2.92 (2.01)
$p = 9$	3.79 (2.54)	2.86 (1.97)
$p = 11$	3.84 (2.56)	2.89 (1.99)

5 Conclusions and Remarks

In this work, a novel missing power load data detection and recovery solution are developed based on the improved CPCAE. For data detection, the missing generation process is firstly accurately modeled to generate the missing mask and then combine the absolute difference sequence and the linear correlation as the metric for the detection of missing segments under different SNR scenarios. In addition, the detected missing mask is further divided into three grades and then reshape the origin data with one dimension into the matrices with two dimensions to enhance the data management. The proposed solution is extensively evaluated through simulation experiments and the numerical results demonstrated the proposed detection and recovery model perform well under different missing situations.

Acknowledgments. This work is supported by the Science and Technology project of the State Grid Zhejiang Electric Power Co., Ltd "Research on basic supporting technology for Deepening Application of power marketing big data".

References

1. Qingyang, L., Long, Y., Wang, J., Xiao, C., Song, Y., Li, H.: Application analysis of big data technology in energy internet. In: International Conference on Intelligent Computing, Automation and Systems (ICICAS) (2020)
2. Dash, S.K., Ray, P.K.: Power quality improvement utilizing PV fed unified power quality conditioner based on UV-PI and PR-R controller. CPSS Trans. Power Electron. Appl. **3**(3), 243–253 (2018)
3. Dong, H., Yuan, S., Han, Z., Ding, X., Ma, S., Han, X.: A comprehensive strategy for power quality improvement of multi-inverter-based microgrid with mixed loads. IEEE Access **6**, 30903–30916 (2018)
4. Shah, P., Hussain, I., Singh, B., Chandra, A., Al-Haddad, K.: GI-based control scheme for single-stage grid interfaced SECS for power quality improvement. IEEE Trans. Ind. Appl. **55**(1), 869–881 (2019)
5. Kumar, N., Singh, B., Panigrahi, B.K.: Framework of gradient descent least squares regression-based NN structure for power quality improvement in PV-integrated low-voltage weak grid system. IEEE Trans. Industr. Electron. **66**(12), 9724–9733 (2019)
6. Miller, J., Parast, M.M.: Learning by applying: the case of the Malcolm Baldrige National Quality Award. IEEE Trans. Eng. Manage. **66**(3), 337–353 (2019)
7. Uprety, S., Cao, C., Gu, Y., Shao, X., Blonski, S., Zhang, B.: Calibration improvements in S-NPP VIIRS DNB sensor data record using version 2 reprocessing. IEEE Trans. Geosci. Remote Sens. **57**(12), 9602–9611 (2019)
8. Neece, M.R., Wulf, S.A.: Selecting excellence: a data-driven process to hire top performance engineers with greater speed and certainty. IEEE Eng. Manag. Rev. **47**(1), 11–16 (2019)
9. Pourramezan, R., Karimi, H., Mahseredjian, J., Paolone, M.: Real-time processing and quality improvement of synchrophasor data. IEEE Trans. Smart Grid **11**(4), 3313–3324 (2020)
10. Borghetti, A., Bottura, R., Barbiroli, M., Nucci, C.A.: Synchrophasors-based distributed secondary voltage/VAR control via cellular network. IEEE Trans. Smart Grid **8**(1), 262–274 (2017)
11. Derviškadić, P.R., Pignati, M., Paolone, M.: Architecture and experimental validation of a low-latency phasor data concentrator. IEEE Trans. Smart Grid **9**(4), 2885–2893 (2018)

Deep Learning-Based Cybersecurity Situation Assessment Method in Big Data Environment

Yan Hu[1], Jian He[2(✉)], and Amar Jain[3]

[1] Department of Continuing Education, Luoding Open University, Luoding 527200, China
[2] Department of Information Engineering, Luoding Polytechnic, Luoding 527200, China
hjy071231@163.com
[3] Madhyanchal Professional University, Ratibad, Bhopal, India

Abstract. In order to better evaluate the security situation of cyberspace, this paper will study a deep learning evaluation method based on big data environment. In this paper, the basic idea of cyberspace security situation assessment is firstly proposed, and then the algorithm principle of assessment method is introduced. Through the study in this paper, the application of deep learning network space security situation assessment method in the big data environment is effective, and the effectiveness is better than traditional methods.

Keywords: Big data · Deep learning · Cyberspace security situation assessment

1 Introduction

Big data is a huge data integration body with massive internal data levels, which can provide powerful data support for various systems. Therefore, many application systems with excellent functions have been developed in the environment of big data, including deep learning system. Deep learning system is mainly based on the neural network structure, analysis of data from big data, analysis the relationship between the available data, the correlation degree and so on depth information, and every time after analysis of the results will be saved, so later in the system application in the same data, the system can be directly, which can identify and further work. This condition, because of deep learning system application range is very wide, so it is also used in the field of cyber security situation assessment, main is to use the system analysis of the learning process of the cyberspace security situation in the related data, it can know the security situation in the process of learning, according to the direction can be judged whether the safety level of cyberspace.

Cyberspace security situation assessment is the modern network users for network security and puts forward demand, namely since spread in our country network, network security issues have existed, although modern people have strong consciousness of network safety protection, use a variety of network security protection measures to protect their network, but because the network security hidden danger in constantly update and upgrade, Makes this problem cannot effect a radical cure, so deal with security update, upgrade in time is the best way to understand its updates, upgrades, determine whether

J. H. Abawajy et al. (Eds.): ICATCI 2022, LNDECT 169, pp. 675–684, 2023.
https://doi.org/10.1007/978-3-031-28893-7_82

the network has the defects of insufficient safety levels, once finding defects is the first time, strengthen the safety protection, for this purpose they need to get network space safety situation assessment service. And space of traditional network security situation assessment service generally has some problems, such as inaccurate evaluation results, the evaluation process is not comprehensive, etc., the reason is that the traditional evaluation method for the service lack the support of big data and study the depth of the big data environment system can effectively solve this problem, make the evaluation results more accurate, more comprehensive evaluation process. However, in order to better promote the relevant methods, it is necessary to carry out research.

The cyberspace security situation is a new concept emerging in the context of network globalization. That is, before 2017, the network was popularized in all countries in the world. At this stage, various security problems often occurred in the network, which posed a serious threat to the property and privacy of network users. Therefore, all countries actively carried out internal network security construction and management, Even put forward new laws and regulations at the judicial level. However, these efforts did not achieve ideal results at that time. In fact, after 2017, various network security problems were still at an endless level, and the problem was still deteriorating at the macro level. There were examples of transnational extortion. The number and type of various malicious software were also constantly updated, indicating that the problem had not been thoroughly addressed. In the face of this situation, people think that we can't just deal with the problems that have occurred, that is, we should solve the problems from the perspective of prevention, so that the related network security problems can not pose a threat. This idea has been widely recognized by people, and also triggered a new round of research. Therefore, with the deepening of research, the relevant fields have put forward a two-level framework for the prevention of network security problems, the first of which is from a macro perspective, The security situation assessment of a specific cyberspace shows that the security level of the cyberspace is low. The lower the security level, the higher the probability, frequency, and degree of cyber attacks on the cyberspace. Therefore, the cyberspace can be regarded as a key target, and then enter the second layer. That is, a more complete and effective prevention system is established against the main reasons for the low security level of the cyberspace, So as to improve its macro level of network security.

It can be seen that the cyberspace security situation is a necessary prerequisite for carrying out network security protection and thoroughly solving problems, which needs to be assessed. However, because of the complexity of the cyberspace structure, a large amount of data must be involved in assessing its security situation. Therefore, a specific big data body, called the cyberspace security situation big data, was born in the cyberspace security situation assessment. This big data body is the same as other big data bodies, but also has the characteristics of huge data volume, various data types, and complicated relationships between different types of data, and iterative data blocks for updating each type of data. To accurately assess the security situation of cyberspace, it is necessary to collect all these data, and then conduct a comprehensive analysis. It is required to establish all data relationships in order to obtain accurate assessment results. It shows that the assessment of cyberspace security situation is a very complex work, which can not be completed by traditional technical means and manual work, but must use more advanced methods, which is also the significance of this study.

In view of the above problems, people have not found a way to process big data for a long time. For example, some early technical means can quickly process massive data, but they cannot analyze the data, or the analysis results are not accurate enough. The reason is that these technologies do not have the function of artificial intelligence and can only be analyzed according to the predetermined logic, which cannot cover all data relationships in big data, Although artificial has intelligent function, it has defects in work efficiency and can't handle massive data. This phenomenon continues until the intelligent technology and artificial neural network model are mature. The reason is that the combination of the two can replace manual processing of massive data and make up for the artificial defects. At the same time, the technical system after the combination of the two can be analyzed through a kind of thinking logic similar to the function of artificial intelligence. The results obtained are no different from the results obtained under human thinking, and even better than the results obtained by human, It shows that the technical system can help people to deal with the problems in big data processing, and represents that cyberspace security situation assessment can be realized. The intelligent technology in the technical system is also an automation technology in essence, but compared with the conventional automation technology, the former has an intelligent module, which can carry an artificial neural network model. Therefore, under the role of the model, the intelligent technology can process data through intelligent logic, such as in-depth learning of data, and constantly evaluate the security situation of cyberspace according to the learning results. The artificial neural network model is the source of intelligent logic. It is a simulation model constructed by imitating the human neural structure and the conduction form of neuron nodes under different thinking modes, which can endow the technical system with intelligent logic and realize deep learning. Artificial neural networks have many forms in the current development, including feedforward, feedback, convolution, etc. Different forms of artificial neural networks have different application conditions, and need to make a reasonable choice according to the actual situation, otherwise the quality of results cannot be guaranteed. In network space security situation assessment, feedforward neural networks are generally recommended. This neural network model has the characteristics of logical forward without repetition, it conforms to the problem solving logic of cyberspace security situation assessment.

2 The Basic Idea of Cyberspace Security Situation Assessment

The cyberspace security situation is a new concept emerging in the context of network globalization. That is, before 2017, the network was popularized in all countries in the world. At this stage, various security problems often occurred in the network, which posed a serious threat to the property and privacy of network users. Therefore, all countries actively carried out internal network security construction and management, Even put forward new laws and regulations at the judicial level. However, these efforts did not achieve ideal results at that time. In fact, after 2017, various network security problems were still at an endless level, and the problem was still deteriorating at the macro level. There were examples of transnational extortion. The number and type of various malicious software were also constantly updated, indicating that the problem had not been thoroughly addressed. In the face of this situation, people think that we can't just deal with the problems that have occurred, that is, we should solve the problems from the perspective of prevention, so that the related network security problems can not pose a threat. This idea has been widely recognized by people, and also triggered a new round of research. Therefore, with the deepening of research, the relevant fields have put forward a two-level framework for the prevention of network security problems, the first of which is from a macro perspective, The security situation assessment of a specific cyberspace shows that the security level of the cyberspace is low. The lower the security level, the higher the probability, frequency, and degree of cyber attacks on the cyberspace. Therefore, the cyberspace can be regarded as a key target, and then enter the second layer. That is, a more complete and effective prevention system is established against the main reasons for the low security level of the cyberspace, So as to improve its macro level of network security.

It can be seen that the cyberspace security situation is a necessary prerequisite for carrying out network security protection and thoroughly solving problems, which needs to be assessed. However, because of the complexity of the cyberspace structure, a large amount of data must be involved in assessing its security situation. Therefore, a specific big data body, called the cyberspace security situation big data, was born in the cyberspace security situation assessment. This big data body is the same as other big data bodies, but also has the characteristics of huge data volume, various data types, and complicated relationships between different types of data, and iterative data blocks for updating each type of data. To accurately assess the security situation of cyberspace, it is necessary to collect all these data, and then conduct a comprehensive analysis. It is required to establish all data relationships in order to obtain accurate assessment results. It shows that the assessment of cyberspace security situation is a very complex work, which can not be completed by traditional technical means and manual work, but must use more advanced methods, which is also the significance of this study.

In view of the above problems, people have not found a way to process big data for a long time. For example, some early technical means can quickly process massive data, but they cannot analyze the data, or the analysis results are not accurate enough. The reason is that these technologies do not have the function of artificial intelligence and can only be analyzed according to the predetermined logic, which cannot cover all data relationships in big data, Although artificial has intelligent function, it has defects in work efficiency and can't handle massive data. This phenomenon continues until the intelligent technology and artificial neural network model are mature. The reason is that the combination of the two can replace manual processing of massive data and make up for the artificial defects. At the same time, the technical system after the combination of the two can be analyzed through a kind of thinking logic similar to the function of artificial intelligence. The results obtained are no different from the results obtained under human thinking, and even better than the results obtained by human, It shows that the technical system can help people to deal with the problems in big data processing, and represents that cyberspace security situation assessment can be realized. The intelligent technology in the technical system is also an automation technology in essence, but compared with the conventional automation technology, the former has an intelligent module, which can carry an artificial neural network model. Therefore, under the role of the model, the intelligent technology can process data through intelligent logic, such as in-depth learning of data, and constantly evaluate the security situation of cyberspace according to the learning results. The artificial neural network model is the source of intelligent logic. It is a simulation model constructed by imitating the human neural structure and the conduction form of neuron nodes under different thinking modes, which can endow the technical system with intelligent logic and realize deep learning. Artificial neural networks have many forms in the current development, including feedforward, feedback, convolution, etc. Different forms of artificial neural networks have different application conditions, and need to make a reasonable choice according to the actual situation, otherwise the quality of results cannot be guaranteed. In network space security situation assessment, feedforward neural networks are generally recommended. This neural network model has the characteristics of logical forward without repetition, it conforms to the problem solving logic of cyberspace security situation assessment. Figure 1 shows the influence mechanism of hidden dangers in cyberspace.

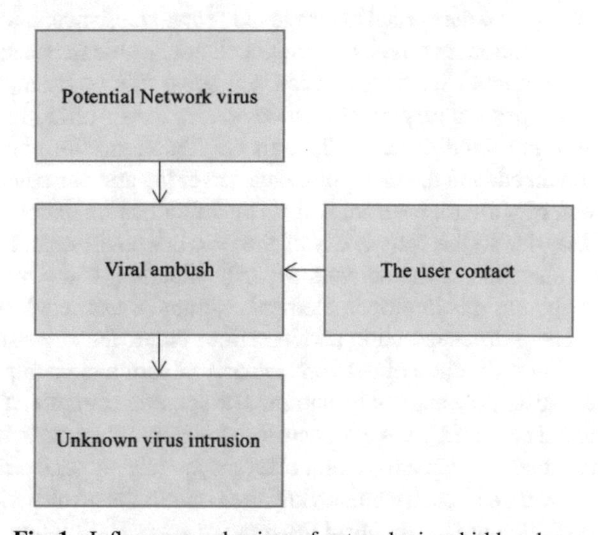

Fig. 1. Influence mechanism of network virus hidden danger

Figure 2 is the basic idea of cyberspace security situation.

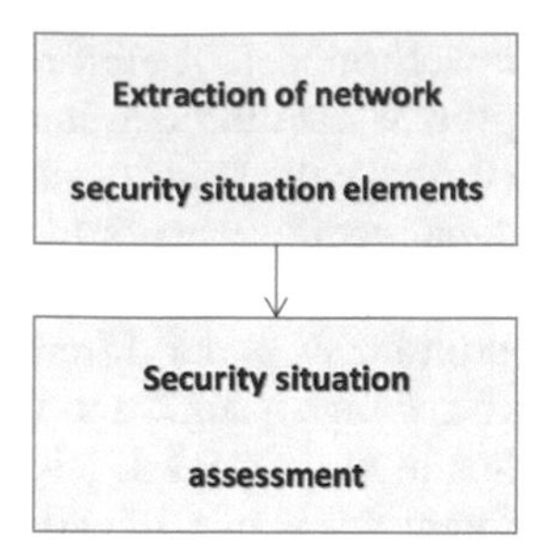

Fig. 2. Basic idea of cyberspace security situation

In Fig. 1, the basic idea of cyberspace security situation is not complicated, which is divided into two parts: element extraction and security assessment: First, because the internal environment of high complexity of modern network space, lead to related diversified network data attributes, such as data source, duplicate records, etc., so the network data with higher dimensional heterogeneous characteristics, internal contains a lot of redundant information, in this case, people will not be able to use the data directly, Network security situational factor extraction can solve this problem, make the data can be applied to network security situation assessment, the so-called security situation elements extraction, is a kind of can collect all the figures on the impact of network environment has changed, and the method of data screening and classification, can be in the complex system of network environment on the initialization data for processing. It should be noted that due to the complexity of the network environment itself, it is also difficult to extract elements of the network security situation. If the method is chosen incorrectly or the operation is improper, elements cannot be extracted well, which will affect the subsequent assessment work, and in serious cases, the assessment work may not be carried out. The extraction of network security situation elements is divided into two stages, namely discovery and confirmation. The specific contents of the two stages are shown in Table 1. Second, network security situation assessment is based on the extracted security situation elements, using relevant models and data analysis work to judge and analyze the data of each element, and then output the results, and compare the results with security standards, to confirm the trend of security situation. Under normal circumstances, the main consideration in network security situation assessment are the elements of network safety record history, suspected attacked records, etc., each element can be divided into a number of indicators, such as network safety record history can be divided into successful safety records, failure safety records, etc., and establish a perfect index system is the basic condition to guarantee the quality of the evaluation results. Network security situation assessment is a dynamic process, the basic idea of which is to integrate all the influencing factors, and then make judgment according to the implicit relationship between each influencing factor. This process goes round and round, and the result obtained each time is called the "final security situation value". According to this value, the security situation in cyberspace can be accurately judged according to the established rules [1–3]. Table 2 shows the rules for judging the security situation in cyberspace.

Table 1. Two stages and description of the extraction of security situation elements in cyberspace

Phase	Content
Found	Detect abnormal behavior
Confirm	Identify the source and attack type, and classify them

Table 2. Network space security situation judgment rules

The final security situation value exceeded the standard	Good security posture
The final security situation value is equal to the standard	General security situation
The final security situation value was below standard	Poor security posture

3 Algorithm Principle of Cyberspace Security Situation Assessment Method

3.1 Machine Learning

The algorithms involved in the cyber space security situation assessment method mainly serve machine learning, that is, machine learning is a kind of machine system to operate according to the pre-set budget logic. In the operation, the deep information of data will be mined through relevant algorithms. Therefore, machine learning is also called "deep learning" [4]. Machine learning is unable to operate out of thin air, must satisfy the four prerequisites before operation, the machine system must have the data, data conversion model, the loss function and train parameters, in the midst of network security situation assessment, to meet the requirements of the four prerequisites for four work, data preprocessing, feature extraction, the attribute reduction, model training, Including data pretreatment is mainly to obtain the data cleaning, integration, etc., especially in the concrete is mainly to extract data security attributes, and manage the attributes, attribute reduction is mainly due to security features of each data attribute is too complex, there are part of the property may be unnecessary attributes, too much to deal directly lead to convergence problems, so for data security feature dimension reduction or attribute integration, makes the attribute reduction and model training is mainly in the related algorithm, on the basis of reference algorithm for training, training every time can make the parameters of assessment model was optimized, until the model parameter optimization to termination conditions, in such a model on the evaluation of the results will be more accurate [5, 6]. Table 3 shows the main algorithms used in machine learning.

Table 3. Main algorithms used in machine learning

The algorithm name	Algorithm description
Dimension reduction algorithm	The main function is to eliminate the redundant attributes of network data, achieve the goal of attribute dimension reduction, and reduce the complexity of the whole calculation process. The main algorithm types include PCA, PLS, etc
Clustering algorithm	In accordance with the principle of similarity within classes and maximum gap between classes, the distance vector between sample points is calculated, and the sample points are classified according to the magnitude of vector-value, generally referred to as the K-means algorithm
Classification algorithm	It is necessary to assign labels to the data training classification model first, and then calculate the label model. The calculation results can be used for the classification of unknown network sample data. At the same time, the data characteristics are proposed, which generally refers to the naive Bayes algorithm

3.2 Neural Network

Neural network is a network security situation assessment model, its itself is a mimic human neural network structure of the structure, internal contains a large number, and a variable number of data nodes, each node is equal to the human neurons, also known as the neuron nodes, and according to the weight to differentiate between neuron node connection path, it is used to learn the implicit information in the input data and classify the input data according to the implicit information. The classification result is very accurate and the purpose of information processing is achieved after classification. Can see that the neural network model is mainly composed of a number of neuron node, each node is a neural network data processing unit, arbitrary neurons possess the output data to other nodes of neurons into the function of the input data, therefore, can be repeated training, to find the relationship between the data in the project, each neuron node and know the correlation between the data items, relationship represents the possible changes of data items in the future, which can judge whether the changes are good or bad, while correlation is an important basis for subsequent data information classification. In actual application of neural network is often seen as a big data processing tool, the reason is that other data processing tools other than the neural network on the functionality, performance, has certain limitation, in the face of big data in large scale and complex data relationships, these data processing tools do not work, it is concluded that the results are often poorly, even the phenomenon of there is not functioning, thus far only can neural network to deal with large data, but as a large data processing tool, neural network operation process is relatively long, which contains a lot of training, and each time can let the weight of neural network model is worth to a reasonable adjustment, the result accuracy improvement [7–10].

In order to meet the requirements of big data application in different fields, neural network models have been developed in related fields, and feedforward neural network, feedback neural network, self-organizing neural network and so on have been proposed. In the security situation assessment of network space, feedforward neural network model is the most suitable one. Type feedforward neural network is a typical one-way network, multilayer structure is mainly composed of input layer, hidden layer and output layer, each layer with a layer of only one-way relationship, and each neuron in each layer, there is no link between nodes, so that in the process of information transmission, the previous layer of output data can be converted to a layer of input data, the final result is obtained by the output layer. Therefore, in the network space security situation assessment, only the initial data as input data into the feedforward neural network model, can get accurate results. It should be noted that when using feedforward neural network for network space security situation assessment, attention must be paid to the selection of activation function, because the wrong selection will lead to inaccurate output results. In general, the commonly used activation functions of feedforward neural network include Relu function, Sigmoid function and Tanh function, among which Relu is the most common and is a typical piecewise linear function. Relu is characterized as zero when the function is negative, and the positive value remains unchanged, so it can play a unilateral inhibition effect. It is very suitable for network space security situation assessment based on feedforward neural network. Formula (1) is the expression of Relu function.

$$f(x) = \max(0, x) \tag{1}$$

3.3 Deep Learning

Deep learning type is a feed-forward neural network implemented by a function, this function will be in bank of feedforward neural network's input layer, and output layer support, through the way of training step by step to learn data, compared with the previous people through iterative algorithm of machine learning function, deep learning to avoid the gradient diffusion modes, This advantage can avoid the generalization of assessment results in the security situation assessment of cyberspace, that is, the security risks that do not belong to the local cyberspace are included in the calculation, so that the originally safe cyberspace is assessed as an unsafe space. At present, deep learning in cyberspace security situation assessment is the main application way of "bottom-up" unsupervised learning, it is a progressive upward from the grass-roots level of the study way, each layer will this layer parameters in the model training, until the termination of this layer parameters conditions will enter the next layer, eventually into the top overall is terminated, the entire process without supervision. This learning method is mainly used for analysis of network space safety data, understand the data attributes, can make the system which can identify the data, i.e. whether the existence of the data is helpful for safety, if is helpful for safety, so the existence of the data attributes will play in the assessment points, points effect conversely, such already can get accurate evaluation results.

4 Conclusion

To sum up, since the methods used in traditional cyber space security situation assessment have defects, it is suggested that the majority of network users actively introduce deep learning methods based on big data environment to deal with the problems. The actual application process of this method may be more complex than the traditional method, but as long as the program setting, the system can run itself, so the convenience is better than the traditional method, and the results obtained by this method are more accurate under the support of the algorithm, neural network model and other elements.

Acknowledgement. This work was supported by Scientific Research Project of Guangdong Provincial Education Department under grant No. 2019GKTSCX132.

References

1. Ozkan, B.E., Bulkan, S.: Hidden risks to cyberspace security from obsolete COTS software. In: 2019 11th International Conference on Cyber Conflict (CyCon) (2019)
2. Rahman, M.: CERTIFICATE OF APPRECIATION Mr. MOIDUR RAHMAN As the keynote speaker on International Seminar on "Network Security and Cryptography" organized by the Network Security and Cryptography (2021)
3. Ashoor, A.S., Kazem, A., Gore, S.: An interactive network security for evaluating linear regression models in cancer mortality analysis and self-correlation of errors by using Durbin-Watson tests in Babylon/Iraq. J. Phys. Conf. Ser. **1804**(1), 012127 (2021)
4. Einy, S., Oz, C., Navaei, Y.D.: The anomaly-and signature-based IDS for network security using hybrid inference systems. Math. Probl. Eng. **2021**(9), 1–10 (2021)
5. Majeed, A., ur Rasool, R., Ahmad, F., Alam, M., Javaid, N.: Near-miss situation based visual analysis of SIEM rules for real time network security monitoring. J. Ambient Intell. Humaniz. Comput. **10**(4), 1509–1526 (2018). https://doi.org/10.1007/s12652-018-0936-7
6. Sethi, T., Mathew, R.: A study on advancement in honeypot based network security model. In: 2021 Third International Conference on Intelligent Communication Technologies and Virtual Mobile Networks (ICICV) (2021)
7. Burbank, J.L.: Security in cognitive radio networks: the required evolution in approaches to wireless network security. In: 2008 3rd International Conference on Cognitive Radio Oriented Wireless Networks and Communications (CrownCom 2008). IEEE (2008)
8. Popovic, D.H., Greatbanks, J.A., Begovic, M., et al.: Placement of distributed generators and reclosers for distribution network security and reliability. Int. J. Electr. Power Energy Syst. **27**(5/6), 398–408 (2005)
9. Hilker, M.: Next challenges in bringing artificial immune systems to production in network security. Prenat. Diagn. **7**(4), 239–244 (2008)
10. Roy, S., Ellis, C., Shiva, S., et al.: A survey of game theory as applied to network security. In: 2010 43rd Hawaii International Conference on System Sciences. IEEE (2010)

Emission Characteristics of Methanol/Diesel Dual Fuel Engine Under Electric Control System Development

Defeng Zhou(✉)

School of Automotive and Traffic Engineering, Jiangsu University, Zhenjiang 212013, China
ezsfdk@126.com

Abstract. In order to fully understand the emission characteristics of methanol/diesel dual-fuel engine under the development of electronic control system, relevant research will be carried out in this paper. Firstly, the development method of methanol/diesel dual fuel engine was proposed, and then the emission test was carried out to understand the emission characteristics of the engine. Through this study, the emission characteristics of methanol/diesel dual-fuel engine under the development of electronic control system are excellent, and it has advantages in many aspects such as economic benefits and environmental protection.

Keywords: Electric control system · Methanol/diesel dual fuel engine · Emission characteristics

1 Introduction

Since the rise of the automotive industry, car gradually into the life of people, the traditional automobile main use of energy are oil, the oil belongs to the non-renewable energy, the high popularity of the modern car, the world's oil reserves are decreasing of speed of development couldn't keep up with supply needs completely, so all countries in the world began to actively seek energy transformation, especially in China, as a lean-oil country. As early as 1993, China has become a net importer of oil, and its annual financial expenditure is very large, so it is urgent for China to transform its energy resources. Application of petroleum energy, meanwhile, although can provide strong driving force for the car, make people travel more convenient, the urban communication more convenient, but also brought serious environmental pollution phenomenon, such as the fog haze with vehicle emission in recent years have inalienable relations, therefore both in terms of economic efficiency, or the point of view of environmental protection, Our country needs to realize energy transformation as soon as possible.

On this basis, domestic related fields have carried out researches from multiple perspectives and proposed many new energy sources that can replace petroleum, including methanol energy. Methanol based on energy, the internal combustion engine can't continue to use, must be dedicated to develop methanol power engine, the engine was

J. H. Abawajy et al. (Eds.): ICATCI 2022, LNDECT 169, pp. 685–692, 2023.
https://doi.org/10.1007/978-3-031-28893-7_83

originally called methanol fuel engine, but the power of the engine at this stage of testing development efficiency of the weak, and the problem still cannot solve directly under the existing technical level, Therefore, the researchers improved the methanol fuel engine, followed the traditional diesel fuel engine system, and developed the methanol/diesel dual-fuel engine, with the help of which to make up for the defects of the methanol fuel engine. And methanol/diesel dual fuel engine proposed time is relatively short, the economic benefit, environmental protection, etc. It remains to be verified, which need to methanol/diesel dual fuel engine related research, therefore, because engine emission characteristics is to validate an important index of economic efficiency, environmental protection, so you can, taking the research of this feature.

2 Development Method of Methanol/Diesel Dual Fuel Engine

2.1 Introduction to Research Background

In the past, the main generator set used by people was diesel engine set, which is still the mainstream in the modern engine industry. The methanol/diesel dual fuel engine is a new type of engine equipment emerging in recent years, which has the advantages of low emissions and low cost, so it has been widely concerned in a short time. Under this condition, people began to compare the diesel engine set and methanol/diesel dual fuel engine set, hoping to understand their advantages and disadvantages. According to this idea, the relevant fields explained that methanol is the main fuel of methanol/diesel dual fuel engine compared with diesel engine, which makes the cost of methanol/diesel dual fuel engine in use lower, At the same time, methanol as a fuel can provide strong power for other units. The power is not different from that of diesel engines, even higher than that of diesel engines. In addition, methanol is a gas fuel, so methanol/diesel dual fuel engine has good emission performance and low pollution. At the same time, methanol can be stored in liquid form. The storage method is very simple, and there is no need to invest too much cost or resources. Therefore, methanol/diesel dual fuel engine has more advantages than diesel engine.

Taking the K6JTA12.8 methanol/diesel dual fuel engine as an example, the engine mainly uses vermicular graphite cast iron cylinder block and cylinder head. The materials have greatly improved the strength of the engine, which is more than three times higher than that of the diesel engine. This ensures the sustainable operation capability of the engine, that is, the overall reliability. At the same time, the engine is also equipped with diamond coated pistons, piston rings and cylinder liners, which greatly reduces the friction coefficient during engine operation and the anti-wear ability of the engine itself. Compared with traditional engines such as diesel engines, methanol/diesel dual fuel engines have better friction coefficient and anti-wear ability. In addition, according to the data test, the service life of the methanol/diesel dual fuel engine is 1.5 million km/25000 h, which is much longer than that of the diesel engine. The DMCC3.0 electronically controlled methanol system used internally can improve the power performance of the engine by about 10% and reduce the temperature by 15%. Even at an altitude of 4000 m, its power attenuation can be kept within 10%, which is impossible for traditional diesel engines.

It can be seen that the methanol/diesel dual fuel engine has many advantages, which indicates that it is worth popularizing. However, because the engine has been launched for a short time, many people do not know about it. The relevant fields have a vague understanding of its emission characteristics, that is, the advantages of methanol/diesel dual fuel engine mostly come from methanol, which is a fuel. The use of methanol as a fuel has changed the fuel system in traditional engine cognition, It will also change the engine emissions, which is a topic of concern for many people. Under this condition, most people do not know enough about the emission characteristics of methanol/diesel dual fuel engines, which will limit the promotion of methanol/diesel dual fuel engines. Therefore, in order to let more people know about methanol/diesel dual fuel engines and promote the promotion of methanol/diesel dual fuel engines, it is necessary to study their emission characteristics.

2.2 Main Ideas

In order to test the emission characteristics of methanol/diesel dual-fuel engine, it is necessary to design the engine before the test. In order to ensure the orderly design work, it is necessary to clarify the main design ideas before the design. The development ideas of methanol/diesel dual-fuel engine in this paper are shown in Fig. 1.

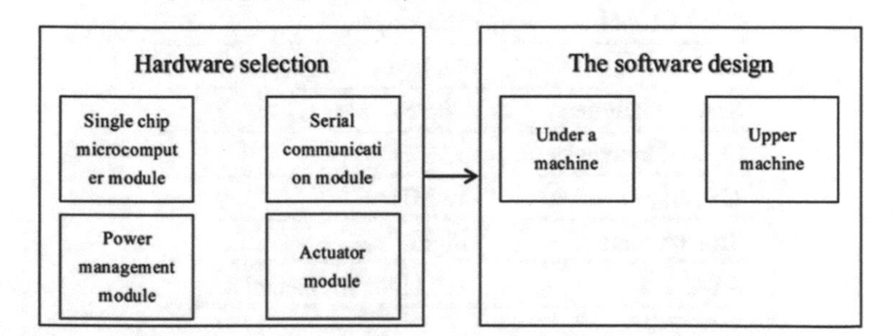

Fig. 1. Development idea of methanol/diesel dual fuel engine

According to Fig. 1, the development ideas of methanol/diesel dual-fuel engine in this paper are divided into two sections, namely hardware selection and software design. Hardware selection, according to the actual needs including single chip module, power management module, serial communication module, actuator driver module four points; Software design, including the next computer software design, upper computer software design two steps [1–3].

2.3 Development Methods

According to the main ideas, the specific development method of methanol/diesel dual fuel engine will be discussed below.

Hardware Selection

First, SCM module: Because methanol/diesel dual fuel engine kernel system belongs to the electronic control system, so according to the actual need, must choose the control unit of electric control system of MCU module as the core, it has to accept, the sensor signal processing, at the same time, according to the preset standard process to control actuators, prompting actuator movement, finally achieved the purpose, Therefore SCM module is very important. Considering the subsequent methanol/diesel dual fuel engine emission characteristics of test requirements, the paper used in the development of the engine, MCU module needs to meet the requirements of some basic requirements, namely the microcontroller chip must have the characteristics of the operation speed, high control precision, at the same time because of the hard to avoid in practical working condition of electromagnetic interference, Therefore, SCM also has a strong ability to resist electromagnetic interference. Combined with these requirements, this paper compares the common microcontroller on the market, and finally chooses ATmega128 microcontroller, which not only meets the above requirements, but also has the advantages of high integration and strong reliability [4–6]. Table 1 shows the configuration of ATmega128 microcontroller.

Table 1. ATmega128 SCM configuration

Project	Configuration
The built-in flash	128 k bytes
SRAM	4 k bytes
The I/O interface	Three
Operating register	32
Crystal frequency	16 MHz
Timer/counter	Eight
ADC	Ten digits in channel 8

PS: More detailed configurations are displayed. Table 1 *shows only the main configurations.*

Second, the power management module: the power management module is an important component of the electronic control system, mainly responsible for the power supply management of the control unit in the system, has the important significance of ensuring the safety and stability of the control unit, so the power management module is also one of the key points of hardware selection. According to the actual work needs, the power management module in engine development must be able to guarantee the operation of the control unit, and ensure that other modules such as single chip microcomputer can provide power support with minimal interference. From this point of view, because this paper chooses ATmega128 microcontroller, according to the power demand of the microcontroller, the power management module needs to provide 5 V DC power supply. In view of this point, this paper chooses LM2576S-5.0 switching buck regulator as the power management module of the engine electronic control system [7–9]. Table 2 shows the configuration of LM2576S-5.0 switch-type step-down regulator.

Table 2. Lm2576S-5.0 switching buck regulator configuration

Project	Configuration
To drive the load	3 A
Dc power control limit	5 V
Voltage error	±4%
Operating register	32
Oscillator frequency error	±10%
Other configuration	Internal voltage regulator, oscillator, reset switch, thermal turn-off, current limiting device, etc.

Third, serial port communication module: For the safety of the engine emission characteristics experimental test, the experimental test in this paper will strictly limit the proximity of personnel to the engine under test state, but this also leads to the manual cannot directly observe and inspect the engine injection characteristics, so the test data can only be collected by technical means. From this point of view, it is necessary to use the serial communication module to solve the problem. The module mainly connects the methanol/diesel dual-fuel engine with the software system, so that the test data can be obtained, and the manual judgment can be made on the equipment side according to the data. It should be noted that the main function of serial communication module is to connect the engine and software system, but it will also affect the test results, such as interference of serial communication module, may lead to data loss, so it should be carefully selected [10]. The serial communication module selected in this paper is MAX3232 chip, as shown in Fig. 2.

Fig. 2. max3232 chip

Fourth, the actuator driver module: the actuator driver module is closely connected with the lower machine of the software system, and is responsible for executing the instructions issued by the lower machine. Therefore, the selection of the actuator driver module must fit the design idea of the lower machine. From this point of view, the pre-selected schemes of this paper are IR2103S, IRF3415 and diode. After comparison, the IR2101 driver chip is selected, which has the characteristics of fast speed, high voltage and high power, and can drive methanol/diesel dual-fuel engine well. At the same time, the chip has a fast response speed, indicating its strong controllability. The engine injection can be controlled according to the actual situation in order to better judge its emission characteristics.

Software System Design

The software system design is mainly divided into two steps: software design of lower computer and software design of upper computer. The details of each step are as follows.

Software Design of the Lower Computer

Because the main function of SCM in the test is to accept the processed camshaft position sensor signal, and at the same time use its timing/counter function to control the square wave generated when the engine methanol injection discharge, so the software design of the lower computer should support these functions. From this perspective, the software of singlechip processor system design of the basic logic is: first, through the monolithic integrated circuit after the camshaft position sensor signal collection and processing, the main method for down along the trigger, logic when MCU falling edge is not collected signals when continuous operation, and to the collected signals, once will interrupt, timer/counter function in operational; Second, when the signal acquisition process is interrupted and the timing/counter function starts to operate, the falling edge signal will continue to be emitted. If the falling edge signal is found again at this stage, the timing/counter function stops, and the signal acquisition process is repeated until there is no falling edge signal emitted. Third, if the time recorded by the timing/counter function is less than the last time, the difference is 50%, then the rising edge signal will be sent through the single chip microcomputer, and the serial port communication module will send the injection discharge instruction to promote the engine injection discharge. The signal generated in the process will be collected by the sensor. The collected signal will be converted into electronic format by the sensor and then returned to the terminal system through the serial communication module. Fourth, because the terminal system itself is a computer, and can not read the electronic format signal, so in the communication process installed a signal format conversion device, can convert the electronic format information into digital format, for reading, observation. Figure 3 Software design flow of single-chip microcomputer system.

Software Design of Upper Computer

PC system is a terminal system of experimental test, the function is to transmit and receive a serial port communication module of signal, read the data, mainly for the emissions of engine speed and sprayed pure and alcohol injection pulse width when the return value, because the data can be read continuously, so the PC software design, through the programming technology to develop the function of data curve drawing, visualization, This function can draw the corresponding curve table from the data continuously transmitted, and then show the curve table to the manual, and the curve table will change with the data change, easy to refer to the manual. In addition, considering

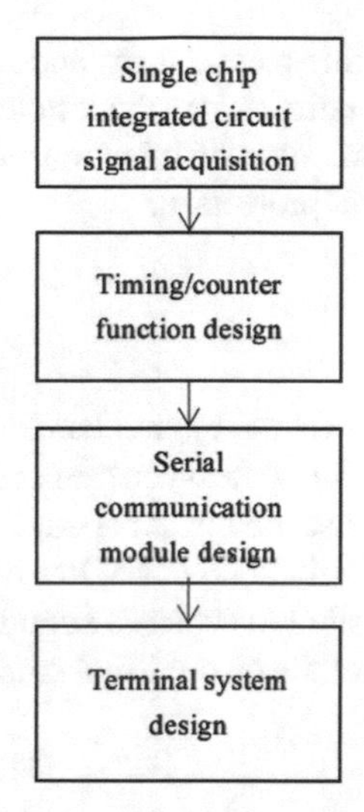

Fig. 3. SCM system software design process

the subjectivity of manual access, this paper processed the data visualization module of the upper computer system, and used VB software to construct a more simple interface of man-machine information interaction.

3 Engine Emission Characteristics Test

3.1 Experimental Methods

It is not complicated to test the emission characteristics of methanol/diesel dual-fuel engine. We only need to confirm the relevant indicators and conduct experiments. We can make a judgment according to the experimental test data. The rotational speed represents its power conversion rate (representing economic benefit), and the return value of injection time and injection pulse width represents methanol emission (representing environmental protection).

3.2 Experimental Results

Through the above development method, the developed methanol/diesel dual-fuel engine was put into the laboratory for testing, and the data of the two indicators were obtained. Then, according to the comparison method, the methanol/diesel dual-fuel engine was compared with the internal combustion engine, and the results were obtained: First of all, in terms of speed, the test data of methanol/diesel dual-fuel engine is 1 min/2200–2500R, and the test data of internal combustion engine is 1 min/1800–2100R, indicating that under the same conditions, the methanol/diesel dual-fuel engine has faster speed, higher power conversion rate and better economic benefits. Secondly, in terms of the combined index of alcohol injection time and alcohol injection pulse width return value, the test data of methanol/diesel dual-fuel engine were 3.5 ms and 3 ms at 1000 r/min, respectively, and those of internal combustion engine were 2.4 ms and 2 ms at 1000 r/min, respectively. It shows that under the same conditions, methanol/diesel dual-fuel engine has higher positive injection time and return value of injection pulse width, lower formaldehyde emission and better environmental protection.

4 Conclusion

In conclusion, methanol/diesel dual fuel engine has more advantages than the internal combustion engine, advantage mainly reflects on the economic benefit and environmental protection, at the same time, because the former change the energy system, car broke the situation, automobile energy non-oil not favour of low cost and renewable formaldehyde to replace, although it is not possible to separate support auto operation needs, But the value is already clear, so methanol/diesel dual fuel engines should be promoted.

References

1. Elumalai, P.V., Parthasarathy, M., Hariharan, V., Jayakar, J., Mohammed Iqbal, S.: Evaluation of water emulsion in biodiesel for engine performance and emission characteristics. J. Therm. Anal. Calorim. **147**(6), 4285–4301 (2021). https://doi.org/10.1007/s10973-021-10825-z
2. Masuk, N.I., Mostakim, K., Kanka, S.D.: Performance and emission characteristic analysis of a gasoline engine utilizing different types of alternative fuels: a comprehensive review. Energy Fuels **35**(6), 4644–4669 (2021)
3. Mathivanan, R., Mathiyazhagan, K., Raghu, L.: Performance and emission characteristics of a turbocharged MPFI engine. J. Sci. Ind. Res. **78**(4), 236–241 (2019)
4. Kumaravel, T.S., Arthanarisamy, M.: Experimental investigation on engine performance, emission, and combustion characteristics of a DI CI engine using tyre pyrolysis oil and diesel blends doped with nanoparticles. Environ. Prog. Sustain. Energy **39**(2), e13321.1-e13321.7 (2020)
5. Moulali, P., Prasad, T.H., Prasad, B.D.: Influence of EGR and Inlet temperature on combustion and emission characteristics of HCCI engine with micro algae oil. J. Sci. Ind. Res. **78**(5), 317–322 (2019)
6. Ramesha, D.K., Vidyasagar, H.N., Trilok, G., et al.: Study of performance, combustion and emission characteristics of DI diesel engine fuelled with neem biodiesel with carbon nano tube as additive. Appl. Mech. Mater. **895**, 237–243 (2019)
7. Kapusuz, M., Gurbuz, M., Ozcan, H.: The effects of photocatalytic activity with titanium dioxide on the performance and emission characteristics of SI engine. Appl. Therm. Eng. **188**, 116600 (2021)
8. Baek, S., Cho, J., Kim, K., Myung, C.-L., Park, S.: Effect of engine control parameters on combustion and particle number emission characteristics from a SIDI engine fueled with gasoline-ethanol blends. J. Mech. Sci. Technol. **35**(3), 1289–1300 (2021). https://doi.org/10.1007/s12206-021-0240-x
9. Hoang, A.T.: Combustion behavior, performance and emission characteristics of diesel engine fuelled with biodiesel containing cerium oxide nanoparticles: a review. Fuel Process. Technol. **218**(1), 106840 (2021)
10. Sathyamurthy, R., Balaji, D., Gorjian, S., et al.: Performance, combustion and emission characteristics of a DI-CI diesel engine fueled with corn oil methyl ester biodiesel blends. Sustain. Energy Technol. Assess. **43**, 100981 (2021)

International Trade Early Warning Analysis Based on CART Algorithm

Xinhui Feng(✉) and Changcai Qin

Yantai University, Yantai 264005, China
1095678516@qq.com

Abstract. In order to use CART algorithm to do a good job in international trade early warning, this paper will carry out relevant research. This paper first introduces the basic concept and model of CART algorithm, then puts forward the key points of international trade early warning, and finally analyzes an international trade case by using the algorithm. Through research, CART algorithm has a good warning function in international trade, and its calculation results are accurate and accurate warning can be achieved.

Keywords: CART algorithm · International trade · Trade alert · Trade warning

1 Introduction

Under the background of internationalization, the trade between China and other countries more frequently, it promoted the economic development of our country, but also increase the influence of the various risks in international trade and the frequent international trade makes incidence increased risk, monomer trade risk extent promoted, so if you can't effectively avoid risk, it is easy to cause the losses in the trade in our country, If the loss occurs frequently, it will threaten the national economic development. Based on this, it is hoped that effectively avoid risk in the international trade, so as to achieve this purpose related organizations launched trade data analysis work, the work in the early indeed played a very good effect, but as the growth of the trade frequency constant, lead to growing trade data, trade data analysis staff found itself has not directly for data analysis, Due to huge trade level as well as the complicated relationships between data outside the confines of human capabilities, the artificial work mode trade data analysis can't effectively, and the results can also be difficult to secure quality, this kind of phenomenon makes people realize the need by technical means to perform the work, so begin to pay close attention to all kinds of algorithms, CART algorithm in the process, get the attention of the people, for now the algorithm is one of the few can meet the demand of the international trade data analysis, to help people avoid risk algorithms, so the people around the algorithm proposed the concept of international trade alert, purpose is based on the algorithm to establish early warning system, people can rely on the system at risk before the outbreak of the prevention and control in time, Therefore, in order to give full play to CART algorithm and do a good job in international trade early warning, it is necessary to carry out relevant research.

J. H. Abawajy et al. (Eds.): ICATCI 2022, LNDECT 169, pp. 693–701, 2023.
https://doi.org/10.1007/978-3-031-28893-7_84

2 Overview and Related Discussion

The essence of international trade early warning is to assess the state of international trade, in which the state specifically refers to the risk state in the trade under its own economic conditions, that is, there are various risks in international trade. For example, some risks will erupt in a short period of time, but will only cause one impact, and the impact is not strong, but there are also some short-term risks that will cause secondary or even multiple impacts, and the impact is uncertain, or the risks are secondary The influence will not break out in a short time, although it may vary from time to time. In the face of all kinds of risks, the country, as a trade subject, naturally wants to avoid risks as much as possible, but simply avoiding risks means that the country may give up some interests. Therefore, the basic idea of modern countries in international trade is to bear certain risks within an acceptable range in order to maximize benefits. On this basis, if the risk is within the acceptable range of the country, it means that the impact of the risk is low and can be ignored. Such risks are not recognized as risks in the modern perspective. On the contrary, if the impact of the risk may exceed the acceptable range of the country, it means that the impact of the risk is high and can not be ignored. Generally, the country will choose to avoid or adjust trade to reduce the impact of the risk.

From this perspective, because the time of risk outbreak is uncertain, countries need to predict risks in advance to ensure their participation in trade activities under the safest conditions, so they need to conduct international trade early warning. The main content of international trade early warning is to analyze trade related information and data, understand the types of possible risks, the probability of various risks, the size of risk influence, etc., and then make accurate judgments based on the situation. However, the previous international trade early warning work did not show its due value, because the early warning methods adopted by people were too extensive, and most of them were dominated manually. With the development of the times, people have paid more attention to international trade early warning, and also realized that there are problems with the previous methods, so they began to seek more effective early warning methods. CART algorithm is one of the most respected early warning methods.

In fact, people realized the importance of early warning of international trade very early, and began to look for accurate early warning methods. At that time, the mainstream methods were generally divided into two categories. One was supervision, which was not a mathematical algorithm. It was a method to judge the state of international trade through information supervision and logical analysis of information, and then conduct early warning. This method was first proposed, But the practical application value is not high, because on the basis of separating from mathematical methods, there are still many vague concepts in the results of information logic analysis. For example, when judging the state, good or bad results will be given, and how good or bad the specific results can not be accurately explained, which makes it difficult for people to make accurate judgments. Therefore, such algorithms were eliminated early. The second is a series of mathematical algorithms, Among them, the most representative is the decision tree algorithm, and CART algorithm is a decision algorithm. Compared with other similar algorithms, CART algorithm is better at solving classification and regression problems.

As a decision algorithm, CART algorithm is different from other similar algorithms, that is, most other algorithms make decision analysis based on the hidden information behind the data. However, CART algorithm introduces probability theory and statistics theory, and gives decision suggestions based on probability and statistics results, which is one of the reasons why CART algorithm can be used in both classification and regression. CART algorithm is often used to compare with C4.5 decision tree algorithm. The former has the function of binary partition of data feature space, while the latter does not. Therefore, there is an essential difference between the decision trees generated by CART algorithm and C4.5 decision tree algorithm. That is, the decision tree generated by CART algorithm is a typical binary tree, but the decision tree generated by C4.5 decision tree algorithm is a unitary decision tree. From a mathematical point of view, the binary tree can segment the scalar attributes and continuous attributes of data. On this basis, the CART algorithm can be used for binary segmentation, and then the input sample results can be predicted according to the input eigenvalues. Finally, the external output can be achieved.

CART algorithm is not perfect, on the contrary, it also has its own advantages and disadvantages. In terms of advantages, CART algorithm is simple in calculation and application, easy to understand the results, and highly interpretable. It can be used to process samples with certain missing attributes, as well as irrelevant features. It is

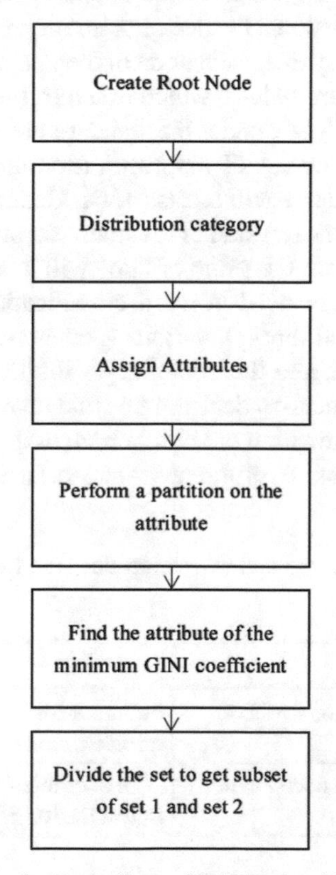

Fig. 1. Basic flow of CART algorithm

extremely efficient in processing. Even if the amount of data is large, it can also give high-quality results in a short time. Disadvantages: CART algorithm does not support online learning, and the decision tree structure is too fixed. Once new data samples are added to the dataset, the decision tree model will be reestablished. Therefore, CART algorithm often repeats this process in practical applications, which will be more cumbersome. In addition, CART algorithm is prone to over fitting during operation, because the decision tree generated by CART algorithm can only classify data with known attributes well to avoid over fitting. However, if the data attributes are unknown, or partially unknown, CART algorithm will have over fitting problems, or even the phenomenon that the operation cannot continue. Figure 1 Basic flow of CART algorithm.

3 Basic Concept and Model of CART Algorithm

3.1 Basic Concepts

CART algorithm is a decision tree algorithm. Different from other decision tree algorithms, the decision tree generated by this algorithm has two functions, namely classification and regression. CART algorithm of the whole operation process can be divided into three steps [1], feature selection, the decision tree generation, the decision tree pruning, the decision tree is a binary tree, each node in the tree's characteristics is carried out in accordance with the standard values, which assumes that a standard first, more than the standard node values are "yes", under the standard node values for "no", this is also one of the characteristics of the CART algorithm (non-numeric values) [2, 3]. On the basis of the node value, each node will be classified. Generally, the node with the value of "Yes" will be classified to the left side of the main rod of the decision tree to form the left branch, while the node with the value of "no" will be classified to the right side to form the left branch. After the branch is formed, classification decision tree or regression decision tree can be generated through various models. After the generation, pruning can operate the decision tree, and then obtain specific results. It is worth noting that although the models of classification decision tree and regression decision tree in CART algorithm are different, the formation process of both of them is a recursive process, but the criteria used in this process are different, as shown in Table 1 [4–6].

Table 1. Classification decision tree and regression decision tree generation criteria of CART algorithm

Project	Guidelines
Classification decision tree	Square error minimization criterion
Regression decision tree	Gini coefficient minimization criterion

3.2 Model

Macro level, CART algorithm the two kinds of decision tree generation model is different, but in this article, the main facing the issues of international trade alert, this problem is a typical regression problems, namely the international trade alert need to analyze the various data, understand the development of the data represented by the event, during constantly return to the trade security index to compare, if the development of events represented by data exceeds the safety standard of a certain index in a certain development process, it indicates that related risks may occur, thus playing an early warning role. Therefore, this paper only introduces the model of regression decision tree.

Regression decision tree model of CART algorithm: First assumes that the input and output variables of the algorithm, in which the output variables as continuous variables, then it is concluded that the given training data set, in the choice of data set value of a variable as main stem of the decision tree, as a standard to distinguish the nodes, main stem so on both side of main stem form after the decision tree is defined as two different areas, among them as "is" area on the left side of the area, "No" area is on the right and is completed in accordance with the main rod standards to find the optimal segmentation variables and optimal segmentation points, a simultaneous nodes, logic is divided into "if the data node in the left main stem towards the direction of the area, then the node values as is, instead of no", the resulting regression decision tree, see formula (1) [7, 8].

$$c_m = aue(y_i | x_i \in R_m), m = 1, 2 \tag{1}$$

In the formula, c_m is the regression decision tree output in the region R_m, equivalent to the mean value of the corresponding output y_i of all input instances x_i on R_m, aue is the criterion coefficient, and m is the standard value of the main rod.

Besides, the return of the formula (1) the decision tree for the classical decision tree, although it can be directly used for early warning analysis of international trade, but the operating process is complicated, and there may be low quality as a result, therefore, in order to avoid problems, also can be optimized for a number of least-squares regression tree, i.e. for each R_m area, a repeat of the classic regression decision tree generation process, this can divide the entire input space into several regions, all of which are expressed as R_1, R_2, … R_n, and then define the output on each region (c_m in Formula 1) to obtain the least squares regression tree (the least squares regression tree is the regression decision tree within each region of the classical decision tree).

4 Key Points of International Trade Early Warning

4.1 Quantity of Traded Products

Trade products in the number of international trade, one of the main points of the early warning due to trade products closely related to the risk of dumping common in international trade, the international trade is the basic feature of foreign products into the domestic market, and to sell in the domestic market, many modern countries are interacting trade through this way, on the one hand to open its own home market, meet the foreign products, on the other hand into foreign market sales of its products, and this way of trade in nature are not unreasonable, but the actual operation of some countries will be a lot of its products on the domestic market to the foreign country, there are many similar products in native and foreign market, such as fruits and vegetables, so for the products in the local market, from will be competing with foreign similar products, if the quantity of similar foreign products is too big, will occupy the local market large area, make no native products, sales channels, and over time leads to the local market channels of foreign goods, weaken the local product productivity, economic power, to some extent, pose a threat to a country's degree of independence, this is the classic dumping risk. In the face of dumping risk, the most important thing people should do is to control the quantity of trade products, so as to avoid foreign products occupying excessive local market share and to protect the advantages and sales channels of local similar products. It shows that the quantity of trade products is one of the key points of international trade early warning [9, 10].

4.2 Unit Price of Trade Products

In fact, the trade and general trading, there is a big difference in the basic logic is both sides open their markets, and funded by the national products introduced each other, let local social organizations in the local market selling products abroad, different from ordinary trade "gang get the cash on delivery" logic, and on the logic of trade, trade products unit price became a key points of risk early warning, namely trade products unit price refers to the country to introduce each other when the unit price of each product, the unit price is equivalent to the cost for the local market, the pricing of products in the domestic market sales is established on the basis of cost of unit price, so when the trade products unit price is too high, will lead to higher price of product sales, consumers may be more than the local market economy, and because of some products is needed for life, so consumers will give local market cause large economic pressure, the social economic risk, in this case, you must to control the unit price of trade products, at least to safeguard consumer economy trade products unit price in the local market inherit the range of the mean, In this way, on the one hand, products can meet people's needs and on the other hand, they will not cause social and economic risks.

4.3 Trade Cycle

International trade occurs among countries, and each country has different trade motivations. For example, some countries choose to trade with other countries because there is a large domestic shortage of a certain product, so they hope to deal with the product shortage through trade. However, this motivation is temporary. That there is a big gap in the country in other countries products at the same time, introduced through trade, will try to develop their own capacity related products, native products once capacity can satisfy the domestic needs, it would probably not choose to trade, so hope that through this kind of motivation can be defined as a countries trade to solve their own. This case will focus on the trade cycle, that is, in the face of their own, the shorter the natural cycle of countries want to trade, the better, but also other countries may have on the capacity limit, cannot provide enough products in the short term, so long phenomenon easy to cause the trade cycle, and once the trade cycle is long, countries will not be able to solve the urgent need, It is easy to cause a variety of social risks, so the country needs to control the trade cycle and carefully choose trade partners, which is the basic strategy to prevent the outbreak of risks.

5 Construction of International Trade Risk Early Warning System Based on CART Algorithm

5.1 Case Overview

With A and B of dumping incident in international trade between countries, for example, of which the core index for trade product quantity, namely A country because of the natural environment is relatively bad, so its agricultural production capacity is insufficient, some fruits and vegetables product import demand, but also because of its unique natural environment A countries have some unique fruits and vegetables, can satisfy the internal demand, however, country B is a large agricultural country with high production capacity of fruits, vegetables and agricultural products, but relatively single varieties. Therefore, both sides can meet each other's needs, so they have reached the intention of trade cooperation. In this process, the huge production capacity of agricultural products in country B means that it may cause dumping on the local market of ordinary fruits, fruits and vegetables in country A. Therefore, country A needs to carry out early warning analysis on international trade risks and establish an early warning system. Table 2 shows the gap and productivity data of common fruits and vegetables in country A, and Table 3 shows the productivity data of common fruits and vegetables in country B.

Table 2. A national gap and capacity data of ordinary melons, fruits and vegetables (2018)

Category	Gaps in the data	Capacity
Vegetables	23.7% material of 500 t/year	1322 t/year
Fruit	17.4% material of 430 t/year	1037 t/year

Table 3. B productivity data of common melons, fruits and vegetables in China (2018)

Category	Capacity data
Vegetables	2371 t/year, surplus 600 t/year
Fruit	3822 t/year, surplus 1000 t/year

5.2 Decision Tree Analysis

According to the above methods, the CART algorithm establishes an international trade risk warning system for country A. The decision tree in the system is A regression decision tree, including two least squares regression decision trees, which can carry out regression analysis on vegetables and fruits respectively. The basic process of analysis is as follows: First, prune the tree after the establishment of the decision tree. The rule is to prune the sub-tree at the bottom of the decision tree to make the decision tree smaller, which can improve the accuracy of the analysis results of the decision tree. Second, the pruning process, any node, for example, after the loss function of the node set, using inequality is calculated, if the results show that the loss function of a node and the whole tree loss function is the same, and the subtree of that node represents the total node number is lower than other subtree node number of the mean, cut off the subtree; Third, choose A country any kind purpose gap data based on node, the same as the B country class purpose capacity data to run the algorithm, known country B products accounted for in the local market, in A country rules as long as the B country home market accounts for less than 20%, dumping risk will happen, and vice risk occurs, so according to this method to have early warning. In addition, this analysis method is also used in the early warning analysis of the unit price of traded products and the trade cycle. The basic process remains unchanged and only data items need to be re-selected, so there is no further elaboration here. Table 4 shows the analysis results.

Table 4. Decision tree analysis results

Category	B country market share	The results of
Vegetables	27.3%	There are risks
Fruit	36.7%	There are risks

6 Conclusion

In conclusion, CART algorithm can play a contextual role in international trade warning, and the operation method is not complicated. The only thing that needs to be paid attention to is the setting of risk standards. If the risk standards are reasonable, the algorithm can be used to avoid trade risks such as dumping.

References

1. Zulfikar, W.B., Taufik, I., Atmadja, A.R., et al.: A classification model for student exchange using CART algorithm. IOP Conf. Ser. Mater. Sci. Eng. **1098**(3), 032054 (2021). (6pp)
2. García, V.J., Márquez, C.O., Isenhart, T.M., et al.: Evaluating the conservation state of the páramo ecosystem: an object-based image analysis and CART algorithm approach for central Ecuador. Heliyon **5**(10), e02701 (2019)
3. Reshetnikov, A.G., Ulyanov, S.V.: Quantum algorithm of imperfect KB self-organization. Pt II: robotic control with remote knowledge base exchange. Adv. Artif. Intell. **3**(2), 27 (2021)
4. Gonzalez, A.: From Arabic Manuscripts to Benin Sculptures: Abraham S. Yahuda and the International Trade of Religious and Cultural Objects, 1902–1944. Seminario online de estudios judíos (2021)

5. Penelope, M., Richard, S., Phillip, B., et al.: Corporate power and the international trade regime as drivers of NCD policy non-decisions: a realist review. Health Policy Plan. **5**, 5 (2021)
6. Chen, Z.: On the influence of international relations on international trade. Mod. Econ. Manag. Forum **3**(1), 30–33 (2022)
7. Cushman, D.O.: The effects of real exchange rate risk on international trade. J. Int. Econ. **15**(1–2), 45–63 (1983)
8. Krishna, P., Senses, M.Z.: International trade and labor income risk in the United States. Soc. Sci. Electron. Publ. **17**, 5–36 (2010)
9. Federspiel, J.J., Stearns, S.C., Domburg, R., et al.: Risk-benefit trade-offs in revascularisation choices. Eurointerv. J. Europcr Collab. Work. Group Interv. Cardiol. Eur. Soc. Cardiol. **6**(8), 936 (2011)
10. Sun, H., Yang, J., Xu, X.: Research on the application of blockchain technology in international trade risk management. J. Asian Res. **4**(3), 36 (2020)

Power Enterprise Talent Selection Method Based on FCM Algorithm

Shurong Zhu(✉)

Jiangxi Electric Vocational & Technical College, Nanchang 330032, China
zsr07008898@126.com

Abstract. In the development of electric power enterprises, talent selection has always been a key issue, based on the current electric power enterprises in the talent selection process is too old, and the lack of talent comprehensive quality, future training and other aspects of consideration. Based on this, this paper proposes a fuzzy c-means algorithm, namely the FCM algorithm, this algorithm can be applied to electric power enterprises in personnel selection, selection methods and functions of the innovation, through the profile coefficient and clustering algorithm, including selection of personnel to provide credibility for the electric power enterprise, enhance the level of overall staff and hope to provide reference for the electric power enterprise.

Keywords: Electric power enterprise · Talent selection · FCM algorithm · Data-processing

1 Introduction

In China's current social development, electric power enterprises are also constantly carrying out various reform measures, in order to meet the requirements of modern society from all aspects. But after the relevant research shows that the current electric power enterprises still exist deficiencies in personnel selection, was not in keeping with The Times, so in order to strengthen the talent selection level, to promote the quality of electric power enterprise overall team, FCM algorithm can be applied to personnel selection work, through the intuitive data to judge of talent, promote the overall development of the electric power enterprise.

2 Algorithm Significance

Because electricity is the necessary energy for modern people to live and work, electric power enterprises, as managers and providers of electric power, should shoulder the important responsibility of ensuring people's electricity quality, which is also related to the overall operation and development of society. To achieve this, electric power enterprises need a large number of high-quality talents to support their work. This demand is more prominent in the context of the diversified development of modern power supply

J. H. Abawajy et al. (Eds.): ICATCI 2022, LNDECT 169, pp. 702–710, 2023.
https://doi.org/10.1007/978-3-031-28893-7_85

modes. Under this condition, power enterprises have to face the problem of talent selection. This problem is not complicated on the surface, but under the influence of some factors, it is actually very complex, and its complexity even exceeds the scope of human capacity. For example, although the demand for talent in power enterprises is large, the number of talents to be recruited is larger, tens of times more than the demand, or even hundreds of times. Therefore, we adhere to the principle of fair selection, Electric power enterprises need to analyze the specific situation of each talent, and then make a comparison, which involves a huge amount of data and complex information relations, making it impossible to handle manually.

In the face of this situation, modern power enterprises believe that it is necessary to establish an efficient and fast talent selection system to solve the problem, and one of the most important links in the design of this system is the selection of selection methods, that is, many power enterprises will use intelligent technology to process huge data in the talent selection system, and establish a talent selection index system according to their own needs, For example, an electric power enterprise has put forward five requirements for talent selection indicators: one is the professional ability of the post, which requires talents to have corresponding professional qualification certificates to prove their professional ability level; The second is work experience, which requires talents to have more than one year of work experience in the same position, so as to quickly integrate into the job; The third is the professional quality, which requires that the talents have not committed any violations or other acts that damage the interests of others, the enterprise collective and other aspects in their previous work experience, and that they should be dutiful and have a certified attitude to ensure that the value of talents can be fully played; The fourth is other technical requirements, which require talents to master other relevant technologies beyond their professional abilities. This requirement is not mandatory, nor does it specify what technologies and technologies talents need to master and what level they want to reach. However, if talents are more excellent in this aspect, they will be more likely to be favored by enterprises, representing that enterprises value compound talents; The fifth is innovation quality, which is also non mandatory and has no specific restrictions, but it represents the value of talents and can help talents to stand out from the competition. Innovation quality generally refers to the level of innovation consciousness and innovation ability of talents. According to these five requirements, the enterprise has selected 1000 reserve talents in the one month talent recruitment, and is ready to continue to observe for four weeks to select. However, in this process, all talents will generate at least thousands of pieces of information every day, which is sufficient to prove that the magnitude of data is huge, and enterprises find that they cannot make accurate decisions on talent selection manually.

In response to this situation, many enterprises will choose to apply intelligent technology, big data mining technology, etc. These technologies are indeed excellent tools to solve similar problems, but the core of these technologies lies in mathematical algorithms, such as intelligent technology, which is characterized by intelligent logic. This logic enables the technical system to analyze data in the human mode of thinking and keep learning, while intelligent logic comes from artificial neural networks, The network is an algorithm driven model. According to this requirement, reasonable selection of mathematical algorithms is the key to solving the problem of talent selection in power enterprises. There are many algorithms applicable to this, such as FCM algorithm.

FCM algorithm is a data clustering analysis algorithm, but different from similar algorithms, the clustering process of this algorithm is based on the optimization of the objective function, that is, through clustering, the objective function can be continuously optimized, and finally the ideal results can be obtained. In this algorithm, the clustering result is the degree of membership of each data point in the cluster center. If the degree of membership of a data point in a cluster center is the highest globally, it means that the relevant data of the data point belongs to the cluster center. The results of this algorithm will express the degree of membership by explicit values, and can show the degree of membership of a data point in several data centers, which is a feature that other similar algorithms do not have.

At present, people regard FCM algorithm as an unsupervised fuzzy clustering method, because it interferes with the time process in an unordered manner. FCM algorithm is very suitable for the same type of problems in talent selection, but the disadvantage of this algorithm in application lies in the need for reference samples, that is, some parameters need to be set first in the application of the algorithm. If the parameter selection is not reasonable, it will lead to errors in the final results. At the same time, when the data sample set is large or there are many data feature items, the algorithm efficiency will be significantly reduced. Figure 1 FCM algorithm flow.

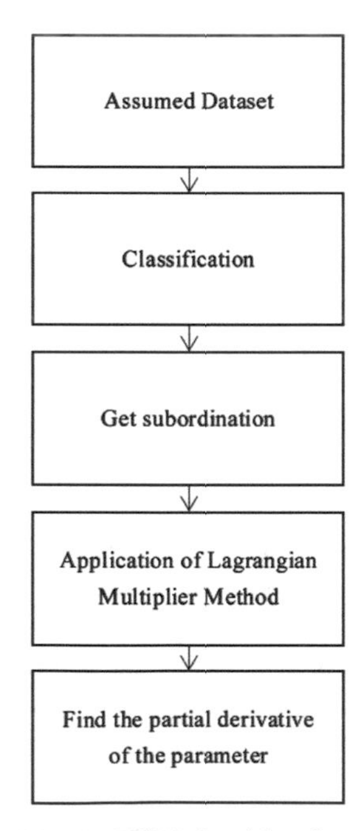

Fig. 1. FCM algorithm flow

3 FCM Algorithm Introduction

FCM algorithm is put forward by scholars, as early as in 1965, this way can to some abstract or physical individual, in accordance with certain rules of divided operation to strengthen its accuracy, for such calculation is also used in the traditional fuzzy clustering estimation, but the FCM algorithm has a high sensitivity and accuracy, can be through the C - average arithmetic.

For example, the site selection of a communication base station should not conflict with the existing base station, but should also solve the existing problems. Therefore, the positive and negative impacts of each site selection should be considered comprehensively to calculate the most reasonable site selection, which can be realized by FCM algorithm, or a certain area should be guarded by a certain army. At present, it is necessary to deploy another unit around the unit, so how to plan the deployment and enhance its deployment effect can also be calculated by FCM algorithm, which has strong applicability [1].

Generally, FCM algorithm needs to set a C-mean value, collect various sample data at the same time, and list the set. For example, n is the number of sequences in the sample set, and C is the number of clusters, so the sequence matrix can be used: $\partial = \{\partial_1, \partial_2, \partial_3 \ldots\ldots \partial_n\}$

$$X = \begin{cases} x_{11} \;\; x_{12} \ldots\ldots x_{1n} \\ x_{21} \;\; x_{22} \ldots\ldots x_{2n} \\ x_{c1} \;\; x_{c2} \ldots\ldots x_{cn} \end{cases}$$

where represents the c membership degree of the NTH element in the sequence to the class, and should satisfy the following formula x_{cn} [2]:

$$x_{cn}\ [0, 1].c = 1, 2, 3, \ldots\ldots, c, n$$

$$\sum_{n=1}^{c} x_{cn} = 1, n = 1, 2, 3, \ldots\ldots, c$$

Through FCM algorithm, the formula can be expressed as:

$$\text{Min}J_n(U, V) = \sum\nolimits_{i=0}^{n} \sum\nolimits_{i=1}^{n} x_{cn} u_{cn}^2$$

which is the distance between data points and clustering prototype measurement, can reflect the sample set, and the differences between Min and smoothing factor and fuzzy parameters selection of n, will directly influence the effect of clustering algorithm, which can then through the algorithm measured the abstract or physical quantities and the influence of the differences between a more substantial and accurately to achieve the purpose of a test $u_{cn}^2 x_{cn} J_n$ [3].

4 Importance of Talent Selection in Electric Power Enterprises

The talent selection of electric power enterprises is very important, which is directly related to the overall development path of electric power enterprises. Talent is the core of development of different enterprises, in the current competitive market environment, the electric power enterprises need to very high standards of talent resources construction, through the talent selection to employ the most suitable for the electric power enterprise staff, the personnel will be in a certain extent, directly affect the future development of the electric power enterprise, Electric power enterprises can provide better service for the society and obtain higher economic benefits only by relying on more high-quality and high-quality talent team [4].

Talent recruitment validity, still can save the cost for the electric power enterprise, for example in the process of talent selection, the electric power enterprise can absorb to conform to the requirements of The Times and their own development needs of high-quality talent, can also reduce the spending on ability training, so as to have more money to their own development projects, to lay a solid foundation for itself. In addition, the electric power enterprise in the process of talent selection, can also through new into talents for the enterprise to provide advanced thoughts, for enterprise to bring the technical innovation, add vigor, especially high technology talents to join, can also raise the competitive power of the electric power enterprise with its own position, side to bring more value for the enterprise, Therefore, the talent selection of electric power enterprises is extremely important [5].

5 Existing Problems of Talent Selection in Electric Power Enterprises

Electric power industry as one of China's main economic pillar, its importance is self-evident, so each enterprise to talented person's thirst for degree is higher, but the current situation, still have a lot of electric power enterprise does not attach importance to talent selection this important process, lack of, the introduction of new mechanisms are lack of understanding of the new market, Combined with advantages in the field of electric power enterprises in our country the status, related personnel salary is higher, so adverse events often occur, for example by the temptation of the electric power enterprise welfare, high treatment, a lot of people through bribery, unwritten rules involved in the selection process, disturb the order of the talent market, a lot of internal staff also more value relations and interests, Ignored for horizontal considerations, such as the qualification, quality, caused by various social sectors, makes the electric power enterprise talent loss with fairness and the selection of principle, it is difficult to bring good for the electric power enterprise development, has caused such as openings, as well as the power and responsibility is not clear, lack of personnel quality, for the electric power enterprise caused varying degrees of damage.

In addition, many power enterprises in the process of talent selection by using the way of selecting are unreasonable, the lack of understanding of the new age, still use the traditional, old-fashioned ways of selecting talents, team worker aging problem is increasingly intensified, seniority, and so on and so forth deepening, the electric power enterprise's social status also creates the opposite effect, As a result, it is difficult for young people to have the opportunity to enter enterprises. As time passes, electric power enterprises also begin to pay little attention to the introduction and training of young people.

In addition, the electric power enterprise itself is relatively high treatment, many talents after entering the enterprise, due to high treatment and welfare, also give up the idea of promotion, content with the status quo, and the enterprise will also reduce the emphasis on training ability, resulting in signs of deterioration of talent ability, neglect the training of technical and management ability talents. However, electric power enterprises also lack the means of post assessment or pre-job test for employees on the job. On the one hand, it is difficult to ensure whether the recruited employees are competent for the post, and on the other hand, it also leads to the undesirable phenomenon of uneven personnel quality in various posts, which is difficult to ensure the future development of enterprises [6].

6 Application Measures of FCM Algorithm in Talent Selection of Electric Power Enterprises

6.1 Establish an Evaluation Index System

FCM algorithm can be fully used in the talent selection of electric power enterprises. The FCM algorithm system needs to be supported and improved by extremely precise data, which can greatly improve the accuracy of talent selection results. In addition, FCM algorithm itself can gather various categories of abstract or physical properties, so it can build indicators suitable for talent selection in power enterprises.

Talent selection is a systematic work, the talent selection work of the current society does not allow power enterprises to rely on traditional means, should be judged on various indicators, and establish an authoritative and should meet the needs of the current society for talent, so as to truly achieve the purpose of selection. The principle of comparability should also be satisfied in the construction of the index system. All indicators should be compared and identified in the form of quantification, and the rationality of the index system framework should be strengthened based on scientific nature. By FCM algorithm, the index system of building should be the comprehensive qualities of a good show, including but not limited to talented person's job performance, scientific research, the basic quality, etc., still should according to the job requirements of the electric power enterprise itself, fine processing of the indicators, such as system of each indicator to rate (Table 1) [7].

Table 1. Index score weight ratio

	A	B	C	D
The index name	Basic quality	Work performance	Scientific research achievements	Scientific research project
Index proportion	15	20	40	25

Each level indicators are made up of different secondary indexes, and each segment of the secondary indexes also can score by proportion, and hired experts are measured, and the secondary indicators can be subdivided into other three indicators, finally by FCM algorithm combining with the subdivision of the indexes for comprehensive evaluation, The accumulated score is the final score of talent selection, so as to conduct selection intuitively through quantitative form.

6.2 Quantitative Data Processing

In the process of electric power enterprise talent selection, should also work of sample collection, such as selection of personnel selected 100 to choose object, its various data acquisition, and through the relevant experts to review, relevant experts should to discuss when evaluating, and ensure the unity of the scoring criteria, strengthen the reasonable and effective results, FCM algorithm can be used to collect all indicators in the domain, for example:

$$\mathrm{Y} = \{y_1, y_2, y_3 \ldots\ldots y_n\}$$

Each element can be represented by FCM algorithm, for example y_i [8]:

$$y_i = \{y_{i1}, y_{i2}, y_{i3} \ldots y_{in}\} \ (1, 2, 3\ldots, \mathrm{i} = \mathrm{n})$$

Put each set into the algorithm, and then show it in the form of matrix to get the initial matrix:

$$\mathrm{Y} = \{y_1, y_2, y_3 \ldots\ldots y_n\} = \begin{pmatrix} y_{11}, y_{12} & \cdots & y_{1i} \\ \vdots & \ddots & \vdots \\ y_{n1}, y_{n2} & \cdots & y_{ni} \end{pmatrix}$$

However, in practical work, each group of data or all kinds of indicators contain different dimensions, and in order to facilitate the comparability principle of data between different dimensions, each indicator should be dimensionally eliminated, such as range conversion, such as [9]:

$$y^1{}_{ni} = \frac{y_{ni} - min\{y_{ni}\}}{max\{y_{ni}\} - min\{y_{ni}\}}, 0 \le y^1{}_{ni} \le 1$$

Only the basic value of each sample data should be included, and the maximum and minimum value of the value should be taken into consideration comprehensively. Finally, each data should be presented digitally, so as to realize the quantification of all indicators and comprehensive scores of talents in the process of talent selection (see Table 2).

6.3 FCM Clustering Algorithm Was Used to Discriminate

After quantifying indicators, FCM algorithm can also cluster the employee population and analyze the clustering evaluation results. For example, talents are assigned to several clusters, and the effectiveness of each cluster depends on the difference between

Table 2. All indicators and comprehensive scores of talents

	Basic quality	Work performance	Scientific research achievements	Scientific research project
y_1	0.7532	0.7864	0.6846	0.1353
y_2	0.7683	0.7985	0.6981	0.0000
y_3	0.6123	0.5891	0.7524	0.3453
.	.	.	.	.
y_{100}	0.6282	0.5628	0.6244	0.2548

the minimum and maximum distance within the cluster, through the sample contour coefficient calculation formula [10]:

$$X(a) = \frac{m(i) - n(i)}{\max\{n(i), m(i)\}}$$

Among them as samples I average distance from all points in the cluster, for all the points in the cluster of adjacent distance, can be subdivided by calculating the average of each cluster, and as a contour coefficient, its value is close to 1, the more can reflect fits, will arrange a worker to the matching degree of the current position, whereas the closer to 1, then the worker does not apply to the current position, thus it can be seen, FCM algorithm can play an indispensable role in talent selection of electric power enterprises, and can make up for the shortcomings of traditional methods.

7 Conclusion

In a word, in the current talent selection work of electric power enterprises, we should clearly recognize the shortcomings of the original selection method, and should actively update the modernization concept, through the FCM algorithm for talent selection work. The use of the algorithm can be all the talent in all kinds of professional index quantitative analysis, through the data up to the enterprise, also can be calculated, the resilience of the algorithm for talents measure talent can be competent for a certain position, so as to improve the effectiveness of the electric power enterprise talent selection work, solve the traditional problems.

References

1. Venkatraj, V., Dixit, M.K.: Challenges in implementing data-driven approaches for building life cycle energy assessment. Renew. Energy Sustain. Energy Rev. **21**(1), 1–8 (2010)
2. Gouda, S.K., Mehta, A.K.: Software cost estimation model based on fuzzy C-means and improved self adaptive differential evolution algorithm. Int. J. Inf. Technol. **14**, 2171–2182 (2022). https://doi.org/10.1007/s41870-022-00882-4
3. Zhang, M., Zhu, L., Sun, Y., Niu, D., Liu, J.: Computed tomography of ground glass nodule image based on fuzzy C-means clustering algorithm to predict invasion of pulmonary adenocarcinoma. J. Radiat. Res. Appl. Sci. **15**(1), 152–158 (2022)

4. Maryam, M.K., Reza, S.A., Arash, D., Maryam, F.: Optimization of fuzzy c-means (FCM). IEEE Trans. Pattern Anal. Mach. Intell. **26**(1), 1–8 (2008)
5. Xu, R.: Fuzzy C-means clustering image segmentation algorithm based on hidden Markov model. Mob. Netw. Appl. **27**, 946–954 (2022)
6. Guo, H., Song, Y., Wang, L., Zhao, J., Marignetti, F.: Predicting the eddy current loss of a large nuclear power turbo generator using a fuzzy c-means deep Gaussian process regression model. Appl. Soft Comput. J. **116**, 108328 (2022)
7. Ferdinando, D.M., Salvatore, S.: A novel quantum inspired genetic algorithm to initialize cluster centers in fuzzy C-means. Expert Syst. Appl. **191**, 116340 (2022)
8. Pang, Y., Shi, M., Zhang, L., Song, X., Sun, W.: PR-FCM: a polynomial regression-based fuzzy C-means algorithm for attribute-related data. Inf. Sci. **585**, 2015 (2022)
9. Amin, G.O., Mahdi, H., Bahareh, A., Ali, B.M.: CGFFCM: Cluster-weight and Group-local Feature-weight learning in Fuzzy C-Means clustering algorithm for color image. J. Appl. Soft Comput. **113**(PB), 317–322 (2021)
10. Valliappa, C., et al.: Hybrid-based bat optimization with fuzzy C-means algorithm for breast cancer analysis. Int. J. Noncommun. Dis. **6**(5), 62 (2021)

Power Information System Network Security Method Based on Big Data

Huan Xu[1](✉), Fen Liu[1], Zhiyong Zha[1], Hao Feng[1], Fei Long[1], Xian Luo[1], and Wen Liu[2]

[1] State Grid Information and Communication Branch of Hubei Electric Power Co., Ltd, Wuhan 430077, China
453287272@qq.com

[2] State Grid Wuhan Power Supply Company, Wuhan 430071, China

Abstract. In order to pass the big data to ensure the safety of electric power information system network, this article will conduct research, mainly discusses the basic concept of big data, and then points out that the electric power information system of network security status, finally construct the big data network security protection system, introduces the application of this system in network security protection system. Through this study, a protection system with stronger network security and protection ability can be established with the help of big data, which can change the status quo of network security of power information system and solve relevant problems.

Keywords: Big data · Electric power information system · Network Security Protection

1 Introduction

As power information construction work of the evolving, power information system has become the indispensable tool in power work, is now fully integrated into the work, but in people gradually found in the application of the system application there are a lot of security risks, the risk through the system not only affect the power work, also can be used in some illegal activities, therefore, how to protect the network security of power information system has become the focus of people's attention. Aiming at this problem, the modern electric power enterprises generally established the related network security protection system, and the system of the protection ability is uneven, the overall existence insufficiency, and can't cope with all the network security risk, and the emergence of large data to the electric power information system network security protection development opportunity, if can play a role of big data, can strengthen the protection, to further ensure network security, it is necessary to carry out relevant research.

The opportunities brought by big data to the development of network security protection of power information system are mainly as follows: Past system is the main reasons of the lack of strength of network security protection technology co, LTD risk analysis data scattered difficult to analysis, and the big data can be used as a kind of advanced

J. H. Abawajy et al. (Eds.): ICATCI 2022, LNDECT 169, pp. 711–719, 2023.
https://doi.org/10.1007/978-3-031-28893-7_86

system first technology to use of the construction of the network security protection system, to strengthen protection system in the process of the attack or virus, such as the ability to identify the risk prompt protection to increase. Secondly, it can provide strong data support for safety protection ability, and promote the system to learn independently. As time goes by, its protection ability will become higher and higher. You can see that the big data completely change the pattern of the development of electric power information system network security protection, away from the traditional reliance on artificial or traditional technology, can make the system more powerful protection, therefore the electric power enterprise should seize the opportunity, actively seek reform, it will cause power guarantees related work.

Big data is a data body composed of various data, and any data can be a part of this data body at macro level. Therefore, big data generally has the characteristics of huge data magnitude. On the micro level, big data is divided into the concepts of industry big data and enterprise big data, which refer to the big data composed of a certain industry or other data. This kind of big data specially serves relevant industries and enterprises. All internal data are related to industry and enterprise operation, and each data has a complex relationship. If we can clarify the relationship, we can deeply understand the relevant issues in the industry and enterprise operation. The results can help people deal with problems and make more accurate predictions about the future, indicating that big data has good application value in any industry or enterprise [1, 2].

2 Background

The main characteristics of big data are huge data magnitude and complex relationships between data. These two points are the embodiment of the value of big data and the difficulty of big data application. That is, huge data all come from related events and represent the whole development process of events. The relationship between data can reveal the state and development direction of events at different stages, so it can be used for event prediction or in-depth analysis, Therefore, relying on big data can help people better understand the nature of events, so as to make accurate judgments and decisions. However, the huge amount of data and complex data relationships pose a huge challenge to manual and traditional technical means. On the one hand, manual data processing has efficiency limitations, which are due to the limitations of manual capabilities and cannot be changed. However, the huge amount of data will lead to manual processing of all data in a short time, so the efficiency of manual work does not meet the real-time requirements of data applications, If an enterprise wants to know its own development status so that it can make the next development plan, but the labor efficiency is low, and the enterprise cannot understand its own situation in a short time, which will slow down the development of the enterprise. On the other hand, as a technical tool, traditional technical means do not have efficiency problems and can quickly process massive data. However, the significance of applying big data lies in understanding the nature of things, which requires in-depth and accurate analysis of data. However, traditional technical means do not have this ability and can basically only be used for data processing and cannot be used for data analysis, even if they can be used for data analysis, The results obtained are often inaccurate, incomplete, not deep and other defects. This structure is not enough to help people make accurate judgments on things, and the quality of decision-making cannot be guaranteed.

It can be seen from the above that although big data has good application value, if it cannot be used correctly, its value cannot be brought into play. Therefore, modern people think that we can solve problems with some advanced technical means. The most representative technical means is intelligent technology, which has intelligent logic under the support of neural network, and can carry out independent learning and in-depth mining according to data, The representative technology system has a capability similar to human intelligence, and as a technical tool, it also has the characteristics of high efficiency in its work, which shows that intelligent technology can comprehensively and deeply analyze big data while giving consideration to efficiency, and the results given are naturally accurate, complete and in-depth, and manual judgment and decision-making can only be made according to the final results.

It is worth mentioning that intelligent technology itself is an automation technology framework, which is not very different from other automation technology frameworks. But because of the intervention of neural networks, intelligent technology is essentially different from other automation technology frameworks, because the most prominent feature of intelligent technology is intelligent logic, which comes from neural networks. The neural network itself is a model built by imitating the brain nerve conduction process under different thinking modes of people, so it can be endowed with intelligent technology and intelligent logic. This model has a variety of forms. At present, there are two common types: feedforward type and feedback type. The feedforward neural network has the feature that the logic does not repeat forward, which means it is more suitable for solving some linear logic problems. Part of the work of network security protection of the power information system belongs to this kind of problem. For example, to protect the system network security, you must first know what security risks exist. On the premise of understanding the security risks, you can naturally make targeted protection decisions, which is a typical linear logic problem. On the contrary, the feedback type has the opposite characteristics. Its logic is often repeated, so it is more suitable for solving some nonlinear logic problems. Some work in the network security protection of the power information system also belongs to this type of problem. For example, to confirm which security hazards can be protected by the current security protection system, it is necessary to connect the security protection system with each security hazard, and also consider the relationship between each security hazard, In this way, iterative logic is formed. It can be seen that different neural network models are applicable to different problems, and there are different problems in the security protection of power information system. Therefore, different neural network driven intelligent technologies should be adopted in big data applications, so as to better play the role of big data. Figure 1 and Fig. 2 show the topology of two neural networks.

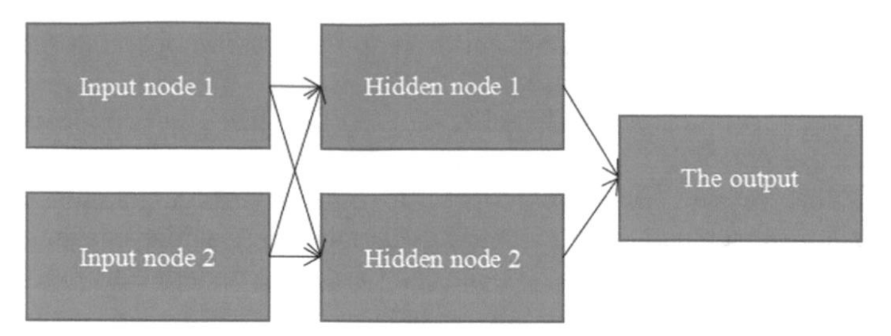

Fig. 1. Topology of feedforward neural network

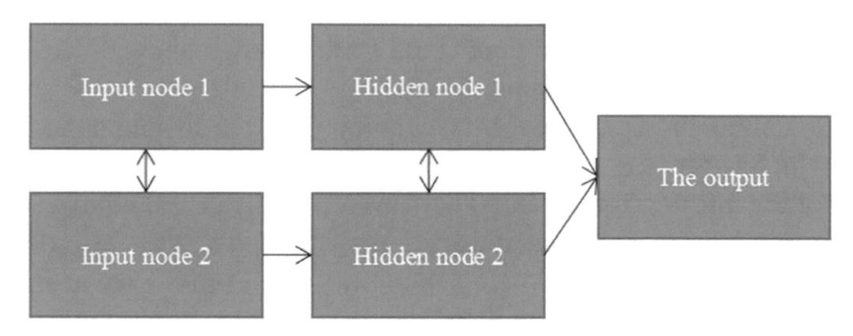

Fig. 2. Feedback neural network topology

3 Power Information System Network Security Problems

3.1 Main Sources of Network Security Threats

On the whole, the network security threats of power information system come from the external and internal environment of the network, as follows: First, the threat in the external environment is mainly refers to the malicious invasion, virus, which the former is to point to by "hacker" unconventional technology cracked the network safe protection system, break through the boundary of internal network invasion of a behavior, once the invasion success of internal network information, such as data may be its theft, listening, and then the information or data used in illegal way. Virus usually refers to a malicious program lurking in the external public network. When internal personnel browse the network, they will invade the network through various induced behaviors, such as inducing internal personnel to download programs with viruses. After downloading, the virus will realize the invasion, and the influence caused by the invasion is similar to that of malicious invasion. Second, the internal environment of the threat is primarily virus in form, but the threat formation mechanism is different, usually because of some internal personnel caused by the improper operation, such as internal personnel were planted on the device the external equipment, the equipment stored in virus, plant can lead to equipment after infected by the virus, then the virus will be through the equipment for the entire network intrusion. These two kinds of security threat sources have been widely paid attention to by electric power enterprises and are important objects of network security protection of electric power information system [3–5].

3.2 Current Situation of Network Security Protection

In the face of two sources of network security threats, modern electric power enterprises have established corresponding security protection systems, but the system has great defects in protection strength, so there are some problems in the current situation: First, in view of the external network security threats, many electric power enterprises is conducted through the firewall protection, this is a kind of border protection for network security technology, the system first internal network and external public network, the dividing line between the two appear, this line can be regarded as the boundary of the internal network, and then you can install a firewall on the boundary, In this way, any external access request must pass the firewall authentication. If the authentication fails, the firewall cannot access the external access request. Firewall does have a certain effect, but as a traditional security protection technology, it has long had the relevant cracking method, and cracking is not difficult, so the firewall is easy to be broken in the security protection, there is insufficient protection strength. Figure 3 is the principle of firewall protection; Second, in view of the internal network security threats, the electric power enterprise is generally through various virus killing software protection, the enterprise will require staff to use regularly check virus killing software for computer equipment, software will obtain relevant procedure characteristics, in the process of combining the virus itself in the record to determine whether a program for viruses and to determine if the virus, then it will automatically kill, or ask the user's opinion. It can be seen that, first of all, the virus detection and killing software cannot operate in time, so it is impossible to ensure that the internal virus will not affect the power information system completely. Secondly, the protection ability of modern virus checking and killing software depends on the virus library. Modern viruses emerge in endlessly, and their camouflage and anti-checking and killing ability is getting stronger and stronger. However, the record update speed of the virus library is slow, so some new viruses cannot be effectively checked and killed, and even cannot be identified [6–8]. Figure 4 shows the protection principle of virus removal software.

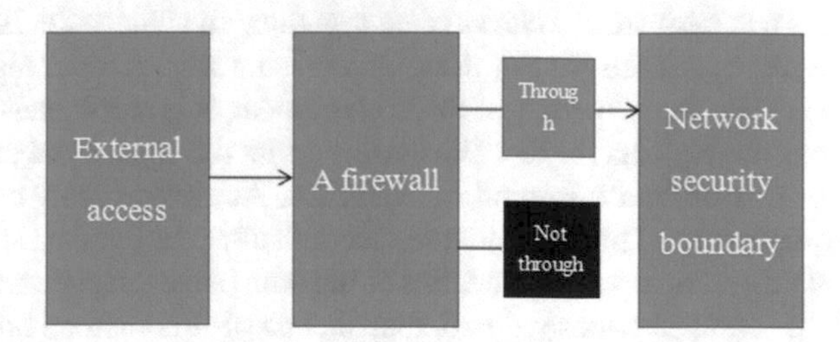

Fig. 3. Firewall protection principle

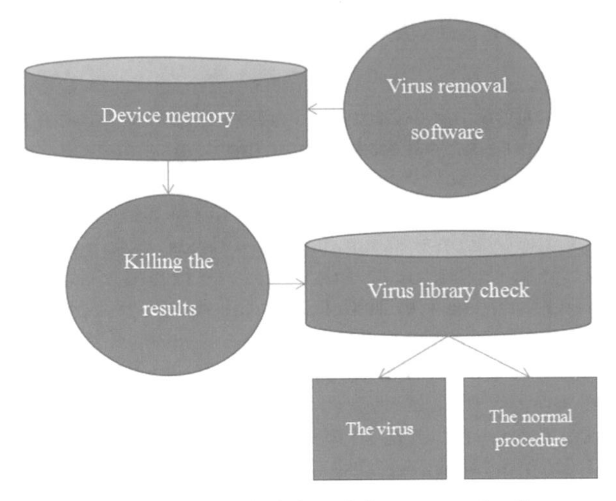

Fig. 4. Protection principles of virus removal software

4 Big Data Power Information System Network Security Protection System Construction Method and Application Strategy

4.1 Construction Method

In order to thoroughly reform the power information system based on big data, other ways and methods should be adopted to construct it, which is analyzed below.

Construction of Application Environment

Because big data can only be applied in a specific environment, it is necessary to establish a big data application environment before the construction of the system network security protection system. The main points are as follows: First, the necessary premise of big data application is to centralize all relevant data together. To achieve this, it is necessary to create a large storage space for big data, which is an important component of big data application environment. In view of this requirement, it is recommended to choose cloud database to store big data [9, 10]. The reason is that the capacity of cloud database is unlimited, which is different from other databases. At present, only cloud database can meet the requirements of big data storage. Second, after the big data storage need to analyze it to play a role, and the characteristics of big data make unusual analysis method is hard to work, so want to choose a particular means of this method are collectively referred to as data mining technology, is a kind of intelligent technology supported by means of data analysis, has the characteristics of fast analysis speed, high accuracy, companies can introduce technology for big data analysis. In addition, data mining technology needs to be supported by algorithms, and the algorithm selected in this paper is k-means algorithm, as shown in Formula (1).

$$E = \sum_{i=1}^{k} \sum_{x \in C_i} \|x - \mu_i\|_2^2 \tag{1}$$

where, k is the data cluster, the number > 1, i is the specific number of k, x is a range of k, and C is the difference between i.

Framework Design

Based on the big data application environment, the basic framework of power information system network security protection system is shown in Fig. 5.

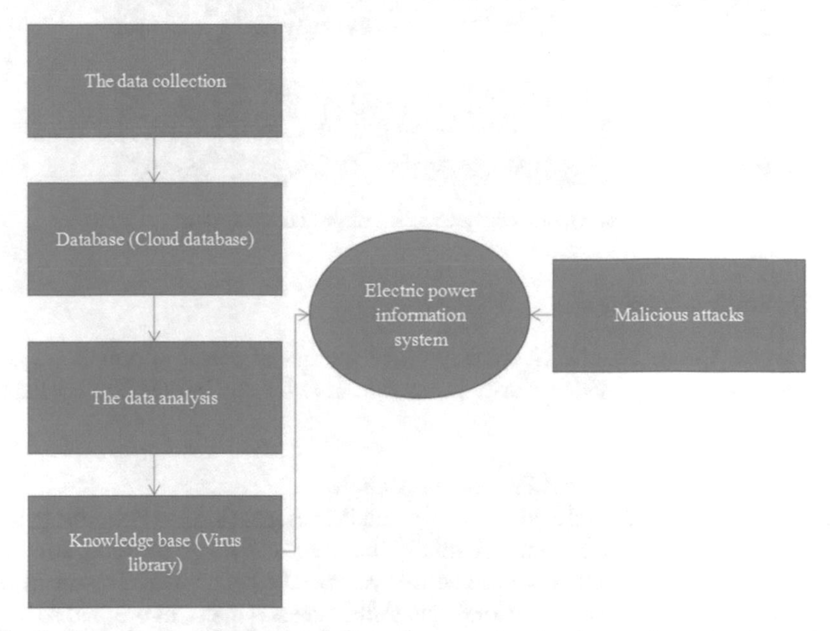

Fig. 5. Basic framework of power information system network security protection system

Figure 5 shows that the protection system is divided into two parts: The first is the protective function of plate, the plate construction of electric power information system safety protection by data acquisition data, big data will be stored in the cloud of database, and then accept the analysis of the data mining technology, the analysis results all import knowledge base, the knowledge base is also a virus, can be used for malicious attacks and similar measures, procedures for identification, because of the large scale data, so recognition is very powerful; Second, is the electric power information system itself, the system has all the work needed to function as the main application system, but also is the main target, malicious attacks when malicious attacks on its launch, plate interceptor will be protection function, the nature of the recording and judgment of the knowledge base and the malicious attacks, if it was malicious attacks on further interception, if unable to identify the nature of the specific, Then it will be treated as "suspected malicious attack", and also further intercept, or notify the administrator to deal with it. At the same time, the program of "suspected malicious attack" will be further analyzed to form new big data, which can be identified next time. Figure 6 shows the operation mechanism of the protection system.

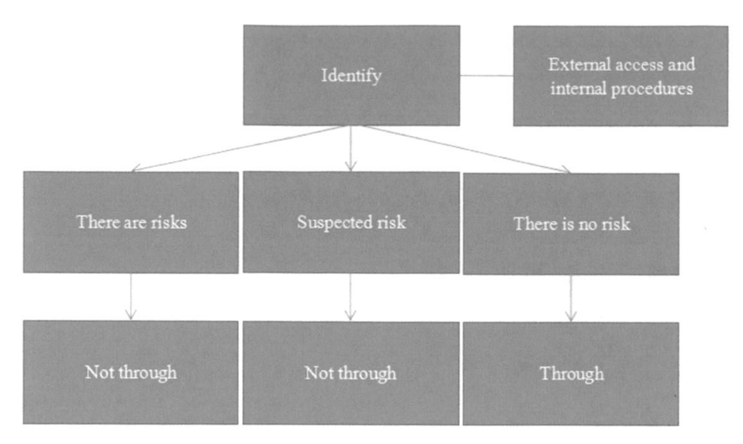

Fig. 6. Operation mechanism of protection system

4.2 Application Policies

In view of the two main network security threat sources of power information system, the above system can provide security protection in two ways in practical application, as follows.

External Environment Threat Prevention Policies
The main threat in the external environment is a malicious invasion and virus, the two may have some differences in form, but essentially the same, can adopt the same strategy to prevent it, namely through the above protective system, whether it is malicious invasion or virus, will be considered as a special kind of external access request, its original definition not featuring on the system logic, but must accept further decision, therefore when encountering malicious invasion or virus transmission system request, the protection system will be to judgment, based on the current knowledge base to store in the process of "knowledge", to be able to identify most of the malicious invasion or viruses, once judged to be malicious invasion or virus, will be refused to visit, and notify the artificial, when it cannot be accurately determined, it will be treated as "suspected malicious attack". The method has been discussed above and will not be repeated here. The protection strategy in addition to using big data to construct a strong knowledge of virus identification data, and improve the safety protection mechanism, at the same time very difficult to break through the protective mechanism, the reason is that the protection system can be supported by large data node array technology to protect themselves, to break through must break through the hundreds of thousands of nodes, the basic impossible, even if it can be done, it takes a lot of time.

Internal Environment Threat Prevention Policy
Internal environment threat to the formation of the channel is more special, protective protective system cannot directly, but can through the protection system on strategy to supervise procedure in all documents, equipment, full backup on the one hand, in case of damaged beyond repair, on the other hand, when any file or program is abnormal, will

be to determine, and then according to the judgment results and protection mechanism for processing, can achieve effective killing.

5 Conclusion

To sum up, the construction of electric power information system does provide a lot of help to electric power enterprises, but it also brings related network security threats. To deal with these risk threats, it is necessary to establish corresponding security protection system, while the traditional system is gradually weakening in practice, so enterprises should actively reform. Reform, big data can provide strong support, strengthen the identification capability of network security protection system, optimize the protection logic and mechanism, and with the aid of related technology, can make protective system protection strengthen ceaselessly, fully meet the requirements of electric power information network security protection system, therefore, enterprises should fully understand the application value of big data, should be introduced in an all-round way.

References

1. Rusdan, M., Manurung, D., Genta, F.K.: Evaluation of wireless network security using information system security assessment framework (ISSAF) (Case Study: PT. Keberlanjutan Strategis Indonesia). Test Eng. Manag. **83**, 15714–15719 (2020)
2. Alaoui, I.E., Gahi, Y.: Network security strategies in big data context. Procedia Comput. Sci. **175**, 730–736 (2020)
3. Opov, V., Emilova, P.: Big data and information security. Conferences of the Department Informatics. Publishing House Science and Economics Varna (2019)
4. Yassir, S., Mostapha, Z., Claude, T., et al.: Workflow scheduling issues and techniques in cloud computing: a systematic literature review. In: Zbakh, M., Essaaidi, M., Manneback, P., Rong, C. (eds.) Cloud Computing and Big Data: Technologies, Applications and Security, vol. 49, pp. 241–263. Springer, Cham (2019). https://doi.org/10.1007/978-3-319-97719-5_16
5. Hasanin, T., Khoshgoftaar, T.M., Leevy, J.L.: A comparison of performance metrics with severely imbalanced network security big data. In: 2019 IEEE 20th International Conference on Information Reuse and Integration for Data Science (IRI). IEEE (2019)
6. Bhosale, S.S., Pujari, V.: Big data application in smart education system. Aayushi Int. Interdiscip. Res. J. **77** (2020). National Seminar on "Trends in Geography, Commerce, IT and Sustainable Development"
7. Kibiwott, K.P., Zhang, F., Kimeli, V.K., et al.: Privacy preservation for ehealth big data in cloud accessed using resource-constrained devices: survey. Int. J. Netw. Secur. **21**(2), 312–325 (2019)
8. Habeeb, R., Nasaruddin, F., Gani, A., et al.: Real-time big data processing for anomaly detection: a survey. Int. J. Inf. Manag. **45**(APR), 289–307 (2019)
9. Udipi, S.: The event data management problem: getting the most from network detection and response. Netw. Secur. **2021**(1), 12–14 (2021)
10. Gupta, D., Rani, R.: Improving malware detection using big data and ensemble learning. Comput. Electr. Eng. **86**, 106729 (2020)

Optimal Care Strategies Based on Single Photon Emission Computed Tomography Examinations

Juan Lin[1](✉), Kun Wu[2], Fenfang Lei[1], and Deepak Kumar Jain[3]

[1] Nursing School of Shaoyang University, Shaoyang 422000, China
270953561@qq.com

[2] The First Affiliated Hospital of Shaoyang University, Shaoyang 422000, China

[3] Chongqing University of Posts and Telecommunications, Chongqing 400065, China

Abstract. Objective: To analyse the choice of care strategy based on single photon emission computed tomography (SPECT) examinations. Methods: The study population was selected from 100 patients who underwent single photon emission computed tomography examination in nuclear medicine in our hospital between January 2019 and December 2021, and they were divided into observation group and control group according to double-blind method, 50 cases in each group, the control group adopted the conventional care mode and the observation group adopted the comprehensive intervention mode, and the clinical compliance, care satisfaction, SAS and SDS scores of patients in both groups were compared after the end of care After the end of care, the clinical compliance, satisfaction, SAS and SDS scores of patients in the two groups were compared to analyse the requirements for the best nursing strategy selection. Results: The results of this study showed that the total clinical compliance of the observation group was 94.0% higher than that of the control group, which was 80.0%; the satisfaction level of the observation group was 96.0% higher than that of the control group, which was 82.0%; the SAS and SDS scores of the observation group were lower than those of the control group. The differences in the indicators between the two groups were statistically significant ($p < 0.05$). Conclusion: Comprehensive interventions during SPECT can improve patients' compliance on the one hand, and improve their emotional and psychological state on the other. This strategy is the best nursing strategy and has the value of promotion and application.

Keywords: Single photon emission computed tomography · Optimal · Care strategies · Nuclear medicine · CT

Nuclear medicine is an emerging discipline in the field of medicine in recent years that has been used to diagnose, treat and study diseases with technology, also known as atomic medicine, and can be integrated with a number of disciplines such as technology and computer technology to diagnose and treat certain difficult conditions. SPECT, short for Single Photon Emission Computed Tomography, is a gamma-ray imaging technique that offers the safety, non-invasiveness and ease of use of nuclear medicine CT technology [1]. However, the test itself is associated with a certain level of nuclear radiation and requires effective nursing interventions to enhance patient compliance during the test. Therefore, this study will also investigate the nursing care strategy during SPECT, aiming

J. H. Abawajy et al. (Eds.): ICATCI 2022, LNDECT 169, pp. 720–729, 2023.
https://doi.org/10.1007/978-3-031-28893-7_87

to provide better quality and protective services to patients. The results of the study are reported below.

1 Analysis of Technical Value of Single Photon Emission Computed Tomography

As a radioisotope CT scanning technology, single photon emission computed tomography (SPECT) has outstanding medical value. In principle, this technology combines radionuclide development technology with CT three-dimensional imaging technology, so that three-dimensional radioisotope distribution map can be obtained, and medical personnel can understand the condition by observing the image. In practical application, the single photon emission computed tomography (SPECT) technology mainly uses ordinary y-rays as the detection object to obtain single photons suspected of internal radioisotopes. The so-called single photons are also called y-photons, which are derived from the decay process of nuclear atomic reactors and have the characteristics of unidirectional radiation.

On the equipment, it is usually considered to use y-photon camera and annular single photon radiation scanner to detect y-rays and develop imaging. Common single photon radionuclides include 99 m technetium, 133 xenon, 201 thallium, etc. In recent years, 81 m krypton and 123 iodine are mostly used. Before detection, the patient is asked to inhale Yang Wu with corresponding radionuclides, and then relevant equipment is used to conduct a circumferential scan of the detection site. The whole process should be conducted longitudinally, and the speed should be as fast as possible, so as to better accept the y photons from different directions in the patient's body. When the y-photon contacts with the sodium iodide crystal on the device probe, the crystal will flash a signal, and the signal will be amplified and become more obvious under the effect of the optical telecommunication tube, and then a series of electronic processing will be carried out to image in the computer. The imaging will rely on computer algorithm to solve the radiation concentration distribution at different points in the fault plane, and then convert it to adjust the image to complete the final imaging, at which time the medical staff can make accurate judgments. In terms of imaging effect, the image is very clear, and various morphological abnormalities can be seen at a glance, such as basal ganglia, thalamus and different cortical regions. It can also show biochemical changes in local visceral blood flow, blood volume, oxygen and glucose metabolism, so it is very helpful for the accuracy of medical judgment.

In the application of technology, no matter what kind of equipment is used for detection, people should master the equipment principle and result recognition method to γ For the camera, there are several detection points on the probe of the device, each of which represents a projection line. The y photons will enter the device through the projection line and finally contact the device. Therefore, the sum of the measured values of y photons on each projection line is equal to the sum of human radioactivity, so that the results can be identified. At the same time, the detection points on each projection line in the equipment application can only be used to detect the radioactivity on a certain fault of the human body. The radioactive output is generally called the fault one-dimensional projection. All projections are first parallel to the detector probe, so the whole is called the parallel beam. The intersection angle between the normal of the detection equipment and the X axis belongs to the observation angle, which enables medical personnel to observe the detection process for control. The y camera belongs to a two position detector and has an additional structure. For example, a parallel hole collimator can be installed on the camera, so that the parallel beam projections of multiple planes can be obtained simultaneously and combined to form a flat film. It should be noted that although the functional development of various devices of this technology has been relatively perfect, in actual work, medical personnel need to understand the structure of the human body in the depth direction, which is impossible for the function of modern equipment. Therefore, medical personnel need to observe from different angles. To know the structure of the human body in the depth direction, they need to observe from different angles.

Single photon emission computed tomography (SPECT) technology has a wide range of applications in the field of modern medicine, mainly in five aspects: first, bone detection, which can achieve bone imaging, and the imaging results can help medical personnel judge the bone metastasis of malignant tumors. The whole process can be divided into disease stages, bone pain evaluation, prognosis judgment, efficacy observation and other links, so that medical personnel can better understand the dangerous parts of pathological fractures; The second is cardiac perfusion tomography and myocardial ischemia diagnosis, that is, this technology can help medical personnel better evaluate the scope of coronary artery disease of patients, and analyze the risk of disease according to the evaluation results, or evaluate the changes of myocardial blood flow perfusion volume caused by coronary stenosis, the function of collateral circulation, myocardial cell vitality, and the prognosis of myocardial infarction, or observe the efficacy, It can also be used to observe the improvement of myocardial ischemia after cardiac bypass surgery and interventional therapy; Third, myocardial infarction can be used for disease diagnosis, such as judging whether myocardial infarction is accompanied by ischemia, myocardial cell survival, etc. It can also be used for the differentiation and diagnosis of other cardiomyopathy and ventricular aneurysm; The fourth thyroid imaging can be used to judge the position and situation of the thyroid, especially for the positioning, diagnosis and differentiation of thyroid nodule function, benign or malignant, thyroid size and weight, and well differentiated thyroid cancer metastasis; Fifthly, the diagnosis of regional cerebral blood flow tomography and ischemic cerebrovascular accident, such as helping to localize the epileptic focus, can enable medical personnel to understand the positive rate of patients in the interval between seizures, judge the blood flow of brain tumors, and identify recurrence and scar after surgery or radiotherapy; The sixth is renal dynamic imaging and renogram examination, which mainly includes understanding the renal artery disease and the blood supply of both kidneys, the judgment of renal function and renal function, the patency of the upper urinary tract and the diagnosis of urinary tract obstruction, monitoring the blood perfusion and function of the transplanted kidney, and the analysis of the impact of diabetes on renal function. In addition, other phenomenal functions also have high application value in medical clinic, such as parathyroid imaging, which can help to diagnose and locate parathyroid adenoma, or adrenal medulla imaging, which can achieve the localization of cell tumor and its metastasis.

In general, the single photon emission computed tomography (SPECT) technology has a high application value, but to give full play to the value of this technology, it is necessary to master its specific application methods, indicating that it is necessary to conduct research. At the same time, technology only plays an auxiliary role in clinical practice, that is, technology can help medical personnel understand specific conditions, but it cannot give strategic suggestions on the best treatment, intervention, nursing, etc. These work still requires medical personnel to make their own judgments, so medical personnel must clearly read and identify the results of this technology, and then consider the best nursing strategy based on their own professional expertise, In order to help patients recover as soon as possible and get rid of pain with the help of strategies. In addition, the so-called optimal nursing strategy does not have a specific concept. It essentially refers to the most suitable and effective nursing strategy for patients, that is, different patients have different personal habits and receptivity in the face of the same disease. Therefore, if a subjective nursing strategy is adopted, it may not be highly coordinated by patients, which will greatly weaken the nursing effect, indicating that this nursing strategy is not the optimal nursing strategy objectively, It also shows that medical personnel should not blindly pursue the nursing effect in the strategy selection, and should ensure that the nursing strategy can be highly coordinated by patients. On this basis, they should choose the best according to the nursing effect. At the same time, in the application of single photon emission computer, some enhancement equipment can be used to ensure the effect. The common enhancement equipment is single photon intensifier, whose structure is shown in Fig. 1.

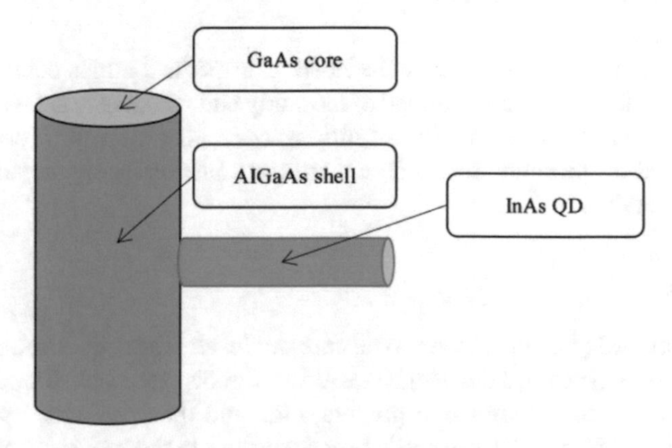

Fig. 1. Single photon intensifier structure

2 Information and Methods

2.1 General Data

One hundred patients who underwent nuclear medicine single photon emission computed tomography examination in our hospital between January 2019 and December 2021 were selected for the study, and they were divided into the observation group and the control group according to the double-blind method, with 50 cases in each group. In the observation group, there were 34 males and 16 females, aged 44–79 years, with a mean age of (59.8 $\pm$ 4.4) years; in the control group, there were 36 males and 14 females, aged 40–78 years, with a mean age of (59.4 $\pm$ 4.7) years. The literacy levels of the patients in both groups are shown in Table 1 below.

Table 1. Comparison of literacy data between the two groups of patients

Group	Number of examples	Below junior high school	Middle school - high school	High school or above
Observation group	50	7	27	16
Control group	50	10	26	14

Inclusion criteria: All patients met the clinical scope of nuclear medicine SPECT, were conscious and had no missing clinical data.
Exclusion criteria: patients with mental illness or cognitive impairment; patients with speech and communication impairment; patients with systemic infectious or immune system diseases; missing clinical data.

The entire study was approved by the hospital's medical ethics committee and all patients and their families were informed of the study and voluntarily signed an informed consent form. There were no significant differences in the general information of all subjects enrolled in the study, so the data were not statistically significant ($P > 0.05$) and were comparable.

2.2 Methods

The control group adopted the conventional care mode, after health education, informing patients of various types of knowledge related to the disease, including the causes of the disease disease, the examination process, etc., and then performed single-photon emission computed tomography examination according to the requirements of medical advice, paying close attention to the patients' vital signs during the examination process and reporting any data abnormalities in a timely manner. The observation organisation uses comprehensive interventions and adopts corresponding care models around the different stages of the examination. See Fig. 2 for the scheme.

Pre-examination
The reason for the need for targeted psychological interventions before the examination is that many patients are relatively unaware of their disease and are prone to negative emotions such as anxiety or depression, which, if not channeled in a timely manner, will inevitably affect the patient's cooperation and compliance during the examination. For example, most patients do not have the necessary understanding of nuclear medicine

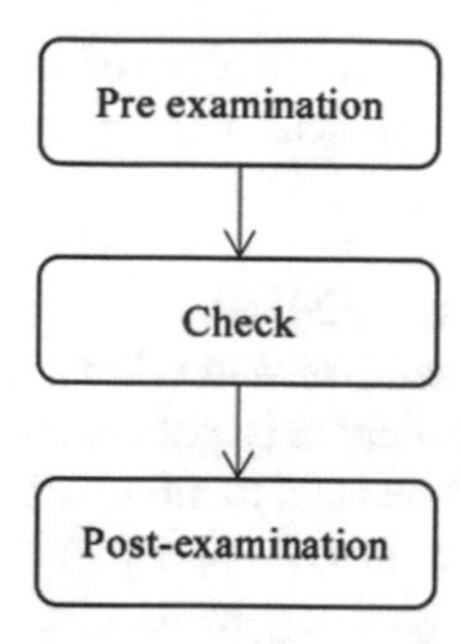

Fig. 2. Solution process

SPECT and believe that the test will have serious nuclear radiation problems, thus generating adverse emotions. The nursing staff should explain the purpose and function of the examination and inform the patient of certain precautions to be taken during the examination.

For example, in the case of adrenal imaging, the use of diuretics or cholesterol-lowering drugs should be prohibited two weeks before the test, and the patient should avoid contrast examinations to avoid delayed excretion and other adverse symptoms; in the case of myocardial perfusion imaging, the use of coronary circulation improving drugs should be reduced or stopped depending on the actual situation; in the case of thyroid imaging, the use of iodine-containing drugs or food containing iodine should also be stopped two weeks before the test. In the case of thyroid imaging, it is also important to stop taking iodine-containing medication or iodine-containing foods two weeks before the test. Patients requiring imaging should be given potassium perchlorate prior to injection, an anti-allergy test prior to injection of labelled antibodies, and dexamethasone; patients requiring cerebral perfusion should be kept visually and audibly closed before and after injection to prevent the psychological impact of equipment noise during the examination.

During the Examination

During the examination, care should be provided to assist the patient to maintain a good body position and mood, and not to move the patient's body as much as possible to avoid any influence on the patient's examination results. Apart from accompanying family members, contact with the patient should be minimised and unnecessary radiation exposure of family members from radioactive drugs should be avoided. In the case of cardiac nuclear examinations, for example, the cardiac nuclear scan takes a long time and the patient's emotional changes and ideological work should always be attended to during the examination, with the patient actively cooperating with the health care staff in the relevant examination.

The drug is injected aseptically to ensure that all the drug is injected into the vein and to prevent damage to the patient's tissues. If the patient absorbs radioactive drugs locally, this can lead directly to complications such as local haematoma phlebitis or even sepsis [3]. If the patient undergoes renal dynamic imaging, the upper arm is elevated and the patient is observed for changes in heart rate and respiration following the injection.

Post-examination

Patients should be kept in isolation for 24 h after the examination to reduce contact with healthy people and should drink plenty of water throughout the isolation period to ensure urination so that the radiopharmaceutical is excreted in the shortest possible time. After the examination, it is important to rest and to monitor the patient's blood count to ensure that it remains normal, and to monitor the patient's ability to sleep and move around. The diet should be light, with no greasy, spicy or stimulating foods and small meals.

2.3 Observation Indicators

After the end of care, the clinical compliance, nursing satisfaction, SAS and SDS scores of the two groups of patients were compared to analyse the requirements for the best nursing strategy selection.

Clinical compliance was categorised as full compliance (complete compliance with all tests as prescribed by the doctor), partial compliance (needing the cooperation of nursing staff or others to carry out all tests), and non-compliance (not cooperating with any tests). The total number of adherence statistics is the number of full and partial adherence.

Satisfaction with nursing care was evaluated according to the hospital's own satisfaction questionnaire, with a score of 100 points, with 80 points or more being very satisfied, 60 points or more being satisfied and 60 points or less being dissatisfied. The overall satisfaction count was for all people with a score of 60 or above.

SAS and SDS scores were assessed according to a self-assessment scale.

2.4 Statistical Methods

All data in the study were analysed using the SPSS 20.0 statistical software after the database had been created using Access, where measurement data were expressed as (mean $\pm$ standard deviation) and count data were expressed as percentages.

3 Results

3.1 Clinical Compliance

The results of this study showed that 94.0% of the total clinical compliance in the observation group was significantly higher than 80.0% in the control group, with statistically significant differences ($p < 0.05$). The detailed data are shown in Table 2 below.

Table 2. Clinical adherence data for both groups (n, %)

Group	Number of examples	Very compliant	Partial compliance	Non-compliance	Total compliance
Observation group	50	28	19	3	47 (94.0)
Control group	50	25	15	10	40 (80.0)
Cardinality	–				5.462
P	–				<0.05

3.2 Satisfaction with Nursing Care

In terms of nursing satisfaction, 96.0% in the observation group was also higher than 82.0% in the control group. The difference in data was statistically significant ($P < 0.05$). Detailed data are shown in Table 3 below.

Table 3. Data on patient satisfaction with care in both groups (n, %)

Group	Number of examples	Very satisfied	Partial satisfied	Not satisfied	Satisfaction with care
Observation group	50	28	19	3	48 (96.0)
Control group	50	25	15	10	41 (820.0)
Cardinality	–				5.577
P	–				<0.05

3.3 Psychological Status Scores

In terms of mental state scores, the observation group's SAS and SDS scores were all lower than those of the control group after care, with statistically significant differences ($p < 0.05$). The detailed data are shown in Table 4 below.

Table 4. Data on post-care psychological status scores for both groups

Group	Number of examples	SAS score	SDS score
Observation group	50	39.4 ± 2.2	36.7 ± 2.0
Control group	50	44.7 ± 2.4	42.0 ± 2.5
t	–	8.812	9.088
P	–	<0.05	<0.05

4 Discussion

The rapid development of clinical nuclear medicine has made this technology play a very important role in the diagnosis and treatment of clinical diseases. Although the phenomenon of SPECT examination is not as accurate as CT examination or MRT examination in terms of fine anatomical structures, it can directly show the location of patients' organs and lesions, determine the metabolic and functional changes of patients, and diagnose the early features of certain diseases [4, 5]. For example, bone imaging can show the morphology and blood supply of the patient's bones, adding tomographic imaging to planar imaging, combining the anatomical specificity of CT with the functional specificity of nuclear medicine, the two complementing each other and showing obvious value in the diagnosis of some orthopaedic diseases; and for example, it can determine the morphology and function of the patient's thyroid gland and analyse whether the patient's thyroid tumour is benign or malignant [6]. For example, it can determine the morphology and function of the thyroid gland and analyse whether the thyroid tumour is benign or malignant [6]. However, it should be noted that the test itself is contaminated with radiation, so attention should be paid to the patient's cooperation and compliance during the test and to actively improve the patient's poor condition. Therefore, the nursing measures adopted in this study were conducted for the three phases of pre-care, during-care and post-care [7–9]. The focus of pre-care is on psychological intervention and psychological reassurance to improve the patient's alertness and the success rate of the examination process; during care, the focus is on the patient's physiological indicators and drug application requirements, and the nursing staff use professional techniques to reduce the patient's pain and unnecessary medical disputes; after care, the focus is on the patient's recovery and nutritional support, etc., so that the radioactive drugs in the body can be excreted in the shortest possible time, which indicates that comprehensive interventions, which can guarantee the safety and cooperation of the patient's examination sessions, are more effective than general care in improving the nurse-patient relationship and relieving the patient's negative emotions [10–12]. The results of the data from this study show that the compliance, satisfaction and SAS and SDS scores of the patients in the observation group were all better than those of the control group, which also confirms the rationality of this method of care.

In conclusion, a comprehensive intervention during SPECT can improve patients' compliance on the one hand, and improve their emotional and psychological state on the other. This strategy is the best nursing strategy and has the value of promotion and application.

References

1. Ratnayake, R., Mundis, G.M., Shahidi, B., et al.: Assessment of impact of single photon emission computed tomography (SPECT) on management of degenerative cervical and lumbar disease: a multi-institution survey of spine surgeons. Spine J. **20**(9), S130–S131 (2020)
2. Hata, H., Kitao, T., Sato, J., et al.: Quantitative bone single photon emission computed tomography analysis of the effects of duration of bisphosphonate administration on the parietal bone. Sci. Rep. **10**, 17461 (2020)
3. Kanzaki, T., Higuchi, T., Takahashi, Y., et al.: Improvement of diagnostic accuracy of Parkinson's disease on I-123-ioflupane single photon emission computed tomography (123I FP-CIT SPECT) using new Japanese normal database. Asia Oceania J. Nucl. Med. Biol. **8**(2), 95–101 (2020)
4. Kitajima, K., Futani, H., Tsuchitani, T., et al.: Quantitative bone single photon emission computed tomography/computed tomography for evaluating response to bisphosphonate treatment in patients with paget's disease of bone. Case Rep. Oncol. **13**(2), 829–834 (2020)
5. Lee, S.J., Park, H.J.: Single photon emission computed tomography (SPECT) or positron emission tomography (PET) imaging for radiotherapy planning in patients with lung cancer: a meta-analysis. Sci. Rep. **10**(1), 14864 (2020)
6. Prosser, A., Tossici-Bolt, L., Kipps, C.M.: The impact of regional 99mTc-HMPAO single-photon-emission computed tomography (SPECT) imaging on clinician diagnostic confidence in a mixed cognitive impairment sample. Clin. Radiol. **75**(9), 714.e7–714.e14 (2020)
7. Mostafa, N.M., Moustafa, S., Hussien, M.T., et al.: Utility of single-photon emission computed tomography/computed tomography in suspected unilateral condylar hyperplasia: a histopathologic validation study. J. Oral Maxillofac. Surg. **79**(5), 1083.e1–1083.e10 (2021)
8. Rao, K., Graves, S., Pollard, J., et al.: Abstract No. 552 Predictive value of voxel-based dosimetry using 99mTc-MAA single-photon emission computed tomography/computed tomography for dose to healthy liver from 90Y radioembolization. J. Vasc. Intervent. Radiol. **32**(5), S151–S152 (2021)
9. Ziolkowska, L., Boruc, A., Sobielarska-Lysiak, D., et al.: Prognostic significance of myocardial ischemia detected by single-photon emission computed tomography in children with hypertrophic cardiomyopathy. Pediatr. Cardiol. **42**(6), 960–968 (2021). https://doi.org/10.1007/s00246-021-02570-9
10. Liu, C.: Exploring the coordination of care for patients undergoing whole-body bone imaging with single-photon emission computed tomography. Chin. Disabil. Med. **28**(4), 2 (2020)
11. Wang, W., Zhu, Y.: Analysis of the application of single photon emission computed tomography/CT imaging in the diagnosis of bone metastases from primary malignant tumors. J. Pract. Med. Technol. **27**(12), 3 (2020)
12. Liu, C., Xiao, B., Huang, L., et al.: Application of 99mTc-methoxyisobutylisocyanide single-photon emission computed tomography combined with diagnostic-grade CT fusion imaging in the identification of benign and malignant lung lesions. Guangxi Med. **42**(24), 4 (2020)

Application of Text Error Correction Algorithm Based on Power Inspection Voice Command Recognition

Ming Li[1], Xiaoling Dong[1](✉), Shuai Gong[1], and Lin Cheng[2]

[1] State Grid Anhui Electric Power Co., Ltd., Information and Communication Branch, Hefei 230061, China
jy_wxh@163.com

[2] Anhui Jiyuan Software Co., Ltd., Hefei 230088, China

Abstract. In order to ensure the accuracy of power inspection voice command recognition and promote the system to make correct actions, this paper will focus on the text error correction algorithm. Firstly, the paper introduces the status quo of voice command recognition in power inspection, and discusses the significance of text error correction algorithm. Secondly, the system is designed around the algorithm. Through this study, text error correction algorithm can accurately correct the voice command of power inspection after recognition, so as to make the command more accurate.

Keywords: Power inspection · Voice command recognition results · Text error correction algorithm

1 Introduction

For maintaining power supply security, stability and other reasons, our country electric power management power inspection work very early, within the scope of the purpose of this work is to understand all of the power equipment, facilities and other related components, once found abnormal will be immediately reported to the and processing, designed to solve the problem quickly, make power supply back to normal, so the work is very important. However, as the grid scale expands, former is given priority to with manual mode of electric power inspection work efficiency and quality of both problems gradually, the reason is that the artificial mode checking work can be influenced by artificial limitations, cannot efficiently, and prone to errors in the work, lead to quality problems, the phenomenon make management aware of the problem, Started with the aid of technology has developed a variety of systems, including the power checking voice command recognition system, the system is to make the inspection personnel using handheld terminal input voice at work, and then through the system to identify voice, understand the voice of instruction meaning, finally according to the instructions which cruise related personnel to help, can work from inspection of a lot of red tape, let work efficiency, quality, so the system has a certain application value. And electric

J. H. Abawajy et al. (Eds.): ICATCI 2022, LNDECT 169, pp. 730–740, 2023.
https://doi.org/10.1007/978-3-031-28893-7_88

power inspection though voice command recognition system has been put into use, but combined use of learned that frequently, voice commands to identify the system error, can't make a correct action result in system, the defects weaken the value of application of the system, therefore needs to be optimized, optimization of Chinese this correction algorithm is a very good choice, it can correct the voice command of power inspection after recognition to avoid the above problems. Therefore, how to apply text error correction algorithm to achieve the purpose after the voice command recognition of power inspection has become a problem worth thinking about, and it is necessary to carry out relevant research.

Power patrol inspection voice command recognition essentially refers to a process in which the machine recognizes the manually input voice. In this process, there are three main reasons for abnormal voice command recognition:

First, there are quality problems in the artificial speech input itself, that is, manual is the source of speech samples, but due to personal factors, the input speech samples may be different from the recognition standards, which will lead to recognition abnormalities. The most common factor interference is accent, which is particularly common in China, that is, China is a country with vast territory and abundant resources, and it is also a multi-ethnic country. Although the common language is Chinese, However, people in different regions will have different accents under the influence of their living environment, which will have an impact on the accuracy of Chinese pronunciation, or even a huge difference in pronunciation. The speech recognition system recognizes speech based on standard Chinese pronunciation, so under the influence of accent, speech cannot be recognized in some cases;

Secondly, after the voice is input to the machine system manually, the voice sample will be transmitted to the terminal or forwarding node through the communication channel. In this process, the voice input link and voice transmission link may be interfered by external factors, especially in the power inspection environment. This phenomenon is very common. That is to say, most of the power inspection environment is outdoors, and the inspectors generally take vehicles to move quickly or work in remote areas of the city. At this time, the voice transmission will be affected under the condition of rapid movement, and the transmission signal will decay rapidly. If the distance of voice transmission is far away, the distance of voice transmission will be extended, so that the voice signal will decay significantly without transmission in place, and the signal will be lost, Even the phenomenon of complete loss. At the same time, there are many electrical equipment in the power inspection link, which will send out certain electrical signals, so the equipment will be interfered by electrical signals when inputting voice, making it unable to accept input voice normally. Any of the above situations will lead to the speech cannot be completely and accurately recognized, and voice command recognition exceptions will occur;

Third, the machine system generally recognizes voice commands according to the corpus. In this way, the machine system may have two situations in the recognition process: (1) there is no corresponding corpus resource in the corpus, which causes the machine system to fail to recognize completely; (2) there is an error in the corpus, which does not correspond to the recognition result, which causes the machine system to fail to recognize completely, or the recognition result to be wrong.

The above three reasons have attracted people's attention since the power patrol inspection speech recognition system was put into use, but the existing research has not found a way to completely solve the related problems, so the current voice command recognition exception problem is unavoidable, and there is a certain incidence in the power patrol inspection process.

In the face of this situation, the existing research has changed its own thinking, thinking that since it is impossible to avoid problems for the time being, we should start from the problems that have occurred to find solutions, so we put forward the view of text error correction. Text error correction is to detect the text generated after the speech recognition of the machine system, and then proofread and correct the errors found in it, so that the results can become closer to the correct answer, or even completely correct, and change the direction to solve the problem. In the past, people realized text error correction by data comparison, that is, the manually input voice will produce different types of data, such as the Chinese character pronunciation in the voice will have a special tone, and the tone will form specific audio data. Therefore, these data will be extracted and compared with the audio data of the standard corpus in the corpus one by one. If the gap between the two is small enough, It means that the relevant corpus in the corpus may be an accurate answer. According to this idea, compare the difference between the audio data of the voice sample and all the audio data of the corpus, and select the corpus with the smallest difference as the correct answer. If the answer is different from the relevant text in the initial text after speech recognition, replace it, and correct the error in this way.

This error correction method is widely used in the past system design, but its actual effect is not significant, because it still cannot solve the influence of artificial accent and other aspects, and it is difficult to correct errors.

In view of this situation, modern research has realized that voice text error correction is a very complex problem, which cannot be handled by simple methods. Therefore, to find a new method, we can consider finding a feasible solution from a mathematical point of view. As a result, many algorithms have attracted people's attention and need to be screened through research.

It is worth mentioning that although relevant research has considered the realization of text error correction through mathematical methods, it does not mean that traditional text error correction methods are eliminated. In fact, relevant research only hopes to optimize traditional text error correction methods through feasible mathematical methods. The optimized text error correction methods can correct and identify text in a large range, and ensure the quality of results, Therefore, this research is mainly carried out on the basis of traditional text error correction methods combined with feasible mathematical methods, aiming to achieve method optimization and provide necessary assistance for power inspection.

2 The Research Background

Power patrol inspection voice command recognition essentially refers to a process in which the machine recognizes the manually input voice. In this process, there are three main reasons for abnormal voice command recognition:

First, there are quality problems in the artificial speech input itself, that is, manual is the source of speech samples, but due to personal factors, the input speech samples may be different from the recognition standards, which will lead to recognition abnormalities. The most common factor interference is accent, which is particularly common in China, that is, China is a country with vast territory and abundant resources, and it is also a multi-ethnic country. Although the common language is Chinese, However, people in different regions will have different accents under the influence of their living environment, which will have an impact on the accuracy of Chinese pronunciation, or even a huge difference in pronunciation. The speech recognition system recognizes speech based on standard Chinese pronunciation, so under the influence of accent, speech cannot be recognized in some cases;

Secondly, after the voice is input to the machine system manually, the voice sample will be transmitted to the terminal or forwarding node through the communication channel. In this process, the voice input link and voice transmission link may be interfered by external factors, especially in the power inspection environment. This phenomenon is very common. That is to say, most of the power inspection environment is outdoors, and the inspectors generally take vehicles to move quickly or work in remote areas of the city. At this time, the voice transmission will be affected under the condition of rapid movement, and the transmission signal will decay rapidly. If the distance of voice transmission is far away, the distance of voice transmission will be extended, so that the voice signal will decay significantly without transmission in place, and the signal will be lost, even the phenomenon of complete loss. At the same time, there are many electrical equipment in the power inspection link, which will send out certain electrical signals, so the equipment will be interfered by electrical signals when inputting voice, making it unable to accept input voice normally. Any of the above situations will lead to the speech cannot be completely and accurately recognized, and voice command recognition exceptions will occur;

Third, the machine system generally recognizes voice commands according to the corpus. In this way, the machine system may have two situations in the recognition process: (1) there is no corresponding corpus resource in the corpus, which causes the machine system to fail to recognize completely; (2) there is an error in the corpus, which does not correspond to the recognition result, which causes the machine system to fail to recognize completely, or the recognition result to be wrong.

The above three reasons have attracted people's attention since the power patrol inspection speech recognition system was put into use, but the existing research has not found a way to completely solve the related problems, so the current voice command recognition exception problem is unavoidable, and there is a certain incidence in the power patrol inspection process.

In the face of this situation, the existing research has changed its own thinking, thinking that since it is impossible to avoid problems for the time being, we should start from the problems that have occurred to find solutions, so we put forward the view of text error correction. Text error correction is to detect the text generated after the speech recognition of the machine system, and then proofread and correct the errors found in it, so that the results can become closer to the correct answer, or even completely correct, and change the direction to solve the problem. In the past, people realized text error correction by data comparison, that is, the manually input voice will produce different types of data, such as the Chinese character pronunciation in the voice will have a special tone, and the tone will form specific audio data. Therefore, these data will be extracted and compared with the audio data of the standard corpus in the corpus one by one. If the gap between the two is small enough, it means that the relevant corpus in the corpus may be an accurate answer. According to this idea, compare the difference between the audio data of the voice sample and all the audio data of the corpus, and select the corpus with the smallest difference as the correct answer. If the answer is different from the relevant text in the initial text after speech recognition, replace it, and correct the error in this way.

This error correction method is widely used in the past system design, but its actual effect is not significant, because it still cannot solve the influence of artificial accent and other aspects, and it is difficult to correct errors.

In view of this situation, modern research has realized that voice text error correction is a very complex problem, which cannot be handled by simple methods. Therefore, to find a new method, we can consider finding a feasible solution from a mathematical point of view. As a result, many algorithms have attracted people's attention and need to be screened through research.

It is worth mentioning that although relevant research has considered the realization of text error correction through mathematical methods, it does not mean that traditional text error correction methods are eliminated. In fact, relevant research only hopes to optimize traditional text error correction methods through feasible mathematical methods. The optimized text error correction methods can correct and identify text in a large range, and ensure the quality of results, Therefore, this research is mainly carried out on the basis of traditional text error correction methods combined with feasible mathematical methods, aiming to achieve method optimization and provide necessary assistance for power inspection.

3 Status Quo and Significance of Voice Command Recognition Algorithm for Power Inspection

3.1 Identification Status

Figure 1 shows the basic process of voice command recognition for power inspection.

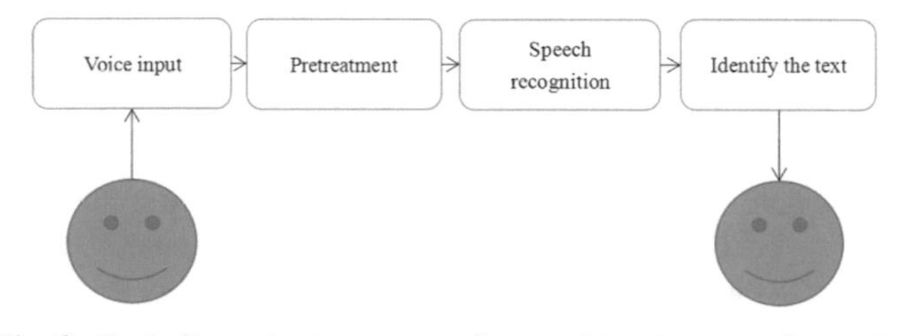

Fig. 1. Basic flow of voice command recognition for power inspection

Combined with Fig. 1, the system works by checking the staff through the handset first login system interface, and then open the voice input function input voice, voice messages will be sent by the handset software program into the system, the system will first to enter voice information preprocessing, again carries on the recognition, identification will be generated after the text, According to the text to understand the instructions of the input speech meaning, will eventually recognize text feedback to inspection personnel, to show in the handheld terminal login system interface, through the inspection after confirmation of the staff can make the right moves, such as inspection staff input power equipment location coordinates "query" voice, will feedback after identification of the same text content, then the staff click "CONFIRM", the system will give the geographical location coordinates of the equipment, so that the inspection staff can get the information support, and do not need to search hard [1–3].

Power inspection on the surface, voice command recognition system of the whole operation process is reasonable, but the practical application of the system has a lot of problems, one of the most important is to identify the given text feedback after wrong, an inspection staff, for example, in the work has repeatedly using the system input 39 times voice did not get the correct feedback, namely the staff during the inspection to one want to query power equipment fault area, so the open system for voice input, the input speech as the "query power equipment fault information, and each time the feedback text errors, such as" query power failures common type "text, there are a lot of time" command is not correct, "or" unable to identify "feedback as a result, after repeating this for 39 times, the staff gave up using the system and manually handled the problem [4–6]. It can be seen that the biggest defect of the system is that there are errors in the text after speech recognition. There are many reasons for such errors, as shown in Table 1.

Table 1. Main causes of text information errors after speech recognition

Cause	Describe
"Accent"	Some inspection staff have accents, and the system cannot accurately recognize the accent, so there will be recognition errors
Interference	In the process of speech input, there may be signal interference, for which the system will reduce the interference through pre-processing, but the degree is not enough (the current technical level is not enough to deal with serious interference), so signal interference will still lead to recognition errors

3.2 Significance of the Algorithm

For power checking voice command recognition system, recognition after the error text greatly limits its value, will continue to use the cause inconvenience, many inspection staff will choose to give up using the system, the meaning of the construction of the system to lose, but also caused the obstructive to the promotion of quality inspection work [7–10]. Look at this point, because the current technical level is limited, people can't directly by technical means to improve the recognition accuracy to solve the problem, so with the help of other methods, there may be error text results for further optimization, it is currently the best idea to solve the problem, which involves is various mathematical algorithms, such as text error correction algorithm. Text error correction algorithm is aimed at correcting the feedback text after input speech recognition by the system. The basic principle is as follows: Combined with the right corpus to semantic match each word in the text, feedback analysis, if the match results show some words with other words in the text within the scope of the semantic matching degree is lower, speaks the words or other words exist error, the two error correction logic is obtained at this moment, one is the correct match the target text, the second is that the correction place matches other words other than the target words. According to the two logics, the correction calculation can get the result of two words matching completely, and then send the two results to the staff, who can understand the instruction by choosing the correct one according to the actual needs. It can be seen that text error correction algorithm is very important for power inspection voice command recognition system, which can solve practical problems and make the system more practical. In addition, the

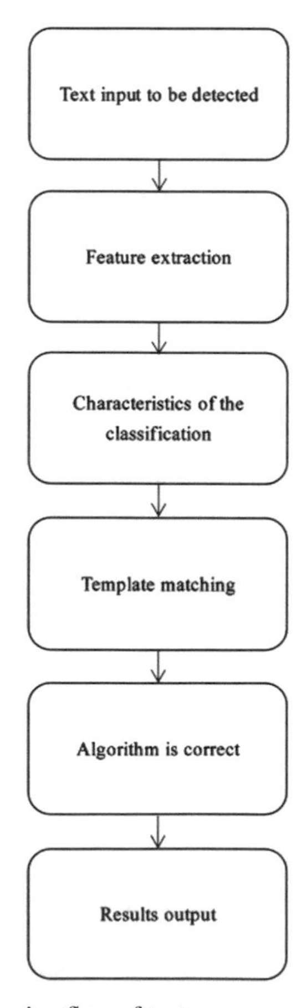

Fig. 2. Application flow of text error correction algorithm

text error correction algorithm does not refer to a particular algorithm, but to a variety of matching algorithms, among which the more representative is the fuzzy matching algorithm, which conforms to the fuzzy characteristics of semantic matching and can get results from two ideas for staff to choose [11]. The model of the algorithm is shown in Formula (1). Figure 2 Application flow of text error correction algorithm.

$$M = \frac{Q_1, Q_2, \ldots.., Q_n}{m_1, m_2, \ldots.., m_n} \tag{1}$$

Where M is the matching value, Q_1,Q_2, ... Q_n and m_1, m_2, ... m_n is the set of corpus and input speech respectively.

4 Design of Text Error Correction System of Voice Command After Recognition Based on Algorithm

4.1 General Framework

Recognize voice commands after text error correction system is a part of power checking voice command recognition system, subsystem of the latter, therefore the system to maintain a connection with the latter, namely as a subsystem, the significance of its existence is to process instructions after identification of the text, and will be processed instruction text feedback sent to the inspection staff, both for the relationship between the two, See Fig. 3 for details. In addition, Fig. 4 shows the architecture of the voice command text error correction system itself.

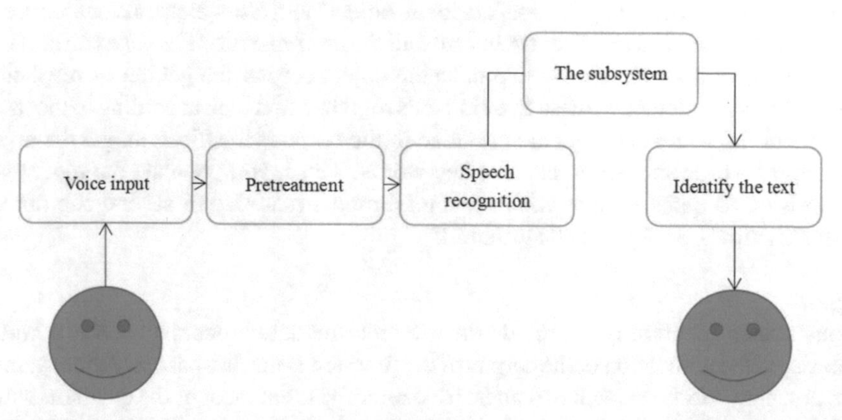

Fig. 3. Relationship between voice command text error correction system and power inspection voice command recognition system after recognition

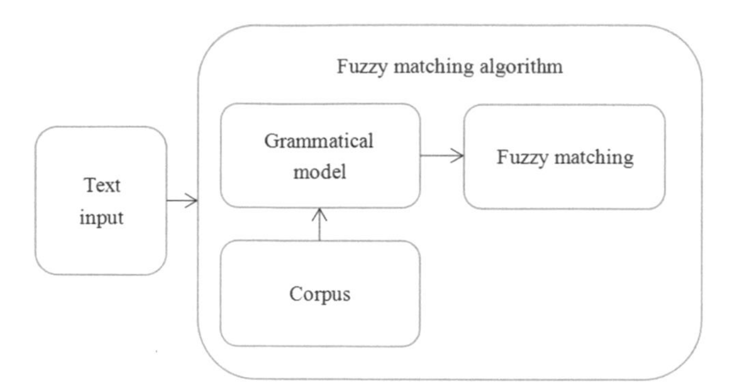

Fig. 4. Architecture of voice command text error correction system

4.2 System Design

Combined with Fig. 3, the design scheme of voice command text error correction system is as follows.

Syntax Model

The grammar model is realized on the basis of corpus, and the corpus in corpus comes from human input, that is, combining the functions required by the system in work, human input relevant corpus, such as "query the location coordinates of power equipment", and then use relevant methods to construct the model: First, the power inspection command word data set and speech recognition text are interpreted to obtain the relevant elements of recognition string, including "action, attribute, object" and other elements, which are the basis of model construction. Second, because all the grammar model is based on the target language object is given priority to rod, so the object corpus can get the corresponding action, attributes, elements, these two elements together, and then according to the model structure of the related corpora are provided in the corpus classification, get the object, attribute keywords and action class of key words; Third, you populate the model with keywords of all types, so that you have a total model made up of several sub-models, namely "Action 1 + object 1 + attribute 1".

Corpus

Corpus's main function is to provide the grammar model corpus, its itself is a kind of database, so the realization of the corpus of the first step is the database selection, namely, there are many kinds of database can be used as corpus, but most of the database will be the defects in practical application, that is because of the limited capacity of the database, And corpus corpus data storage capacity of the demand is higher, so we cannot rely on a single database to store, if all the corpus data is stored at the same time, the related organizations must constantly to expand by way of building physical server database capacity, it can really solve the problem, but will bring huge cost, so do not use. In this case, it is recommended that relevant organizations adopt cloud database, which has a huge data storage capacity and is almost unlimited. Meanwhile, in practical application, the expansion of its capacity is easy to operate and does not cost anything. Therefore, this paper will also choose cloud database as corpus. But it is important to note that leads to the use of cloud database information security problem, the reason is that the cloud database storage capacity is infinite, because the database of data storage in an open Internet, makes the internal storage of corpus will be exposed in the public eye, so in order to avoid the corpus in the cloud database using the information, related organizations need to get security protection correctly, can consider to use firewall technology for cloud database storage space planning, fraction in its unlimited storage capacity as a initial capacity, so you can make the corpus data are protected, if later need to expand capacity, you can temporarily remove firewall, after capacity expansion, restore firewall protection.

Fuzzy Matching

After the completion of the above two steps, fuzzy matching algorithm can be formally adopted to correct the identified text. The specific method is as follows: First according to the Chinese phonetic corpus edit distance, i.e., related studies have shown that text recognition among the many forms of error, the text error phenomenon is the most common Chinese characters, pinyin error such as the "power" as "ceremony" and so on, lead to errors occur, in the face of such situation can edit distance by Chinese phonetic corpus to solve the problem. Edit distance relies mainly on the minimum edit distance in fuzzy matching error correction link, the link of the difference between traditional likelihood matching algorithm, can greatly reduce the error rate of recognition results in the process, the specific process in order to establish identification string, the string matching, and then delete, add, replace operation, such as making recognition and matching chain between the operand is reduced, Select the matching group with the smallest operands as the final result (the size of the operands = the size of the distance between the two), so that the result is the most accurate; The second is the replacement operation, that is, through fuzzy matching, we can know whether there is an error in the text and where the error exists. Then, words and phrases should be replaced according to the two logics, so the replacement operation is generated. This operation step can be realized with the help of AC automata, that is, focusing on two logic, list the pinyin initials of the identification string, get the initial string, and then calculate the length of each initial letter in the initial string, get the initial string length set synchronously, set it as $X_1,X_2,\ldots,X_n$ after completion, the same method is used to calculate the set of input speech $m_1,m_2,\ldots,m_n$ initial string length set, the matching degree can be obtained by comparing the two sets of initial string length sets. Finally, the initial letter with insufficient matching degree in the input speech set is taken as the replacement object, $X_1,X_2,\ldots,X_n$ in the corresponding initial letter filled in, synchronization will replace the corresponding word or phrase, so as to achieve error correction.

5 Conclusion

In conclusion, the power checking voice command recognition system for electricity inspection work have very good help, so that the system is related organization attaches great importance to, but because of the technical level is limited, the system of the current is not very good, which can identify the input speech instruction text often makes mistakes, so the feedback system practicability is not high, needs to be optimized. In view of this, text error correction algorithm can achieve the purpose of system optimization, so relevant organizations should actively use this algorithm to optimize the system.

References

1. Nguyen, Q.-D., Le, D.-A., Phan, N.-M., Zelinka, I.: OCR error correction using correction patterns and self-organizing migrating algorithm. Pattern Anal. Appl. **24**(2), 701–721 (2020). https://doi.org/10.1007/s10044-020-00936-y
2. Yu, Z.: Menzies T.FAST 2: an intelligent assistant for finding relevant papers. Expert Syst. Appl. **120**(APR), 57–71 (2019)
3. Balaha, H.M., Saafan, M.M.: Automatic exam correction framework (AECF) for the MCQs, essays, and equations matching. IEEE Access **PP**(99), 1 (2021)
4. Jain, A., Jain, M., Jain, G., et al.: 'UTTAM': an efficient spelling correction system for Hindi language based on supervised learning. ACM Trans. Asian Lang. Inf. Process. **18**(1), 8.1–8.26 (2019)
5. Ju, R., Wang, R., Yu, P., et al.: Research on IEEE 1588 clock synchronization error correction algorithm based on Kalman filter. J. Phys.: Conf. Ser. **1449**(1), 012109(8pp) (2020)
6. Ma, R., Zhang, S., Ruan, T., et al.: Scanning-position error-correction algorithm in dual-wavelength ptychographic microscopy. Chin. Phys. Lett. **37**(4), 44201 (2020)
7. Elshahaby, H., Rashwan, M.: An end to end system for subtitle text extraction from movie videos. J. Ambient Intell. Humaniz. Comput. **13**, 1–13 (2022). https://doi.org/10.1007/s12652-021-02951-1
8. Huynh, V.N., Hamdi, A., Doucet, A.: When to use OCR post-correction for named entity recognition. In: International Conference on Asian Digital Libraries, ICADL 2020: Digital Libraries at Times of Massive Societal Transition, pp. 33–42 (2020)
9. Chen, Z.: Think twice: a post-processing approach for the Chinese spelling error correction. Appl. Sci. **11**(13), 5832 (2021)
10. Zhong, X., Jin, G.: Application of hamming code based error correction algorithm in quantum key distribution system. In: 2020 IEEE 3rd International Conference on Electronics Technology (ICET), 8–12 May 2020
11. Kalmykov, I.A., Sidorov, N.S., Tyncherov, K.T., et al.: Error correction algorithm developed for special-purpose computing devices based on polynomial modular codes. J. Phys: Conf. Ser. **1661**, 012006 (2020)

Development Measure and Evolution Analysis of Urban Green Economy Based on Particle Swarm Optimization Algorithm

Wenjia Cao(✉)

School of Economics, Changchun University of Finance and Economics, Harbin 130122, China
523971521@qq.com

Abstract. In order to measure and analyze the evolution of urban green economy development, this paper will carry out relevant research based on particle swarm optimization algorithm. Firstly, the paper introduces the basic concept and operation process of PSO, then puts forward the measure and evolution analysis scheme of green economic development, and uses PSO to analyze a city, and puts forward development suggestions. Through the analysis of this paper, particle swarm optimization algorithm has a high application value in the measurement and evolution analysis of urban green economic development, but to give full play to the role of the algorithm, it is necessary to establish a corresponding scheme system, so as to obtain accurate results.

Keywords: Particle swarm optimization · City · A green economy · PSO

1 Introduction

Because early urban development concept lack of attention to urban environmental problems, so all countries in the world in the development of the city there is the phenomenon of excessive access to resources, this kind of phenomenon lasted for a long time, and intensifying, and because of this problem after the line gradually, various countries' the development of the city space smaller, continue not only cannot promote the development of economy, It can even lead to economic setbacks. So at least in the present stage in our country have realized in the future, have begun to make a regulation, put forward the green economy development, it is a market-oriented, based on the traditional industrial economy, for the purpose of the harmony of economy and environment form the new economic development, city industry in pursuit of economic development at the same time, the environmental protection question, and give attention to two or morethings It is necessary to avoid irreversible or excessive impacts on the environment as far as possible, so as to maintain long-term and steady economic development. Thus, under the guidance of green economic development form has a strong role, under the form of the major industry in China are aware of their current development strategy, goal and behavior is not reasonable, and have the relevant reform work, has made certain progress, makes the form of a green economy development is carried out in the city development.

J. H. Abawajy et al. (Eds.): ICATCI 2022, LNDECT 169, pp. 741–749, 2023.
https://doi.org/10.1007/978-3-031-28893-7_89

But due to different cities there are various differences in various aspects, making the city green economy development speed, the situation is different, there are a lot of city in this case all have a green economic development bottleneck, the next step development rare is silly, so in order to continue to promote the development of city green economy, let the green economy to become a normal economic development in our country, Relevant fields believe that people need to adopt corresponding methods to measure and analyze the evolution of urban green economy development, so as to understand the actual demand and clarify the next development direction. Views in related fields have attracted people's attention, but it also brings a new problem, that is, how to carry out measurement and evolution analysis. To solve this problem, some algorithms are needed, and particle swarm algorithm is a good choice, so this paper will carry out relevant research.

2 The Research Background

The development of urban green economy is a relatively broad concept, involving all aspects of urban economic development. For example, the development of urban transportation can well promote urban economic development. In addition, the development of all industries in the city also has the same role. The development of green economy requires that the development of all aspects of the city should not be excessive and should be restricted to a certain extent. On the one hand, it should make full use of natural resources, On the other hand, the demand for natural resources from development activities should not exceed the limit, at least not exceed the regeneration rate of natural resources.

The significance of this is to ensure the long-term and sustainable development of urban economic development, that is, the excessive demand for natural resources in the past in the development of human society. So today, many natural resources have been exhausted, including some renewable resources, and some non renewable resources and hard to regenerate resources have been completely exhausted. Therefore, many urban economic development is in a bottleneck, which has aroused people's vigilance, Recognize that continuing to demand natural resources and pursuing development excessively is equivalent to cutting off the road behind development. Therefore, the follow-up development should avoid this phenomenon and ensure that urban development always has sufficient resource reserves, so as to continue to move forward.

The proposition of the concept of urban green economy development has been widely concerned by people, not only because it can ensure the long-term effectiveness and sustainability of urban economic development, but also because green economy can promote urban economic development on a macro level as a whole, so that it can reach a higher level. That is to say, under the green economy concept, the urban economic development will be restricted to some extent, but the existence of the restrictions will transform the urban economic development, and various industries will change from the previous independent development state to the cooperative development state. In this state, different industries will interweave and integrate, and the development momentum and resources will be more sufficient. At the same time, the capital operation will be more reasonable, which will greatly help the urban economic development.

The most typical case of industrial cooperative development is tourism, that is, tourism is one of the pillar industries in China. In its early development, tourism has already had a cooperative relationship with other industries, such as catering, accommodation and other industries. In the context of urban green economy development, the multi-party cooperative relationship is not only deepened, but also the tourism industry has maintained a cooperative relationship with more other industries, such as retail, cultural industry and so on.

In this context, tourism can better meet consumer demand and play a role in promoting consumer desire, which undoubtedly increases the proportion of cities in the tourism market, effectively speeds up the circulation of urban economy and helps urban economic development. It can be seen that the concept of urban green economy has high value, which is also a major reason why the concept can become the development orientation of modern market economy.

Particle Swarm Optimization (PSO) is a kind of intelligent optimization algorithm, which is famous for imitating the foraging behavior of birds, and has been widely used in different fields for a long time. The main function of particle swarm optimization algorithm is to find the global optimal solution or the unique solution in the global, but its search mechanism is a typical optimization mechanism, that is, people often get different answers when dealing with problems, and each answer is not necessarily correct or optimal. At this time, people need to identify and compare each answer to confirm whether the answer can solve the problem, Or whether it can solve the problem better than other answers.

This process is very complex and requires a specific method as support. Particle swarm optimization has such a function that it can find all solutions in the global, then compare them, and finally output the results. If a solution is closest to the correct answer, it means that it is the current optimal solution. If a solution that is closer is found in subsequent calculations, The original answer will be optimized, or all the suspected solutions will be found in the whole world, and then the problem will be solved around each solution. If the problem cannot be solved according to a certain solution, it means that the suspected solution is not correct and can be deleted. In this way, the problem system will be optimized, and the final remaining solution is the optimal solution.

Compared with other similar algorithms, PSO has five advantages: first, compared with other algorithms, PSO does not need to obtain relevant information about the problem in the operation, but directly solves it through real numbers, which is also the source of its strong universality. That is, other algorithms generally need to analyze the problem manually to find necessary data, parameters, etc. before operation, but PSO does not need to do so, As long as basic data training samples are provided; Secondly, the principle of particle swarm algorithm is very simple, which can be well explained through the bird flock foraging mode, so the algorithm is also easier to implement, and it generally does not need to adjust the parameters after implementation, and can be directly put into use in most cases; Third, the convergence speed of particle swarm optimization algorithm is very fast, and does not occupy too much device memory, which is impossible for other algorithms; Fourthly, the particle movement in the operation of particle swarm optimization algorithm is characterized by leaps and bounds, which makes it easier for the algorithm to find the global optimal solution, which is different from other algorithms that are easily limited to the local optimal solution; Fifthly, there are two learning factors in the particle swarm optimization algorithm, C1 and C2, which can well balance the local search ability and the global search ability. For example, the larger the C1 and C2 values are, the stronger the local search ability is. On the contrary, the stronger the global search ability is. This also shows that the particle swarm optimization algorithm has good adjustability and the convergence process can be guaranteed. It can be seen that these advantages make the particle swarm optimization algorithm get extensive attention, but the algorithm has not been widely used in the measurement and evolution analysis of urban green economy development, so it needs to be studied.

3 Operation Process of Particle Swarm Optimization

3.1 Basic Concepts

Particle swarm optimization (PSO) algorithm is a kind of people inspired by the flock foraging behavior, and the proposed algorithm is proposed in observing birds collective foraging behavior, found that the birds always fly in a regional scale to find food, this presentation on behalf of each bird knows oneself and the distance between the food, so don't out of the range, also explain every bird did not know the specific direction, in this case as each bird fly ceaselessly, the distance between each bird and food will update, then each bird will through their own unique way of communication to inform the whole flock their current and the distance between the food, and then the birds will begin with a bird as the target of nearest with food, narrowing the scope of the flight, the birds flying in the process of behavior into two categories, the first is random flight, is that all birds flying aimlessly in the search scope, this flight behavior will appear before the information update, only the second category is a flying target, namely after the information, all birds will have a clear destination, and then fly to the destination, this kind of flying behavior only appears when the information is updated. The flock foraging behavior occurs obvious iterative rule, namely the initial stage of foraging range is very large, each bird random flight, and then wait for updated information, as the birds have target analysis, start foraging range iterations, the iteration after the completion of the foraging narrow, on behalf of the entire flock and the food went a step further, so cycle and iterated that birds finally can find food [1, 2]. Particle swarm optimization (PSO) is proposed under this law. The author extracted some elements in the foraging behavior of birds and gave them new definitions to create PSO [3]. Table 1 is the algorithm definition of relevant elements in the foraging behavior of flocks.

Table 1. Algorithm definition of relevant elements in foraging behavior of flocks

The flock elements	Algorithm defined
Foraging range	To solve the space
Bird (individual)	Feasible solution
Food	The optimal solution

3.2 Operation Process

Although there are certain differences in the application methods of particle swarm optimization algorithm in different application scenarios, the running process of the algorithm remains unchanged in any scenario. In order to play the role of the algorithm in practice, it is necessary to master its basic running process [4, 5]. Figure 1 shows the running process of particle swarm optimization.

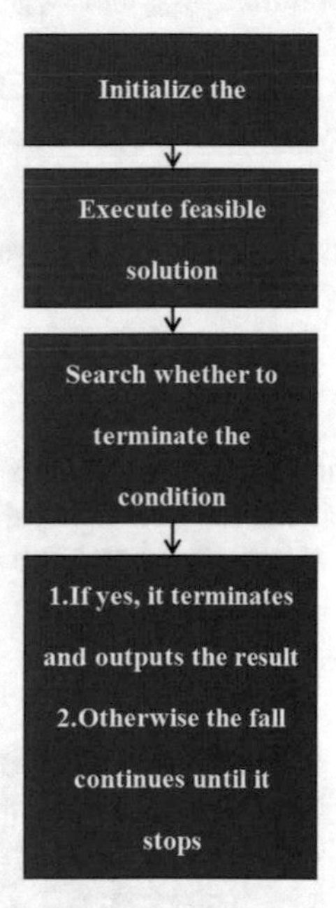

Fig. 1. Running flow of particle swarm optimization

According to Fig. 1 understand that although the particle swarm optimization (pso) algorithm is inspired by the flock foraging behavior of an algorithm program, but the operation process of the algorithm and bird flock foraging behavior still have certain difference, namely the so-called flock foraging behavior of the optimal solution is to food, and there may be more than one food foraging scope, and the food for the birds are the optimal solution, But for particle swarm algorithm the optimal solution is only one, the rest of the solution are time solution, thus directly calculated in accordance with the flock foraging behavior, easily lead to the optimal solution too much or fail to distinguish between the optimal solution of the phenomenon, this kind of situation on behalf of the algorithm convergence is very poor, the output of the results do not have high application value, therefore, in order to avoid this kind of situation, People in the operation of the particle swarm algorithm in the process adds a rule, is the algorithm run a termination conditions, as long as meet the termination conditions, the solution obtained at present is the optimal solution, namely to find the optimal solution of the flock foraging behavior belongs to the global optimal solution, but the algorithm does not seek the global optimal solution, is the pursuit of local optimal solution, The intervention of the termination condition enables the algorithm to pursue the local optimal solution, so that the results can meet the actual demand, but also can avoid the occurrence of bad problems such as algorithm convergence [6–8]. Formula (1) is the operating expression of particle swarm optimization.

$$w^{(t)} = (w_{ini} - w_{end})(G - g)/G + w_{end} \tag{1}$$

where, $w^{(t)}$ is the local optimal solution, w_{ini} and w_{end} are the initial and final weights of w, G is the maximum iteration number, a form of termination condition, and g is the current iteration number. Thus, it can be seen that the algorithm terminates when $G - g = 0$ and the solution is the locally optimal solution after calculation by Formula (1).

4 Evolution Analysis, Analysis Process and Development of Urban Green Economy

4.1 Analysis Scheme

According to the measurement and evolution analysis requirements of urban green economy development, the analysis scheme can be roughly divided into two dimensions, each dimension needs to analyze related indicators, and the index system of each dimension is the same. The two dimensions of the analysis scheme are shown in Fig. 2, and the index system is shown in Table 2.

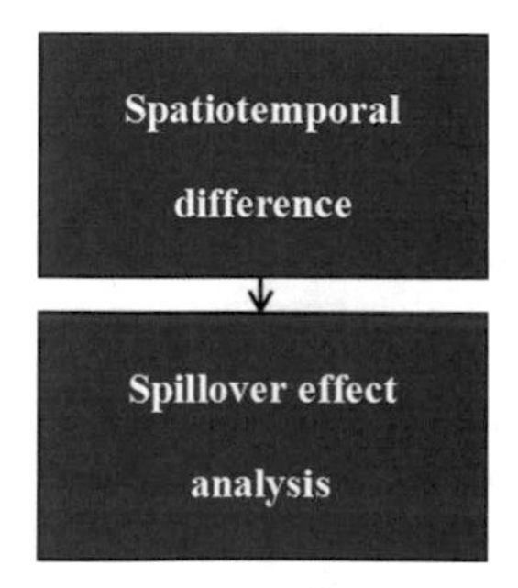

Fig. 2. Analyzes the two dimensions of the scheme

Table 2. Index system of analysis scheme

Level indicators	The secondary indicators
Resource utilization	Energy consumption index, per capita electricity and water consumption
Environmental governance	Carbon dioxide emissions, environmental pollution control input
Quality of growth	Per capita GDP growth rate, residential disposable income
Environmental quality	Forest cover and air pollution index

4.2 Analysis Process

In order to verify the effectiveness of the algorithm and analysis scheme, the following will take the green economic development of a city this year as an example to analyze it from two dimensions.

Analysis of Temporal and Spatial Differences. The analysis of spatio-temporal difference dimension can be divided into two steps, namely, overall difference analysis and regional difference analysis:

First, because the overall variance analysis involves a large amount of data, so the direct use of particle swarm algorithm may not be able to define the termination conditions, so this paper in the process of particle swarm optimization (PSO) into the gini coefficient, namely according to index of all class to convert the gini, available index of each class of the gini coefficient, on the developing trend of each index of gini coefficient of the year, The comprehensive performance of urban green economy development at each level of indicators can be seen. On this basis, the algorithm termination condition is set in this paper, that is, when the Gini coefficient of the first-level index reaches the range of 0.5 ± 0.1 during an iteration of the algorithm, the current iteration number is the maximum iteration number and the algorithm terminates, otherwise the algorithm continues to fall. According to this method, this paper obtained the results, namely, in terms of resource utilization, the algorithm iterated three times, and the Gini coefficient of each time was 0.3, 0.4 and 0.6, indicating that the development of urban green economy showed a rising trend in resource utilization index and performed well. In terms of environmental governance, the algorithm iterated three times, and the Gini coefficient of each time was 0.1, 0.3 and 0.3, indicating that the development of urban green economy was stable in terms of environmental governance index, with a general performance. In terms of growth quality, the algorithm iterated five times, and the Gini coefficient of each time was 0.1, 0.1, 0.3, 0.1, 0.5, indicating that the development of urban green economy showed an unstable trend in growth quality index, with a general performance. In terms of environmental quality, the algorithm iterated twice, and the gini coefficient of each iteration was 0.4 and 0.6, indicating that the development of urban green economy showed a rising trend in environmental quality index and performed well [9].

Secondly, for convenience, this paper constructs the data curve of the urban green economic development index with the help of data visualization software, and then analyzes the regional differences. In the process, A horizontal comparative analysis is conducted first, and it can be seen that there are significant differences between region A and Region B of the city. That is, the indexes of resource utilization, environmental governance, growth quality and environmental quality in Region A are 0.6, 0.71, 0.63 and 0.9 respectively, while those in Region B are 0.51, 0.53, 0.61 and 0.42 respectively. Therefore, region A of the city has A good development trend in horizontal comparison, while region B needs further development. Secondly, A longitudinal comparative analysis was conducted. The main targets were all secondary targets. The results showed that the indexes of all secondary indicators in region A were 0.59, 0.52, 0.82, 0.55, 0.63, 0.67, 0.77 and 0.63, respectively. The indexes of all secondary indexes in region B are 0.4, 0.45, 0.67, 0.5, 0.61, 0.52, 0.33 and 0.42 respectively, indicating that the development level of all secondary indexes in region B is comprehensively weaker than that in Region A.

Spillover Effect Analysis. Spillover effect analysis is also divided into two steps, namely, spatial autocorrelation analysis and spatial measurement of spillover effect:

First, in terms of spatial autocorrelation analysis, found that this year the city's green economy development index showed a trend of floating, floating range is 0.42 to 0.56, then the particle swarm algorithm has obtained the floating range of the optimal solution, according to the optimal solution, this year the city's green economy development index overall is more close to the highest (0.56), This performance shows that the green economy development index of the city has a positive benefit effect, indicating that the overall development trend of the green economy of the city is good, and the economic output capacity is increasing.

Secondly, in terms of spatial measurement of spillover effect, particle swarm optimization algorithm also cannot be directly used in this analysis step, so external help is needed. In this paper, spatial Dubin model is selected for initial measurement, and then particle swarm optimization algorithm is used to obtain the optimal solution. Firstly, based on the spatial dubin model, it is found that the spatial spillover effect value of green economy in this city is 0.331, indicating that the green economy development index of this city has a positive spillover trend, indicating that the green economy development level of a certain region of the city is good, and it can also drive the development of other regions with a lower development level than the city. Secondly, particle swarm optimization algorithm was used to find the optimal solution, and the results showed that the comprehensive index of the city's green economic development was 0.77, but there was a typical U-shaped non-linear relationship between monthly GDP. This was because in the early economic development, the green economic development index would decline with the per capita GDP growth, but it would encounter an inflection point in the future. At this time, it will increase with the per capita GDP growth, and the inflection point index is 0.82. Figure 3 Dubin model construction process

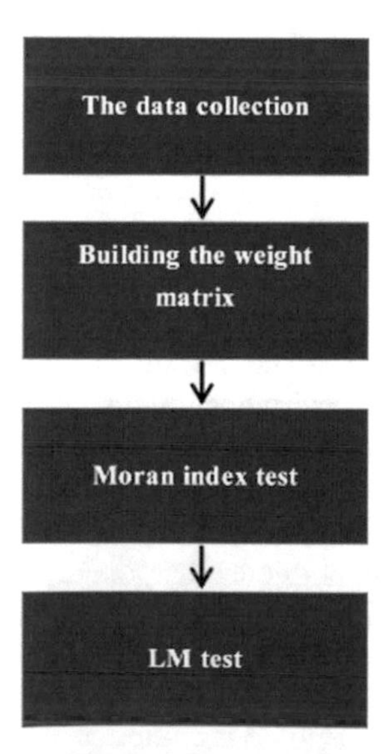

Fig. 3. Dubin model construction process

4.3 Development Suggestions

Through the above analysis, the city green economy development measure and evolution analysis results overall performance is good, the city's comprehensive index of green economic development trend, and is approaching a turning point, when the city green economy development also into another step, but in order to quickly reach the inflection point, suggest that the all-round development of the city focus on the area B, methodically, region A can be used to drive region B.

5 Conclusion

Particle swarm optimization (PSO) has good application value in urban green economy development measurement and evolution analysis. We only need to master the basic flow of the algorithm and then establish corresponding analysis scheme to get accurate results. Therefore, the active use of particle swarm optimization can effectively promote the development of urban green economy and break through the current bottleneck.

References

1. Shi, Y.E.: Particle swarm optimization: developments, applications and resources. Congress on Evolutionary Computation. In: Proceedings of the 2001 Congress on Evolutionary Computation. IEEE (2002)
2. Eberhart, R.C.: Comparing inertia weights and constriction factors in particle swarm optimization. In: Proceedings of the 2000 IEEE Congress on Evolutionary Computation. IEEE, La Jolla, CA (2002)
3. Eberhart, R.C., Shi, Y.: Comparison between genetic algorithms and particle swarm optimization. In: Porto, V.W., Saravanan, N., Waagen, D., Eiben, A.E. (eds.) EP 1998. LNCS, vol. 1447, pp. 611–616. Springer, Heidelberg (1998). https://doi.org/10.1007/BFb0040812
4. Abido, M.A.: Optimal power flow using particle swarm optimization. Int. J. Electr. Power Energy Syst. **24**(7), 563–571 (2002)
5. Nabi, S., Ahmed, M.: PSO-RDAL: particle swarm optimization-based resource-and deadline-aware dynamic load balancer for deadline constrained cloud tasks. J. Supercomput. **78**(4), 4624–4654 (2021)
6. Wang, C.H., Wang, Y., et al.: Path planning of an saucer-type autonomous underwater glider based on adaptive quantum behaved particle swarm optimization. In: 2020 5th International Conference on Automation, Control and Robotics Engineering (CACRE) (2020)
7. Maihemuti, S., Wang, W., Wang, H., et al.: Voltage security operation region calculation based on improved particle swarm optimization and recursive least square hybrid algorithm. J. Modern Power Syst. Clean Energy **9**(1), 138–147 (2021)
8. Talita, A.S., Nataza, O.S., Rustam, Z.: Nave bayes classifier and particle swarm optimization feature selection method for classifying intrusion detection system dataset. J. Phys. Conf. Ser. **1752**(1), 012021 (2021)
9. Ashraf, Z., Malhotra, D., Muhuri, P.K., Lohani, Q.M.D.: Interval type-2 fuzzy vendor managed inventory system and its solution with particle swarm optimization. Int. J. Fuzzy Syst. **23**(7), 2080–2105 (2021). https://doi.org/10.1007/s40815-021-01077-y

Design Scheme Analysis of Text Duplicate Search Optimization Algorithm in Natural Language Processing

Junjie Ma[1](✉), Zhou Li[1], Xuemin Han[1], Wenhe Zhuo[1], and Pengxi Liu[2]

[1] State Grid Anhui Electric Power Co., Ltd., Hefei 230061, China
13162013910@163.com

[2] State Grid Anhui Siji Technology Co., Ltd., Hefei 230088, China

Abstract. In order to optimize text retrieval results in natural language processing, this paper will carry out relevant research. Firstly, the key technologies of the algorithm scheme are analyzed, and then the scheme flow is designed and tested. Through the research of this paper, a more effective text search optimization algorithm for natural language processing is obtained, which can obtain the search results more accurately compared with previous algorithms.

Keywords: Natural language · Processing text · Text rechecking

1 Introduction

Natural language processing (NLD) text duplicate checking is commonly seen in college students' graduation thesis duplicate checking. The purpose is to understand the repetition rate between the paper and previous published data, so as to control plagiarism. This technology is mainly implemented by keyword feature matching algorithm, which can give relatively accurate results. But with the deepening of the application, it was found that the defects of the algorithm is more obvious, that is the language of previous published data in some papers on the "reference", according to the rules of the language should not be included in the check in the reference range, and the current algorithm cannot do this, so lead to students' thesis writing is difficult, can't give full play to their own learning, often need to cooperate with the rules of re-examination to modify the paper repeatedly, causing great trouble to students, but also easy to lead to students through re-examination will change the paper "beyond recognition", affecting the quality of the paper. Began to constantly in the related fields of this background, the natural language processing algorithm to optimize rechecking during the course of the text, has obtained some achievements, but the results have not yet fully put into practical application, the reason is that the optimized algorithm is lack the support of solution cannot be effective in application, thus it is necessary to expand research, aiming to make check weight algorithm to optimize as soon as possible, to provide better services for students or other groups.

J. H. Abawajy et al. (Eds.): ICATCI 2022, LNDECT 169, pp. 750–758, 2023.
https://doi.org/10.1007/978-3-031-28893-7_90

2 Research Meaning

Natural language processing algorithms do not refer to a particular algorithm, but a series of algorithms with language processing functions. Although these algorithms have the same functional types, they are different in function and application conditions in actual use. In terms of function, some natural language processing algorithms focus on semantic recognition, that is, the key to natural language processing is semantic recognition, which is related to the accuracy of processing results. Accurate recognition results can let people know the meaning expressed by natural language samples, so that they can better compare with the model, judge whether there is plagiarism, and calculate the proportion. If the recognition is not accurate enough, Then the judgment result will be inaccurate, which will have a significant negative impact on the natural language sample provider. Therefore, the original intention of the relevant natural language processing algorithm is to accurately identify it. The functional role naturally focuses on this point. While other natural language processing algorithms focus on the distinction of the nature of the recognition results. That is, as mentioned above, there are two phenomena in natural language texts, plagiarism and reference. The two phenomena are very similar in form, and even the latter has a higher plagiarism rate. Therefore, if it is impossible to distinguish the nature of these two phenomena in the recognition results, it is easy to mix them up, making the final results inaccurate. In response to this phenomenon, relevant fields have developed natural language processing algorithms that can distinguish the nature of the results according to two ideas: first, degree discrimination, which is mainly based on the similarity of relevant languages between natural language samples and templates. If the similarity is lower than the standard value, it will be deemed as reference, otherwise, it is defined as plagiarism. This idea has some merit, but the reliability of the recognition results is still questionable, so it was discussed very early; The second is symbol recognition. According to the writing norms of natural language texts, specific symbols need to be added when referring to the language of other articles. Therefore, the author's intention can be known by recognizing this symbol for identification. This idea has received extensive attention and has been put into use. However, because the idea is not mature enough, the application effect is not good, and there is still much room for improvement. Focusing on the defects of various natural language processing algorithms, it is worth thinking about how to establish a more perfect algorithm system, which is also the significance of this study.

3 Key Technologies of Text Duplicate Search Algorithm in Natural Language Processing

3.1 Sample Training Techniques

Word2vec is a common sample training technology in natural language processing text retrieval. Word2vec is a technical means that can train massive texts, as well as a technical means with unsupervised learning mechanism. In application, it can train and learn according to the characteristics of samples, extracting labels and other ways. The basis of Word2vec technology is artificial neural network, the main form of which is feedforward type, as shown in Fig. 1 [1–3]. With the support of the neural network, the technical system can predict the probability of words appearing in the text according to the context of the text, and then perform the fitting calculation similar to vector calculation [4, 5].

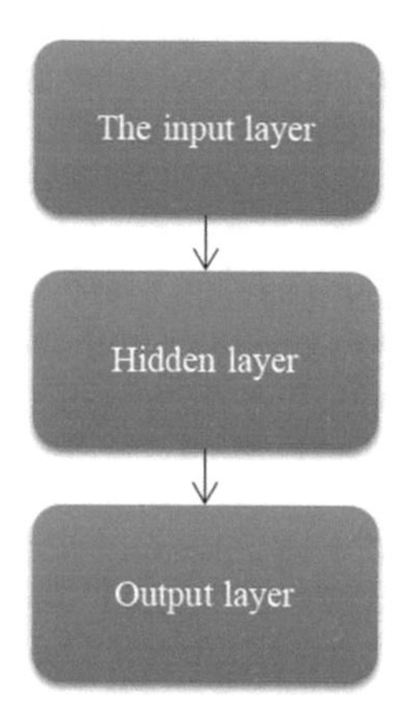

Fig. 1. Topology of feedforward neural network for sample training techniques

Combined with Fig. 1, the neural network is composed of input layer, hidden layer and output layer, the input layer number is 1, contains a number of input nodes, the nodes in the sample is known as the input vector in the training techniques, unlimited number of hidden layer, internal contains a number of hidden nodes, the nodes in sample training is known as the hidden layer vector, Is transformed from each input vector through the weight matrix of the hidden layer. The number of the output layer is 1 and contains 1 output node, which is called the output vector in the sample training technology and is transformed from all the hidden nodes through the weight matrix of the output layer. Combined with the neural network, the training sample of words, words could be started training as the input vector, such as implicit layer under its own weight matrix, to analyze the input vector and analysis in addition to get keywords feature, will combine all the input vector, form the nodes correlation model, then model is transferred to the output layer, Is converted to the result under the weight matrix of the output layer. In addition, the ultimate goal of sample training technology is to find the most appropriate weight matrix, so as to find the output vector matching the input vector through fitting, such as vectorizing the text with any words in any line of the matrix, and selecting the word vector as the parameter to achieve the goal. Then, it can be transferred to the neural network to provide parameter support for training, so that the word vector is constantly updated. In this way, after each training, the sample training results will be higher and higher, indicating that the algorithm has been optimized [6, 7].

3.2 LDA Theme Model

LDA theme originated in the latent semantic indexing model thought, this thought also referred to as the probabilistic latent semantic indexing, it has been widely used in the text theme distribution prediction, thus it can be defined as subject distribution prediction, theme LDA model is under the idea of a kind of can speculate that the text theme distribution model. The principles of the LDA theme model are as follows: Assumption in the text theme with A, B, C word characteristics, these characteristics in the text has the subject representative, namely feature point out where is the theme, and these characteristics under the rules of grammar will be in the text the probability distribution status, such as "I" in the most generally appear in the text in the language of the beginning or end (for example, I ate A meal, Or he hurts me), indicating that the beginning and end are the high probability distribution areas of the theme "I", and the other areas are the low probability distribution areas. Based on this probability distribution rule, the LDA theme model can obtain the concentrated theme of each text, and then the text analysis can obtain the results by extending the theme. Theme it is worth noting that the LDA model analysis results cannot be directly read human, so to make the model into the practical application, also need to use other technical means, such as machine learning technology, the technology under the action of the LDA theme model can directly used in NLP, such as a variety of system tasks, specific operation way is: Combining the LDA topic model with the weight matrix of Word2vec technology, the LDA topic model can be optimized and the intuitive results can be given. Then, NLP topic words can be extracted from the intuitive results for text modeling. This way in natural language processing has a high application value of rechecking during the course of the text, the reason is that the natural language text has strict rules of grammar, so it's easy to grasp its theme distribution probability, and the main word in the text of the overall quantity is far less than the number of other types of words, so rely on theme LDA model for dimension as well as to obtain information implied in the text of the subject, according

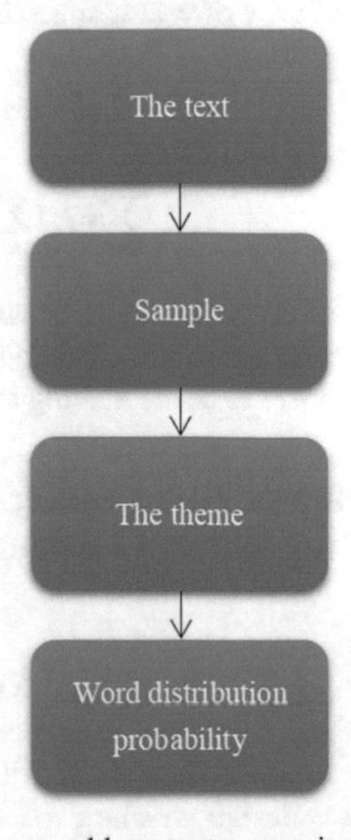

Fig. 2. LDA topic model in natural language processing text retrieval operation flow

to this information, accurate results can be obtained [8]. Figure 2 shows the flow of the LDA topic model in natural language processing text retrieval.

3.3 Similarity Calculation

People in the mathematical point of view as the distance between the text, the text similarity that is text similarity calculation in essence is a subject on the text of the LDA model and text vector calculated the distance between the process, if the distance between the two is smaller, means that the smaller the difference between the two, on the other hand, the greater the so of text in natural language processing, if the distance between the two is small, it means that the repetition rate is high; otherwise, it means that the repetition rate is low. According to this logic, we can not only get the repetition result, but also show that similarity calculation is the key technology in natural language processing text repetition retrieval. In the natural language processing text, similarity calculation mainly replaces the traditional keyword matching algorithm, or is used as the secondary processing technology of the matching algorithm [9, 10]. With the help of similarity calculation, the result of duplicate search can be more accurate, and the phenomenon of reference being duplicate is also effectively avoided. At present, the common method to calculate the similarity of a vector space cosine similarity calculation, edit distance calculation, euclidean distance, hamming distance, etc., these algorithms in practice because of the different variable dimension is sensitive, can't be calculated by ordinary coordinate distance, so the algorithm applies differences between, in this paper, the main choice of vector space cosine similarity calculation method, the algorithm can calculate the difference between two individuals, by measuring the vector to complete the re-check detection, the results are very accurate, and the form is easy to understand. The calculation method of cosine similarity of vector space is shown in Formula (1).

$$sim(x, y) = \cos\theta \frac{\sum_{i=1}^{n} (x_i, y_i)}{\sqrt{\sum_{i=1}^{n} x_i^2}\sqrt{\sum_{i=1}^{n} y_i^2}} \tag{1}$$

Where x and y are feature items, i is a feature in any feature item, θ is the Angle between any two vectors in the vector space. If the calculated result shows that the cosine value is close to 1, the Angle is closer to 0°, indicating the high vector similarity.

4 Flow Design and Testing of Text Duplicate Search Algorithm Optimization Scheme for Natural Language Processing

4.1 Process Design

Using the above three key technologies, the natural language processing text replay algorithm itself has been optimized. However, in order to obtain better replay results, the optimization algorithm alone is not enough, and the replay process needs to be optimized. Therefore, the process design will be carried out below, and Fig. 3 is the process framework.

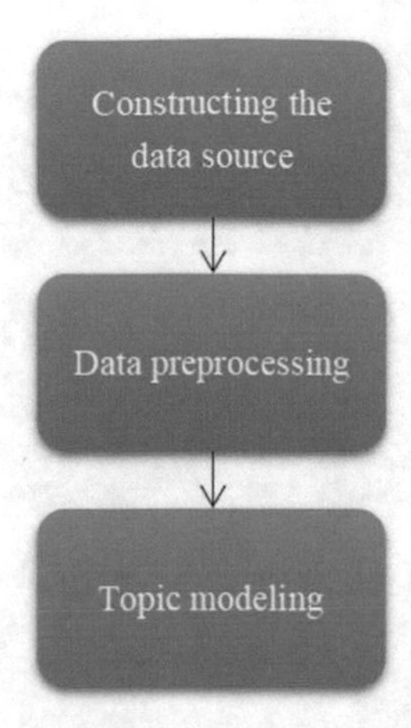

Fig. 3. Optimized recheck flow framework

(1) Constructing data sources. Data source is the basic support of text replay in natural language processing, which directly affects the running process of algorithm and the accuracy of final results. Therefore, the quality of data source should be emphasized in the design of replay process. The construction method of data source can be realized by relying on open source corpus. However, in the face of a variety of open source corpora, reasonable selection should be made in the construction of data source. The ideas can be referred to as follows: Because there is a proportion relation between the corpus of various corpora, the weight ratio of different corpora can be obtained according to this relation. The higher the weight ratio is, the better. The reason is that the higher the weight ratio is, the more complete the corpus is, which is more conducive to the quality of the results. According to this train of thought, in the data source construction should be as far as possible to expand the data source of the corpus data level, that is to maximize the corpus as a data source, this is generally recommended that choose the corpus of college student practice report, compared with other corpora, the corpus is no disadvantage on first in order of magnitude, second language is very targeted, therefore, this corpus can be used to accurately obtain the natural language text that meets the requirements, and then conduct the duplicate search. Figure 4 Data source construction idea.
(2) Data preprocessing. Because different corpora have different data collection rules, and most corpora are not strictly organized in this aspect, it is inevitable that there will be low-quality data in the corpus. Direct re-checking of such data will lead to inaccurate results. Therefore, in order to avoid this phenomenon, data preprocessing is needed. The data preprocessing tools selected in this paper are shown in Table 1. In order to realize the preprocessing process, this paper has carried on the development work: firstly, all the corpus in the corpus is gathered into the text, and then the text is imported into the text database of the local device. After the completion of the pure text processing, a number of. TXT format text is obtained, and all the unified storage. Second to unified save plain text preprocessing, including word processing mainly rely on all kinds of word segmentation algorithm to achieve, common dictionary

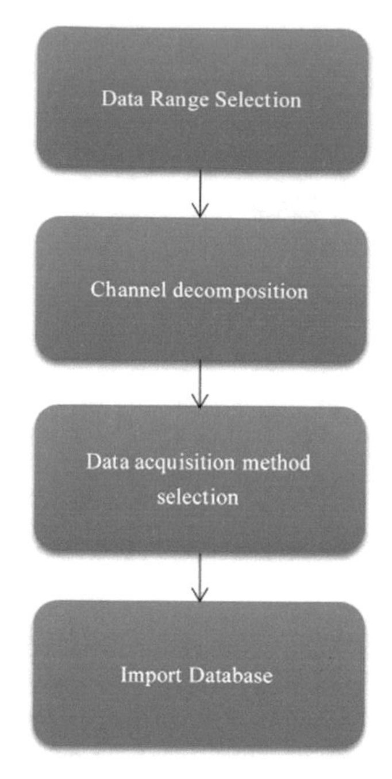

Fig. 4. Data source construction idea

word segmentation algorithm and machine learning word segmentation algorithm, neural network segmentation algorithm (the algorithm has very little to do with this article, the reason will not be description), in processing with Jieba open source tools can use these algorithms for word processing, cleaning tools are used to remove incomplete or other non-conforming words, as well as to remove noise from word data. There are many types of tools available in the current state of the art, such as batch tools, which are available in Python OS libraries.

Table 1. Data preprocessing tools

Tool	Function
Word segmentation tools	To classify words according to their parts of speech
Cleaning tools	Clean out incomplete data and noise in data

(3) Topic modeling. Topic modeling is divided into two phases: In this stage, the relevant data is sorted according to the weight of words, and then the word quantization process is carried out. Word2vec technology is used in the process, that is, the Word2vec tool is obtained by the Gensim library of The Python environment, and then the corpus statements are input into it. Statements can be arranged automatically through the tool to form statement sequence, and then word sequence can be

obtained according to the statement sequence, so that the premise of word quantization is prepared. After the preparation work is completed, the corresponding word list can be put forward from each sentence according to the word segmentation rules to complete word vectorization. The second stage is the stage of vector mapping to the model, that is, the product of the spatial vector and the topic weight of the text subject words is obtained with the help of Word2vec technology, and the binary value is given to the LDA topic model as the importance weight to complete the process of vector mapping to the model.

4.2 Test

On the basis of the above process, the corresponding LDA topic model is obtained. For this model, similarity calculation method is adopted to check the duplicate of 2000 data. At the same time, before the duplicate check, manual control of the repetition rate is carried out to ensure that the overall data repetition rate is kept within the range of 50% ± 1%. The results showed a replication rate of 50.6% for 2,000 pieces of data, in line with expectations.

5 Conclusion

In conclusion, natural language processing text rechecking technology is very important for college students, but because of the defect of previous methods, technology, so the technology is not perfect, in the practical application has brought a lot of trouble to the student, and through the optimized algorithm and scheme, to solve the problem in the past, let the check results more accurate.

References

1. Sakai, H., Iiduka, H.: Riemannian adaptive optimization algorithm and its application to natural language processing. IEEE Trans. Cybern. **99**, 1–12 (2021)
2. Shrestha, H., Dhasarathan, C., Munisamy, S., et al.: Natural language processing based sentimental analysis of Hindi(SAH)Script an optimization approach. Int. J. Speech Technol. **23**(4), 757–766 (2020)
3. Maragheh, H.K., Gharehchopogh, F.S., Majidzadeh, K., et al.: A new hybrid based on long short-term memory network with spotted hyena optimization algorithm for multi-label text classification. Mathematics **10**(3), 1–24 (2022)
4. Mosa, M.A., Anwar, A.S., Hamouda, A.: A survey of multiple types of text summarization with their satellite contents based on swarm intelligence optimization algorithms. Knowl. Based Syst. **163**(6), 518–532 (2019)
5. Kerk, L.C., Lau, G.C., Ahmad, S.N., et al.: A robust algorithm for global optimization problems. J. Phys. Conf. Ser. **1988**(1), 012055 (2021)
6. Meng, X.: Grasshopper optimization algorithm for multi-objective optimization problems. Comput. Rev. **60**(5), 214–214 (2019)
7. Cui, Z., Zhao, L., Zeng, Y., et al.: Novel PIO algorithm with multiple selection strategies for many-objective optimization problems. Complex Syst. Model. Simul. **1**(4), 291–307 (2021)

8. Pietron, M., Karwatowski, M., Wielgosz, M., et al.: Fast compression and optimization of deep learning models for natural language processing. In: 2019 Seventh International Symposium on Computing and Networking Workshops (CANDARW). IEEE (2019)
9. Gullu, M., Polat, H., Cetin, A.: Author identification with chicken swarm optimization algorithm and adaboost approaches. In: 2020 5th International Conference on Computer Science and Engineering (UBMK) (2020)
10. Tsai, A., Ghaoui, L.E.: Sparse optimization for unsupervised extractive summarization of long documents with the frank-wolfe algorithm. In: Proceedings of SustaiNLP: Workshop on Simple and Efficient Natural Language Processing (2020)

Author Index

J. H. Abawajy et al. (Eds.): ICATCI 2022, LNDECT 169, pp. 759–761, 2023.
https://doi.org/10.1007/978-3-031-28893-7

Printed in the United States
by Baker & Taylor Publisher Services